T0201223

MODELING AND ESTIMATION OF STRUCTURAL DAMAGE

MODELING AND ESTIMATION OF STRUCTURAL DAMAGE

Jonathan M. Nichols

Naval Research Laboratory
Washington DC, USA

Kevin D. Murphy

Mechanical Engineering Department
University of Louisville
Kentucky, USA

Library of Congress Cataloging-in-Publication Data applied for.

ISBN: 9781118777053

A catalogue record for this book is available from the British Library.

Set in 10/12pt TimesLTStd by SPi Global, Chennai, India

Printed and bound in Singapore by Markono Print Media Pte Ltd

1   2016

Contents

Preface   **xi**

1   Introduction   **1**
1.1   Users' Guide   1
1.2   Modeling and Estimation Overview   2
1.3   Motivation   4
1.4   Structural Health Monitoring   7
      1.4.1   Data-Driven Approaches   10
      1.4.2   Physics-Based Approach   14
1.5   Organization and Scope   17
      References   18

2   Probability   **21**
2.1   Probability Basics   23
2.2   Probability Distributions   25
2.3   Multivariate Distributions, Conditional Probability, and Independence 28
2.4   Functions of Random Variables   32
2.5   Expectations and Moments   39
2.6   Moment-Generating Functions and Cumulants   43
      References   49

3   Random Processes   **51**
3.1   Properties of a Random Process   54
3.2   Stationarity   57
3.3   Spectral Analysis   61
      3.3.1   Spectral Representation of Deterministic Signals   62
      3.3.2   Spectral Representation of Stochastic Signals   65

	3.3.3	Power Spectral Density	67
	3.3.4	Relationship to Correlation Functions	71
	3.3.5	Higher Order Spectra	74
3.4		Markov Models	81
3.5		Information Theoretics	82
	3.5.1	Mutual Information	85
	3.5.2	Transfer Entropy	87
3.6		Random Process Models for Structural Response Data	91
		References	93

4		**Modeling in Structural Dynamics**	**95**
4.1		Why Build Mathematical Models?	96
4.2		Good Versus Bad Models – An Example	97
4.3		Elements of Modeling	99
	4.3.1	Newton's Laws	101
	4.3.2	Background to Variational Methods	101
	4.3.3	Variational Mechanics	103
	4.3.4	Lagrange's Equations	105
	4.3.5	Hamilton's Principle	108
4.4		Common Challenges	114
	4.4.1	Impact Problems	114
	4.4.2	Stress Singularities and Cracking	117
4.5		Solution Techniques	119
	4.5.1	Analytical Techniques I – Ordinary Differential Equations	119
	4.5.2	Analytical Techniques II – Partial Differential Equations	128
	4.5.3	Local Discretizations	131
	4.5.4	Global Discretizations	132
4.6		Volterra Series for Nonlinear Systems	133
		References	140

5		**Physics-Based Model Examples**	**143**
5.1		Imperfection Modeling in Plates	143
	5.1.1	Cracks as Imperfections	143
	5.1.2	Boundary Imperfections: In-Plane Slippage	145
5.2		Delamination in a Composite Beam	151
5.3		Bolted Joint Degradation: Quasi-static Approach	160
	5.3.1	The Model	161
	5.3.2	Experimental System and Procedure	164
	5.3.3	Results and Discussion	166
5.4		Bolted Joint Degradation: Dynamic Approach	172
5.5		Corrosion Damage	178
5.6		Beam on a Tensionless Foundation	182
	5.6.1	Equilibrium Equations and Their Solutions	184
	5.6.2	Boundary Conditions	185
	5.6.3	Results	187

5.7   Cracked, Axially Moving Wires   189
        5.7.1   Some Useful Concepts from Fracture Mechanics   191
        5.7.2   The Effect of a Crack on the Local Stiffness   193
        5.7.3   Limitations   194
        5.7.4   Equations of Motion   196
        5.7.5   Natural Frequencies and Stability   198
        5.7.6   Results   198
        References   200

6   **Estimating Statistical Properties of Structural Response Data**   **203**
6.1   Estimator Bias and Variance   206
6.2   Method of Maximum Likelihood   209
6.3   Ergodicity   213
6.4   Power Spectral Density and Correlation Functions for LTI Systems   218
        6.4.1   Estimation of Power Spectral Density   218
        6.4.2   Estimation of Correlation Functions   234
6.5   Estimating Higher Order Spectra   240
        6.5.1   Coherence Functions   246
        6.5.2   Bispectral Density Estimation   248
        6.5.3   Analytical Bicoherence for Non-Gaussian Signals   257
        6.5.4   Trispectral Density Function   264
6.6   Estimation of Information Theoretics   275
6.7   Generating Random Processes   284
        6.7.1   Review of Basic Concepts   285
        6.7.2   Data with a Known Covariance and Gaussian Marginal PDF   287
        6.7.3   Data with a Known Covariance and Arbitrary Marginal PDF   290
        6.7.4   Examples   295
6.8   Stationarity Testing   302
        6.8.1   Reverse Arrangement Test   304
        6.8.2   Evolutionary Spectral Testing   306
6.9   Hypothesis Testing and Intervals of Confidence   312
        6.9.1   Detection Strategies   313
        6.9.2   Detector Performance   319
        6.9.3   Intervals of Confidence   327
        References   329

7   **Parameter Estimation for Structural Systems**   **333**
7.1   Method of Maximum Likelihood   336
        7.1.1   Linear Least Squares   338
        7.1.2   Finite Element Model Updating   341
        7.1.3   Modified Differential Evolution for Obtaining MLEs   344
        7.1.4   Structural Damage MLE Example   347
        7.1.5   Estimating Time of Flight for Ultrasonic Applications   352
7.2   Bayesian Estimation   363
        7.2.1   Conjugacy   365
        7.2.2   Using Conjugacy to Assess Algorithm Performance   366

	7.2.3	*Markov Chain Monte Carlo (MCMC) Methods*	374
	7.2.4	*Gibbs Sampling*	379
	7.2.5	*Conditional Conjugacy: Sampling the Noise Variance*	380
	7.2.6	*Beam Example Revisited*	383
	7.2.7	*Population-Based MCMC*	386
7.3	Multimodel Inference		392
	7.3.1	*Model Comparison via AIC*	392
	7.3.2	*Reversible Jump MCMC*	397
	References		400

8	**Detecting Damage-Induced Nonlinearity**		**403**
8.1	Capturing Nonlinearity		407
	8.1.1	*Higher Order Cumulants*	408
	8.1.2	*Higher Order Spectral Coefficients*	410
	8.1.3	*Nonlinear Prediction Error*	412
	8.1.4	*Information Theoretics*	414
8.2	Bolted Joint Revisited		415
	8.2.1	*Composite Joint Experiment*	415
	8.2.2	*Kurtosis Results*	417
	8.2.3	*Spectral Results*	419
8.3	Bispectral Detection: The Single Degree-of-Freedom (SDOF), Gaussian Case		421
	8.3.1	*Bispectral Detection Statistic*	422
	8.3.2	*Test Statistic Distribution*	423
	8.3.3	*Detector Performance*	425
8.4	Bispectral Detection: the General Multi-Degree-of-Freedom (MDOF) Case		429
	8.4.1	*Bicoherence Detection Statistic Distribution*	433
	8.4.2	*Which Bicoherence to Compute?*	434
	8.4.3	*Optimal Input Probability Distribution for Detection*	436
8.5	Application of the HOS to Delamination Detection		438
8.6	Method of Surrogate Data		444
	8.6.1	*Fourier Transform-Based Surrogates*	446
	8.6.2	*AAFT Surrogates*	448
	8.6.3	*IAFFT Surrogates*	449
	8.6.4	*DFT Surrogates*	450
8.7	Numerical Surrogate Examples		451
	8.7.1	*Detection of Bilinear Stiffness*	451
	8.7.2	*Detecting Cubic Stiffness*	456
	8.7.3	*Surrogate Invariance to Ambient Variation*	461
8.8	Surrogate Experiments		464
	8.8.1	*Detection of Rotor – Stator Rub*	465
	8.8.2	*Bolted Joint Degradation with Ocean Wave Excitation*	467
8.9	Surrogates for Nonstationary Data		475
8.10	Chapter Summary		476
	References		478

9   Damage Identification   **481**
9.1   Modeling and Identification of Imperfections in Shell Structures   481
      9.1.1   Modeling of Submerged Shell Structures   482
      9.1.2   Non-Contact Results Using Maximum Likelihood   487
      9.1.3   Bayesian Identification of Dents   492
9.2   Modeling and Identification of Delamination   501
9.3   Modeling and Identification of Cracked Structures   508
      9.3.1   Cracked Plate Model   508
      9.3.2   Crack Parameter Identification   510
      9.3.3   Optimization of Sensor Placement   523
9.4   Modeling and Identification of Corrosion   527
      9.4.1   Experimental Setup   530
      9.4.2   Results and Discussion   532
9.5   Chapter Summary   538
      References   540

10   Decision Making in Condition-Based Maintenance   **543**
10.1 Structured Decision Making   544
10.2 Example: Ship in Transit   545
      10.2.1 Loading Data   547
      10.2.2 Ship "Stringer" Model   552
      10.2.3 Cumulative Fatigue Model   559
10.3 Optimal Transit   562
      10.3.1 Problem Statement   562
      10.3.2 Solutions via Dynamic Programming   563
      10.3.3 Transit Examples   565
10.4 Summary   568
      References   569

Appendix A   Useful Constants and Probability Distributions   **571**

Appendix B   Contour Integration of Spectral Density Functions   **575**
      Reference   580

**Appendix C   Derivation of Terms for the Trispectrum of an MDOF Nonlinear
                Structure**   **581**
C.1   Simplification of $C_{pijk}^{VIII}(\tau_1, \tau_2, \tau_3)$ 582
C.2   Submanifold Terms in the Trispectrum   583
C.3   Complete Trispectrum Expression   585

Index   **587**

Preface

This book is intended as a guide to solving the type of modeling and estimation problems associated with the physics of structural damage. These two topics (modeling and estimation) are intimately related, such that a discussion of one is at least partially incomplete without the other. The job of the model is to understand and predict behavior, in this case the observed behavior of a damaged structure. This model includes both deterministic (physics-driven) and stochastic components (e.g., measurement error). The parameters that describe the model, including damage, are typically unknown *a priori* and must be estimated from observed data. This book provides the readers with both the modeling tools needed to describe structural damage and the estimation tools needed to identify the damage parameters associated with those models.

More general discussions of both structural modeling and estimation theory can be found separately in other places. However, we have found that the modeling and estimation problems that arise in structural damage identification differ sufficiently from those found elsewhere to deserve separate treatment. We have also found that an integrated treatment of these topics is lacking in a single source. Therefore, it is the goal of this book to serve as that single source. That being said, much of the material presented generalizes to other types of modeling and estimation problems faced by researchers in structural dynamics. Readers interested in these more general problems will hopefully also find this a useful guide.

The material that follows was developed over a number of years and was influenced by the thinking of a number of talented individuals. Several of these individuals we feel deserve special mention for both their intellectual contributions and their friendship. First, we thank Dr. James D. Nichols (Jon's father), a Senior Scientist with the U.S. Geological Survey. His contributions to this work were both technical (see chapter on decision theory in particular) and philosophical. Over the past two decades, we have spent much time (over many beers) discussing efficient ways to conduct science. His strong advocacy for model-based, hypothesis-driven science can be seen in every chapter and, to a large extent, sets the tone for the entire book. Proponents of the "data-driven" world-view would be well served to have a pint or two discussing this topic with Jim.

We also thank PierGiovanni Marzocca of Clarkson University. Jon was fortunate enough to work with "Pier" during two summers at the U.S. Naval Research Laboratory (NRL). In addition to being a good collaborator, Pier was the driving force behind much of the Volterra series modeling which appears in several places in this book. Gustavo Rhode of Carnegie Mellon also deserves mention. Gustavo's attention to detail and technical brilliance in the theory of random processes taught us a great deal on how to think about, and write about, this challenging subject. In the same subject area, Frank Bucholtz of NRL also deserves mention for his willingness to host "white board" discussions on probability and statistics. Frank has the rare ability to drill down into the details of a problem, no matter how seemingly trivial the topic might be, to the point where you either become an expert or leave his office convinced you understand nothing whatsoever.

We also owe a large debt of gratitude to Ned Moore (Kevin's former Ph.D. student), now of Central Connecticut State University. Ned was tasked with the practical implementation of many of the system identification techniques described in this book. Chapter 9 would certainly not have existed without his help. In addition, we thank Chris Earls of Cornell University, and his students Chris Stull and Heather Reed, who were behind the bulk of the work on identifying dents in plate structures.

Finally, we would be remiss in not acknowledging Dr. Paul Hess of the Office of Naval Research. For many years now, Paul has been a strong proponent of the model-based view of damage identification. Certainly, without his support a great deal of the material in this book would never have been developed.

On a more personal note, the second author thanks the first author, Jon. It was Jon who opened the door to this modeling and estimation world for me; I'm richer for it. In the process, Jon also introduced me to my wife. So I'm doubly in his debt. And on that note, I also thank my wife Françoise for her unbounded supply of patience and support.

Jon would like to thank his wife Susan and two children, Kirstin and Cassidy, for their patience with this project. Jon would also like to thank his parents, Lois and Jim, for their continued support and advice; this book is dedicated to them.

Jonathan M. Nichols
Crofton, MD, October 2015

1

Introduction

1.1 Users' Guide

Anyone who has done a fair bit of technical writing will likely agree that the best way to truly understand a topic is to try and clearly explain that topic to others. There is no better way to expose one's own technical deficiencies than to sit down and try and describe a subject in writing. This is certainly true of the material presented in this book. In fact, our original intent was not to write a book but rather document what we had learned about modeling and estimation so as to improve our own understanding and to keep from having to "relearn" the material over time.

In particular, we wanted to focus on some of the details of modeling and estimation that are frequently overlooked or implicitly assumed without explanation. Understanding the origins of these assumptions has helped us tremendously in our own research and we hope the book provides a similarly useful reference for others. One of our chief aims is therefore to clearly explain the roots of modeling and estimation for structural response data, tracing the mathematical reasoning back to the originators. So much of what we do in engineering sciences builds on the brilliance of A. Kolmogorov (probability), G. D. Birkhoff (signal processing), N. Wiener (spectral analysis), and J.-L. Lagrange (mechanics), to name a few. Time and time again we have seen that those who are making the most meaningful contributions in their respective fields of study are those who return to these foundations before moving forward.

That being said, there are different ways one can use this book. For example, one could choose to learn the details of probability theory in Chapter 2 or simply proceed to the later, more applied chapters and simply reference back to the mathematics when needed. The same is true for much of Chapter 3. The material of Chapter 6 explains the origins of estimation theory; however, one could move straight to Chapters 7–10 where that material is applied to problems in damage detection and identification. In short, the detail is provided, but it may not be necessary for much of what the reader is trying to accomplish. The idea was to at least give the reader the option of exploring modeling and estimation to whatever depths he or she deems appropriate.

From a structural modeling point of view, the book is well-suited to those who have taken basic undergraduate courses in mechanics of solids and dynamics. In terms of mathematics, the book presumes familiarity with basic calculus operations, series expansions

---

(e.g., Taylor series), as well as differential equations. Familiarity with probability theory and spectral analysis is also a plus, although we have taken great pains to explain these topics carefully and clearly for the interested reader. This likely places the useful starting range of the book somewhere in the later undergraduate years. This is consistent with courses currently being taught in the structural health monitoring (SHM) field at various universities. Our brief survey of such courses places the majority in the junior or senior years, continuing on as part of a graduate program.

1.2   Modeling and Estimation Overview

Most of us who entered into science and engineering disciplines did so because at some level we were fundamentally interested in questions about how things work. Whether the curiosity relates to atmospheric events, cell biology, or (more to the point of this book) why bridges don't fall down, the common link is a desire to understand the world around us. As we have all learned by now, this understanding is achieved through modeling and prediction. We construct models of the phenomenon of interest and predict outcomes. Models that predict well are retained; those that do not are discarded.

The main goal of science is, in fact, to produce useful models of reality so that we may reliably predict outcomes. There is a tremendous power in prediction. It allows us to generalize what we have observed to things that we have not yet observed. Thus, every time we build a bridge with a different design from a previous one, we don't have to worry about whether or not it will collapse. We can sufficiently model this new design and confidently predict its integrity over the intended lifetime. The model further allows us to try a number of different designs and predict their efficacy without having to build and test each architecture.

All models are, by definition, wrong, of course. They are simply abstractions of reality that we find useful for their ability to make predictions. One cannot hope to model *exactly* the observed data, nor would we want to. Increasing model complexity without significantly improving prediction is essentially pointless. As Einstein put it, "It can scarcely be denied that the supreme goal of all theory is to make the irreducible basic elements as simple and as few as possible without having to surrender the adequate representation of a single datum of experience" [1]. This guiding principle of modeling is sometimes referred to as the principle of parsimony and plays a prominent role throughout this book.

In engineering we are taught to derive *deterministic models* by applying some basic physical principles, for example, $F = ma$, and invoking some simplifying assumptions (parsimony!) about our operating regime to yield a set of governing equations. For example, to predict the vibrational response of a cantilevered beam to an initial tip displacement, we could start with Newton's laws, make some simplifying assumptions about the homogeneity of the material comprising the beam, amplitude of the resulting vibrations, and so on, and develop a solution. This solution is expected to be a good predictor of our observed response in the regime defined by our assumptions. There is no need for us to solve the full (nonlinear) governing equations.

However, even with the most sophisticated of models there will always be some remaining error in our predictions. We acknowledge that we cannot describe the exact behavior and instead describe "expected" or "typical" behavior using *probabilistic models*. Sensor noise is often the primary culprit in this type of error. For example, we might attach a resistive strain gage to our cantilevered beam and record the response. We can describe most of what we observe using our aforementioned deterministic model, however we can't predict the exact voltage that will be read because of both residual model error and sensor noise. There are a number of different "noise" mechanisms, however at this stage, it will suffice to say that noise gives rise to observations that we cannot explain with a deterministic model.[1] Instead, we describe the *probability distribution* of the response, that is, predict the values we are likely to observe. It may at first seem quite unsatisfying to have to resort to a (partially) probabilistic description of our data, however probabilistic models are quite powerful and are every bit as useful as deterministic models in describing the world around us. We will demonstrate that so long as we can describe our uncertainty, we can minimize its influence on our ability to predict.

Thus, our observations are to be characterized by both deterministic and probabilistic components. In fact, the key ingredients to any structural estimation problem are (i) a probabilistic model describing the uncertainty in the observed data and (ii) a deterministic structural model (or models) governed by a set of model parameters. Given these two ingredients, we can begin to discuss the subject of estimation. This subject can be loosely defined as the process of extracting our deterministic and probabilistic model parameters given the data we have observed. The subject of estimation is absolutely essential to damage identification as it is through estimation that we connect our model to reality. At the end of the day, we will declare "good" estimates to be the ones that are highly probable. As we will see, there are two fundamentally different viewpoints on how to arrive at "most probable." Once we have our model parameters, our data model is completely specified and we can turn to the task of making predictions and, ultimately, decisions regarding the maintenance of a particular structure.

As implied by the title, our focus is on the modeling and estimation of structural damage. This particular problem poses some unique challenges in both arenas. With regard to the former, the structural damage will alter the model of the pristine structure, often in a nontrivial way. Moreover, the damage model should reduce to the undamaged model in the limiting case that some damage-related parameter goes to zero, that is, the model should predict both healthy and damaged response data. In terms of damage parameter estimation, the problem is similarly challenging. Typically, one would like to identify damage before it becomes large and influences structural performance. However, the smaller the damage the less influence it will have on the observed data, making it more difficult to estimate the associated damage parameters. Special attention is therefore paid to both the estimates *and* the uncertainty in the estimates which, for small damage, can be large. Quantifying this uncertainty is essential to making decisions regarding how the structure is maintained. This relationship is made explicit in the final chapter of the book.

We also cover cases where the goal is to detect the damage presence, not necessarily identify the complete damage state (magnitude, location, orientation, etc.). The approach we will take is still based on the physics of damage, however in this case the problem will be viewed as one of model selection. Specifically, we will consider cases where damage results in a nonlinearity in a structure that is otherwise (when healthy) best described by a linear model. Our job will be to assess the likelihood that our observed data were produced by one of those two models (linear vs. nonlinear). While not as powerful as approaches that identify specific damage-related parameters, model selection can be used successfully in situations where there is a large amount of uncertainty in the detailed physics of the damage. Moreover, we will show that even this simple assumption about the physics of damage divorces the practitioner from

---

[1] The mathematical history of "noise" is actually a fascinating subject summarized nicely in a review article by Cohen [2].

having to rely on basic "change detection" in a structure's response as a damage detection strategy.

1.3   Motivation

So why should we focus on the modeling of structural damage in the first place? After all, the material in this book can be applied toward many other problems in structural dynamics (in fact, the original intent of this work was to provide a general reference in structural system identification). In looking back at our own research and that of many of our colleagues, problems involving "structural damage" were a recurring theme. The motivations for this research are varied and typically include a statement suggesting that an understanding of damage physics is necessary for development of some future "automated" system for monitoring the condition of a structure and making decisions about how to best maintain it (best typically implied to mean, "least costly"). Indeed, there is an increasing recognition in both military and commercial communities that an understanding of damage physics is of paramount importance. Consider, for a moment, three situations where one may want to understand and predict the condition of a structure:

1. Improve safety
2. Reduce maintenance costs
3. Increase operational envelope.

Each of these items is a strong motivating factor for understanding damage physics with large financial and performance incentives.

In the Department of Defense, there are financial pressures to reduce maintenance costs while at the same time increasing the operational envelope of a given asset (e.g., increasing ship speed while reducing the number of repairs). For example, certain classes of ships have experienced wide-spread cracking of deck plates, requiring millions of dollars annually to repair. Figure 1.1 shows two sample cracks, one taken from a top-side view, the other from

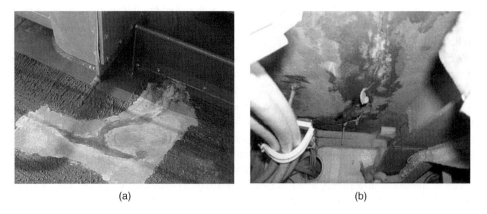

(a)   (b)

Figure 1.1   (a) Top-side view of a recently repaired deck plate crack and (b) view of the crack from inside the ship. This crack would normally be hidden beneath several inches of insulation

beneath the deck plate showing a crack that is normally hidden beneath the insulation. The cause of this cracking has been investigated and is now understood to be due to stress corrosion caused by sensitization of the aluminum alloy used in construction (5456 material). However, in order for this type of cracking to initiate and persist, the material must be sustaining large stresses. It is the origin of these stresses that is still largely unknown (at least at the time of this writing).

In a partial response to this question, one of the deck plates of the affected ship was instrumented with a fiber-optic strain sensing system (see Figure 1.2). The ship then underwent a series of high speed turns during transit, the goal being to test the strain response at the edge of the operational envelope. The strain time-history in Figure 1.2 shows only a minor signal resulting from these maneuvers, measuring $<15\ \mu\epsilon$ (micro-strain) in amplitude. This translated to a stress amplitude of <1 ksi, far below the yield stress for this material (≈ 33 ksi). This is certainly useful information, however it does not offer much in the way of predictive power. All we can say with any certainty is that these particular maneuvers are unlikely to be the source of the cracking.

Clearly, a predictive model that could accurately forecast high stress conditions, crack lengths and locations, and/or plate stiffness would be of much greater value. Ship operators need to understand when a crack has evolved to the point where it is compromising the safety of the crew or of the ship. Should the ship's captain turn around or complete the mission? In the absence of a model, this information is simply not available. In Chapter 10 we address this particular problem in its entirety and show how a model-based approach can be used to make decisions regarding how best to use a maritime asset in transit.

US Army ground vehicles have also been the subject of damage identification efforts. A number of these vehicles were experiencing cracking in the wheel spindle (part of the wheel hub assembly); cracks greater than 0.2 in. meant that the part required replacement [4]. The

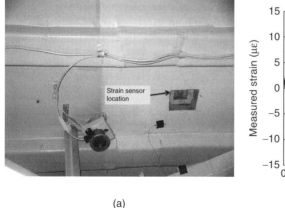

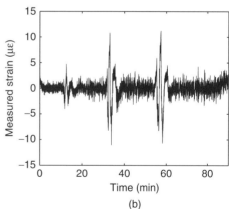

(a)

(b)

Figure 1.2 (a) Fiber-optic strain sensors are affixed to the underside of an aluminum deck plate, located behind the insulation and (b) detrended strain time-history showing the influence of high speed maneuvers (turns) on the measured response of the deck at a particular location. The magnitude of the signal ($<15\ \mu\epsilon$) suggests stresses far below the yield stress of the plate. *Source*: Adapted from [3], Figure 9, reproduced with permission of the Society of Naval Architects and Marine Engineers

Figure 1.3 Wheel spindle crack on US Army ground vehicles (a) proved a challenging problem in damage detection. The spindle (b) is hidden behind the wheel and wheel assembly (c), making it difficult to identify the damage presence without removing the vehicle from service, removing the assembly, and visually inspecting the part. Automated methods for detecting damage in these types of situations have the potential to eliminate costly repairs and downtime. *Source*: Adapted from [5], Figure 1.5, reproduced with permission of John Wiley & Sons

question, of course, is how does one know when the crack has reached the critical length? An inefficient strategy would periodically pull a vehicle out of service, remove the entire wheel assembly, and check for the appearance of a crack. However, removal of an asset from service while in-theater is a costly action to take(in terms of dollars and downtime). The particular vehicle in question is shown in Figure 1.3 along with a depiction of the spindle location behind the wheel (indicated by the black arrow) and a closeup of the spindle itself. In response to this problem, researchers at Purdue University, led by Dr Douglas Adams, developed a simple test for spindle cracks that could be performed *in situ*. On the basis of a finite element model of the component, it was determined that a crack would alter the frequency response of the assembly in a specific manner. The test therefore uses estimates of the frequency response (a subject we discuss at length in Sections 3.3 and 6.4) to detect the crack presence without removing the entire assembly [4].

In civil and commercial domains, similar safety and financial pressures have yielded additional research toward the development of various "monitoring" technologies. Bridges and other components of the civil infrastructure are now being monitored at various sites around the globe for the express purpose of assessing structural integrity. The goal of these installations is typically to monitor peak loads or displacements to confirm they are within normal operating ranges. As an example, consider the strain monitoring system depicted in Figure 1.4 and installed on the I-10 bridge in New Mexico. Performed in 1997, the goal of this installation was to demonstrate the feasibility of such a system for the monitoring of civil structures. In this case, a fiber-optic strain monitoring system developed at the Naval Research Laboratory was used to monitor the strain response of the bridge at various points. Among other things, the system was used to study the peak strains observed as a function of the type of traffic traversing the bridge. Figure 1.5 shows a histogram of strain response data obtained over many days of operation. The histogram clearly shows two distinct peaks, associated with vehicles of different sizes. Car traffic produces smaller strain signals (~ 30 $\mu\epsilon$), while trucks yield larger strains as expected (~ 50 $\mu\epsilon$).

Each of these case studies is an example of what is commonly referred to as "SHM." The next section discusses this field in more detail, describing the basic approaches and philosophies used in tackling this challenging problem. While this book is not meant to be a "SHM" book, it certainly provides tools that are likely to be useful to those in the field. In what follows we therefore attempt to place our work in the context of this more general area of study.

1.4   Structural Health Monitoring

The field of SHM comprises a body of work aimed at the identification of damage for the reasons discussed in the previous section. We should state upfront that the material presented here is not at all meant to be a comprehensive look at the SHM field as it is understood by most practitioners. A good overview of the SHM field, including numerous approaches to damage detection and identification, is given by Farrar and Worden [6] and also Adams [5]. Perhaps the most glaring omission in this book is a discussion of the types of sensors used to acquire structural response data. Data acquisition is certainly an integral part of any SHM system and has been given extensive treatment in numerous references (see, e.g., [7] or Chapter 4 and Appendix B of [5]). While our experimental examples make use of such systems, a detailed discussion of their construction and operation is not provided.

In addition, one can loosely group SHM techniques into "local" versus "global" methods. The former, as one might guess, uses data acquired from localized areas of a structure where damage is presumed to exist. The latter, global approach, is the focus of most of the examples in this work and presumes that the entire structure is being interrogated (e.g., is undergoing vibration) and that we are measuring this response at one or more locations. This represents a more challenging problem as identification requires locating the damage from these observations. However the global approach has the obvious advantage that *a priori* damage location information is not required.

Nonetheless, many of these "local" approaches to the damage identification problem have achieved solid results in a variety of contexts and therefore deserve mention (see, e.g., [8] for an overview). Thermography [9], eddy-current techniques [10], and ultrasound [11] (to name a few) have all been used to identify localized structural damage; none of these are given in-depth treatment in this book. However, this is not to say that the methods developed in

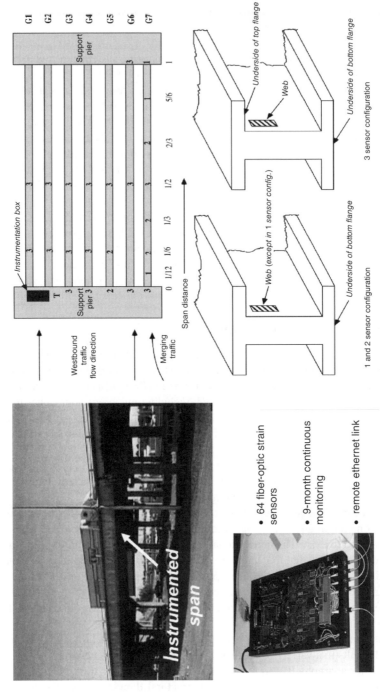

Figure 1.4 Monitoring of the I-10 bridge in New Mexico circa 1997. Shown are the bridge, the fiber-optic strain monitoring system electronics, and a schematic of the bridge span and associated sensor locations

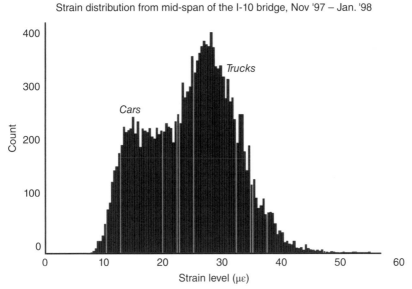

Figure 1.5 Histogram of the bridge strain response showing range of strain response in ~30–50 με range. The two peaks are associated with different types of vehicles, cars and trucks. Such data provide information relevant to the operation and maintenance of structures and are therefore a necessary part of a structural monitoring system

this book are not able to handle such problems. Indeed, the framework we describe, and the estimation tools presented, are 100% applicable to other types of physics and the corresponding measurements. What changes, of course, is the physics-based modeling component of the problem. In thermography, for example, the data consist of a temperature field and its temporal evolution. Specifically, a short duration heat pulse is applied to the structure and an infrared camera captures the subsequent change in the temperature field over time. The lack of thermal conductivity near a delamination site should result in different heat dissipation properties than the surroundings. For example, consider the lingering air munition (LAM) wing shown in Figure 1.6. A series of infrared images were recorded 0.03, 0.5, 0.7, 1, and 1.5 s after the thermal pulse was applied. The damage was a series of delaminations, located on the grid shown in the figure. Initially the damage shows up as black (cooler) dots in the appropriate locations, while after 1.5 s have passed, the delamination sites appear warmer (whiter) than the surroundings. Now, if we have a model that predicts this temperature field for a given type/number of delaminations, much of what follows in this book can be used without modification to estimate the delamination size, location, extent, and so on. We, in fact, consider the delamination problem in later chapters (modeling in Section 5.2 and detection/identification in Sections 8.5 and 9.2), but in the context of vibration-based (as opposed to thermally based) methods.

This work also makes little reference to what is perhaps the primary viewpoint of the SHM field, "statistical pattern recognition" [12]. Using this approach, one acquires training data from a structure, models the acquired data (e.g., neural nets, autoregressive models, etc.) and looks for patterns that are unique to damage presence, location, size, and so on. While the estimation of *data* model parameters is considered in pattern recognition, that approach does

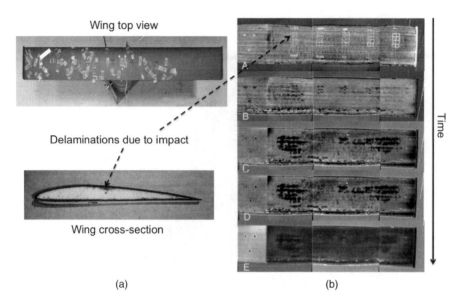

Wing top view

Delaminations due to impact

Wing cross-section

Time

(a)   (b)

Figure 1.6 Thermographic imaging used to detect impact damage in a LAM wing. (a) The top view of the wing and wing cross-section. (b) A temporal progression of the thermal profile is shown. The delaminations are located inside the grid of light squares on the first image. Although difficult to see, the delaminations begin as dark (cold) dots and in the final image show up as light (warm) dots. The images were recorded at times 0.03, 0.5, 0.7,1, and 1.5 s after the thermal "flash." Damage diagnosis using thermal imaging is currently a qualitative exercise with a trained observer making the final determination

not consider the formal estimation of *physical* damage parameters, the subject of this work. We therefore arrive at a bifurcation in the SHM field between data-driven and physics-based approaches to the problem. This dichotomy mirrors often-debated differences among those advocating inductive versus deductive approaches to science.

Induction is the process by which one attempts to infer physical laws from observation alone. Put another way, based on acquired data one attempts to formulate an explanation (model) for the physics that produced those data. By contrast, the deductive approach first postulates a hypothesis (model) to explain the observations without recourse to a particular data set. That hypothesis is then either accepted or discarded depending on how well it explains what we observe. These differing philosophies also give rise to different viewpoints on probabilistic modeling and estimation; these are discussed in the next chapter. In the meantime we discuss both data-driven and physics-based approaches and motivate our preference for the latter in damage identification applications.

1.4.1 Data-Driven Approaches

The idea behind statistical pattern recognition is to observe the system in either a healthy or damaged state, build a model of the data for each state, and then attempt to classify future observations based on these models. At a minimum, one requires data from a healthy structure. Deviations from the corresponding "healthy" model can then be used to detect damage-induced changes. If one is fortunate enough to also possess data from a particular damage state, future data could possibly be used to classify that particular damage state as well. The approach is inherently inductive as one is attempting to gain insight into the underlying damage physics by observing the system.

As an example of this approach to damage detection, consider the bolted joint shown in Figure 1.7. In 2004, the US Navy was interested in automated methods for determining loss of connectivity among bolted composite joints. Composites were being considered as a building material for future vessels and little was understood about the integrity of such connections. In response, a laboratory study was conducted, whereby a composite beam was bolted to a steel frame and subject to vibrations. The strain response of the beam was recorded at multiple points on the structure, including the steel frame. Details of this experiment can be found in Ref. [13].

Using observations from the fully clamped (healthy) beam, two different data-driven models were created: one a nonlinear model, described in detail in Section 8.1.3 and the other a linear, auto-regressive model, described in Section 3.4. Both models are therefore expected to be good predictors of a healthy connection. As the connection degrades and the dynamics across the joint begin to change, one would expect the models to break down and the associated prediction error to increase. Indeed, this is exactly the behavior seen in the results displayed in Figure 1.8. The dark gray region is associated with an interval of confidence for the healthy prediction error values (a topic we discuss in Section 6.9). Prediction error values that fall above this region are indicative of a significant change in the dynamics. Both sets of prediction errors (linear and nonlinear) are displayed as a function of bolt preload, as measured by the instrumented bolts used in this study. Both models indicate a detectable loss of preload somewhere between 500 and 1000 lb.

It is worth recalling the key assumption that went into this diagnosis: that the *only* source of change is, in fact, the joint physics through a weakening of the connection. If this is indeed the case, the approach can clearly detect the damage presence (see again Figure 1.8). However,

Figure 1.7 Composite, bolted joint structure used in experiments for detecting loss of joint integrity. Two data-driven approaches were used to model the "healthy" structural response. The working hypothesis is that a loss of joint integrity will cause deviations from the healthy model response and produce a corresponding rise in model prediction error

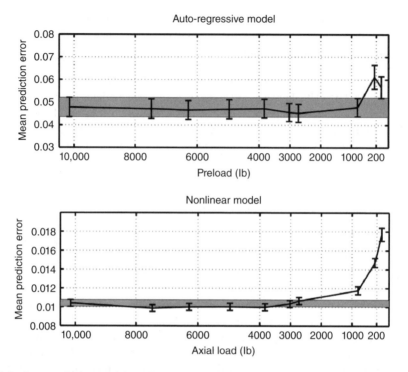

Figure 1.8   Increase in data model prediction error as a function of preload loss in the bolts. Assuming a loss of preload is the only source of change in the measured signals, a loosened connection can be detected at somewhere near the 1000 lb preload point

it is also worth mentioning that the detection algorithm output is in units of "mean prediction error." Beyond detecting the damage presence, these units are not of much use. What we would really like is an algorithm that outputs the pound-force of preload at the joint. Provided that the relationship between prediction error and preload is monotonic, we could do this. For example, fitting the lower plot of Figure 1.8 to a polynomial would allow us to map an observed prediction error to a unique preload. In short, if the relationship between preload and prediction error is monotonic *and* the only source of change in the system is the preload, then we may use this data-driven approach to reliably assess the connectivity of the joint. We have found, however, that in practice these assumptions are not always met, particularly the second.

In fact, the main difficulty with the inductive approach is that what has been learned is a relationship between data and the underlying *conditions* that produced the data. Such a model then allows us to relate newly acquired data to a particular set of conditions. Obviously, the hope is that by "conditions" we mean "state of the structure", however the link between the two may be weak, or simply may not exist. Take, for example, a condition monitoring exercise where the ambient temperature of a healthy structure is fluctuating. Higher temperatures can lower the stiffness of a material and therefore the structure's natural frequencies (the specific relationship between stiffness and natural frequency is discussed in Chapter 4). Of course, structural damage is also frequently modeled as a localized reduction in structural stiffness, for example, due to the presence of a crack. If we had previously learned of a relationship between lower frequencies and structural damage we might very well classify a warm structure as damaged (this specific problem is addressed in Section 8.8.2 using physics-based approaches).

This is a widely recognized challenge to condition monitoring using data-driven methods. One solution is to try and include covariates (sources of variation) in the database used to generate the model. If we are fortunate enough to build a classifier that can incorporate all of the major sources of variability, this can sometimes produce good results. For example, in the work of Sohn *et al.* the data model included temperature as a covariate. The authors were then able to successfully separate the damage-induced changes from those caused by temperature [14]. In short, if one can control, or properly account for all other sources of variability in an observed signal, then a damage detection strategy based on pattern recognition can be very effective.

However, based on our experience with inductive, data-driven approaches we can levy several basic criticisms. First, the resulting model is based on a limited sampling of the world and therefore may or may not have any predictive power when circumstances change. Except for very controlled environments, we can be sure that the observed data will *not* contain all possible sources of variation. The resulting model can only be expected to hold under the conditions for which the data have been collected and therefore may or may not be useful in general.

Along these same lines, we have the challenge of how to use the approach to predict all possible combinations of damage parameters. To accomplish this, one would, in principle, need to acquire data from a structure in all possible damage states (locations, magnitudes, orientations, etc.) and build a classifier that could accurately describe newly acquired data. Thus, one would need to record both damaged and healthy response data under all possible operating conditions (e.g., different loadings, weather, etc.). Considering that most damage parameters (e.g., crack length) are continuous random variables, this data set is infinite, and therefore unobtainable. Even if we were to consider some restricted set of discrete parameter values to train on, the number of acquired data would be immense; moreover, we would never know if we had captured all of the covariates (operating conditions). *In short, a model based on data, not physics, can reliably be expected to predict only that which has been previously observed.*

Secondly, we mention that in modeling the data, one is implicitly modeling all of the uncertainty (e.g., noise, ambient fluctuations, etc.) that come with those data. We would just as soon not use the same model for both the noise and physics, and focus instead on minimizing the influence of the noise on our ability to detect and identify structural damage. This is precisely what methods rooted in damage physics attempt to do, that is, isolate aspects of the observed data unique to the damage via the model while minimizing uncertainty. Our ability to minimize uncertainty is therefore predicated on our model fidelity *and* on our ability to reliably estimate model parameters, a subject we discuss at length in Chapters 6 and 7. Put another way, rather than use a pattern recognition engine to tease out the physics from the noise, why not put in the physics ourselves and let the algorithm minimize the influence of noise or covariates? That being said, it should be stressed that a number of "machine learning" techniques have been developed for SHM purposes that attempt to do precisely that: extract the (hopefully) low-dimensional structure in the data that captures the damage physics [6]. While we believe algorithms attempting to "learn" the physics cannot meet the ultimate goals of the SHM field (for the aforementioned reasons), their widespread popularity and success in certain SHM application areas should not be overlooked.

Finally, on a related note, models based solely on data often ignore what we know about the physics of structures, information that has been developed and refined over hundreds of years. Indeed, we rely heavily on past work in structural mechanics in developing the physics-based approach discussed next. Particularly when one considers the advances in computational power, we can create very detailed models of damage physics that can, with great accuracy, predict observed data.

In short, we believe it will be difficult to achieve the aims of the SHM field without formally incorporating damage physics in the identification process. Nonetheless, the framework we offer in Chapters 2, 3, and 6 is wholly applicable to data-driven approaches. These chapters collectively explain how to estimate parameters in the types of data models frequently used in data-driven approaches. In addition, in Section (6.9), a formal procedure for generating damage test statistics, akin to the frequently cited "feature values" in the pattern recognition literature, is provided as well as a means of assessing type I and type II errors associated with those test statistics. It is therefore our hope that regardless of whether one takes a data-driven or physics-based approach, the ideas and techniques described herein can be of use.

1.4.2   Physics-Based Approach

Our contention is that a firm grasp of damage physics is essential to developing an effective SHM system, however this is only one piece of the puzzle. In this book we also cover estimation and decision making, components we also view as essential to a working condition monitoring system. Our reasons for focusing on modeling the physics, as opposed to the data, are rooted in the belief that the former will prove more useful and practical to the end goal of decision making. This last step is more or less overlooked in the SHM literature, yet is perhaps the most important step of all. Consider the three previously mentioned situations where one may want an automated SHM system (improve safety, reduce maintenance costs, increase operational envelope). In all three, the desired output is a decision. A system designed to improve safety is required to decide whether a given structure is considered "safe." Reducing maintenance costs requires that decisions be made on the part of the maintainer on how to balance usage with the inevitable degradation. Increasing the envelope of operation (e.g., telling a vessel operator he/she can go faster) involves a decision about how much risk one is willing to accept versus the performance gains associated with pushing the bounds of safe usage. For this reason, we feel that a prospective SHM system must include the basic ingredients required to eventually arrive at a decision.

A schematic of this approach to SHM is provided in Figure 1.9 and consists of three basic components. Just as with the data-driven approach, the first component is to collect data relevant to the problem at hand. Ideally, this would consist of both the input data (loading on the structure) and vibration data measured by a collection of sensors (e.g., strain gages, accelerometers, etc.). As we have already mentioned, data acquisition for SHM is the subject of numerous papers and texts and is therefore not discussed here. The second phase, and the primary focus of this work, is the modeling and estimation of the current system state. The former is discussed in Chapters 4 and 5, while the latter is the subject of Chapters 6 and 7. The job of the decision engine is then to provide the owner of the structure with optimal decisions on how to best maintain that structure and represents the final component of the SHM system. To make decisions, one needs to be able to predict how those decisions will influence the state of the

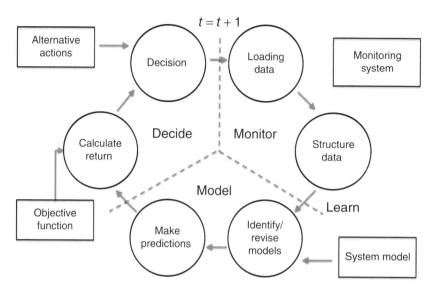

Figure 1.9 Schematic of a physics-based SHM process. Data are collected and a damage model is inferred. On the basis of the identified model, one predicts the future integrity of the structure. These predictions, along with measures of their uncertainty, can then be used to make decisions, the ultimate goal of the system. This book is primarily focused on the second stage that is, the modeling and estimation of the damage, although in the final chapter we explore the topic of decision making. Figure courtesy of M. Runge, US Geological Survey. *Source*: Reproduced from [15], Figure 1, with permission of Elsevier

structure; thus, we need to do more than simply detect damage, we need to identify it and predict its evolution.

Trying to accomplish this without physics-based modeling is challenging at best, for reasons we have articulated in the previous section. This is not to say that physics-based approaches are without challenges. Modeling complex structures is certainly difficult, however we know quite a bit about their behavior; knowledge obtained from over 200 years of work in mechanics. Especially when one considers advances in computational techniques (e.g., finite element method), even complex structural behavior can be accurately predicted. Interestingly, the "structures are too complex" argument is often listed as the prime reason for adopting data-driven approaches in SHM (see discussion surrounding Eq. (5.2) of Ref. [12]). However, this is precisely the situation where physics is of most use. It is *only* through a model that complex vibrations can be uniquely mapped to a system (damage) state. A pattern by construction cannot provide the necessary physical understanding as the number of rational hypothesis (states) that can explain any given phenomenon (pattern) is infinite. This is especially true in complex systems. To once again reference an oft-quoted physicist: "Reason gives the structure to the system; the data of experience and their mutual relations are to correspond exactly to consequences in the theory. On the possibility alone of such a correspondence rests the value and the justification of the whole system, and especially of its fundamental concepts and basic laws. But for this, these latter would simply be free inventions of the human mind which admit of no *a priori* justification either through the nature of the human mind or in any other way at all."[2] [1].

We also stress again that the model need not capture all of the structural complexity to be useful. In fact, we do not *want* to model much of the detail. A good example is presented later in this work where deviations from a true clamped boundary condition are treated by simple addition of a lumped edge stiffness parameter. While we could have developed a far more sophisticated (spatially dependent) clamping model (including the bolt properties, preloads, etc.), this was entirely unnecessary for identifying the damage. Another example occurs in Section 9.4, where a uniform thickness model is used for an experimentally corroded plate. Although the actual corrosion is quantifiably nonuniform, the uniform model still allows us to identify the presence and extent of corrosion-induced degradation.

It should also be mentioned that physics-based approaches do typically require at least some training data to ensure that the predictions are accurate (i.e., "model validation"). We have found that it is sufficient to record one or two data sets (preferably from a healthy structure) and then use those data sets to refine model parameters and to see if we are properly capturing the relevant physics. A good example of this process occurs in a case study described in Section 9.3.2 and mentioned in the previous paragraph. In trying to model the physics of a clamped plate, our initial data showed that our ability to capture the boundary physics was inadequate. We therefore had to refine the form of the boundary model to improve our predictions using the acquired data. It should be stressed, however, that this refinement process took minimal resources as only a few data sets were required. Once the adjustment was made, the model could easily discern changes in the response signal due to boundary effects and other previously unobserved sources of change from those involving structural damage.

In SHM, models allow us to view the problem as one of estimation: given the observed data, directly estimate the state of damage. The end result of this approach is precisely what we want: an estimate of the damage state in real physical units (as opposed to "feature" values) and a means of quantifying the uncertainty in that estimate. This information is necessary to make optimal decisions regarding maintenance, the purported goal of SHM. Secondly, the aforementioned operational variability (e.g., temperature fluctuations) is no longer an issue as we never have to compare it to a previously recorded data set or pattern. One simply records the data and performs the estimation directly. In addition, we have at our disposal a host of estimation tools, developed over the past century, that allow us to make claims of optimality regarding our estimate (i.e., given the noise in the system, what is the best possible damage estimate we can make). Such claims are more difficult to make using a pattern recognition approach to the problem.

Thus, we either accept the challenge of physics-based modeling or that of recording a tremendously large (technically infinite) data set and attempting to infer the physics based on patterns alone. The former approach leverages centuries of prior work and provides optimal damage parameter estimates with little reliance on prerecorded data. The latter does not produce physical parameter estimates, cannot make claims of optimality, or identify those phenomena that have not been previously observed. In short, we feel the challenge of modeling damage in structures, particularly with the present-day computational power, is far more tractable and useful than is the data-driven approach of extracting features that are unique to a given damage state (size, location, orientation, etc.).

---

[2] Another relevant passage from the same essay " … any attempt logically to derive the basic concepts and laws of mechanics from the ultimate data of experience is doomed to failure" [1].

1.5   Organization and Scope

A note on how the book is organized. It should be clear at this point that the problem posed by damage identification is one that draws on numerous disciplines. We require tools from probabilistic modeling, deterministic modeling (mechanics), detection and estimation theory, and, ultimately, decision theory. Even when taken separately, these subjects are quite involved as indicated by the numerous volumes devoted to their treatment. Taken together they represent a challenge for a prospective author.

We have chosen to focus heavily on material in statistical signal processing and estimation theory. Whether one takes a data-driven or physics-based approach to the SHM problem, a thorough understanding of how to model and describe a temporal sequence of observations is absolutely essential. After all, it is hard to envision an SHM technique that is not based on this type of acquired data. In the absence of this understanding, it is fairly easy to wind up with a damage detection algorithm that is not detecting damage but rather some other effect (estimation error, input vibrations that are nonlinear, etc.). Moreover, by providing these details we explain a number of important aspects to signal processing and system identification, many of which are often implicitly assumed but seldom discussed. For example,

- Why does the analysis of structural response data focus so heavily on the power spectral density function? covariance? (*Answer:* Jointly Gaussian data model & Isserlis' theorem, pages 56 and section 3.3.4.)
- From the perspective of mean-square-error in a parameter estimate, what is the worst possible probability model for the contaminating noise? (*Answer:* Normal or Gaussian distribution, page 362.)
- Why does the Fourier transform of a stationary stochastic signal $x(t)$ not exist but the Fourier transform of the associated auto-covariance $R_{XX}(\tau)$ does? (*Answer:* For structural response data, $x(t)$ is not absolutely integrable while $R_{XX}(\tau)$ is, pages 63 and 72.)
- Why do we use the mean-square-error as a quantity to be minimized in estimation? Why not, for example, mean-cubed error? (*Answer:* This choice depends entirely on the signal noise model, in fact, sometimes we should *not* be using mean-square-error!, page 336, 362.)
- Is the discrete Fourier transform a good approximation to the continuous Fourier integral? Under what conditions is this true? (*Answer:* Yes, provided long time-sequences and a jointly Gaussian noise model, page 228.)
- When estimating probabilities of correct classification, we often simply add up the number we got correct and divide by the total number of trials. Is this a good estimator? (*Answer:* Yes, under a very specific probability model. Other estimators can yield even more information, see, for example, classification example in section 7.2.2.)

Each of these questions is to us a fascinating line of inquiry, some of which have caused us some difficulty over the years in our attempts to answer larger questions about system identification (in particular the identification of structural damage). As a result, this book is at least partially dedicated to shedding light on aspects of signal processing and estimation that are sometimes overlooked in the SHM field. Our hope is that by providing clear explanations and some mathematical rigor we can spare the reader the time spent digging for answers to such questions. In the end we have found that effort spent trying to understand how to model, process, and predict structural response data is well worth it, regardless of the specific field of inquiry.

The approach we have taken is to therefore provide a thorough treatment of the material we have used in our applications at the exclusion of other, related material. For example, auto-regressive models are ubiquitous in signal processing and system identification. However, we have not used them in our particular approaches to the modeling and identification of structural damage; hence, little of this material appears here. Conversely, we make heavy use of the higher order spectra in damage detection; hence, a great deal of information on these quantities is provided.

The basic outlines of the book are as follows. A thorough description of the probabilistic modeling tools is provided in Chapters 2 and 3 followed by a detailed description of the basic principles used in modeling damage mechanics (Chapters 4 and 5). Taken together, these chapters provide both the deterministic and probabilistic models we require to describe our data. Next, we focus on the problem of estimation as it pertains to the statistical properties of structural response data. Understanding these properties is essential to understanding our approaches to the damage detection problem. Chapter 6 therefore develops and then applies these estimators to the output of specific structural models. The subject of structural parameter estimation is then described in Chapter 7. At this point, we have all of the tools needed to model, detect, and identify structural damage. Chapter 8 presents several examples of the detection problem, while Chapter 9 tackles the identification of specific damage-related parameters. We conclude Chapter 10 with some work designed to take what we know about the structure and make decisions regarding optimal structural maintenance. In the end, we will have provided the reader with a bottom-to-top approach to the types of modeling and estimation problems that he/she is likely to encounter in studying the physics of structural damage.

References

[1] A. Einstein, On the method of theoretical physics, Philosophy of Science 1 (2) (1934) 163–169.

[2] L. Cohen, The history of noise, IEEE Signal Processing Magazine 22 (6) (2005) 20–45.

[3] J. M. Nichols, M. Seaver, S. T. Trickey, K. Scandell, L. W. Salvino, E. Aktaş, Real-time strain monitoring of a navy vessel during open water transit, Journal of Ship Research 54 (4) (2010) 225–230.

[4] S. Ackers, R. Evans, T. Johnson, H. Kess, J. White, D. E. Adams, P. Brown, Crack detection in a wheel end spindle using wave propagation via modal impacts and piezo actuation, in: T. Kundu (Ed.), Proceedings of the SPIE, Health Monitoring and Smart Nondestructive Evaluation of Structural and Biological Systems V, Vol. 6177, SPIE, Bellingham WA, USA, 2006, pp. 1–13.

[5] D. E. Adams, Health Monitoring of Structural Materials and Components: Methods With Applications, John Wiley & Sons, West Sussex, 2007.

[6] C. R. Farrar, K. Worden, Structural Health Monitoring: A Machine Learning Perspective, John Wiley & Sons, Inc., New York, 2012.

[7] A. Guemes, D. Balageas, C.-P. Fritzen, Structural Health Monitoring, John Wiley & Sons, Inc., Newport Beach, CA, 2008.

[8] J. E. Doherty, Nondestructive evaluation, in: A. S. Kobayashi (Ed.), Handbook on Experimental Mechanics, Prentice-Hall, Englewood Cliffs, NJ, 1987.

[9] J. K. C. Shih, R. Delpak, C. W. Hu, P. Plassmann, A. Wawrzynek, M. Kogut, Thermographic nondestructive testing damage detection for metals and cementitious materials, Imaging Science Journal 48 (2000) 33–43.

[10] H. T. Banks, M. L. Joyner, B. Wincheski, W. P. Winfree, Real time computational algorithms for eddy-current based damage detection, Inverse Problems 18 (2002) 795–823.

[11] D. Tuzzeo, F. L. di Scalea, Non-contact air-coupled ultrasonic guided waves for detection of hidden corrosion in aluminum plates, Journal of Research in Nondestructive Evaluation 13 (2002) 61–78.

[12] K. Worden, C. R. Farrar, G. Manson, G. Park, The fundamental axioms of structural health monitoring, Proceedings of the Royal Society of London A 463 (2007) 1639–1664.

[13] J. M. Nichols, C. J. Nichols, M. D. Todd, M. Seaver, S. T. Trickey, L. N. Virgin, Use of data-driven phase space models in assessing the strength of a bolted connection in a composite beam, Smart Materials and Structures 13 (2004) 241–250.

[14] H. Sohn, K. Worden, C. R. Farrar, Statistical damage classification under changing environmental and operational conditions, Journal of Intelligent Material Systems and Structures 13 (2002) 561–574.

[15] J. M. Nichols, P. L. Fackler, K. Pacifici, K. D. Murphy, J. D. Nichols, Reducing fatigue damage for ships in transit through structured decision making, Marine Structures 38 (2014) 18–43.

2

Probability

The subject of probability is all too often a source of confusion. Part of the problem is that probability theory can be a very nonintuitive subject, where the mathematics often disagrees with our heuristic understanding. Another issue is that probability theory is either not taught or is given scant treatment in certain science and engineering curricula. Unfortunately, our lack of familiarity with the subject often leads to incorrect statements when assigning probability to our experimental results. The goal of this section is to introduce probability, provide the necessary mathematical framework for assigning probability, and to clarify the interpretation of probabilistic statements. Special attention is given to establishing a clear, understandable notation that is used consistently throughout this book. While the notation may at first seem cumbersome, it eliminates much of the ambiguity that can sometimes accompany probabilistic models. Although we devote a good deal of attention to this topic, we feel it is necessary for the reader to have a thorough understanding of probabilistic models. This understanding is absolutely essential for performing estimation and therefore for the identification of structural damage.

Simply put, *probability theory provides us with a mathematical model for uncertainty*. As with any model, the end goal is prediction. Scientists and engineers are frequently interested in using probability to predict the outcomes of an experiment. Let's say we are interested in the ultimate strength of type 1018 annealed steel for a particular design application. We might want to make statements such as "given a certain set of conditions (e.g., room temperature), there is a 90% probability that the ultimate strength lies in the range $325 \leq S_u \leq 357$ MPa." [1] Clearly, this is a useful statement to be able to make. It says that under the specified conditions, we predict with 90% certainty that the true value of S_u is somewhere in the interval $[325, 357]$ MPa, that is, we have used a probabilistic model. Probability theory gives us the tools needed to arrive at such a model. In this book we focus on making probabilistic statements about structural model parameters, particularly those that relate to structural damage.

Perhaps not surprisingly, there are two primary schools of thought on probability; this dichotomy mirrors in some ways the "data-driven" versus "model-based" discussion in the previous chapter. Probability can be appropriately viewed as an empirical construct, equal

---

[1] This is not typically what is done. Usually, we are simply given a number for S_u in a table; how this number was obtained is largely left a mystery.

Modeling and Estimation of Structural Damage, First Edition. Jonathan M. Nichols and Kevin D. Murphy.
© 2016 John Wiley & Sons, Ltd. Published 2016 by John Wiley & Sons, Ltd.

to the number of occurrences m of a particular experimental outcome divided by the total number of experiments N. Our intuition tells us that as the number of experiments becomes large, the probability of that outcome should be close to the ratio m/N. This is certainly a valid interpretation and, in fact, the axioms of probability described in the next section can be deduced from this definition. We can immediately see how this approach could give rise to useful results in structural parameter estimation.

Consider the frequently used approach to developing confidence intervals for S_u in our annealed steel example. The typical approach would be to repeat an experiment some number of times and quantify our uncertainty based on the resulting spread of outcomes. Thus, we might perform an experiment to determine S_u on 100 different samples and generate a confidence interval based on the central 90 values (discard the highest and lowest five outcomes for S_u). Let's say those 90 values spanned the interval $[325, 357]$ MPa. We might be tempted to make the desired probabilistic statement, that is, that there is a 90% probability that $325 \leq S_u \leq 357$ MPa. Unfortunately, this would not be technically accurate. All we can say is that using this specific procedure for determining S_u, we would expect 90% of the values to be in the range $[325, 357]$ Mpa. This range doesn't necessarily say anything about the true underlying value of S_u, but instead speaks of the variability in the process of acquiring the 100 separate outcomes. Put another way, intervals formed through replication have to do with the machinery used to produce the outcomes and not necessarily the probability of the outcome. This general approach to probabilistic modeling is often referred to as a *frequentist* approach, so named because inference is developed on the basis of repetition, that is, the frequency with which results in certain intervals are obtained from prior observations. Note also that by defining probability this way, one can only model experiments where previous data are available.

Instead, consider a different viewpoint, one that views probability as a model that may be specified in the absence of prior data. For example, if we were testing a material similar in molecular composition and treatment to our annealed steel, we might predict that this new material would have an ultimate strength lying in a similar range. Even without conducting the experiment, we can make this prediction based on what we know about the physics of the problem. We certainly wouldn't say that the probability of S_u for this new material lying in the range $[325, 357]$ is zero simply because we had not previously conducted any trials ($m = N = 0$)! This alternate viewpoint relies exclusively on the axiomatic definition of probability outlined in the next section. Applying this viewpoint to modeling and estimation problems is often labeled a "Bayesian" approach, after an original advocate, Thomas Bayes. Even without access to much data, we may still use this approach to develop credible intervals (Bayesian intervals of confidence) for the parameters of interest.

Our objective here is not to assign right or wrong to these two views of probability, but rather to clearly differentiate between them, highlighting strengths/weaknesses, and show how they both can be used to draw useful inference about structural damage. In fact, a fair question might be who cares? Does it matter whether our statements about S_u were formed using a true probabilistic model or were obtained solely through replication? We demonstrate in later chapters that this distinction can sometimes be important and not simply a matter of philosophical debate. However, both viewpoints lead to powerful methods for drawing inferences from observed data. We therefore develop both frequentist and Bayesian estimation approaches and use them in the experimental examples provided in later chapters. Both require the fundamental understanding of probability we attempt to provide in Section 2.1.

Finally, before discussing probability in more detail, it is important to also understand statistics and how they relate to probability. Statistics are descriptive rather than predictive. In the same way that engineers use "stiffness" as a parameter in many deterministic structural models (see Chapter 4), statistics can be thought of as parameters that describe a probabilistic model. For example, we may use the estimated variance of a set of observations to quantify the variability arising from fluctuations in sensor readings, ambient conditions, and so on. The variance is a parameter that typically describes the width of a probability model and therefore speaks of how uncertain we are about the outcome. Statistics are extremely important in describing our probabilistic model and will play an important role in identifying properties of our structural response data that are indicative of structural damage (see Chapter 8). A number of important statistics associated with common probability models are therefore also described in this chapter.

2.1 Probability Basics

The modern mathematical framework for developing probabilistic models was put forward in the 1930s by Kolmogorov. In an early monograph, appropriately titled "Foundations of Probability" [1], Kolmogorov lays out the axioms for the theory of probability. The resulting formalism is powerful and underlies all of the approaches to estimation discussed in this book. Although this section contains a few abstract concepts to which we seldom make direct reference, we feel that an understanding of the origins of probabilistic models is essential to understanding their predictive power. Our goal here is to introduce the reader to the foundations of probability theory and present it in the context of the types of problems faced by engineers. More complete treatments may be found in Rosenthal [2] or the original manuscript of Kolmogorov [1].

Simply stated, probability is used to model an experiment whose outcome is uncertain. This "experiment" could be the roll of a dice or, more likely in our case, the reading of a voltage from a sensor. We will henceforth denote the observed outcome of the experiment with a lower case letter, for example, "x." This is the data we will be collecting from the structure of interest.

We seek to predict the likelihood (probability) of certain experimental outcomes. For example, what is the probability of reading a voltage in the range $[0, 1]$V given a signal that is bounded by a digitizer to lie on the interval $[-5, 5]$V? To answer this question, we need to assign a probability to the subset $\sigma_1 = [0, 1] \subset [-5, 5]$. The mathematics of probability therefore require us to assign a number to a set of points that is a subset of the possible space of experimental outcomes. To do this, we require ideas and proofs from measure theory, which is a branch of mathematics that essentially deals with quantifying the "size" of sets of points. Our measure of "size" in this context is "probability," and hence mathematicians will often refer to the "probability measure." With this discussion in mind, we can turn to the formalism required to assign probability to an experimental outcome.

Our probabilistic model for any observation will be defined on a *probability space*, specified by the triplet $(\xi, \sigma, P(\sigma_i \in \sigma))$. The sample space is denoted by the nonempty set ξ and contains *all possible outcomes* of the experiment (i.e., what values can x possibly attain?). If we consider a single roll of a standard dice, the possible outcomes are $\xi = \{1, 2, 3, 4, 5, 6\}$. In the voltage example, the set might contain the real numbers in a certain voltage range, for example, $\xi = [-5, 5]$V.

The set $\boldsymbol{\sigma}$ is a set of *events*, or subsets of ξ, and we denote the ith such event $\sigma_i \subset \xi$. These are the subsets to which we would like to assign probability. It does not have to include *all* possible subsets, however we require of $\boldsymbol{\sigma}$ that it include both the null set $\emptyset = \{\}$ and ξ (all outcomes) as possibilities. It will also be required that

$$\text{if } \sigma_i \in \boldsymbol{\sigma}, \sigma_i^C \in \boldsymbol{\sigma} \tag{2.1a}$$

$$\sigma_1 \cup \sigma_2 \cup \cdots \in \boldsymbol{\sigma} \tag{2.1b}$$

$$\sigma_1 \cap \sigma_2 \cap \cdots \in \boldsymbol{\sigma} \tag{2.1c}$$

where σ_i^C is the complement of σ_i defined as $\sigma_i^C \equiv \xi - \sigma_i$. That is to say, for a collection of subsets $\sigma_i \in \boldsymbol{\sigma}$, then $\boldsymbol{\sigma}$ is closed under the formation of complements, countable unions, and countable intersections. These requirements are in place to ensure the existence of limits in developing probability theory [2]. The collection of sets obeying these properties is referred to as a σ-algebra. In keeping with the dice example, one possible σ-algebra is $\boldsymbol{\sigma} = \{\sigma_1 = \{2, 4, 6\}, \sigma_2 = \{1, 3, 5\}, \sigma_3 = \{1, 2, 3, 4, 5, 6\}, \sigma_4 = \emptyset\}$. In our voltage example, we might choose $\boldsymbol{\sigma} = \{\sigma_1 = [-5, 0), \sigma_2 = [0, 5], \sigma_3 = [-5, 5], \sigma_4 = \emptyset\}$. We will ultimately be interested in the probability that our sensor output lies in one or more of these subsets (ranges), that is, we seek to predict the probability of making a particular measurement.

Thus, the final element in our probability space, $P(\sigma_i)$, is a real-valued function that assigns a probability to each of the events $\sigma_i \in \boldsymbol{\sigma}$. The function $P(\cdot)$ is also required to obey several constraints

$$P(\sigma_i) \geq 0 \tag{2.2a}$$

$$P(\xi) = 1 \tag{2.2b}$$

$$P\left(\sum_i \sigma_i\right) = \sum_i P(\sigma_i) \text{ for } \sigma_i \subset \xi \quad \text{and} \quad \sigma_i \cap \sigma_j = \emptyset, i \neq j \tag{2.2c}$$

The first two statements match our intuition about assigning probability: the function is strictly nonnegative and the event containing all possible outcomes is assigned a probability of 1 (one of them *has* to occur in our experiment). The last statement says that $P(\cdot)$ is *countably additive*, that is, adding the probabilities of disjoint subsets is the same as taking the proba- bility of the sum (union) of those subsets. Again, this requirement exists to ensure that limits may be taken [2]. For the discrete situation, it is easy to construct $\boldsymbol{\sigma}$ and $P(\cdot)$ as there are clearly a countable number of subsets of ξ. For the sample σ-algebra given for our dice example, we have $P(\sigma_1) = 0.5, P(\sigma_2) = 0.5, P(\sigma_3) = 1, P(\sigma_4) = 0$. Similarly, in our voltage example (assuming all voltages are equally likely), we would have $P(\sigma_1) = 0.5, P(\sigma_2) = 0.5$, $P(\sigma_3) = 1$, and $P(\sigma_4) = 0$. However, in this more practical example, where the subsets are intervals of real numbers, there is a potential dilemma in creating a valid σ-algebra and in assigning probability.

We could have, for example, chosen $\boldsymbol{\sigma}$ to be comprised of one or more singleton subsets, where each $\sigma_i = \{s\}$ and "s" takes on a unique value in the specified range, that is, $s \in [-5, 5]$. The probability assigned to any *specific* value in the continuum is $P(\sigma_i = s) = 0$. Thus, if our set of events were comprised solely of singleton subsets $\boldsymbol{\sigma} = \bigcup_{s \in \xi} \{s\}$, we would not only have an *uncountably infinite* number of subsets but we would also have $P\left(\sum \sigma_i\right) = 1 \neq \sum P(\sigma_i) = 0$. This clearly violates our requirements for a probability space (see Eq. 2.2c). Fortunately, it turns out that one can always define $P(\cdot)$ on a specific kind of σ-algebra, the Borel algebra (defined as the smallest σ-algebra containing all open sets on the interval of interest [2]), and that this probability measure will possess the desired properties (including 2.2c). Moreover, the "extension theorem," given by Kolmogorov, states that we can always uniquely "extend" a given σ-algebra to a Borel algebra; hence, we can always define a unique probability measure to our experimental outcome, regardless of whether we are dealing with discrete or continuous random variables [1]. All this means is that technically we should refer to the probability measure as being with respect to the Borel algebra associated with our σ-algebra. This is certainly good news as we are often modeling our data as continuous random variables.[2] We will never have to explicitly construct such a probability space, however we will always implicitly rely on the fact that one exists.

Before concluding this section we need to introduce the concept of a *random variable*. These are the quantities that we will be using to model the specific measurements we make in an experiment. A random variable is, in general, a real-valued function $X(s \in \xi)$, defined everywhere on the set of possible outcomes. For example one might wish to consider $X(s) = a + bs$ for real constants a, b. However, in this book we always assume the identity $X(s \in \xi) = s$, that is to say, *the random variable is defined directly as an element of the set of possible outcomes our experiment could produce*. For this reason we will denote our random variable as simply X, with the implicit understanding that it is a function (albeit a simple one). Throughout this book, random variables are "upper case" letters, corresponding to the lower case "observation" it is being used to model, for example, the measurement x is modeled with X.

We may then consider an event σ associated with our random variable and the probability assigned to that event. For example, we might consider a measured voltage that can take any value $X \in \xi = [-5 \, \text{V}, 5 \, \text{V}]$ with uniform probability. Using the formalism of this section, we may assign a probability to the event $\sigma = [0, 5]$, which, for this example, is $P(0 \leq X \leq 5 \, \text{V}) = 0.5$. This is to be read as the probability that a number contained in our set of possible outcomes lies in the range $[0, 5 \, \text{V}]$. Because we specified uniform probability in the problem statement, we know that there is exactly a 50% chance that a member of this particular σ lies in the second half of the total voltage range. Armed with a proper probability space and random variables defined on that space, we are in a position to begin to form the kinds of probabilistic models required in later chapters.

2.2 Probability Distributions

We have just shown that regardless of whether we consider a discrete or a continuous random variable, we can always construct a proper probability space and define a random variable X, assumed to exist over the space of possible outcomes of our experiment, any one of which is denoted x. We have also shown that we can assign a probability to the event that our random variable falls in a certain range of values. Specifically, we assign the probability $P(a \leq X \leq b)$ to the event $\sigma \in [a, b]$ for real constants a, b.

---

[2] Of course, in practice this isn't quite accurate. The acquired voltage is the result of a digitization process so that, in fact, we are always working with discrete random variables where the formation of a probability space is trivial. However, our mathematical models will often treat the observations as continuous random variables, and hence the importance of Kolmogorov's work on the topic.

Using this construction, we can define the *cumulative distribution function* (CDF) associated with our random variable

$$P_X(x) \equiv P(X \leq x) \tag{2.3}$$

which is read as the probability that an element of the possible outcomes is less than the value "*x*" (our measurement). Before discussing the properties of this function, consider first the notation used in defining (2.3). The subscript is used throughout this book to denote that the CDF is defined for random variable X. By defining (2.3) as a function of the experimental outcome, we are implicitly assigning probability to the family of events defined by the set $\sigma(x) = [-\infty, x]$. With the function (2.3) we can also express a probability assigned to an arbitrary range as $P(a \leq X \leq b) = P_X(b) - P_X(a)$. The function $P_X(x)$ is sufficient to assign probability to any event on a real-number axis, thus (2.3) is a very compact way of representing a complete probabilistic model [1].

The CDF therefore plays a central role in the analysis of random variables. The function will always be limited by $P_X(-\infty) = 0$, $P_X(\infty) = 1$. We also have from axiom (2.2c)

$$P_X(x + dx) = P(X \leq x) + P(x < X \leq x + dx)$$
$$= P_X(x) + P(x < X \leq x + dx) \geq P_X(x) \tag{2.4}$$

so that the function is always monotonically increasing. Example CDFs for our discrete (dice roll) and continuous (voltage reading) examples are shown in Figure 2.1.

The data we acquire in structural vibration problems will nearly always be modeled as continuous random variables (a notable exception is the example given in Section 7.2.1 involving the multinomial distribution). Because a continuous random variable will always have a continuous CDF, we can guarantee the existence of its derivative [3]

$$p_X(x) \equiv \lim_{\Delta x \to 0} \frac{P(x < X \leq x + \Delta x)}{\Delta x} = dP_X(x)/dx \tag{2.5}$$

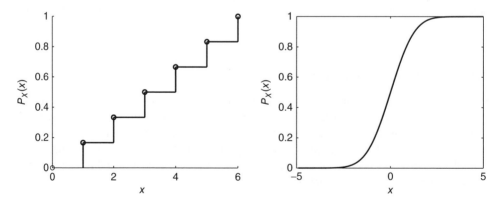

Figure 2.1 Cumulative distribution functions associated with the discrete random variable $X \in \{1, 2, 3, 4, 5, 6\}$ (used to model the roll of a fair dice) and a continuous random variable $X \in [-5, 5]$ (used to model a voltage measurement). The specific functional form of this latter probability model is given by substituting Eqn. 2.8 into Eqn. 2.6.

which is referred to as the *probability density function* or PDF. The PDF and CDF of a random variable are therefore related via

$$P_X(x) = \int_{-\infty}^{x} p_X(u)du. \tag{2.6}$$

Because $P_X(x)$ is monotonically increasing, we have $p_X(x) \geq 0$. The PDF of a random variable is a function with which we have some familiarity. This function can provide us with the likelihood of observing our random variable in a given interval $[a, b]$

$$\int_{a}^{b} p_X(x)dx = P(a \leq X \leq b). \tag{2.7}$$

Most of us have dealt with the normal or Gaussian PDF shown in Figure 2.2. This is the PDF associated with the CDF of Figure 2.1, and is given by

$$p_X(x) = \frac{1}{\sqrt{2\pi\sigma_X^2}} e^{-\frac{1}{2\sigma_X^2}(x-\mu_X)^2} \tag{2.8}$$

where μ_X, σ_X^2 are parameters characterizing this distribution, referred to as the mean and the variance, respectively. The values of the normal distribution near $x = \mu_X$ have a higher probability of occurrence than those near the extremes, while σ_X controls the width or "spread" of the distribution. Occasionally, we will use the notation $X \sim N(\mu_X, \sigma_X^2)$ to mean that the random variable X is normally distributed with mean μ_X and variance σ_X^2. A table listing other common distributions (and some of their properties) is provided in Appendix A. We should note that our example, Figure 2.2b, actually depicts a truncated Gaussian distribution. The Gaussian distribution has infinite support, however in our example we have restricted the random variable to the range $[-5, 5]$, as would be the case in an experiment. It is also worth mentioning that the PDF of a random variable is not unitless. The PDF will have units of $[x]^{-1}$ owing to

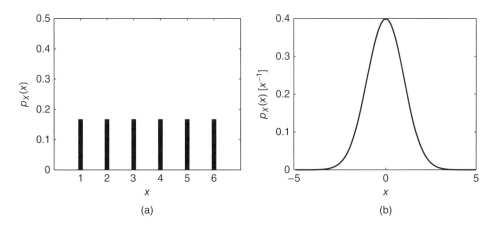

Figure 2.2 Probability mass (a) and distribution (b) functions associated with the discrete and continuous CDFs shown in Figure 2.1

the differential element "dx" in Eq. (2.5). Thus, to obtain a probability from a PDF one must integrate over the interval (i.e., the event range) of interest as in Eq. (2.7).

An analogous quantity to (2.5) can be defined for discrete distributions. The difference, of course, is that for discrete distributions it makes no sense to look at the probability of finding an observation in an interval $[x, x + dx]$, as the support of such a model is a discrete set of points $X = \{x_1, x_2, \cdots\}$. Thus, for a discrete random variable we instead define a *probability mass function* (PMF)

$$p_X(x) \equiv P(X = x) \tag{2.9}$$

which, unlike the PDF, is unitless. Here, as always, the laws of probability must apply; thus, $p_X(x) \geq 0 \; \forall \; x$ and $\sum_X p_X(x) = 1$. In analogy with Eq. (2.6), for a discrete random variable we have for the CDF

$$P_X(x_M) \equiv \sum_{m=1}^{M} p_X(x_m). \tag{2.10}$$

In the dice example, described by the CDF in Figure 2.1, the PMF is simply the constant $p_X(x) = \frac{1}{6}$ for $x = 1, 2, \cdots, 6$. An important example of detection and identification problems involving discrete PMFs is discussed in detail in Section (7.2.2).

2.3 Multivariate Distributions, Conditional Probability, and Independence

The discussion in the previous section can be easily extended to the multivariate case. Consider the two random variables X, Y, each representing a set of outcomes from different experiments.[3] Analogous to Eq. (2.3), we may define the *joint* CDF

$$P_{XY}(x, y) = P(X \leq x, Y \leq y) \tag{2.11}$$

that is, the probability that the random variable X is less than or equal to x and also that the random variable Y is less than y. In the case of continuous random variables, we may further define a multivariate PDF

$$p_{XY}(x, y) = \frac{\partial^2 P_{XY}(x, y)}{\partial x \, \partial y} \tag{2.12}$$

from which we may write, in analogy to Eq. (2.7),

$$P(a < X \leq b, c < Y \leq d) = \int_a^b \int_c^d p_{XY}(x, y) dy \, dx \tag{2.13}$$

which is the probability of the events $X \in (a, b]$ *and* $Y \in (c, d]$ occurring.

In the chapters to come, a multivariate (joint) PDF is often given, but the interest is more in the PDF associated with a subset of the constituent random variables. For example, we

---

[3] In this book, different "experiments" always refers to different data points acquired from the same physical experimental setup.

may be given $p_{XY}(x, y)$ but we want an expression for $p_X(x)$. To see how to relate these two distributions, we begin by noting that the laws of probability dictate

$$P_{XY}(x, \infty) = P(-\infty < X \le x, -\infty < Y \le \infty) = P_X(x)$$

$$P_{XY}(\infty, y) = P(-\infty < X \le \infty, -\infty < Y \le y) = P_Y(y) \quad (2.14)$$

owing to the fact that the event containing the set of all possible outcomes is assigned a probability of 1 (Eq. 2.2b). We additionally have that

$$p_X(x) = dP_X(x)/dx$$

$$= \frac{d}{dx} \left\{ \int_{-\infty}^{x} \int_{-\infty}^{\infty} p_{XY}(u, v) dv \, du \right\}$$

$$= \int_{-\infty}^{\infty} p_{XY}(x, v) dv$$

$$p_Y(y) = \int_{-\infty}^{\infty} p_{XY}(u, y) du. \quad (2.15)$$

These are the so-called *marginal distributions* associated with the joint density $p_{XY}(x, y)$. In short, we can always form a marginal density for a joint distribution of random variables by integrating out (with limits $-\infty, +\infty$) one of the random variables. It is worth pointing out that the notation for a marginal distribution is identical to that used when specifying any distribution for a single random variable. Whether $p_X(x)$ is called a "marginal distribution" or just a "distribution" depends entirely on the ancestry of the function (whether it was derived from a joint model) and not on anything having to do with the properties of the function.

It will also prove useful to define a notation for *conditional probability*, that is, the probability assigned to one event given knowledge of another. This concept is most easily illustrated with an abstract experiment; we will give a more practical example later. Consider a bag of 10 colored marbles, 4 red, 6 green. Model consecutive draws from this bag with random variables X and Y. Table 2.1 shows the joint probability associated with the four possibilities, red–red, red–green, green–red, green–green.

The key here is to note that after the first draw, whether it be red or green, the number of remaining marbles is decreased by one. With this in mind, the probability of choosing a green marble, $Y = green$, given that we have chosen a red in the first draw, $X = red$ is seen to be

Table 2.1 Probabilities associated with all possible outcomes of two consecutive draws from a bag known to contain four red marbles and six green

	Red	Green
Red	$\frac{4}{10} \times \frac{3}{9} = \frac{2}{15}$	$\frac{4}{10} \times \frac{6}{9} = \frac{4}{15}$
Green	$\frac{6}{10} \times \frac{4}{9} = \frac{4}{15}$	$\frac{6}{10} \times \frac{6}{9} = \frac{2}{5}$

given by

$$p_Y(\text{green}|\text{red}) = p_{XY}(\text{red}, \text{green})/p_X(\text{red}) = \frac{4}{15} \div \frac{4}{10} = \frac{2}{3}$$

where we have introduced a notation for conditional probability. We will henceforth use $y|x$ in the argument of $p_Y(\cdot)$ to denote the outcome of the experiment modeled by random variable Y as conditional on knowledge of X. Put another way, knowledge of the first experiment has changed our prediction for the second.

As a second example, consider consecutive rolls of the dice and let $X \in \{1, 2, 3, 4, 5, 6\}$ and $Y \in \{1, 2, 3, 4, 5, 6\}$ Assuming a fair dice, we have $p_X(x) = p_Y(y) = 1/6$ and $p_{XY}(x, y) = 1/36$ for any observed values of x and y. This yields $p_X(x|y) = 1/6$, that is to say, given that we have already observed a particular value y, then our probability of observing the value x is still $1/6$. This is just a mathematical statement that our previous roll has no bearing whatsoever on our current roll and ultimately stems from our equiprobability assumption on the outcomes of the dice. The laws of conditional probability can similarly be applied to case of continuous random variables [3], so that, in general, we may write

$$p_X(x|y) = \frac{p_{XY}(x, y)}{p_Y(y)}. \tag{2.16}$$

This is an extremely useful mathematical statement that shows up periodically in this book. It turns out that Bayes' rule (see Section 7.2), on which Bayesian estimation is based, is just a simple restatement of the law of conditional probability.

Finally, we briefly discuss what it means for two random variables to be independent. Random variables X and Y are said to be independent if

$$P_{XY}(x, y) = P_X(x)P_Y(y)$$

$$p_{XY}(x, y) = p_X(x)p_Y(y) \tag{2.17}$$

that is to say, the joint distribution and joint density factor. The conditional probability example involving consecutive rolls of a dice demonstrated this property. The assumption of independence can greatly simplify our probability models, as demonstrated in later chapters. This assumption will allow us to form the joint distribution of any number of random variables given the distribution of the constituent random variables.

We have now discussed probability for the occurrence of single and of multiple events. In addition, we have discussed how to form marginal distributions from joint distributions, conditional probability, and what it means for the random variables to be independent. Before we conclude this section, let's summarize with a quick example. Assume we have acquired a voltage from two different sensors using a data acquisition system. Our system is set so that sensor 1 registers a voltage on the interval $[-5, 5]$V and the sensor 2 voltage is on the interval $[-10, 10]$V. From Section 2.1 we already know how to form a proper probability space and assign probabilities to various events on that space. Now we wish to treat the acquired samples as random variables $X \in [-5, 5]$ and $Y \in [-10, 10]$. We know, of course, that $P(-5 \leq X \leq 5) = P(-10 \leq Y \leq 10) = 1$. We also know that if we treat these random variables as continuous, we may form $p_X(x), p_Y(y), p_{XY}(x, y)$. The actual form of these PDFs will depend on how we choose to model the random variables, that is, we have to assume something about the likelihood of finding values in certain subintervals of our space of voltages. Specific models for $p_X(x), p_Y(y), p_{XY}(x, y)$ are discussed subsequently. In addition, if we are given $p_{XY}(x, y)$, we can always find the marginal distributions $p_X(x) = \int_{-10}^{10} p_{XY}(x, v) dv$ and $p_Y(y) = \int_{-5}^{5} p_{XY}(u, y) du$. We might also choose to look at the distribution for random variable X given that we have already observed $-5 \le Y \le 5$ V, that is, look at $p_X(x|y) = p_{XY}(x, y)/p_Y(y)$. Finally, if we choose to model these acquired voltages as independent, we may write $p_{XY}(x, y) = p_X(x) p_Y(y)$.

Everything we have just discussed for two random variables extends directly to sequences of random variables. Consider the N random variables $\mathbf{X} \equiv (X_1, X_2, \cdots, X_N)$ which we will use to model the observations $\mathbf{x} = (x_1, x_2, \cdots, x_N)$. In our notation we will henceforth use boldface type to denote a vector. Just as with the $N = 2$ case, we would like to make probabilistic statements about joint events. In fact, our focus in Chapter 3 is on the joint CDF

$$P_{\mathbf{X}}(X_1 \le x_1, X_2 \le x_2, \cdots, X_N \le x_N) \tag{2.18}$$

and joint PDF

$$p_{\mathbf{X}}(x_1, x_2, \cdots, x_N) \tag{2.19}$$

where the latter can be expressed $p_{\mathbf{X}}(x_1, x_2, \cdots, x_N) = \frac{\partial^N P_{\mathbf{X}}(X_1, X_2, \cdots, X_N)}{\partial x_1 \partial x_2 \cdots \partial x_N}$ in cases where the CDF is differentiable. Equation (2.19) is a model that describes the probability of having observed the particular sequence $x_1, x_2, \cdots, x_N$ and will play a prominent role in estimating damage-related parameters. Using this notation, in analogy to Eqs. (2.15), we can define the marginal density associated with the ith random variable in the sequence

$$p_{X_i}(x_i) = \int_{\mathbb{R}^{N-1}} p_{X_1 X_2 \cdots X_N}(x_1, x_2, \cdots, x_N) dx_1 dx_2 \cdots dx_{i-1} dx_{i+1} \cdots dx_N \tag{2.20}$$

where the notation $\mathbb{R}^{N-1}$ denotes the multidimensional (with infinite limits) integral over all variables other than x_i (see Eq. 3.76 for a more detailed explanation of this notation).

Finally, we might also consider two vectors of random variables $\mathbf{X}$ and $\mathbf{Y}$. The marginal density for the vector $\mathbf{X}$ given the joint density $P_{\mathbf{XY}}(\mathbf{X}, \mathbf{Y})$ is simply

$$p_{\mathbf{X}}(\mathbf{x}) = \int_{-\infty}^{\infty} p_{\mathbf{XY}}(\mathbf{x}, \mathbf{y}) d\mathbf{y} \tag{2.21}$$

(which is itself a joint density). Likewise, statements of conditional probability and independence can be written

$$p_{\mathbf{X}}(\mathbf{x}|\mathbf{y}) = p_{\mathbf{XY}}(\mathbf{x}, \mathbf{y})/p_{\mathbf{Y}}(\mathbf{y}) \tag{2.22}$$

and

$$p_{\mathbf{XY}}(\mathbf{x}, \mathbf{y}) = p_{\mathbf{X}}(\mathbf{x}) p_{\mathbf{Y}}(\mathbf{y}) \tag{2.23}$$

respectively.

2.4   Functions of Random Variables

Given a continuous random variable X with PDF $p_X(x)$, we seek an expression for the PDF of a new random variable $Y = h(X)$. This is a common situation in signal processing and will be of particular use in developing techniques for damage detection. The transformation of random variables will play a role in (among other places) testing for nonstationarity (Section 6.8.2), developing test statistic distributions in damage detection (Section 8.3), and in establishing the surrogate data approach used in detecting damage-induced nonlinearities in structures (Section 8.6).

First, consider the case where $h(\cdot)$ is a real, one-to-one, continuous function of its argument. For infinitesimal increments $\Delta x, \Delta y$, we may write $P(x < X \leq x + \Delta x) = P(y < Y \leq y + \Delta y)$, that is to say, the probability of finding the random variable X in the interval $[x, x + \Delta x]$ is the same as that for Y on $[y, y + \Delta y]$ for small $\Delta x, \Delta y$. In terms of the PDFs of X and Y, we may therefore write

$$p_Y(y)\Delta y \approx P(y < Y \leq y + \Delta y) = P(x < X \leq x + \Delta x) \approx p_X(x)\Delta x \qquad (2.24)$$

so that

$$p_Y(y) = \lim_{\Delta x, \Delta y \to 0} p_X(x)/|(\Delta y/\Delta x)|$$

$$= p_X(x = h^{-1}(y)) \left| \frac{dy\,(x = h^{-1}(y))}{dx} \right|^{-1}$$

$$= p_X(h^{-1}(y))|h'(h^{-1}(y))|^{-1}. \qquad (2.25)$$

The absolute value in the denominator ensures that the result is nonnegative, as required of a PDF. In addition, because the left-hand side is a function of the random variable Y, the right-hand side must be evaluated at $x = h^{-1}(y)$. Equation (2.25) says that for a one-to-one function, knowing both the derivative and the inverse is sufficient to construct the PDF for Y given the PDF for X. For other well-behaved continuous functions, for example, $y = x^2$, the situation is only slightly more complicated. It can be shown that for continuous functions that are not one-to-one, that is, multiple values in the domain map to a single value in the range, Eq. (2.25) becomes

$$p_Y(y) = \sum_{k=1}^{n_x} p_X(h_k^{-1}(y))|h'(h_k^{-1}(y))|^{-1} \qquad (2.26)$$

where n_x is the number of regions in the domain that maps to a single value in the range and $x = h_k^{-1}(y)$ are the inverse mappings for each branch of the function.

As a useful example, we may consider the general case where we measure a random variable Y that is a polynomial function of a standard Gaussian random variable X, $p_X(x) = \frac{1}{\sqrt{2\pi\sigma_X^2}}e^{-x^2/(2\sigma_X^2)}$ for real, non-zero constant σ_X. Specifically, we consider the transformation $y = a_0 + a_1 x + a_2 x^2$. Clearly, $n_x = 2$ for this quadratic equation and we can write both branches of the inverse as

$$
x = \begin{cases} h_1^{-1}(y) = -\dfrac{a_1 + \sqrt{a_1^2 + 4a_2 y - 4a_0 a_2}}{2a_2} & \text{if} \quad x \le -\dfrac{a_1}{2a_2} \\[3mm] h_2^{-1}(y) = -\dfrac{a_1 - \sqrt{a_1^2 + 4a_2 y - 4a_0 a_2}}{2a_2} & \text{if} \quad x > -\dfrac{a_1}{2a_2} \end{cases}
$$

Note that the inverse is only real-valued for $y \ge \dfrac{-a_1^2 + 4a_0 a_2}{4a_2}$ and our solution will therefore be restricted to this domain. Figure 2.3 shows both branches of the inverse transformation for the parameters $a_o = 1.177, a_1 = 0.663, a_2 = 0.079$. The expression for $p_Y(y)$ also requires the derivative

$$
h'(x) = a_1 + 2xa_2
$$

evaluated along both branches of our function, giving

$$
h'_{1,2}(x = h_{1,2}^{-1}(y)) = \pm\sqrt{a_1^2 + 4ya_2 - 4a_0 a_2} \; .
$$

Thus, our final expression for the transformed density is given by

$$
p_Y(y) = \frac{1}{\sqrt{2\pi\sigma_X^2}\left|a_1^2 + 4ya_2 - 4a_0 a_2\right|} \left\{ e^{-\dfrac{\left(a_1 - \sqrt{a_1^2 + 4ya_2 - 4a_0 a_2}\right)^2}{8a_2^2\sigma_X^2}} + e^{-\dfrac{\left(a_1 + \sqrt{a_1^2 + 4ya_2 - 4a_0 a_2}\right)^2}{8a_2^2\sigma_X^2}} \right\}
$$

$$(2.27)$$

Figure 2.3 shows this density plotted for the parameters $a_o = 1.177, a_1 = 0.663, a_2 = 0.079, \sigma_X = 1$ over the valid range of the random variable Y.

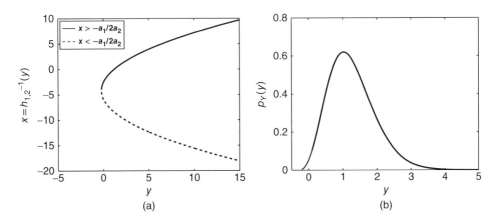

Figure 2.3 Inverse transformations $h_{1,2}^{-1}(y)$ used in deriving $p_Y(y)$ given that $p_X(x) \sim N(0, \sigma_X^2)$ and $Y = a_0 + a_1 X + a_2 X^2$. For this figure, the parameter values $a_0 = 1.177, a_1 = 0.663, a_2 = 0.079$ were used. This transformation exists (is real-valued) only on the range $y \in [\frac{-a_1^2 + 4a_0 a_2}{4a_2}, \infty]$, and hence this range is the support of $p_Y(y)$, shown in (b)

Equation (2.27) actually represents an important class of distributions, that is, those distributions that can be formed as static (doesn't depend on time), nonlinear transformations of a Gaussian distribution. A large number of distributions can be well-approximated using this strategy. In fact, the example parameters were chosen so as to approximate a Rayleigh distribution. Still higher order polynomial transformations can produce other, more exotic looking distributions (see, e.g., [4]). A similar transformation will prove useful in Section 6.5.3 in deriving the nonlinear bicoherence function and later in Section 8.4.3 in deriving higher order statistical properties of a nonlinear system response in a damage detection application.

We may extend the given approach to vectors of random variables, in which case Eq. (2.25) becomes

$$p_{\mathbf{Y}}(\mathbf{y}) = p_{\mathbf{X}}(\mathbf{h}^{-1}(\mathbf{y}))|\mathbf{J}(\mathbf{h}^{-1}(\mathbf{y}))|^{-1} \tag{2.28}$$

where $\mathbf{J}$ is the Jacobian of the transformation, defined as the matrix

$$\mathbf{J} \equiv J_{ij} = \frac{\partial y_i}{\partial x_j} \tag{2.29}$$

for $i, j = 1 \ldots K$ random variables and where $|\cdot|$ takes the determinant.

As a second example, assume we are interested in a random variable Y that can be expressed as a linear combination of two independent, normally distributed random variables, i.e., $Y_1 = h_1(X_1, X_2) = a_1X_1 + a_2X_2$ for a_1, a_2 real constants and $X_1 \sim N(\mu_1, \sigma_1), X_2 \sim N(\mu_2, \sigma_2)$. In this case, the solution isn't immediately obvious as we are combining two random variables into a single random variable; therefore, we cannot directly apply Eq. (2.28). However, we may define the auxiliary random variable $Y_2 = h_2(X_1, X_2) = a_1X_1 - a_2X_2$ and solve for the joint distribution

$$p_{Y_1Y_2}(y_1, y_2) = p_{X_1X_2}(h_1^{-1}(x_1, x_2), h_2^{-1}(x_1, x_2)) \begin{vmatrix} \frac{\partial Y_1}{\partial X_1}\big|_{h_1^{-1}(X_1)} & \frac{\partial Y_1}{\partial X_2}\big|_{h_1^{-1}(X_2)} \\ \frac{\partial Y_2}{\partial X_1}\big|_{h_2^{-1}(X_1)} & \frac{\partial Y_2}{\partial X_2}\big|_{h_2^{-1}(X_2)} \end{vmatrix}^{-1}$$

$$= p_{X_1}((y_1 + y_2)/2a_1)p_{X_2}((y_1 - y_2)/2a_2) \left| \begin{pmatrix} a_1 & a_2 \\ a_1 & -a_2 \end{pmatrix} \right|^{-1}$$

$$= \frac{1}{4a_1a_2\pi\sigma_1\sigma_2} e^{-\frac{(y_1+y_2-2a_1\mu_1)^2}{8a_1^2\sigma_1^2} - \frac{(y_2-y_1+2a_2\mu_2)^2}{8a_2^2\sigma_2^2}}. \tag{2.30}$$

Note that in the second line we have factored the joint PDF $p_{X_1X_2}(x_1, x_2) = p_{X_1}(x_1)p_{X_2}(x_2)$ following the assumption that the random variables X_1, X_2 are independent. Note also that we solved for both X_1, X_2 in terms of Y_1, Y_2 before substitution. However, what we are interested in is the marginal distribution

$$p_{Y_1}(y_1) = \int_{Y_2} p_{Y_1 Y_2}(y_1, y_2) dy_2$$

$$= \frac{1}{a_1 a_2 \sigma_1 \sigma_2 \sqrt{2\pi} \sqrt{\frac{1}{a_1^2 \sigma_1^2} + \frac{1}{a_2^2 \sigma_2^2}}} e^{-\frac{(y_1 - a_1 \mu_1 - a_2 \mu_2)^2}{2\left(a_1^2 \sigma_1^2 + a_2^2 \sigma_2^2\right)}}$$

$$= \frac{1}{\sqrt{2\pi} \sqrt{a_1^2 \sigma_1^2 + a_2^2 \sigma_2^2}} e^{-\frac{(y_1 - a_1 \mu_1 - a_2 \mu_2)^2}{2\left(a_1^2 \sigma_1^2 + a_2^2 \sigma_2^2\right)}}. \tag{2.31}$$

This is a useful expression that occurs frequently in engineering applications. Often times, sources of error, say manufacturing tolerances on multiple components of a piece of machinery, are assumed normally distributed with a given mean (usually zero) and variance. The manufacturer may be interested in the alignment error for the entire machine given the alignment error in the components. He/she may also wish to add safety factors $a, b > 1$ before determining the final error distribution. In this case, the total error might be modeled as the weighted sum of the individual error components, which is precisely what we have modeled in Eq. (2.31). The result is that the total error is also normally distributed but with variance $\sigma_{tot}^2 = a_1^2 \sigma_1^2 + a_2^2 \sigma_2^2$. This result extends to the sum of K normally distributed random variables, that is, $\sigma_{tot}^2 = \sum_{k=1}^{K} a_k^2 \sigma_k^2$. This is a common way of combining errors, however it is *only* valid under the assumptions that (i) the total error can be expressed as a linear combination of the constituent errors, that (ii) the underlying PDF for each of the constituent errors is Gaussian, and (iii) that those errors are independent. Incidentally, we have just demonstrated that a linear combination of independent, normally distributed random variables will also be normally distributed. This fact is seen to be useful in later chapters (see, e.g., 6.7).

This represents a common situation: we are given the PDFs of two random variables and seek the distribution of their sum. Other common PDFs of interest include the difference, product, and quotient of random variables. To use Eq. (2.28), we had to define an auxiliary random variable ($Y_2 = X_1 - X_2$), find the joint distribution $p_Y(\mathbf{y})$, and then take the marginal distribution by integrating over Y_2, leaving us with $p_{Y_1}(y_1)$. There exists a slightly different approach to this problem, however, that yields a simpler form for the transformation. Recall that for $Y = X_1 + X_2$, we have the equality

$$P_Y(y) = P(Y < y) = P(X_1 + X_2 < y). \tag{2.32}$$

The right-hand side could also have been written in terms of the joint PDF of X_1, X_2 as

$$P_Y(y) = \int_{-\infty}^{\infty} \left[\int_{-\infty}^{y - x_1} p_{X_1 X_2}(x_1, x_2) dx_2 \right] dx_1. \tag{2.33}$$

We also have by definition

$$p_Y(y) = \frac{d}{dy}P_Y(y) = \frac{d}{dy}\int_{-\infty}^{\infty}\left[\int_{-\infty}^{y-x_1}p_{X_1X_2}(x_1,x_2)dx_2\right]dx_1$$

$$= \int_{-\infty}^{\infty}\frac{d}{dy}\left[\int_{-\infty}^{y-x_1}p_{X_1X_2}(x_1,x_2)dx_2\right]dx_1$$

$$= \int_{-\infty}^{\infty}p_{X_1X_2}(x_1,y-x_1)dx_1. \tag{2.34}$$

If the random variables X_1, X_2 are further assumed independent, we have

$$p_Y(y) = \int_{-\infty}^{\infty}p_{X_1}(x_1)p_{X_2}(y-x_1)dx_1. \tag{2.35}$$

Returning to the previous example, we seek the sum of the random variables a_1X_1 and a_2X_2 given that both X_1, X_2 are normally distributed with mean and standard deviation $\mu_{1,2}, \sigma_{1,2}$, respectively. First, it is easy to show using Eq. (2.25) that if $X \sim N(\mu_X, \sigma_X^2)$, then $aX \sim N(a\mu_X, a^2\sigma_X^2)$ (see Section 2.6 for an alternate derivation). Next, we may use Eq. (2.35) to solve for the PDF of the sum

$$p_Y(y) = \int_{-\infty}^{\infty}\frac{1}{a_1\sigma_1\sqrt{2\pi}}e^{-\frac{(x_1-a_1\mu_1)^2}{2a_1^2\sigma_1^2}}\frac{1}{a_2\sigma_2\sqrt{2\pi}}e^{-\frac{(y-x_1-a_2\mu_2)^2}{2a_2^2\sigma_2^2}}dx_1 \tag{2.36}$$

Carrying out this integral returns the same result as Eq. (2.31). Using this same approach, we can derive the PDFs for a number of other common cases as well. Specifically, for independent random variables X_1, X_2, we have [5]:

Find $p_Y(y)$ given $Y = X_1 + X_2$ and $p_{X_1}(x_1), p_{X_2}(x_2)$.

$$p_Y(y) = \int_{-\infty}^{\infty}p_{X_1}(x_1)p_{X_2}(y-x_1)dx_1 \tag{2.37}$$

Find $p_Y(y)$ given $Y = X_2 - X_1$ and $p_{X_1}(x_1), p_{X_2}(x_2)$.

$$p_Y(y) = \int_{-\infty}^{\infty}p_{X_1}(x_1)p_{X_2}(y+x_1)dx_1 \tag{2.38}$$

Find $p_Y(y)$ given $Y = X_1 \times X_2$ and $p_{X_1}(x_1), p_{X_2}(x_2)$.

$$p_Y(y) = \int_{-\infty}^{\infty}p_{X_1}(x_1)p_{X_2}(y/x_1)\frac{1}{|x_1|}dx_1 \tag{2.39}$$

Find $p_Y(y)$ given $Y = X_2/X_1$ and $p_{X_1}(x_1), p_{X_2}(x_2)$.

$$p_Y(y) = \int_{-\infty}^{\infty}p_{X_1}(x_1)p_{X_2}(yx_1)|x_1|dx_1 \tag{2.40}$$

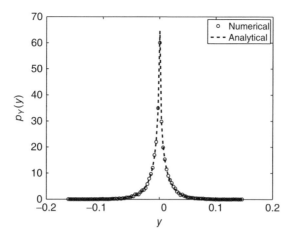

Figure 2.4 Analytical expression (2.42) plotted along with the estimated PDF obtained via Monte Carlo simulation. Parameters were chosen as $\sigma_1 = 0.1, \sigma_2 = 0.2$

Each of these integrals is assumed to extend over the range of Y, which, of course, depends functionally on the range of X_1, X_2. These formulas can be surprisingly difficult to implement in practice for even seemingly simple cases. For example, consider the product $Y = X_1 X_2$, where $X_1 \sim N(0, \sigma_1), X_2 \sim N(0, \sigma_2)$. Finding $p_Y(y)$ requires the evaluation of Eq. (2.39)

$$p_Y(y) = \int_{-\infty}^{\infty} \frac{1}{2\pi\sigma_1\sigma_2|x_1|} e^{-\frac{x_1^2}{2\sigma_1^2} - \frac{(y/x_1)^2}{2\sigma_2^2}} dx_1. \tag{2.41}$$

Carrying out the integration results in

$$p_Y(y) = \frac{K_0\left(\frac{|y|}{\sigma_1\sigma_2}\right)}{\sigma_1\sigma_2} \tag{2.42}$$

where $K_n(v)$ is the modified Bessel function of the second kind of order n and argument v. This PDF given by (2.42) is depicted in Figure 2.4. Also shown are the results of a Monte Carlo simulation, whereby 15,000 independent draws of both X_1, X_2 (with standard deviations $\sigma_1 = 0.1, \sigma_2 = 0.2$, respectively) were multiplied to give $Y = X_1 X_2$. The results were histogrammed and normalized by the histogram bin width and number of data to give the numerically estimated PDF shown in the figure (see Section 6.3 for a discussion of histogram estimators of a PDF).

Before leaving this section, we consider one other transformation of a continuous random variable X that is of some practical utility in the simulation of random processes. Denote the CDF for this variable as $P_X(x)$. It can be shown ([6], page 27) that the observation

$$y = P_X(x) \tag{2.43}$$

has a uniform distribution on the interval $[0, 1]$. This makes some intuitive sense as the fraction of the domain of a CDF occupied by a random variable is inversely proportional to its probability of occurrence. That is to say, more likely values of X fall in a narrower domain

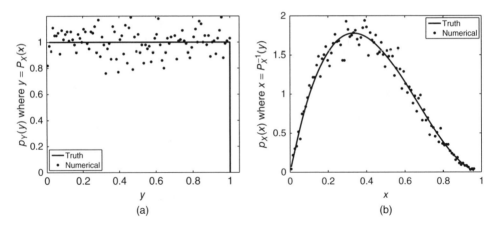

Figure 2.5 (a) $N = 10,000$ values of a Beta-distributed random variable X are mapped to the uniformly distributed random variable Y by applying the transformation $y = P_X(x)$, where $P_X(\cdot)$ is the Beta CDF with parameters $\lambda = 2, \eta = 3$. As predicted by Eq. (2.43), the values y are uniformly distributed, as seen from their normalized histogram. (b) Analytical expression for the Beta distribution along with a normalized histogram of the $N = 10,000$ draws $x = P_X^{-1}(y)$, where each y is taken from a uniform distribution. This technique can be used to generate random numbers from any continuous probability distribution for which we can write an inverse CDF

of the CDF function (see, e.g., Figure 2.1), that is, where the derivative (the PDF) is largest. Conversely, values of X that are unlikely occupy a large domain (the tails of the distribution). Equation (2.43) therefore takes the few low probability values on the domain and concentrates them in a fairly narrow range. At the same time, it takes the many high probability values in the narrow domain and spreads them out over the corresponding range. The mapping therefore has the effect of "boosting" draws from the tail while diluting the high probability values.

The effects of this transformation can be seen in Figure 2.5a, which shows $N = 10,000$ values of a random variable X drawn from the Beta PDF with parameters $\lambda = 2, \eta = 3$ (see Table A.2) and transformed according to 2.43 using the corresponding Beta CDF. Clearly, the resulting values are well approximated by the uniform distribution, as predicted. Perhaps of more practical utility is the fact that (2.43) also implies that if

$$x = P_X^{-1}(y), Y \sim U(0, 1) \tag{2.44}$$

then X will have the CDF $P_X(\cdot)$. This gives us an excellent prescription for generating random numbers from any probability distribution for which we can write down an expression for the CDF. For example, if we want N values of a continuous random variable with PDF $p_X(x)$, we simply generate N values, y, from the uniform distribution and then take for each realization $x = P_X^{-1}(y)$. As an example, we generate $N = 10,000$ values of a uniform distribution and use these values to generate values from the same Beta distribution (with parameters $\lambda = 2, \eta = 3$) by applying the inverse Beta CDF. The results are shown in Figure 2.5. This particular transformation is used in Chapter 6 in the generation of random processes with particular joint statistical properties.

2.5 Expectations and Moments

We now turn our intention to the different properties that can be used to describe a probability distribution. Perhaps the most common descriptor is the expected value of a distribution. Given a continuous random variable X and the associated probability distribution $p_X(x)$, define the expected value as

$$E[X] = \int_X x p_X(x) dx \qquad (2.45)$$

where the integral is taken over the support of the random variable and is denoted as $\int_X$. Thus, we can think of the expected value operator as a weighted average of the random variable X, where the weightings are given by the values of the PDF. For example, if we carry out the integral in Eq. (2.45) for the normal distribution of Eq. (2.8), we get $E[X] = \mu_X$. We can similarly define expectations (average properties) for products of different random variables

$$E[XY] = \int_Y \int_X xy p_{XY}(x, y) dx\, dy$$

$$E[XYZ] = \int_Z \int_Y \int_X xyz p_{XYZ}(x, y, z) dx\, dy\, dz$$

$$\vdots \qquad (2.46)$$

or in the case of powers of a single random variable

$$E[X^2] = \int_X x^2 p_X(x) dx$$

$$E[X^3] = \int_X x^3 p_X(x) dx$$

$$\vdots \qquad (2.47)$$

Expectations are useful descriptors of a given probability distribution. They can provide information about the spread of a distribution, how asymmetric a distribution is, and so on. The definition can be further generalized to consider functions of random variables. For example, given a random variable X with probability distribution $p_X(x)$, we may write for the expectation of $g(X)$

$$E[g(X)] = \int_X g(x) p_X(x) dx. \qquad (2.48)$$

Assume that we have a random variable X that we use to model the error in a given sensor measurement (i.e., noise). Further assume that the error has a zero mean, that is, $E[X] = 0$. Regardless of the specific PDF used to model this error, we can get a sense for the error magnitude by studying $E[X^2]$. In fact, many approaches to estimation are specifically designed to minimize this quantity, as we will see later. To take another more specific example, assume again that the random variable X is normally distributed but with a nonzero mean (Eq. 2.8). In this case, carrying out the integrals given by the first and second lines of Eq. (2.47), we get $E[X^2] = \mu_X^2 + \sigma_X^2$ and $E[X^3] = \mu_X^3 + 3\mu_X\sigma_X^2$. However, if X is first transformed to be zero-mean, these expectations become $E[(X - \mu_X)^2] = \sigma_X^2$ and $E[(X - \mu_X)^3] = 0$, respectively.

With these expression in mind, we may also define the *central moments* associated with a random variable

$$E[(X - E[X])^n] = \int_X (x - \mu_X)^n p_X(x)dx \tag{2.49}$$

For example, taking $n = 2$, it is common to define

$$\sigma_X^2 \equiv E[(X - E[X])^2] \tag{2.50}$$

as the variance of a given distribution. In keeping with the parameterization of the normal distribution, the mean and variance of a random variable X will be denoted μ_X, σ_X^2, respectively, regardless of the distribution. The first and second (central) moments of a number of other commonly found distributions are found in Appendix A.

These illustrative examples have focused on properties of a single random variable, for example, $E[X^2], E[X^3]$. However, Eqs. (2.46) are more general, allowing us to focus on properties of joint distributions among two or more different random variables. In particular, the property $E[(X - \mu_X)(Y - \mu_Y)]$ is of some importance to signal processing and estimation as it quantifies dependency between random variables X and Y. To explore this property further, consider a common model for the joint distribution of two random variables

$$p_{XY}(x,y) = \frac{1}{2\pi\sigma_X\sigma_Y\sqrt{1-\rho_{XY}^2}} e^{-\frac{1}{2(1-\rho_{XY}^2)}\left[\frac{(x-\mu_X)^2}{\sigma_X^2} + \frac{(y-\mu_Y)^2}{\sigma_Y^2} - \frac{2\rho_{XY}(x-\mu_X)(y-\mu_Y)}{\sigma_X\sigma_Y}\right]} \tag{2.51}$$

where μ_X, μ_Y are the mean values associated with X, Y, respectively, and σ_X^2, σ_Y^2 are the corresponding variances. Equation (2.51) is the two-dimensional extension of the normal distribution and can be used to model the case where both X and Y are normally distributed (possibly dependent) random variables and where we are interested in the probability that $x < X \le x + dx$ and $y < Y \le y + dy$.

The parameter $-1 < \rho_{XY} < 1$ is referred to as the *correlation coefficient* and can be obtained by taking the expectation $E[(X - \mu_X)(Y - \mu_Y)]$ (with respect to 2.51) and rearranging the result, giving

$$\rho_{XY} = \frac{E[(X - \mu_X)(Y - \mu_Y)]}{\sigma_X\sigma_Y}. \tag{2.52}$$

As we see in the next section, it is *only* for the joint normal distribution that ρ_{XY} is a complete description of statistical dependency between the two random variables. For other distributions, ρ_{XY} may partially capture dependency, but other expectations may be required to fully understand the relationship between X and Y.

The numerator in Eq. (2.52) is referred to as the *covariance* between random variables X and Y and can be simplified as follows:

$$C_{XY} \equiv E[(X - E[X])(Y - E[Y])]$$
$$= E[XY - E[X]Y - E[Y]X + E[X]E[Y]]$$
$$= E[XY] - E[X]E[Y]. \tag{2.53}$$

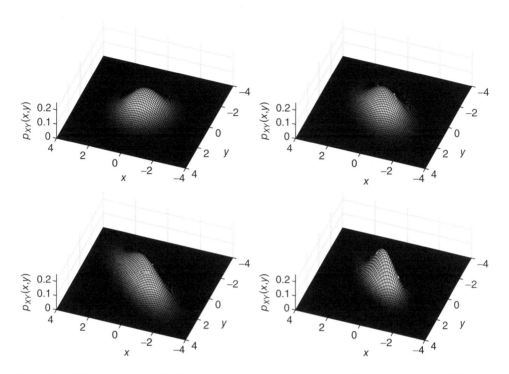

Figure 2.6 Bivariate normal distribution ($\mu_X = \mu_Y = 0$, $\sigma_X = \sigma_Y = 1$) with $\rho_{XY} = 0$, $\rho_{XY} = 0.25$, $\rho_{XY} = 0.5$, and $\rho_{XY} = 0.75$

The correlation coefficient is therefore the normalized covariance and provides a measure of linear dependence of one normally distributed random variable on the other. To see why this is so, consider a graphical depiction of the joint distribution (2.51). Figure 2.6 plots the zero mean bivariate normal distribution ($\sigma_Y = \sigma_Y = 1$) for various values of ρ_{XY}. For $\rho_{XY} = 0$, we are left with the relationship $p_{XY}(x, y) = p_X(x)p_Y(y)$ and by Eq. (2.17) the two variables are statistically independent. As the magnitude of the correlation coefficient is increased ($\rho_{XY} \rightarrow \pm 1$), the distribution collapses toward a line in the x, y plane defined by the equation

$$y(x) = \mu_Y + Sign[\rho_{XY}]\frac{\sigma_Y}{\sigma_X}(x - \mu_X). \tag{2.54}$$

This can be seen by examining the argument in the exponent of the joint distribution (2.51).

The shape of the joint PDF is clearly governed by $\left[\frac{(x-\mu_X)^2}{\sigma_X^2} + \frac{(y-\mu_Y)^2}{\sigma_Y^2} - \frac{2\rho_{XY}(x-\mu_X)(y-\mu_Y)}{\sigma_X\sigma_Y}\right]$. Set-

ting this argument equal to zero gives the maximum of the distribution, which can be solved for y when $\rho_{XY} = \pm 1$, yielding Eq. (2.54). What this means is that for perfect correlation the value $Y = y$ is completely specified by the value $X = x$. Moreover, we see that the relationship between X and Y is linear. Finally, it can be shown (substituting Eq. 2.51 into Eq. 2.15) that the marginal distributions associated with the bivariate normal are

$$p_X(x) = \frac{1}{\sqrt{2\pi\sigma_X^2}} e^{-\frac{1}{2\sigma_X^2}(x-\mu_X)^2}$$

$$p_Y(y) = \frac{1}{\sqrt{2\pi\sigma_Y^2}} e^{-\frac{1}{2\sigma_Y^2}(y-\mu_Y)^2} \tag{2.55}$$

for X and Y, respectively. This holds regardless of the value taken by ρ_{XY}. However, in the case of perfect correlation we can again make use of Eq. (2.25) and arrive at $p_Y(y)$ from $p_X(x)$ using the linear relationship given by Eq. (2.54).

It should also be noted that, in general, for linearly related random variables the output distribution $p_Y(y)$ will have the same functional form as the input distribution $p_X(x)$. To see why this is so, consider random variables Y, X, where $Y = mX + b$ and m, b are scalar constants. Considering Eq. (2.25), we have $h^{-1}(y) = x = (y - b)/m$ and $|h'(x)| = |m|$. The transformed density is therefore

$$p_Y(y) = \frac{1}{|m|} p_X((y-b)/m) \tag{2.56}$$

which is of the same functional form as the input distribution, but with a shifted and rescaled argument. This general property will become important when developing the method of surrogate data to detect the presence of structural damage (see Section 8.6).

Although this discussion has focused on the joint normal distribution, the covariance and correlation coefficients are universal descriptors of linear dependency among two or more random variables. Regardless of the form of the distribution, it is common to use the notation ρ_{XY}, C_{XY} for the correlation coefficient and covariance, respectively. However, it is important to remember that for these quantities to be *complete* descriptors of dependency, one must consider the jointly Gaussian model. However, as seen subsequently (see Section 3, Theorem 3.1.1), the utility of these statistics extends well beyond the simple case of two random variables.

We should also stress that the dependency being quantified by ρ_{XY}, C_{XY} is a linear dependency, regardless of the form of the joint PDF. In situations where nonlinear dependencies exist, two random variables can still be completely uncorrelated. Such examples abound in the literature. A common one is to take X as a zero mean, normally distributed random variable and take $Y = X^2$. Thus, Y is a quadratically nonlinear function of X and, hence, completely dependent on its value. However,

$$\begin{aligned} C_{XY} &= E(XY) - E(X)E(Y) \\ &= E[X^3] - E[X]E[X^2] \\ &= E(X^3) \\ &= 0 \end{aligned} \tag{2.57}$$

as was shown earlier in Section (2.5). The nonlinear (quadratic) dependency is *not* captured by ρ_{XY}, yet clearly Y depends on X directly. In fact, taking Y to be *any* even-ordered polynomial function of X will also result in zero covariance.

2.6 Moment-Generating Functions and Cumulants

Before we end our discussion of basic statistical properties, there are two more functions that require an introduction as both are used in later chapters (specifically Sections 3.3.5, 6.5, and 8.3). These are the so-called moment-generating function, its generalization the characteristic function, and cumulants. As the name implies, the former is a function from which all moments associated with a PDF (or PMF) can be derived. Consider the case of a continuous random variable with PDF $p_X(x)$. Define the moment-generating function

$$M_X(t) = \int_{-\infty}^{\infty} p_X(x)e^{tx}\, dx \qquad (2.58)$$

which can be seen by inspection (using Eq. 2.48) to be $E[e^{tX}]$. Now, if the argument e^{tX} is expanded in an $n-$term power series about $t = 0$, we have

$$M_X(t) = E[e^{tX}] = 1 + tE[X] + \frac{t^2}{2}E[X^2] + \frac{t^3}{6}E[X^3] + \cdots + \frac{t^n}{n!}E[X^n] \qquad (2.59)$$

Thus, if we knew $M_X(t)$, we could expand in a power series in t and retain the coefficients as the moments $E[X], E[X^2], \cdots$ Alternatively, we could simply note that

$$\frac{d^n M_X(t)}{dt^n} = \int_{-\infty}^{\infty} x^n p_X(x)e^{tx}\, dx \qquad (2.60)$$

which, for $t = 0$, provides another mechanism by which $E[X^n]$ can be determined. Specifically,

$$E[X^n] = \frac{d^n M_X(t)}{dt}\Big|_{t=0}. \qquad (2.61)$$

In principle, any order moment could be computed from this function. However, for some distributions all moments do not exist, hence by (2.61) neither will $M_X(0)$. It can also be seen from Eq. (2.59) that as $t \to \infty$ the moments must vanish, or the series will not converge.

For these reasons it makes more sense to define the *characteristic function* for the random variable X as

$$\phi_X(it) = \int_{-\infty}^{\infty} e^{itx} p_X(x)dx \qquad (2.62)$$

which fixes the problems of the moment-generating function. It can be shown [7] that this expression exists and will not diverge for any distribution, while remaining useful in the sense that it still can be used to define all moments of a probability distribution; for example,

$$E[X^n] = \frac{d^n \phi_X(it)}{dt^n}\Big|_{t=0}. \qquad (2.63)$$

Clearly, Eq. (2.62) is convenient in the sense that it allows one to derive all moments of a given distribution from a single function. However, the utility of Eq. (2.62) extends far beyond mathematical efficiency. The characteristic function is frequently used to simplify derivations involving transformations of random variables. First, note that Eq. (2.62) can be inverted to solve for $p_X(x)$ via

$$p_X(x) = \frac{1}{2\pi} \int_{-\infty}^{\infty} \phi_X(it)e^{-itx}\, dt. \qquad (2.64)$$

We see in Section 3.3 (Eq. 3.35) that $p_X(x)$, $\phi_X(it)$ form what is referred to as a Fourier transform pair, although they are defined with a slightly different convention than traditional Fourier analysis. By definition, the characteristic function is actually the complex conjugate of the Fourier transform of $p_X(x)$.

Now, return to the example in Section (2.4) where the goal was to find the PDF of the linear combination of two random variables. Specifically, let's consider the problem of finding $p_Y(y)$, where $Y = a_1 X_1 + a_2 X_2$ and both X_1, X_2 are normally distributed with mean and variance $\mu_{1,2}, \sigma_{1,2}^2$, respectively. First, we may use the characteristic function to find the distribution of the random variables $\tilde{X}_1 = a_1 X_1$ and $\tilde{X}_2 = a_2 X_2$. Defining $\phi_{\tilde{X}_1}(it) = E[e^{ita_1 X_1}]$ where the expectation is with reference to the PDF of X_1, we have

$$\phi_{\tilde{X}_1}(it) = \int_{-\infty}^{\infty} p_{X_1}(x_1) e^{ita_1 x_1} \, dx_1$$

$$= e^{ia_1\mu_1 - \frac{1}{2}a_1^2 t^2 \sigma_1^2} \tag{2.65}$$

so that taking the Fourier transform yields

$$p_{\tilde{X}_1}(\tilde{x}_1) = \frac{1}{2\pi} \int_{-\infty}^{\infty} \phi_{\tilde{X}_1}(it) e^{-it\tilde{x}_1} \, dt$$

$$= \frac{1}{\sqrt{2\pi}a_1\sigma_1} e^{-\frac{(x-a_1\mu_1)^2}{2a_1^2\sigma_1^2}} \tag{2.66}$$

This establishes the previously derived result (Section 2.4) that if $X \sim \mathcal{N}(\mu, \sigma^2)$, then $aX \sim \mathcal{N}(a\mu, a^2\sigma^2)$. Now we turn our attention to finding the sum of two independent random variables with distributions $\tilde{X}_1 \sim \mathcal{N}(a_1\mu_1, a_1^2\sigma_1^2)$, $\tilde{X}_2 \sim \mathcal{N}(a_2\mu_2, a_2^2\sigma_2^2)$. It was shown that Eq. (2.37) provided a means of finding this distribution via

$$p_Y(y) = \int_{-\infty}^{\infty} p_{\tilde{X}_1}(\tilde{x}_1) p_{\tilde{X}_2}(y - \tilde{x}_1) d\tilde{x}_1 \tag{2.67}$$

We point out that the form of this expression is that of a *convolution* between two functions, a concept discussed at length in Chapter 4. What this means is that we can write the characteristic function for the random variable Y as the product

$$\phi_Y^*(it) = \phi_{\tilde{X}_1}^*(it)\phi_{\tilde{X}_2}^*(it) \tag{2.68}$$

We therefore have that the integral (2.67) can be represented as inverse Fourier transform of the product of characteristic functions

$$p_Y(y) = \frac{1}{2\pi} \int_{-\infty}^{\infty} \phi_{\tilde{X}_1}^*(it)\phi_{\tilde{X}_2}^*(it) e^{ity} \, dt \tag{2.69}$$

where * denotes the complex conjugate. Writing

$$\phi_{\tilde{X}_1}^*(it) = \int_{-\infty}^{\infty} p_{\tilde{X}_1}(\tilde{x}_1) e^{-it\tilde{x}_1} \, d\tilde{x}_1$$

$$\phi_{\tilde{X}_2}^*(it) = \int_{-\infty}^{\infty} p_{\tilde{X}_2}(\tilde{x}_2) e^{-it\tilde{x}_2} \, d\tilde{x}_2 \tag{2.70}$$

we have that

$$
\begin{aligned}
p_Y(y) &= \frac{1}{2\pi} \int_{-\infty}^{\infty} \phi_{\tilde{X}_1}^*(it_1)\phi_{\tilde{X}_2}^*(it_2)e^{ity} \, dt \\
&= \frac{1}{2\pi} \int_{-\infty}^{\infty} \left[e^{-ia_1 t\mu_1 - \frac{1}{2}a_1^2 t^2 \sigma_1^2 - ia_2 t\mu_2 - \frac{1}{2}a_2^2 t^2 \sigma_1^2} \right] e^{ity} \, dt \\
&= \frac{1}{\sqrt{2\pi}\sqrt{a_1^2\sigma_1^2 + a_2^2\sigma_2^2}} e^{-\frac{(y-a_1\mu_1 - a_2\mu_2)^2}{2(a_1^2\sigma_1^2 + a_2^2\sigma_2^2))}}
\end{aligned}
\tag{2.71}
$$

which is the desired distribution, derived earlier (Section 2.4, Eq. 2.31). In general, the distribution for the sum of two random variables is the inverse Fourier transform of the product of the complex conjugate of the two constituent characteristic functions. This result can be generalized to produce one of the most useful results in statistics.

Consider a series of independent, identically distributed random variables $X_n, n = 1 \cdots N$ with mean μ_X and variance σ_X^2. We may standardize these random variables to zero mean, unit variance via the transformation $Y_n = \frac{X_n - \mu_X}{\sigma_X}$. Now consider the random variable $Z = \frac{1}{\sqrt{N}} \sum_{n=1}^{N} Y_n = \frac{1}{\sqrt{N}} \sum_{n=1}^{N} (X_n - \mu_X)/\sigma_X$, that is, a normalized sum of zero mean, unit variance random variables. The normalization constant ensures that $E[Z^2] = 1$. Note that to this point we have made no assumptions about the PDF for the X.

Denote the characteristic function associated with the Y_n as $\phi_Y(it)$, so that for the random variable $Y_n/\sqrt{N}$ we have the characteristic function $\phi_Y(it/\sqrt{N})$. We can use the abovementioned convolution result to write for the characteristic function of Z,

$$
\begin{aligned}
\phi_Z^*(it) &= \prod_{n=1}^{N} \phi_Y^*\left(it/\sqrt{N}\right) \\
&= [\phi_Y^*\left(it/\sqrt{N}\right)]^N
\end{aligned}
\tag{2.72}
$$

Now, we know from the properties of a characteristic function that $d\phi_Z(it)/dt = E[Z] = 0$ and that $d^2\phi_Z(it)/dt^2 = E[Z^2] = 1$. Thus, if we expand the characteristic function for Z as a Taylor series, and let t become small, we have

$$
\begin{aligned}
\phi_Z^*(it) &= \left[1 + \left(\frac{it}{\sqrt{N}}\right) E[Z] + \frac{(it)^2}{2N} E[Z^2] + \frac{(it)^3}{6N^{3/2}} E[Z^3] + \cdots \right]^N \\
&= \left[1 - \frac{t^2}{2N} + \frac{(it)^3}{6N^{3/2}} E[Z^3] + \cdots \right]^N
\end{aligned}
\tag{2.73}
$$

In the limit of large N, this function becomes

$$
\phi_Z^*(it) = \lim_{N \to \infty} \left[1 - \frac{t^2}{2N} + h.o.t \right]^N = e^{-t^2/2}
\tag{2.74}
$$

where $h.o.t$ denotes a higher order term which vanishes in the limit. Thus, for large N we are left with the characteristic function for the Gaussian distribution, which means

$$
p_Z(z) \approx \frac{1}{\sqrt{2\pi}} e^{-\frac{1}{2}z^2}.
\tag{2.75}
$$

This is referred to as the *central limit theorem* and it means that regardless of the PDF for X, if the random variable of interest can be described as the summation of a large number of independent random variables with arbitrary probability model, the resulting distribution will be normal. The only conditions of significance in this derivation are that of independence among the values and that the second derivative of the PDF for X exists (this actually excludes Cauchy random variables from the application of this theorem, as the Cauchy distribution does not possess a finite variance). The central limit theorem is invoked later in Section 6.9 when it is time to form intervals for our test statistics used for detecting damage. Many such statistics are describable as the summation of a large number of random variables, hence we are justified in assuming a normal probability model.

As these examples have shown, there are numerous ways of simplifying expressions involving random variables. In this chapter alone we have shown three different approaches to deriving the PDF of $Y = a_1 X_1 + a_2 X_2$. The approach one ultimately uses in practice will be problem-specific. We rely heavily on this material in later chapters (particularly Chapter 8), when it comes time to quantify detectors of structural damage.

It also proves useful in later chapters to introduce *cumulants* and the associated *cumulant-generating function*

$$K(it) = \log \ (E[e^{itX}]).   \tag{2.76}$$

The utility of such a function becomes clear upon simplification. First, we denote $g(it) = E[e^{itX}] - 1$ and expand

$$K(it) = \log \ (1 + g(it))$$

$$= \sum_{n=1}^{\infty} \frac{g(it)^n (-1)^{n+1}}{n}$$

$$= \sum_{n=1}^{\infty} \frac{(-1)^n [-g(it)]^n (-1)^{n+1}}{n}$$

$$= - \sum_{n=1}^{\infty} \frac{-g(it)^n}{n}   \tag{2.77}$$

whereby substituting back in for $g(it)$ gives

$$K(it) = - \sum_{n=1}^{\infty} \frac{(1 - E[e^{itX}])^n}{n}.   \tag{2.78}$$

However, we have already provided the series expansion for $E[e^{itX}]$ which, upon substitution into the above, yields

$$K(it) = - \sum_{n=1}^{\infty} \frac{1}{n} \left(- \sum_{m=1}^{\infty} E[X^m] \frac{(it)^m}{m!} \right)^n.   \tag{2.79}$$

Collecting the coefficients of $(it)^j / j!$ in the expansion, the first five terms are given by

$$K(it) \approx \sum_{j=1}^{5} k_j \left(\frac{(it)^j}{j!} \right)$$

$$= E[X]t$$
$$+ (E[X^2] - E[X]^2)(it)^2/2!$$
$$+ (E[X^3] - 3E[X]E[X^2] + 2E[X]^3)(it)^3/3!$$
$$+ (E[X^4] - 4E[X]E[X^3] - 3E[X^2]^2 + 12E[X]^2E[X^2] - 6E[X]^4)(it)^4/4!$$
$$+ (E[X^5] - 5E[X]E[X^4] - 10E[X^2]E[X^3] + 20E[X]^2E[X^3] + 30E[X]E[X^2]^2$$
$$- 60E[X]^3E[X^2] + 24E[X]^5)(it)^5/5. \tag{2.80}$$

The coefficients k_j are the *cumulants* associated with the random variable X, the first five of which can be more compactly written

$$k_1 = E[X] \equiv \mu_X$$
$$k_2 = E[(X - \mu_X)^2] \equiv \sigma_X^2$$
$$k_3 = E[(X - \mu_X)^3]$$
$$k_4 = E[(X - \mu_X)^4] - 3E[(X - \mu_X)^2]^2$$
$$k_5 = E[(X - \mu_X)^5] - 10E[(X - \mu_X)^2]E[(X - \mu_X)^3]. \tag{2.81}$$

The first thing to note is that while the moment-generating function provides expectations of random variables raised to different powers, the cumulant- generating function produces mean-centered expectations. Secondly, for zero-mean random variables we note that the first three moments are equal to the first three cumulants. However, for cumulants of order four and higher, we see that additional terms are produced. It will be shown later that these terms are very important when studying the higher order statistical properties of structural response data (see Section 8.3). This is due to the fact that for a normally distributed random variable, these extra terms cause the associated cumulants to vanish. In fact, for the normal distribution, all cumulants greater than second order vanish (see Theorem 3.1.1)! In Section 5.4, for example, we show how the fourth cumulant can be used to detect damage-induced deviations from a Gaussian structural response.

Finally, we note that this discussion extends to joint probability distributions as well. That is to say, we can define a characteristic function for the vector of random variables $\mathbf{X} \equiv (X_1, X_2, \cdots)$ as $E[e^{i\mathbf{X}^T \mathbf{t}}]$ such that all joint moments can be obtained in analogy to Eq. (2.63). For example, consider the vector of independent random variables $\mathbf{X} \equiv (X_1, X_2)$, where $X_1 \sim p_{X_1}(x_1)$ and $X_2 \sim p_{X_2}(x_2)$. The joint characteristic function is

$$\phi_{\mathbf{X}}(it) = \int_{-\infty}^{\infty} e^{i\mathbf{x}^T \mathbf{t}} p_{X_1}(x_1)p_{X_2}(x_2)dx_1 \ dx_2 \tag{2.82}$$

From this function we can derive the joint moment

$$E[X_1 X_2] = \frac{\partial}{\partial it_1} \frac{\partial}{\partial it_2} \phi(it)\Big|_{\substack{t_1=0 \\ t_2=0}}$$
$$= \int_{-\infty}^{\infty} x_1 x_2 p_{X_1}(x_1)p_{X_2}(x_2)dx_1 \ dx_2 \tag{2.83}$$

which is the result stated in Eq. (2.46) for independent random variables (joint distribution factors). Thus, in the multivariate case, we can, in general, write

$$E[X_1 X_2 \cdots X_N] = \frac{\partial}{\partial it_1} \frac{\partial}{\partial it_2} \cdots \frac{\partial}{\partial it_N} \phi(it)|_{t_1=t_2=\cdots=t_N=0} \qquad (2.84)$$

for the expectation of the product of N independent random variables. For random variables that are not independent, one replaces the product of individual PDFs in (2.82) with the joint PDF. We may similarly consider the multivariate cumulant- generating function

$$K(it) = E[\log (\phi_X(it))] \qquad (2.85)$$

where upon expanding as a series, the first four mean-centered joint cumulants can be obtained as

$$k_{X_1} = E[X_1] = \mu_{X_1}$$
$$k_{X_1 X_2} = E[(X_1 - \mu_{X_1})(X_2 - \mu_{X_2})]$$
$$k_{X_1 X_2 X_3} = E[(X_1 - \mu_{X_1})(X_2 - \mu_{X_2})(X_3 - \mu_{X_3})]$$
$$k_{X_1 X_2 X_3 X_4} = E[(X_1 - \mu_{X_1})(X_2 - \mu_{X_2})(X_3 - \mu_{X_3})(X_4 - \mu_{X_4})]$$
$$- E[(X_1 - \mu_{X_1})(X_2 - \mu_{X_2})]E[(X_3 - \mu_{X_3})(X_4 - \mu_{X_4})]$$
$$- E[(X_1 - \mu_{X_1})(X_3 - \mu_{X_3})]E[(X_2 - \mu_{X_2})(X_4 - \mu_{X_4})]$$
$$- E[(X_1 - \mu_{X_1})(X_4 - \mu_{X_4})]E[(X_2 - \mu_{X_2})(X_3 - \mu_{X_3})] \qquad (2.86)$$

where, again, we see that the first three joint cumulants are equal to the first three central moments, with higher cumulants showing additional terms. In a later chapter, the importance of defining higher order spectral properties in terms of joint cumulants (as opposed to joint moments) is demonstrated. It is also shown that joint cumulants are extremely important in the detection of damage-induced nonlinearity in a structure given a sequence of observations of the structure's response.

To this point, our discussion of probability has gradually been building toward the multi-variate case where $N > 2$ random variables are being considered. In fact, we now possess the basic tools needed to construct probabilistic models for entire sequences of observations, such as those produced by one or more sensors located on a structure. The next chapter addresses this topic in some detail.

We will continue to focus on the joint normal distribution as the model for the observed data. This is not just a matter of analytical convenience. The joint normal is an excellent model for most sources of sensor noise, which typically gives rise to the measurement uncertainty we are trying to minimize when it comes time to estimate parameters. Ironically, although the joint normal is probably the simplest joint distribution to handle analytically, it is the most difficult to overcome when estimating parameters! It has been demonstrated (see, e.g., [8, 9]) that among a very large number (possibly all) distributions, the joint normal leads to the largest theoretical minimum variance in the estimated parameters (the so-called Cramér–Rao lower bound discussed in Section 7.1). The fact that the most useful probability model with the simplest analytical form also represents the worst possible case from an estimation perspective proves once again that there is no free lunch in science.

References

[1] A. N. Kolmogorov, Foundations of Probability, Chelsea Publishing Co., New York, 1956.

[2] J. S. Rosenthal, A First Look at Rigorous Probability Theory, World Scientific Publishing, Singapore, 2000.

[3] R. N. McDonough, A. D. Whalen, Detection of Signals in Noise, 2nd ed., Academic Press, San Diego, CA, 1995.

[4] J. M. Nichols, C. C. Olson, J. V. Michalowicz, F. Bucholtz, The bispectrum and bicoherence for quadratically non-linear systems subject to non-Gaussian inputs, IEEE Transactions on Signal Processing 57 (10) (2009) 38793890.

[5] A. M. Mood, F. A. Graybill, D. C. Boes, Introduction to the Theory of Statistics, 3rd ed., McGraw-Hill, New York, 1974.

[6] L. Devroye, Non-Uniform Random Variate Generation, Springer-Verlag, New York, 1986.

[7] M. B. Priestly, Spectral Analysis and Time Series, Probability and Mathematical Statistics, Elsevier Academic Press, London, 1981.

[8] T.-H. Li, K.-S. Song, Estimation of the parameters of sinusoidal signals in non-Gaussian noise, IEEE Transactions on Signal Processing 57 (1) (2009) 6272.

[9] P. Stoica, P. Babu, The Gaussian data assumption leads to the largest Cramèr–Rao bound, IEEE Signal Processing Magazine 28 (2011) 132133.

3

Random Processes

As was mentioned in the introduction, one of our primary goals is to accurately model structural vibration data. Typically, these data take the form of a temporal sequence of points acquired with one or more sensors. For example, let's say we turn on a strain sensor mounted on a bridge girder and measure the ambient structural vibration due to traffic. The sensor has a fixed sampling rate and we record N temporally spaced observations $x(t_1), x(t_2), \ldots, x(t_N)$, where the sample times may be related to their discrete time indices through the sampling interval, i.e., $t_n = n\Delta_t$. Note we sometimes use the notation $x(n) \equiv x(t_n)$, $n = 1 \cdots N$ to denote a discrete sequence of observations and provide the sampling interval as a separate piece of information. So while in Chapter 2 the "experiments" we were modeling consisted of a single observation (e.g., a single voltage reading), here the experiment consists of recording the entire N-sequence.

Owing to the uncertainty inherent in any measurement process, we cannot predict the exact sequence of values and so we model each observation as a random variable, e.g., $x(t_1)$ is modeled with $X(t_1)$, $x(t_2)$ with $X(t_2)$, and so on. Our model for the entire observed signal is therefore the vector $\mathbf{X} \equiv \{X(t_1), X(t_2), \ldots, X(t_N)\}$. The experiment that produced the sequence of observations will be referred to as a *random process* and the vector $\mathbf{X}$, the random process model. In the language in Chapter 2, $\mathbf{X}$ is a collection of real-valued functions, each defined on the space of possible outcomes. Typically, these outcomes are an N-dimensional product of closed intervals. A ± 5 V digitizer, for example, would yield $\mathbf{X} \in \{[-5\text{ V}, 5\text{ V}] \times [-5\text{ V}, 5\text{ V}] \times, \underset{N\text{-}times}{\cdots}, \times [-5\text{ V}, 5\text{ V}]\}$, which covers the space of all possible N-vectors recorded by the digitizer. Our σ-algebra for each experiment is assumed to contain a finite collection of subsets on this range, the Borel sets, including the null set and the set of all possible outcomes. Thus, using the framework we established in Chapter 2 we can define for the sequence a joint CDF, $P_{\mathbf{X}}(\mathbf{x}) \equiv P(X(1) \le x(1), X(2) \le x(2), \ldots, X(N) \le x(N))$, and for continuous random variables, a joint probability density function (PDF), $p_{\mathbf{X}}(\mathbf{x})$.

Each observed sequence of N digitized values (i.e., each vector outcome) will henceforth be referred to as a *realization* of our experiment. Every time we turn on our sensor and record N values, we are collecting a different realization, which is one possibility from the set of N-vectors we could have observed. If we were to run the experiment K different times, we observe the data $\mathbf{x}^{(k)}, k = 1 \cdots K$, each of which is expected to follow our random process model $\mathbf{X}$. Note that the notation we have established here is fairly general. For example, each

Modeling and Estimation of Structural Damage, First Edition. Jonathan M. Nichols and Kevin D. Murphy.
© 2016 John Wiley & Sons, Ltd. Published 2016 by John Wiley & Sons, Ltd.

of the random variables that comprise $\mathbf{X}$ could be defined on a different space of outcomes and have entirely different probability models. The more typical case is to assume that each of the constituent random variables follows the same model, a point that is discussed in Section 3.2. Random process models are useful in that they allow us to make predictions about the statistical properties of the observations (e.g., what values are the sequence of observations most likely to take?). In addition, as shown in Chapter 7, without a model of the uncertainty we cannot make definitive statements about our ability to identify the deterministic component of the observations.

Our reason for focusing on discrete sequences of observations is simply because that is the type of data we will always be working with in practice. However, it is important mathematically to see that the notation can be conceptually extended to experimental outcomes that are continuous functions of time. In the continuous case, we assume that the analog waveform $x(t)$ can be modeled as a continuous random process, $X(t)$, where the set of possible experimental outcomes is now a continuous, real-valued function of time. In the formal construction in Chapter 2, this means that for a given experimental outcome $s(t) \in \xi(t)$ we define the random variable $X(s(t)) = s(t)$. That is to say, just as before, the random process is defined directly as an element in the set of possible outcomes (signals). In shorthand, it is common to use the notation $X(t)$. As in the discrete case, each experimental outcome yields a different realization $x^{(k)}(t), k = 1 \cdots K$, each of which can be modeled with $X(t)$.

From an estimation standpoint, thinking of the experimental outcome as a collection of N discrete random variables is more natural, as this is what a typical sensor gives us. However, in developing certain functions (particularly those that involve the taking of limits) of our random process, it will be useful to allow the sampling interval to shrink to zero and think of the random process as continuous. We illustrate the situation graphically in Figure 3.1. Shown here are five different realizations of a random process $x^{(k)}(t), k = 1 \cdots 5$. Each time the experiment is run, a different continuous time output occurs. Our data collection system then samples the process at times t_n. In Figure 3.1, sample times t_1, t_2, t_3 are called out; these locate the observations $x(t_1), x(t_2), x(t_3)$ in each realization. In estimation we will be concerned with the joint PDF $p_{\mathbf{X}}(x(t_1), x(t_2), x(t_3))$ associated with these observations. The joint PDF is a probabilistic model that holds for any realization we happen to observe, that is to say it provides the likelihood of observing a given sequence. The specific form of the PDF used to model the observations is chosen by the practitioner. As with any other model, this choice is made on the basis of a combination of experience and knowledge of the underlying physics. Specific choices of probabilistic models are provided in subsequent sections.

As the number of observations N in a given realization becomes large, there are an increasing number of possibilities for our model. In the most general case, each observation could be modeled as a random variable from a *different* probability distribution, i.e., $p_{X(1)}(x(1)) \neq p_{X(2)}(x(2)) \cdots \neq p_{X(N)}(x(N))$. This basic concept is illustrated pictorially in Figure 3.2. In this example, we observe a sequence of $N = 6$ values $x(1), x(2), \ldots, x(6)$. The first three values are modeled as independent draws from a normal distribution with zero mean and unit variance, while the second three values in the series are modeled as independent draws from a chi-squared distribution with two degrees of freedom (denoted $\chi^2(2)$; see Appendix A). The assumption of independence allows us to write the joint probability model

$$p_{\mathbf{X}}(x(1), \ldots, x(6)) = \frac{1}{(2\pi)^{3/2}} e^{-\frac{1}{2}\sum_{n=1}^{3} x(n)^2} \times \frac{1}{2} e^{-\frac{1}{2}\sum_{n=4}^{6} x(n)} \tag{3.1}$$

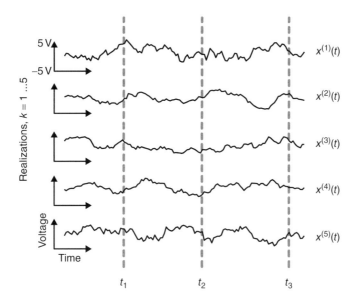

Figure 3.1   Different continuous time random processes $x^{(k)}(t)$. Each of these can be modeled as a different realization of a random process $X(t)$. We can also consider specific sampling times, e.g., t_1, t_2, t_3 and the associated joint PDF $p_{\mathbf{X}}(x(t_1), x(t_2), x(t_3))$. Each time we record a new realization, the likelihood of observing the sequence of values $x(t_1), x(t_2), x(t_3)$ is governed by this probabilistic model

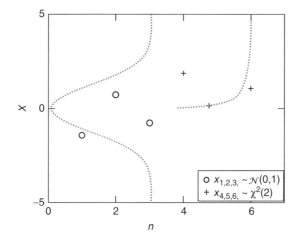

Figure 3.2   Time series consisting of $N = 6$ observations. The first three observations are modeled as independent draws from a standard normal distribution (zero mean, unit variance). The second three observations are modeled as independent draws from a χ^2 distribution with two degrees of freedom (see Appendix A for details on this distribution). The PDFs associated with these two distributions are overlaid on the plot

Admittedly, this is a strange model, however it is not hard to conceive of a situation where this might occur. For example, a sensor may be used to measure both magnitude and phase of an acoustic field. During ambient vibration, both the real and imaginary parts of the measured signal may be modeled as normally distributed noise. During the first three observations, we record only the real part of the signal. However, after the first three observations, the data acquisition system is "toggled" to record the magnitude squared of the next three observations (i.e., sum of real and imaginary parts squared). In this case, our model would be accurate, as it can be shown that the magnitude squared of a complex normally distributed random variable is a chi-squared distributed random variable with two degrees of freedom [1] (we could also have quickly arrived at this result ourselves using the methods in Section 2.4).

A brief note on notation. In the remainder of this book, we often refer to a random variable at a specific discrete time (e.g., $X(1)$); thus, the PDF would read $p_{X(1)}(x(1))$. Rather than carry the cumbersome subscript $X(1)$, we simply use $p_X(x(1))$, because in the vast majority of cases the space of possible outcomes is the same for each observation, that is, $X(1) = X(2) = \cdots = X(N) \equiv X$. The joint PDF for observations $x(1), x(2)$ will therefore read $p_X(x(1), x(2))$. Similarly, in defining properties of a continuous time random process, we will use $p_X(x(t))$ instead of $p_{X(t)}(x(t))$ to denote the PDF of the random variable X at time t. Again, this notation implicitly assumes the same probabilistic model at each point in time. In the rare event that the random variables *are* modeled with different distributions, we will simply retain the temporal indices.

The more common situation occurs when modeling joint properties of random variables from *different* random processes. For example, if we have sensors on M spatial locations on a structure, we will record the (multiple) time series $x_i(t), i = 1 \cdots M$. In this case, it makes sense to define the joint probability models $p_{X_i X_j}(x_i(t), x_j(t))$ for any $i, j \in [1, M]$. When warranted, we will also use subscripts to describe joint statistical properties of a particular random process, e.g., the covariance $C_{X_i X_j}(t) = E[(X_i(t) - \mu_{X_i}(t))(X_j(t) - \mu_{X_j}(t))]$. It will always be clear from the context which random variable(s) are being used to describe which observations. This notation will be particularly useful in deriving higher order spectral (HOS) properties (see Section 3.3.5).

3.1   Properties of a Random Process

Just as we defined expectations of random variables, we can define expectations of random processes. The only difference is that our random variable is now indexed by time. For example, we define the mean

$$\mu_X(t) \equiv E[X(t)] = \int_{-\infty}^{\infty} x(t) p_X(x(t)) dx(t) \tag{3.2}$$

as the expected value of our random process at time t. It is important to keep in mind what this expectation represents when it comes to random processes. This expectation refers to the average value of "X" at time "t." Imagine that we repeated an experiment over and over again (i.e., looked at different realizations) and at $t = 2$ s (relative to the start of each experiment) recorded $x^{(k)}(t = 2)$ for each realization k. If we then looked at all of the $x^{(k)}(2)$ values ($k = 1 \cdots K$), adding them up and dividing by K would be a good estimate of $\mu_X(2)$. This implies (correctly) that at each point in time, our random process $x(t)$ could have a different mean. Of course, this is seldom the case in practice as we almost always model the mean of a random process as time invariant, e.g., $\mu_X(t) = \mu_X$. Nonetheless, the expectations of different properties of a temporal random process are always defined with respect to an ensemble of measurements at particular points in time.

In general, to *completely* describe a discrete random process consisting of N observations, one would have to specify all of the joint moments

$$E[X(t)]$$

$$E[X(t_1)X(t_2)]$$

$$E[X(t_1)X(t_2)X(t_3)]$$

$$\vdots \tag{3.3}$$

As we have already shown, the entire collection of joint moments can be obtained from the joint characteristic function (2.82). Studying all possible expectations of products of the random variables comprising a random process is impractical. Rather, we tend to focus on the first few joint moments as they convey much of the information needed to accurately model a random process, particularly the mean given by Eq. (3.2), and the autocorrelation function

$$R_{XX}(t_n, t_m) \equiv E[X(t_n)X(t_m)] = \int_{-\infty}^{\infty} \int_{-\infty}^{\infty} x(t_n)x(t_m)p_{\mathbf{X}}(x(t_n), x(t_m))dx(t_n)dx(t_m). \tag{3.4}$$

These two descriptors can be further combined to form the *autocovariance* function

$$C_{XX}(t_n, t_m) = E[(X(t_n) - \mu_X(t_n))(X(t_m) - \mu_X(t_m))]$$

$$= R_{XX}(t_n, t_m) - \mu_X(t_n)\mu_X(t_m) \tag{3.5}$$

where we see that, in the special case of a zero-mean random process, the autocorrelation and autocovariance functions are equivalent.

Given a sequence of observations recorded at times $t_1, \ldots, t_N$, Eq. (3.5) defines the $N \times N$ *covariance matrix*

$$\mathbf{C}_{\mathbf{XX}} =$$

$$\begin{bmatrix} E[(X(t_1) - \mu_X(t_1))(X(t_1) - \mu_X(t_1))] & E[(X(t_1) - \mu_X(t_1))(X(t_2) - \mu_X(t_2))] & \cdots \\ E[(X(t_2) - \mu_X(t_2))(X(t_1) - \mu_X(t_1))] & E[(X(t_2) - \mu_X(t_2))(X(t_2) - \mu_X(t_2))] & \cdots \\ \vdots & \vdots & \ddots \\ E[(X(t_N) - \mu_X(t_N))(X(t_1) - \mu_X(t_1))] & E[(X(t_N) - \mu_X(t_N))(X(t_2) - \mu_X(t_2))] & \cdots \end{bmatrix}$$

$$\begin{matrix} E[(X(t_1) - \mu_X(t_1))(X(t_N) - \mu_X(t_N))] \\ E[(X(t_2) - \mu_X(t_2))(X(t_N) - \mu_X(t_N))] \\ \vdots \\ E[(X(t_N) - \mu_X(t_N))(X(t_N) - \mu_X(t_N))] \end{matrix} \Bigg]. \tag{3.6}$$

This matrix holds a special place in signal and image processing as it defines the PDF for a jointly Gaussian sequence, $\mathbf{x} = (x(t_1), \ldots, x(t_N))$. That is to say, if we let $\mathbf{C}_{\mathbf{XX}}$ be defined as (3.6) and $\mu_X = \mu_X(t_n), n = 1 \cdots N$, our model for the observed sequence is written

$$p_{\mathbf{X}}(\mathbf{x}) = \frac{1}{2\pi^{N/2}|\mathbf{C}_{\mathbf{XX}}|^{1/2}} e^{[-\frac{1}{2}(\mathbf{x} - \mu_X)^T \mathbf{C}_{\mathbf{XX}}^{-1}(\mathbf{x} - \mu_X)]} \tag{3.7}$$

where $|\cdot|$ takes the determinant of a matrix. Equation (3.7) is the N-dimensional extension of the bivariate normal distribution already discussed in Chapter 2, Eq. (2.5.1). To see this, set $N = 2$ and recall that by definition $\sigma_{X_1}^2 = E[(X(t_1) - \mu_X(t_1))^2]$, $\sigma_{X_2}^2 = E[(X(t_2) - \mu_X(t_2))^2]$, and $\rho_{X_1 X_2} = E[(X(t_1) - \mu_X(t_1))(X(t_2) - \mu_X(t_2))]/\sigma_{X_1}\sigma_{X_2}$. Using these definitions we see that

$$\begin{vmatrix} E[(X(t_1) - \mu_X(t_1))(X(t_1) - \mu_X(t_1))] & E[(X(t_1) - \mu_X(t_1))(X(t_2) - \mu_X(t_2))] \\ E[(X(t_2) - \mu_X(t_2))(X(t_1) - \mu_X(t_1))] & E[(X(t_2) - \mu_X(t_2))(X(t_2) - \mu_X(t_2))] \end{vmatrix}$$

$$= \sigma_{X_1}^2 \sigma_{X_2}^2 - E[(X(t_1) - \mu_X(t_1))(X(t_2) - \mu_X(t_2))]^2$$

$$= \sigma_{X_1}^2 \sigma_{X_2}^2 (1 - \rho_{X_1 X_2}^2) \tag{3.8}$$

and

$$\mathbf{C_{XX}}^{-1} = \begin{bmatrix} \dfrac{1}{\sigma_{X_1}^2 (1-\rho_{X_1 X_2}^2)} & -\dfrac{\rho_{X_1 X_2}}{(1-\rho_{X_1 X_2}^2)\sigma_{X_1}\sigma_{X_2}} \\ -\dfrac{\rho_{X_1 X_2}}{(1-\rho_{X_1 X_2}^2)\sigma_{X_1}\sigma_{X_2}} & \dfrac{1}{\sigma_{X_2}^2 (1-\rho_{X_1 X_2}^2)} \end{bmatrix} \tag{3.9}$$

from which the argument in the exponent of (3.7) becomes

$$-\frac{1}{2}(\mathbf{x} - \mu_{\mathbf{X}})^T \mathbf{C_{XX}}^{-1}(\mathbf{x} - \mu_{\mathbf{X}})$$

$$= -\frac{1}{2(1 - \rho_{X_1 X_2}^2)} \left(\frac{(x_1 - \mu_{X_1})^2}{\sigma_{X_1}^2} + \frac{(x_2 - \mu_{X_2})^2}{\sigma_{X_2}^2} - \frac{2\rho_{X_1 X_2}(x_1 - \mu_{X_1})(x_2 - \mu_{X_2})}{\sigma_{X_1}\sigma_{X_2}} \right) \tag{3.10}$$

thus, we again arrive at Eq. (2.51). It can therefore be stated that the covariance matrix completely defines a joint Gaussian PDF.

The fact that the autocovariance is often taken as a complete description of a random process is not just a matter of convenience. Many of the random processes encountered in practice are accurately modeled by a joint normal distribution, that is, each observation is a normally distributed random variable. Given that the covariance matrix specifies a joint normal distribution, it is perhaps not surprising that *all* of the central moments of a joint Gaussian distribution can be expressed as a function of its entries. Specifically, it was shown by Isserlis [2] that the expectation of the product of $2N$ normally distributed, zero-mean random variables reduces to the sum of products of two normally distributed random variables. Formally:

Theorem 3.1.1 ((*Isserlis' theorem*)) *If* $X_1, X_2, \dots, X_{2N+1}$ $(N = 1, 2, \cdots)$ *are each Gaussian random variables with* $E[X_i] = \mu_i$ *and* $E[X_i^2] = \sigma_i^2$*, then*

$$E[(X_1 - \mu_{X_1})(X_2 - \mu_{X_2}) \cdots (X_{2N} - \mu_{X_{2N}})] = \sum \prod E[(X_i - \mu_{X_i})(X_j - \mu_{X_j})]$$

$$E[(X_1 - \mu_{X_1})(X_2 - \mu_{X_2}) \cdots (X_{2N+1} - \mu_{X_{2N+1}})] = 0 \tag{3.11}$$

where the notation $\sum \prod$ means summing over all distinct ways of grouping the $2N$ random variables into pairs. It turns out that there are $\frac{(2N)!}{2^N N!}$ such pairs. For example, for a random variable $X \sim N(\mu_X, \sigma_X^2)$, $E[(X - \mu_X)^4] = 3E[(X - \mu_X)^2] = 3\sigma_X^4$, as there are three distinct ways of creating pairs from a group of four random variables ($N = 2$). Returning to the expression for the fourth-order cumulant (Eq. 2.81), we see then that $k_4 = E[(X - \mu_X)^4] - 3\sigma_X^4 = 0$, thus proving (at least partially) our earlier assertion that all cumulants of greater than order two vanish for a normally distributed random variable. The theorem also states that all odd-order moments associated with a jointly Gaussian distribution are zero.

Equation (3.11) turns out to be an extremely useful expression in studying random vibration problems, as it gives the expectation of products of $2N$ normally distributed random variables using linear combinations of the products of two zero-mean, normally distributed random variables (see, e.g., [3, 4]). This means that the autocovariances $C_{XX}(t_n, t_m) = E[(X(t_n) - \mu_X(t_n))(X(t_m) - \mu_X(t_m))], n, m \in [1, N]$ associated with the random processes $X(t_n), Y(t_m)$ are sufficient to describe *all* joint moments, a fact that has motivated extensive treatment of the covariance function in the literature. Note, however, that this theorem *only* holds for a jointly Gaussian (or jointly "mixed-Gaussian" [5]) random process. Other joint distributions will require other joint expectations to fully characterize the process.

In fact, in the analysis of damaged structural response data, it becomes advantageous to look at the *higher order* correlation functions, e.g.,

$$C_{XXX}(t_n, t_m, t_p) = E[(X(t_n) - \mu_X(t_n))(X(t_m) - \mu_X(t_m))(X(t_p) - \mu_X(t_p))]$$

$$C_{XXXX}(t_n, t_m, t_p, t_q) = E[(X(t_n) - \mu_X(t_n))(X(t_m) - \mu_X(t_m))$$

$$\times (X(t_p) - \mu_X(t_p))(X(t_q) - \mu_X(t_q))]. \qquad (3.12)$$

As shown later, these functions can play a very useful role in the detection and identification of nonlinearity in structural systems as they are the building blocks for HOS analysis. More specifically, they can be used to define the higher order joint cumulants for a random process (see Section 2.6)

$$k_{XXX}(t_n, t_m, t_p) = C_{XXX}(t_n, t_m, t_p)$$

$$= E[(X(t_n) - \mu_X(t_n))(X(t_m) - \mu_X(t_m))(X(t_p) - \mu_X(t_p))]$$

$$k_{XXX}(t_n, t_m, t_p, t_q) = C_{XXXX}(t_n, t_m, t_p, t_q) - C_{XX}(t_n, t_m)C_{XX}(t_p, t_q)$$

$$- C_{XX}(t_n, t_p)C_{XX}(t_m, t_q) - C_{XX}(t_n, t_q)C_{XX}(t_m, t_p) \qquad (3.13)$$

which define the HOS. There is no theoretical reason why one wouldn't consider still higher order correlation functions (e.g., $C_{XXXXX}(t_n, t_m, t_p, t_q, t_r)$). However, it turns out that the main difficulty in working with higher order statistics is the problem being addressed by this book: estimation. In short, the amount of data required to estimate these functions from data is prohibitively large once one gets beyond a fourth-order correlation function. The estimation of the HOS is discussed in some detail in Section 6.5.

3.2   Stationarity

In the above-mentioned development, we have left open the possibility that at each point in time the observations $x(t_n)$ are being modeled with a different probability distribution. While there are certainly cases when this might be desirable, in this book we focus largely on *stationary* random processes (the exceptions being the discussion on evolutionary spectra presented in Section 6.8.2 and the identification of damage based on impulse response data in Chapter 9).

The assumption of stationarity is a common one, and it means that the joint probability distribution we are using to model our random process is time-invariant. Mathematically speaking, a stationary random process is one in which

$$p_{\mathbf{X}}(x(t_1), x(t_2), \ldots, x(t_N)) = p_{\mathbf{X}}(x(t_1 + \tau), x(t_2 + \tau), \ldots, x(t_N + \tau)) \qquad (3.14)$$

where τ is a measure of time delay. Equation (3.14) states that the joint PDF of the random process is invariant to the time at which the measurement is recorded. Returning to the example of Figure 3.2, if we assume that all observations beyond the first three remained chi-squared distributed for all time, all subsequent realizations of $N = 6$ observations would be stationary, jointly chi-squared distributed. If we instead assumed that the sequence (three Gaussian-distributed random variables followed by three chi-squared random variables) repeated for all time, the random process would be nonstationary as the joint distribution associated with any realization of $N = 6$ observations would absolutely depend on where in the sequence those six observations were taken. However, we should note that in the case just described, the sequence *would* be stationary with a certain periodicity. Every six observations, the joint distribution would be the same (three Gaussian, three chi-squared). This is an example of a *cyclostationary* random process, that is to say, one in which the joint probability distribution governing the observations is periodically stationary.

A distinction sometimes drawn in the literature is between *strictly* stationary and *weakly* stationary. The definition of strict stationarity has already been given (3.14). However, in some situations we may relax this constraint and instead only require that the first two moments (mean, variance) be the same for each observation. A time series for which each observation is drawn from a PDF with the same mean and variance (but not necessarily higher moments) is referred to as weakly stationary. For such a random process, we have the mathematical relationships

$$E[X(t_n)] = E[X(t_n + \tau)]$$
$$E[(X(t_n) - E[X(t_n)])(X(t_m) - E[X(t_n)])] =$$
$$E[(X(t_n + \tau) - E[X(t_n)])(X(t_m + \tau) - E[X(t_n)])] \qquad (3.15)$$

which state that both mean and variance are invariant to temporal locations n, m. For the mean of the random process, this means that

$$\mu_X(t_n) = \int_{-\infty}^{\infty} x(t_n) p_X(t_n) dx(t_n)$$
$$= \text{const.} \equiv \mu_X \qquad (3.16)$$

because the probability density $p_X(t_n)$ no longer depends on the time t_n. When discussing joint moments, this means that the expectation depends only on the *relative separation* between observations. That is to say, the joint expectation $E[X(t_n)X(t_m)]$ is a function of the integer delay $\tau = (t_m - t_n)$ only. More specifically, for a stationary random process we may substitute

$t_m = t_n + \tau$ in (3.4), giving

$$R_{XX}(t_n, t_n + \tau) = E[X(t_n)X(t_n + \tau)]$$

$$= \int_{-\infty}^{\infty} \int_{-\infty}^{\infty} x(t_n)x(t_n + \tau)p_{\mathbf{X}}(x(t_n), x(t_n + \tau))dx(t_n)dx(t_n + \tau)$$

$$\equiv R_{XX}(\tau) \tag{3.17}$$

defined as the autocorrelation function for the stationary random process $\mathbf{X}$. The absolute time t_n is irrelevant because by the assumption of stationarity, the joint PDF is invariant to its value. By extension we also therefore have the stationary autocovariance function

$$C_{XX}(\tau) = R_{XX}(\tau) - \mu_X^2. \tag{3.18}$$

Therefore, each row of the full covariance matrix (Eq. 3.6) is determined entirely by (3.18). Also, for a stationary random process, we can see that $E[X(t_m)X(t_n)] \rightarrow E[X(t_n)X(t_n + \tau)] = [X(t_m)X(t_m - \tau)]$ and therefore

$$R_{XX}(\tau) = R_{XX}(-\tau)$$

$$C_{XX}(\tau) = C_{XX}(-\tau) \tag{3.19}$$

that is, both the stationary autocorrelation and autocovariance are even functions. Recall also that Isserlis' theorem does not require stationarity, only that the underlying joint PDF be Gaussian. The mean and variance associated with each observation can be different. However, if the random process is stationary, we can say that knowledge of $C_{XX}(\tau)$ completely characterizes the joint distribution of the observations. This is why in many cases one needs to only verify that a time sequence is weakly stationary (as opposed to strictly stationary), because in the weakly stationary case only first and second joint moments are required to specify the entire probabilistic model.

In general, we see that the assumption of stationarity removes one of the independent "time" dimensions in a correlation function. The same holds true for higher joint moments, for example, the functions defined in Eqs. (3.12) become

$$C_{XXX}(t_n, t_m, t_p) \rightarrow C_{XXX}(\tau_1, \tau_2)$$

$$C_{XXXX}(t_n, t_m, t_p, t_q) \rightarrow C_{XXXX}(\tau_1, \tau_2, \tau_3) \tag{3.20}$$

and are functions of the time delays $\tau_1 = t_m - t_n, \tau_2 = t_p - t_n, \tau_3 = t_q - t_n$ only. Again, this is only true for a stationary random process and proves useful in the upcoming section on HOS analysis (see Section 3.3.5). By extension, the cumulants introduced in Section 2.6 can also, for stationary data, be written as a function of time delays, e.g., (3.13) becomes

$$k_{XXX}(t_n, t_m, t_p) \rightarrow k_{XXX}(\tau_1, \tau_2)$$

$$k_{XXXX}(t_n, t_m, t_p, t_q) \rightarrow k_{XXXX}(\tau_1, \tau_2, \tau_3) \tag{3.21}$$

Studying the stationary autocovariance is a good way to understand how the observations $x(t_n)$ and $x(t_n + \tau)$ are related. Certainly for $\tau = 0$ one recovers $E[(X - \mu_X)(X - \mu_X)] = C_{XX}(0) = \sigma_X^2$, that is, the variance of the random variable X (which, again, is independent of the discrete time n for a stationary time-series). The variance σ_X^2 is, in fact, the maximum of the stationary autocovariance, thus a common normalization scheme gives

$$\rho_{\mathbf{XX}}(\tau) = \frac{C_{\mathbf{XX}}(\tau)}{\sigma_X^2} \tag{3.22}$$

resulting in the normalized autocovariance function which is bounded on the interval [$0 \leq |\rho(\tau)| \leq 1$]. Note this is just the extension (temporally) of the correlation coefficient for the bivariate normal distribution presented earlier (Eq. 2.52).

This discussion is easily extended to the covariance between two different random processes, $\mathbf{X}, \mathbf{Y}$. Such a situation commonly arises when multiple sensors are placed on a structure and we are interested in relationships among those data (see Section 9.3.3 on sensor placement study). We define the *cross-correlation* function for two stationary random processes $\mathbf{X}, \mathbf{Y}$ as

$$R_{XY}(t_n, t_n + \tau) = E[X(t_n)Y(t_n + \tau)]$$

$$= \int_{-\infty}^{\infty} \int_{-\infty}^{\infty} x(t_n)y(t_n + \tau)p_{\mathbf{XY}}(x(t_n), y(t_n + \tau))dx(t_n)dy(t_n + \tau)$$

$$\equiv R_{XY}(\tau) = R_{YX}(-\tau) \tag{3.23}$$

and the corresponding autocovariance

$$C_{XY}(\tau) = R_{XY}(\tau) - \mu_X \mu_Y$$

$$= C_{YX}(-\tau) \tag{3.24}$$

as well as the cross-correlation coefficient

$$\rho_{XY}(\tau) = \frac{C_{XY}(\tau)}{\sqrt{\sigma_X^2 \sigma_Y^2}}$$

$$= \rho_{XY}(-\tau). \tag{3.25}$$

As we have frequently noted, these functions do not depend on the absolute time index t_n because of the assumption of stationarity on the joint probability density.

Finally, we mention yet another assumption frequently invoked in the analysis of structural response data. It is common to assume that the noise observed by a sensor is modeled by a random process which is (i) joint normally distributed, (ii) stationary, and (iii) where each value is statistically independent of the next. These last two assumptions are so frequently used that the observations in such a process are referred to as *independent, identically distributed* or i.i.d. The "identically distributed," of course, refers to stationarity, while the assumption of "independence" allows the joint distribution to factor. Recalling from Chapter 2, Eq. (2.17), for a random process with N independent observations we may write

$$p_{\mathbf{X}}(\mathbf{x}) = \prod_{n=1}^{N} p_X(x(t_n)). \tag{3.26}$$

The *only* way for this to be true for a jointly Gaussian distribution is for the off-diagonal terms in the covariance matrix to be identically zero. In other words, if each member of the sequence is drawn independently from the same Gaussian distribution, we have $C_{\mathbf{XX}}(\tau) = 0 \ \forall \ |\tau| > 0$. This is also true for higher order statistics. For example, in a jointly non-Gaussian model, the i.i.d. assumption implies that $E[X(t_n)X(t_n + \tau_1)X(t_n + \tau_2)] = 0, \forall \ |\tau_1|, |\tau_2| > 0$.

To summarize this section: if the time series being analyzed are modeled as jointly Gaussian random processes, *all* of the information about the time series is carried by the autocovariance function (3.5) or (alternatively) the correlation coefficient. If we further restrict the random process to be at least weakly stationary, the mean value is no longer a function of time and the auto-covariance is a one-dimensional function of the temporal separation between observations. In this case, Isserlis' theorem tells us that Eq. (3.18) defines the entire joint PDF. Both the assumptions made here (jointly Gaussian and stationary) are typically made in practice, sometimes implicitly, and often in cases where they may not be warranted. If the noise model being considered is *not* Gaussian and/or the underlying probability distribution is changing in time, then Eq. (3.18) may not be a sufficient descriptor of the random process. While non-Gaussian random process models may certainly prove more challenging to handle analytically, the framework we have developed allows us to handle these cases as will be shown in later sections.

3.3 Spectral Analysis

Spectral analysis comprises a very large, and very well-studied body of material, the origins of which date back to the early 19th century. The breadth of the subject is supported by its usefulness across a variety of disciplines ranging from the very abstract to the very applied. What began as a mathematical technique for analyzing deterministic, periodic functions has been subsequently extended to realizations of a random process, precisely the type of data we are most interested in. We follow this same progression here, beginning with a discussion of the spectral properties of deterministic signals and then generalizing to random process models. In the process we discuss, in detail, the conditions under which spectral analysis is possible.

In short, spectral analysis allows us to model a given signal as a superposition of some (possibly infinite) number of sine and cosine functions whose arguments are a product of frequency and time. Fourier analysis can be used to identify the amplitudes of those frequencies. For structural systems, the measured data are often dominated by a small number of "modes" (see Chapter 4), and hence a sinusoidal model for such data makes sense. Moreover, structural damage can be expected to alter the modal properties, and hence the historical usage of modal analysis in the damage detection field.

Of particular importance in our development are the spectral properties of the autocorrelation function of a random process. As we have already shown, the autocorrelation offers a complete description of the statistical properties of a jointly normal, stationary random process. Moreover, for structural systems undergoing random vibration, $R_{XX}(\tau)$ is often a decaying periodic function comprised largely of only a few frequencies (see Section 6.4.2) and is therefore ideally suited to spectral analysis. The distribution of Fourier amplitudes associated with these frequencies will be referred to as the *power spectral density* or PSD function. As it is related to $R_{XX}(\tau)$ by a deterministic transformation (the Fourier transform), the PSD too offers a complete description of a stationary, jointly normal random process.

The relationship between correlation and signal power is an important one that we leverage throughout this book. In fact, the same procedure we use to derive this relationship will also allow us to derive relationships between higher order stationary correlation functions (e.g., 3.20) and their associated spectral functions, the so-called higher order spectra. As we will

see, however, the spectral analysis of random processes leads directly to some thorny technical issues which we feel the reader should be aware of. We attempt to point out these issues
and provide solutions as concisely as possible, presenting only the material required in subsequent chapters. For those interested in the details of spectral analysis we recommend the
comprehensive work of Priestly [6].

3.3.1   Spectral Representation of Deterministic Signals

At the heart of spectral analysis is the Fourier series, named for the French mathematician Jean
Baptiste Joseph Fourier (1768–1830). For a *deterministic, periodic* function $x(t)$ with period
$2T$, we may write [6]

$$x(t) = \frac{1}{2}a(0) + \sum_{n=1}^{\infty}[a(n)\cos(\pi t/T) + b(n)\sin(\pi t/T)] \tag{3.27}$$

where

$$a(n) = \frac{1}{T}\int_{-T}^{T} x(t)\cos\left(\frac{\pi nt}{T}\right)dt \tag{3.28}$$

and

$$b(n) = \frac{1}{T}\int_{-T}^{T} x(t)\sin\left(\frac{\pi nt}{T}\right)dt. \tag{3.29}$$

Thus, we describe our signal as a sum of sines and cosines at discrete frequencies $f_n = n/2T$
cycles per second, or Hertz, denoted Hz. This holds for any $2T$ periodic function, including
those with discontinuities (e.g., sawtooth wave), provided that 1) $\int_{-T}^{T}|x(t)|dt < \infty$ and 2) $x(t)$
is "well-behaved," possessing a finite number of maxima, minima, and discontinuities on the
interval $(-T, T)$.

The Fourier series representation can be made more compact by using a complex exponential form to rewrite (3.27) as

$$x(t) = \sum_{n=-\infty}^{\infty} X(n)e^{2\pi i nt/2T} \tag{3.30}$$

where by Eqs. (3.28, 3.29) the complex coefficients $X(n) = a(n) + ib(n)$ are [6]

$$X(n) = \frac{1}{2T}\int_{-T}^{T} x(t)e^{-2\pi i nt/2T}\,dt$$

or

$$X(f_n) = \frac{1}{2T}\int_{-T}^{T} x(t)e^{-2\pi i f_n t}\,dt. \tag{3.31}$$

In our notation, we always use a capital letter to represent the Fourier coefficients associated
with the "lower case" signal, e.g., $X(f)$ are the Fourier coefficients used to model $x(t)$. However,
to this point in the book we have also used capital letters to represent random variables. We
explicitly note here that when the argument of the upper case quantity is a frequency, we are

always talking about Fourier coefficients. It will always be clear from the context whether we are referring to a random variable or a Fourier coefficient.

Equation (3.30) can also be used to model *deterministic, nonperiodic* signals provided we are willing to restrict the definition of the signal to the finite region $-T < t < T$ and then assume that the signal repeats itself *ad infinitum*. That is to say, for $t \in [-T, T]$ assume

$$x(t + 2Tp) = x(t), \quad p = \cdots, -2, -1, 0, 1, 2, \cdots \tag{3.32}$$

so that $x(t)$ is by definition $2T$ periodic. So long as we define our signal this way, we can use (3.27) to model our nonperiodic signal just as we did for the periodic signal. Moreover, the existence of (3.31) is still guaranteed for well-behaved signals, again so long as the signal is absolutely integrable on the pertinent range, i.e., $\int_{-T}^{T} |x(t)| dt < \infty$. So far, however, we still have placed two very important restrictions on our analysis, namely, that (i) the signal is deterministic and (ii) the signal is periodic (either naturally or by restricting the range and assuming infinite repetition), and therefore comprises a discrete set of frequencies. Neither of these assumptions holds for random process models. We proceed to address this second restriction.

Returning to (3.31), we note that in the last step we have made the replacement $f_n = n/2T$, defined as the frequency associated with the nth complex amplitude $X(f_n)$. The frequency spacing is therefore dictated by the fundamental signal period and is given by $\Delta f = 1/2T$. Thus, we could have alternatively written (3.30) as

$$x(t) = \sum_{n=-\infty}^{\infty} \left(\int_{-T}^{T} x(t) e^{-2\pi i f_n t} \, dt \right) e^{2\pi i f_n t} \Delta f. \tag{3.33}$$

It seems quite natural to let $T \to \infty$ so that the discrete set of frequencies becomes continuous, thereby giving us the well-known Fourier transform integral pair

$$X(f) = \int_{-\infty}^{\infty} x(t) e^{-2\pi i f t} \, dt$$

$$x(t) = \int_{-\infty}^{\infty} X(f) e^{2\pi i f t} \, df. \tag{3.34}$$

In some instances, it will prove notationally convenient to define this pair in terms of the angular frequency variable $\omega = 2\pi f$. In this case, the Fourier transform pair becomes

$$X(\omega) = \int_{-\infty}^{\infty} x(t) e^{-i\omega t} \, dt$$

$$x(t) = \frac{1}{2\pi} \int_{-\infty}^{\infty} X(\omega) e^{i\omega t} \, d\omega. \tag{3.35}$$

Both formulations are used in this book, depending on the application. By (3.34) the signal is now assumed to be modeled over a continuous range of frequencies. We have apparently resolved the restriction on the signal being $2T$ periodic by simply allowing $T \to \infty$. However, one important distinction with the discrete representation (3.30) is that the absolute integrability condition for the existence of the transform becomes [6]

$$\int_{-\infty}^{\infty} |x(t)| dt < \infty \tag{3.36}$$

whereas in our prior analysis we only had to satisfy $\int_{-T}^{T} |x(t)|dt < \infty$ for $2T$ periodic signals. By allowing $T \to \infty$, we attain the familiar integral representation (3.34), however in the process, we have almost guaranteed that for the signals we care about such a representation will not exist! It is almost certainly *not* the case that a typical realization of a random process will obey (3.36). Imagine, for example, that $x(t)$ is a sequence of i.i.d. random variables from any probability distribution. Such a random process is comprised of a nondecaying, infinite sequence of values such that (3.36) is not met. However, as we have stressed, random process models are used almost exclusively to model structural response data. A spectral theory that can accommodate such models is therefore a necessity in practice and not simply a matter of mathematical rigor.

The development of such a theory will be partially aided by first considering the choice we must make in representing $x(t)$. Do we assume a discrete set of frequencies and use (3.30)? Or do we choose to model the signal over a continuous range of frequencies using (3.34) or (3.35)? It turns out there is a compact representation that can be used to represent *both* types of signals and that this representation will be crucial to our eventual goal of handling stochastic signals.

Consider first our continuous inverse Fourier transform as the limit of the Riemann summation $\lim_{\Delta_f \to 0} \sum_{n=1}^{N} e^{i2\pi f_n t} X(f_n) \Delta_f$ over the range $[f_1, f_N]$ in uniform increments of $\Delta_f = f_{n+1} - f_n$. From basic calculus, we know that this approximation holds so long as $X(f)$ is a well-behaved, continuous function. If $X(f)$ possesses sudden "jumps" or discontinuities, this definition of an integral breaks down. For example, if the signal $x(t)$ is periodic, then $X(f)$ will be comprised of a finite number of discrete spectral lines. We clearly cannot use the traditional integral representation for periodic signals which is why, of course, we began our analysis with the discrete summation (3.30).

Nonetheless, we can still combine (3.30) and (3.34), if we are willing to consider a more general definition of the integral. For a function $Z(f)$, define the Riemann–Stieljes integral over the range $[a, b]$ with respect to the deterministic function $g(f)$ as

$$\int_a^b g(f)dZ(f) = \lim_{\Delta_f \to 0} \sum_{n=1}^{N} g(f)(Z(f_n) - Z(f_{n-1})) \tag{3.37}$$

where $a \le f_1 < f_2, \dots, < f_N \le b$ and where we have defined $\Delta_f = f_n - f_{n-1}$. The strength of this representation is that the integral exists even when the function $Z(f)$ is nondifferentiable, i.e., when $dZ(f)$ is "spikey." The main requirement is that the increments be of "bounded variation," meaning $\sum_{n=2}^{N} |Z(f_n) - Z(f_{n-1})| < \infty$. However, in the event that $Z(f)$ is continuous, we may recover the traditional definition of the integral by writing $\int_a^b g(f)\frac{dZ(f)}{df}df = \int_a^b g(f)X(f)df$, where we have let $X(f) \equiv dZ(f)/df$ (our reasons for defining $X(f)$ this way will be clear shortly). In this case, the existence of the transform is governed by (3.36).

In spectral analysis we take $g(f) \equiv e^{i2\pi ft}$, extend the limits, and write the *Fourier–Stieltjes transform*

$$x(t) = \int_{-\infty}^{\infty} e^{i2\pi ft} \, dZ(f). \tag{3.38}$$

This representation is valid for deterministic periodic signals and a-periodic signals that are absolutely integrable and therefore combines both discrete and continuous Fourier representations. For example, consider $x(t)$ periodic so that $dZ(f)$ consists of unit amplitude spikes at frequencies $f_n, n = 1, 2, \cdots$ In this case, we see that the corresponding function $Z(f)$ is given by

$$Z(f) = \begin{cases} 0 : & f < f_1 \\ 1 : & f_1 \leq f < f_2 \\ 2 : & f_2 \leq f < f_3 \\ \cdots : & \cdots \end{cases} \tag{3.39}$$

that is, a monotonically increasing series of "steps." The advantage of the representation (3.38) is that we can write the inverse Fourier transform of a discontinuous spectral function $dZ(f)$ as a continuous transform. Moreover, if $Z(f)$ is differentiable we can always rewrite the Fourier–Stieltjes transform

$$x(t) = \int_{-\infty}^{\infty} e^{i2\pi ft} dZ(f) = \int_{-\infty}^{\infty} e^{i2\pi ft} \frac{dZ(f)}{df} df$$

$$= \int_{-\infty}^{\infty} e^{i2\pi ft} X(f) df \tag{3.40}$$

and therefore recover the continuous Fourier representation with $X(f) = dZ(f)/df$ (the existence of which is predicated on 3.36). Thus, we have a general, compact spectral representation for a broad class of deterministic signals.

The function $Z(f)$ is of an interesting character in the sense that the derivative $dZ(f)/df$ behaves like the continuous Fourier transform, while the increments $Z(f_{n+1}) - Z(f_n)$ behave like discrete Fourier coefficients $X(n)$. This representation is consistent, as it must be, from the perspective of units as well. If the signal $x(t)$ has units of volts, (V), we know the continuous Fourier transform $X(f)$ has units (V · s) or, alternatively, Volts per unit frequency. The signal $Z(f)$ clearly has units of (V) as well so that $X(f) = dZ(f)/df$ is (again) (V · s). However, in the periodic case, $dZ(f)$ is a voltage difference, and hence the inverse Fourier transform will also possess units of volts. What is interesting, however, is that while the units are consistent, the magnitude of $dZ(f)$ changes drastically depending on the type of signal being modeled. For periodic signals $dZ(f) \sim O(1)$, while for a continuous spectral representation $dZ(f) \sim O(df)$ [6]. We return to this point when discussing the spectral representation of stochastic signals.

This would seem to be quite a bit of work to go through simply to come up with a unified spectral representation. After all, we usually know which representation (discrete sum or continuous integral) would be most appropriate for a given application. As we see next, however, the usefulness of (3.38) extends far beyond the ability to combine (3.30) and (3.34). It turns out that the Fourier–Stieltjes representation is the key to the spectral analysis of random process data.

3.3.2 Spectral Representation of Stochastic Signals

Consider now $x(t)$ to be a realization of a random process $X(t)$, defined for all time $-\infty < t < \infty$. In the previous section we discussed one problem with the spectral representation of such signals, namely, conformity to (3.36). A straightforward solution from both a practical and mathematical point of view is to simply restrict the definition of $x(t)$ such that

$$x_R(t) = \begin{cases} x(t) : & -T < t < T \\ 0 : & \text{otherwise} \end{cases}. \tag{3.41}$$

Thus, by definition, any well-behaved random process will obey (3.36). Therefore, we can almost always define the Fourier transform pair $x_R(t), X_R(f)$ where

$$X_R(f) \equiv \int_{-T}^{T} x_R(t)e^{-2\pi ift} \, dt. \tag{3.42}$$

The restriction to the interval $(-T, T)$ is not really problematic in practice as our observed data are finite and therefore estimates of $X_R(f)$ are always restricted to a finite domain. Indeed, we take precisely this "windowing" approach when estimating spectral properties in Section 6.4.1. However, this discrepancy becomes important as the theory to which we compare our estimates often demands that the random process extend infinitely in time. Moreover, the existence of the transformation is not the only issue we face here. We are proposing to take an integral transformation of a sequence of random variables, which would seem to imply that the Fourier representation, $X(f)$, is also a random process!

Fortunately, there exists a rigorous mathematical treatment of spectral analysis for random processes that solves the above-described problems. Detailed treatment of this topic can be found (originally) in Wiener [7] and subsequently in Yaglom [8] and Priestly [6]. To present this theory, we first note that for sequences of random variables, even basic calculus operations such as "integration" and "differentiation" require a different treatment. Differentiation, for example, requires evaluating differences in function values at ever closer intervals $df \to 0$, i.e., we take a limit. If the elements of the function are random variables, the concept of a single "limiting value" does not make sense. One solution is to change what we mean by a "limit" for random processes and require that the derivatives converge in expectation, that is to say, $\lim_{df \to 0} E\left[\left\{ X'(f) - \frac{X(f+df)-X(f)}{df} \right\}^2 \right] \to 0$. Convergence in mean-square value is one practically useful way to define the existence of derivatives $X'(f) \equiv dX(f)/df$ of a random process.

Similarly, we typically define the integral of $X(f)$ as the limit of a Riemann summation, as we showed in the previous section. Just as with differentiation, for random processes we can only really view the existence of the integral in the sense of expectation, i.e., for $X(f)$ continuous, $E[\{|X(f) - \int X'(f)df\}^2| \to 0$ as $df \to 0$. The question is whether we can develop a spectral model for which convergence in this sense is guaranteed.

It turns out that if we take the Riemann–Stieljes approach of the previous section, we can, in fact, demonstrate convergence of the resulting representation in the mean-square sense. Specifically, we replace the traditional Fourier integral with the Fourier–Stieltjes integral

$$x(t) = \int_{-\infty}^{\infty} e^{2\pi ift} dZ(f) \tag{3.43}$$

where the function $Z(f)$ is a complex random process and therefore so too is $dZ(f) = Z(f + df) - Z(f)$ (again, because we are modeling $x(t)$ as a random process, it is required that the spectral model $Z(f)$ also be a random process). At each frequency f, each of the frequency increments $dZ(f)$ is therefore treated as a random variable. As such, we can speak about the statistical properties of $dZ(f)$, for example, the mean $E[dZ(f)]$. The statistical properties of the random process $dZ(f)$ are discussed in the next section. Perhaps the most remarkable aspect of the representation (3.43) is that it can be proved that in expectation

$$E\left[|X(t) - \int_{-\infty}^{\infty} e^{2\pi ift} dZ(f)|^2 \right] = 0 \tag{3.44}$$

for random processes $X(t)$ that are not absolutely integrable [8]. The representation therefore converges in a mean-square sense even though we retain the infinite limits on the integral. The function $Z(f)$ therefore behaves in many ways as a general integral of $X(f)$. We further note (as we did in Section 3.3.1) that the properties of $Z(f)$ are defined such that one can also evaluate (3.43) for periodic (where $dZ(f)$ is comprised of discrete spectral lines) or deterministic a-periodic functions in addition to random processes. Hence, (3.43) is an extremely general and practical spectral model for our data. This representation was pioneered by Weiner [7] and remains the standard spectral analysis tool for random processes at present.[1] It is also of interest to note, that in the case of random processes, the coefficients are of order $dZ(f) \sim O(\sqrt{df})$, a fact we return to in the next section [6].

We elaborate on this formulation in the next section as it also solves a major challenge in defining the PSD for a random process. As this book deals almost exclusively with the analysis of data modeled as a random process, the representation (3.43) is of great importance. As we mentioned previously, the quantity $dZ(f)/df$ effectively plays the role of the Fourier transform $X(f)$. However, in the practical cases we are interested in, the function $Z(f)$ is *not* differentiable, and hence the replacement $dZ(f) \to X(f)df$ is usually improper [6, 8]. As a final point, we note that we could also have used the angular frequency variable in the representation, i.e.,

$$x(t) = \int_{-\infty}^{\infty} e^{i\omega} dZ(\omega) \qquad (3.45)$$

where the random process $dZ(\omega)$ retains the same interpretation as $dZ(f)$. Depending on the application, we use both types of frequency representations in this book.

3.3.3   Power Spectral Density

One of the most useful aspects of Fourier analysis is that it provides insight into which frequencies or frequency ranges contain most of the signal energy. Structural vibration data, for example, often have energy concentrated in just a few frequency bands (structural "modes," see Chapter 4). In this section, we make explicit the concepts of signal "energy" and "power" as they play a central role in later chapters. For example, in Chapter 9, we develop an estimation approach that works by matching the power/frequency distribution between signal and model. The PSD also plays a crucial role in the development of "surrogate" data sets used in Chapter 8.

Consider first an expression for the total signal "energy"[2]

$$E_X = \int_{-\infty}^{\infty} x^2(t)dt. \qquad (3.46)$$

---

[1] In the cited work, Weiner remarks about two prior efforts using traditional Fourier methods applied to random processes: "In both cases one is astonished by the skill with which the authors use clumsy and unsuitable tools to obtain the right results, and one is led to admire the unfailing heuristic insight of the true physicist." Quite a "compliment"!
[2] "Energy" is in quotes because the units of E_X are not units of energy, but rather carry units $[x]^2 \cdot [time]$. However, they are directly proportional to energy via an appropriately chosen constant; for example, if $x(t)$ is in volts, the energy dissipated across a resistor ω (Ohms) is simply E_X/ω.

Assuming the existence of the Fourier transform so that $x(t) = \int_{-\infty}^{\infty} X(f)e^{i2\pi ft}\, df$, we may write

$$
\begin{aligned}
E_X &= \int_{-\infty}^{\infty} x^2(t)dt = \int_{-\infty}^{\infty} x(t) \left(\int_{-\infty}^{\infty} X(f)e^{2\pi ift}\, df \right) dt \\
&= \int_{-\infty}^{\infty} X(f) \left(\int_{-\infty}^{\infty} x(t)e^{2\pi ift}\, dt \right) df \\
&= \int_{-\infty}^{\infty} X(f)X^*(f)df \\
&= \int_{-\infty}^{\infty} |X(f)|^2 df.
\end{aligned}
\tag{3.47}
$$

This is known as Parseval's relation and states that the total signal energy is equal to the integral of the magnitude squared of the Fourier transform. It is worth pausing to consider the units in the given expression. Let $[x]$ denote the units of $x(t)$. The left-hand side of (3.47) has units $[x]^2 \cdot s$. The quantity $X(f)$ has units $[x] \cdot s$ and therefore $|X(f)|^2$ has units $[x]^2 \cdot s^2$. The frequency increment df goes as s^{-1} so that the right-hand side of (3.47) also has units $[x]^2 \cdot s^2 \cdot s^{-1} = [x]^2 \cdot s$ as it must. As part of this exercise, we see the term $|X(f)|^2$ has units proportional to signal energy multiplied by time, or, alternatively, signal energy per unit frequency. The quantity inside the integral, $|X(f)|^2$, is therefore an energy density function and we may regard $|X(f)|^2 df$ as proportional to the portion of the signal energy found in the frequency interval $[f, f + df]$.

Here again, however, we see the problem associated with the infinite limits on the integrals. A realization of a random process (extending infinitely in time) would seem to possess infinite energy unless, of course, we restrict the limits on our random process as was done in (3.42). Thus, the quantity $|X_R(f)|^2 df$ will have the interpretation as the fraction of energy in the signal $x_R(t)$ in frequency interval $[f, f + df]$. However, we still have the problem in the limiting case that $T \to \infty$. A simple yet practical solution is to study a different quantity!

It is common practice to focus on the signal *power* density in the $[-T, T]$ interval and take the limit

$$
\lim_{T \to \infty} \frac{|X_R(f)|^2}{2T}.
\tag{3.48}
$$

Signal power is a much better behaved function mathematically (tends to converge as $T \to \infty$ for most random processes) and is therefore preferred to signal energy. Finally, we note that expression (3.48) is defined for a given realization of our random process, which tells us little about "expected" behavior. Considering this function in expectation leads to the *PSD* function

$$
S_{XX}(f) = \lim_{T \to \infty} E\left[\frac{|X_R(f)|^2}{2T} \right].
\tag{3.49}
$$

This quantity plays a central role in the analysis of random processes and plays a crucial role in later chapters. Had we used the angular frequency ω in the derivation, the expression would read

$$
S_{XX}(\omega) = \lim_{T \to \infty} \left(\frac{1}{2\pi} \right) E\left[\frac{|X_R(\omega)|^2}{2T} \right]
\tag{3.50}
$$

As we will show, this quantity is easily estimated from observed data. Because we are *always* dealing with a finite sequence of observations, limiting the signal to $[-T, T]$ is not at all an issue when it comes to estimating (3.49), as we see in Section 6.4.1.

Nonetheless, we will later be required to mathematically derive the PSD for random processes that exist over the entire real number line and cannot be restricted to the $[-T, T]$ interval. For this case, we rely on the spectral representation (3.43) discussed earlier. Using this representation, it will be shown (see Wiener [7] and also Priestly [6]) that for a random process $x(t)$ with spectral representation (3.43), we may alternatively write

$$S_{XX}(f)df = E[dZ(f)dZ^*(f)]$$

$$= E[|dZ(f)|^2]. \tag{3.51}$$

In fact, we show how to arrive at this expression in Section 3.3.5. From Eq. (3.51) we have that the total signal power in the frequency increment $[f, f + df]$ is equal to the expected magnitude of the Fourier–Stieltjes representation of the random process. We can understand the power of the Fourier–Stieltjes representation by again considering units. As we have mentioned, $dZ(f) \sim O(\sqrt{df})$ hence the produce $E[dZ(f)dZ^*(f)] \sim O(df)$, so that both the left- and right-hand side of (3.51) are of the same order. Thus, we have a stable definition of the PSD for all stationary random processes, even under the limiting operation of $df \to 0$. Using this formulation, it can also be shown (see development of Eq. 3.73) that if $x(t)$ is zero mean, then the random process $dZ(f)$ has the properties

$$E[dZ(f)] = 0$$

$$E[dZ(f)dZ^*(f')] = 0 \quad \forall f \neq f' \tag{3.52}$$

that is to say, the random variables $dZ(f)$ and $dZ(f')$ are uncorrelated and zero mean. Another interesting property of $dZ(f)$ is that for periodic functions we had $dZ(f) \sim O(1)$, while for a-periodic signals $dZ(f) \sim O(df)$. Hence, the spectral coefficients associated with a random process lie somewhere between the periodic and a-periodic signal coefficients in terms of their magnitude which, again, is $O(\sqrt{df})$. It is also noted here that nothing changes in Eqs. (3.51) or (3.52) in replacing f with ω as the frequency variable.

In practice, we tend to work with (3.49) in forming PSD estimates $\hat{S}_{XX}(f)$ from our sensor observations. However, it is important to keep in mind (3.51) when developing analytical results concerning the spectral analysis of random processes. A good example of when this representation is important occurs in Chapter 10 in developing a method for generating realizations of random processes. Using the more elegant Fourier–Stieltjes approach eliminates mistakes in units that can easily occur using the traditional Fourier analysis. This representation will also be used in deriving the HOS density functions in Section 3.3.5 and in developing a test for nonstationarity in Section 6.8.2.

As an additional note on PSD, while we can define (3.49) for a signal with infinite energy (random processes, periodic signals) a signal that obeys $\int_{-\infty}^{\infty} x^2(t)dt < \infty$ will, by Parseval's relationship, be bounded in the frequency domain and hence as $T \to \infty$ will have zero power. For this class of signals, it makes sense to use the energy density, however in this work we always assume the relevant quantity for the frequency domain description of a random process to be power, i.e., Eq. (3.49).

To conclude this section, we note that it is also common in signal processing to study the relationship between multiple signals. Although it lacks the physical interpretation of signal energy, we can consider the quantity

$$E_{XY} = \int_{-\infty}^{\infty} x(t)y(t)\, dt \tag{3.53}$$

which, in analogy to Eq. (3.47), reduces to

$$
\begin{aligned}
E_{XY} &= \int_{-\infty}^{\infty} x(t)y(t)dt = \int_{-\infty}^{\infty} x(t) \left(\int_{-\infty}^{\infty} Y(f)e^{2\pi ift}\, df \right) dt \\
&= \int_{-\infty}^{\infty} Y(f) \left(\int_{-\infty}^{\infty} x(t)e^{2\pi ift}\, dt \right) df \\
&= \int_{-\infty}^{\infty} Y(f)X(-f)df = \int_{-\infty}^{\infty} Y(f)X^*(f)df
\end{aligned}
\tag{3.54}
$$

Following the same logic surrounding Eq. (3.47) leads naturally to the definition of the cross-spectral density function for restricted random processes $x_R(t), y_R(t)$

$$S_{XY}(f) = \lim_{T\to\infty} \frac{E[Y_R(f)X_R^*(f)]}{2T} \tag{3.55}$$

where by convention the conjugated argument appears second in the expectation. Thus, we could also write

$$S_{YX}(f) = \lim_{T\to\infty} \frac{E[X_R(f)Y_R^*(f)]}{2T}. \tag{3.56}$$

Alternatively, using the Fourier–Stieltjes representation for each signal, we could have written

$$S_{XY}(f)df = E[dZ_Y(f)dZ_X^*(f)]. \tag{3.57}$$

Had we used instead the angular frequency variable, we would have the corresponding relationships

$$S_{XY}(\omega) = \lim_{T\to\infty} \left(\frac{1}{2\pi} \right) \frac{E[Y_R(\omega)X_R^*(\omega)]}{2T}$$

$$S_{XY}(\omega)d\omega = E[dZ_Y(\omega)dZ_X^*(\omega)]. \tag{3.58}$$

The analysis in this section has, to this point, made no assumptions about whether the random process $X(t)$ is real or complex valued. We will obviously always be interested in the former situation, in which case more can be said about the relationship between the PSD functions for negative and positive frequency variables. Specifically, it can be seen from the given development that the following properties hold for both the auto- and cross-spectral densities of a real-valued random process

$$S_{XX}(f) = S_{XX}(-f) = S_{XX}^*(f)$$

$$S_{XY}(f) = S_{YX}(-f) = S_{XY}^*(-f) \tag{3.59}$$

regardless of whether we use f or ω as the frequency variable. Next, it is shown how the spectral densities are related to the joint statistical properties of stationary random processes.

3.3.4   Relationship to Correlation Functions

Given the above-mentioned background material, we are now in a position to explore the spectral properties of correlation functions. As we saw in Section 3.1, random processes are described by their statistical properties. More specifically, one typically looks at expectations of joint moments of a random process. For stationary, jointly Gaussian random processes, it was shown that the autocovariance function offered a complete probabilistic description. However, the autocovariance (or autocorrelation) functions are seldom displayed as a function of time-delay, i.e., it is not typical to plot $R_{XX}(\tau)$ versus τ. Rather, it is far more common to look at the Fourier transform of $R_{XX}(\tau)$. The reason is that for structural systems, $R_{XX}(\tau)$ can often be treated as a periodic function, comprised of one or more dominant tones. As we have just shown, the Fourier transform can be used to highlight the frequencies and amplitudes of those tones.

It turns out there is a useful relationship between the autocorrelation function for stationary processes and the PSD of the process. The origins of the relationship trace to the work of Wiener [7] and the Russian mathematician Aleksandr Khinchin. Consider two random processes $x_R(t), y_R(t)$, each defined as in (3.41) and possessing the corresponding Fourier representation (3.42). As before, the restriction to the interval $[-T, T]$ is made to bound the energy associated with our signals. Using these definitions, define the correlation function

$$\tilde{R}_{XY}(\tau|T) = \frac{1}{2T} \int_{-\infty}^{\infty} x_R(t) y_R(t+\tau) dt \qquad (3.60)$$

which is similar in appearance to the cross-correlation for stationary processes $R_{XY}(\tau)$, as defined in Eq. (3.23). Consider now the Fourier transform of this function

$$\int_{-\infty}^{\infty} \tilde{R}_{XY}(\tau|T) e^{-2\pi i f \tau} \, d\tau = \frac{1}{2T} \int_{-\infty}^{\infty} \int_{-\infty}^{\infty} x_R(t) y_R(t+\tau) e^{-2\pi f \tau} \, dt \, d\tau$$

$$= \frac{1}{2T} \int_{-\infty}^{\infty} \int_{-\infty}^{\infty} x_R(t) e^{i2\pi ft} y_R(t+\tau) e^{-i2\pi f(t+\tau)} \, dt \, d\tau$$

$$= \frac{1}{2T} Y_R(f) X_R^*(f) \qquad (3.61)$$

whereby taking the limit of the expected value

$$\lim_{T\to\infty} \int_{-\infty}^{\infty} E[\tilde{R}_{XY}(\tau|T)] e^{-2\pi i f \tau} \, d\tau = \lim_{T\to\infty} \frac{1}{2T} E[Y_R(f) X_R^*(f)]$$

$$= S_{XY}(f) \qquad (3.62)$$

Of course, this is not yet the desired result. The function (3.60) is quite different from the cross-correlation $R_{XY}(\tau)$ (3.23).

To understand how the two are related, we write

$$E[\tilde{R}_{XY}(\tau|T)] = \frac{1}{2T} \int_{-\infty}^{\infty} E[X_R(t) Y_R(t+\tau)] dt$$

$$= \frac{1}{2T} \int_{-\infty}^{\infty} R_{XY}(\tau) dt$$

$$= \frac{1}{2T} \int_{-T}^{T-|\tau|} R_{XY}(\tau)dt \tag{3.63}$$

where in the last line we have made use of the fact that the signals $x_R(t)$, $y_R(t)$ are defined to be zero outside the interval $[-T, T]$. Carrying out the integration yields

$$E[\tilde{R}_{XY}(\tau|T)] = R_{XY}(\tau) \left(1 - \frac{|\tau|}{2T} \right) \tag{3.64}$$

and, thus, we arrive at

$$S_{XY}(f) = \lim_{T \to \infty} \int_{-\infty}^{\infty} R_{XY}(\tau) \left(1 - \frac{|\tau|}{2T} \right) e^{-2\pi f \tau} \, d\tau$$

$$= \int_{-\infty}^{\infty} R_{XY}(\tau) e^{-2\pi i f \tau} \, d\tau. \tag{3.65}$$

Considering the case of a single stationary random process $x(t)$ with autocorrelation $R_{XX}(\tau)$, we can similarly write

$$S_{XX}(f) = \int_{-\infty}^{\infty} R_{XX}(\tau) e^{-2\pi i f \tau} \, d\tau. \tag{3.66}$$

Had we used the angular frequency variable, the relationships (3.65, 3.66) would have become

$$S_{XY}(\omega) = \frac{1}{2\pi} \int_{-\infty}^{\infty} R_{XY}(\tau) e^{-i\omega\tau} d\tau$$

$$S_{XX}(\omega) = \frac{1}{2\pi} \int_{-\infty}^{\infty} R_{XX}(\tau) e^{-i\omega\tau} d\tau \tag{3.67}$$

respectively. Moreover, we see that for this relationship to hold, it is sufficient for the auto- or cross-correlation functions to be absolutely integrable [6] i.e.,

$$\int_{-\infty}^{\infty} |R_{XY}(\tau)| d\tau < \infty \tag{3.68}$$

so that Eq. (3.65) is dominated by $R_{XY}(\tau)$ (as opposed to $|\tau|/2T$). Now the given analysis also implies, of course, that

$$R_{XY}(\tau) = \int_{-\infty}^{\infty} S_{XY}(f) e^{i2\pi f \tau} \, df \tag{3.69}$$

that is to say, the cross-correlation is the inverse Fourier transform of the cross-spectral density function. This has come to be known as the Wiener–Khinchin theorem in honor of Norbert Wiener and Alexandre Khinchin, whose independent research led to its development.

However, the result is even more general than what we have so far presented. Recall that we began this analysis with the restricted signals $x_R(t)$, $y_R(t)$ but would like to extend the analysis to random processes of arbitrary, possibly infinite length. It turns out that for most structural systems, the response auto- and cross-correlation functions do, in fact, decay to zero as $\tau \to \infty$ so that the existence condition is met even for signals with infinite support. This is true even if the structure's response is appropriately modeled as a random process. Nonetheless, we know from our previous discussion that we may instead consider the more general Fourier–Stieltjes representation

$$R_{XY}(\tau) = \int_{-\infty}^{\infty} e^{i2\pi f\tau}\, dH_{XY}(f) \tag{3.70}$$

which we expect to be valid for almost any type of spectral function $H_{XY}(f) = \int_{-\infty}^{f} S_{XY}(f)df$, even if it possesses discontinuities. Because of its central importance in statistical signal processing, we give the complete result here for the autocorrelation and autospectral densities

Theorem 3.3.1 (*Wiener–Khinchin theorem*) *Given a stationary, zero-mean random process $X(t)$ with continuous spectral representation $x(t) = \int_{-\infty}^{\infty} e^{i2\pi ft}dZ(f)$, the autocorrelation function can be written*

$$R_{XX}(\tau) = \int_{-\infty}^{\infty} e^{i2\pi f\tau}\, dH_{XX}(f) \tag{3.71}$$

where $dH_{XX}(f) = E[dZ(f)dZ^*(f)]$

This result is proved to hold in the mean-square sense discussed earlier in the context of Eq. (3.43), and convergence is thus guaranteed. A similar theorem holds for the cross-spectral density (3.65). When the function $H_{XX}(f)$ is differentiable, we have $dH_{XX}(f)/df = S_{XX}(f)$ and therefore we recover

$$R_{XX}(\tau) = \int_{-\infty}^{\infty} e^{i2\pi f\tau}S_{XX}(f)df \tag{3.72}$$

which is the same as (3.66). Throughout this book we assume the existence condition (3.68) is met and that $H_{XX}(f)$ is differentiable, hence the Fourier transform pair (3.72) and (3.66) may be used.

We can actually use (3.71) to demonstrate one of the key properties of the random process $dZ(f)$ given by Eq. (3.52). In particular, we write

$$R_{XX}(\tau) = E[X(t)X(t+\tau)]$$

$$= E\left[\int_{-\infty}^{\infty}\int_{-\infty}^{\infty} dZ(f)e^{i2\pi ft}dZ(f')e^{i2\pi f'(t+\tau)}\right]$$

$$= \int_{-\infty}^{\infty}\int_{-\infty}^{\infty} e^{i(f+f')t}e^{i2\pi f'\tau}E[dZ(f)dZ(f')] \tag{3.73}$$

which we know to be a function of τ only for stationary data. Thus, for all $f \neq -f'$ we have that $E[dZ(f)dZ(f')] = 0$ and the random variables $dZ(f)$ and $dZ(f')$ are uncorrelated.

We may also use this relationship to develop an important result that appears later in the book. For the special case of an independent, identically distributed (i.i.d.) random process **X**, we have that $R_{XX}(0) = \sigma_X^2$ and $R_{XX}(\tau \neq 0) = 0$. In this case, we can also see (from Eq. 3.66, for example) that $S_{XX}(f)$ will not be a function of frequency, but rather a constant P. In this special case,

$$\sigma_X^2 = \int_{-\infty}^{\infty} P\, df \tag{3.74}$$

leaving us with infinite variance. As we have pointed out, in practice we never have a truly unlimited bandwidth; rather, we are always restricted to an interval $f \in \left[\frac{-1}{2\Delta}, \frac{1}{2\Delta}\right]$, where Δ is usually the interval with which we have sampled the random process. In this case, the above becomes

$$\sigma_X^2 = P \int_{\frac{-1}{2\Delta}}^{\frac{1}{2\Delta}} df$$

$$= \frac{P}{\Delta} \tag{3.75}$$

hence $P = \sigma_X^2 \Delta$. This will always be the case of practical interest and for an i.i.d. random process allows us to relate the PSD to process variance σ_X^2. This relationship is used later in Chapter 8. Note that had we used angular frequency variable ω, the relationship would instead have been $P = \sigma_X^2 \Delta/(2\pi)$.

To summarize this section, in the analysis of random process data it is advantageous to consider the signal power in different frequency bands. If we are willing to consider truncated signals of the form (3.41), we may use the standard Fourier representation and consider the PSD given by Eq. (3.49). If we, instead, are required to consider a random process with infinite support, we use the Fourier–Stieltjes representation in which case the PSD is given by the expression (3.51). Moreover, we can relate the PSD in a straightforward manner to the auto- and cross-correlation functions via the Wiener–Khinchin theorem. This particular relationship plays a large role in later chapters. For example, in Section (8.6) we use the relationships (3.65, 3.66) in the generation of surrogate data sets for detecting damage-induced nonlinearity. Later, in Section 6.5 they are used to derive analytical expressions for the HOS properties of structural response data, which can be used to identify structural damage. A brief introduction to the HOS is given next.

3.3.5 Higher Order Spectra

The preceding section has demonstrated the relationship between PSD and the autocorrelation function. Specifically, it was shown that if the data are assumed stationary, the two quantities are related via Fourier transform. In this section we extend these results to higher order statistical properties, for example, third-order correlation function. Before looking at this topic, however, we introduce some notations for this section. Rather than use a different letter, for example, X, Y, Z, we adopt a more compact notation and identify different random processes using subscripts, e.g., $X_i(t)$. In this way, we can index an arbitrarily large number of random processes without requiring an arbitrarily large alphabet. It will also prove convenient to work in angular frequency ω (as opposed to frequency f, in Hz) so as to prevent us from having to carry around 2π in the exponentials. Finally, we frequently encounter multidimensional integrals with infinite limits. For this reason we introduce the shorthand

$$\int_{\mathbb{R}^N} \equiv \int_{-\infty}^{\infty} \int_{-\infty}^{\infty} \cdots N \text{times} \cdots \int_{-\infty}^{\infty} \tag{3.76}$$

which we invoke when the number of integrals becomes large (>2).

To begin the discussion, assume two random variables $X_i(t_1), X_j(t_2)$ and the existence of spectral representations $x_i(t_1) - \mu_{X_i} = \int_{-\infty}^{\infty} dZ_i(\omega_1)e^{i\omega_1 t_1}$ and $x_j(t_2) - \mu_{X_j} = \int_{-\infty}^{\infty} dZ_j(\omega_2)e^{i\omega_2 t_2}$.

From our previous section, we know that we may write the stationary cross-correlation function as

$$C_{X_iX_j}(\tau) = E[(X_i(t_1) - \mu_{X_i})(X_j(t_1 + \tau) - \mu_{X_j})]$$

$$= \int_{-\infty}^{\infty} S_{X_iX_j}(\omega)e^{i\omega\tau}\,d\omega. \tag{3.77}$$

However, we could have also started from the more general, nonstationary covariance function and substituted in the spectral representation to give

$$C_{X_iX_j}(t_1, t_2) = E[(X_i(t_1) - \mu_{X_i})(X_j(t_2) - \mu_{X_j})]$$

$$= E\left[\int_{-\infty}^{\infty}\int_{-\infty}^{\infty} dZ_i(\omega_1)e^{i\omega_1 t_1}dZ_j(\omega_2)e^{i\omega_2 t_2}\right]$$

$$= \int_{-\infty}^{\infty}\int_{-\infty}^{\infty} e^{i(\omega_1 t_1 + \omega_2 t_2)}E[dZ_i(\omega_1)dZ_j(\omega_2)] \tag{3.78}$$

If we then invoked stationarity, i.e., let $t_2 = t_1 + \tau$ where the absolute index t_1 is irrelevant, we have that

$$C_{X_iX_j}(t_1, t_1 + \tau) = C_{X_iX_j}(\tau)$$

$$= \int_{-\infty}^{\infty}\int_{-\infty}^{\infty} e^{i(\omega_1 t_1 + \omega_2 t_1)}e^{i\omega_2 \tau}E[dZ_i(\omega_1)dZ_j(\omega_2)]$$

$$= \int_{-\infty}^{\infty} \{\delta(\omega_1 + \omega_2)E[dZ_i(\omega_1)dZ_j(\omega_2)]\}e^{i\omega_2 \tau} \tag{3.79}$$

where, in the last step, we have noted that because the left-hand side is independent of t_1, the *only* place the right-hand side exists is when $\omega_1 = -\omega_2$. Equating (3.77) and (3.79) then yields

$$S_{X_iX_j}(\omega_2)d\omega_2 = E[dZ_j(\omega_2)dZ_i(-\omega_2)]$$

$$= E[dZ_j(\omega_2)dZ_i^*(\omega_2)]. \tag{3.80}$$

The subscript on the frequency variable is arbitrary, of course, and we can simply denote this function $S_{X_iX_j}(\omega)$. What this analysis shows is that we can interpret the previously discussed stationary cross-spectral density of Eq. (3.57) as the more general spectral function $E[dZ_i(\omega_1)dZ_j(\omega_2)]$ restricted to the line $\omega_1 = -\omega_2$. Note that combining Eqs. (3.79) and (3.80) we have provided an alternative proof of the Weiner–Khinchin theorem, as our derivation has shown that

$$S_{X_iX_j}(\omega) = \frac{1}{2\pi}\int_{-\infty}^{\infty} C_{X_iX_j}(\tau)e^{-i\omega\tau}\,d\tau \tag{3.81}$$

(the factor of 2π results from our definition of the forward and inverse Fourier transforms).

This same basic approach can also allow us to look at the spectral representation of other, higher order correlation functions. Consider the general case of different observations $x_i(t_1), x_j(t_2), x_k(t_3)$ with spectral representations $x_i(t_1) - \mu_{X_i} = \int_{-\infty}^{\infty} e^{i\omega_1 t_1}dZ_i(\omega_1)$,

$x_j(t_2) - \mu_{X_j} = \int_{-\infty}^{\infty} e^{i\omega_2 t_2} dZ_j(\omega_2)$, $x_k(t_3) - \mu_{X_k} = \int_{-\infty}^{\infty} e^{i\omega_3 t_3} dZ_k(\omega_3)$. The general, third-order moment is defined as

$$C_{X_i X_j X_k}(t_1, t_2, t_3) = E[(X_i(t_1) - \mu_i)(X_j(t_2) - \mu_j)(X_k(t_3) - \mu_k)]. \tag{3.82}$$

Just as we did for the PSD, we can substitute the spectral representation into the above, resulting in

$$C_{X_i X_j X_k}(t_1, t_2, t_3) =$$

$$E\left[\int_{\mathbb{R}} dZ_i(\omega_1) e^{i\omega_1 t_1} \int_{\mathbb{R}} dZ_j(\omega_2) e^{i\omega_2 t_2} \int_{\mathbb{R}} dZ_k(\omega_3) e^{i\omega_3 t_3}\right]$$

$$= \int_{\mathbb{R}^3} E[dZ_i(\omega_1) dZ_j(\omega_2) dZ_k(\omega_3)] e^{i(\omega_1 t_1 + \omega_2 t_2 + \omega_3 t_3)} \tag{3.83}$$

As before, we invoke stationarity, in which case we may define $\tau_1 = t_2 - t_1$ and $\tau_2 = t_3 - t_1$ and write

$$C_{X_i X_j X_k}(\tau_1, \tau_2) = E[(X_i(t_1) - \mu_{X_i})(X_j(t_1 + \tau_1) - \mu_{X_j})(X_k(t_1 + \tau_2) - \mu_{X_k})]. \tag{3.84}$$

The absolute time index t_1 is arbitrary and will be denoted simply $t_1 \rightarrow t$ in the following development. In the stationary case, we then see that (3.83) becomes

$$C_{X_i X_j X_k}(\tau_1, \tau_2) =$$

$$\int_{\mathbb{R}^2} \int_R E[dZ_i(\omega_1) dZ_j(\omega_2) dZ_k(\omega_3)] e^{i(\omega_1 + \omega_2 + \omega_3)t} e^{i(\omega_2 \tau_1 + \omega_3 \tau_2)}. \tag{3.85}$$

Because this function is stationary and does not depend on time "t," we know this function only exists for $\omega_1 = -\omega_2 - \omega_3$; thus, we could have also written (3.85) as

$$C_{X_i X_j X_k}(\tau_1, \tau_2) = \int_{\mathbb{R}^2} \left\{ \int_R \delta(\omega_1 + \omega_2 + \omega_3) E[dZ_i(\omega_1) dZ_j(\omega_2) dZ_k(\omega_3)] \right\} e^{i(\omega_2 \tau_1 + \omega_3 \tau_2)}$$

$$= \int_{\mathbb{R}^2} \{E[dZ_j(\omega_2) dZ_k(\omega_3) dZ_i(-\omega_2 - \omega_3)]\} e^{i(\omega_2 \tau_1 + \omega_3 \tau_2)} \tag{3.86}$$

where the term in braces $\{\cdot\}$ is simply the double Fourier–Stieltjes transform of $C_{X_i X_j X_k}(\tau_1, \tau_2)$ with reference to τ_1, τ_2. In analogy to the PSD, we may therefore define the bispectral density function

$$B_{X_i X_j X_k}(\omega_2, \omega_3) d\omega_2 d\omega_3 = E[dZ_j(\omega_2) dZ_k(\omega_3) dZ_i(-\omega_2 - \omega_3)] \tag{3.87}$$

which can be alternatively written (considering 3.87 and 3.86) as

$$B_{X_i X_j X_k}(\omega_2, \omega_3) = \frac{1}{4\pi^2} \int_{\mathbb{R}^2} C_{X_i X_j X_k}(\tau_1, \tau_2) e^{-i(\omega_2 + \omega_3)t} d\tau_1 d\tau_2. \tag{3.88}$$

Just as we defined the PSD as the Fourier transform of the autocovariance function, we define the bispectral density as the double Fourier transform of the third joint moment. The prefactor $\frac{1}{4\pi^2}$ is again included for consistency with how we have defined the inverse Fourier transform. Had we used frequency variable f instead of ω, this prefactor would not be present.

Under the assumption of stationarity, the full bispectral density is restricted to the subspace $\omega_1 = -\omega_2 - \omega_3$ and is therefore a function of two frequency variables only. As with the PSD, we notice in the derivation that the first argument of $C_{X_i X_j X_k}(\tau_1, \tau_2)$ leads to a term with negative frequency arguments. This will have implications later, when it is time to compare estimate to theory. The estimator will be based on the right-hand side of (3.87) as shown in Section 6.5, which has a different ordering in the arguments than the subscripts on the left-hand side of (3.87). Our convention in this book is such that the order of the arguments in the estimator is i, j, k. That is to say, we write

$$B_{X_k X_i X_j}(\omega_1, \omega_2) d\omega_1 d\omega_2 = E[dZ_i(\omega_1) dZ_j(\omega_2) dZ_k(-\omega_1 - \omega_2)] \tag{3.89}$$

so that the random processes on the right-hand side are in the order i, j, k, with k (the last argument) assigned the negative frequencies. This convention can certainly be altered, so long as the corresponding change is made to the left-hand side. We have also noted that the choice of subscripts on the frequencies is arbitrary, and hence we use the more conventional ω_1, ω_2 (as opposed to ω_2, ω_3).

This discussion pertains to a continuous random process $X(t)$ extending infinitely in time, a luxury afforded us by the Fourier–Stieltjes representation. However, we can also consider the case where our structural response data are restricted to the interval $[-T, T]$, as will always be the case with our acquired data. Just as we did for the PSD function, such a restriction suggests the definition

$$B_{X_k X_i X_j}(\omega_1, \omega_2) = \lim_{T \to \infty} \left(\frac{1}{4\pi^2}\right) E\left[\frac{X_i(\omega_1) X_j(\omega_2) X_k(-\omega_1 - \omega_2)}{2T}\right] \tag{3.90}$$

where $X_i(\omega)$ is the traditional Fourier representation of $x_i(t)$. Recall that when we defined the PSD, we arrived at expression (3.49) in a somewhat heuristic manner. We did so by normalizing the expected value of signal energy by the signal duration to get a better behaved function in the limit $T \to \infty$; the result was an expression with units of power.

A different way to arrive at (3.90) can be realized by considering a single realization of each of the constituent random processes, defined only on the $-T, T$ interval. Following [9] (specifically Eqs. 3.1 and 3.2 of the cited work), a suggested approximation to the needed cumulant is

$$\tilde{C}_{X_i X_j X_k}(\tau_1, \tau_2) \approx \frac{1}{2T} \int_{-T}^{T} x_i(t) x_j(t + \tau_1) x_k(t + \tau_2) dt \tag{3.91}$$

which can be substituted into Eq. (3.88) to give

$$B_{X_i X_j X_k}(\omega_1, \omega_2) = \frac{1}{4\pi^2(2T)} \int_{-T+\tau_1}^{T-\tau_1} \int_{-T+\tau_2}^{T-\tau_2} \int_{-T}^{T} x_i(t) x_j(t + \tau_1) x_k(t + \tau_2)$$
$$\times e^{-i(\omega_1 \tau_1 + \omega_2 \tau_2)} e^{i(-\omega_1 - \omega_2)t} e^{i(\omega_1 + \omega_2)t} \, d\tau_1 \, d\tau_2 \, dt$$

$$\approx \frac{1}{4\pi^2(2T)} X_i(-\omega_1 - \omega_2) X_j(\omega_1) X_k(\omega_2) \tag{3.92}$$

in the limit as $T \to \infty$. If we further consider the expectation over multiple realizations, we then recover (3.90). When it comes to estimating the bispectral density, we will use (3.90) as opposed to (3.87).

It is also worth pointing out that, in analogy to Eq. (3.75), we can define the constant bispectrum for an i.i.d. random process, but one that has been artificially band-limited by a finite sampling interval $1/\Delta$ as

$$B_{X_k X_i X_j}(\omega_1, \omega_2) = \text{const.} = \frac{k_3 \Delta^2}{4\pi^2} \tag{3.93}$$

where k_3 is the third cumulant (see Eq. 2.81). Like Eq. (3.75), Eq. (3.93) proves useful later in Chapter 8.

From this discussion it is tempting to want to define the nth-order HOS as the n-dimensional Fourier transform of the joint moment function $C_{X_1 X_2 \cdots X_n}(t_1, t_2, \ldots, t_n) = E\left[\prod_{i=1}^{n}(x(t_i) - \mu_{X_i}(t_i))\right]$. For the PSD and bispectral densities, this is in fact the case. However, in general, the HOS are defined in terms of joint *cumulants* (introduced in Section 2.6) as opposed to joint central moments. That is to say, the HOS are formally the n–dimensional Fourier transform of the nth-order cumulants, the first five of which are given in Eqs. (2.86), i.e.,

$$\text{HOS}_{X_1 X_2 \cdots X_n}(\omega_1, \omega_2, \ldots, \omega_n)$$
$$= \left(\frac{1}{2\pi}\right)^n \int_{\mathbb{R}^n} k_{X_1, X_2, \ldots, X_n}(t_1, t_2, \ldots, t_n) e^{-i(\omega_1 t_1 + \omega_2 t_2 + \cdots + \omega_n t_n)} dt_1 \, dt_2 \cdots dt_n \tag{3.94}$$

As a direct consequence of Isserlis' theorem (3.11) and the definition of cumulants (2.86), the HOS have the useful property that they vanish for order $n > 2$ provided that the random processes X_i are joint normally distributed. This property proves key in using the HOS in Section 8.3 to identify damage-induced nonlinearity in structures.

While the HOS are not formally the Fourier transforms of joint moments, we can see that for $n > 3$ the joint cumulants *can* be represented as functions of joint moments. We also know that under the assumption of stationarity, the nth-order moment spectra reduces to a function of $n - 1$ frequency variables, where $\omega_1 = -\omega_2 - \omega_3 - \cdots - \omega_n$. Thus, we define the joint moment spectra

$$S_{X_1 X_2 \cdots X_n}(\omega_1, \omega_2, \ldots, \omega_{n-1}) =$$
$$\left(\frac{1}{2\pi}\right)^{n-1} \int_{R^{n-1}} E[(X_1(t) - \mu_{X_1})(X_2(t+\tau_1) - \mu_{X_2}) \cdots (X_n(t+\tau_{n-1}) - \mu_{X_N})]$$
$$\times e^{-i(\omega_2 \tau_1 + \omega_3 \tau_2 + \cdots + \omega_n \tau_{n-1})} d\tau_1 \, d\tau_2 \cdots d\tau_{n-1} \tag{3.95}$$

which for $n = 2$ and $n = 3$ become the power and bispectral densities, respectively. Note that on the basis of our previous development we could have also defined the moment spectra in terms of the Fourier–Stieltjes representation as

$$S_{X_1 X_2 \cdots X_n}(\omega_2, \cdots \omega_n) d\omega_2 \cdots d\omega_n = E\left[dZ_1\left(-\sum_{q=2}^{n} \omega_q\right) dZ_2(\omega_2) \cdots dZ_n(\omega_n)\right] \tag{3.96}$$

or, using our preferred ordering scheme

$$S_{X_n X_1 \cdots X_{n-1}}(\omega_1, \cdots \omega_{n-1}) d\omega_1 \cdots d\omega_{n-1} = E\left[dZ_1(\omega_1) dZ_2(\omega_2) \cdots dZ_n\left(-\sum_{q=1}^{n-1} \omega_q\right)\right]. \tag{3.97}$$

Armed with this notation, we can consider HOS beyond the bispectrum. HOS for $n > 4$ are not typically examined because of the difficulty associated with their estimation (the reasons for which become apparent in Section 6.5). However, the trispectrum, $n = 4$, is sometimes studied and is used later in this book in damage detection applications. Through examination of the $n = 4$th order cumulant, we may define the general (nonstationary) trispectrum as [10]

$$
T_{X_i X_j X_k X_p}(\omega_1, \omega_2, \omega_3, \omega_4) = \left(\frac{1}{16\pi^4} \right)
$$

$$
\int_{\mathbb{R}^4} k_{X_i X_j X_k X_p}(t_1, t_2, t_3, t_4) e^{-i(\omega_1 t_1 + \omega_2 t_2 + \omega_3 t_3 + \omega_4 t_4)} dt_1 \, dt_2 \, dt_3 \, dt_4
$$

$$
= \left(\frac{1}{16\pi^4} \right) \int_{\mathbb{R}^4} C_{X_i X_j X_k X_p}(t_1, t_2, t_3, t_4) e^{-i(\omega_1 t_1 + \omega_2 t_2 + \omega_3 t_3 + \omega_4 t_4)} dt_1 \, dt_2 \, dt_3 \, dt_4
$$

$$
- \left(\frac{1}{16\pi^4} \right) \int_{\mathbb{R}^2} C_{X_i X_j}(t_1, t_2) e^{-i(\omega_1 t_1 + \omega_2 t_2)} dt_1 \, dt_2 \int_{\mathbb{R}^2} C_{X_k X_p}(t_3, t_4) e^{-i(\omega_3 t_3 + \omega_4 t_4)} dt_3 \, dt_4
$$

$$
- \left(\frac{1}{16\pi^4} \right) \int_{\mathbb{R}^2} C_{X_i X_k}(t_1, t_3) e^{-i(\omega_1 t_1 + \omega_3 t_3)} dt_1 \, dt_3 \int_{\mathbb{R}^2} C_{X_j X_p}(t_2, t_4) e^{-i(\omega_2 t_2 + \omega_4 t_4)} dt_2 \, dt_4
$$

$$
- \left(\frac{1}{16\pi^4} \right) \int_{\mathbb{R}^2} C_{X_i X_p}(t_1, t_4) e^{-i(\omega_1 t_1 + \omega_4 t_4)} dt_1 \, dt_4 \int_{\mathbb{R}^2} C_{X_j X_k}(t_2, t_3) e^{-i(\omega_2 t_2 + \omega_3 t_3)} dt_2 \, dt_3 \quad (3.98)
$$

associated with random processes $X_i(t)$, $X_j(t)$, $X_k(t)$, and $X_p(t)$. As with the power spectrum and bispectrum, these processes can be used to model data acquired from a structure at four different locations. For brevity, we assume in what follows that we are dealing with mean-centered random processes, i.e., allow $X_i(t) \rightarrow X_i(t) - \mu_{X_i}$.

Under the assumption that the data are stationary, the four terms in Eq. (3.98) become independent of the absolute time index and reduce to functions of the moment spectra defined by Eq. (3.95). Defining $\tau_1 = t_2 - t_1$, $\tau_2 = t_3 - t_1$, $\tau_3 = t_4 - t_1$ as measures of time delay, the fourth moment becomes $C_{X_i X_j X_k X_p}(t_1, t_2, t_3, t_4) = E[X_i(t_1) X_j(t_1 + \tau_1) X_k(t_1 + \tau_2) X_p(t_1 + \tau_3)]$. Making this change in the first term in (3.98) leads to

$$
\left(\frac{1}{16\pi^4} \right) \int_{\mathbb{R}^4} E[X_i(t_1) X_j(t_1 + \tau_1) X_k(t_1 + \tau_2) X_p(t_1 + \tau_3)] e^{-i(\omega_1 + \omega_2 + \omega_3 + \omega_4) t_1}
$$

$$
\times e^{-i(\omega_2 \tau_1 + \omega_3 \tau_2 + \omega_4 \tau_3)} dt_1 \, d\tau_1 \, d\tau_2 \, d\tau_3
$$

$$
= \left(\frac{1}{16\pi^4} \right) \int_{\mathbb{R}^3} E[X_i(t_1) X_j(t_1 + \tau_1) X_k(t_1 + \tau_2) X_p(t_1 + \tau_3)] e^{-i(\omega_2 \tau_1 + \omega_3 \tau_2 + \omega_4 \tau_3)} d\tau_1 \, d\tau_2 \, d\tau_3
$$

$$
\times \int_{\mathbb{R}} e^{-i(\omega_1 + \omega_2 + \omega_3 + \omega_4) t_1} dt_1
$$

$$
= \delta(\omega_1 + \omega_2 + \omega_3 + \omega_4) \left(\frac{1}{8\pi^3} \right) \int_{\mathbb{R}^3} E[X_i(t_1) X_j(t_1 + \tau_1) X_k(t_1 + \tau_2) X_p(t_1 + \tau_3)]
$$

$$
\times e^{-i(\omega_2 \tau_1 + \omega_3 \tau_2 + \omega_4 \tau_3)} d\tau_1 \, d\tau_2 \, d\tau_3 \quad (3.99)
$$

where we have separated the integral involving t_1 because the expectation is not a function of t_1 due to stationarity. This term can be immediately recognized as being in the form of our generalized moment spectrum (3.95). Hence, if we choose the ordering scheme defined in

Eq. (3.97), we can finally rewrite Eq. (3.99) as

$$\delta(\omega_1 + \omega_2 + \omega_3 + \omega_4) S_{X_p X_i X_j X_k}(\omega_1, \omega_2, \omega_3). \tag{3.100}$$

We have also explicitly noted that the fourth-order stationary moment spectrum is only supported (i.e., is only nonzero) on the frequency plane defined by $\omega_4 = -\omega_1 - \omega_2 - \omega_3$. For the power spectrum and bispectrum, this is not typically noted explicitly, however for reasons that will become apparent shortly, we do so for the trispectrum.

In addition to this term, the trispectrum has three other terms that are the products of Fourier transforms of second moments. We have already shown at the beginning of this section how to simplify these terms under the assumption of stationarity. Defining $\tau_1 = t_2 - t_1$, the first of the second moment terms in Eq. (3.98) becomes

$$\int_{\mathbb{R}^2} C_{X_i X_j}(t_1, t_1 + \tau_1) e^{-i(\omega_1 t_1 + \omega_2(t_1 + \tau_1))} dt_1 \, d\tau_1$$

$$= \int_{\mathbb{R}} C_{X_i X_j}(\tau_1) e^{-i\omega_2 \tau_1} d\tau_1 \int_{\mathbb{R}} e^{-i(\omega_1 + \omega_2)t_1} dt_1$$

$$= 4\pi^2 \delta(\omega_1 + \omega_2) S_{X_i X_j}(\omega_2) = 4\pi^2 \delta(\omega_1 + \omega_2) S_{X_j X_i}(\omega_1) \tag{3.101}$$

where we have used Eqs. (3.79) and (3.80) in the simplification. Similarly, defining $\tau_3 = t_4 - t_3$,

$$\int_{\mathbb{R}^2} C_{X_k X_p}(t_3, t_3 + \tau_3) e^{-i(\omega_3 t_1 + \omega_4(t_3 + \tau_3))} dt_3 \, d\tau_3$$

$$= \int_{\mathbb{R}} C_{X_k X_p}(\tau_3) e^{-i\omega_4 \tau_3} d\tau_3 \int_{\mathbb{R}} e^{-i(\omega_3 + \omega_4)t_3} dt_3$$

$$= 4\pi^2 \delta(\omega_3 + \omega_4) S_{X_k X_p}(\omega_4)$$

$$= 4\pi^2 \delta(\omega_3 + \omega_4) S_{X_p X_k}(\omega_3)$$

$$= 4\pi^2 \delta(\omega_3 - \omega_1 - \omega_2 - \omega_3) S_{X_p X_k}(\omega_3)$$

$$= 4\pi^2 \delta(\omega_1 + \omega_2) S_{X_p X_k}(\omega_3) \tag{3.102}$$

so that the first term involving the product of second-order moment spectra becomes

$$-\left(\frac{1}{16\pi^4}\right) \int_{\mathbb{R}^2} C_{X_i X_j}(t_1, t_2) e^{-i(\omega_1 t_1 + \omega_2 t_2)} dt_1 \, dt_2 \int_{\mathbb{R}^2} C_{X_k X_p}(t_3, t_4) e^{-i(\omega_3 t_3 + \omega_4 t_4)} dt_3 \, dt_4$$

$$= \delta(\omega_1 + \omega_2) S_{X_j X_i}(\omega_1) S_{X_p X_k}(\omega_3). \tag{3.103}$$

The other terms of the trispectrum involving products of integrals of second moments possess nonvanishing support on the planes $\omega_1 + \omega_3$ and $\omega_2 + \omega_3$, so that the complete trispectrum for multivariate, stationary data is given by the expression

$$T_{X_i X_j X_k X_p}(\omega_1, \omega_2, \omega_3) = S_{X_p X_i X_j X_k}(\omega_1, \omega_2, \omega_3) - \delta(\omega_1 + \omega_2) S_{X_j X_i}(\omega_1) S_{X_p X_k}(\omega_3)$$

$$- \delta(\omega_1 + \omega_3) S_{X_k X_i}(\omega_1) S_{X_p X_j}(\omega_2) - \delta(\omega_2 + \omega_3) S_{X_p X_i}(\omega_1) S_{X_k X_j}(\omega_2) \tag{3.104}$$

What is interesting about the trispectral density is that different terms exist on different sub-manifolds (defined by the various delta functions). The main term is defined everywhere in the $\omega_1, \omega_2, \omega_3$ plane, however the terms involving products of second moments are defined only on the planes defined by, for example, $\omega_1 = -\omega_2$. As we show in Section 6.5.4 and in Appendix C, however, these terms are important in testing for damage-induced nonlinearities and must be retained.

Continuing the analogy to the bispectral density, using the Fourier–Stieltjes representation we could have also defined the main trispectral density term as

$$S_{X_p X_i X_j X_k}(\omega_1, \omega_2, \omega_3) d\omega_1 \, d\omega_2 \, d\omega_3 = E[dZ_i(\omega_1)dZ_j(\omega_2)dZ_k(\omega_3)dZ_p(-\omega_1 - \omega_2 - \omega_3)]$$

(3.105)

or, for our usual restriction to the $[-T, T]$ interval, as

$$S_{X_p X_i X_j X_k}(\omega_1, \omega_2, \omega_3) = \lim_{T \to \infty} \left(\frac{1}{8\pi^3} \right) \frac{E[X_i(\omega_1)X_j(\omega_2)X_k(\omega_3)X_p(-\omega_1-\omega_2-\omega_3)]}{2T}.$$

(3.106)

The rest of the terms required of the trispectral density are the familiar PSDs, which can also be defined using either the Fourier–Stieltjes representation (as in 3.57) or the limiting Fourier representation (3.55). It is this latter form that is of most use in developing estimators for the HOS in Section 6.5.

One final point concerns the use of the terms "bispectrum" and "trispectrum" and "bispectral density" and "trispectral density." By definition, Eq. (3.94) are density functions, expressed in units of $[x]^3/Hz^2$ and $[x]^4/Hz^3$ for the bispectrum and trispectrum, respectively. It is more common, however, to ignore the sampling interval in the estimate (to be discussed) giving the trispectrum (for example) units $[x]^4$. Whether one uses the "spectrum" or "spectral density" is largely a matter of preference, so long as proper units are retained when estimating these quantities from observed data. Analytically, however, it makes more sense to be true to the definition and derive the expressions as true density functions. However, throughout this book we retain bispectrum/trispectrum as they are the more commonly used terms.

3.4 Markov Models

One class of probability model of particular relevance to the study of random processes is the Markov model. Simply put, a Markov model is one in which the probability of being in a given state at time t is conditional on the state at $t - \tau$ but *not* on the system state for times $< t - \tau$, i.e., the system has no memory beyond the past time horizon τ. For a sequence of observations, we may write this mathematically as

$$p_X(x(t)|x(t - \tau_1), x(t - \tau_2), \dots, x(t - \tau_N))$$
$$= p_X(x(t)|x(t - \tau_1), x(t - \tau_2), \dots, x(t - \tau_N), x(t - \tau_N - 1), \cdots). \quad (3.107)$$

This is a very general description of the dynamics of a random process, stating only that the system state at time t is conditional on states no further than $t - \tau$ seconds in the past. Markov models play a large role in the upcoming chapters on estimation (see specifically Section 7.2.3) and in the description of certain information-theoretic quantities described in the next section.

Although not discussed here, Eq. (3.107) is the general description of the familiar autoregressive (AR) model for a discrete sequence of observations $x(t_1), x(t_2), \ldots, x(t_N)$

$$x(t_{n+1}) = a_1 x(t_n) + a_2 x(t_{n-1}) + \cdots + a_M x(t_{n-M+1}) + \eta_{t_n} \qquad (3.108)$$

whereby observations of a random process are modeled as a linear combination of past values specified by the order M and the weights a_i, $i = 1 \cdots M$. The values η_{t_n} are usually taken as additive Gaussian noise. Clearly, the probability of being in a given state at time t_{n+1} is conditional on linear combinations of past states. We mention that AR models have been used successfully to detect damage in structures in several cases (see, for example, [11, 12]); the interested reader is referred to the cited works for additional information.

3.5 Information Theoretics

To this point, we have discussed a number of statistical properties of probabilistic models. In later chapters, we use these properties to draw inferences about the state of structural damage. So far the focus has been on particular expectations of products of the random variables comprising our structural response data, for example, covariance. If we know enough about the physics of the damage, we might reliably predict which of these statistical moments carries the most information about the damage presence and then use it as a damage detection statistic. Indeed, the HOS discussed in Section 8.3 is used for just that purpose. However, in some instances it is not clear which statistical properties are of most use in identifying structural damage. In these situations, it turns out to be useful to develop measures that reflect the entire joint distribution of a signal.

One way to capture general relationships among random variables (or random processes) is to quantify the information content. The conceptual framework for quantifying information for discrete random variables was put forth by Shannon [13]. Shannon's paper laid out the arguments for using the scalar quantity entropy,

$$H_X = -\sum_{n=1}^{N} p_X(x(n))\log_2(p_X(x(n))), \qquad (3.109)$$

as a measure of information contained in the probability mass function $p_X(x(n))$ associated with a discrete random variable X which can achieve one of N possible states $(x(1), x(2), \ldots, x(N))$. Because the PMF is unitless, so too is the discrete signal entropy (3.109). Nonetheless, it is common to assign the units of "bits," assuming base 2 logarithms are used in the definition, or "nats" (natural units) if one uses base e. One can similarly define the joint entropy among two PMFs

$$H_{XY} = -\sum_{n=1}^{N}\sum_{m=1}^{N} p_{XY}(x(n), y(m))\log_2(p_{XY}(x(n), y(m))). \qquad (3.110)$$

Shannon's original focus was communication theory and understanding the limits on signal compression and transmission. Since then, however, the formalism has grown to include other applications and even other functions comprising the field of information theory. One of the first extensions is to consider the continuous analogy to (3.109) and define the continuous or *differential* entropy

$$h_X = -\int_X p_X(x)\log_2(p_X(x))dx \tag{3.111}$$

for PDF $p_X(x)$. The limits of the integral are understood to extend over the support of $p_X(x)$, which is our space of possible outcomes X. Similarly, for the joint entropy we can define

$$h_{XY} = -\int_{XY} p_{XY}(x, y)\log_2(p_{XY}(x, y))dx\, dy. \tag{3.112}$$

Differential entropy differs in several nontrivial ways from its discrete counterpart, for example differential entropy can be negative, whereas discrete entropy cannot. These differences are discussed in detail in Ref. [14], but they are not critical to the way entropy is used here.

The differential entropy is a scalar quantity that is a function of the entire PDF associated with the random variable(s). As such, changes to any of the statistical properties associated with that random variable will change the entropy. It is also worth noting that the functional form of the entropy is that of an expectation of a function. Recall from Chapter 2 that the expectation of a function is given by Eq. (2.48). If we substitute $f(x) = \log_2(p_X(x))$, we can see that the entropy is also expressible as

$$h_X = -E[\log_2(p_X(x))]. \tag{3.113}$$

We also note that in some references the entropy is referred to as a measure of uncertainty. This is actually a more appropriate interpretation as the PDFs (or PMFs) are models of uncertainty.

In our applications we are interested in the entropy associated with the random processes, $\mathbf{X}, \mathbf{Y}$ which are used to model our usual discrete sequences of observations $\mathbf{x} = (x(t_1), x(t_2), \ldots, x(t_N))$ and $\mathbf{y} = (y(t_1), y(t_2), \ldots, y(t_N))$, respectively. Here, as is often the case, the random variables are indexed by sampling times $t_1, t_2, \ldots, t_N$, each assumed to be separated by the constant sampling interval Δ_t. Given such a collection, we are interested in both the single and joint differential entropies,

$$h_{\mathbf{X}} = -\int_{\mathbb{R}^N} p_{\mathbf{X}}(\mathbf{x})\log_2(p_{\mathbf{X}}(\mathbf{x}))d\mathbf{x}$$

$$h_{\mathbf{Y}} = -\int_{\mathbb{R}^N} p_{\mathbf{Y}}(\mathbf{y})\log_2(p_{\mathbf{Y}}(\mathbf{y}))d\mathbf{y}$$

$$h_{\mathbf{XY}} = -\int_{\mathbb{R}^{2N}} p_{\mathbf{XY}}(\mathbf{x}, \mathbf{y})\log_2(p_{\mathbf{XY}}(\mathbf{x}, \mathbf{y}))d\mathbf{x}\, d\mathbf{y} \tag{3.114}$$

where the notation $\int_{\mathbb{R}^N}$ is again the N-dimensional integral extending over each observation.

Before describing the uses of differential entropy, it is first instructive to consider the case where an expression may be obtained analytically. The only multivariate probability distribution that readily admits an analytical solution for the joint, differential entropies is the jointly

Gaussian distribution. The Gaussian models for both the individual and joint data vectors are

$$p_{\mathbf{X}}(\mathbf{x}) = \frac{1}{(2\pi)^{N/2}|C_{\mathbf{XX}}|^{1/2}} e^{-\frac{1}{2}(\mathbf{x}^T C_{\mathbf{XX}}^{-1}\mathbf{x})}$$

$$p_{\mathbf{Y}}(\mathbf{y}) = \frac{1}{(2\pi)^{N/2}|C_{\mathbf{YY}}|^{1/2}} e^{-\frac{1}{2}(\mathbf{y}^T C_{\mathbf{YY}}^{-1}\mathbf{y})}$$

$$p_{\mathbf{XY}}(\mathbf{x}, \mathbf{y}) = \frac{1}{(2\pi)^{N}|C_{\mathbf{XY}}|^{1/2}} e^{-\frac{1}{2}(\mathbf{x}^T C_{\mathbf{XY}}^{-1}\mathbf{y})} \tag{3.115}$$

where $C_{\mathbf{XX}}$, $C_{\mathbf{YY}}$ are the $N \times N$ covariance matrices associated with the random processes $\mathbf{X}$, $\mathbf{Y}$, respectively, and $|\cdot|$ takes the determinant. The matrix $C_{\mathbf{XY}}$ is the $2N \times 2N$ covariance matrix associated with the joint data vector $[\mathbf{x}, \mathbf{y}]$.

If the joint distribution is given by Eq. (3.115), the entropy for the random process $\mathbf{X}$ (for example) is

$$h_{\mathbf{X}} = -E[\log \ (p_{\mathbf{X}}(\mathbf{x}))]$$

$$= E[\log \ (2\pi^{N/2}|C_{\mathbf{XX}}|^{1/2})] + E\left[\frac{1}{2}(\mathbf{x} - \mu_{\mathbf{X}})^T C_{\mathbf{XX}}^{-1}(\mathbf{x} - \mu_{\mathbf{X}})\right]$$

$$= \frac{1}{2}\log_2((2\pi)^N) + \frac{1}{2}|C_{\mathbf{XX}}| + \frac{1}{2}E[(\mathbf{x} - \mu_{\mathbf{X}})^T C_{\mathbf{XX}}^{-1}(\mathbf{x} - \mu_{\mathbf{X}})]$$

$$= \frac{1}{2}\log_2((2\pi)^N) + \frac{1}{2}|C_{\mathbf{XX}}| + \frac{N}{2}$$

$$= \frac{1}{2}\log \ ((2\pi e)^N |C_{\mathbf{XX}}|) \tag{3.116}$$

and similarly for $\mathbf{Y}$

$$h_{\mathbf{Y}} = \frac{1}{2}\log \ ((2\pi e)^N |C_{\mathbf{YY}}|) \tag{3.117}$$

The joint entropy is likewise governed by $C_{\mathbf{XY}}$ and is given by

$$h_{\mathbf{XY}} = \frac{1}{2}\log \ ((2\pi e)^{2N} |C_{\mathbf{XY}}|). \tag{3.118}$$

Not surprisingly, for two jointly Gaussian random processes, *all* of the information is captured by the covariance matrices.

The differential entropy by itself lacks a straightforward physical interpretation for a random process. For example, what does it mean to compute the joint differential entropy for a time series of observations recorded from a structure? Aside from mapping the joint distribution to a scalar value, it is unclear why one would be interested in such a quantity in the study of structural dynamics.

It turns out that differential entropy lies at the heart of several important information-theoretic quantities that have been used to great effect in the study of dynamical systems. Specifically, we will consider two such quantities: the average mutual information function and the transfer entropy (TE). The mutual information can be thought of as a generalization of the cross-correlation, providing a measure of statistical dependence among random processes. This measure has seen use in structural dynamics as a method of damage detection [15, 16], and in choosing optimal sensor placement in a structural monitoring application [17]. Both applications are considered later in Chapters 8 and 9. The mutual information also lies at the heart of many techniques in model selection, a topic that is discussed in the context of damage identification in Section 7.3.

TE is another information-theoretic quantity that provides a slightly different definition of statistical dependency, using conditional probability to define what it means for one random process to provide information about another. This definition is based on Markov models, defined in Section 3.4, of the constituent random processes. This measure was originally designed to quantify information transport in dynamical systems [18] and was later extended to continuous random variables [19]. Since its introduction, the TE has also been applied to studying the output of structural systems for damage detection, an application that is also considered in Chapter 8 [15, 20]. Both the mutual information and TE functions are described in the following sections for general probability models, and in the specific case of the joint Gaussian probability model.

3.5.1 Mutual Information

Again, assume we can monitor two random processes $X(t), Y(t)$ by recording their values at N discrete points in time giving the time series $x(t_n), y(t_m)$, $n, m = 1 \cdots N$. These could be the sensor responses at two points on a structure, for example. Each sequence can be modeled with a joint PDF, e.g., $p_X(\mathbf{x})$. We know from Chapter 2 that if two sequences of random variables are statistically independent (see Section 2.3, Eqs. 2.17 and 2.23), then $p_{XY}(\mathbf{x}, \mathbf{y}) = p_X(\mathbf{x})p_Y(\mathbf{y})$, that is, the joint PDF factors. To quantify the degree to which the random processes are independent, one may compute the *average mutual information* function

$$I_{XY} = \int_X \int_Y p_{XY}(\mathbf{x}, \mathbf{y}) \log_2 \left(\frac{p_{XY}(\mathbf{x}, \mathbf{y})}{p_X(\mathbf{x})p_Y(\mathbf{y})} \right) d\mathbf{x} \, d\mathbf{y} \qquad (3.119)$$

which, like the joint entropy, is a scalar quantity [21]. The mutual information function is non-negative and it is assumed in the case of zero probability that $0 \log_2[0] = 0$. By the argument of the logarithm we see that if the random processes are statistically independent, the mutual information is zero and rising to a maximum if one of the random variables uniquely specifies the other. The expression (3.119) is sometimes referred to as a "distance" between two probabilistic models, in this case between a coupled random process model and one in which the constituent processes are independent. Technically, the measure I_{XY} does not satisfy the mathematical requirements of a distance, yet is referred to in later chapters as such, but with "distance" in quotes.

It can be noticed that by definition we could have also written the mutual information

$$I_{XY} = E\left[\log_2 \left(\frac{p_{XY}(\mathbf{x}, \mathbf{y})}{p_X(\mathbf{x})p_Y(\mathbf{y})} \right) \right] \qquad (3.120)$$

a form that is of some use when it comes to estimating this quantity from observation (see Section 6.6). This form also plays a part in demonstrating the important role that mutual information has when estimating signal parameters in the presence of non-Gaussian noise (see Section 7.1.5).

Making use of the fact that $\int_X \int_Y p_{XY}(\mathbf{x}, \mathbf{y}) \log_2(p_X(\mathbf{x})) d\mathbf{y} \, d\mathbf{x} = \int_X p_X(\mathbf{x}) \log (p_X(\mathbf{x})) d\mathbf{x}$, we may re-write (3.119) as

$$
\begin{aligned}
I_{\mathbf{XY}} = &\int_X \int_Y p_{XY}(\mathbf{x}, \mathbf{y}) \log_2(p_{XY}(\mathbf{x}, \mathbf{y})) d\mathbf{x} \, d\mathbf{y} \\
&- \int_X p_X(\mathbf{x}) \log_2(p_X(\mathbf{x})) d\mathbf{x} \\
&- \int_Y p_Y(\mathbf{y}) \log_2(p_Y(\mathbf{y})) d\mathbf{y} \\
= &\, h_{\mathbf{X}} + h_{\mathbf{Y}} - h_{\mathbf{XY}}
\end{aligned}
\tag{3.121}
$$

which is immediately recognized as a sum of joint differential entropies. Thus, if we consider both random processes to be jointly Gaussian distributed, we immediately see that the mutual information function becomes

$$
\begin{aligned}
I_{\mathbf{XY}} &= -\frac{1}{2} \log \left(\frac{(2\pi e)^{2N} |\mathbf{C_{XY}}|}{\left((2\pi e)^N |\mathbf{C_{XX}}| (2\pi e)^N |\mathbf{C_{YY}}| \right)} \right) \\
&= -\frac{1}{2} \log \left(\frac{|\mathbf{C_{XY}}|}{|\mathbf{C_{XX}}| \|\mathbf{C_{YY}}\|} \right).
\end{aligned}
\tag{3.122}
$$

Equation (3.122) holds for both stationary and nonstationary Gaussian sequences. Unfortunately, the needed determinants do not admit simple analytical expressions for structural response data. Moreover, in the non-Gaussian case, the estimation of high-dimensional integrals required of the entropy expressions (3.114) is a daunting task as seen in Chapter 6.

However, as we have often done, we can simplify the expressions considerably if we are willing to assume the random processes are stationary. Let each observation in the signals be modeled by the PDFs $p_X(x(t_n))$, $p_Y(y(t_m))$ and a joint probability density $p_{XY}(x(t_n), y(t_m))$. The mutual information function between these two observations is

$$
I_{XY}(t_n, t_m) = \int_X \int_Y p_{XY}(x(t_n), y(t_m)) \log_2 \left[\frac{p_{XY}(x(t_n), y(t_m))}{p_X(x(t_n)) p_Y(y(t_m))} \right] dx(t_n) dy(t_m)
\tag{3.123}
$$

Introducing the time delay $t_m = t_n + \tau$, Eq. (3.123) for any two observations of a stationary random process becomes

$$
\begin{aligned}
I_{XY}(\tau) = &\int_X \int_Y p_{XY}(x(t_n), y(t_n + \tau)) \log_2 \left[\frac{p_{XY}(x(t_n), y(t_n + \tau))}{p_X(x(t_n)) p_Y(y(t_n + \tau))} \right] dx(t_n) dy(t_n + \tau) \\
= &\int \int p_{XY}(x(t_n), y(t_n + \tau)) \log_2 [p_{XY}(x(t_n), y(t_n + \tau))] dx \, dy \\
&- \int p_X(x(t_n)) \log_2 [p_X(x(t_n))] dx \\
&- \int p_Y(y(t_n + \tau)) \log_2 [p_Y(y(t_n + \tau))] dy \\
= &\, h_X + h_Y - h_{XY}(\tau).
\end{aligned}
\tag{3.124}
$$

The resulting quantity is no longer a function of the absolute time index t_n, but of the temporal separation between t_n and t_m. This particular form of the mutual information most easily lends itself to analytical and computational treatment. Equation (3.124) has the feel of a stationary correlation function. This is not an unreasonable interpretation, in fact the mutual information can be thought of as a general measure of correlation between two random variables that is not restricted to second-order statistical properties. Moreover, the estimation problem is greatly simplified as the constituent entropies are now one- and two-dimensional integrals.

Returning to (3.122), we see that if the observations $x(t_n), y(t_n + \tau)$ are jointly Gaussian distributed we have

$$|C_{XX}| = \sigma_X^2$$
$$|C_{YY}| = \sigma_Y^2 \qquad (3.125)$$

and

$$|C_{XY}| = \begin{vmatrix} E[(X(t_n) - \mu_X)(X(t_n) - \mu_X)] & E[(X(t_n) - \mu_X)(Y(t_n + \tau) - \mu_Y)] \\ E[(Y(t_n + \tau) - \mu_Y)(X(t_n) - \mu_X)] & E[(Y(t_n + \tau) - \mu_Y)(Y(t_n + \tau) - \mu_Y)] \end{vmatrix}$$
$$= \sigma_X^2 \sigma_Y^2 - E[(X(t_n) - \mu_X)(Y(t_n + \tau) - \mu_Y)]^2$$
$$= \sigma_X^2 \sigma_Y^2 (1 - \rho_{XY}^2(\tau)) \qquad (3.126)$$

where we have used the definition of the cross-correlation coefficient (see Eq. 2.52). The mutual information then becomes

$$I_{XY}(\tau) = -\frac{1}{2}\log_2(1 - \rho_{XY}^2(\tau)) \qquad (3.127)$$

which tells us that the mutual information between two Gaussian distributed random variables is simply a function of the cross-correlation coefficient. In fact, one may view the general time-delayed mutual information function (3.124) as a *nonlinear* cross-correlation function capable of capturing both second-order (linear) and higher order (nonlinear) correlations among time series data. The algorithm required for estimating Eq. (3.124) is provided in Chapter 6. In Chapter 8, we use the function (3.124) for the detection of structural damage.

3.5.2 Transfer Entropy

One potential drawback of using the mutual information function to quantify coupling is that it does not consider the dynamics of the underlying processes explicitly. Rather, one introduces a time-delay in one random process with respect to the other and observes the corresponding fluctuations in statistical properties. While the resulting quantity certainly captures the dynamical properties of the two systems under study, it is not a formal measure of dynamical dependence. A different metric, the transfer entropy, was introduced by Schreiber [18] and incorporates the dynamic nature of the random processes directly. The TE metric was designed specifically to look at information transport and has been used already in examining physiological coupling [19], financial time series [22], ecological and climate dynamics [23, 24], chemical systems [25], and more to the point of this book, the output of structural systems [16, 26].

Recall from Section 3.4 that a Markov model is a probabilistic model that uses conditional probability to describe the dynamics of a random process. For example, we might specify $p_X(x(t_n)|x(t_{n-1}))$ as the probability of observing the value $x(t_n)$ given that we have already observed $x(t_{n-1})$. The idea that knowledge of past observations changes the likelihood of future observations is certainly common in dynamical systems. A dynamical system whose output is a repeating sequence of $010101\cdots$ is equally likely to be in state 0 or state 1 (probability 0.5) if the system is observed at a randomly chosen time. However, if we know the value at $t_1 = 0$, the value $t_2 = 1$ is known with probability 1. This concept lies at the heart of the Pth order Markov model, which by definition obeys (see Eq. 3.107, Chapter 2)

$$p_X(x(t_{n+1})|x(t_n), x(t_{n-1}), x(t_{n-2}), \ldots, x(t_{n-P}))$$
$$= p_X(x(t_{n+1})|x(t_n), x(t_{n-1}), x(t_{n-2}), \ldots, x(t_{n-P}), x(t_{n-P-1}), \cdots)$$
$$\equiv p_X(x(t_n)^{(1)}|\mathbf{x}(t_n)^{(P)}). \tag{3.128}$$

That is to say, the probability of the random variable attaining the value $x(t_{n+1})$ is conditional on the previous P values only. In the same way that we have denoted the time lags with a superscript, e.g., P, we denote the unit time step advance as $p_X(x(t_{n+1})) \equiv p_X(x(t_n)^{(1)})$.

Armed with this notation, we consider the work of Kaiser and Schreiber [19] and define the continuous TE between random processes $X(t)$ and $Y(t)$

$$TE_{Y\to X} = \int_{\mathbb{R}^{P+Q+1}} p_X(x(t_n)^{(1)}|\mathbf{x}^{(P)}(t_n), \mathbf{y}^{(Q)}(t_n))$$
$$\times \log_2\left(\frac{p_X(x(t_n)^{(1)}|\mathbf{x}^{(P)}(t_n), \mathbf{y}^{(Q)}(t_n))}{p_X(x(t_n)^{(1)}|\mathbf{x}^{(P)})}\right) dx(t_n^{(1)}) d\mathbf{x}(t_n)^{(P)} d\mathbf{y}(t_n)^{(Q)} \tag{3.129}$$

where the $P + Q + 1$-dimensional integral extends over the support of the random variables. By definition, this measure quantifies the ability of the random process $\mathbf{Y}$ to predict the dynamics of the random process $\mathbf{X}$. To see why, we can examine the argument of the logarithm. In the event that the dynamics of $\mathbf{X}$ are *not* coupled to $\mathbf{Y}$, one has the Markov model in the denominator of (3.129) as a good predictor of future values of the random process $\mathbf{X}$. However, should $\mathbf{Y}$ carry information about the transition probabilities of $\mathbf{X}$, the numerator is a better model. The TE is effectively mapping the difference between these hypotheses to the scalar $TE_{Y\to X}$. Thus, in the same way that mutual information measures deviations from the hypothesis of independence, TE measures deviations from the hypothesis that the dynamics of $\mathbf{X}$ can be described entirely by its own past history and that no new information is gained by considering the dynamics of system $\mathbf{Y}$.

As with the mutual information, the estimation of (3.129) is greatly aided by assuming that the processes are stationary and ergodic, with each observation coming from the same underlying probability distribution. In this case, the absolute time index t_n is of no consequence and may be therefore dropped from the notation. With the assumption of stationarity, we may use the law of conditional probability (2.16) and expand Eq. (3.129) as

$$TE_{Y\to X} = \int_{\mathbb{R}^{P+Q+1}} p_{X^{(1)}\mathbf{XY}}(x^{(1)}, \mathbf{x}^{(P)}, \mathbf{y}^{(Q)}) \log_2(p_{X^{(1)}\mathbf{XY}}(x^{(1)}, \mathbf{x}^{(P)}, \mathbf{y}^{(Q)}))$$
$$\times dx^{(1)} d\mathbf{x}^{(P)} d\mathbf{y}^{(Q)}$$

$$-\int_{\mathbb{R}^{P+Q}} p_{\mathbf{XY}}(\mathbf{x}^{(P)}, \mathbf{y}^{(Q)}) \log_2(p_{\mathbf{XY}}(\mathbf{x}^{(P)}, \mathbf{y}^{(Q)})) d\mathbf{x}^{(P)} d\mathbf{y}^{(Q)}$$

$$-\int_{\mathbb{R}^{P+1}} p_{X^{(1)}\mathbf{X}}(x^{(1)}, \mathbf{x}^{(P)}) \log_2(p_{X^{(1)}\mathbf{X}}(x^{(1)}, \mathbf{x}^{(P)})) dx^{(1)} d\mathbf{x}^{(P)}$$

$$+\int_{\mathbb{R}^{P}} p_{\mathbf{X}}(\mathbf{x}^{(P)}) \log_2(p_{\mathbf{X}}(\mathbf{x}^{(P)})) d\mathbf{x}^{(P)}$$

$$= -h_{X^{(1)}\mathbf{X}^{(P)}\mathbf{Y}^{(Q)}} + h_{\mathbf{X}^{(P)}\mathbf{Y}^{(Q)}} + h_{X^{(1)}\mathbf{X}^{(P)}} - h_{\mathbf{X}^{(P)}} \tag{3.130}$$

where the terms $h_{\mathbf{X}^{(M)}} = -\int_{\mathbb{R}^{M}} p_{\mathbf{X}}(\mathbf{x}) \log_2(p(\mathbf{x})) d\mathbf{x}$, are the differential entropies associated with an M-dimensional random variable $\mathbf{X}$. As with the mutual information, we see that the TE is comprised of differential entropies.

It is instructive to study this quantity first in the case where an analytical solution is possible. Again assuming that both random processes $\mathbf{X}$ and $\mathbf{Y}$ are jointly Gaussian distributed (see 3.115), we may follow the same procedure as was used to derive (3.116), which ultimately yields

$$TE_{Y \to X} = \frac{1}{2} \log_2 \left(\frac{|C_{\mathbf{X}^{(P)}\mathbf{Y}^{(Q)}}||C_{X^{(1)}\mathbf{X}^{(P)}}|}{|C_{X^{(1)}\mathbf{X}^{(P)}\mathbf{Y}^{(Q)}}||C_{\mathbf{X}^{(P)}}|} \right). \tag{3.131}$$

For P, Q large the needed determinants become difficult to compute. We therefore employ a simplification to the model that retains the spirit of the TE, but makes an analytical solution more tractable. In our approach, we set $P = Q = 1$, i.e., both random processes are assumed to follow a first-order Markov model. However, we allow the time interval between the random processes to vary, just as we did for the mutual information. Specifically, we model $X(t)$ as the first-order Markov model $p_X(x(t_n + \Delta_t)|x(t_n))$ and use the TE to consider the alternative $p_X(x(t_n + \Delta_t)|x(t_n), y(t_n + \tau))$ (note that $t_{n+1} = t_n + \Delta_t$). Although we are only using first-order Markov models, by varying the time delay τ we can explore whether the random variable $Y(t_n + \tau)$ carries information about the transition probability $p_X(x(t_n + \Delta_t)|x(t_n))$. Should consideration of $y(t_n + \tau)$ provide no additional knowledge about the dynamics of $X(t)$, the TE will be zero, rising to some positive value should $y(t_n + \tau)$ carry information about future values of $X(t)$ not possessed in $x(t_n)$. In what follows, we refer to this particular form of the TE as the time-delayed transfer entropy, or, TDTE. In this simplified situation, the needed covariance matrices are (using the overbar to denote mean, or average value)

$$C_{XY}(\tau) = \begin{bmatrix} E[(X(t_n) - \bar{X})^2] & E[(X(t_n) - \bar{X})(Y(t_n + \tau) - \bar{Y})] \\ E[(Y(t_n + \tau) - \bar{Y})(X(t_n) - \bar{X})] & E[(Y(t_n + \tau) - \bar{Y})^2] \end{bmatrix}$$

$$C_{X^{(1)}XY}(\tau) = \begin{bmatrix} E[(X(t_n + \Delta_t) - \bar{X})^2] & E[(X(t_n + \Delta_t) - \bar{X})(X(t_n) - \bar{X})] \\ E[(X(t_n) - \bar{X})(X(t_n + \Delta_t) - \bar{X})] & E[(X(t_n) - \bar{X})^2] \\ E[(Y(t_n + \tau) - \bar{Y})(X(t_n + \Delta_t) - \bar{X})] & E[(Y(t_n + \tau) - \bar{Y})(X(t_n) - \bar{X})] \end{bmatrix}$$

$$\begin{bmatrix} E[(X(t_n + \Delta_t) - \bar{X})(Y(t_n + \tau) - \bar{Y})] \\ E[(X(t_n) - \bar{X})(Y(t_n + \tau) - \bar{Y})] \\ E[(Y(t_n + \tau) - \bar{Y})^2] \end{bmatrix}$$

$$C_{X^{(1)}X} = \begin{bmatrix} E[(X(t_n + \Delta_t) - \bar{X})^2] & E[(X(t_n + \Delta_t) - \bar{X})(X(t_n) - \bar{X})] \\ E[(X(t_n) - \bar{X})(X(t_n + \Delta_t) - \bar{X})] & E[(X(t_n) - \bar{X})^2] \end{bmatrix} \tag{3.132}$$

and $C_{XX} = E[(X(t_n) - \bar{X})^2] \equiv \sigma_X^2$ is simply the variance of each observation in the random process **X**. The assumption of stationarity also allows us to write $E[(X(t_n + \Delta_t) - \bar{X})^2] = \sigma_X^2$ and $E[(Y(n + \tau) - \bar{Y})^2] = \sigma_Y^2$. Making these substitutions into (3.131) yields the expression

$$TE_{Y \to X}(\tau)$$

$$= \frac{1}{2} \log_2 \left[\frac{\left(1 - \rho_{XX}^2(\Delta_t)\right)\left(1 - \rho_{XY}^2(\tau)\right)}{1 - \rho_{XY}^2(\tau) - \rho_{XY}^2(\tau - \Delta_t) - \rho_{XX}^2(\Delta_t) + 2\rho_{XX}(\Delta_t)\rho_{XY}(\tau)\rho_{XY}(\tau - \Delta_t)} \right]$$

(3.133)

where we have defined particular expectations in the covariance matrices in terms of the cross-correlation coefficient $\rho_{XY}(\tau) \equiv E[(X(t_n) - \bar{X})(Y(t_n + \tau) - \bar{Y})]/\sigma_X \sigma_Y$. Note that the covariance matrices are positive-definite matrices and that the determinant of a positive-definite matrix is positive [26]. Thus, the quantity inside the logarithm will always be positive and the logarithm will exist. In fact, it turns out that by construction the argument of the logarithm in (3.133) is always ≥ 1, so that the TE is always positive.

Now, the hypothesis that the TE was designed to test is whether past values of the process $Y(t)$ carry information about the transition probabilities of the second process $X(t)$. Thus, if we are to keep with the original intent of the measure, we would only consider $\tau < 0$. However, this restriction is only necessary if one implicitly assumes a nonzero TE means $Y(t)$ is *influencing* the transition $p_X(x(t_n + \Delta_t)|x(t_n + \tau))$ as opposed to simply carrying additional information *about* the transition. Again, this latter statement is a more accurate depiction of what the TE is really measuring and we have found it useful to consider both negative and positive delays τ in trying to understand coupling among system components.

It is also interesting to note the bounds of this function. Certainly, for constant-valued signals we have $\rho_{XX}(\Delta_t) = \rho_{XY}(\tau) = 0 \; \forall \; \tau$ and the TE is zero for any choice of time scales τ defining the Markov processes. Knowledge of $Y(t)$ does not aid in forecasting $X(t)$ simply because the transition probability in going from $x(t_n)$ to $x(t_n + \Delta_t)$ is already unity. Likewise, if there is no coupling between system components we have $\rho_{XY}(\tau) = 0$ and the TDTE becomes $TE_{Y \to X}(\tau) = \frac{1}{2} \log_2 \left[\frac{1 - \rho_{XX}^2(\Delta_t)}{1 - \rho_{XX}^2(\Delta_t)} \right] = 0$. At the other extreme, for *perfectly* coupled systems, i.e., $X = Y$, consider $\tau \to 0$. In this case, we have $\rho_{XY}^2(\tau) \to 1$, and $\rho_{XY}(\tau - \Delta_t) \to \rho_{XX}(-\Delta_t) = \rho_{XX}(\Delta_t)$ (in this last expression we have noted the symmetry of the function $\rho_{XX}(\tau)$ with reference to the time delay). The TE then becomes

$$TE_{Y \to X}(0) = \frac{1}{2} \log_2 \left[\frac{0}{0} \right] \to 0$$

(3.134)

and the random process $Y(t)$ at $\tau = 0$ is again seen to carry no information about the dynamics of $X(t)$. However, in this special case, the zero TE is simply due to the fact that $p_X(x(t_n + \Delta_t)|x(t_n)) = p_X(x(t_n + \Delta_t)|x(t_n), x(t_n))$. This example highlights the care that must be taken in interpreting the TE. In particular, we must not interpret the measure to quantify the amount of information carried about the dynamics of one random process by another. Rather, the TDTE measures the *additional* information provided by one random process about the dynamics of another. In the perfectly coupled case, there is no added information content in $Y(t)$ not already contained in $X(t)$. Hence, the interpretation as a measure of information exchange (the original intent of the measure) is on firm ground, however interpreting the TDTE as a general measure of coupling is not always valid.

We should also point out that the average mutual information function can resolve this ambiguity if one wishes to use the TDTE as a diagnostic of dynamical coupling. For two Gaussian random processes, the mutual information was just shown to be $I_{XY}(\tau) = -\frac{1}{2}\log_2[1 - \rho_{XY}^2(\tau)]$ (Eq. 3.127). Hence, for perfect coupling $I_{XY}(0) \to \infty$, whereas for uncoupled systems $I_{XY}(0) \to 0$. Computing both time-delayed mutual information and TEs permits stronger inference about the nature of dynamical coupling.

However, for our purposes we are simply interested in measures that will attain different values depending on whether the random processes are generated by a linear or nonlinear system. Because the TE and mutual information functions are defined in terms of the entire joint probability distribution function associated with those random processes, they will be sensitive to the presence of higher order statistics in a nonlinear system output. Because structural damage is frequently associated with the presence of a nonlinearity, both measures can be used to create effective detectors. Use of information-theoretics as damage detection statistics is discussed at length in Chapter 8 and demonstrated to correctly diagnose the presence of damage in an experimental rotary system in Section 8.8.1.

3.6   Random Process Models for Structural Response Data

This chapter has focused on the probabilistic modeling of sequences of observations. The reason for this, as is apparent from the subsequent chapters, is that such sequences are precisely the type of data we will be dealing with in practice. To identify the properties of a structure (e.g., size of a crack), we need to observe the structure. These observations may be collected via laser vibrometer, accelerometers, strain gages, or linear variable differential transformers (LVDT; measures displacement), to name a few. When we make these measurements, we have (or should have!) some deterministic model that predicts what we will observe. However, there will certainly be aspects of the problem we cannot accurately predict; these are modeled probabilistically. We therefore use the tools developed in this chapter to model those aspects of physics we cannot predict with a deterministic model.

The obvious place such uncertainty crops up is sensor noise, which is often appropriately modeled as a joint normally distributed random process. Another source of uncertainty is the source of structural vibrations. Oftentimes, we model the input (loading) to a structure as a random process from a particular joint probability distribution, again often assumed normally distributed. Wind loading on a building, traffic loading on a bridge, or engine rattling on a shock absorber are all instances where the input to the structure is appropriately modeled as a random process. Thus, we have uncertainty in both the structural inputs and in the structural response, which must be accounted for in our analysis.

This chapter has also discussed different properties we might use to describe a random process. These are of particular value in Chapter 8, which focuses on damage detection. For example, we see several examples where HOS analysis of a structure's response to random vibration can be used to detect the presence of damage-induced nonlinearities. We also see how to use what we have learned about spectral density functions to create "surrogate" structural response data (see Section (8.6)), again with the goal being to detect the presence of damage. Probabilistic models are also the key to quantifying the performance of a given damage identification algorithm, as discussed in Section 6.9.

Finally, probabilistic models form the core of our approach to estimating parameters associated with structural damage. Thus, before we conclude this chapter we provide a preview of what is to come regarding estimation (our eventual goal). To develop accurate techniques for parameter estimation, a description (model) of the observed structural response data is required. In structural dynamics, random process models are typically comprised of two parts: a deterministic component that is the result of our physics-based model and a stochastic component describing the uncertainty present in experimental measurements (e.g., sensor noise). For example, consider the physics of a simple, one-degree-of-freedom structure subject to harmonic excitation. The deterministic model is

$$m\ddot{x}(t) + c\dot{x}(t) + kx(t) = F\sin(\omega t) \tag{3.135}$$

where m, c, k, F, ω are physical system parameters with units kg, kg/s, kg/s^2, kg $\cdot$ m/s^2, rad/s, respectively. The acceleration response, $\ddot{x}(t)$ in m/s^2, is well known and is given by Rao [27]

$$\ddot{x}(t) = A(\omega)\sin(\omega t + \phi) \tag{3.136}$$

where $A(\omega)$ is a frequency-dependent amplitude (to be discussed in Section 4.5). Now let's assume that we have an accelerometer mounted on the structure and that we sample N points of the response with a sampling interval of Δ_t seconds. We might model the measured response

$$\ddot{y}(n) = \ddot{x}(n) + \eta(n)$$
$$= A(\omega)\sin(\omega\, n\Delta_t + \pi) + \eta(n), n = 1 \cdots N \tag{3.137}$$

where each of the $\eta(n)$ are taken as independent, normally distributed random variables, H(n), with zero mean and variance σ^2. This is a frequently used additive model for sensor noise. The discrete time index n indicates data collected at sampling times $t_n = n\Delta_t$, $n = 1 \cdots N$. As mentioned earlier, this model is insufficient to predict *exactly* the values $\ddot{y}(n)$ because the $\eta(n)$ are chosen randomly from a normal distribution with mean zero and standard deviation, σ. However, we *can* make useful statements about the probability of having observed the sequence $\ddot{y}(n)$ using a random process model.

If each $\eta(n)$ is normally distributed with zero mean, we know that the quantity $\eta(n) + \ddot{x}(n)$ is still normally distributed, but with nonzero mean $\ddot{x}(n)$. This allows us to write for any particular n

$$p_H(\ddot{y}(n)) = \frac{1}{\sqrt{2\pi\sigma^2}}e^{-\frac{1}{2\sigma^2}(\ddot{y}(n)-\ddot{x}(n))^2}. \tag{3.138}$$

If we further assume each noise value is independent, i.e., $p_{HH}(\eta(n), \eta(m)) = p_H(\eta(n))p_H(\eta(m))$, $\forall\, n \neq m$ we then have for the entire observed sequence

$$p_{\mathbf{H}}(\ddot{\mathbf{y}}) = \frac{1}{(2\pi\sigma^2)^{N/2}}e^{-\frac{1}{2\sigma^2}\sum\limits_{n=1}^{N}(\ddot{y}(n)-\ddot{x}(n))^2}. \tag{3.139}$$

This particular expression shows up numerous times in this book as it is an extremely useful random process model. It is typical in structural dynamics problems for the deterministic component to play the role of the mean value of the (usually Gaussian) noise distribution. It is important to note that the form of the random process model is in practice governed by the form of the noise. If we had an additive, independent chi-squared noise model (see Appendix A), $p_{\mathbf{H}}(\ddot{\mathbf{y}})$ would have been expressed as the product of N chi-squared random variables. The ability to probabilistically describe a sequence of observations is absolutely essential to the process of estimation. Indeed, we rely heavily on the material of this section in developing methods for estimating the model parameters (m, c, k) in Chapter 7. However, before proceeding we clearly need the tools to develop the physics-based portion of Eq. (3.139), $\ddot{x}(n)$. The next two chapters develop several approaches to providing such models.

References

[1] R. N. McDonough, A. D. Whalen, Detection of Signals in Noise, 2nd ed., Academic Press, San Diego, CA, 1995.

[2] L. Isserlis, On a formula for the product-moment coefficient of any order of a normal frequency distribution in any number of variables, Biometrika 12 (1/2) (1918) 134–139.

[3] J. M. Nichols, C. C. Olson, J. V. Michalowicz, F. Bucholtz, The bispectrum and bicoherence for quadratically nonlinear systems subject to non-Gaussian inputs, IEEE Transactions on Signal Processing 57 (10) (2009) 3879–3890.

[4] J. M. Nichols, P. Marzocca, A. Milanese, The trispectrum for Gaussian driven, multiple degree-of-freedom, non-linear structures, International Journal of Non-Linear Mechanics 44 (2009) 404–416.

[5] J. V. Michalowicz, J. M. Nichols, F. Bucholtz, C. C. Olson, A general Isserlis theorem for mixed-Gaussian random variables, Statistics and Probability Letters 81 (8) (2011) 1233–1240.

[6] M. B. Priestly, Spectral Analysis and Time Series, Probability and Mathematical Statistics, Elsevier Academic Press, London, 1981.

[7] N. Wiener, Generalized harmonic analysis, Acta Mathematica 55 (1) (1930) 117–258.

[8] A. M. Yaglom, An Introduction to the Theory of Stationary Random Functions, Dover Publications, New York, 1962.

[9] M. Rosenblatt, J. W. V. Ness, Estimation of the bispectrum, Annals of Mathematical Statistics 36 (4) (1965) 1120–1136.

[10] M. J. Hinich, Higher order cumulants and cumulant spectra, Circuits, Systems, and Signal Processing 13 (4) (1994) 391–402.

[11] J. S. Owen, B. J. Eccles, B. S. Choo, M. A. Woodings, The application of auto-regressive time series modelling for the time-frequency analysis of civil engineering structures, Engineering Structures 23 (2001) 521–536.

[12] H. Sohn, K. Worden, C. R. Farrar, Statistical damage classification under changing environmental and operational conditions, Journal of Intelligent Material Systems and Structures 13 (2002) 561–574.

[13] C. E. Shannon, Communication in the presence of noise, Proceedings of the IEEE (reprinted from "Proceedings of the IRE 37(1), 10–21, 1949") 86 (2).

[14] J. V. Michalowicz, J. M. Nichols, F. Bucholtz, Handbook of Differential Entropy, CRC Press, Boca Raton, FL, 2014.

[15] J. M. Nichols, Examining structural dynamics using information flow, Probabilistic Engineering Mechanics 21 (2006) 420–433.

[16] J. M. Nichols, M. Seaver, S. T. Trickey, L. W. Salvino, D. L. Pecora, Detecting impact damage in experimental composite structures: an information-theoretic approach, Smart Materials and Structures 15 (2006) 424–434.

[17] I. Trendafilova, W. Heylen, H. V. Brussel, Measurement point selection in damage detection using the mutual information concept, Smart Materials and Structures 10 (3) (2001) 528–533.

[18] T. Schreiber, Measuring information transfer, Physical Review Letters 85 (2000) 461.

[19] A. Kaiser, T. Schreiber, Information transfer in continuous processes, Physica D 166 (2002) 43–62.

[20] J. M. Nichols, S. T. Trickey, M. Seaver, Detecting damage-induced nonlinearities in structures using information theory, Journal of Sound and Vibration 297 (2006) 1–16.

[21] S. Kullback, Information Theory and Statistics, Dover Publications, 1968.

[22] R. Marschinski, H. Kantz, Analysing the information flow between financial time series, European Physical Journal B 30 (2002) 275–281.

[23] L. J. Moniz, E. G. Cooch, S. P. Ellner, J. D. Nichols, J. M. Nichols, Application of information theory methods to food web reconstruction, Ecological Modelling 208 (2007) 145–158.

[24] L. J. Moniz, J. D. Nichols, J. M. Nichols, Mapping the information landscape: discerning peaks and valleys for ecological monitoring, Journal of Biological Physics 33 (2007) 171–181.

[25] M. Bauer, J. W. Cox, M. H. Caveness, J. J. Downs, N. F. Thornhill, Finding the direction of disturbance propagation in a chemical process using transfer entropy, IEEE Transactions on Control Systems Technology 15 (1) (2007) 12–21.

[26] S. M. Kay, Fundamentals of Statistical Signal Processing: Estimation Theory, Vol. I, Prentice Hall, New Jersey, 1993.

[27] S. S. Rao, Mechanical Vibrations, 3rd ed., Addison-Wesley, New York, 1995.

4

Modeling in Structural Dynamics

Now that we have developed rigorous tools that quantitatively describe probabilistic functions and random processes, we turn our attention to the development of deterministic models, whose output describes the behavior of the physical structure under consideration. This description, when combined with the material of the last three chapters, completes our model of structural response data, thereby allowing for the estimation of unknown structural parameters (Chapter 7), and the quantification of damage in structural systems (Chapters 8 and 9).

As we mentioned in the introduction, a physics-based model is a mathematical description of some aspect of a system's behavior (e.g., amplitude of vibration, strain at a particular location, etc.) and is obtained using the physical properties of the system (e.g., geometry, materials, etc.) and some underlying, governing physical law. The job of the model is to predict the behavior we expect to observe. For example, we might want a model to predict the first natural frequency of vibration as a function of the length of a crack. An overview of the model development process is shown in Figure 4.1 and is briefly discussed. Our emphasis on a physics-based approach to modeling (as opposed to data modeling) is that we hope to relate what we measure in practice (strain, displacement, etc.), to a physical quantity of interest, in particular, structural damage. Moreover, because physical laws tend to be differential in nature, the resulting model equations are often ordinary differential equations (ODEs) or partial differential equations (PDEs), which have to be solved to obtain the desired system behavior, such as a strain field, a response frequency, or a reaction force.

Interestingly, Figure 4.1 can be viewed in terms of disciplines. The left-hand side is all driven by physics and engineering. In other words, we have to develop a quantitative description of the relevant physical characteristics of a system (material behavior, geometry, etc.) and then combine them with known physical laws; this leads to the governing equations. The center, by contrast, is actually a mathematical problem: how do you solve the equations that emerge from the underlying physics? These solutions, the right-hand side, describe the system physical response. A key step involved in the right-hand side, which is not pictured, is validation. Once the response is obtained, we should turn back to the physics to see if the result makes physical sense. Validating the answer typically involves comparing the obtained solution to that of a simplified problem or, when possible, with experimental data. To accomplish the validation step, we require the tools of estimation, the subject of Chapter 7.

Modeling and Estimation of Structural Damage, First Edition. Jonathan M. Nichols and Kevin D. Murphy.
© 2016 John Wiley & Sons, Ltd. Published 2016 by John Wiley & Sons, Ltd.

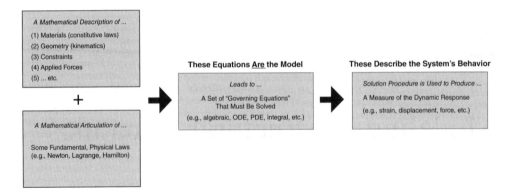

Figure 4.1   A general overview of the modeling process

4.1   Why Build Mathematical Models?

Why build models? It's a fair enough question. After all, good models take some expertise to develop. Solutions have to be obtained – sometimes at great numerical cost – and then validated. So why spend all of this time and energy? What's the upside?

In the context of parameter estimation, and damage estimation specifically, models are absolutely essential. These models *contain* the unknown parameters (e.g., a crack length, a delamination depth, a corrosion level, etc.) that characterize the extent to which the structure is damaged. So without the model there is no parametric description of the damage. In other words, there would be no damage parameters to estimate.

But in a wider sense, models provide physical insight. And this is invaluable for a design engineer who wants to promote (curtail) desirable (undesirable) behavior in components or systems. As an example of the latter, consider resonance. If you're driving a beam-like system near its fundamental frequency, resonance will result. A designer may wish to increase the fundamental frequency out of this operating range. On the basis of the simple Euler–Bernoulli (EB) beam theory, the ith frequency is $\omega_i = \beta_i^2 \sqrt{\frac{EI}{mL^4}}$ [1]. So the designer is left with four options (presuming a change in boundary conditions is not an option): (i) increase the modulus E, (ii) increase the area moment of inertia I, (iii) decrease the mass per unit length m, or (iv) decrease the length L. And this formula provides both qualitative insight and quantitative details on how to get this job done. For example, to increase the frequency you should decrease L (qualitative information) *and*, because $\omega_i \propto 1/L^2$ (quantitative information), you know how much L must change to produce the desired change in ω_i. For more complex problems, where only numerical results exist, a parametric study (a plot) involving the appropriate normalized parameters serves the same purpose as a formula. In the absence of a model, a designer would have only two choices. He/she could guess on the basis of intuition, which is a rudderless endeavor. Or he/she could build a physical prototype, test, and redesign iteratively until an acceptable design is achieved. This is an extremely time-consuming and expensive option.

So the short answer to the original question is this: anyone who needs to develop a fundamental insight into how certain system parameters impact the behavior of a system should care about modeling. It is fast and inexpensive in comparison to every other option.

4.2   Good Versus Bad Models – An Example

As mentioned, a *physical system model* is a mathematical description/representation of a system, component, or process that incorporates some of the relevant, underlying physics of the problem. And because of the rigor and exactness of science – physics, in this case – people can get lulled into the notion that there is only one "right" model and all others must be "wrong." But that's not true. Let's consider three scenarios.

First, there is the possibility that someone may misapply a physical principle. This, of course, leads to a "wrong" model because you're violating the underlying physics. But these gross errors are usually easy to identify, simply because they do not have predictive power, that is, the model will do a poor job describing the observed data.

Second, a model may not replicate some observed behavior because a crucial piece of physics, which is responsible for the behavior, has been omitted. This doesn't mean that the physics that was incorporated is wrong, per se. The model may still work under certain limited circumstances, but is entirely unable to reproduce the system's behavior in others. For example, consider taking strain measurements during a pull-test in an environment where the temperature during the test can vary by as much as 50 °F (no temperature compensation made to the measurements). To make sense of the data, one must include a temperature-dependent constitutive model or the model is doomed. A temperature-independent model could be created. But it won't be able to mimic the data. This isn't a gross error, as in the first case. It is an error in judgment; a modeler has to understand the application, know what parameters are central to the problem, and foresee how the model will be used.

The third possibility is that a model will do well under certain circumstances and then less well under others (but still offer some predictive power). This is the most common scenario. Under these circumstances, it's not a binary outcome (good/bad). Instead, there's a sliding scale with varying *degrees of goodness*. Circumstances matter. A helpful analogy is baking a cake. At sea level, one recipe might be very successful. That same recipe may not turn out well at high altitudes. And at intermediate elevations, the recipe achieves an intermediate level of success. If you think of gradually moving from sea level to higher elevations, the performance of the original recipe steadily diminishes. There isn't a magic elevation below which the recipe is great and above which it absolutely fails. The degradation is gradual. Models behave similarly. The good news is that when a model begins to perform poorly, it can just be modified to accommodate new circumstances (just as at high elevations a cake recipe can be augmented to perform better at altitude – by adding less baking powder and more egg whites, for example). The moral is that, insofar as possible, one must be mindful of the circumstances under which a model is being used and know the efficacy of the model in those cases.

Let's highlight this idea with a specific, physical problem. Consider the lateral deformation $w(x)$ of a pinned–pinned beam, subjected to a static, half-sine wave, distributed load $q(x) = q \sin(\pi x/L)$. Ultimately, we would want to find the deflection of the beam. But for now, we'll just focus on a description of the internal strain energy, as that gives us a measure of the *goodness* of the model; the strain energy is directly related to the stiffness, which impacts the deflection (for details on how this leads to the deflection, see Hamilton's principle in Section 4.3.5). The simplest model is the EB beam; this includes only energy due to bending. But to quantify the fidelity of the EB model, a second model is needed for comparison. This second model will be the more complex Timoshenko (T) beam, which contains energy due to bending *and* shear. *Note*: If you're just interested in the results and not the details of the two models, skip the next two paragraphs. We provide the physical principles and details that underly this brief example subsequently.

In the EB model, it's assumed that only the normal stresses – arising from bending – contribute significantly to the strain energy; shear is ignored. The net axial elongation $u = u(x, y)$ at any location is given by

$$u_{total} = u_s + u_b$$
$$= u_s - y\frac{\partial w}{\partial x} \tag{4.1}$$

where u_s is the uniform stretching that occurs across the cross-section due to axial loads, y is the elevation above/below the neutral axis, and $w(x)$ is the lateral deformation of the neutral axis. The deformation of all points (x, y) in the continuum are described by u_{total} and w. The local linear strain in the x-direction is given by $\epsilon_{11} = \frac{\partial u_{total}}{\partial x}$:

$$\epsilon_{11} = u_{s,x} - yw_{,xx} \tag{4.2}$$

where the comma denotes differentiation with respect to the variable that follows. The associated strain energy is

$$U_b = \frac{1}{2} \int \int \int \sigma_{11}\epsilon_{11} \, d\Omega \tag{4.3}$$

where $d\Omega$ is an infinitesimal volume element. Assuming that the material acts as a 1-D linear elastic continuum (i.e., $\sigma_{11} = E\epsilon_{11}$), the energy becomes

$$U_b = \frac{1}{2}b \int_0^L \int_{-h/2}^{h/2} (E\epsilon_{11}) \, \epsilon_{11} \, dx \, dy \tag{4.4}$$

where h is the thickness, b is the depth, and L is the length. Substituting in the expression for the strain (Eq. 4.2) and integrating through the thickness gives the traditional EB result. If we focus just on the bending contributions (i.e., ignoring strains induced by axial loads), the strain energy is

$$U_b = \frac{EI}{2} \int_0^L (w_{,xx})^2 \, dx. \tag{4.5}$$

This strain energy, along with the external work done by the applied lateral load $q(x)$, can be combined into Hamilton's principle (see Section 4.3.5) to arrive at an ODE governing the static system: $(EIw_{,xx})_{,xx} - q = 0$. The solution to this equation *is* the lateral deflection $w = w(x)$. For the pinned–pinned case, this turns out to be $w(x) = \frac{q_o}{EI}\left(\frac{L}{\pi}\right)^4 \sin\left(\frac{\pi x}{L}\right)$. Substituting this into the bending energy yields

$$U_b = \frac{q_o^2 L^5}{4\pi^2 EI} \tag{4.6}$$

That does it for the EB strain energy. Throughout this discussion we neglected any shear stress contributions to the strain energy. Of course, the EB model can make shear stress predictions. These shear stresses are simply not included in the strain energy calculation, which affects the predicted deflections.

For a second model, we do include shear terms in the overall energy. The shear stress τ_{12} is found via equilibrium: $\partial\sigma_{11}/\partial x + \partial\tau_{12}/\partial y = 0$, with the normal stress being $\sigma_{11} = My/I$.

These two can be combined and integrated with respect to y; the boundary conditions are applied to ensure that the shear is zero at the free surfaces $y = \pm h/2$. The shear stress distribution results:

$$\tau_{12} = -\frac{\partial M}{\partial x}\left(\frac{1}{2I}\right)\left(y^2 - \frac{h^2}{4}\right). \tag{4.7}$$

The bending moment is $M = EIw_{,xx}$, meaning that the shear is related to the third spatial derivative of w. Much like the stretching energy, the shear strain energy is the spatially integrated product of the stress and strain:

$$U_s = \frac{1}{2}\int\int\int \tau_{12}\gamma_{12}d\Omega = \frac{1}{2}\int\int\int \tau_{12}^2/G\, d\Omega. \tag{4.8}$$

We continue with the assumption of linear elastic behavior: $\tau = G\gamma$ with $G = \frac{E}{1+v}$ being the shear modulus and γ being the shear strain. Combining Eq. (4.7) with our constitutive model, the strain energy may be integrated and yields

$$U_s = \frac{q_o^2 L^5(1+v)}{20\pi^2 EI}\left(\frac{h}{L}\right)^2. \tag{4.9}$$

The total strain energy for the T model is the sum of U_b and U_s.

So we've developed two expressions: one for the bending energy U_b and the other for the shear energy U_s. How well does the EB (U_b only) model work? Consider Figure 4.2a. This shows the strain energy ratio U_s/U_b as a function of the thickness to length ratio h/L. At small values of h/L, which correspond to long, thin beams, $U_s/U_b \to 0$. In other words, these beams are dominated by bending; shear effects are not important. This is underscored by Figure 4.2b, which shows the lateral deflection for $h/L = 0.05$. The deflections for the EB beam (bending only) and the T beam (including shear) are virtually indistinguishable. Of course, the EB model is much simpler and, hence, preferable. At intermediate thickness-to-length ratios ($h/L \approx 0.38$), the bending and shear contributions are comparable. And for larger values of h/L (i.e., short, thick beams), shear begins to dominate the energy. For example, consider the case $h/L = 0.5$. Figure 4.2c shows the deflection of the two beam models; the EB model is entirely too flexible and grossly overpredicts the deflection. The more complex T (shear) beam theory has to be used to make reasonable, quantitative predictions.

Figure 4.2a clearly represents this sliding *scale of goodness*, alluded to earlier. The EB model does well at small values of h/L, but gradually gets worse as h/L is increased. It doesn't suddenly and catastrophically fail at some critical h/L. It's up to the user to decide how critical it is that the model be accurate, that is, up to what value of h/L is the EB result acceptable? If it is ultimately decided that accuracy requirement isn't met with the simpler approach, then the more complex model is needed.

This example also goes to the heart of the modeling challenge: one must display good judgment of what physical features play a critical role in the physics – under the given circumstances – and what can be readily ignored.

4.3   Elements of Modeling

Building a successful structural model is challenging because, despite the roadmap of Figure 4.1, each problem contains its own unique twists that complicate the process.

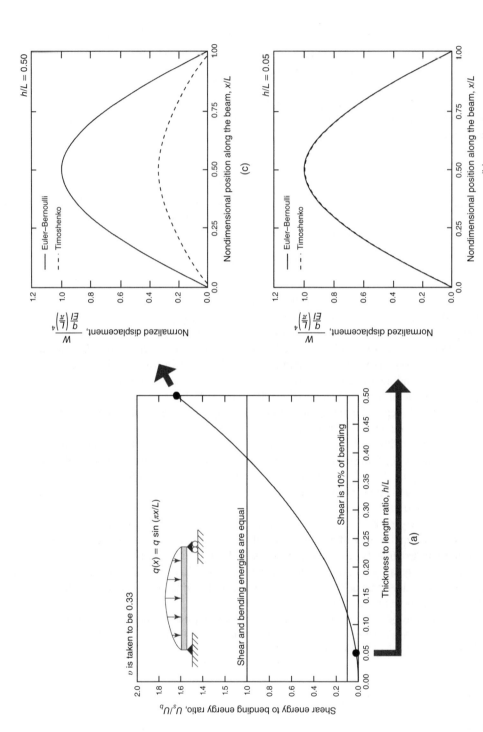

Figure 4.2 (a) The relative importance of the bending and strain energies, as a function of the normalized beam thickness. (b) The beam deflection based on the EB and T models for $h/L = 0.05$; the deflections are comparable. (c) The beam deflection based on the EB and T models for $h/L = 0.5$; the EB theory significantly overpredicts the deflection

These can include complex geometries, boundary conditions (contact conditions), material properties, constraints, defects, and so on. The real challenge is to discern what physical features constitute a "must have" versus a "nice to have" – and recognize that superfluous items add unnecessary complexity and increase simulation time. And then, of course, the modeler has to research related problems, borrowing liberally from the literature, and then develop a model to accommodate the particular problem at hand.

In what follows, we lay out some specifics on the fundamental laws and their applications. This includes some discussion of materials, geometry, constraints, and so on. Much of this takes its cue from the left-hand side of Figure 4.1. This is followed by a short discussion of the mathematical solutions to the governing equations. In Chapter 5, we look at a number of specific research problems, to demonstrate some of the challenges that arise. Many of these models are motivated by the Bayesian structural health monitoring (SHM) approach that is developed in Chapters 7, 9.

4.3.1   Newton's Laws

Newton's laws are fairly straightforward and are discussed in a variety of undergraduate and graduate textbooks on two- and three-dimensional dynamics, respectively [2–5]. Newton's laws are helpful in that they help develop a good physical feel for a problem. After all, conceptually dismembering a system in a free body diagram – exposing all participating forces – gets you into the nitty gritty of the system. But it has its drawbacks. The free body diagrams can be difficult to draw in 3-D, all terms are vector quantities, and constraint forces always appear. In addition, the kinematics involve accelerations, which are more complicated than, say, velocities. Given these issues, Newton isn't always the easiest way forward. We spend more time discussing – and using – energy-based methods. Energy methods eliminate the aforementioned issues and, as a side benefit, produce the expressions for the boundary conditions. Understanding the energy-based approach requires us first to understand the basics of variational calculus, the key points of which are outlined next.

4.3.2   Background to Variational Methods

Imagine having an integral I of an unknown function $y(x)$. The goal of the variational calculus is to determine the function $y(x)$ that extremizes the integral I. This may seem a little bizarre at first. But consider Figure 4.3a and this simple question: what function $y(x)$ goes through a and b and has the shortest arc length? An infinitesimal bit of arc length is $ds = \sqrt{dx^2 + dy^2} = \sqrt{1 + (y')^2}dx$. And the total arc length is the sum of all of these infinitesimal bits: $I = \int_a^b \sqrt{1 + (y')^2}dx$. So this problem seeks the function $y(x)$ that renders I a minimum. The answer is (of course) a straight line: $y(x) = mx + c$. But this example highlights the broader problem: find an unknown function that extremizes some specified integral.

Now let's look at the problem in a larger context: find the function $y(x)$ that renders $I = \int_a^b f(y, y', x)dx$ stationary. To solve this, we will take a cue from traditional calculus and Figure 4.3c. In traditional calculus, to extremize (render stationary) $f(x)$, we presume the

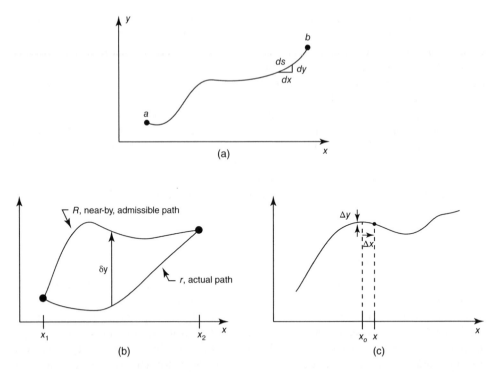

Figure 4.3 (a) Finding the shortest distance between points a and b. (b) The variation is the difference between the actual solution and an imagined, nearby path. (c) The standard calculus approach to finding an extremum

answer is x_o and look at values of the function in the neighborhood of x_o and construct a Taylor series:

$$f(x) = f(x_o) + \left.\frac{df}{dx}\right|_{x_o}(x - x_o) + \frac{1}{2}\left.\frac{d^2f}{dx^2}\right|_{x_o}(x - x_o)^2 + \cdots$$

$$df(x) = \left.\frac{df}{dx}\right|_{x_o} dx + \text{H.O.T.} \tag{4.10}$$

where $dx = (x - x_o)$ and H.O.T. stands for higher order terms (quadratic and above, in this case). Extremum are found when $df = 0$ for an arbitrary, nonzero dx. This gives the standard first-order requirement that $df/dx = 0$.

In the calculus of variations, we presume we have the function $y(x)$ that extremizes the integral I, as shown in Figure 4.3b. We consider functions near $y(x)$ that differ by some small *variation*, δy: $Y = y + \delta y$. Then we construct a Taylor series just as we do in traditional calculus:

$$I(Y, Y', x) = I(y, y', x) + \left.\frac{\partial I}{\partial y}\right|_{x_o}(Y - y) + \frac{1}{2}\left.\frac{\partial I}{\partial y'}\right|_{x_o}(Y' - y') + \text{H.O.T.}$$

$$\delta I = \left.\frac{\partial I}{\partial y}\right|_y \delta y + \frac{1}{2}\left.\frac{\partial I}{\partial y'}\right|_y \delta y' + \text{H.O.T.} \tag{4.11}$$

Stationary values are obtained when $\delta I = 0$. Note that taking the variation of a function involves differentiation. For example, if $g = g(q_1, q_2)$, then $\delta g = (\partial g/\partial q_1)\delta q_1 + (\partial g/\partial q_2)\delta q_2$. If we neglect the higher order terms and use the fact that I is a known integral of an unknown function y, then we find

$$\delta I = \frac{\partial}{\partial y}\left[\int_a^b f(y, y', x)dx\right]\delta y + \frac{\partial}{\partial y'}\left[\int_a^b f(y, y', x)dx\right]\delta y' = 0.$$

Presuming that the order of integration and differentiation is interchangeable, this becomes

$$\delta I = \int_a^b \left[\frac{\partial f}{\partial y}\delta y + \frac{\partial f}{\partial y'}\delta y'\right]dx = 0.$$

The second term in the integral can be integrated by parts once (to convert the $\delta y'$ to δy) to give:

$$\delta I = \int_a^b \left[\frac{\partial f}{\partial y}\delta y\right]dx + \frac{\partial f}{\partial y'}\delta y\Big|_a^b - \int_a^b \left[\frac{\partial}{\partial x}\left(\frac{\partial f}{\partial y'}\right)\delta y\right]dx = 0.$$

But from Figure 4.3b, the boundary terms are clearly zero because the variation at the endpoints are zero: $\delta y(a) = \delta y(b) = 0$. And so this expression reduces to

$$\delta I = \int_a^b \left[\frac{\partial f}{\partial y} - \frac{\partial}{\partial x}\left(\frac{\partial f}{\partial y'}\right)\right]\delta y \, dx = 0.$$

And the only way for this expression to be satisfied for an arbitrary variation δy is to require that the integrand be zero:

$$\frac{\partial}{\partial x}\left(\frac{\partial f}{\partial y'}\right) - \frac{\partial f}{\partial y} = 0. \tag{4.12}$$

Here's the takeaway: to find the function y that extremizes the integral $I = \int_a^b f(y, y', x)dx$, one must satisfy the differential equation, Eq. (4.12). This, of course, is only valid when $f = f(y, y', x)$. There's no reason f couldn't also depend on, say, the second derivative of y. That situation would simply require more integration by parts and is not conceptually different. But the case where $f = f(y, y', x)$ is of particular interest in dynamics because energy formulations usually require the displacement y (potential energy and external work), the velocity y' (kinetic energy), and the independent variable time t (i.e., $t = x$). So this sort of problem shows up with some regularity.

4.3.3   Variational Mechanics

Hamilton's principle and Lagrange's equations are integrated forms of Newton's second law. They contain all of the same physics and, hence, are equivalent to Newton. But they tend to simplify the process of analysis; as a result, they are generally preferable. It's also worth pointing out that Hamilton and Lagrange are scalar, energy-based relationships, as opposed to the vector equations of Newton.

To arrive at Hamilton, we begin with Newton's law for a collection of particles:

$$\sum_{i=1}^N \mathbf{F}_i = \sum_{i=1}^N m_i\ddot{\mathbf{r}}_i \tag{4.13}$$

where N represents the total number of particles in the system, F_i is the net force on the ith particle, m_i is the mass of the ith particle, $\mathbf{r}_i(t)$ is the position (or trajectory) of the ith in an inertial frame, and the dots refer to differentiation with respect to time. Proceeding along a *variational* line of reasoning, we consider another candidate trajectory $\mathbf{R}(t)$ that is "nearby" the actual trajectory and is geometrically admissible, that is, it satisfies some specified end conditions at t_1 and t_2. These two trajectories differ by a small *variation*: $\delta\mathbf{r} = \mathbf{R} - \mathbf{r}$. If Newton's law is dotted with this virtual displacement, one gets the following scalar equation:

$$\sum_{i=1}^{N} \mathbf{F}_i \cdot \delta\mathbf{r}_i - \sum_{i=1}^{N} m_i\ddot{\mathbf{r}}_i \cdot \delta\mathbf{r}_i = 0. \tag{4.14}$$

The first term is the virtual work δW done by the actual forces $\mathbf{F}_i$ and the virtual displacements $\delta\mathbf{r}_i$. The second term can be simplified using the chain rule (rearranged):

$$\ddot{\mathbf{r}} \cdot \delta\mathbf{r}_i = \frac{d}{dt}(\dot{\mathbf{r}}_i \cdot \delta\mathbf{r}_i) - \delta\left(\frac{1}{2}\dot{\mathbf{r}}_i \cdot \dot{\mathbf{r}}_i\right).$$

Using this expression in Eq. (4.14) yields

$$\delta W - \sum_{i=1}^{N} m_i\left[\frac{d}{dt}\left(\dot{\mathbf{r}}_i \cdot \delta\mathbf{r}_i\right) - \delta\left(\frac{1}{2}\dot{\mathbf{r}}_i \cdot \dot{\mathbf{r}}_i\right)\right] = 0. \tag{4.15}$$

Once m_i is multiplied through, the last term becomes the variation of the kinetic energy for the ith particle. And the sum over all the particles is the variation of the net kinetic energy δT. This can be recast as

$$\delta W + \delta T = \sum_{i=1}^{N} m_i\frac{d}{dt}(\dot{\mathbf{r}}_i \cdot \delta\mathbf{r}_i). \tag{4.16}$$

The right-hand side can be simplified if we integrate over the time considered $t \in [t_1, t_2]$

$$\int_{t_1}^{t_2} [\delta W + \delta T]dt = \int_{t_1}^{t_2}\left[\sum_{i=1}^{N} m_i\frac{d}{dt}\left(\dot{\mathbf{r}}_i \cdot \delta\mathbf{r}_i\right)\right]dt. \tag{4.17}$$

We will assume the order of the integration and the sum on the right-hand side can be interchanged. The integrand on the right side is simplified to $m_i d(\dot{\mathbf{r}}_i \cdot \delta\mathbf{r}_i)$. The right-hand side then becomes

$$\sum_{i=1}^{N} m_i\left[\int_{t_1}^{t_2} d\left(\dot{\mathbf{r}}_i \cdot \delta\mathbf{r}_i\right)\right] = \sum_{i=1}^{N} m_i\left[\dot{\mathbf{r}}_i \cdot \delta\mathbf{r}_i\big|_{t_1}^{t_2}\right] = 0$$

because the variation $\delta\mathbf{r}_i = 0$ is always zero at the end points of the interval t_1 and t_2, per Figure 4.2b. In other words, Eq. (4.17) becomes

$$\delta \int_{t_1}^{t_2} (T + W)\, dt = 0. \tag{4.18}$$

This is Hamilton's principle. In other words, this equation seeks to find a trajectory (a function) that extremizes what is referred to as the *action integral*: $\int_{t_1}^{t_2}(T + W)dt$. Of course, the work can be broken down into conservative (W_c) and nonconservative (W_{nc}) work. The former is derivable from a scalar potential and is path-independent, while the latter is not. In fact, the conservative work is just the negative of the scalar potential: $W_c = -V$. And so Hamilton's principle may be written as

$$\delta \int_{t_1}^{t_2} (T - V + W_{nc}) \, dt = 0. \tag{4.19}$$

In the next two sections, we highlight a few simple examples to show the utility of this result. More in-depth, research-level problems are explored in Chapter 5.

4.3.4   Lagrange's Equations

Lagrange's equations are particularly useful in describing discrete systems, that is, systems described by a discrete number of generalized coordinates. We'll assume that there are N of these coordinates, given by q_i. For the moment, let's also assume that the system is conservative, such that $W_{nc} = 0$. If we define $\mathcal{L} = T - V$ (noting that typically $T = T(\dot{q}_i(t))$ and $V = V(q_i(t))$), then Eq. (4.19) is reduced to $\delta \int_{t_1}^{t_2} \mathcal{L}(q_i, \dot{q}_i, t) dt = 0$, which is the form of our original variational problem. So the motion may be obtained by solving the following set of differential equations

$$\frac{\partial}{\partial t} \left(\frac{\partial \mathcal{L}}{\partial \dot{q}_i} \right) - \frac{\partial \mathcal{L}}{\partial q_i} = 0, \quad i = 1, 2, \ldots, N \tag{4.20}$$

for the $q_i(t)$'s. Relaxing the condition that $W_{nc} = 0$ changes the problem only slightly. In this case, the governing differential equations are

$$\frac{\partial}{\partial t} \left(\frac{\partial \mathcal{L}}{\partial \dot{q}_i} \right) - \frac{\partial \mathcal{L}}{\partial q_i} = Q_i, \quad i = 1, 2, \ldots, N \tag{4.21}$$

where Q_i is found by taking a variation of the nonconservative work term: $\delta W_{nc} = Q_i \delta q_i$ (sum implied).

In what follows we show a few simple examples to flesh out the process of using Lagrange's equations to get the governing equations for a system.

Example: Consider a simple mass-spring system shown in Figure 4.4a. We could displace the system from equilibrium, draw a free body diagram, and then apply Newton's second law, giving the familiar result: $m\ddot{x} + kx = 0$. To instead use Eq. (4.20), we set $q \equiv x$ and form the spring potential $V = \frac{1}{2}kx^2$ and the kinetic energy $T = \frac{1}{2}m\dot{x}^2$. The Lagrangian is $\mathcal{L} = T - V$. And the result is

$$\frac{\partial}{\partial t} \left(\frac{\partial \mathcal{L}}{\partial \dot{x}} \right) = \frac{\partial}{\partial t}(m\dot{x}) = m\ddot{x}$$

and

$$\frac{\partial \mathcal{L}}{\partial x} = -kx$$

So the resulting differential equation is

$$m\ddot{x} + kx = 0 \tag{4.22}$$

which agrees with the traditional equation obtained via Newton. Equation (4.22) is one of the most commonly used expressions in structural dynamics and is used throughout this book

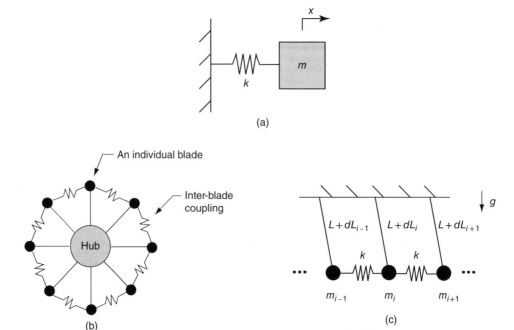

(a)

(b)

(c)

Figure 4.4   (a) Standard one degree-of-freedom (DOF) oscillator. (b) An extremely simple gas turbine model. (c) A series of interconnected pendula with the symmetry slightly broken

as the simplest relevant structural model. However, any real structural system will contain a dissipative term, that is, damping. While other forms of damping exist (e.g., Coulomb), a viscous damping model, where the dissipative force is proportional to the mass velocity, $\dot{x}$ is the most common.

To add this element of realism to the above-described framework, the nonconservative work, W_{nc}, would be nonzero. To demonstrate how this is incorporated, let's consider the case of a damper being placed in parallel to the spring in Figure 4.4a. Let's also assume that the damping force is linear and viscous in nature: $F_d = -c\dot{x}$ (negative because the force opposes the velocity). In this case, the work may be expressed as $W_{nc} = F \cdot x = (-c\dot{x})x$. The first variation of the work is: $\delta W_{nc} = Q\delta q = (-c\dot{x})\delta x$. By inspection, we conclude $Q = -c\dot{x}$. The earlier mass-spring equation is only changed by the presence of Q on the right-hand side of Eq. 4.21. The result is

$$m\ddot{x} + kx = -c\dot{x}$$

or

$$m\ddot{x} + c\dot{x} + kx = 0 \tag{4.23}$$

which agrees with the standard mass-spring-damper result found by a Newtonian approach. Of course, this is just the simplest form of dissipation in structural models; however, it is (i) an easy model to implement (parsimonious) and (ii) captures the essential behavior observed (at least qualitatively) for many structural systems. More complex forms of dissipation may be added if the situation warrants it. But the general process outlined here, using W_{nc}, remains valid.

Example: Models for gas turbine blades often consist of a number of flexible bodies mounted on a central hub, as shown schematically in Figure 4.4b. Cyclic symmetry is commonly assumed, where every blade is identical to its neighbor and there is uniform spacing. In cases where the blades differ slightly – perhaps due to a crack or loose manufacturing tolerances – that symmetry is broken and undesirable behavior such as mode localization can occur [6]. If you simply want to study mode localization, then the full turbine model (including blade extension bending, torsion, etc.) isn't necessary. A very simple model will suffice; the fundamental ingredients include a periodic, interconnected structure where the symmetry is (slightly) broken. One simple model consists of a series of pendula interconnected by springs. Here the symmetry is broken by the pendula having slightly different lengths. L is the nominal length and ΔL_i is the deviation from that nominal value. The angular displacement of the ith pendula is given in generalized coordinates by $q_i \equiv \theta_i$, with θ_i being measured "up" from the vertical. The net kinetic energy is

$$T = \sum_{k=1}^{N} m(L_k \dot{\theta}_k)^2$$

where $L_k = L + \Delta L_k$. The potential is

$$V = \sum_{k=1}^{N} m_k g L_k \left[1 - \cos(\theta_k)\right] + \sum_{k=1}^{N-1} \frac{1}{2} k[L_k \sin(\theta_k) - L_{k+1} \sin(\theta_{k+1})]^2.$$

Applying Lagrange's equations (Eq. 4.20) and assuming small motion (i.e., linearizing by making $\sin(\theta_k) \approx \theta_k$ and $\cos(\theta_k) \approx 1$), the following matrix differential equation results:

$$\mathbf{M}\ddot{\mathbf{x}} + \mathbf{K}\mathbf{x} = 0 \tag{4.24}$$

where $\mathbf{x} = \{\theta_1, \theta_2, \ldots, \theta_N\}^t$ and the mass and the stiffness matrices are

$$\mathbf{M} = mL \begin{bmatrix} (1+\eta_1) & 0 & 0 & 0 & \cdots \\ 0 & (1+\eta_2) & 0 & 0 & \cdots \\ 0 & 0 & (1+\eta_3) & 0 & \cdots \\ \vdots & \vdots & \vdots & \vdots & \ddots \end{bmatrix}$$

and

$$\mathbf{K} = kL \begin{bmatrix} (1+\eta_1) & -(1+\eta_2) & 0 & 0 & 0 & \cdots \\ -(1+\eta_1) & 2(1+\eta_2) & -(1+\eta_3) & 0 & 0 & \cdots \\ 0 & -(1+\eta_2) & 2(1+\eta_3) & -(1+\eta_4) & 0 & \cdots \\ 0 & 0 & -(1+\eta_3) & 2(1+\eta_4) & -(1+\eta_5) & \cdots \\ \vdots & \vdots & \vdots & \vdots & \vdots & \ddots \end{bmatrix}$$

$$+ mgL \begin{bmatrix} (1+\eta_1) & 0 & 0 & 0 & \cdots \\ 0 & (1+\eta_2) & 0 & 0 & \cdots \\ 0 & 0 & (1+\eta_3) & 0 & \cdots \\ \vdots & \vdots & \vdots & \vdots & \ddots \end{bmatrix}$$

Here $\eta_i = \Delta L_i / L$. The governing matrix equations are coupled through the stiffness because of our choice of generalized coordinates. The emphasis here is to find the governing equations of motion – and to provide a simple example of modeling choices – all of which was motivated by the localization problem. But if one wanted to continue with the mode localization problem, a periodic solution would be assumed: $\mathbf{x} = \mathbf{A}e^{i\omega t}$. This leads to the free vibration eigenvalue problem; under certain detunings (ΔL's), localization can be found in the modes' shapes (eigenvectors). We will tackle the eigenvalue problem shortly in Section 4.5.1.2.

Of course, other physical complications can arise. The problem may also include external excitation, leading to nonconservative work terms and a nonzero right-hand side. Or, if we reconsider the system of Figure 4.4b, the central hub may have a nonzero angular velocity Ω, producing a skew-symmetric gyroscopic matrix. We can also consider nonconservative damping forces as we did in the single-degree-of-freedom (SDOF) example (4.23). Following such a procedure frequently leads to a dissipative matrix $\mathbf{C}$ that possesses the same mathematical structure as either the mass or damping matrices. That is to say, we can write

$$\mathbf{C} = \alpha\mathbf{M} + \beta\mathbf{K} \tag{4.25}$$

where α, β are real-valued, usually small, constants. In this case, (4.24) becomes

$$\mathbf{M\ddot{x} + C\dot{x} + Kx = 0}. \tag{4.26}$$

Again, this choice is motivated by a combination of simplicity and its usefulness in describing observed structural response data. Analytically, there are also advantages to this form of damping, as is highlighted in Section 4.5.1. Practical implementation of this general approach to modeling dissipative forces appears in Sections 5.4, 8.5, 8.7.1, 8.7.2, 9.2, 9.3, 9.4, and 10.2.2.

The main point is that Lagrange's equations are the perfect tool for dealing with problems with a finite number of discrete generalized coordinates. The result is a set of matrix equations in the unknown coordinates that can be solved for either analytically or numerically. Collectively, the coordinates describe the structures' response and give us the needed model predictions at structural locations where we will record experimental data.

4.3.5   Hamilton's Principle

Lagrange (Eq. 4.21) is the ideal solution to Hamilton's principle (Eq. 4.19) when the problem involves a set of discrete, time-dependent variables: $q_i(t)$. But it is less helpful for spatially extended systems, whose displacement variable depends on time *and* space. A simple example is a beam, where the lateral deflection depends on (i) where you look on the beam and (ii) the instant of time under consideration. In other words, $w = w(x, t)$. In these cases, it is advantageous to use Hamilton's principle directly because, as it turns out, you get both the governing equation(s) *and* the boundary conditions. This will become clearer through a few examples.

Example: As a first example, let's consider the motion of the flexible bar shown in Figure 4.5a. This bar could undergo a number of motions: it could elongate (x-direction), it could bend (y-direction), or it could twist (rotation about the x-axis). But one can imagine scenarios where the geometry, neighboring geometries, or external loads restrict some of these motions. For example, if the bar is mounted on a bearing at $x = L$ (Figure 4.5b), then bending will be severely restricted and perhaps may be ignored. In this case, a system model only has to capture axial and torsional motion – greatly simplifying the analysis. The point is that these modeling

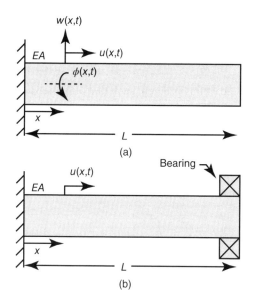

w(x,t)

EA

u(x,t)

$\phi(x,t)$

x

L

(a)

u(x,t)

Bearing

EA

x

L

(b)

Figure 4.5 (a) A bar undergoing a combined axial-bending-torsion deformation. (b) The bearing at $x = L$ restricts bending

choices must be based on a thorough understanding of how the system will be used and how it is likely to respond. As a means of demonstrating Hamilton's principle (Eq. 4.19), let's consider the system of Figure 4.5b and assume only axial stretching ($u = u(x,t)$) is relevant. The goal is to obtain the governing equation and boundary conditions. The net strain energy is the integrated effect of all of the infinitesimal energies: $U = V = \frac{1}{2}\sigma_{11}\epsilon_{11}\,dV$. We assume a uniform cross-section ($dV = A\,dx$) and that the deformations are small such that $\epsilon_{11} = \frac{\partial u}{\partial x}$. Lastly, we assume that the bar behaves like a linear elastic continuum. The constitutive model is $\sigma_{11} = E\epsilon_{11}$. Combining all of this leads to

$$U = \frac{1}{2}\int_0^L (E\epsilon_{11})\epsilon_{11}\,A\,dx = \frac{EA}{2}\int_0^L u_{,x}^2\,dx$$

To use this potential energy term in Eq. (4.19), we must take the first variation with respect to the unknown ($u_{,x}$) and this takes the form of a first-order Taylor series: $\delta U = (\partial U/\partial u_{,x})\delta u_{,x}$:

$$\delta U = EA\int_0^L u_{,x}\,\delta u_{,x}\,dx$$

The kinetic energy is the integrated effect of all of the kinetic energies of each infinitesimal material segment: $dT = \bar{m}\dot{u}^2\,dx$, where $\bar{m}$ is the mass per unit length (assumed constant). So the kinetic energy is

$$T = \frac{1}{2}\int_0^L \bar{m}\dot{u}^2\,dx$$

and its first variation ($\delta T = (\partial T/\partial\dot{u})\delta\dot{u}$) is

$$\delta T = \int_0^L \bar{m}\dot{u}\,\delta\dot{u}\,dx$$

Taking the difference $(\delta T - \delta U)$ and substituting into Eq. (4.19) leads to

$$\int_{t_1}^{t_2} \left[\int_0^L \left(\bar{m}\ddot{u}\, \delta\dot{u} - EAu_{,x}\, \delta u_{,x} \right)\, dx \right] dt = 0.$$

Ideally, we would like to recast this problem such that only the variation of u appears (not its derivatives) inside the time and space integral – the reason will become clear in a moment. To accomplish this, we can integrate the second term in the integral by parts:

$$\int_{t_1}^{t_2} \left[-EAu_{,x}\delta u \big|_0^L + \int_0^L \left(\bar{m}\ddot{u}\, \delta\dot{u} + EAu_{,xx}\, \delta u \right)\, dx \right] dt = 0.$$

If we swap the order of the space and time integral, we can integrate the $\delta\dot{u}$ term as well. This gives

$$\int_{t_1}^{t_2} \left[-EAu_{,x}\delta u \big|_0^L + \int_0^L \left(-\bar{m}\ddot{u}\delta u + EAu_{,xx}\, \delta u \right)\, dx \right] dt = 0$$

Note that we've zeroed out the boundary term $\bar{m}\dot{u}\,\delta u\big|_{t_1}^{t_2}$ because the variations at the endpoints are defined to be zero: $\delta u(t_1) = \delta u(t_2) = 0$ (see Figure 4.5b). We've also taken the liberty of switching the order of the integrals back. Notice that the space integral now contains only variations of u, which may be factored out (this is why we did the integration by parts ... to factor this out!),

$$\int_{t_1}^{t_2} \left[-EAu_{,x}(L)\delta u(L) + EAu_{,x}(0)\delta u(0) + \int_0^L \left(-\bar{m}\ddot{u} + EAu_{,xx} \right)\, \delta u\, dx \right] dt = 0.$$

Obviously, the overall integral is zero. But this necessitates that each term is individually zero. Focusing on the space integral, we notice that because δu is arbitrary, the bracketed term must be zero.

$$-\bar{m}\ddot{u} + EAu_{,xx} = 0.$$

This *is* the governing PDE of motion; it's the classic wave equation. The boundary terms give options. Either

$$EAu_{,x}(0) = 0 \quad \text{or} \quad \delta u(0) = 0.$$

The first is a statement of no axial force at zero (which is not the physical case drawn) and the second is a statement of no displacement at zero (which is the physical case drawn). The other term provides the option of either

$$EAu_{,x}(L) = 0 \quad \text{or} \quad \delta u(L) = 0.$$

Again, the first states that the axial force at $x = L$ is zero, which is true. The second says that the deflection is zero (untrue). Thus, the power of Hamilton becomes evident. For continuous systems, Hamilton's principle gives us the governing PDE *and* the appropriate boundary conditions. Of course, this particular problem was relatively straightforward and the boundary conditions could have been intuited. But more complex boundaries are not so obvious, as we see in the next example.

Example: Consider the system shown in Figure 4.6. It consists of a bending, linear elastic beam with a concentrated end mass. The end mass is connected to an inextensible cord that

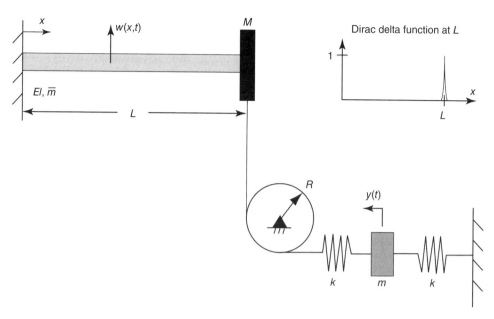

Figure 4.6 A dynamical system consisting of a bending beam (top left) with a concentrated end mass M connected to a single point mass m via springs

wraps around a pulley and then connects to the spring/point mass system at the bottom right. The displacement of the point mass is a discrete variable $y(t)$, but the deflection of the beam is a continuous function of space and time $w(x,t)$. And it's the distributed beam that makes Hamilton's principle a more attractive approach than Lagrange's equations. Also, note that we need to ascertain (and satisfy) boundary conditions as well – a complication that doesn't arise in purely discrete problems.

To use Eq. (4.19), we need to develop expressions for the kinetic (T) and potential (V) energy. Nonconservative (path-dependent) effects are not present, so $W_{nc} = 0$. The kinetic energy consists of three terms:

$$T = \frac{1}{2} \int_0^L [\bar{m}\dot{w}^2 + M\delta(x - L)\dot{w}^2]\, dx + \frac{1}{2}m\dot{y}^2$$

The first term is the kinetic energy of the beam, obtained by integrating the kinetic energy of each infinitesimal beam segment of mass/length $\bar{m}$ and length dx. The second term accounts for the motion of the large mass M, which is described by the motion $w(L,t)$. To place this mass effect at the end of the beam, the kinetic energy is multiplied by a Dirac delta function centered at $x = L$ (shown as an inset in the figure). Thus, the product $M\delta(x - L)$ makes no contribution along the length of the beam until one reaches $x = L$, when its value becomes M. The third, nonintegrated term is the kinetic energy of the point mass m. The first variation of the kinetic energy is obtained by a first-order Taylor series in the unknown variables that appear (in this case, $\dot{w}$ and $\dot{y}$): $\delta T = (\partial T/\partial \dot{w})\delta\dot{w} + (\partial T/\partial \dot{y})\delta\dot{y}$. This gives

$$\delta T = \int_0^L [\bar{m}\dot{w}\,\delta\dot{w} + M\delta(x - L)\dot{w}\delta\dot{w}]\, dx + m\dot{y}\,\delta\dot{y}$$

Care must be taken not to confuse the notation. There is a δ associated with the Dirac function and a δ associated with the variation process – and they have completely different meanings.

The potential energy is

$$V = \frac{1}{2} \int_0^L [EIw_{,xx}^2 + k\delta(x-L)(w-y)^2] \, dt + \frac{1}{2}ky^2$$

The first term is the integrated energy stored in the deformed beam, and this is taken directly from Eq. (4.5). The second term in the integral is the energy stored in the spring between the end mass M and the discrete mass m. Again, a Dirac delta function is used to locate the effect of the spring at $x = L$. The last term represents the energy stored in the spring to the right of the discrete mass m. The first variation of V is obtained by a Taylor series in the unknown variables that appear (in this case $w_{,xx}$, w and y): $\delta V = (\partial V/\partial w_{,xx})\delta w_{,xx} + (\partial V/\partial w)\delta w + (\partial V/\partial y)\delta y$. The result is

$$\delta V = \int_0^L [EIw_{,xx}\delta w_{,xx} + k\delta(x-L)(w-y)(\delta w - \delta y)] \, dx + ky \, \delta y.$$

The expressions for δT and δV can be subtracted and used in Hamilton's principle, Eq. (4.19).

$$\int_{t_1}^{t_2} \left[\int_0^L [\bar{m}\dot{w}\delta\dot{w} + M\delta(x-L)\dot{w}\,\delta\dot{w}] \, dx + m\dot{y}\,\delta\dot{y} \right.$$
$$\left. - \int_0^L [EIw_{,xx}\delta w_{,xx} + k\delta(x-L)(w-y)(\delta w - \delta y)] \, dx - ky\,\delta y. \right] dt = 0. \quad (4.27)$$

This expression is littered with a number of variations: $\delta\dot{w}$, $\delta\dot{y}$, $\delta w_{,xx}$, δw and δy. The objective at this point is to integrate the equations such that only variations of the displacement variables (δw and δy) appear – much like what was done in the prior example. To integrate the first two terms that involve $\delta\dot{w}$, the order of the integrals is interchanged and the terms are integrated by parts with respect to time. The result is

$$[\bar{m}\dot{w} + M\delta(x-L)\dot{w}] \, \delta w \big|_{t_1}^{t_2} - \int_{t_1}^{t_2} [\bar{m}\ddot{w} + M\delta(x-L)\ddot{w}] \, \delta w \, dt.$$

But we know that the variations are zero at the endpoints: $\delta w(t_1) = \delta w(t_2) = 0$ (see Figure 4.3b). So this eliminates the endpoint terms. The third term in Eq. (4.27), $m\dot{y}\,\delta\dot{y}$, behaves in a similar manner. If we reswap the order of integration (so that the time integral is on the outside again), we can eliminate the variation $\delta w_{,xx}$. This requires us to integrate by parts twice, both times with respect to space. This term becomes

$$-EIw_{,xx}\delta w_{,x}\big|_0^L + EIw_{,xxx}\delta w\big|_0^L - \int_0^L EIw_{,xxxx} \, \delta w \, dx.$$

The last two terms in Eq. (4.27) involve δw and δy and need not be integrated. Reassembling all of this information into Eq. (4.27) leads to

$$\int_{t_1}^{t_2} \left[\int_0^L \{-\bar{m}\ddot{w} - EIw_{,xxxx}\} \, \delta w \, dx + \{-m\ddot{y} - 2ky + kw(L)\} \, \delta y \right.$$
$$+ \{-M\ddot{w}(L) + EIw_{,xxx}(L) + k(y - w(L))\} \, \delta w(L)$$
$$\left. + \{-EIw_{,xx}(L)\} \, \delta w_{,x}(L) + \{EIw_{,xx}(0)\} \, \delta w_{,x}(0) + \{EIw_{,xxx}(0)\} \, \delta w(0) \right] dt = 0.$$

This gives us a wealth of information. We'll look the bracketed terms, that is, { - }, one by one. But note that the entire time integral is zero, which means that each individual term should be zero. The first term multiplies δw, which is arbitrary. So the term in the brackets must be zero:

$$\bar{m}\ddot{w} + EIw_{,xxxx} = 0. \tag{4.28}$$

This is the governing PDE for the beam motion. The second term multiplies δy, which is also arbitrary. So this bracketed term must also be zero:

$$m\ddot{y} + 2ky - kw(L) = 0.$$

This is the ODE for the discrete mass m, which is clearly (and expectedly) coupled to the motion of the end of the beam. The third term is the first boundary condition at $x = L$ and presents us with a choice. Either

$$M\ddot{w}(L) - EIw_{,xxx}(L) - k(y - w(L)) = 0 \quad \text{or} \quad \delta w(L) = 0.$$

The left expression is a statement of Newton's second law applied at $x = L$. It contains a shear force (third derivative), a spring force, and the inertia/kinematic effect, $M\ddot{w}$. The right expression states that the displacement at $x = L$ is zero: $\delta w = 0$ (but this is clearly not the physical problem posed). So the left expression is the operative boundary condition here. The fourth term is the second boundary condition and again offers a choice. Either

$$EIw_{,xx}(L) = 0 \quad \text{or} \quad \delta w_{,x}(L) = 0.$$

The left (right) expression states that there is no moment (slope) at $x = L$. The physical problem clearly doesn't impose a slope restriction and, hence, the moment must be zero. The fifth term is the third boundary condition:

$$EIw_{,xx}(0) = 0 \quad \text{or} \quad \delta w_{,x}(0) = 0.$$

These are the same options as the last set – but simply applied at $x = 0$. But in this case (where the left end is built-in), there is a slope restriction and we conclude that there must be no slope: $\delta w_{,x}(0) = 0$. The sixth – and last – term provides the fourth boundary condition:

$$EIw_{,xxx}(0) = 0 \quad \text{or} \quad \delta w(0) = 0.$$

The left (right) expression states that there is no shear (displacement) at $x = 0$. But given that the left end is cantilevered, the displacement condition must be in effect (in fact, a nonzero shear must exist to provide a restraint force, preventing lateral displacements at $x = 0$).

In summary, you have to express the system's energy in terms of spatial integrals (because it's spatially extended) and then take variations with respect to the unknown variables. These are substituted into Hamilton's principle and integration by parts is carried out to eliminate all of the derivatives of the variations (e.g., $\delta w_{,x}$ or $\delta \dot{w}$) inside the spatial integral(s). This process should be fairly mechanical. But the payoff is tremendous. Once complete, the governing equations of motion (PDEs and/or ODEs) emerge as well as any boundary condition. And these boundary conditions are the bonus; neither Lagrange nor Newton will provide these, requiring that the modeler intuit them somehow. And in complex problems, this can be difficult.

4.4 Common Challenges

Structural damage often leads to modeling difficulties in the form of discontinuities. In what follows, we discuss a few of the more common problems encountered and how they can be addressed.

4.4.1 Impact Problems

Impacts can be defined as any event in which there is a discontinuity in the stiffness. Some impact events are highly desirable; shot peening, for example, is used to create local compressive surface stresses that inhibit crack growth and tend to increase the number of cycles to failure in high-cycle components. But, typically, impacts are not intentionally engineered into a system. They arise out of imperfections and can have undesirable effects, such as excessive noise or wear. For example, Figure 4.7a shows a loosely bolted joint. Rattling may result between any of the components (the bolt, the washer, the two components A and B, or the nut). Figure 4.7b is a schematic of a delaminated beam, where the upper laminate may "slap" into its neighbor under periodic loading. These, and other impact systems, are often modeled as a SDOF impact oscillator, Figure 4.7c [7, 8]; Figure 4.7d shows a typical contact force-displacement relationship. Figure 4.7e shows another continuous system that may involve impacts; a buckled beam in contact with a tensionless, linear elastic foundation.

All impact problems suffer from the stiffness discontinuity that takes place at the moment of impact (Figure 4.7d); in essence, it's the challenge of capturing the impact event temporally. As a ubiquitous problem for impacting systems, this is our focus. But it's also worth mentioning that in spatially extended systems, there may be a spatial component to the impact event. For example, the compressed beam of Figure 4.7e may have some portions impacting the substrate, while others are not. As such, it's not reducible to a simple, single impact problem (Figure 4.7c). Instead, the restoring force per unit length is $F(x,t) = k[w(x,t)]w(x,t)$, with the stiffness being a highly nonlinear function of the displacement w:

$$k[w(x,t)] = \begin{cases} 0, & w(x,t) \geq 0 \\ K, & w(x,t) < 0 \end{cases} \tag{4.29}$$

which is a spatially pointwise function. This significantly complicates the problem, as one must ascertain the *regions of contact/no-contact* at every instant. This is typically done via a finite element discretization (see Section 4.5.3). The downside is that a finite element solution can be computationally intensive. However, there are many circumstances under which a simplified model is appropriate. Consider Figure 4.7b. If the size of the delamination zone is small compared to the length of the overall structure, then it's reasonable to model the entire delamination as impacting the substrate simultaneously. This removes the complex spatial dependency, leaving only the temporal complexity.

There are two "most common" impact models. The first is the coefficient of restitution model. This model uses the conservation of momentum to determine the rebound velocity as a function of the inbound velocity: $v_r^n = e v_i^n$, where e is the coefficient of restitution, which is a material property of the interacting bodies, and v_r^n is the normal component of the rebound velocity and v_i^n is the normal component of the inbound velocity. The strength of this approach is that it has experimental data behind it (to get e). But it also assumes that the

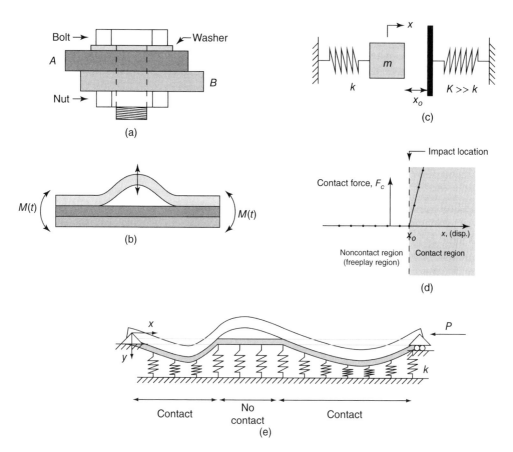

Figure 4.7   Several systems that may display impacts: (a) a loosely bolted joint, (b) a delaminated composite structure, (c) the arch-type single-degree-of-freedom (SDOF) problem. (d) The force-displacement relationship at impact. (e) A bucked beam in contact with a tensionless foundation

structure rebounds instantly. This isn't an issue if the natural time scales of the problem are large, because the actual contact times are usually quite small. In other cases, the time spent in contact may be nonnegligible. In these cases, a model more like Figure 4.7c would be in order. This consists of two distinct oscillators: one is in the free-play region with frequency $\omega_n = \sqrt{k/m}$ and the other is in the contact region with a frequency $\omega_n = \sqrt{(k+K)/m}$. In an otherwise linear structure, the problem of the instant of impact, that is, when the impact happens, is not known *a-priori*.

One approach is to solve for the motion analytically in both the free-play and contacting regions and match them at the unknown instant if impact. This leads to a nonlinear equation for the impact time that must be solved numerically [7]. Once solved, the result is a piece-wise description of the motion. Of course, numerical integration of the governing equations is another alternative; in fact, this may be necessary if any of the spring stiffnesses are nonlinear. The difficulty here is determining *when* the impacting condition is met – so that the wall stiffness can be switched from 0 to some value K N/m.

More specifically, assume we wish to study a simple spring-mass damper (Eq. 4.23) and implement the stiffness change (4.29). To numerically integrate we place the system in first-order form, whereby we define the mass displacement $w_1 \equiv w(t)$ and the velocity $w_2 \equiv \dot{w}_1(t)$. In this case, the second-order system can be written

$$\frac{dw_1}{dt} = w_2$$

$$\frac{dw_2}{dt} = -k\,[w_1]\,w_1 - cw_2 + F(t)$$

$$\frac{dw_3}{dt} = 1 \qquad\qquad\qquad (4.30)$$

where k follows Eq. (4.29), c is the familiar damping constant, and $F(t)$ is an externally applied force. Our inclusion of $w_3(t)$ as an auxiliary "time" variable will become clear shortly.

As indicated by Eq. (4.29), assume our numerical integration routine produces a sequence of solution values: $\{w_1(n), \dot{w}_1(n), t_n\}$, at discrete times $t_n = n\Delta_t$ where Δ_t is the time step of the integrator. We seek to simulate the system such that $k = 0$ for $w_1(n) > 0$ and $k = K$ for $w_1(n) < 0$. One approach is to use a conventional numerical integration routine (e.g. Runge–Kutta) and *step through* the impacting condition, return to the system state before the condition being met, then take a smaller step (i.e., shrink Δ_t), or interpolate, and hope to land at the exact point of impact. This process is usually iterative, using successively smaller time steps until an error tolerance $w_1(n) - 0 < \epsilon$ is met, at which point the stiffness is set to 0 or K (depending on direction). This process, while effective, can significantly slow computation and will leave residual error ϵ in the point at which the stiffness change is implemented.

As an alternative, we can use a simple, yet effective, numerical trick, first made available by Henon [9]. The idea is to monitor the integration until the impacting condition is met, just as is done in the iterative method. However, at this point, one recasts the problem such that "time" appears as a dependent variable [9] in the equations of motion, while displacement becomes the independent variable. This can be done by dividing through (4.30) by dw_1/dt, giving

$$\frac{dw_1}{dw_1} = 1$$

$$\frac{dw_2}{dw_1} = \frac{1}{w_2}(-k\,[w_1]\,w_1(t) - cw_2(t) + F(t))$$

$$\frac{dw_3}{dw_1} = \frac{1}{w_2} \qquad\qquad\qquad (4.31)$$

Of course, we know exactly the spatial point at which the impact will occur ($w_1 = 0$), so we can form $\Delta_{w_1} = 0 - w_1$ and integrate (4.31) with this backward step size – exactly to the point of first contact. Because w_3 is the "time" variable, we can record the value of w_3 at the end of this "back step." At this point, we can make the appropriate adjustment in stiffness value (i.e., implement 4.29), switch back to system (4.30), and continue integrating. If we wish to maintain a constant time step throughout the simulation, we can do so by setting the first "post-transition" time step so that Δ_t is conserved. Note that (4.30) and (4.31) can be coded identically, but for a constant $C_1 = 1$ or $C_1 = 1/w_2$ premultiplying each of the equations.

Hence, in implementation we simply decide whether we are integrating in time or space and set C_1 accordingly.

For multiple degree-of-freedom (MDOF) systems, the procedure is identical. Here, however, instead of monitoring for the transition $w_1 > 0$, it becomes the *relative* mass displacements that are compared when deciding which value to use for stiffness. We have used this approach frequently when simulating data from systems with bilinear stiffness characteristics and do so later in Section 8.7.1.

4.4.2 Stress Singularities and Cracking

Although many of kinds of damage are considered in this book (corrosion, dents, etc.), cracks usually stand front and center in the field of SHM. In this section, we discuss some of the basics associated with crack modeling. For academic studies, the simplest cracked structure model is shown in Figure 4.8a. It consists of a bending beam with a crack located at x_o along its span. Ideally, the beam would be pretensioned sufficiently such that, as it undergoes the bending vibrations, the two open faces of the crack never close; this is referred to as a breathing crack. One well-recognized model for a beam with a symmetric crack was developed by Christides and Barr [10]. They derived a continuous EB theory that modified the uncracked beam's local stress field by superimposing an (empirical) exponentially decaying stress that mimics the effect of a crack. Because the stress is finite at the crack tip, this model actually treats the crack more like a notch. But, if calibrated properly, this model certainly captures the behavior of the system.

A simpler (more cartoonish) model for this system is shown in Figure 4.8b. Two continuous beams are connected by a torsional spring. This single torsional spring represents the reduced

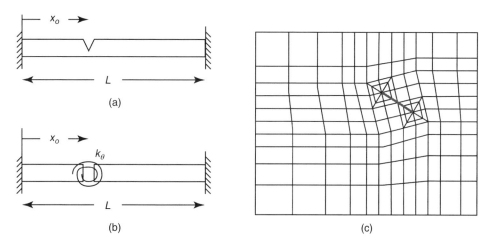

(a)

(b)

(c)

Figure 4.8 (a) A beam model with a crack at $x = x_o$. (b) This model is one way of handling the crack problem: placing an effective torsional spring in place of the crack. (c) A finite element discretization of a plate with a crack (gray bold line) using crack singularity elements provides a slightly more sophisticated approach that is used in Section 9.3

bending stiffness in the system and should be equivalent to Figure 4.8a. The question then turns to choosing an appropriate value for k_θ. This is done via fracture mechanics. The local change in compliance due to the crack is defined as

$$\Delta c_{ij} = \frac{\partial^2}{\partial P_i \, \partial P_j} \int_0^a G(\bar{a}) \, d\bar{a}$$

where G is the energy release rate (energy released per unit crack advance), a is the crack depth, and P_i is the axial load ($i = 1$) or the bending load $i = 2$. For plain strain problems, $G = \frac{1-\nu^2}{E}(K_I^b + K_I^s)^2$ and the stress intensity factors in bending and stretching (K_I^b and K_I^s) are available in the tables of Ref. [11]. The local compliance matrix becomes

$$[C] = \begin{bmatrix} \left(\frac{L}{EA}\right)_o + \Delta c_{11} & \Delta c_{12} \\ \Delta c_{21} & \left(\frac{L^3}{EI}\right)_o + \Delta c_{22} \end{bmatrix}$$

which can be inverted to get the stiffness terms. If one ignores the bending/stretching coupling (off diagonal terms), the bending and stretching rigidities may be expressed: $EA = (EA)_o - \Delta K_{11} L \delta(x - x_o)$ and $EI = (EI)_o - \Delta K_{22} L^3 \delta(x - x_o)$. And these expressions may be used in writing the strain energy for use in Hamilton's principle (Eq. 4.19). More details are provided in Refs [12] and [13].

A central assumption was that the aforementioned models were for breathing cracks. If the beam is not under tension or a large (out of the page) bending moment is applied, then the crack may close, invalidating the models. Moreover, this closing event may be dynamic, in which case the system becomes an impact oscillator as well [14].

Outside of these relatively simple geometries, however, analytic methods aren't terribly effective. This is, in part, due to more complicated boundaries encountered in real problems (sometimes with complex two- or three-dimensional edges). But the crack itself poses problems; the stress must be infinite at the crack tip and decay as you move away. This is what motivated Christides and Barr [10] to introduce the decaying exponential functions near the crack tip in their beam problems. The most effective – and flexible – solution to this problem is to use the finite element method (FEM) to solve the dynamic crack problem. The downside is that each solution is specific to the particular geometry used and broader trends are hard to tease out as there is no mathematical function describing the behavior.

Many FEM models of cracked plates have extremely dense meshes near the crack tips because the stress changes so quickly near the tip (as $\sigma \to \infty$). Convergence therefore necessitates many grid points, resulting in a very large increase in the computational cost. However, if we use what we know about the physics of cracks, we can design specialized "singularity elements" to ease the computational burden. Specifically, we can take advantage of the asymptotic limit of the stress near the crack tip.

For an infinitely sharp crack in a linear elastic material, the stress singularity varies as $\sigma(r) \propto 1/\sqrt{r}$, where r is the distance from the crack tip in any direction. Rather than use a lot of elements near the crack tip, we use elements that are specially designed to give a $1/\sqrt{r}$ stress distribution near the crack tip [15, 16]. To achieve this, eight-noded Mindlin elements are used throughout the bulk of the structure (noncracked portion). But near the crack tip, the elements are augmented; the three nodes on one side are collocated, producing a triangular element. This new collocated node is placed on the crack tip. The nodes on the adjacent sides are then moved to the one-quarter position (three-quarters away from the collocated node). The shape function for this new, augmented element is exactly $1/\sqrt{r}$, producing the desired stress field! So a lot of elements are not needed to capture the behavior; one simple element does it. A sample of this is shown in Figure 4.8c. The mesh is made up almost entirely of rectangular elements – except near the crack tips, where these augmented triangular elements are used. The effect is a "pinwheel" around the crack tips. But the mesh is remarkably sparse and converges to produce highly accurate results [17]. This approach to modeling is used in Section 9.3 to identify the crack length, location, and orientation in an experimental system.

4.5   Solution Techniques

For a moment, let's revisit Figure 4.1. In broad brushstrokes, this lays out a strategy for building a dynamic model for a structure. To this point, the focus has been on the left and middle (incorporating the necessary physical law, material behavior, geometry, forces, etc.), which generates the governing equations of motion. Because these are dynamic problems, involving inertia ma, the result is necessarily a set of ordinary or PDEs – with the possibility of algebraic constraints being coupled in, as well. For this brief discussion, we restrict ourselves to ODEs and PDEs.

4.5.1   Analytical Techniques I – Ordinary Differential Equations

4.5.1.1   Single DOF Systems

One of the most important problems that occurs over and over again is the SDOF oscillator. Earlier, we showed how one could derive the equations of motion of such a system using energy methods (Eqs. 4.22 and 4.23). Because of its ubiquity, we focus on this system, and then extend the one-dimensional solution approach to more realistic, MDOF system models. The standard forced single DOF system has the form:

$$m\ddot{x} + c\dot{x} + kx = F\cos(\omega t)$$

or

$$\ddot{x} + 2\zeta\omega\dot{x} + \omega_n^2 x = \frac{1}{m}F\cos(\omega t) \tag{4.32}$$

where $\omega_n = \sqrt{k/m}$ and $\zeta = \frac{c}{2\sqrt{km}} = \frac{c}{2\omega_n m}$ are referred to as the natural frequency and damping ratio, respectively. Equation (4.32) has all of the basic components required of a structural model. Inertia, in the form of the mass m, a restoring force k (if we push on the structure, it tends to bounce back elastically), and a dissipative force, c (left to its own devices, a vibrating structure will eventually return to a state of rest). The forcing is seen to be a harmonic signal with a phase that is effectively determined by the time at which we "start" the model. This may at first seem strange given that most structures are not subject to perfectly periodic forcing *in situ*. However, as we know from Chapter 3, the frequency domain description (thinking of a signal as combinations of sines and cosines) can be used to describe periodic, nonperiodic, and even nondeterministic signals. Thus, the model (4.32) is actually a prelude to a very general forcing model that we can use to gain a great deal of insight into the behavior of a structure in what follows.

The problem (4.32) is a straightforward, constant coefficient, second-order, nonhomogeneous differential equation with a standard solution (see Ref. [18]):

$$x(t) = e^{-\zeta\omega_n t}\left[x(0)\cos\left(\omega_n\sqrt{1-\zeta^2}\,t\right) + \frac{\dot{x}(0)+\zeta\omega_n x(0)}{\omega_n\sqrt{1-\zeta^2}}\sin\left(\omega_n\sqrt{1-\zeta^2}\,t\right)\right]$$

$$+\frac{1/k}{\sqrt{(1-(\omega/\omega_n)^2)^2+(2\zeta\omega/\omega_n)^2}}(F\cos(\omega t - \phi)) \tag{4.33}$$

where $x(0), \dot{x}(0)$ are the initial position and velocity, respectively, and $\phi = \tan^{-1}\left(\frac{2\zeta\omega/\omega_n}{1-(\omega/\omega_n)^2}\right)$. The first line captures the homogeneous or so-called "free decay" portion of the solution. That is to say, in the absence of a forcing function, and subject to initial displacement or velocity, the response will be of an exponentially decaying, oscillatory form. The second portion of the solution is that which persists because of the application of the (in this model, harmonic) forcing. You can arrive at Eq. (4.33) by any number of solution techniques: the method of undetermined coefficients, Laplace transforms, power series, variation of parameters, and so on. Next, we briefly review the first of these approaches.

The first step is to note that for this class of differential equations (linear, second order) the solution is comprised of both the homogeneous portion (where we set the right-hand side, or forcing, equal to zero) and the nonhomogeneous, where the forcing is included. For the former, one assumes a solution of the form $x(t) = Ae^{\lambda t}$, where A is a complex constant coefficient. Substituting this into Eq. (4.33), one finds $\lambda^2 + 2\zeta\omega_n\lambda + \omega_n = 0$ from which we have $\lambda_{1,2} = \left(-\zeta \pm i\sqrt{1-\zeta^2}\right)\omega_n = -\zeta\omega_n \pm i\omega_d$. Here we have defined $\omega_d \equiv \omega_n\sqrt{1-\zeta^2}$. In this book, and most texts on structural vibration, we are always interested in the so-called "lightly damped" case where $\zeta \ll 1$. The reason, quite simply, is that most physical systems possess dissipative terms that produce a much smaller force than the restoring term. We therefore only consider the $\zeta < 1$ case in what follows and simply note that the other cases are handled in other references (e.g., Ref. [19], Chapter 4). Because we have two valid roots, the solution must be written as: $x(t) = e^{-\zeta\omega_n t}(A_1 e^{i\omega_d t} + A_2 e^{-i\omega_d t})$. Again, A_1, A_2 are in general complex.

At this point, we can convert the exponentials to trigonometric functions and write

$$x(t) = e^{-\zeta\omega_n t}\left((A_1+A_2)\cos(\omega_d t) + i(A_1-A_2)\sin(\omega_d t)\right)$$

$$= e^{-\zeta\omega_n t}\left(A_1'\cos(\omega_d t) + A_2'\sin(\omega_d t)\right)$$

$$= Xe^{-\zeta\omega_n t}\cos(\omega_d t + \phi) \tag{4.34}$$

where we have introduced the combined coefficients A_1', A_2' which we constrain to be real. Note that in going from the first to the second line of Eq. (4.34), there was some sleight of hand; we went from a complex description to an all-real description. The only way that the constants A_1' and A_2' can be real is if A_1, A_2 are complex conjugates of one another. This, of course, must be the case because the solution is required to be real. The final form is obtained by noting that $a\cos(c) + b\sin(c) = \sqrt{a^2+b^2}\cos(c+\phi)$ with $\phi = \tan^{-1}(-b/a)$. Thus, we can write $X = \sqrt{A_1'^2 + A_2'^2}$ and $\phi = \tan^{-1}(-A_2'/A_1')$. From the second line of Eq. (4.34), we can use the initial position and velocity, $x(0), \dot{x}(0)$ to solve for A_1', A_2'. The result is the first portion of Eq. (4.33).

The second, nonhomogeneous portion of the solution involves the forcing functions and will therefore produce a nonzero $x(t)$ for as long as the forcing is applied. This portion of the solution is sometimes referred to as the "steady state" solution as it is this part that will dominate (persist) as time becomes large and the first term decays to zero. In fact, for most of the problems in this book, the focus will be exclusively on the steady state solution as it is typically the long-term structural response we are interested in.

The first thing to note is that the general forcing model can be written as the real part of a complex exponential: $F\cos(\Omega t) = \text{Re}\{Fe^{i\Omega t}\}$. Alternatively, should we prefer for our forcing function $F\sin(\Omega t)$ we could use $\text{Im}\{Fe^{i\Omega t}\}$. In short, by considering the exponential form of the harmonic forcing function, we can effectively solve for both types of forcing simultaneously.

For the nonhomogeneous portion of Eq. (4.32), we use the alternate form

$$\ddot{x} + 2\zeta\omega_n\dot{x} + \omega_n^2 x = \frac{1}{m}Fe^{i\omega t} \tag{4.35}$$

which has the general solution $x(t) = X(i\omega)e^{i\omega t}$ [19]. That is to say, if the forcing is harmonic with frequency ω, so too will be the response. The only question concerns the complex amplitude of the response at this frequency. Substituting this solution into (4.35) we obtain

$$X(i\omega) = \frac{F/m}{-\omega^2 + i2\zeta\omega\omega_n + \omega_n^2}$$

$$= \frac{F/k}{1 + i2\zeta\omega/\omega_n - (\omega/\omega_n)^2}$$

$$= FH(i\omega) \tag{4.36}$$

where the complex function $H(i\omega)$ is referred to as the frequency response function (FRF) of the system. As written, this function is seen to quantify the ratio of displacement amplitude to input force, that is,

$$H(i\omega) = \frac{X(i\omega)}{F} \tag{4.37}$$

Thus, we see that the units of $H(i\omega)$ are m/N. In some books, frequency response is defined as the dimensionless ratio given by the denominator in the second line of Eq. (4.36). We can write this function $G(i\omega) = kH(i\omega) = 1/(1 + i2\zeta\omega/\omega_n - (\omega/\omega_n)^2) = \frac{kX(i\omega)}{F}$ so that $G(i\omega)$ is seen to quantify the ratio of restoring force to input force as a function of frequency ω. In this book, however, our analyses will be cast in terms of $H(i\omega)$.

We also note that depending on the parameterization used, $H(i\omega)$ can be written as

$$H(i\omega) = \frac{1/k}{1 + i2\zeta\omega/\omega_n - (\omega/\omega_n)^2} = G(i\omega)/k$$

$$= \frac{1/m}{-\omega^2 + i2\zeta\omega\omega_n + \omega_n^2}$$

$$= \frac{1}{-m\omega^2 + i2c\omega + k} \tag{4.38}$$

In our later sections on Volterra series and higher order spectral analysis (see Sections 4.6 and 6.5), it is the last of these expressions that we typically use.

Finally, to arrive at the forced portion of the response in Eq. (4.33), we again note that any complex number can be written in terms of a magnitude and phase so that

$$H(i\omega) = |H(i\omega)|e^{i\phi} \tag{4.39}$$

where $|H(i\omega)| = \frac{1/k}{[(1-\omega^2/\omega_n^2)^2+(2\zeta\omega/\omega_n)^2]^{1/2}}$ and $\phi = \tan^{-1}\left(\frac{2\zeta\omega/\omega_n}{1-\omega^2/\omega_n^2}\right)$. Substituting back in for $x(t)$ gives

$$x(t) = X(i\omega)e^{i\omega t}$$
$$= F|H(i\omega)|e^{i\omega t+\phi} \tag{4.40}$$

whereupon taking the real (cosine) portion yields the nonhomogeneous portion of Eq. (4.33). Also note that if we want either the velocity or acceleration response ($\dot{x}(t), \ddot{x}(t)$), we simply differentiate (4.40) with respect to time yielding

$$\dot{x}(t) = Fi\omega|H(i\omega)|e^{i\omega t+\phi}$$
$$\ddot{x}(t) = -F\omega^2|H(i\omega)|e^{i\omega t+\phi} \tag{4.41}$$

Thus, we have that the velocity and acceleration FRFs are given by $H_v(i\omega) = i\omega H(i\omega)$ and $H_a(i\omega) = -\omega^2 H(i\omega)$, respectively. These relationships are used later in Sections 6.4.1 and 6.5.2.

The solution, Eq. (4.33), is important because it serves as a *building block solution*, meaning that you build solutions for other (more physically meaningful) problems using this one. Here are two common examples. First, if the physical problem involves a more complex but still periodic excitation, then you could express that excitation using the Fourier series of Section 3.3.1 to give

$$m\ddot{x} + c\dot{x} + kx = \sum_{j=1}^{\infty} F(i\omega_j)e^{i\omega_j t} \tag{4.42}$$

where $F(i\omega_j)$ is now a complex, frequency-dependent coefficient (as in Eq. (3.30)). The homogeneous part of the solution, an exponential decay with frequency $\left(\omega_n\sqrt{1-\zeta^2}\right)$, remains the same, while the nonhomogeneous part is solved via superposition. Because this is a linear system, the solution to a superposition of sinusoids is the superposition of the solution to a single sinusoid which we have just derived. Therefore, if we are interested in the long-term (forced) response only, the solution is governed by the last two terms of Eq. (4.33) and becomes

$$x(t) = \sum_{j=1}^{\infty} H(i\omega_j)F(i\omega_j)e^{i\omega_j t} \tag{4.43}$$

Our solution is therefore comprised of a sum of sinusoids with varying amplitude at each frequency. The amplitude is governed by the structural properties of the oscillator along with the frequencies associated with the excitation.

The form of the solution Eq. (4.43) bears a good deal of similarity to the expressions used in the development of Fourier analysis in Section 3.3.1. In fact, we used Section 3.3.1 to discuss at length the conditions under which the infinite Fourier series can be replaced with the Fourier integral. If we therefore consider the forcing to be over a continuous (rather than discrete) set of frequencies, it becomes a function of the continuous forcing variable ω (as opposed to ω_j, $j = 1, \ldots, \infty$) and the sum is replaced by an integral (again, keeping in mind the mathematical implications of doing this, Section 3.3.1). Given the development in going from Eq. (3.27) to Eq. (3.30), we see that the Eq. (4.43) is of the form of the inverse Fourier transform of the function $H(\omega)F(\omega)$ (see Eq. 3.35), that is,

$$x(t) = \int_{-\infty}^{\infty} H(i\omega)F(i\omega)e^{i\omega t}\, d\omega. \tag{4.44}$$

Hence, if we take the Fourier transform of both sides, we arrive at the equivalent frequency domain solution to (4.33),

$$X(i\omega) = H(i\omega)F(i\omega). \tag{4.45}$$

Equation (4.45) relates the frequency content of the input to that of the response under the action of the structure. In some instances it makes more sense to model the system response in the frequency domain, as opposed to the time domain (particularly for steady state dynamics). We make use of this frequency domain relationship between input and output in later chapters (e.g., 6.4.1).

Now consider a different approach to generating a general forcing model. One powerful tool that we make use of later (see Section 4.6) is the unit impulse response, obtained by examining the response of Eq. (4.32) to a unit amplitude pulse. Conceptually, these impulses are sharp spikes, occurring at time $t = \theta$ over infinitesimal time spans $\Delta\theta$; it's a unit impulse because the area under the impulse, $\delta(t - \theta)\Delta\theta$, is 1. In fact, the Dirac delta function is used simply to center the force at the right instant of time $t = \theta$. The value of the function at that instant is $F(\theta)\delta(t - \theta)\Delta\theta$. $F(\theta)$ scales the unit force and the last two terms have a value of one, with a noted time shift to the right of θ. Now imagine that we describe an arbitrary forcing $F(t)$ as a series of scaled unit impulses. The *net* force at time t is the summed effect of all of the prior forces: $F(t) = \sum F(\theta)\delta(t - \theta)\Delta\theta$.

Now consider the response. If a unit impulse is applied to a system at $t = \theta$, the response is given by the *unit impulse response function* times the impulse duration, $h(t - \theta)\Delta\theta$. The function $h(\cdot)$ will differ from system to system. If an impulse with arbitrary magnitude is applied (i.e., it's a nonunit impulse) and the system is linear and time-invariant, then the response simply scales with the magnitude of the impulse:

$$\Delta x(t) = F(\theta)h(t - \theta)\Delta\theta.$$

Of course, this is for a single impulse and we know we want to describe the excitation $F(t)$ as a series of scaled unit impulses. We can find this solution by superposition. Summing the effects of each impulse gives

$$x(t) = \int_0^t F(\theta)h(t - \theta)d\theta. \tag{4.46}$$

This is known as the *convolution integral* or the *impulse response* solution. The chore is to find the unit impulse response function $h(t - \theta)$ for a particular system and then carry out the integration of Eq. (4.46). However, the real utility of (4.46) can be seen through a simple manipulation.

First, we consider *causal* systems, meaning $h(t < 0) = 0$, that is to say the system cannot respond until the forcing begins. We can also easily imagine that $F(\theta < 0) = 0$, that is, until time $t = 0$ there is no forcing. This means the limits on the integration (4.46) can be stretched over the entire real number line without influencing the result. Now, take the Fourier transform of both sides of (4.46) giving

$$X(i\omega) = \int_{-\infty}^{\infty} \int_{-\infty}^{\infty} F(\theta)h(t - \theta)e^{-i\omega t} \, d\theta \, dt \tag{4.47}$$

With the change of variables, $\lambda = t - \theta$ we see that $dt = d\lambda$ and $t = \lambda + \theta$ so that

$$X(i\omega) = \int_{-\infty}^{\infty} \int_{-\infty}^{\infty} F(\theta)h(\lambda)e^{-i\omega\lambda}e^{-i\omega\theta} \, d\theta \, d\lambda$$

$$= \int_{-\infty}^{\infty} F(\theta)e^{-i\omega\theta} d\theta \int_{-\infty}^{\infty} h(\lambda)e^{-i\omega\lambda} \, d\lambda$$

$$= H(i\omega)F(i\omega) \tag{4.48}$$

which is the same relationship we arrived at using the Fourier integral approach to modeling the excitation, Eq. (4.45). In this case, we considered a general forcing function as being comprised as an infinite number of scaled impulses. In the former case we modeled the excitation with the Fourier integral, an approach we know from Chapter 3 can model harmonic inputs or even random processes. Both approaches should therefore be capable of modeling any type of forcing function we require, hence the fact that both forcing models yield the same response model is not entirely surprising. The importance of the frequency-domain model equation (4.45) cannot be understated, and it is used time and time again in this book. The ubiquity of the relationship is the motivating force behind the extensive treatment of Fourier analysis in Section 3.3.

In fact, the impulse response $h(t)$ is seldom used directly; methods by which it is estimated are typically based on the relationship (4.45). It is also frequently the case that a frequency-domain description of the forcing is easy to come by. Thus, to determine $x(t)$, one can simply take the inverse Fourier transform of the product $H(i\omega)F(i\omega)$. Moreover, when it comes to looking at a *nonlinear* system response, the techniques we use are based on a frequency domain description of the signals of interest and are described as combinations of both $H(i\omega)$ and $F(i\omega)$ (see Section 4.6). A final word on notation for this section: we have explicitly noted that both the frequency response and forcing are functions of the complex argument $i\omega$. This is merely to drive home the point that both are complex valued quantities. However, in many of the following chapters we implicitly acknowledge the complex nature of frequency-dependent functions and leave them as functions of real frequency ω only.

4.5.1.2 MDOF Systems

Given the development and results of our SDOF system, we can now consider a fully coupled, harmonically forced, MDOF system described by

$$\mathbf{M}\ddot{\mathbf{x}} + \mathbf{C}\dot{\mathbf{x}} + \mathbf{K}\mathbf{x} = \mathbf{F}e^{i\omega t}. \tag{4.49}$$

Equation (4.49) is also linear, but now consists of N equations in the N response variables (and their time derivatives) specified by the vector $\mathbf{x}$. Each of the matrices $\mathbf{M}, \mathbf{C}, \mathbf{K}$ is therefore an $N \times N$ matrix with constant entries. The forcing is still harmonic for the same reasons as the previous section, only now there is the possibility of the excitation being applied at any or all DOFs. However, the far more common case is for the excitation to be applied at a particular point, hence the amplitude vector **F** is usually taken as zero everywhere except for the forcing location. This is the situation we focus on in our modeling efforts.

As with the SDOF system, the MDOF case will have both homogeneous and nonhomogeneous solutions. The solution proceeds by first assuming the undamped case, **C** = 0. As in the SDOF case, we begin by assuming a solution of the form

$$\mathbf{x} = \mathbf{u}e^{\lambda t} \tag{4.50}$$

and substituting into Eq. (4.49) with no forcing (**F** = 0). The result is the eigenvalue problem

$$[\lambda^2 \mathbf{M} + \mathbf{K}]\mathbf{u} = 0. \tag{4.51}$$

The solution is given by setting the determinant $|\lambda^2 \mathbf{M} + \mathbf{K}| = 0$ (see again Ref. [19]), which can be shown to be satisfied by N different *eigenvalues* $\lambda_n^2 = -\omega_n^2, n = 1 \cdots N$. The associated *eigenvectors* $\mathbf{u}_n, n = 1 \cdots N$ are a set of orthogonal vectors (provided **M** and **K** are real and symmetric) that give the relative amplitudes of each of the frequency components ω_n^2 that make up the response. Thus, the eigenvalues (frequencies) and eigenvectors define N solutions in the form of Eq. (4.50). Of course, there are two possibilities for λ in Eq. (4.50). Because $\lambda_n^2 = -\omega_n^2$ we have that $\lambda_n = \pm i\omega_n$. Just as with the SDOF system, this means our undamped solution is written $\mathbf{x} = \sum_{n=1}^{N}(A_{1,n}e^{i\omega_n t} + A_{2,n}e^{-i\omega_n t}) \mathbf{u}_n$, where $A_{1,n}, A_{2,n}$ are constants obtained via the initial conditions. We know from our prior discussion that we could alternatively write the solution $\mathbf{x} = \sum_{n=1}^{N} X_n \cos(\omega_n t + \phi_n) \mathbf{u}_n$. The difference between this and the SDOF case is that the final solution for **x** is a linear combination of the SDOF case where the weights are given by $\mathbf{u}_n$.

Two remaining issues, however, are (i) the inclusion of a dissipative force, that is, damping and (ii) how to decide on a proper normalization scheme for **u** given that this vector contains *relative* amplitudes of solution (i.e., there are an infinite number of solutions because a constant times this vector is also a solution). To solve the latter issue, a useful normalization scheme prescribes that $\mathbf{u}_n^T \mathbf{M} \mathbf{u}_n = 1$, which is referred to as the "mass-normalized" approach. Returning to the eigenvalue problem and premultiplying by $\mathbf{u}_n^T$ we get

$$-\omega_n^2 \mathbf{u}_n^T \mathbf{M} \mathbf{u}_n + \mathbf{u}_n^T \mathbf{K} \mathbf{u}_n = 0 \tag{4.52}$$

so that by this normalization scheme, $\mathbf{u}_n^T \mathbf{K} \mathbf{u}_n = \omega_n^2$. To handle the full, damped problem Eq. (4.49), it is common to model the damping as a linear combination of mass and stiffness as we showed with Eq. (4.25), that is, $\mathbf{C} = \alpha \mathbf{M} + \beta \mathbf{K}$. In this case, the term $\mathbf{C}\dot{\mathbf{x}}$ will, after premultiplying by $\mathbf{u}_n^T$ as we did for the other terms, become $\mathbf{u}_n^T(\alpha\mathbf{M} + \beta\mathbf{K})\mathbf{u}_m = (\alpha + \beta\omega_n^2)\delta_{mn}$. In other words, it becomes a diagonal matrix. If we then let $2\zeta_n\omega_n = (\alpha + \beta\omega_n^2)$ we return to the familiar damping model for our single DOF system. While the genesis of such a damping model is clearly analytical convenience, it turns out to be a reasonable description of dissipative forces in real structural systems and is therefore frequently used. To use the model, one chooses the α, β values and then sets ζ_n accordingly. In fact, we will use precisely this approach in developing an analytical expression for the auto- and cross-correlation functions in Section 6.4.2.

Given this normalization scheme and damping model, we can return to the MDOF problem. First, however, it is useful to introduce the modal matrix: $\mathbf{Q} = [\mathbf{u}_1, \mathbf{u}_2, \cdots]$, where $\mathbf{u}_n$ is the nth

mass-normalized eigenvector. This allows us to write the solution to Eqn. 4.49 compactly as $\mathbf{x} = \mathbf{Q}\mathbf{z}$. Substituting this solution into Eq. (4.49) and premultiplying both sides by $\mathbf{Q}^T$ then decouples the entire set of equations leaving

$$\ddot{z}_n + 2\zeta_n\omega_n\dot{z}_n + \omega_n^2 z_n = \sum_{m=1}^{N} Q_{mn}F_m e^{i\omega t}, \quad i = 1, 2, \ldots, N$$

$$= \mathbf{Q}^T\mathbf{F}e^{i\omega t} \tag{4.53}$$

where F_m is the mth amplitude in the harmonic forcing vector. This brings us back to the single DOF oscillator of Eq. (4.42) with a slightly modified forcing function and described by a different set of (modal) coordinates, $\mathbf{z}$.

For the homogeneous solution to the damped problem, each $z_n(t)$ is clearly obtained by considering the first portion of Eq. (4.33). After solving for each of the $z_n(t)$, we then have $\mathbf{x} = \mathbf{Q}\mathbf{z}$ as the solution to the model Eqn. 4.49. For the forced problem, the solution for each DOF "n" is therefore given by Eq. (4.40), with the change in forcing magnitude from F to the vector $\tilde{\mathbf{F}} = \mathbf{Q}^T\mathbf{F}$ and with ω_n^2 in the denominator instead of the stiffness parameter k (a consequence of the mass normalization). That is to say, for the nonhomogeneous solution we have

$$\mathbf{x} = \sum_{n=1}^{N} \frac{\tilde{F}_n^T}{\omega_n^2}|H_n(i\omega)|e^{i(\omega t + \phi_n)}\mathbf{u}_n \tag{4.54}$$

where the frequency response and phase for the nth mode have already been derived in developing Eq. (4.40). We make use of this expression in Section 6.4.2, for example, when it comes to deriving analytical expressions for commonly used correlation functions.

What we have shown is that we can recast the more complex problem as a collection of single DOF oscillators, to which we know the solution(s). In a sense, we trick the problem into looking like something that we already know how to solve – the single DOF problem. This is a common tack. However, as we show next, this is not the only approach to solving the problem (Eq. 4.49). In some cases, particularly those with relatively few DOFs, we can tackle the problem directly without requiring a change to modal coordinates $\mathbf{z}$. The following approach, illustrated by example, is one we use later in developing the Volterra series models for nonlinear system response (Section 4.6).

Assume that we have a two-DOF linear system with constant coefficient matrices

$$[M] = \begin{bmatrix} m_{11} & m_{12} \\ m_{21} & m_{22} \end{bmatrix}$$

$$[C] = \begin{bmatrix} c_{11} & c_{12} \\ c_{21} & c_{22} \end{bmatrix}$$

$$[K] = \begin{bmatrix} k_{11} & k_{12} \\ k_{21} & k_{22} \end{bmatrix} \tag{4.55}$$

Equation (4.49) is then a set of two, coupled, ordinary differential equations. In this case, both the forcing and response are now vectors, that is, $\ddot{\mathbf{x}}(t) \equiv (\ddot{x}_1(t), \ddot{x}_2(t))$ and $\mathbf{F}(t) \equiv (F_1(t), F_2(t))$. Again, we assume harmonic forcing. Given that we have shown how *any* type of forcing can be described as a linear combination of harmonics, this is a very general forcing model.

Just as with any linear system, superposition holds. Hence, we may consider the response of the system to each forcing function independently and then simply sum the result. That is to say, we assume a solution of the same form we did in Eq. (4.40) for each DOF separately. To illustrate, consider harmonic forcing of unit amplitude

$$\mathbf{F}(t) = \left\{ \begin{matrix} e^{i\omega t} \\ 0 \end{matrix} \right\} \tag{4.56}$$

which will, in general, lead to the harmonic response

$$\mathbf{x}(t) = \left\{ \begin{matrix} H_{11}(i\omega)e^{i\omega t} \\ H_{21}(i\omega)e^{i\omega t} \end{matrix} \right\} \tag{4.57}$$

where the notation $H_{ij}(i\Omega)$ denotes the frequency-dependent response amplitude at DOF i to (unit) input at DOF j. Note that rather than assume the general form $\mathbf{x}(t) = X(i\omega)e^{i\omega t}$ we have anticipated that, for unit amplitude forcing, the response amplitude will be the frequency-dependent ratios $H_{ij}(i\omega)$. Thus, we can write the two equations (after canceling $e^{i\omega t}$ on both sides)

$$-m_{11}\omega^2 H_{11}(i\omega) - m_{12}\omega^2 H_{21}(i\omega) + i\omega c_{11}H_{11}(i\omega) + i\omega c_{12}H_{21}(i\omega)$$
$$+ k_{11}H_{11}(i\omega) + k_{12}H_{21}(i\omega) = 1$$
$$-m_{21}\omega^2 H_{11}(i\omega) - m_{22}\omega^2 H_{21}(i\omega) + i\omega c_{21}H_{11}(i\omega) + i\omega c_{22}H_{21}(i\omega)$$
$$+ k_{21}H_{11}(i\omega) + k_{22}H_{21}(i\omega) = 0. \tag{4.58}$$

If we repeat this process with the forcing vector,

$$\mathbf{F}(t) = \left\{ \begin{matrix} 0 \\ e^{i\omega t} \end{matrix} \right\} \tag{4.59}$$

we get two more equations with unknowns $H_{12}(i\omega)$ and $H_{22}(i\omega)$. In matrix form, the four equations can be written

$$\begin{bmatrix} (-\omega^2 m_{11} + i\omega c_{11} + k_{11})H_{11}(i\omega) + (-\omega^2 m_{12} + i\omega c_{12} + k_{12})H_{21}(i\omega) \\ (-\omega^2 m_{21} + i\omega c_{21} + k_{21})H_{11}(i\omega) + (-\omega^2 m_{22} + i\omega c_{22} + k_{22})H_{21}(i\omega) \end{bmatrix}$$

$$\left. \begin{matrix} (-\omega^2 m_{11} + i\omega c_{11} + k_{11})H_{12}(i\omega) + (-\omega^2 m_{12} + i\omega c_{12} + k_{12})H_{22}(i\omega) \\ (-\omega^2 m_{21} + i\omega c_{21} + k_{21})H_{12}(i\omega) + (-\omega^2 m_{22} + i\omega c_{22} + k_{22})H_{22}(i\omega) \end{matrix} \right] = \begin{bmatrix} 1 & 0 \\ 0 & 1 \end{bmatrix} \tag{4.60}$$

which can be conveniently factored

$$\begin{bmatrix} -\omega^2 m_{11} + i\omega c_{11} + k_{11} & -\omega^2 m_{12} + i\omega c_{12} + k_{12} \\ -\omega^2 m_{21} + i\omega c_{21} + k_{21} & -\omega^2 m_{22} + i\omega c_{22} + k_{22} \end{bmatrix} \times \begin{bmatrix} H_{11}(i\omega) & H_{12}(i\omega) \\ H_{21}(i\omega) & H_{22}(i\omega) \end{bmatrix} = \begin{bmatrix} 1 & 0 \\ 0 & 1 \end{bmatrix} \tag{4.61}$$

In matrix form we may write this as $\mathbf{Z}(i\omega)\mathbf{H}(i\omega) = \mathbf{I}$. Simple matrix inversion yields each of the needed transfer functions:

$$\mathbf{H}(i\omega) = \mathbf{Z}^{-1}(i\omega) \tag{4.62}$$

Such inversions can be challenging and are most easily accomplished using symbolic manipulation software. Hence, the final solutions for $x_1(t), x_2(t)$ can be written as

$$x_1(t) = H_{11}(i\omega)e^{i\omega t} + H_{12}(i\omega t)e^{i\omega t}$$

$$x_2(t) = H_{21}(i\omega)e^{i\omega t} + H_{22}(i\omega t)e^{i\omega t}. \qquad (4.63)$$

Had we instead used a frequency-dependent forcing with nonunit amplitude, that is, $\mathbf{F}(t) = (A(i\omega)e^{i\omega t}, B(i\omega)e^{i\omega t})^T$, the derivation would have proceeded in the same manner, yielding for the response

$$x_1(t) = A(i\omega)H_{11}(i\omega)e^{i\omega t} + B(i\omega)H_{12}(i\omega t)e^{i\omega t}$$

$$x_2(t) = A(i\omega)H_{21}(i\omega)e^{i\omega t} + B(i\omega)H_{22}(i\omega t)e^{i\omega t}. \qquad (4.64)$$

The procedure we have just described extends trivially to higher DOF systems. One assumes a harmonic forcing applied independently at each DOF, writes the DOF $\times$ DOF matrix equation, and solves for the unknown transfer functions via matrix inversion. As we see in Section 4.6, this basic procedure is referred to as harmonic probing in the context of nonlinear time-invariant systems.

While the above-described approaches are common in the analysis of lumped parameter systems (and are relied on heavily in later chapters), perhaps an even more powerful and far-reaching technique is the power series expansion. This handles all of the standard constant coefficient differential equations that come up, such as Eq. (4.32) (remember that $\sin(t)$ and $\cos(t)$ are just shorthand notations for two specific infinite series). Series solutions are also adept at handling all kinds of nonconstant coefficient ODEs. These include – but are not limited to – Mathieu's equation [20], Frobenius' equation, Legendre's equation, Bessel's equation, and most general Sturm-Liouville-type problems which come up in mathematical physics [21]. And despite the fact that most nonlinear problems don't have exact analytic solutions, certain ones do. For example, certain forms of the nonlinear Schrödinger equation and the elastica problem may be solved in terms of elliptic integrals, which involve series expansions [22, 23]. For the sake of brevity, we won't cover any of the specific techniques, largely because there are so many and they vary so widely in detail. Instead, we encourage the interested reader to see the references cited herein (and elsewhere).

4.5.2 Analytical Techniques II – Partial Differential Equations

In structural problems, the kinds of partial differential equations that emerge tend to be either hyperbolic second order (i.e., wave equation-like) or fourth order (beam-like). For the time being, let's focus on the wave equation problem, which governs the linear transverse vibrations of strings, axial vibrations of bars, torsional vibrations of bars, stretching vibrations of membranes, and propagation of acoustic waves [24]. The governing equation is

$$\nabla^2 w = \frac{1}{c^2}\frac{\partial^2 w}{\partial t^2} \qquad (4.65)$$

which reduces to

$$\frac{\partial^2 w}{\partial x^2} = \frac{1}{c^2}\frac{\partial^2 w}{\partial t^2} \qquad (4.66)$$

in one spatial dimension. A one-dimensional problem is a vibrating string, where the motion depends on one spatial variable x only. Two analytical methods are commonly used to solve this problem. The first is the D'Alembert or traveling wave solution. Here, any disturbance in the media creates two mirror disturbances (which sum up to the original disturbance at $t = 0$) and travel in opposite directions as time evolves. This can be shown by introducing the change in variables $\xi = x - ct$ and $\eta = x + ct$. Using the chain rule Eq. (4.66) can be recast in terms of ξ and η (rather than x and y): $\frac{\partial^2 w}{\partial \xi \partial \eta} = 0$. Integrating twice leads to the solution:

$$w = \psi(\eta) + \phi(\xi) = \psi(x + ct) + \phi(x - ct).$$

These functions represent the left (ψ) and right (ϕ) traveling disturbances [25]. This is shown schematically in Figure 4.9, where the initial spike disturbance is divided in two with half traveling to the right and half to the left at speed c. In an infinite medium, that's the end of the problem. But in finite problems, the physics of the boundaries impacts how the solution proceeds. In general, some amount of the wave is reflected at the boundary (possibly at some phase relative to the incident wave, depending on damping) and the remainder transmitted through the medium [26]. This can also be done in three dimensions, using vector representations for the wave fronts. While this process is helpful from a conceptual standpoint, there are more efficient and quantitative analytical descriptions of the solution to the wave equation. Moreover, these other techniques are readily extendable to fourth-order, beam-like problems.

The second approach is a separation of variables approach; let's consider the vibrations of a string. Here we assume that the solution $w = w(x, t)$ may be split into two constituent parts, with one being a function of x only and the other being a function of t only: $w(x, t) = W(x)T(t)$. Substituting this into Eq. (4.66), taking the derivatives, and dividing by the product $W(x)T(t)$, we find:

$$c^2 \frac{W''}{W} = \frac{\ddot{T}}{T}$$

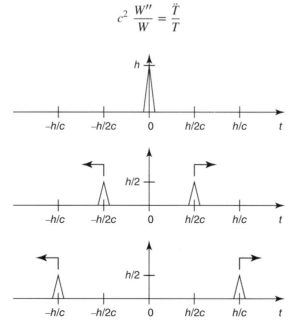

Figure 4.9   A 1-D traveling wave representation of the solution to the wave equation

Because the right and left sides are functions of different independent variables, the only way they can be equal is if they are both equal to the same constant. With foresight (to the time equation), we will call this constant $-\omega^2$. The time and space equations become:

$$\ddot{T} + \omega^2 T = 0$$

and

$$W'' + \left(\frac{\omega}{c}\right)^2 W = 0$$

Interestingly, this separation process has reduced the PDE into two, second-order ODEs. And they both have the same form as Eq. (4.32) – our building block problem from the previous section. The solution to the time equation is simply: $T(t) = A\cos(\omega t) + B\sin(\omega t)$. It is harmonic in time. The spatial part is more interesting though. It is also harmonic: $W(x) = D\cos(\omega x/c) + E\sin(\omega x/c)$, but the constants D and E are found via the boundary conditions. For simplicity, let's assume the ends of the string are fixed: $W(0) = W(L) = 0$. This leads to $D = 0$ and $E\sin(\omega L/c) = 0$. To avoid the trivial solution $E \neq 0$ so $\sin(\omega L/c) = 0$ or, in other words,

$$\omega_i = \frac{i\pi c}{L} \qquad i = 1, 2, \dots, \infty$$

So there are an infinite number of these *natural frequencies* ("natural" because it's how the string naturally vibrates if left to its own devices); in mathematical terms, these are the eigenvalues, referenced in Eq. (4.51). This process should make sense in that the frequency comes directly from the boundary conditions. If the boundary conditions change, so do the frequencies. Consider replacing the fixed ends with very weak springs. We know – just from physical intuition – that the system would vibrate very differently, that is, at a very different set of frequencies. The only portion of the spatial solution that survives is the part associated with E. These functions are the mode shapes (eigenfunctions) and satisfy orthogonality [21]: $\Psi_i(x) = \sin(\omega_i x/c) = \sin(i\pi x/L)$.

To construct the total (product) solution $w(x,t) = W(x)T(t)$, it's useful to remember that both T and W contain ω_i – and that *all* of these frequencies must be retained. Having said that, the total solution takes the form of a Fourier series:

$$w(x,t) = \sum_{i=1}^{\infty} [A_i \cos(\omega_i t) + B_i \sin(\omega_i t)]\Psi_i(x)$$

and the constants A_i and B_i are found by applying the given initial conditions ($w(x,0)$ and $\dot{w}(x,0)$) and making use of the spatial orthogonality of the mode shapes [21]. Specifically, to find an arbitrary coefficient, say A_j (the jth A coefficient), you would apply the initial displacement condition, multiply by $\Psi_j(x)$, and integrate over the spatial domain (from 0 to L). The result is

$$A_j = \frac{\int_0^L w(x,0)\Psi_j(x)\,dx}{\int_0^L \Psi_j^2\,dx}$$

In a similar way, you can find

$$B_j = \frac{\int_0^L \dot{w}(x,0)\Psi_j(x)\,dx}{\int_0^L \omega_j \Psi_j^2\,dx}$$

The strength of this method is that it gives you an analytic solution to the problem that can be easily evaluated numerically, that is, you can quickly calculate the solution at any time t and at any location x. This method also provides physical insight in terms of the makeup of the natural frequencies and mode shapes for a given problem.

This discussion has dealt almost entirely with problems with one spatial dimension x. But there's no need for such a restrictive perspective. Equation (4.65) permits up to four independent variables: three spatial variables and time. The string is an example of a 1-D problem. A vibrating circular membrane is an example of a 2-D problem; but in this case the spatial variables are the polar coordinates (r, θ) and lead to Bessel's equation. The advantage is that polar coordinates greatly simplify the description of the boundary conditions. An example of a 3-D problem is the probability distribution for an electron's position, given as a solution to Schrödinger's equation. These distributions are the s, p, d, and f orbital shapes (eigenfunctions). The separation solution process for the 2-D and 3-D problems are identical to the 1-D case and doesn't bear repeating.

Vibration problems involving bending (e.g., EB or T beams) are only slightly different. In these cases, there are four spatial derivatives. In the case of the EB beam, the separation process yields the following spatial equation:

$$W^{IV} + \frac{\bar{m}\omega^2}{EI} W = 0$$

which has a solution form: $W(x) = A\cos(\beta x) + B\sin(\beta x) + D\cosh(\beta x) + E\sinh(\beta x)$. Here $\beta = \sqrt[4]{\bar{m}\omega^2/(EI)}$. Again, applying the four boundary conditions leads to the natural frequencies and orthogonal mode shapes. The product solution, including all of the frequencies and modes, is then constructed in the form of a Fourier series with the initial conditions (and orthogonality) supplying the unknown constants.

4.5.3   Local Discretizations

Analytical solutions to certain problems simply don't exist. Two common culprits undermining analytical solutions are complex boundaries (e.g., complex shapes that don't lend themselves to a single type of coordinate system) and nonlinearity (either material or geometric) in the governing equation. And even when analytical solutions do exist, they can get cumbersome, making numerical solutions preferable. Whatever the circumstances, numerical solutions to a governing PDE + boundary conditions are often needed.

One approach is to break the spatial domain into small, discrete chunks (hence the term *discretization*) – and then solve the governing equations at the discrete points and interpolate the behavior in between points. This resolves the spatial problem, which can be solved iteratively in time to solve the entire problem. Local discretization is ideal for complex boundaries with odd twists and turns because the discrete points can be wrapped around the edges, as needed. The finite element method (FEM) is one such approach and a sample discretization for a flat rectangular plate is shown in Figure 4.8. In fact, the elements are based on a variational formulation, which circumvents the integration by parts described in Section 4.3.5. No attempt is be made here to cover this vast subject; the interested reader is referred to Refs [27, 28]. But here are two simple suggestions that will help ensure that your simulations are producing physically meaningful results.

First, be sure to conduct a simple grid convergence study. It may sound like a trivial first step, however this simple validation is essential to making sure that the results obtained are due to the physics of the problem and not numerical error. Start with a very coarse grid, run the simulation, and obtain whatever measure of the response is required for the problem at hand. Next, make the grid more dense and rerun the simulation. Continue to do this until the change in your answer is sufficiently small (an engineering judgment). Such studies were performed prior to solving many of the estimation problems that appear in Chapter 9.

Second, don't solve your entire problem right out of the gate. Instead, sneak up on it by solving a problem that you know the answer to. Here's a simple example. Say you want to know the natural frequency of a rotating helicopter blade in operation. You could create a FEM simulation of the actual geometry, have it rotate, and then solve the free vibration problem. You'll get answers, but you won't know if your code is giving reliable answers. Instead, simplify the problem. Remove the twist from the blade and make it a beam with a rectangular cross-section. While you're at it, set the rotation speed to zero, $\Omega = 0$. What you have now is a cantilever beam – for which there are analytical solutions. Run your simulations and make sure they agree with the analytical solutions. Once correct, you could introduce a nonzero rotation speed. And, while you may not have an exact answer for a rotating beam, it's clear that the beam should elongate and stiffen under rotation. So the frequencies should increase; the code should reflect that as well. Finally, when you go to the true geometry, you won't have analytical solutions to guide you (that's why you're doing the FEM solution in the first place). But you will be reasonably confident that you are using your code correctly and that your trends are correct.

4.5.4   Global Discretizations

In local discretizations, the governing equations are satisfied pointwise and interpolation functions are used to smooth out the behavior between evaluation points. By contrast, global discretization methods attempt to satisfy the governing equation across the entire body simultaneously – but do so in a modal context. Let's look at a really simple example to flesh this out. Consider the small amplitude axial vibrations of an elastic rod. This is exactly the problem we solved using Hamilton's principle in Section 4.3.5. We can define the mass and stiffness operators as: $M[u] = \bar{m}\ddot{u}$ and $L[u] = -EAu_{,xx}$. The differential equation becomes

$$M[u] + L[u] = 0.$$

The fixed-free boundary conditions are $u(0, t) = u_{,x}(L, t) = 0$. From here, we take a cue from the separation of variables approach and create an *assumed* separable solution: $u_a(x, t) = \sum a_i(t)\psi_i(x)$, where ψ_i is some set of spatial functions known to satisfy the boundary conditions. From here, we create an error function ϵ:

$$\epsilon = M[u_a] + L[u_a].$$

Obviously, if $\epsilon = 0$, then our assumed solution is the exact solution. But for an arbitrary $\psi_i(x)$, this won't happen. Instead, we require that the weighted error integrated over the domain be zero:

$$\int_0^L \epsilon\psi_j(x)\, dx = 0 \qquad j = 1, 2, \ldots, \infty$$

This is referred to as Galerkin's method. It effectively removes all of the spatial dependence out of the problem and leaves just an ODE (or a set of ODEs) in time, which must be solved. To round out the problem a bit more, we will consider a single term ($j = 1$ only) expansion with $\psi(x) = Lx - x^2/2$. Note that this choice automatically satisfies both boundary conditions. Substituting this into the error function gives

$$\epsilon = m\ddot{a}\left(Lx - \frac{1}{2}x^2\right) - EAa(1)$$

Now, we must drive the weighted error to zero:

$$\int_0^L \left[m\ddot{a}\left(Lx - \frac{1}{2}x^2\right) - EAa\right]\left(Lx - \frac{1}{2}x^2\right)dx = 0$$

Rearranging this, we find

$$\left[\int_0^L m\left(Lx - \frac{1}{2}x^2\right)^2 dx\right]\ddot{a} + \left[\int_0^L -EA\left(Lx - \frac{1}{2}x^2\right)dx\right]a = 0$$

which simplifies to the standard building block differential equation (again):

$$M\ddot{a} + Ka = 0.$$

Here $M = \int_0^L m(Lx - \frac{1}{2}x^2)^2 dx$ and $K = \int_0^L -EA(Lx - \frac{1}{2}x^2)dx$. If a multiterm expansion is used, then the result is a matrix set of ordinary differential equations. The solution to these equations may be analytical (Section 4.5.1) or numerical.

Galerkin's method, the dominant global discretization technique, is global because the expansion functions ψ_i span the spatial domain and stand in contrast to local techniques. The biggest problem is to identify an acceptable set of expansion functions. A common approach in nonlinear problems is to use the mode shapes to the linear problem as the expansion functions, because they are guaranteed to satisfy the boundary conditions. This approach will be used to derive the governing equations of the structure used in the main example of Chapter 10.

4.6 Volterra Series for Nonlinear Systems

As we have shown with Eq. (4.46), the behavior of a causal linear system with memory can be described by the well-known convolution integral $y(t) = \int h(\theta)x(t - \theta)d\theta$, where $x(t)$ is the input, $y(t)$ is the output, and $h(\theta)$ is the impulse response of the system. While many systems encountered in the engineering sciences are well-approximated by such a model, there are also situations where we would like an analytical approach to modeling a *nonlinear* system response. It turns out that for a causal, weakly nonlinear system with finite memory, the time-domain response may be modeled using a Volterra series model [29]

$$y(t) = y_1(t) + y_2(t) + \cdots + y_N(t)$$
$$= \int_{\mathbb{R}} h_1(\theta)x(t - \theta)d\theta + \int_{\mathbb{R}^2} h_2(\theta_1, \theta_2)x(t - \theta_1)x(t - \theta_2)d\theta_1 d\theta_2 + \cdots$$
$$+ \int_{\mathbb{R}^N} h(\theta_1, \theta_2, \ldots, \theta_N)\prod_{n=1}^N x(t - \theta_n)d\theta_1 d\theta_2 \cdots d\theta_N. \tag{4.67}$$

The Volterra series model therefore relates the output to linear combinations of powers of the input through a series of multidimensional integrals. The weightings associated with the Nth input power are specified by the Nth-order Volterra kernel $h(\theta_1, \theta_2, \dots, \theta_N)$. It is worth remarking that $h(\theta_1, \theta_2, \dots, \theta_N) > 0$ for any $\theta_n < 0$ $n \in [1, N]$ in order for the system to be causal. Existence and uniqueness of the kernels, as well as convergence of the series equation (4.67), are discussed in Ref. [29].

In general, the Volterra series will converge for smoothly nonlinear systems and for a range of system parameters. Such conditions are satisfied by a large number of mildly nonlinear systems, as has been witnessed by the rich literature on the subject.

Analogous expressions to Eqs. (4.67) can be written in the frequency domain resorting to higher order frequency response functions(HOFRFs), where the nth-order FRF is defined as the nth-order Fourier transform of the correspondent time-domain kernel representation:

$$H_1(\omega_1) = \int_{\mathbb{R}} h_1(\theta) e^{-i\omega_1 \theta} \, d\theta$$

$$H_2(\omega_1, \omega_2) = \int_{\mathbb{R}^2} h_2(\theta_1, \theta_2) e^{-i\omega_1 \theta_1} e^{-i\omega_2 \theta_2} \, d\theta_1 \, d\theta_2$$

$$\vdots$$

$$H_N(\omega_1, \omega_2, \dots, \omega_N) = \int_{\mathbb{R}^N} h_N(\theta_1, \theta_2, \dots, \theta_N) e^{-i\omega_1 \theta_1} e^{-i\omega_2 \theta_2} \cdots e^{-i\omega_N \theta_N} \, d\theta_1 \, d\theta_2 \cdots d\theta_N.$$
$$(4.68)$$

As in the linear case, these complex functions carry the same information about the system as their time-domain counterparts, but they are often easier to work with, both experimentally and analytically. If these HOFRFs are available, the model given by Eqs. (4.67) can be obtained, as will be shown in detail for the second-order component $H_2(\omega_1, \omega_2)$. To further the development, it is useful to introduce an auxiliary function of two temporal variables with the property $y_{(2)}(t, t) = y_2(t)$ [29]. By then letting $t = t_1 = t_2$, the second-order term in the Volterra series can be rewritten as

$$y_2(t) = y_{(2)}(t_1, t_2) = \int_{\mathbb{R}^2} h_2(\theta_1, \theta_2) x(t_1 - \theta_1) x(t_2 - \theta_2) d\theta_1 \, d\theta_2 \qquad (4.69)$$

whereupon taking the double Fourier transform yields

$$Y_{(2)}(\omega_1, \omega_2) = \int_{\mathbb{R}^2} y_{(2)}(t_1, t_2) e^{-i\omega_1 t_1} e^{-i\omega_2 t_2}$$

$$= \int_{\mathbb{R}^4} h_2(\theta_1, \theta_2) x(t_1 - \theta_1) x(t_2 - \theta_2) e^{-i\omega_1 t_1} e^{-i\omega_2 t_2} \, d\theta_1 \, d\theta_2 \, dt_1 \, dt_2 \quad (4.70)$$

Making the change of variables $\sigma_1 \equiv t_1 - \theta_1$ and $\sigma_2 = t_2 - \theta_2$, the integral is easily simplified as

$$Y_{(2)}(\omega_1, \omega_2) = \int_{\mathbb{R}^2} h_2(\theta_1, \theta_2) e^{-i\omega_1 \theta_1} e^{-i\omega_2 \theta_2} \, d\theta_1 \, d\theta_2 \int_{\mathbb{R}} x(\sigma_1) e^{-i\omega_1 \sigma_1} d\sigma_1 \int_{\mathbb{R}} x(\sigma_2) e^{-i\omega_2 \sigma_2} \, d\sigma_2$$

$$= H_2(\omega_1, \omega_2) X(\omega_1) X(\omega_2) \qquad (4.71)$$

To recover $y_2(t)$, we allow $t_1 = t_2 = t$ and define $\omega = \omega_1 + \omega_2$ so that

$$y_2(t) = \int_{\mathbb{R}^2} Y_{(2)}(\omega_1, \omega_2) e^{i(\omega_1 + \omega_2)t} \, d\omega_1 \, d\omega_2$$

$$= \int_{\mathbb{R}} \int_{\mathbb{R}} Y_{(2)}(\omega_1, \omega - \omega_1) d\omega_1 e^{i\omega t} \, d\omega \qquad (4.72)$$

and so by definition, $Y_2(\omega) = \int_{\mathbb{R}} Y_{(2)}(\omega_1, \omega - \omega_1) d\omega_1$ which, by Eq. (4.71), gives

$$Y_2(\omega) = \int_{\mathbb{R}} H_2(\omega_1, \omega - \omega_1) X(\omega_1) X(\omega - \omega_1) d\omega_1. \qquad (4.73)$$

When the system contains higher order nonlinearities, expressions analogous to Eq. (4.73) can be written taking into consideration HOFRFs. In this example, we have related the input $X(\omega)$ to the second-order output $Y_2(\omega)$. Thus, if a two-term Volterra model had been used the frequency response would be $Y(\omega) = Y_1(\omega) + Y_2(\omega)$, where $Y_1(\omega) = H_1(\omega)X(\omega)$ and $Y_2(\omega)$ is given by Eq. (4.73). In general, the Volterra series models can describe the system response in either the time or frequency domain.

It is one thing to write down the form of the Volterra model but to this point we have not discussed how one might actually obtain $H_2(\omega_1, \omega_2)$, for example, from a given dynamical system. To identify the Volterra kernels, we can make use of the *harmonic probing* technique [30]. Harmonic probing is simply an extension of the usual procedure for obtaining the linear transfer function and is based on the fact that if the input consists of a sum of sinusoids at frequencies $\omega_1, \omega_2, \ldots, \omega_N$, the nth-order FRF $H_N(\omega_1, \omega_2, \ldots, \omega_N)$ provides, in magnitude and phase, the output component at the sum of the forcing frequencies $\sum_{n=1}^{N} \omega_n$. This easily follows from Eq. (4.73) for second-order Volterra models, although it is valid also when higher kernels appear (higher order models) [31]. This observation offers both a direct way of measuring the nth-order FRF experimentally and a way of obtaining it analytically from the governing equations. The expressions needed to derive the nth-order kernel are therefore obtained by substituting a sum of "n" sinusoids into the frequency domain Volterra model, for example, Eq. (4.73), and solving for $y(t)$. To get the nth-order FRF for a particular system, one simply uses a sum of "N" sinusoids as the input and the associated $y(t)$ for the output, substitutes into the governing equations, equates terms at the sum frequency, and solves for the nth-order FRF. This entire procedure is described in detail in Refs [30, 31]. The input/output relationships for a nonlinear Volterra model can be described either in the time or frequency domains. Once the HOFRFs are known, it is possible to describe the output to any time-dependent excitation, not only to harmonic. The derivation used here proceeds in the frequency domain, as it is the auto-bispectrum that is the main object of this investigation. However, as remarked in Ref. [32], the development can also proceed using Laplace or z-transform and their corresponding domains.

To illustrate the procedure, consider an SDOF with quadratic nonlinearities in both the restoring force and dissipative force

$$m\ddot{y}(t) + c\dot{y}(t) + ky(t) + c_{non}\dot{y}^2(t) + k_{non}y^2(t) = x(t) \qquad (4.74)$$

where $m, c, k, c_{non}, k_{non}$ are real-valued, positive-valued parameters and $x(t)$ is the input signal. To obtain the first-order (linear) Volterra kernel we use the familiar harmonic input $x(t) = e^{i\omega_1 t}$, in which case the output takes the form $y(t) = H_1(\omega_1)e^{i\omega_1 t}$. Substituting into the governing equations yields

$$-m\omega_1^2 H_1(\omega_1)e^{i\omega_1 t} + i\omega_1 c H_1(\omega_1)e^{i\omega_1 t} + k H_1(\omega_1)e^{i\omega_1 t}$$
$$- H_1^2(\omega_1)e^{2i\omega_1 t} + H_1^2(\omega_1)e^{2i\omega_1 t} = e^{i\omega_1 t} \tag{4.75}$$

Equating coefficients of $e^{i\omega_1 t}$, we have

$$H_1(\omega_1) = \frac{1}{-m\omega_1^2 + i\omega_1 c + k} \tag{4.76}$$

which is the familiar linear transfer function. In the same way, the second-order FRF can be determined by applying a dual-tone input $x(t) = e^{i\omega_1 t} + e^{i\omega_2 t}$, resulting in an output of the form:

$$y(t) = H_1(\omega_1)e^{i\omega_1 t} + H_1(\omega_2)e^{i\omega_2 t} + H_2(\omega_1, \omega_1)e^{i2\omega_1 t} + H_2(\omega_2, \omega_2)e^{i2\omega_2 t}$$
$$+ H_2(\omega_1, \omega_2)e^{i(\omega_1 + \omega_2)t} + H_2(\omega_2, \omega_1)e^{i(\omega_2 + \omega_1)t} \tag{4.77}$$

By placing Eq. (4.77) and its derivatives into Eq. (4.74) and equating terms at the sum frequency, the second-order FRF can be obtained as

$$H_2(\omega_1, \omega_2) = (-k_2 + c_2\omega_1\omega_2)H_1(\omega_1)H_1(\omega_2)H_1(\omega_1 + \omega_2). \tag{4.78}$$

Clearly, the second-order kernel does not exist if the system is linear ($k_2 = c_2 = 0$). To obtain the third-order kernel we follow the same procedure, applying a three-tone input $x(t) = e^{i\omega_1 t} + e^{i\omega_2 t} + e^{i\omega_3 t}$, assuming $y(t) = H_3(\omega_1, \omega_2, \omega_3)e^{i(\omega_1 + \omega_2 + \omega_3)t}$, substituting into the governing equations, and equating coefficients of $e^{i(\omega_1 + \omega_2 + \omega_3)t}$ to solve for $H_3(\omega_1, \omega_2, \omega_3)$. The result is

$$H_3(\omega_1, \omega_2, \omega_3) = \frac{1}{3}H_1(\omega_1 + \omega_2 + \omega_3)\{[-3k_3 + 3ic_3\omega_1\omega_2\omega_3]H_1(\omega_1)H_1(\omega_2)H_1(\omega_3)$$
$$+ [-2k_2 + 2c_2(\omega_1 + \omega_2)\omega_3]H_1(\omega_3)H_2(\omega_1, \omega_2)$$
$$+ [-2k_2 + 2c_2(\omega_1 + \omega_3)]H_1(\omega_2)H_2(\omega_1, \omega_3)$$
$$+ [-2k_2 + 2c_2(\omega_2 + \omega_3)\omega_1]H_1(\omega_1)H_2(\omega_2, \omega_3)\} \tag{4.79}$$

Note that in some instances we are interested in relating the input $x(t)$ to the output $\dot{y}(t)$ or $\ddot{y}(t)$ (as opposed to $y(t)$). We already know from earlier in this chapter that the linear kernels become

$$H_{1,v}(\omega) = i\omega H_{1,y} = \frac{i\omega}{-m\omega^2 + ic\omega + k} \quad \text{(velocity)}$$

$$H_{1,a}(\omega) = -\omega^2 H_{1,y} = \frac{-\omega^2}{-m\omega^2 + ic\omega + k} \quad \text{(acceleration)} \tag{4.80}$$

for the velocity and acceleration responses, respectively (we are denoting the displacement response $H_{1,y}$). From these expressions it is straightforward to calculate

$$H_{2,v}(\omega_1, \omega_2) = i(\omega_1 + \omega_2)(-k_2 + c_2\omega_1\omega_2)H_{1,y}(\omega_1)H_{1,y}(\omega_2)H_{1,y}(\omega_1 + \omega_2) \quad \text{(velocity)}$$

$$H_{2,a}(\omega_1, \omega_2) = -(\omega_1 + \omega_2)^2(-k_2 + c_2\omega_1\omega_2)H_{1,y}(\omega_1)H_{1,y}(\omega_2)H_{1,y}(\omega_1 + \omega_2) \quad \text{(acceleration)} \tag{4.81}$$

Similarly, for the third Volterra kernel, we obtain $H_{3,v}$ and $H_{3,a}$ by premultiplying Eq. (4.79) by $i(\omega_1 + \omega_2 + \omega_3)$ and $-(\omega_1^2 + \omega_2^2 + \omega_3^2)$, respectively.

It appears that even for a quadratic nonlinear system where $k_3 = c_3 = 0$, the third-order FRF still exists. Similar considerations can be extended to higher order kernels. This implies that the output terms $y_3(t), y_4(t)$ and so forth of Eq. (4.67) are present, although for the series to be convergent they need to bring less and less contribution to the response. In general, when using a Volterra series representation, the number of terms to be retained depends on the inherent nonlinearities, and on the excitation level that clearly brings higher order terms into play. To verify the accuracy of the procedure, an approximated output response can be found using a truncated series, which can be compared with the solution obtained from a numerical scheme, when the nonlinear governing equation is available [29, 33]. From the previous equations for the first HOFRFs, some important properties can be inferred:

- HOFRFs contain products of linear FRFs;
- HOFRFs are independent of the input to the system, as it is in the linear case;
- HOFRFs are symmetric, because they are insensitive to any frequency variable permutation, for example, $H_2(\omega_1, \omega_2) = H_2(\omega_2, \omega_1)$;
- the complex conjugate of HOFRFs is simply obtained by changing the sign of the frequency variables, for example, $H_2^*(\omega_1, \omega_2) = H_2(-\omega_1, -\omega_2)$. This property, which follows also directly from Eqs. (4.68), can be exploited when absolute values need to be taken, for example, $|H_n(\omega_1, \omega_2, \ldots, \omega_n)|^2 = H_n(\omega_1, \omega_2, \ldots, \omega_n)H_n(-\omega_1, -\omega_2, \ldots, -\omega_n)$ for a general nth-order FRF.

The Volterra models we have discussed so far allow us to describe single-input, single-output structural systems. Seldom are we presented with such a simple model in practice. The more common scenario is for a forced structure to be modeled as an MDOF system. Fortunately, the Volterra series model still applies in this case. Consider an MDOF structural system described by the general expression

$$
m_s \ddot{y}_s(t) + \sum_{i=1}^{Order} [c_i^{s,s} \dot{y}_s^i(t) + k_i^{s,s} y_s^i(t)]
$$

$$
- \sum_{i=1}^{Order} \sum_{\substack{p=1 \\ p \neq s}}^{Ndof} [c_i^{s,p}(\dot{y}_p(t) - \dot{y}_s(t))^i + k_i^{s,p}(y_p(t) - y_s(t))^i] = \delta(r-s)Ax_r(t) \qquad (4.82)
$$

HOFRFs for a class of MDOF systems can be computed using a straightforward extension of the SDOF method. The original reference by Gifford and Tomlinson [32] describes the needed approach. Here, the procedure is illustrated for a two-DOF system, however it is applicable to systems of arbitrary complexity. The case to be considered, which will suffice to exhibit the general approach, is that of a two-DOF system where mass m_1 is connected to the ground and mass m_2 is connected only to mass m_1. In addition, the nonlinearities are present only in the connection between mass m_1 and mass m_2; all other springs and dampers are taken as linear (see Figure 8.25 and the surrounding discussion for the complete system description). This is only to simplify the exposition; in principle, the nonlinearities can be located in any position of the structure, and there is no reason why several grounded or ungrounded springs may not be taken to be nonlinear. However, such problems typically require computer-assisted algebra and are therefore not particularly informative as illustrative examples.

The equations of motion for this example system are

$$m_1\ddot{y}_1 + c_1^{1,1}\dot{y}_1 + k_1^{1,1}y_1 + c_1^{1,2}(\dot{y}_1 - \dot{y}_2)$$

$$+ k_1^{1,2}(y_1 - y_2) + c_2^{1,1}\dot{y}_1^2 + k_2^{1,1}y_1^2 - c_2^{1,2}(\dot{y}_1 - \dot{y}_2)^2 - k_2^{1,2}(y_1 - y_2)^2 = x_1(t) \qquad (4.83)$$

$$m_2\ddot{y}_2 + c_1^{2,2}\dot{y}_2 + k_1^{2,2}y_2 + c_1^{1,2}(\dot{y}_2 - \dot{y}_1)$$

$$+ k_1^{1,2}(y_2 - y_1) + c_2^{2,2}\dot{y}_2^2 + k_2^{2,2}y_2^2 + c_2^{1,2}(\dot{y}_2 - \dot{y}_1)^2 + k_2^{1,2}(y_2 - y_1)^2 = x_2(t) \qquad (4.84)$$

where $x_1(t)$ and $x_2(t)$ are the forcing functions. Again, in this work we are considering the case where only a single input is active (i.e., either $x_1(t) = 0$ or $x_2(t) = 0$). The multiple input case requires a more sophisticated approach for determining the Volterra kernels [30]

We first define the linear mass, damping, and stiffness matrices as

$$\mathbf{m} = \begin{bmatrix} m_1 & 0 \\ 0 & m_2 \end{bmatrix}$$

$$\mathbf{c} = \begin{bmatrix} c_1^{1,1} + c_1^{1,2} & -c_1^{1,2} \\ -c_1^{1,2} & c_1^{2,2} + c_1^{1,2} \end{bmatrix}$$

$$\mathbf{k} = \begin{bmatrix} k_1^{1,1} + k_1^{1,2} & -k_1^{1,2} \\ -k_1^{1,2} & k_1^{2,2} + k_1^{1,2} \end{bmatrix}.$$

To evaluate the first-order kernels (or FRFs), the system is excited at DOF r by the single input $x_r(t) = Xe^{i\omega t}$. Substituting into Eq. (4.84) and equating coefficients of $Xe^{i\omega t}$, the first Volterra kernel transform matrix can be obtained as

$$\mathbf{H}_1(\omega) = \begin{bmatrix} H_1^{1,1}(\omega) & H_1^{2,1}(\omega) \\ H_1^{1,2}(\omega) & H_1^{2,2}(\omega) \end{bmatrix},$$

where matrix $\mathbf{H}_1$ is structured such that every column corresponds to a different forcing location (e.g., column 1 represents $r = 1$, while column 2 represents $r = 2$). For this particular example, the procedure yields

$$\mathbf{H}_1(\omega) = \frac{1}{\mathbf{p}(\omega)} \begin{bmatrix} (k_1^{2,2} + k_1^{1,2}) + i\omega(c_1^{2,2} + c_1^{1,2}) - m_2\omega^2 & k_1^{1,2} + i\omega c_1^{1,2} \\ k_1^{1,2} + i\omega c_1^{1,2} & (k_1^{1,1} + k_1^{1,2}) + i\omega(c_1^{1,1} + c_1^{2,2}) - m_1\omega^2 \end{bmatrix}$$

$$(4.85)$$

where

$$\mathbf{p}(\omega) = det(\mathbf{k} + i\omega\mathbf{c} - \omega^2\mathbf{m}) \qquad (4.86)$$

and "det" takes the determinant. Clearly, the first-order Volterra kernels are independent of the nonlinear terms in Eq. (4.84). In addition, Eq. (4.85) shows that the H_1^r functions have the same poles but different zeroes. This is an important piece of information because the structure of these kernels will reveal features in the autobispectral density function, used later in Chapters 6 and 8 for damage detection applications.

To evaluate the second-order kernels, the system is again excited at point r by a single input comprised of two tones $x_r(t) = X_1 e^{i\omega_1 t} + X_2 e^{i\omega_2 t}$. Substituting into the equations of motion and equating coefficients of $X_1 X_2 e^{i(\omega_1+\omega_2)t}$, the Volterra kernel transform matrix can be obtained:

$$\mathbf{H_2}(\omega_1,\omega_2) = \begin{bmatrix} H_2^{1,1}(\omega_1,\omega_2) & H_2^{2,1}(\omega_1,\omega_2) \\ H_2^{1,2}(\omega_1,\omega_2) & H_2^{2,2}(\omega_1,\omega_2) \end{bmatrix}$$

$$= \mathbf{H_1}(\omega_1+\omega_2) \begin{bmatrix} T_2^{1,1}(\omega_1,\omega_2) & T_2^{2,1}(\omega_1,\omega_2) \\ T_2^{1,2}(\omega_1,\omega_2) & T_2^{2,2}(\omega_1,\omega_2) \end{bmatrix}$$

where the quantities $T_2^{i,j}$ are defined to be

$$\mathbf{T_2^r} = \begin{bmatrix} T_2^{r,1}(\omega_1,\omega_2) \\ T_2^{r,2}(\omega_1,\omega_2) \end{bmatrix}$$

$$= \begin{bmatrix} (\omega_1\omega_2 c_2^{1,1} - k_2^{1,1})H_1^{r,1}(\omega_1)H_1^{r,1}(\omega_2) - (\omega_1\omega_2 c_2^{1,2} - k_2^{1,2}) \\ \times [H_1^{r,2}(\omega_1) - H_1^{r,1}(\omega_1)][H_1^{r,2}(\omega_2) - H_1^{r,1}(\omega_2)] \\ (\omega_1\omega_2 c_2^{2,2} - k_2^{2,2})H_1^{r,2}(\omega_1)H_1^{r,2}(\omega_2) + (\omega_1\omega_2 c_2^{1,2} - k_2^{1,2}) \\ \times [H_1^{r,2}(\omega_1) - H_1^{r,1}(\omega_1)][H_1^{r,2}(\omega_2) - H_1^{r,1}(\omega_2)] \end{bmatrix}$$

The procedure can be extended in a straightforward manner to include any DOF and to any kernel representation. As indicated in Ref. [32], the second-order Volterra kernel transform for the SISO system consisting of a force input at point r an output of displacement measured at point s is given by

$$H_2^{r,s}(\omega_1,\omega_2) = \sum_{m=1}^{M} (\omega_1\omega_2 c_2^{1,m} - k_2^{1,m})H_1^{s,m}(\omega_1)H_1^{r,m}(\omega_2)H_1^{r,m}(\omega_1+\omega_2)$$

$$- \sum_{m=1}^{M-1} \sum_{n=m+1}^{M} (\omega_1\omega_2 c_2^{m,n} - k_2^{m,n})[H_1^{s,m}(\omega_1+\omega_2) - H_1^{s,n}(\omega_1+\omega_2)]$$

$$[H_1^{r,m}(\omega_1) - H_1^{r,n}(\omega_1)][H_1^{r,m}(\omega_2) - H_1^{r,n}(\omega_2)]. \tag{4.87}$$

The second-order Volterra kernel is expressed by a set of first-order kernels that are computed from just the linear part of the equations of motion, and the coefficients of the quadratic polynomial functions representing the stiffness and damping terms in the structure. These kernels are used later to give an analytical expression for the autobispectral density function computed for an arbitrary combination of system response data. This procedure is used extensively in later chapters concerning the identification of damage-induced nonlinearities in

MDOF systems. Specifically, we are interested in how the nonlinearity influences the statistical properties of the structures response. By using the Volterra series approach, we are able to make analytical predictions and ultimately determine the expected damage detection probabilities.

References

[1] D. Young, R. Felgar, Tables of Characteristic Functions Representing Normal Modes of Vibration of a Beam, University of Texas Publications, 1949.

[2] J. L. Meriam, L. G. Kraige, Engineering Mechanics, Vol. 2, John Wiley & Sons, Inc., New York, 1992.

[3] D. T. Greenwood, Principles of Dynamics, 2nd ed., Prentice-Hall, Englewood Cliffs, NJ, 1988.

[4] M. Beatty, Principles of Engineering Mechanics: Kinematics - The Geometry of Motion, Vol. 1, Springer-Verlag, New York, 1986.

[5] M. Beatty, Principles of Engineering Mechanics: Dynamics - The Analysis of Motion, Vol. 2, Springer-Verlag, New York, 2006.

[6] C. Pierre, E. H. Dowell, Localization of vibrations by structural irregularity, Journal of Sound and Vibration 114 (1987) 549–564.

[7] S. W. Shaw, P. J. Holmes, A periodically forced piecewise linear oscillator, Journal of Sound and Vibration 90 (1983) 129–155.

[8] P. V. Bayly, L. N. Virgin, An experimental study of an impacting pendulum, Journal of Sound and Vibration 164 (1993) 364–374.

[9] M. Henon, On the numerical computation of Poincare maps, Physica D 5 (1982) 412–414.

[10] S. Christides, A. D. S. Barr, One dimensional theory of crack Euler-Bernoulli beams, International Journal of Mechanical Sciences 26 (11) (1984) 637–648.

[11] H. Tada, P. Paris, G. Irwin, The Stress Analysis of Cracks Handbook, Del Research Corp., Hellertown, PA, 1972.

[12] K. D. Murphy, Y. Zhang, Vibration and stability of a cracked translating beam, Journal of Sound and Vibration 237 (2000) 319–335.

[13] T. G. Chondros, A. D. Dimarogonas, J. Yao, A consistent cracked bar vibration theory, Journal of Sound and Vibration 200 (3) (1997) 303–313.

[14] Y. C. Chu, M.-H. H. Shen, Analysis of forced bilinear oscillators and the applications to cracked beam dynamics, AIAA Journal 30 (10) (1992) 2512–2519.

[15] R. S. Barsoum, On the use of isoparametric finite elements in linear fracture mechanics, International Journal for Numerical Methods in Engineering 10 (1976) 25–37.

[16] R. D. Henshell, K. G. Shaw, Crack tip finite elements are unnecessary, International Journal for Numerical Methods in Engineering 9 (1975) 495–509.

[17] E. Z. Moore, J. M. Nichols, K. D. Murphy, Crack identification in a freely vibrating plate using Bayesian parameter estimation, Mechanical Systems and Signal Processing 25 (6) (2011) 2125–2134.

[18] S. S. Rao, Mechanical Vibrations, 3rd ed., Addison-Wesley, New York, 1995.

[19] L. Meirovitch, Introduction to Dynamics and Control, John Wiley & Sons, Inc., 1985.

[20] N. W. McLachlan, Theory and Application of Mathieu Functions, Dover Publications, New York, 1964.

[21] E. Kreyszig, Advanced Engineering Mathematics, John Wiley & Sons, Inc., New York, 2006.

[22] H. T. Davis, Introduction to Nonlinear Differential and Integral Equations, Dover Publications, New York, 1962.

[23] F. Bowman, Introduction to Elliptic Functions with Applications, Dover Publications, New York,1961.

[24] L. Meirovitch, Analytical Methods in Vibrations, MacMillan Publishing, New York, 1967.

[25] K. F. Graff, Wave Propagation in Elastic Solids, Dover Publications, New York, 1975.

[26] L. E. Kinsler, A. R. Frey, A. B. Coppens, J. V. Sanders, Fundamentals of Acoustics, John Wiley & Sons, Inc., New York, 2000.

[27] K. J. Bathe, Finite Element Procedures in Engineering Analysis, Prentice-Hall, New York, 1982.

[28] T. J. Oden, J. N. Reddy, An Introduction to the Mathematical Theory of Finite Elements, Dover Publications, New York, 2011.

[29] M. Schetzen, The Volterra and Wiener Theories of Nonlinear Systems, John Wiley & Sons, Inc., New York, 1980.

[30] K. Worden, G. R. Tomlinson, Nonlinearity in Structural Dynamics, Institute of Physics Publishing, Bristol, Philadelphia, PA, 2001.

[31] E. Bedrosian, S. O. Rice, The output properties of volterra systems (nonlinear systems with memory) driven by harmonic and Gaussian inputs, Proceedings of the IEEE 59 (12) (1971) 1688–1707.

[32] S. J. Gifford, G. R. Tomlinson, Recent advances in the application of functional series to non-linear structures, Journal of Sound and Vibration 135 (2) (1989) 289–317.

[33] S. Boyd, L. O. Chua, C. A. Desoer, Analytical foundations of Volterra series, IMA Journal of Mathematical Control and Information 1 (1984) 243–282.

5

Physics-Based Model Examples

In the previous chapter we reviewed the mechanics of modeling structural systems and, in particular, structural damage. This chapter takes this material and begins applying it to specific example systems, both numerical and experimental. The sections in this chapter therefore serve as representative case studies for the types of damage frequently encountered in practice.

5.1 Imperfection Modeling in Plates

Plate structures are one of the most common types of engineering structures encountered. These versatile structures are exposed to a multitude of different environments, loading scenarios, boundary conditions, and so on. They're also likely to be damaged in countless ways. And, of course, damage is the underlying motivation for our modeling efforts. But, rather than key in on *damage* per se, we will look at a number of modeling challenges associated with plate *imperfections* – with damage (e.g., a crack) being one category of imperfection.

5.1.1 Cracks as Imperfections

The behavior of cracks in an otherwise continuous media is a challenge because of the crack tip stress singularity that was briefly touched on in Section 4.4.2. In short, the stress field approaches infinity as you move radially in toward the crack tip. This can be seen clearly through the William's solution [1], which describes radial variation of the stress field in a two-dimensional infinite or semi-infinite continua: $\sigma(r) \propto 1/\sqrt{r}$. At the crack tip, the stress is infinite; as you move away from the crack tip, the stress evanesces. But the real challenge is geometry; if the crack tip is near a boundary (or other complicating geometric features), the stress field may distort considerably from this $1/\sqrt{r}$ trend in the vicinity of the boundary. Therefore, instead of spending an inordinate amount of time creating a complicated analytical model, most engineers turn to a local discretization technique such as the finite element method (FEM) of Section 4.5.3. This approach accommodates unusual geometries/boundaries while providing approximate (yet accurate) solutions to the stress field. The dilemma lies in capturing the stress singularity. This requires a finer and finer mesh – with more stresses (at

Modeling and Estimation of Structural Damage, First Edition. Jonathan M. Nichols and Kevin D. Murphy.
© 2016 John Wiley & Sons, Ltd. Published 2016 by John Wiley & Sons, Ltd.

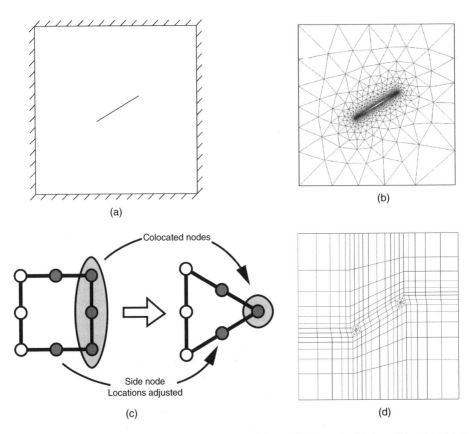

Figure 5.1 (a) An idealized, fully clamped plate containing a single crack. (b) A traditional FEM discretization of the cracked plate. Note the number of elements around the crack. (c) The process of creating a specialty element to sit at the crack tip, capturing the $1/\sqrt{r}$ stress distribution. (d) The FEM discretization using these specialty elements around the crack tips; note how few elements are used

each node) to calculate. As an example, consider Figure 5.1. Figure 5.1a shows the cracked plate geometry and Figure 5.1b shows a typical crack-related FEM mesh; note the density of the mesh around the crack, which increases simulation time. This is particularly burdensome for dynamic problems, where the solution at each nodal point has to be integrated in time. So what can be done?

One good solution was introduced in Refs [2] and [3]. These authors took a standard Mindlin (eight-noded quad) element, shown schematically in Figure 5.1c, and augmented it. How? To begin with, they colocated three nodes at one point – this point will sit at the crack tip. Next, they moved the adjacent side nodes to sit two-thirds of the distance from the colocated node. This process leaves an element for which the stress will naturally vary with $1/\sqrt{r}$, where r is the distance from the colocated node! In other words, you don't need a lot of elements to get the correct asymptotic behavior near the crack tip; it's already built into the element. Figure 5.1d shows a mesh using standard Midline elements everywhere – except at the crack tip. Surrounding the crack tip, in a pin-wheel geometry, are these crack-tip elements. The result is a very sparse mesh that is perfect for dynamic analysis. A convergence study [4] comparing meshes with and without quarter-point elements showed that the use of quarter-point elements reduced the required number of degrees of freedom (DOFs) by 80%. And, because solution run time increases (roughly) quadratically with the DOF count, the savings in model complexity should lead to approximately a 96% savings in run time when compared to the unmodified mesh.

5.1.2   Boundary Imperfections: In-Plane Slippage

By far the most common (fixed) plate boundary is the clamped condition, which prescribes no lateral displacement and no slope. Moreover, there is no in-plane motion at the boundary. However, when there are large out-of-plane deformations, the in-plane stresses can be con-sidered – enough to overcome the clamping friction and the plate can actually slip in-plane, relieving the stress. This is common in thermally loaded plates; the elevated temperature con-tributes directly to in-plane compressive stresses. The question becomes: how can you model this imperfect clamping?

Large deflection plate theory requires some description of the mid-plane stretching. This can be accomplished by satisfying Newton's laws, an elastic constitutive (stress–strain) model, and a first-order nonlinear strain-displacement relation. The unknowns are the two in-plane (u, v) displacements and the one lateral (w) displacement; see Figure 5.2a. The problem can be recast to remove the unknown in-plane displacements u and v – leaving only the out-of-plane displacement w as the unknown. This is accomplished by satisfying the compatibility rela-tionships [5], which are geometric constraint equations. Think of compatibility this way – a prescribed displacement field produces a unique strain field through differentiation: $\epsilon_{ij} = \frac{\partial u_i}{\partial x_j}$. But a prescribed strain field cannot uniquely specify the displacement field because there are nine strains but only three displacements (i.e., the problem is overconstrained). Once all of these elements are combined, the compatibility problem can be expressed in terms of the fol-lowing nonlinear plate equation [6, 7]:

$$\bar{\nabla}^4 w - \left(\frac{a}{b}\right)^2 \left[\frac{\partial^2 F}{\partial \eta^2} \left(\frac{\partial^2 w}{\partial \xi^2} + \frac{\partial^2 w_o}{\partial \xi^2} \right) + \frac{\partial^2 F}{\partial \xi^2} \left(\frac{\partial^2 w}{\partial \eta^2} + \frac{\partial^2 w_o}{\partial \eta^2} \right) \right.$$
$$\left. - 2 \frac{\partial^2 F}{\partial \xi \partial \eta} \left(\frac{\partial^2 w}{\partial \xi \partial \eta} + \frac{\partial^2 w_o}{\partial \xi \partial \eta} \right) \right] + \frac{\partial^2 w}{\partial \tau^2} + \beta \frac{\partial w}{\partial \tau} + \Delta P = 0 \qquad (5.1)$$

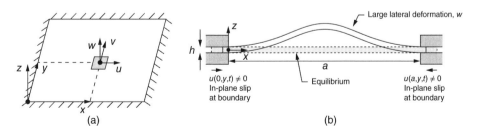

Figure 5.2   (a) An idealized, fully clamped plate. The displacements of an arbitrary material element are given by $\{u, v, w\}$ in the $\{x, y, z\}$ directions, respectively. (b) An edge-on view of the deformed plate showing slippage at the boundary

where $w(x, y, t)$ is the lateral deflection, $w_o(x, y)$ is any initial geometric imperfection, and F is the Airy stress function (more on that in a moment). Also, ξ and η are the normalized spatial coordinates, τ is the normalized time, and $\bar{\nabla}^4$ is the normalized biharmonic operator:

$$\xi = \frac{x}{a}, \; 0 \le \xi \le 1, \quad \eta = \frac{y}{b}, \; 0 \le \eta \le 1$$

$$\tau = t \sqrt{\frac{D}{\rho h a^4}}$$

and

$$\bar{\nabla}^4 = \left(\frac{1}{a^4}\right)\left[\frac{\partial^4}{\partial \xi^4} + 2\left(\frac{a}{b}\right)^2 \frac{\partial^4}{\partial \xi^2 \eta^2} + \left(\frac{a}{b}\right)^4 \frac{\partial^4}{\partial \eta^4}\right]$$

where a and b are the edge lengths in the x and y directions, respectively, h is the plate thickness, D is the flexural rigidity, and ρ is the material density.

Now back to the Airy stress function F. This function acts much like a traditional potential function; recall that in conservative force problems we have $\mathbf{F} = \nabla U$. Similarly, in 2-D irrotational flows we have $\mathbf{v} = \nabla\Phi$. In other words, the force vector $\mathbf{F}$ (the velocity vector $\mathbf{v}$) is derivable from a scalar potential function U (Φ). Similarly, the derivatives of the Airy stress function gives the in-plane stress, σ_{ij}. Specifically, the stresses are related to the second derivative of the Airy stress function:

$$\left(\frac{D}{b^2}\right)\frac{\partial^2 F}{\partial \eta^2} = N_\xi - \frac{Eh}{1-v}\alpha\Delta T$$

$$\left(\frac{D}{a^2}\right)\frac{\partial^2 F}{\partial \xi^2} = N_\eta - \frac{Eh}{1-v}\alpha\Delta T \tag{5.2}$$

where E is the modulus and N_ξ, N_η are the in-plane tensile (if positive) loads per unit length due to the strain field, α is the coefficient of thermal expansion, and ΔT is the temperature rise above ambient. In other words, *the Airy stress function* (actually, its second derivatives) *directly gives the in-plane stresses*. Finally, the problem formulation is completed by the compatibility equation, which relates the in-plane stresses to the lateral displacement and takes the form:

$$\bar{\nabla}^4 F = 12(1-v^2)\left(\frac{a}{b}\right)^2 \left[\left(\frac{\partial^2 w}{\partial \xi \partial \eta}\right)^2 + 2\frac{\partial^2 w}{\partial \xi \partial \eta}\frac{\partial^2 w_o}{\partial \xi \partial \eta} - \frac{\partial^2 w}{\partial \xi^2}\frac{\partial^2 w}{\partial \eta^2} - \frac{\partial^2 w}{\partial \xi^2}\frac{\partial^2 w_o}{\partial \eta^2} - \frac{\partial^2 w}{\partial \eta^2}\frac{\partial^2 w_o}{\partial \xi^2}\right].$$
$$\tag{5.3}$$

The idea is to develop a solution for $w = w(\xi, \eta, \tau)$. In the process, we also obtain an expression for F. The in-plane slipping condition is built into the solution procedure. So it's instructive to go through the process.

The solution procedure is outlined in Ref. [8] and the references contained therein. To begin, an assumed displacement field is proposed of the form:

$$w(\xi, \eta, \tau) = \sum_i \sum_j a_{ij}(\tau)\Psi_i(\xi)\Phi_j(\eta)$$

$$w_o(\xi, \eta) = \tilde{a}_{ij}(\tau)\tilde{\Psi}_i(\xi)\tilde{\Phi}_j(\eta) \tag{5.4}$$

where Ψ_i and Φ_i are assumed mode shapes, which must be comparison function as required by Galerkin's procedure (see section 4.5.4). In other words, they must exactly satisfy the out-of-plane boundary conditions. Here, the "cosine" functions are used:

$$\Psi_i(\xi) = \cos([i-1]\pi\xi) - \cos([i+1]\pi\xi)$$

$$\Phi_j(\eta) = \cos([j-1]\pi\eta) - \cos([j+1]\pi\eta) \tag{5.5}$$

Note that the $a_{ij}(\tau)$'s are the unknowns, while the amplitude of the initial geometric imperfection $\tilde{a}_{ij}$ is assumed known.

The first step is to determine the particular solution F_p for the compatibility equation, Eq. (5.3), in terms of the unknown amplitudes a_{ij}. And then we'll get the homogeneous solution F_h before moving on to the amplitudes a_{ij}. Substituting Eqs. (5.4) into Eq. (5.3) gives

$$\bar{\nabla}^4 F = 12(1-v^2)\left(\frac{a}{b}\right)^2 \left[\sum_i \sum_j \sum_k \sum_l a_{ij}(\tau)a_{kl}(\tau) \left[\Psi_i'\Phi_j'\Psi_k'\Phi_l' - \Psi_i''\Phi_j\Psi_k\Phi_l'' \right] \right.$$

$$\left. + \sum_i \sum_j a_{ij}(\tau)\tilde{a}_{mn} \left[2\Phi_i'\Psi_j'\tilde{\Phi}_m'\tilde{\Psi}_n' - \Psi_i''\Phi_j\tilde{\Psi}_m\tilde{\Phi}_n'' - \Phi_i\Psi_j''\tilde{\Phi}_m''\tilde{\Psi}_n \right] \right] \tag{5.6}$$

where the primes denote differentiation with respect to that function's independent variable (i.e., $\Psi' = \partial\Psi/\partial\xi$ and $\Phi' = \partial\Phi/\partial\eta$). The right-hand side can be viewed as being comprised of two parts: one involving the initial imperfection and the other not. With this in mind, F_p is expanded in terms of two unknown functions F_{ijkl}^I and F_{ij}^{II}, as follows

$$F_p(\xi, \eta, \tau) = \sum_i \sum_j \sum_k \sum_l a_{ij}(\tau)a_{kl}(\tau)F_{ijkl}^I(\xi, \eta) + \sum_i \sum_j a_{ij}(\tau)\tilde{a}_{mn}F_{ij}^{II}(\xi, \eta). \tag{5.7}$$

This is used to reduce Eq. (5.6) to two fourth-order partial differential equations (PDEs) (for F_{ijkl}^I and F_{ij}^{II}), which are dependent on space only. Specifically,

$$\bar{\nabla}^4 F_{ijkl}^I = 12(1-v^2)\left(\frac{a}{b}\right)^2 \left[\Psi_i'\Phi_j'\Psi_k'\Phi_l' - \Psi_i''\Phi_j\Psi_k\Phi_l'' \right]$$

and

$$\bar{\nabla}^4 F_{ij}^{II} = 12(1-v^2)\left(\frac{a}{b}\right)^2 \left[2\Phi_i'\Psi_j'\tilde{\Phi}_m'\tilde{\Psi}_n' - \Psi_i''\Phi_j\tilde{\Psi}_m\tilde{\Phi}_n'' - \Phi_i\Psi_j''\tilde{\Phi}_m''\tilde{\Psi}_n \right]. \tag{5.8}$$

To obtain solutions to these equations, F_{ijkl}^I and F_{ij}^{II} are first expanded as series in terms of the assumed mode shapes Ψ and Φ.

$$F_{ijkl}^I = \sum_p \sum_q \mathcal{F}_{qpijkl}^I \Psi_p(\xi)\Phi_q(\eta)$$

$$F_{ij}^{II} = \sum_p \sum_q \mathcal{F}_{qpij}^{II} \Psi_p(\xi)\Phi_q(\eta). \tag{5.9}$$

These expressions can be substituted into Eq. (5.8) and Galerkin's method is applied by multiplying the resulting equations by Ψ_r, Φ_s and integrating over the domain. This results in a set of linear algebraic equations for the expansion coefficients $\mathcal{F}_{qpijkl}^I$ and $\mathcal{F}_{qpij}^{II}$, which can be solved in a straightforward manner. Once these tensor coefficients are obtained, the particular solution is complete (through Eqs. 5.7 and 5.9).

Because the particular solution was expanded in terms of the modes, it should be evident that F_p does *not* contribute at the boundaries, that is, $F^I = F^{II} = 0$ on the edges. In other words, the particular solution only defines the in-plane stress field on the inside of the domain and the boundary stresses are completely defined by the homogeneous solution F_h. The homogeneous form of Eq. (5.3) is simply $\bar{\nabla}^4 F = 0$. A solution to this equation is

$$F_h = \frac{1}{2} \left(\bar{N}_\xi \eta^2 + \bar{N}_\eta \xi^2 \right) \tag{5.10}$$

where $\bar{N}_\eta$ and $\bar{N}_\xi$ are the in-plane stresses at the edges. To compute these stresses, they are assumed to be proportional to the "average" in-plane edge displacement. This is entirely analogous to Hooke's law and, as a result, can be thought of as a distributed spring along the edge of the plate; this is how we allow slipping at the edges. Specifically,

$$\bar{N}_\xi = -\left(\frac{a}{h}\right) K_\xi \int_0^1 \Delta_\xi d\eta$$

$$\bar{N}_\eta = -\left(\frac{b}{h}\right) K_\eta \int_0^1 \Delta_\eta d\xi \tag{5.11}$$

where K_ξ and K_η are the equivalent in-plane boundary stiffnesses in the ξ and η directions, respectively. Further, Δ_ξ and Δ_η are the stretching of the plate: $\Delta_\xi = u(1, \eta) - u(0, \eta)$ and $\Delta_\eta = v(\xi, 1) - v(\xi, 0)$. Appropriate expressions for u and v come from combining the nondimensional stress–strain relations

$$\frac{D}{a^2} \frac{\partial^2 F}{\partial \xi^2} = \frac{Eh}{1 - v^2}(\epsilon_\eta + v\epsilon_\xi) - \frac{Eh}{1 - v}\alpha\Delta T$$

$$\frac{D}{b^2} \frac{\partial^2 F}{\partial \eta^2} = \frac{Eh}{1 - v^2}(\epsilon_\xi + v\epsilon_\eta) - \frac{Eh}{1 - v}\alpha\Delta T \tag{5.12}$$

with the nondimensional, nonlinear strain-displacement relations

$$\epsilon_\xi = \frac{\partial u}{\partial \xi} + \frac{1}{2}\left(\frac{h}{a}\right)^2 \left(\frac{\partial w}{\partial \xi}\right)^2 + \left(\frac{h}{a}\right)^2 \frac{\partial w}{\partial \xi} \frac{\partial w_o}{\partial \xi}$$

$$\epsilon_\eta = \frac{\partial v}{\partial \eta} + \frac{1}{2}\left(\frac{h}{b}\right)^2 \left(\frac{\partial w}{\partial \eta}\right)^2 + \left(\frac{h}{b}\right)^2 \frac{\partial w}{\partial \eta} \frac{\partial w_o}{\partial \eta}. \tag{5.13}$$

By substituting Eqs. (5.13) in Eqs. (5.12), the terms $\partial u/\partial \xi$ and $\partial v/\partial \eta$ can be isolated and integrated to obtain: $u = \int_0^1 (\partial u/\partial \xi)d\xi$ and $v = \int_0^1 (\partial v/\partial \eta)d\eta$. These expressions for u and v, along with Eqs. (5.10) and (5.11), complete the homogeneous solution F_h. The total solution to the compatibility equation (Eq. 5.3) is the Airy stress function: $F = F_p + F_h$.

The last step is to substitute our Airy stress function F and the assumed displacement field (Eq. 5.4) into the equation of motion, Eq. (5.1). Galerkin's procedure is applied by multiplying the resulting equations by $\Psi_r\Phi_s$ and integrating over the domain. This removes the spatial dependence and gives $r \times s$ nonlinear, coupled ordinary differential equations (ODEs) in the modal coefficients, $a_{ij}(\tau)$. The rsth equation takes the form

$$\sum_i \sum_j S_{ijrs}\ddot{a}_{ij} + \beta \sum_i \sum_j S_{ijrs}\dot{a}_{ij} + \sum_i \sum_j \sum_k \sum_l \sum_p \sum_q B_{ilklpqrs}a_{ij}a_{kl}a_{pq}$$

$$+ \sum_i \sum_j \sum_k \sum_l C_{ilklrs}a_{ij}a_{kl} + \sum_i \sum_j D_{ilrs}a_{ij} + E_{rs} = \Delta P_{rs} \qquad (5.14)$$

where the S coefficients are modal mass terms; β is a proportional damping coefficient; B, C, D, and E are stiffness coefficients; and ΔP is an excitation term.

Of course, the question remains: how do you quantify the boundary stiffnesses K_ξ and K_η, such that the amount of boundary slip matches up with those seen in any given experimental test rig? Assessing these values – and the initial imperfection amplitude $\tilde{a}_{mn}$ – may be done via Southwell's technique.

Southwell [9] suggested that the initial imperfection and the buckling load of a plate could be measured experimentally from load-displacement measurements in the linear range of the structure. Consider a one-mode, linear model for a heated plate. In the presence of an initial imperfection, the center deflection will grow with temperature and, as ΔT_{cr} is approached, become unbounded (in a linear sense). This behavior can be approximated by

$$a_{11} = \frac{\tilde{a}_{11}}{(1 - \Delta T/\Delta T_{cr})}. \qquad (5.15)$$

Again, a_{11} is the absolute first-mode displacement and $\tilde{a}_{11}$ is the constant initial imperfection. The displacement relative to the imperfect state is

$$a_{11}^* = a_{11} - \tilde{a}_{11}. \qquad (5.16)$$

This is actually more practical because all experimental readings will be taken from the imperfect state. Using this expression to eliminate a_{11} from Eq. (5.15) and rearranging gives

$$a_{11}^* = \Delta T_{cr} \left(\frac{a_{11}^*}{\Delta T}\right) - \tilde{a}_{11}. \qquad (5.17)$$

By applying known ΔT's and measuring the associated a_{11}^*'s, ΔT_{cr} and $\tilde{a}_{11}$ may be obtained by plotting the straight line of Eq. (5.17). Of course, this presumes that deformations are limited to the linear range of the plate. Finally, K_ξ and K_η can then be obtained by adjusting their values in the model until $\Delta T_{cr}^{exp} = \Delta T_{cr}^{theory}$. In this way, the in-plane boundary stiffness may be obtained indirectly via the buckling temperature measurement.

To demonstrate the utility of the model, consider a plate with dimensions 381 mm × 304.8 mm × 3.185 mm (15 in. × 12 in. × 0.125 in., aspect ratio = $\phi = 1.25$). It is AISI 321 stainless steel with $E = 193$ GPa, $\nu = 0.33$, $\alpha = 16.7 \times 10^{-6}$ m/m/°K, and $\rho = 8$ g/cm^3. The plate is mounted on a test rig at the thermal acoustic fatigue apparatus (TAFA) at the NASA Langley Research Center [10]; see Figure 5.3a.

At the back of the specimen area, a set of 10 quartz lamp units provide a radiant thermal load. Using all 10 lamps, this system is capable of generating a maximum heat flux of 511 kJ/(m^2 s), which would heat a thin plate to over 1095 °C. Of course, to create in-plane stresses the mounting frame must not expand with the plate (*note*: If the plate and frame expanded together, there

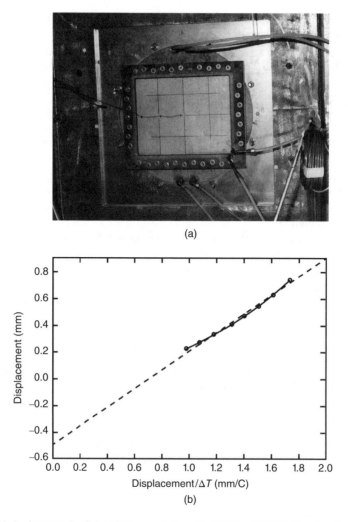

(a)

(b)

Figure 5.3 (a) A photograph of the plate mounted on the TAFA at NASA Langley Research Center and (b) the Southwell plot generated for the 3.125 mm plate. The best fit (dashed) line gives the buckling temperature and the initial imperfection amplitude

would be no in-plane stress generated). To avoid thermal expansion of the frame, several steps were taken. Min-K, an insulating blanket material, was placed on the back wall of the test rig surrounding the plate to minimize thermal conduction through the wall to the support frame. Zircar, a ceramic insulation material, was placed between the plate and the frame to minimize conduction to the frame. Finally, a water channel was mounted on the inside edge of the frame to provide continuous cooling to the frame. The plate and frame were instrumented with thermocouples and four high-temperature strain gages; see Ref. [8] for details.

The lamps were turned on low and the plate was brought up to a steady-state temperature and then the bending strain was measured. The temperature was increased and the measurement repeated. Bending strain data was converted to displacement data using one mode and the relations

$$\epsilon_\xi = -z \left(\frac{h}{a}\right) \frac{\partial^2 w}{\partial \xi^2}$$

$$\epsilon_\eta = -z \left(\frac{h}{b}\right) \frac{\partial^2 w}{\partial \eta^2} \qquad (5.18)$$

At each ΔT, four bending strains were recorded. Hence, the problem of solving for $\tilde{a}_{11}$ and ΔT_{cr} using Eq. (5.17) was overconstrained but was determined in a least-squares sense.

Figure 5.3b shows a Southwell plot for the 3.175 mm thick plate, along with a least-squares (dashed) line. On the basis of these results, the initial imperfection was $\tilde{a}_{11} = 0.038$, which corresponds to a static center deflection of 0.483 mm (i.e., 15% of the thickness). The buckling temperature is $\Delta T_{cr} = 86.1°C$. By adjusting the in-plane boundary stiffnesses in the model, we could get $\Delta T_{cr}^{exp} = \Delta T_{cr}^{theory}$. The stiffnesses that accomplish this are: $K_\xi = K_\eta = 450$. This is equivalent to having a uniformly distributed spring along the face of the plate's edge, with a stiffness of 8.225×10^{-3} (N/mm)/(mm^2). This relatively weak spring means that there is a good deal of movement at the boundary. This is likely due to the poor gripping ability of the ceramic insulative material (Zircar), which was placed between the plate and the frame to limit heat transfer to the frame.

Thus far, we've determined the model parameters. But we haven't seen if the model (using those parameters) renders meaningful results. In other words, can the model successfully predict the system behavior? To assess this, we consider the behavior of the natural frequencies as a function of temperature. To determine the frequencies experimentally, the temperature was held constant and a low level, broadband acoustic input was applied to the plate. The frequency response function (FRF) was obtained from the strain response; peaks in the FRF correspond to natural frequencies. The temperature was incremented and the process repeated. The model frequencies were obtained by finding numerically the (possibly nonlinear) equilibrium position of the plate and linearizing Eq. (5.14) about that position. The frequencies of this linearized system are found by solving the standard free vibration eigenvalue problem [8].

Figure 5.4 shows a comparison of the measured frequencies (circles) and the model frequencies (dashed line), using the values $\tilde{a}_{11} = 0.038$ and $K_\xi = K_\eta = 450$. It is clear that the model makes good quantitative predictions about the behavior of the frequencies even well into the post-buckled regime ($\Delta T / \Delta T_{cr} > 1$), where the Southwell data (upon which our values of $\tilde{a}_{11}$, K_ξ and K_η were based) was not applicable. Using the methods introduced in this section, we have shown how to estimate the amplitude of an initial imperfection in a plate and how to use that knowledge to describe the vibrational behavior of the plate when the boundaries are not quite fully clamped.

5.2   Delamination in a Composite Beam

Composite materials have become an essential part of many structures and are primarily valued for their high strength-to-weight ratio. Although advances are continually being made in material construction, failure is still an issue for composites. One of the primary failure mechanisms is delamination, whereby two or more of the composite layers separate. The results are reduced component strength and higher potential for failure. In addition, delamination failures may not be localized; once the delamination is initiated, it may propagate throughout the structure either under a static load or through a fatigue scenario.

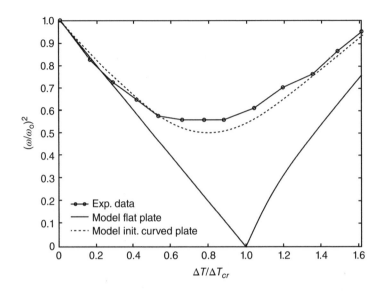

Figure 5.4 This shows the model predicted natural frequencies for the plate with $K_\xi = K_\eta = 450$. Also shown are the measured frequencies

This section is concerned with developing a model for a delaminated composite structure and validating this model with experimental data. Previous efforts have modeled the delamination as a nonlinearity in the composite structure. For example, the work of Schwarts-Givli *et al.* used finite elements to consider both contact and geometric nonlinearities in modeling damaged sandwich panels [12]. Another nonlinear model, this time for delaminated beams, was also given in Ref. [13]. In another work, Hunt *et al.* [14] considered a static, low-dimensional model of an axially compressed, delaminated strut. Under this axial load, the delamination could buckle locally, resulting in a nonlinear response.

The model developed here is along the lines of that described in Ref. [14], with a different loading geometry and with certain symmetry assumptions removed. Specifically, there is a lateral load and the delamination isn't centered on the beam but, rather, located at an arbitrary position along its length. However, there is no initial lateral deformation (curvature) in the beam, retaining all prebuckled lateral symmetry. This model is intentionally low-dimensional because (i) simpler models tend to be faster (simulation-wise) than high-dimensional finite element analysis (FEA) models and (ii) it is usually easier to attribute response phenomenon to specific physical features of the system. Best of all, this model is clearly able to capture the important physics of a delamination in a thin laminate beam subjected to an applied lateral load. In what follows we describe the model and an associated experiment aimed at verifying model predictions.

The system under consideration is shown schematically in Figure 5.5. It consists of a beam that is cantilevered at the left end ($x = 0$), with a delamination existing in the region $x_a < x < x_b$. The beam has thickness h and the delamination exists at a depth of ah from the top of the beam where $a < 1$. There is a lateral load with magnitude P applied at x_o. With zero applied load, regions 2 and 3 (the delaminated portion of the beam) are in direct contact. Regions 1, 4, and 5 are always intact. As the load is increased, the upper (lower) portion

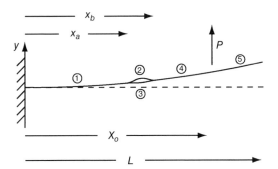

Figure 5.5 A schematic of the laterally loaded beam with a small delamination. The delamination exists in the range $x_a < x < x_b$ and has an upper portion (region 2) and a lower portion (region 3); both are exaggerated for the sake of clarity *Source*: Reproduced from [15], Figure 1, with permission of Elsevier

experiences compression (tension) until the delamination buckles at some critical load. This forms a *blister* on the surface, as shown schematically in the figure.

From a modeling perspective (see also Refs [14] and [16]), the delamination is nothing more than a site where there is no appreciable interlaminate bond, allowing separation of the two mating surfaces. The intact regions, 1, 4 and 5, are therefore modeled as linear elastic Euler–Bernoulli beams. The delaminated regions, 2 and 3, are modeled as geometrically nonlinear elastic Euler–Bernoulli beams because the buckled geometry requires substantial midline stretching, which is a nonlinear effect. The deformed equilibrium geometry of the system is obtained by minimizing the total strain energy less the external work done, $U - W$, a procedure we described in Section 4.3.4.

As a starting point, consider the undamaged beam, where all of the laminates are thoroughly bonded. Assuming that the tip deflection is not too large, the deformation of the beam may be described using the linear Euler–Bernoulli beam theory. The deflections to the left and right of the load are

$$w_L(x) = \frac{P}{6EI} (3x_o x^2 - x^3) \tag{5.19}$$

and

$$w_R(x) = \frac{Px_o^2}{6EI} (3x - x_o) \tag{5.20}$$

respectively. These describe the lateral (y direction) deformation of the neutral axis (NA). In this linear theory, there is no axial deformation (i.e., axial strain) of the NA; it is assumed that there is a linear distribution of strain through the thickness.

Now consider the beam with the delamination. For clarity, Figure 5.6a shows the geometry of the unbuckled beam at the transition from region 1 to regions 2 and 3. The upper portion has thickness ah and the lower portion has thickness $(1 - a)h$.

The deformation of each portion of the beam will be described by the deformation of its own NA. $w_i(x_i)$ is the deflection of the ith NA, where x_i is a local axial coordinate measured from the left-most end of the ith region; see Figure 5.5. The assumed deflection functions are

$$w_1(x_1) = w_L(x)$$

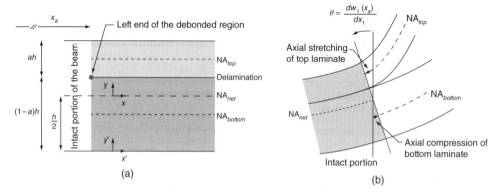

Figure 5.6 (a) A schematic showing the undeformed locations of the net neutral axis (NA) for the entire beam and the local NAs in segments 2 and 3. (b) A schematic showing the additional axial deflection of the local NAs, produced from the finite rotation at the left end of the delaminated region. A similar scenario exists on the right end *Source*: Reproduced from [15], Figure 2, with permission of Elsevier

$$w_2(x_2) = w_L(x_2 + x_a) + q_2\,\Psi(x_2) + \frac{1}{2}(1-a)h$$

$$w_3(x_3) = w_L(x_3 + x_a) + q_3\,\Psi(x_3) - \frac{ah}{2} \qquad (5.21)$$

$$w_4(x_4) = w_L(x_4 + x_b)$$

$$w_5(x_5) = w_R(x_5 + x_o)$$

where $w_L(x)$ and $w_R(x)$ are given by Eqs. (5.19) and (5.20), respectively. The function $\Psi(x)$ has been chosen to be $\Psi(x) = \cosh(\beta x) - \cos(\beta x) - \alpha(\sinh(\beta x) - \sin(\beta x))$. This is the first vibration mode shape of a clamped–clamped beam (zero displacement and zero slope at the ends) with $\beta = 4.73$ and $\alpha = 0.9825$ [17]. Of course, any function satisfying the geometric boundary conditions would be admissible. The constants q_2 and q_3 are unknown. This deformation field, Eq. (5.21), is simply that of the linear beam with a *blister shape* (Ψ) superposed in regions 2 and 3. Therefore, q_2 and q_3 are the relative amplitudes of the NAs of the upper and lower laminate. For example, if the delamination is prebuckled, $q_2 = q_3 = 0$ and the linear solution is recovered plus a static shift in regions 2 and 3, because w_2 and w_3 describe the NA of their respective portions; see Figure 5.6. If post-buckled, $q_2 \neq 0$ and $q_3 \neq 0$ and the deformation is the linear solution with a blister in regions 2 and 3 superimposed. And because Ψ has zero displacement and zero slope at the ends, geometric continuity is assured at the edges of the delaminated regions, that is, the displacement and slope are continuous at $x = x_a$ and $x = x_b$.

To determine these deformations, an expression for the strain energy must be developed (again see Section 4.3.4). Assuming that each laminate experiences pure bending about its own NA, the stretching and bending energies uncouple. In this case, the bending energy is

$$U_{bending} = \sum_{i=1}^{5} \frac{EI_i}{2} \int_0^{L_i} \left(\frac{\partial^2 w_i}{\partial x_i^2}\right)^2 dx_i \qquad (5.22)$$

where E is the elastic modulus and $I_1 = I_4 = I_5 = \frac{bh^3}{12}$, $I_2 = \frac{b(ah)^3}{12}$ and $I_3 = \frac{b((1-a)h)^3}{12}$ are the area moments of inertia. From Figure 5.5, the interval lengths are $L_1 = x_a$, $L_2 = L_3 = x_b - x_a$, $L_4 = x_o - x_b$, $L_5 = L - x_o$. There is no implied sum on the repeated index.

The linear bending theory that governs regions 1, 4, and 5 does not include axial stretching of the NAs. However, in regions 2 and 3 there will be considerable midline stretching during buckling. This axial stretching of the neutral axes is a nonlinear effect and may be expressed to first order as $u_i = \frac{1}{2} \int_0^{L_i} \left(\frac{\partial w_i}{\partial x_i} \right)^2 dx_i$. Again, there is no implied sum on the repeated index. In addition to this midline stretching, there is additional stretching/compression at the ends of this interval. This is shown in Figure 5.6b for the left end of the delamination. The nonzero slopes at the left and right ends produce additional axial deformation: stretching (compression) of the NA in region 2 (3) at the left end. The situation is reversed at the right end As such, the total axial stretching of the two NAs are

$$\delta_2 = \frac{1}{2} \int_0^{L_2} \left(\frac{\partial w_2}{\partial x_2} \right)^2 dx_2 + \frac{1}{2}(1-a)h\frac{\partial w_1}{\partial x_1}(x_a) - \frac{1}{2}(1-a)h\frac{\partial w_4}{\partial x_4}(x_b) \qquad (5.23)$$

and

$$\delta_3 = \frac{1}{2} \int_0^{L_3} \left(\frac{\partial w_3}{\partial x_3} \right)^2 dx_3 - \frac{ah}{2}\frac{\partial w_1}{\partial x_1}(x_a) + \frac{ah}{2}\frac{\partial w_4}{\partial x_4}(x_b) \qquad (5.24)$$

In addition to the traditional nonlinear stretching, given by (5.23) and (5.24), these expressions contain slope terms that arise because the centerlines of the laminates rotate (about the net NA) at the ends of the delamination region; these slope terms couple the stretching with the bending. Also, these centerlines are then presumed to act as the local NAs (with zero bending strain at the axis and a linear distribution above and below it). To highlight the need for coupling the bending and stretching, consider the *prebuckled beam* under a load P. The strain at the centerline of the upper laminate may be calculated by considering the beam to be thoroughly intact. In this case, the centerline of the upper laminate has a bending strain but no stretching strain:

$$\epsilon = -\frac{y}{\rho} = -y w_{L,xx} = -\left[\frac{h(1-a)}{2} \right] w_{L,xx} \qquad (5.25)$$

where $y = h(1-a)/2$ is the distance from the net NA of the entire beam to the centerline of the upper laminate and $1/\rho = \kappa = w_{L,xx}$ is the curvature of the intact beam.

Alternatively, the strain could be found using the present model, where the upper laminate is presumed to bend about its own centerline. One may be tempted to say that the strain is zero: $\epsilon = -y/\rho = 0$, because $y = 0$ at the centerline of the upper laminate. This result, however, does not be consistent with the intact model, Eq. (5.25). The remedy is to recognize that the bending and stretching are not decoupled. The total strain at the centerline of the upper laminate is

$$\epsilon = -\frac{y}{\rho} + \frac{\delta_2}{x_b - x_a} \qquad (5.26)$$

The first term is zero, because $y = 0$ at the centerline. But the second term is nonzero (even in this prebuckled state) because of the nonzero rotations at the end of the delamination interval, as shown in Figure 5.6. It is relatively easy to show that Eqs. (5.25) and (5.26) yield the same strain. This helps validate the present model and underscores the importance of these coupling terms.

With these expressions, the stretching energy becomes

$$U_s = \frac{1}{2}K_2\delta_2^2 + \frac{1}{2}K_3\delta_3^2 \tag{5.27}$$

where $K_2 = \frac{(ah)bE}{L_2}$ and $K_3 = \frac{(1-a)hbE}{L_3}$.

The total strain energy of the system is $U = U_b + U_s$. Equilibrium may be determined by minimizing the difference between the strain energy and the external work done by P with respect to our unknowns q_2 and q_3:

$$\frac{\partial[U - Pw_5(0)]}{\partial q_2} = 0 \tag{5.28}$$

and

$$\frac{\partial[U - Pw_5(0)]}{\partial q_3} = 0. \tag{5.29}$$

This renders a set of nonlinear, algebraic equations that may be solved numerically, using a multidimensional Newton–Raphson algorithm, for the unknowns q_2 and q_3. In a moment, these buckling amplitudes are examined as a function of the applied load P.

To test the model predictions, an experimental system that allows the strain versus load relationship for a delaminated composite beam to be measured was developed. The test setup is shown schematically in Figure 5.7. The beam consisted of 16 plies of carbon-fiber cross-ply laminates, each one measuring 250 mm in length and 60 mm in width. The total thickness of the beam is 2.25 mm. A delamination two plies deep ($a = 0.125$) and 20 mm long was built into the beam during construction. The delaminated portion spanned the entire width of the beam.

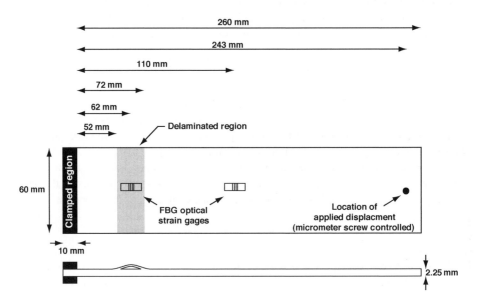

Figure 5.7 A schematic of the experimental setup *Source*: Reproduced from [15], Figure 3, with permission of Elsevier

The strain response was measured at two locations on the top surface of the beam using two fiber Bragg grating (FBG) strain sensors. One sensor was in the center of the delamination and the other was 48 mm away, toward the free end. The optical hardware used to demodulate the FBG signals (i.e., recover the strain-induced optical wavelength shifts) along with the associated data acquisition system are described in Ref. [18]. All data were acquired at a rate of 2 kHz for roughly 2.5 s (5000 samples) and then averaged to give a single strain value. The data were not filtered.

The beam was loaded using a screw-controlled micrometer, located 243 mm from the clamped end. This allowed the user to "dial in" a prescribed displacement. Because the objective was to examine the strain-load response, a load cell was placed between the micrometer and the beam such that the applied load could be monitored. A picture of the entire experimental setup is shown in Figure 5.8a. Also shown in Figure 5.8b is a close-up picture of the delaminated portion of the beam under loading. The top two plies can clearly be seen separating from the rest of the beam. Visually, this is the same behavior shown in Figure 5.5 and it will be demonstrated that the experimentally measured strain response associated with the delamination also matches the proposed model.

The first test focused on measuring the modulus E, because this value is not otherwise available (because of anisotropy in each ply, the modulus will depend on the particular lay-up of the beam). The test involved applying several known displacements, dialed in via the micrometer, and measuring the load P. The load versus deflection relationship for an isotropic, linear elastic solid is: $P = \left[\frac{3EI}{L^3}\right]\delta$ where δ is the displacement. Using the measured data with this relationship permits an estimate of an equivalent E. This value is used in the model.

The second set of tests involved using the micrometer to dial in a displacement, measuring the applied load, and recording the strain on the top surface of the delamination. Strain data were recorded for increasing loads and decreasing loads well into the buckled region.

Figure 5.9 shows the lateral deflection centerline of the upper laminate at the center of the delamination: $w_2(x_2 = (x_b - x_a)/2)$. At small loads, the delamination is not buckled ($q_2 = 0$) and the deformation is simply $w_2 = w_L((x_a + x_b)/2) + \frac{1}{2}(1 - a)h$, which is linear in P. As the lateral load is increased beyond 7N, the upper laminate buckles and q_2 becomes nonzero. This is clearly shown in the inset. The net deflection w_2 jumps suddenly at this load as

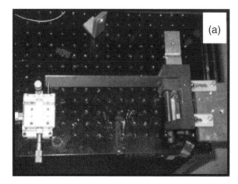
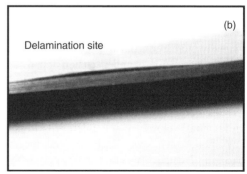

Delamination site

Figure 5.8 (a) The experimental setup and (b) a close up of the delaminated beam *Source*: Reproduced from [15], Figure 4, with permission of Elsevier

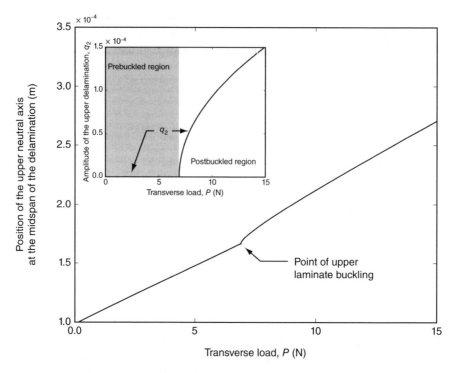

Figure 5.9 The position of the centerline of the upper laminate at the midspan of the delamination as a function of the applied lateral load. The deflection is linear with the load P (see Eq. 5.19) until the upper laminate buckles and q_2 becomes nonzero; see the insert *Source*: Reproduced from [15], Figure 5, with permission of Elsevier

buckling occurs. Note that the symmetry in the relative displacement (q_2) is not immediately evident in the absolute displacement (w_2). As the load is increased still further, the deflection continues to grow, although not in a linear manner.

Figure 5.10 shows the static strain response of the delamination ϵ as a function of the load P. The experimental data show that as the load is increased (squares), the strain becomes increasingly negative. This says that as the load increases, the upper portion of the beam experiences more and more compressive strain (and, hence, compressive stress). Finally, as the load is gradually increased from 7 to 8.3 N, the delamination buckles upward. Once buckled, the top of the delamination is under tension and, as a result, the strain jumps to a positive value. As the load is increased further, the tensile strain continues to increase.

As the load is decreased quasi-statically (triangles), the process is undone. The tensile strain decreases until the delamination unbuckles; region 2 of the beam comes back into contact with region 3. However, it is interesting to note that this transition occurs between 6.85 and 5.6 N, that is, at a lower load level than the snap-to-buckling that occurred in the increasing load data. This clearly demonstrates that there is hysteresis – a direction dependence – to the buckling/unbuckling phenomenon.

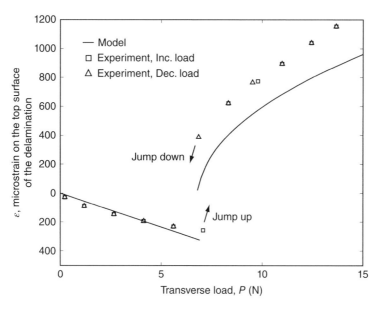

Figure 5.10 The strain response of the upper laminate as a function of the applied lateral load P. The model is given by the solid line and the squares and triangles are the experimental data for increasing and decreasing load, respectively *Source*: Reproduced from [15], Figure 6, with permission of Elsevier

The analytical results show similar behavior. At low loads, the strain is negative and the delamination is unbuckled. Agreement in this regime is excellent. At larger loads, the model predicts buckling of the delamination through a pitchfork bifurcation in q_2. The strains jump suddenly from negative (compressive) to positive (tensile) as the laminate snaps through. However, the postbuckled strain values are somewhat lower than those observed experimentally, that is, the model is stiffer than the experiment. This could occur for a variety of reasons. For example, the experimental strains reported are not actually point-strains but represent an averaged strain over the length of the sensor. Also, with the anisotropic nature of the material – and the fact that the modulus was measured only in the prebuckled region – the assumption of an equivalent isotropic modulus (from prebuckled data) may not be entirely valid. In addition, the model only includes bending and stretching energies while ignoring shear contributions to the energy, rendering the model more stiff. Nonetheless, the model does a reasonable job of capturing the response both quantitatively and qualitatively.

A noticeable shortcoming of the present model is that it does *not* display the hysteresis that was observed experimentally. This is explained through symmetry. Consider a flat plate under an axial compressive load. A perfectly flat plate will buckle via a pitchfork bifurcation, which is symmetric and without hysteresis. If the plate has an imperfection (say, a small initial curvature), then the pitchfork is replaced with a saddle-node bifurcation, which experiences hysteresis in the displacement-load curve [19]. In the present model, all of the five elements are assumed to be perfectly flat when unloaded. This is, of course, not physically realistic for the experiment. Consequently, the addition of a small geometric imperfection would break the symmetry of the system and, in all likelihood, reproduce the hysteresis observed experimentally.

This example clearly shows the power of understanding the mechanics of structural damage. The model used in this example possesses just two DOFs (q_2 and q_3) and is reasonably easy to implement, yet clearly captures the physics of the damage. Moreover, because of its low dimensionality, the model is computationally fast; the one thousand points used to create the analytical curves in Figure 5.10 were generated almost instantaneously via the Newton–Raphson algorithm. The true benefit of such models is that they allow the practitioner to focus on the properties of the structural response that are *unique to the damage*. As we stressed in the introduction, it is *only* through these types of modeling efforts that one is likely to develop a system capable of differentiating damage-induced behavior from other sources of variability in observed response data. This particular model has shown that delamination produces both quadratic and cubic nonlinearities local to the damage area. In Chapter 8 we return to this model and the problem of detecting the damage using vibrational response data.

5.3 Bolted Joint Degradation: Quasi-static Approach

One of the more common failure mechanisms for structural systems is a loss of connectivity of threaded fastener assemblies. Threaded fasteners are popular because of advantages such as the ability to develop a clamping force and the ease with which they may be disassembled for maintenance. It has been well-documented that such fasteners loosen under shock, vibration, or thermal loading (as we just saw in section 5.1.2), and a recent comprehensive discussion of these effects is given in Ref. [20]. In fact, a combined finite element and experimental study of dynamic shear loading-induced loosening and showed that the minimum load required to initiate loosening is lower than previously reported [21].

Because of the highly localized nature of bolt loosening and failure, most approaches in this field have involved two- and three-dimensional finite element formulations, for example, [22]. These approaches are well-suited to studying the fundamental nature of the problem and guiding the design process. From a vibration domain structural monitoring perspective, the question is whether these loosening effects may be detected indirectly through vibration measurements and modal analysis of the vibration. In vibration-based SHM, the majority of the literature has, in fact, considered features derived from a modal analysis of the structure, for example, resonant frequency shifts, mode shape shifts, modal damping changes, flexibility, and so on. The literature expanse in this area is too vast to cite here, even representatively. A relatively recent good summary of various applications of modal-based methods to a wide variety of structural assessment scenarios may be found in Ref. [23].

In this section, we construct a model and a corresponding experiment to determine whether modal analysis is an appropriate tool for such a problem. A beam is bolted to supports at its boundaries, but springs encase the bolts such that greater control over the clamping force is retained for study purposes. A transition from fully clamped to fully free is implemented at one boundary, and a modal analysis is performed at various steps along the way. We show that resonant frequencies and mode shapes are relatively insensitive to clamping force changes over wide ranges with a narrow region of sudden transition. This lack of "smoothness" in transition suggests that modal properties may not be ideal features to use in bolted joint health monitoring, especially from a prognostic point of view.

5.3.1   The Model

5.3.1.1   Elastic Boundary Constraints

The overall system to be modeled is a thin aluminum beam. In Section 4.3.5 (see Eq. 4.28), we derived the equations of motion for such a beam using energy methods. The resulting equations can be nondimensionalized, giving an expression for the free vibration of the beam

$$w_{xxxx} + w_{tt} = 0 \qquad (5.30)$$

where $w(x, t)$ is the vertical displacement nondimensionalized by beam length L, x is the axial position coordinate along the beam nondimensionalized by beam length L, and t is time nondimensionalized by $\sqrt{\rho A L^4 / EI}$, where ρ is the beam mass per unit volume, A is the cross-sectional area, E is Young's modulus, and I is the area moment of inertia. All properties are assumed uniform along the beam.

For a beam whose edges are fastened by bolts, the exact boundary conditions are very complicated, as we discussed, for example, in section 5.1.2. The main purpose in developing a model in this section is to capture the simplest relevant behavior describing the loss of clamping force on the beam boundary as the fastener loosens, without regard to the exact mechanism of loosening or localized details of the process evolution. We propose that a quasi-linear model utilizing generalized elastic boundary constraints captures this behavior and matches well with experimental observations (to be presented later). Indeed, elastic constraint boundary conditions could be construed as the most general linear boundary condition, as specific familiar boundary conditions such as "clamped" or "free" may be easily derived from limiting cases of these general elastic constraints. Variations on elastic-type formulations have been used in the literature previously for other purposes, for example, [24]. Figure 5.11 shows the boundary of a beam subject to general linear elastic constrains in shear, denoted by the K_V spring, and in moment, denoted by the rotational K_M spring.

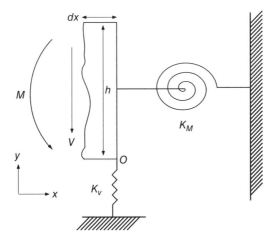

Figure 5.11   Elastic edge constraints for a Euler–Bernoulli beam *Source*: Reproduced from [25], Figure 1, with permission of Elsevier

A static moment and y-direction force balance acting on the typical edge element shown is

$$\sum F_{y\text{-}direction} = 0 = -V + K_V Lw$$

$$\sum M_O = 0 = M - K_M w_x - VL \, dx \tag{5.31}$$

where forces in moments in other planes are ignored and the usual linearity assumptions implicit in the Euler–Bernoulli theory are maintained for model consistency. As $dx \to 0$ in the limit, the general boundary conditions for the moment M and shear force V, applicable at the boundaries, are thus

$$M = K_M w_x$$

$$V = -K_V Lw \tag{5.32}$$

For Euler–Bernoulli beams, the internal bending moment and shear force are given by

$$M = -\frac{EI}{L} w_{xx}$$

$$V = -\frac{EI}{L^2} w_{xxx}. \tag{5.33}$$

Equating Eqs. (5.32) and (5.33), the elastic boundary constraints may be reexpressed as

$$w_{xx} = -\alpha w_x$$

$$w_{xxx} = \beta w \tag{5.34}$$

where $\alpha = K_M L/EI$ and $\beta = K_V L^3/EI$. As both K_M and K_V approach zero in the limiting case $(\alpha, \beta \to 0)$, the classical free edge boundary condition is obtained. As these values approach infinity $(\alpha, \beta \to \infty)$, the classical clamped edge boundary condition is obtained.

The values of α and β in a purely linear formulation would be constants, as they reflect the stiffness values of the respective springs. Bursi and Jaspart [22] showed how the clamping force and clamping moment present in a fastener are strongly nonlinear functions of fastener axial strain during transition from the elastic to the inelastic regime. The slopes of the force/displacement and moment/rotation curves presented reflect the effective stiffness present, and the dominant characteristic of these curves is that the slope is initially extremely large and goes to zero as the forces and moments saturate during initial plastification. This nonlinearity may be readily incorporated into the current model by allowing α and β to be modified by an appropriate function that trends to reflect the elastic to plastic transition of a fastener assembly. We propose that this function, denoted by K, be of the form

$$K(f) = \tanh\left(\kappa(1-f)\right)\left(p + (1-p)\tanh\left(\kappa\frac{f}{1-f}\right)\right) \tag{5.35}$$

where κ and p are tuning parameters, and f is a measure of the ratio of the clamping force (or moment) to the maximum clamping force (or moment). A series of these functions for $p = 2$ and various κ is shown in Figure 5.12. The p parameter primarily adjusts the magnitude of the stiffness function at $f = 1$, which corresponds to the maximum clamping force or moment possible (both theoretically infinite in the idealized "fully clamped" boundary condition). The κ parameter adjusts the rate at which the stiffness changes, particularly at extremes

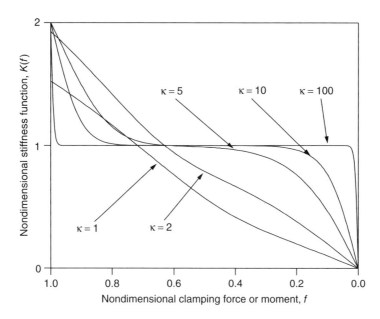

Figure 5.12   Proposed nonlinear stiffness function $K(f)$ used to describe the clamping force transition
Source: Reproduced from [25], Figure 2, with permission of Elsevier

in clamping force near the minimum ($f = 0$) or the maximum ($f = 1$). It should be noted that the horizontal axis is plotted in reverse, to indicate that normally a fastened joint begins life at maximum clamping force and moment and loses that functionality over time, as discussed previously.

It is also noted that the stiffness function, particularly at larger κ, is flat for a large spectrum of nonextreme clamping loads. In the experiment to be discussed shortly, springs were used in place of fasteners so that better control of the clamping force was retained. The springs themselves are fairly linear over a wide region of compression, but as the extremes of a fully compressed state or a fully noncompressed state are approached, the springs behave nonlinearly with rapid effective stiffness changes over a very short distance. This behavior exactly mimics the function K, particularly for larger κ: the springs behave linearly over a wide region, ($K = 1$ in that region), with rapid effective stiffness increases and decreases at full compression and no compression. For perfectly idealized linear springs, K would be unity for all clamping forces or moments. Thus, the expressions for the boundary conditions in Equations (5.34) could be modified to give

$$w_{xx} = -K_\alpha \alpha w_x$$

$$w_{xxx} = K_\beta \beta w \tag{5.36}$$

where the nominal stiffnesses α and β have been modified by corresponding functions K_α and K_β to reflect the nonlinear stiffness effects. In this reformulation, α and β are now interpreted as constants that scale the flat region of K, and all stiffness variation is contained within K. These boundary conditions, with the new nonlinear stiffness function, are now not easily satisfied, because the function K varies as w changes during vibration (i.e., the nonlinear stiffness function depends on the displacement w itself). However, if the vibrations are assumed small, then no appreciable stiffness change occurs during the vibration except at the very extremes of the spring compression range. Thus, we have a "quasi-linear" boundary condition in the sense that for a given initial, static spring compression, K is assumed constant.

5.3.1.2   Solution Results

Now that the boundary condition model has been established, the problem may be solved by usual modal expansion methods (see section 4.5.1). In the experiment, the springs were controlled at only the $x = 1$ boundary, the springs at $x = 0$ were left fully compressed such that the idealized clamped boundary condition applies. Assuming that $w(x, t) = \phi(x)e^{ik^2 t}$, Eqs. (5.30) and (5.36) are reduced to the eigenvalue problem

$$\phi_{xxxx} = k^4 \phi$$

$$\phi(0) = 0$$

$$\phi_x(0) = 0$$

$$\phi_{xx}(1) = -K_\alpha \alpha \phi_x(1)$$

$$\phi_{xxx}(1) = K_\beta \beta \phi_{xx}(1) \tag{5.37}$$

The solution to this problem is

$$\phi(x) = \sin kx - \sinh kx + C(\cos kx - \cosh kx)$$

$$C = \frac{k^3(\cos k + \cosh k) + K_\beta \beta(\sin k - \sinh k)}{k^3(\sin k - \sinh k) + K_\beta \beta(\cosh k - \cos k)} \tag{5.38}$$

where the values of k are taken to be the roots of the characteristic equation

$$K_\alpha \alpha K_\beta \beta + k^4 + k(K_\beta \beta - k^2 K_\alpha \alpha) \cos k \sinh k$$

$$+ \cosh k((k^4 - K_\alpha \alpha K_\beta \beta) \cos k + k(K_\beta \beta + k^2 K_\alpha \alpha) \sin k) = 0. \tag{5.39}$$

It is readily shown that under the idealized limiting cases of fully clamped ($K_\alpha \alpha, K_\beta \beta \to \infty$) and fully free ($K_\alpha \alpha, K_\beta \beta \to 0$) boundary constraints, the characteristic equation reduces to the familiar expressions

$$1 - \cos k \cosh k = 0$$

$$1 + \cos k \cosh k = 0 \tag{5.40}$$

respectively.

5.3.2   *Experimental System and Procedure*

A physical representation of the model is shown in Figure 5.13. An aluminum beam measuring 5×10^{-1} m in length, 5×10^{-2} m in width, and 3.175×10^{-3} m in thickness was attached to rigid supports at both ends with a bolt-and-spring assembly, as shown. Each spring had a

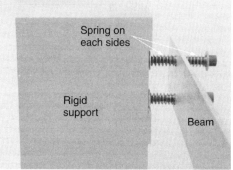

Figure 5.13 Experimental setup for the experiment used in testing model predictions

linear spring constant of 64.6 N/m and relaxed length of 2.0 cm. Four nominally identical spring-and-bolt assemblies were used at each boundary of the beam. On one boundary, all the springs were tightened to a fully compressed condition (nominal spring length of 0.55 cm) such that the idealized clamped boundary condition was realized. Tightening or loosening the bolts on the other boundary results in compression or elongation of the springs on both sides of the beam and thus serves as a means of controlling the clamping force. Thirteen different compression lengths (clamping force levels) were tested, from the fully compressed state to a final idealized "free" condition, where the springs and bolt were completely removed.

Impact modal testing was performed at each clamping force level, using a multiple-input/single-output (MISO) approach. The beam was divided into equal segments of 5 cm giving a total of 10 points, including the spring-constrained end. The eighth point, located 40 cm from the clamped end, was used to measure the beam's response to impact excitation, while the remaining nine points served as the locations at which the impact was applied. The response location was chosen so as to avoid taking measurements at a node for the first five predicted modes. Excitation was provided by means of an Endevco modal hammer, and response data were acquired using an Ometron VH300+™ single-point laser Doppler vibrometer (LDV). Both the excitation and response signals were recorded simultaneously using the PULSE™ multianalyzer system from Bruel and Kjaer at a sampling rate of 2048 Hz. The beam was struck 10 times at each of the impact locations, and the results were averaged in PULSE to obtain one frequency response function (FRF) using the H_1 estimator for each of the nine excitation/response pairs. As a result of the averaging process, the bandwidth of the sampled data was reduced to 800 Hz. This testing procedure was repeated at each of the clamping force levels for a total of $13 \times 9 = 117$ frequency response measurements. For each FRF, a peak detection algorithm was used to track the first five resonant frequencies. Although frequency information alone could have been extracted with a single-input/single-output (SISO) approach, the current method is required for obtaining mode shape data. Mode shape estimates were obtained by means of an eigensystem realization algorithm [26]. In this procedure, each of the nine FRFs was used to build a state-space model of the dynamics, and the eigenvectors of the state matrix may be scaled to obtain the mode shape estimates. In what follows, the mode shapes were scaled such that the maximum nodal excursion was unity.

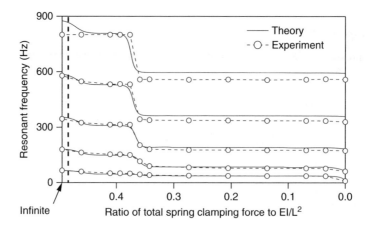

Figure 5.14 The trend in the first five identified resonant frequencies as the clamping force is varied (theory and experiment) *Source*: Reproduced from [25], Figure 4, with permission of Elsevier

5.3.3   Results and Discussion

The first five resonances identified by this experimental procedure, along with the corresponding resonances predicted by the model, are shown in Figure 5.14 under decreasing clamping force. Similarly, the first five mode shapes are shown in Figures 5.15–5.19. For the springs and beam used, $\beta = 1.71 \times 10^1$. Little moment resistance was observed by the bolt-and-spring assembly except near the full compression state, so α was made arbitrarily small (order 10^{-7}); in this way, the moment resistance effects will only be significant near the fully compressed state. For both stiffness functions K_α and K_β, the tuning parameters were chosen to be $p = 2 \times 10^8$ and $\kappa = 25$ to give a large clamping force and moment at the full compression level and considerably rapid transition at the stiffness "boundary layer" extremes (full compression and no compression, or no springs at all).

The experimental and predicted frequency and mode shape data match very well across the full clamping force range. Figure 5.14 is plotted with decreasing nondimensionalized clamping force from left to right, again to imply the normal transition of fastener joints from maximum clamping to progressively loosened conditions. The lack of agreement in the fifth resonance at large clamping loads is due to hardware limitations; the 800 Hz Nyquist frequency in the experiment was not sufficient to identify that resonance at the higher clamping levels. As the clamping loads are reduced, the frequencies do not change significantly and then rapidly drop as a critical clamping load is reached. After further load reduction, the frequencies again do not change appreciably, even after the clamping load is completely removed (a free boundary condition). Very similar behavior may be observed in all the identified mode shapes, where again the fifth mode could not be identified at larger clamping forces because of Nyquist limitations. In the mode figures, the clamping force level is indicated by a number in the lower left corner, with "1" meaning fully clamped and "13" meaning fully free.

The sudden transition of the modal parameters at some critical clamping force, with relatively insensitive fluctuation both pre- and post-critical, suggests that resonances and mode shapes may not be ideal candidates for vibration-based SHM features. Ideal features in this application should typically track linearly or nearly so with the damage scenario

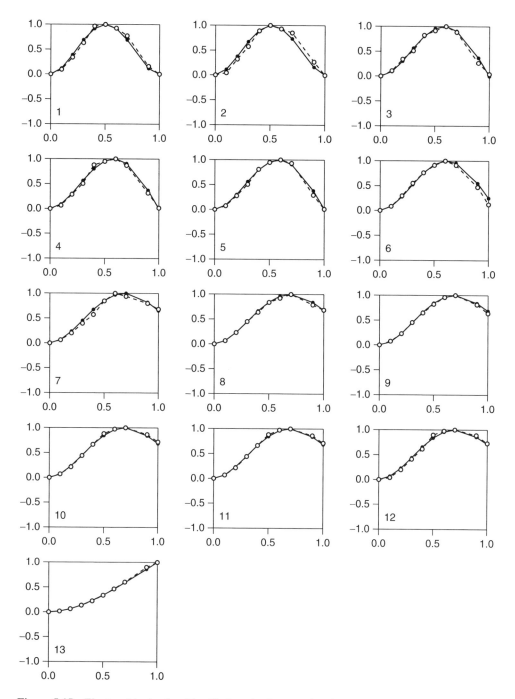

Figure 5.15 The trend in the first identified mode shape as the clamping force is varied (solid lines with filled circles is theory and dashed lines with unfilled circles is experiment) *Source:* Reproduced from [25], Figure 5, with permission of Elsevier

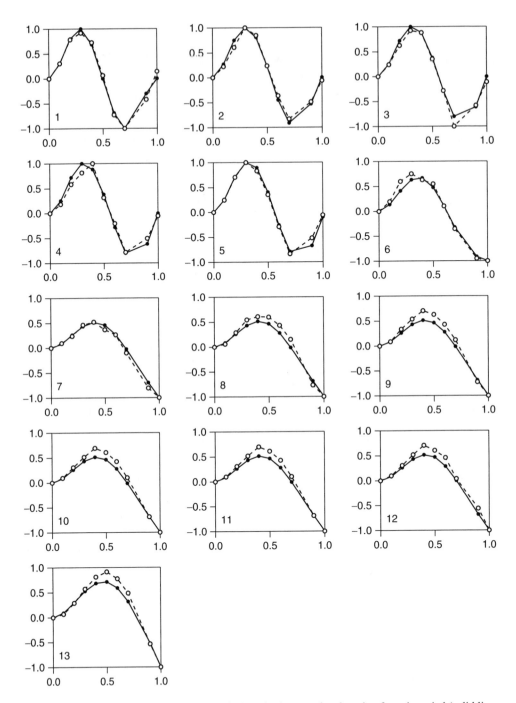

Figure 5.16 The trend in the second identified mode shape as the clamping force is varied (solid lines with filled circles is theory and dashed lines with unfilled circles is experiment) *Source*: Reproduced from [25], Figure 6, with permission of Elsevier

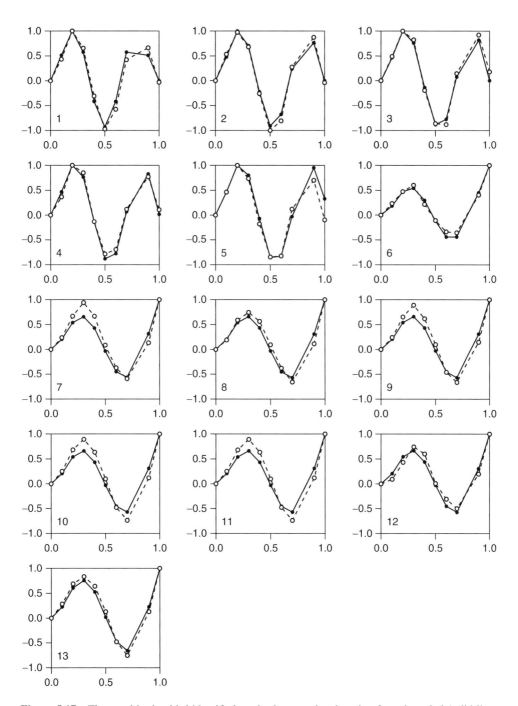

Figure 5.17 The trend in the third identified mode shape as the clamping force is varied (solid lines with filled circles is theory and dashed lines with unfilled circles is experiment) *Source*: Reproduced from [25], Figure 7, with permission of Elsevier

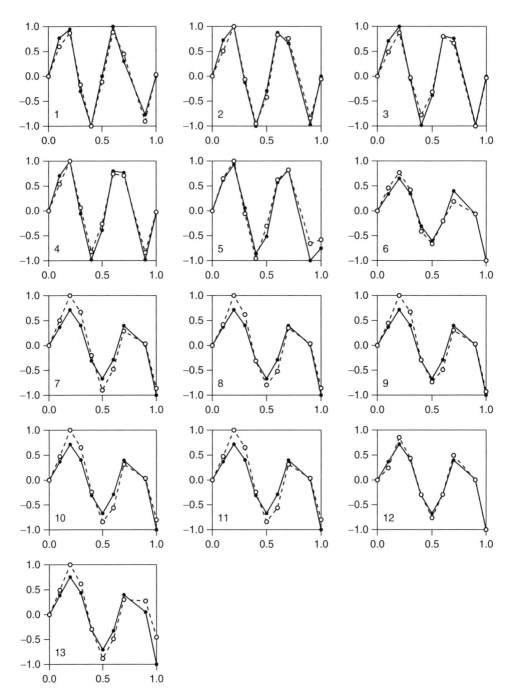

Figure 5.18   The trend in the fourth identified mode shape as the clamping force is varied (solid lines with filled circles is theory and dashed lines with unfilled circles is experiment) *Source*: Reproduced from [25], Figure 8, with permission of Elsevier

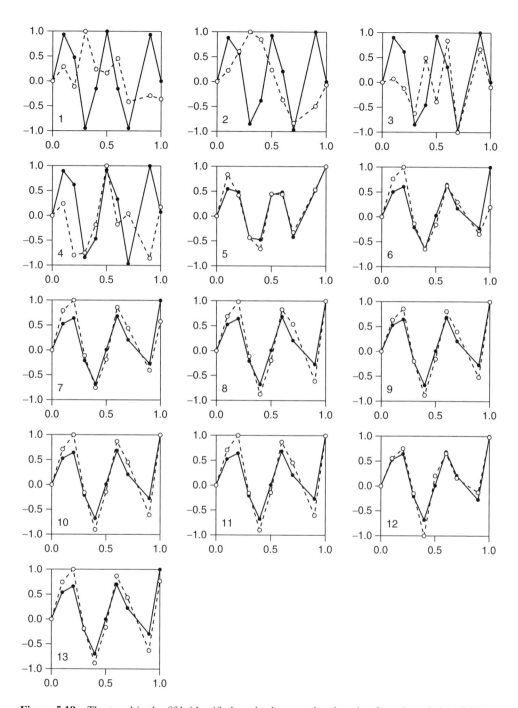

Figure 5.19 The trend in the fifth identified mode shape as the clamping force is varied (solid lines with filled circles is theory and dashed lines with unfilled circles is experiment) *Source*: Reproduced from [25], Figure 9, with permission of Elsevier

(clamping force loss in this study), so that the inverse problem (using the modal properties to classify the joint clamp force) is tractable. The resonant frequencies and mode shapes at the fully clamped condition are sufficiently close to the same properties measured just before the large shift, meaning that there is little ability to classify the data according to the clamping force. This would be especially true in a less controlled environment, where the modal properties may be significantly more "noisy" and less robustly identifiable. Finally, the existence of a sudden shift in the modal properties leaves little recourse for building in safety factors into any prognostic capability that may be developed.

5.4   Bolted Joint Degradation: Dynamic Approach

The approach we just described was quasi-static in the sense that the analysis linearized about local regions of a globally nonlinear stiffness function located at the joint. The benefit of this approach is that it afforded us analytical solutions that could be used to track modal properties as a function of the clamping force. However, we may also consider a loosened joint as a fully nonlinear vibrations problem. The downside to such an approach is that we will have to resort to numerical modeling in place of our analytical solution. On the positive side, such an approach allows us to consider the more practical situation whereby the beam is subject to random vibrations.

Here we construct a finite element model beam with clearance nonlinearity at one boundary (the local discretization approach mentioned in section 4.5.3). Figure 5.20 shows a schematic of the model, with the numbering of the nodes and the DOFs. The structure is discretized in $n - 1$ beam plane elements, resulting in n nodes and $2n$ DOFs, because every node admits a translation and a rotation. An odd number of nodes can be chosen, so that there is a node at the midspan that coincides with the location of the external force.

The equations of motion for this finite element system can be written in terms of the node displacement vector $\mathbf{x}$ as

$$\mathbf{M}\ddot{\mathbf{x}} + \mathbf{C}\dot{\mathbf{x}} + \mathbf{K}\mathbf{x} + k_c \left(x_{end} - sign(x_{end}) \frac{\delta_g}{2} \right) \mathbf{a} = \mathbf{F} \qquad (5.41)$$

where we have that

- $\mathbf{x}$ are the displacements, organized in a $2n \times 1$ vector
- $\mathbf{M}, \quad \mathbf{C}, \quad \mathbf{K}$ are the $2n \times 2n$ mass, damping, and stiffness matrices
- x_{end} is the translational DOF at the right end, equal to x_{2n-1}
- δ_g is the size of the gap nonlinearity (see Fig. 5.20) centered about the undeformed beam configuration
- k_c is a constant, assuming a value of 0 if $|x_{end}| < \delta_g/2$, and a large value for $|x_{end}| > \delta_g/2$
- $\mathbf{a}$ is $2n \times 1$ vector consisting of all zeros and a single "1" at the end node and is simply used to select the DOF at which the gap is located
- $\mathbf{F}$ is a $2n \times 1$ vector of the external forces

As in the previous section, the only difference between this and a standard Euler beam model is the manner in which the "loose" boundary condition is implemented. Rather than consider a smooth function as before, the boundary in this model is implemented as a true discontinuity in

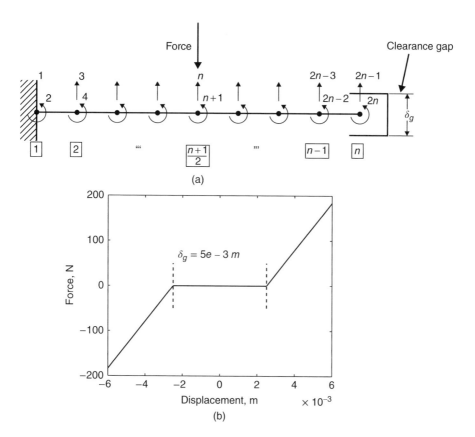

Figure 5.20 Finite element beam model (a) along with the stiffness profile assigned the end element. The stiffness discontinuity coincides with the quasi-static model of the previous section with $\kappa \to \infty$

Source: Adapted from [27], Figure 1, reproduced with permission of SAGE

stiffness. Specifically, there exists a "dead-band" in displacement over which the end boundary condition is free, that is, there is no constraint on displacement. Outside of this band there exists a positive stiffness of a very large value, simulating a nearly clamped boundary outside the dead-band region. The idea behind such a model is to capture the general behavior, whereby the end of the beam reaches the bolt head followed by free downward motion until an impact with the lower portion of the clamped structure occurs. This stiffness behavior is captured in Figure 5.20, which shows the force/displacement relationship for a beam with a gap between bolt and beam of $\delta_g = 0.005\,m$. We note that this is not a collision model. If that were the case, the force/displacement relationship would have infinite slope outside the dead-band region and we would have to use some form of impact model to describe the collision between beam and bolt head.

The mass, damping, and stiffness matrices are formed using standard Euler–Bernoulli beam elements. We further prescribe the modal damping model of Eq. (4.25). Recall that the motivation behind this approach was to diagonalize the damping matrix to decompose the MDOF problem into a sequence of SDOF problems. In the current, nonlinear example, we have to rely on numerical integration and the need for this specific form is of diminished importance. However, the idea of applying a standard damping factor $2\zeta\omega_j$ proportional to the "jth" frequency is still a good one. Given this functional form for the damping, we can write the classical (undiagonalized) damping matrix

$$C = M\left(\sum_{j=1}^{2n} 2\zeta\omega_j u_j u_j^T\right) M \tag{5.42}$$

where u_j are the eigenvectors of the matrix $M^{-1}K$ and ω_j are the associated natural frequencies (see the eigenvalue problem, Eqn. 4.51). The dimensionless damping ratio ζ is assumed constant for all modes in what follows. In this particular version of the model, the forcing is applied at the midpoint of the beam and consists of a filtered, independent, identically distributed (i.i.d.) sequence of Gaussian distributed values.

Simulating the Eqs. (5.41) can be challenging as most structural systems are referred to as "stiff" systems. As we know from our modeling, a stiff system will possess high natural frequencies, implying that fast time scales are required to capture the response. If the time step used in an explicit numerical integration scheme (e.g., the familiar Runge–Kutta algorithm) is too large, numerical instabilities can cause the solution to diverge from the true solution, that is, the solution will often "blow up." A brute force approach is to retain an explicit solver, but use a small time step, Δ_t. This comes at the cost of having to integrate for a very long time to cover a meaningful sampling period.

Fortunately, alternative integration schemes exist. It is common practice in simulating structural systems to turn to *implicit* integration methods; to this end, a number of possibilities are available. We have found that using a constant (possibly large) Δ_t, together with a Newmark integration scheme, produces excellent results. For the Newmark scheme, an effective dynamic stiffness K^* and an effective dynamic load F^* can be defined as

$$K^* = \frac{1}{\alpha\Delta_t^2}M + \frac{\beta}{\alpha\Delta_t}C + K$$

$$F^* = F_{ext} + M\left(\frac{1}{\alpha\Delta_t^2}x_i + \frac{1}{\alpha\Delta_t}\dot{x}_i + \left(\frac{1}{2\alpha} - 1\right)\ddot{x}_i\right)$$

$$+ C\left(\frac{\beta}{\alpha\Delta_t}x_i + \left(\frac{\beta}{\alpha} - 1\right)\dot{x}_i + \left(\frac{\beta}{\alpha} - 2\right)\frac{\Delta_t}{2}\ddot{x}_i\right) \tag{5.43}$$

where

$$F_{ext} = F - k_c\left(x_{end} - \text{sign}(x_{end})\frac{\delta_g}{2}\right)a \tag{5.44}$$

and with the subscript i denoting the discrete time index, for example, $t = i\Delta_t$. Herein, the values $\alpha = 1/4$ and $\beta = 1/2$, corresponding to constant mean acceleration, are used. Using this approach, the acceleration and velocity are readily computed at each time step. The solution marches in time, solving the linear system

$$K^* x_{i+1} = F^* \tag{5.45}$$

Furthermore, because the effects of the variable constraint appear in the external load, as in Eq. (5.44), the matrix K^* is constant. The matrix inversion required of Eq. (5.45) is therefore performed only once at discrete time $i = 0$, thus the cost of the algorithm at each subsequent time step is almost entirely due to the matrix multiplication $\mathbf{K}^{*-1}\mathbf{F}^*$.

It is also worth remembering that when simulating the response of a structure to stochastic loading we have to be careful about how the excitation is handled. Often, it is the case that the input loading is specified via power spectral density (PSD) function $S_{FF}(f)$ in units of N^2/Hz. Our job is to simulate an input signal with units N that is consistent with this PSD. Because we are simulating the response to a stochastic input, our system of equations (5.41) is technically a stochastic differential equation (SDE). The theory behind SDEs is challenging and we do not attempt to cover it here. Rather, we give a prescription for designing the input loading based on a specific type of input, namely, a loading time series that is a jointly Gaussian random process. Assume that each member of the loading signal is drawn independently from a Gaussian distribution with zero mean and unit variance, that is, $F(i) \sim \mathcal{N}(0, 1)$. It was shown in Chapter 3, Eqn. 3.75, (see also Chapter 6, Eqn. 6.75) that the PSD for such an input is a constant and is given by

$$S_{FF}(f) \approx \sigma^2 \Delta_t \qquad (5.46)$$

hence we can set the variance of our input values to be simply $\sigma^2 = S_{FF}(f)/\Delta_t$ to create a realization $F(i), i = 1 \cdots N$ of the input random process to use in the simulation.

In anticipation of experiments detailed in Section 8.2, consider a uniform beam of length $L = 1$ (m), width $b = 0.02$ (m) and height $h = 0.005$ (m). The material is assigned a density $\rho = 7800 \, [kg/m^3]$ and an elastic modulus $E = 210 \times 10^9 \, [N/m^2]$. A total of $n = 21$ elements is used for the discretization. For this problem, this number of elements provides a good balance between computational burden and model accuracy (both of which increase with n). Given these parameters, we can use the methods of Chapter 4 to show that the natural frequencies are given by

$$f_i = \frac{\lambda_i^2}{2\pi} \sqrt{\frac{EI}{\rho bh L^4}} \qquad (5.47)$$

where the λ_i will depend on the boundary conditions. For a cantilevered beam, the λ_i are the roots of the equation $\cos(\lambda_i)\cosh(\lambda_i) = 1$, while for a clamped–clamped beam they must satisfy $\cos(\lambda_i)\cosh(\lambda_i) = -1$ (see Eqn. 5.40). As we showed in the previous section, a loose joint is effectively a transition between these two cases and essentially results in shifting mode i of the cantilevered case to mode $i - 1$ for the clamped–clamped case. For example, the first four frequencies in the cantilevered case are $4.19, 26.26, 73.54$, and 144.11 Hz. If the beam is, instead, clamped at both ends, the frequencies are $26.67, 73.51, 144.11$, and 238.22.

As we have already mentioned, this approach views a bolted joint as a nonlinear stiffness constraining the clamped ends. The large stiffness used to constrain the translation at the left end (and the right one, when the displacement exceeds the gap) is set to $100 \times bh^3 E/(L/(n-1))^3$, which is 100 times the value of the stiffness K_{11} of the fixed/free structure. Analogously, $100 \times bh^3 E/(L/n-1)^3$ is used to constrain the rotation at the left end, corresponding to 100 times the value K_{22} of the fixed/free structure. For these simulations, the gap opening is set to $\delta_{gap} = 0.005$ (m) and a modal damping (for all modes) equal to $\zeta = 0.04$ is used.

The external excitation is Gaussian distributed, with zero mean. A fifth-order Butterworth filter, with cutoff frequency at 500 Hz, is used to filter the otherwise white signal: this results in an input force spectrum almost flat up to 500 Hz, with value $0.02 \, N^2/Hz$. This excitation

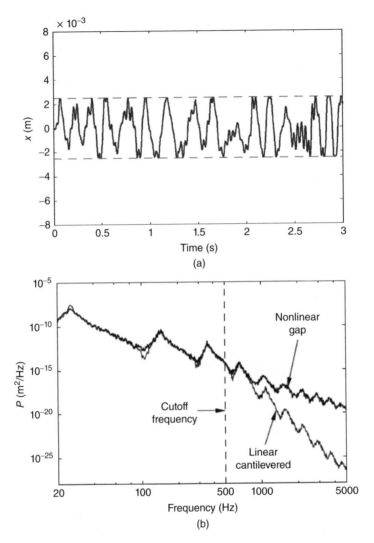

Figure 5.21 Both time (a) and frequency (b) domain descriptions of the bolted joint response. The hard limit imposed by the gap prevents the temporal response from exceeding the ± 0.0025 *m* limits. The frequency domain implication is that higher frequencies become greatly emphasized over the linear, fixed-free case. As we show in Section (8.2), this amplified frequency content can be used as a damage-detection statistic

profile is comparable to that used in the experimental setup presented in 8.2. The integration is carried out with a time step $1e - 6$ (s). Every 50 time steps, the displacement, velocity, and acceleration outputs are recorded, resulting in a sampling performed via decimation with $\Delta_t = 5e - 5$ (s).

A typical time-series response is shown in Figure 5.21. The hard limits to displacement at ± 2.5 mm are clearly visible in the response. Figure 5.21 also reports the PSD for the displacement response of both the linear and nonlinear (damaged) model. The definition used for the PSD in this work, along with the estimation approach taken, can be found later in Section 6.4. The estimation for both cases is performed using the Welch algorithm, as implemented in the Signal Processing Toolbox of MATLAB [17], with a Hanning window and a 50% overlap between records. The spectrum of the linear case is in very good agreement with the one that can be obtained analytically, using the formulae for beam theory. The two spectra of the linear and nonlinear (because of the gap) cases are very similar until the cutoff frequency of 500 Hz. However, the linear response drops in accordance with the input spectrum above 500 Hz, while the nonlinear one does not. This implies that even though the frequency content of the excitation above the cutoff frequency is low, the displacement response of the beam with the gap has frequency content in that band.

The ability to create frequencies in the output that are not present in the input is well known for nonlinear systems [28]. Also, a slight shift of the peaks can be noted from the spectrum of the cantilevered beam to the one with the gap. Such a shift is to the right, as expected, because, in a way, the gap acts as a "stiffening spring." In fact, this introduces a stiffness that, although not active all the times, increases the frequencies of the response.

The change in spectral properties is a frequently used identifier of damage-induced non-linearity. In fact, the added frequency content resulting from the nonlinearity will be used as a damage detection statistic in the experimental case study of section 8.2.3. A less familiar approach is to look at the probability distribution of the response variable. For a linear system, we show at the beginning of Chapter 8 that a Gaussian input will lead to a Gaussian output. A nonlinear system, however, can alter the probability distribution of the input such that deviations from the normality scan be used to detect the nonlinearity. As a preview of the results of Section 8.2, Figure 5.22 shows the predicted change in statistical properties of the beam

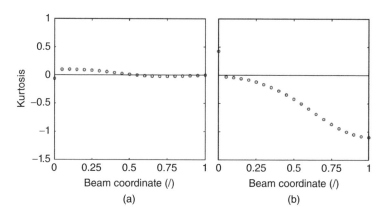

Figure 5.22   Estimated kurtosis from the model response at different points along the length of the beam. For a linear structure, a jointly Gaussian input should yield a jointly Gaussian response, while a nonlinear structure will almost certainly yield a response with a non-Gaussian probability distribution. As demonstrated in Section 2.6, the kurtosis should be zero for a Gaussian random variable and nonzero for certain types of non-Gaussianity. The damaged structure clearly exhibits nonzero kurtosis, partic-ularly near the damage site ($l = 1$) *Source*: Adapted from [27], Figure 3, reproduced with permission of SAGE

displacement response (specifically the kurtosis) due to the presence of damage-induced non-linearity. Because a Gaussian input is provided, if the system is linear, a Gaussian output is expected, while non-Gaussianity would be an indication of nonlinearity. A simple measure of non-Gaussianity is the *kurtosis*, defined here as

$$k \equiv \frac{E[(Y - \mu_X)^4]}{E[(Y - \mu_Y)^2]^2} - 3$$

$$= \frac{k_4}{E[(Y - \mu_Y)^2]^2}$$

$$= \frac{k_4}{\sigma_Y^4} \tag{5.48}$$

where y is the response, modeled as a random variable Y with μ_Y as its mean and σ_Y as its standard deviation. Notice that this is simply the normalized fourth cumulant k_4 introduced in Section 2.6. Recall from that section that all cumulants vanish for a Gaussian distribution, hence for a Gaussian-excited structure $k = 0$. This is confirmed in Figure 5.22, for the cantilever (linear) beam. On the other hand, the figure also shows that when a gap is introduced in the structure, the kurtosis becomes different from zero, with larger values around the location of the nonlinearity. On the basis of this model, it is expected that the loosening joint will result in response data that exhibit these common characteristics of nonlinearity in both the time and frequency domains. Once we have developed a better understanding of the response data and, more importantly, how to estimate characteristics of the response from experimental data, we revisit this problem in Section 8.2.

5.5   Corrosion Damage

Corrosion, and the structural degradation that accompanies it, is a recurring problem for any (partially) metallic structure that is exposed to the elements. Whether it is an offshore oil rig or a component in a battery, corrosion undermines a structure's integrity and limits its useable lifetime. Of course, corrosion is not confined to a simple redox mechanism; it can depend sensitively on the precise chemistry involved in the base metal and the surrounding fluid, the flow velocity and geometry of the surrounding fluid, and a host of other factors. Uniform corrosion, which is the variety considered here, is typically associated with high-velocity fluid flows.

The present model was developed to quantify the extent of corrosion damage, which is presumed uniform, in a homogeneous, isotropic metal plate exposed to a corrosive salt water environment [29]. The damage parameter is the effective, or average, thickness of the plate. As a result, the model cannot successfully identify highly localized phenomena, such as pitting corrosion or intergranular corrosion.

The model of the plate used here is adapted from the model described in Ref. [30] and is appropriate for describing the behavior of thick or thin plates subject to random vibration. To derive this model, we note that the energy functional Π for a plate can be expressed in terms of the maximum strain energy $U_{\max}$ and the maximum kinetic energy $T_{\max}$:

$$\Pi = U_{\max} - T_{\max} \tag{5.49}$$

The potential energy and kinetic energy can be calculated using transverse deflection and rotations of the cross-sections. The functions ψ_x and ψ_y are the mode shapes for the rotations of the plate about the x and y axis, and Z is the lateral displacement mode shape of the plate. It is assumed that these functions are separable, and so can be expressed using the expansion technique described in Section 4.5.4 as

$$Z(\xi,\gamma) = \sum_{i=1}^{S}\sum_{k=1}^{T} c_{ik}X_i(\xi)Y_k(\gamma) \tag{5.50a}$$

$$\psi_x(\xi,\gamma) = \sum_{i=1}^{S}\sum_{k=1}^{T} d_{ik}\phi_i(\xi)Y_k(\gamma)/a \tag{5.50b}$$

$$\psi_y(\xi,\gamma) = \sum_{i=1}^{S}\sum_{k=1}^{T} e_{ik}X_i(\xi)\alpha_k(\gamma)/b. \tag{5.50c}$$

In Eq. (5.50), $Y_k(\eta)$, $X_i(\xi)$, $\phi_i(\xi)$, and $\alpha_k(\gamma)$ are the admissible functions for displacement and rotation of the plate in the x and y directions, $\xi = x/a$ and $\gamma = y/b$ are nondimensional coordinates, and c_{ik}, d_{ik} and e_{ik} are unknown vectors of constants which can be calculated by solving the eigenproblem associated with Π. The eigenproblem can be constructed by using Eq. (5.50) to represent the strain energy and kinetic energy in Eq. (5.49). We then use the variational approach outlined in Sections 4.3.4 and 4.3.5 and set the first variation of the energy functional Π to zero, which yields:

$$([K] - \Omega^2[M])\begin{Bmatrix} c_{ik} \\ d_{ik} \\ e_{ik} \end{Bmatrix} = \{0\}. \tag{5.51}$$

In Equation (5.51), Ω is the vector of normalized natural frequencies of the plate, and (K) and (M) are the stiffness and mass matrices of the system, respectively.

The parameters S and T in Eq. (5.50) set the number of admissible functions used, with larger values increasing the accuracy of the analysis at the expense of increasing computation time. On the basis of the results of a convergence study, S and T were both set to 3.

A unit strip width section taken parallel to an edge of a Mindlin plate is equivalent to a uniform Timoshenko beam from a modeling perspective. The static solution to a Timoshenko beam under an arbitrary load is therefore an excellent candidate for use as the admissible functions of Eq. (5.51). To satisfy the elastic boundary conditions, in the x direction the admissible functions therefore take the form:

$$X_i(\xi) = C_0^i + C_1^i\xi + C_2^i\xi^2/2 + C_3^i(\xi^3/6 - R_x\xi) + [R_x(i\pi)^2 + 1]\sin(i\pi\xi), i = 1\cdots 3 \tag{5.52}$$

$$\phi_i(\xi) = C_4^i + C_5^i\xi + C_6^i\xi^2/2 + i\pi\cos(i\pi\xi), i = 1\cdots 3 \tag{5.53}$$

where $R_x = (1 - v^2)D/Gh\kappa_x a^2$, v is Poisson's ratio, D is the flexural rigidity of the plate, G is the elastic modulus of shear, h is the plate thickness, and κ is the shear correction factor. The beam functions in the y direction, $Y_k(\gamma)$ and $\alpha_k(\gamma)$ are defined in a similar manner. The constants C_t^i are obtained using the boundary conditions of the plate:

$$M(0) = -k_{x0}^r\psi(0), M(1) = k_{x1}^r\psi(1), V(0) = k_{x0}^r z(0), V(1) = -k_{x1}^r z(1) \tag{5.54}$$

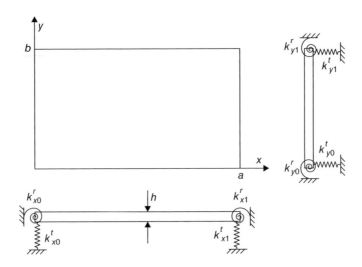

Figure 5.23   Schematic representation of the plate geometry, including elastic boundary supports

where k^r and k^t are the rotational and translational stiffnesses along the edge of the plate, as shown in Figure 5.23. Also, ψ and z are the beam rotations and displacements, respectively. We have found through past modeling efforts that the inclusion of these boundary stiffness parameters is necessary to accurately model the behavior of the plate, and is therefore also important for damage estimation.

In the limits of zero and infinite torsional stiffness, one obtains the simply supported and fully clamped boundary conditions, respectively. Of course, the actual plate is somewhere in between these conditions. To achieve this, the translational stiffnesses k^t_{x0}, k^t_{x1}, k^t_{y0}, and k^t_{y1} are all set to 1×10^9 N/m, a value suggested in Ref. [30] as approximating a fully restrained condition. The torsional spring stiffnesses carry a great deal more uncertainty and will therefore be treated as unknowns to be estimated; we will do this for an experimental corrosion example in Chapter 9.

Although the specific torsional stiffness values are unknown, they can be assumed to be uniform around all four edges of the plate. When normalized by their edge length and flexural rigidity, we have that $\tau = \frac{k^r_{x0} a}{D} = \frac{k^r_{x1} a}{D} = \frac{k^r_{y0} b}{D} = \frac{k^r_{y1} b}{D}$; this differs from Ref. [30], where they are all allowed to vary independently. This simplification reduces the complexity of the model, yet clearly preserves the physics of the test plate and rig, as demonstrated in our results. That said, it is a simple matter to include each stiffness as a separate parameter in the estimation algorithm should the physics dictate separate treatment.

The mass and stiffness matrices of the plate vibration eigenproblem were constructed of algebraic expressions using symbolic math software based on Eqs (5.49)–(5.54). This allowed for very rapid evaluation of the mode shapes and natural frequencies of the plate, as the stiffness and mass matrices are only size 27×27. A finite element model with similar accuracy would have mass and stiffness and mass matrices at least an order of magnitude larger. Again, we stress the importance of a fast, efficient forward model as the estimation techniques used require tens of thousands of model evaluations to estimate the complete parameter distribution (see Chapter 7 on approaches to estimating structural damage).

Once the mode shapes and natural frequencies are determined, it is a simple matter to calculate the displacements. From these, the strain time series at any particular point on the plate may be found using the definition of strain in Mindlin plates:

$$\epsilon_x = z\frac{\partial \Psi_y}{\partial x}, \epsilon_y = z\frac{\partial \Psi_x}{\partial y}, \epsilon_{xy} = z\left(\frac{\partial \Psi_y}{\partial y} - \frac{\partial \Psi_x}{\partial x}\right). \tag{5.55}$$

Here, Ψ_x and Ψ_y are the rotation of the cross-section of the plate at any point, and at any time

$$\Psi_x(x, y, t) = \sum_{j=1}^{S} A_j \psi_x(x, y) e^{i\omega_j t}/a$$

$$\Psi_y(x, y, t) = \sum_{j=1}^{T} A_j \psi_y(x, y) e^{i\omega_j t}/b \tag{5.56}$$

where the constants A_j are found via analytic methods and are related to the initial conditions of the plate, and $S = T = 3$ based on the convergence study mentioned earlier. Given this construction, the strain field at a particular location (x, y) can be written

$$w(x, y) = \epsilon_x \cos^2\phi + \epsilon_y \sin^2\phi + \epsilon_{xy} \sin\phi \cos\phi \tag{5.57}$$

given the orientation of the gage on the plate, γ, as well as the position of the sensor (x, y) and the time of interest t.

At this point, we can specify the geometry of the plate and the material properties. The resulting model unknowns are the torsional boundary stiffness k^t and the plate thickness h (which varies based on the level of corrosion). The point of developing this model was to provide a means to estimate the level of corrosion damage, as measured by the reduction in the average plate thickness. The details of the estimation process and a formal experimental example of corrosion identification are given later in Section 9.4. However, first we offer a brief experimental justification of the model.

Consider as a test specimen, a 0.76 m by 0.60 m (30 in. by 24 in. 5:4 aspect ratio) 6061-T6 aluminum plate. The thickness is normalized by the original thickness so that the extent of the corrosion damage is clear. Of course, no plate has truly uniform thickness. So the thickness measured (used to validate the estimated values) is the average values of nine separate measurements taken around the perimeter of the plate. An initial set of vibration data was acquired using a new (i.e., uncorroded) plate; this was done in the test rig shown in Figure 5.24a. To acquire the strain response data, the plate was mounted into the test rig. The plate was clamped to a bolted fixture, where each of the 28, 12 mm bolts were tightened to 122 N-m (90 ft-lbs). Lock washers were used on each fastener. To induce corrosion, a saltwater bath was created; see Figure 5.24b. The saltwater had the same salt concentration as common seawater (3.5).

The plate was tested first in the pristine, or healthy, condition. It was struck four separate times and the vibration response data (measured as strain vs. time) for each sensor was used to infer the thickness of the plate. The inference process provided very good estimates of the thickness and the torsional stiffnesses [29]. Figure 5.24c shows the experimental strain response and model predicted strain response (using the estimated parameters). There is generally very good agreement between the experimentally recorded strain data and the model predicted response, as shown. This level of agreement continues as damage accrues, provided

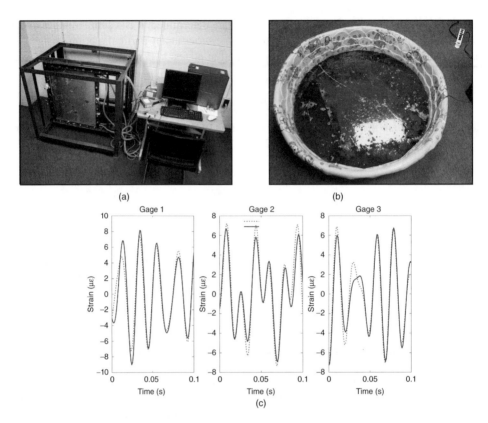

Figure 5.24 (a) The experimental test setup and (b) the accelerated corrosion environment during used. The final plots (c) show model and experimental strain response data measured at different spatial locations. The model developed in this section provides accurate predictions of the experimental plate response.

the level of noise in the experimental strain signal remains low. This predictive capability combined with its relative simplicity and the low simulation time (because it is largely analytic) make this an attractive model. Again, this experiment is elaborated on in Section 9.4.

5.6   Beam on a Tensionless Foundation

A common characteristic of many damage scenarios is undesired contact between components of a structure. One common approach to modeling contact behavior is to replace one of the contacting components (presumably the less important one) with an elastic foundation. This approach has the benefit of incorporating the flexibility of the secondary component while greatly simplifying the analysis. This approach has been used repeatedly in the study of beams and plates in contact with other bodies; see the early Refs [31] and [32].

The most common beam-foundation model allows for both compressive and tensile stresses to exist across the interface between the beam and the foundation. In other words, if a downward lateral load is applied to a beam resting on a foundation, the beam will be compressed into

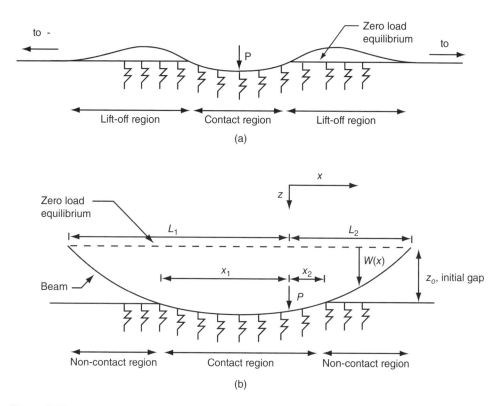

Figure 5.25 (a) A schematic of the infinite point loaded beam on a tensionless foundation. (b) A schematic of the finite beam with an eccentric point load and a finite gap

the foundation. If the direction of the load is reversed, the beam and the foundation are pulled up, creating tension in the foundation. However, in applications where adhesion between the beam and the substrate is not assured, this latter behavior is physically unrealistic. Instead, a more appropriate model would be a foundation which reacts to compressive stresses but is incapable of experiencing tension. Such foundations are termed *one-way* or *tensionless* foundations.

Most studies have looked at an infinite (or semi-infinite) beam resting on a tensionless, elastic foundation [33–36]. However, as it turns out, many of the results presented occur because of the "infinite beam" assumption. More specifically, they stem from the symmetry that naturally arises; because the beam is infinite, any applied point load must be centered on the beam. So what are these results? To explain, consider the schematic beam-foundation system shown in Figure 5.25a. As a load is applied, a portion of the beam is compressed into the substrate; outside the contact zone the outer edges "lift off" from its zero load equilibrium position. This liftoff condition must always occur, regardless of the magnitude of the applied load. This liftoff condition produces another interesting feature. As greater loads are applied, the central portion of the beam plunges deeper into the foundation. However, the outer edges of the plate experience greater liftoff. The result is that the contact area remains constant. Moreover, the contact area is independent of the magnitude of the applied load. All of these general conclusions arise from the infinite geometry of the beam and manifest themselves mathematically in the symmetry of the solution to the equilibrium problem.

The system under consideration here differs in three fundamental ways from the infinite, symmetric problem described in the early literature. First of all, the beam is finite. This requires specifying some appropriate boundary conditions which, of course, influence the results. We will focus on the pinned–pinned beam. Second, the finite span permits an off-center point load, which breaks the symmetry of the system. Finally, in all cases other than the free–free case, the boundary conditions enable a nonzero gap to exist between the beam and the foundation. All three of these features are built into the mathematical model, which produces different phenomena than the prior systems.

Consider the beam-foundation system shown in Figure 5.25b. It consists of a linear elastic Euler–Bernoulli beam and a linear, tensionless foundation. The beam has length L and flexural rigidity EI. An external load P is applied to the beam and the origin of the coordinate system is centered at this load. The distance to the left (right) end of the beam is L_1 (L_2). The contact zone spans the region $-X_1 < X < X_2$. The transverse deflection of the beam is given by $W(x)$. A finite gap Z_o may exist between the foundation and the no-load equilibrium of the beam (given by the dashed line). The substrate is a tensionless Winkler foundation with modulus k. Of course, the beam will separate from the foundation once the displacement is less than the gap size, $W < Z_o$.

5.6.1   Equilibrium Equations and Their Solutions

The transverse deflection of the NA $W(x)$ is broken into three distinct regions:

$$W(x) = \begin{cases} W_1 & -L_1 < X < -X_1 \\ W_2 & -X_1 < X < X_2 \\ W_3 & X_2 < X < L_2 \end{cases}$$

These correspond to the deflections in the left noncontact region ($W_1 < Z_o$), the contact region ($W_2 \geq Z_o$), and the right noncontact region ($W_3 < Z_o$). Similarly, the equilibrium equation is divided into three parts:

$$EI\frac{d^4 W_1}{dX^4} = 0 \quad -L_1 < X < -X_1 \tag{5.58}$$

$$EI\frac{d^4 W_2}{dX^4} + k(W_2 - Z_o) = P\delta(X) \quad -X_1 < X < X_2 \tag{5.59}$$

$$EI\frac{d^4 W_3}{dX^4} = 0 \quad X_2 < X < L_2 \tag{5.60}$$

where $\delta(X)$ is Dirac delta function. For convenience, we introduce the quantity $\beta^4 = \frac{k}{4EI}$ along with the following quantities:

$$\xi_1 = \beta X_1, \quad \xi_2 = \beta X_2, \quad l_1 = \beta L_1, \quad l_2 = \beta L_2, \quad l = \beta L, \quad z_o = \beta Z_o$$

$$w = \beta W, \quad \xi = \beta X, \quad F = \frac{P}{4\beta^2 EI}$$

Introducing these quantities into the equilibrium equations (Eqs. 5.58–5.60) produces the following nondimensional equations:

$$\frac{d^4 w_1}{d\xi^4} = 0 \qquad -l_1 < \xi < -\xi_1 \tag{5.61}$$

$$\frac{d^4 w_2}{d\xi^4} + w_2 - z_o = F\delta(\xi) \qquad -\xi_1 < \xi < \xi_2 \tag{5.62}$$

$$\frac{d^4 w_3}{d\xi^4} = 0 \qquad \xi_2 < \xi < l_2. \tag{5.63}$$

The total solutions to Equations (5.61)–(5.63) are found via power series expansion and may be simplified to the following:

$$w_1(\xi) = A_1 \xi^3 + B_1 \xi^2 + C_1 \xi + D_1 \tag{5.64}$$

$$w_2(\xi) = A_2 \cosh(\xi) \sin(\xi) + B_2 \cosh(\xi) \cos(\xi) + C_2 \sinh(\xi) \sin(\xi)$$

$$+ D_2 \sinh(\xi) \cos(\xi) - \frac{F}{2} \sinh |\xi| + \frac{F}{2} \cosh(\xi) \sin |\xi| + z_o \tag{5.65}$$

$$w_3(\xi) = A_3 \xi^3 + B_3 \xi^2 + C_3 \xi + D_3. \tag{5.66}$$

Here, A_i, B_i, C_i, and D_i ($i = 1, 2, 3$) are unknown constants (there are 12). In addition to these constants, the size of the contact zone, given by ξ_1 and ξ_2, is also unknown. Hence, there are a total of 14 unknown constants.

5.6.2   Boundary Conditions

To determine these 14 constants, there must be an equal number of boundary/matching conditions. Let's begin with the matching conditions that occur at the edge of the contact. At $\xi = -\xi_1, \xi_2$, the geometric boundary conditions require continuity of the displacement and slope. These are expressed as

$$w_1(-\xi_1) = w_2(-\xi_1), \qquad \frac{dw_1(-\xi_1)}{d\xi} = \frac{dw_2(-\xi_1)}{d\xi} \tag{5.67}$$

$$w_2(\xi_2) = w_3(\xi_2), \qquad \frac{dw_2(\xi_2)}{d\xi} = \frac{dw_3(\xi_2)}{d\xi}. \tag{5.68}$$

There are also four natural boundary conditions at $\xi = -\xi_1, \xi_2$. These require continuity of the bending moment and shear force. These are

$$\frac{d^2 w_1(-\xi_1)}{d\xi^2} = \frac{d^2 w_2(-\xi_1)}{d\xi^2}, \qquad \frac{d^3 w_1(-\xi_1)}{d\xi^3} = \frac{d^3 w_2(-\xi_1)}{d\xi^3} \tag{5.69}$$

$$\frac{d^2 w_2(\xi_2)}{d\xi^2} = \frac{d^2 w_3(\xi_2)}{d\xi^2}, \qquad \frac{d^3 w_2(\xi_2)}{d\xi^3} = \frac{d^3 w_3(\xi_2)}{d\xi^3}. \tag{5.70}$$

At the edge of the contact zone, it is evident that the displacement must also equal the gap size. So two additional conditions may be written:

$$w_2(-\xi_1) = z_o \qquad w_2(\xi_2) = z_o. \tag{5.71}$$

Equations (5.67)–(5.71) give 10 conditions. Four additional conditions arise from the boundaries at $\xi = -l_1, l_2$. For the pinned–pinned beam, these are

$$w_1(-l_1) = 0, \quad \frac{d^2 w_1(-l_1)}{d\xi^2} = 0, \quad w_3(l_2) = 0, \quad \frac{d^2 w_3(l_2)}{d\xi^2} = 0. \tag{5.72}$$

This leaves 14 boundary/matching conditions to determine the 14 unknown constants, A_i, B_i, C_i, D_i ($i = 1, 2, 3$), ξ_1 and ξ_2. The boundary conditions are applied to the solutions given in Eqs. (5.64)–(5.66) so that the unknowns may be determined. At first glance, this may appear to be a simple, linear boundary value problem. However, because ξ_1 and ξ_2 appear in the argument to the solutions, the problem is nonlinear. Hence, a multidimensional Newton–Raphson algorithm [37] is used to obtain the roots to this problem numerically.

At this juncture, it is worth pointing out some of the similarities and differences from the infinite beam case presented by Weitsman [33]. In that work, Weitsman focused only on the deflection solution in the contacting regime (i.e., w_2). Moreover, because of the symmetry associated with the infinite domain, only even terms are retained in the solution to Eqs. (5.65). In other words, the terms $\cosh(\xi)\sin(\xi)$ and $\sinh(\xi)\cos(\xi)$ are omitted. This leaves:

$$w_2(x) = B_2 \cosh(\xi)\cos(\xi) + C_2 \sinh(\xi)\sin(\xi) - \frac{F}{2}\sinh|\xi| + \frac{F}{2}\cosh(\xi)\sin|\xi|. \tag{5.73}$$

Note that z_o is presumed to be zero. In this case, there are only three unknowns: B_2, C_2, $\xi_1 = \xi_2$ (this last equality is assured by symmetry). These constants are determined through the matching conditions:

$$w_2(\xi_1) = 0, \quad \frac{d^2 w_2(\xi_1)}{d\xi^2} = 0, \quad \frac{d^3 w_2(\xi_1)}{d\xi^2} = 0. \tag{5.74}$$

The first condition is clear (because $z_o = 0$) and the latter two stem from the fact that there is neither a moment nor a shear force throughout regions 1 and 3 (the left and right sides of the beam, outside the contact zone).

What immediate conclusions may be drawn from this problem and how does it differ from the finite problem? First of all, the infinite problem has only 3 unknowns compared to the 14 for the finite case. The most obvious physical result is that the infinite problem is inherently symmetric. This is evident from the solution form for w_2 and the fact that ξ_1 is presumed equal to ξ_2. What is less obvious is that, after applying the boundary conditions, the contact length, $2\xi_1$, is found to be constant and, hence, is independent of the applied load, see Ref. [33]. This constant contact length feature leads to a third phenomenon. Namely, as the load is increased, the contacting portion of the beam plunges deeper into the foundation, but, to enforce the constant contact length requirement, the beam must *lift off* from the foundation (shown schematically in Figure 5.25a). And because the constant contact length is assured for any load magnitude, the liftoff phenomenon must also occur for *any* applied load. By contrast, the finite beam problem need not be symmetric, as one would expect if the load were off-center. In addition, the contact length is not constant but depends on a number of things. Once this constant contact length requirement is relaxed, the liftoff condition is no longer assured. Of course, given the fact that the finite beam solution contains all of the same terms as the infinite solution (and then some), liftoff is still a possibility – but it is not a requirement.

5.6.3 Results

In the case of a pinned–pinned beam, two issues are considered. The load may be either symmetric or asymmetric and the gap size may be nonzero. For the moment, let's fix the gap size at $z_o = 0$ and focus on the symmetric and asymmetric loading cases. Figure 5.26a shows the contact length as a function of the beam length for the cases $l_1 = 0.5l$, $0.6l$. These results (with $z_o = 0$) were found to be independent of the load magnitude. For short beams, the contact length of the symmetric case grows linearly with slope one. Again, this indicates that the entire beam is in contact with the foundation. This persists until $l = 6.187$. As the beam length is increased further, the contact radius actually begins to shrink. This beam length ($l = 6.187$), separating the regions of growing versus shrinking contact length, is the critical length for the zero gap separation case. The shrinking contact is accomplished by the liftoff phenomenon, which is initiated at the critical beam length. As the beam length is increased further, the contact length decreases and asymptotically approaches π.

The asymmetric loading case in Figure 5.26a (for $l_1 = 0.6l$) begins at zero and has a contact length that initially grows linearly with the beam length. In this regime, the entire beam is in contact with the foundation. As the length is increased, the contact curve shows two peaks – the contact length grows, shrinks, grows again, and then shrinks again before it asymptotically approaches π. To explain this behavior, consider Figure 5.26b, which shows the behavior of the two distances ξ_1 and ξ_2 that define the edges of contact. The symmetric case is shown as well, for comparison. As before, the left-side contact point initially grows as $\xi_1 = 0.6l$. This side of the contact peaks before the symmetric case and then shrinks gradually to the asymptotic value of $\pi/2$. The shrinking contact area (after the peak) occurs because the left side begins to lift off the foundation. The right-side contact initially grows more slowly at $\xi_2 = 0.4l$ and peaks after the symmetric case. The subsequent drop in the contact length is due to right-side liftoff. If these two functions are summed, they form the asymmetric case shown in Figure 5.26a. The first peak is attributed to the peak in ξ_1. The drop in contact results from the drop in ξ_1 after its peak (ξ_2 is increasing but not fast enough to offset the drop in ξ_1). The second peak is due to ξ_2. Finally, Figure 5.27 shows the deflection under two different loads, $F = 0.1, 0.2$, for a beam length of $l = 30$. Here, both sides have clearly lifted off. Moreover, the contact length, which is near π, is independent of the applied load.

Now consider the effect of a finite gap separation, z_o. Figure 5.28a shows the contact behavior of the symmetric beam both with and without a gap separation. The solid curve shows the contact length with $z_o = 0$ and a load of $F = 0.1$ and 0.5; they are coincident, meaning that the contact length is load-independent in the zero gap case. As before, the curve begins at the origin and is nonmonotonic. Now consider a small, nonzero gap separation of $z_o = 0.05$. The contact length is given by the dashed (dash-dotted) line for a load of $F = 0.1$ ($F = 0.5$). These two curves quite clearly show that the contact length is *load-dependent* when the gap size is nonzero. Regarding the three cases shown in Figure 5.28a, three other observations may be made. First, the contact length curve does not begin at the origin as do all the other (zero gap) cases. Contact cannot be initiated until the center point deflection is equal to the gap size – and short beams are too stiff to have this much deflection under a small load. So contact is not initiated until the center point deflection becomes (in dimensional terms) $W(\frac{l}{2}) = Z_o = \frac{PL^3}{48EI}$. With this expression, the appropriate parameter values may be used to determine the critical beam length associated with initial contact. For example, in the case of $F = 0.1$ and $z_o = 0.05$, this value is $l = 1.816$. Second, the curves may or may not increase monotonically, depending

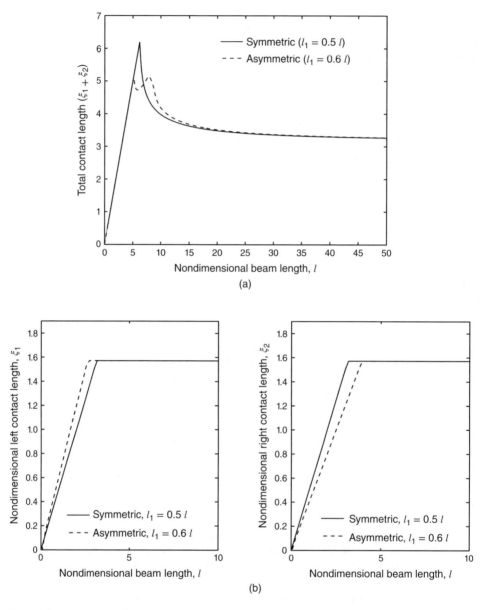

Figure 5.26 (a) The total contact length versus the beam length for the symmetric and asymmetric loading. (b) The left- and right-side contact length versus total beam length for the symmetric and asymmetric case

on the load level. Third, as the load is increased or as the gap separation decreases, the contact length curve will approach the zero gap result. This suggests, at least heuristically, that the quantity F/z_o would indicate the relative proximity of the nonzero gap curve to the zero gap curve, large values of F/z_o being closer to the zero gap case. This observation also explains

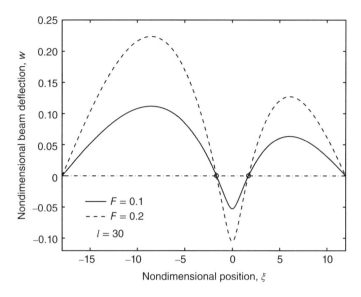

Figure 5.27 Deflection curves showing liftoff for the asymmetrically loaded beam under two loads. Here, $l = 30$ and $z_o = 0$. *Note*: the contact length is the same

why the zero gap case is independent of the load F ($F/z_o = \infty$, regardless of the load and the curves converge).

Finally, the asymmetric loading case is shown in Figure 5.28b, where $l_1 = 0.6l$. The no-gap case shows the double peak associated with uneven liftoff, as described previously. The introduction of the gap produces behavior similar to the symmetrically loaded case. However, at short beam lengths, the system is sufficiently stiff such that the offset load point does not come into contact with the foundation.

The static model presented here is one possible means of describing tensionless contact. And the model results demonstrate that the tensionless contact problem shows richer and varied behavior than was previously anticipated.

5.7 Cracked, Axially Moving Wires

Axially moving systems are common in engineering applications. Examples include bandsaws, belt drives, fiber drawing systems, web handling devices, and so on. In each of these examples, the material system is transported along its major axis at a prescribed *axial transport speed, c*. In the course of the operation, damage can accumulate in the translating media. The particular problem under consideration here is motivated by the wire electrodischarge machining (EDM) operation.

Figure 5.29a and b shows how the wire EDM system works. A thin wire is brought into close proximity to a grounded, conductive workpiece. Sparks are discharged across the wire to the workpiece; in the process, material is ablated from the workpiece – effectively *cutting* it. Of course, this process also ablates the wire. To prevent wire rupture, new wire is continuously

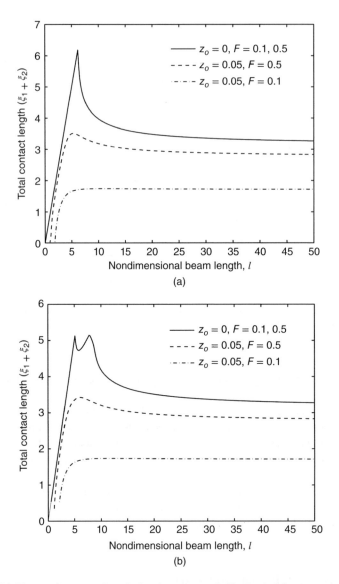

Figure 5.28 (a) The total contact length for the symmetrically loaded beam versus beam length. (b) The total contact length for the asymmetrically loaded beam versus beam length

fed into (pulled out of) the kerf from a supply (take-up) reel. As a result, the wire has a net axial transport speed, c. Figure 5.29c shows an actual workpiece being cut via wire EDM. The goal here is to develop a useful model to describe the vibrations of an EDM wire, because any vibrations degrade the accuracy of the cut. In particular, we will focus on identifying the natural frequencies, because these correspond to resonant conditions in the forced problem. We will also identify the limitations of this model.

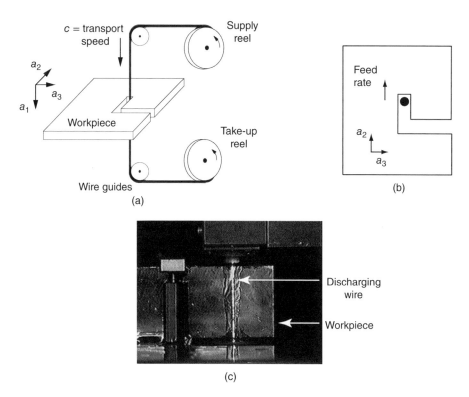

Figure 5.29 The wire electrodischarge machining (EDM) system: (a) a 3-D schematic of the system operation, (b) a 2-D representation of how the cut progresses, (c) a photograph of an actual EDM cutting operation

5.7.1 Some Useful Concepts from Fracture Mechanics

To begin, consider Figure 5.30a; this shows a crack in a linearly elastic material. Points in the material are identified by the polar coordinates r and θ, centered at the crack tip. It is well established that the stress field associated with a crack decays as $1/\sqrt{r}$ and varies with θ [38]. This is usually expressed in the form

$$\sigma_{ij} = \frac{K_I}{\sqrt{r}} f_{ij}(\theta)$$

where K_I is the stress intensity factor and is a function of the applied load and the size or depth of the crack. If a combined load is applied, the resultant stress is simply the sum of the individual stresses because the system is linear. For this study, the primary focus is on combined bending and axial stretching of a long, thin beam. In this case, shear is ignored and the net axial stress is

$$\sigma = \sigma^b + \sigma^s = \frac{1}{\sqrt{\pi r}} \left[K_I^b f_b(\theta) + K_I^s f_s(\theta) \right]. \tag{5.75}$$

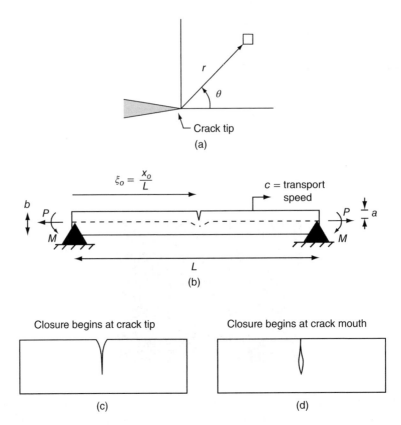

Figure 5.30 (a) A schematic showing the stress at an arbitrary material element at (r, θ) away from the crack tip; (b) the model of the EDM wire as a cracked, translating beam; (c) the first mechanism for crack closure – closure begins at the crack tip; (d) the second mechanism for crack closure – closure begins at the mouth

If the crack is allowed to propagate, new surface area is continually generated. This disruption of the continuum requires that some of the internal strain energy of the material be dissipated. This converted energy is usually described in terms of an energy release rate, G. Of course, the amount of energy dissipated depends on *how* the crack is propagated. Typically, cracks propagate in one of three ways: Mode I – crack opening, Mode II – sliding (in-plane shear), and Mode III – tearing (antiplane shear), see Ref. [38]. For the geometry under consideration, only Mode I deformations are relevant.

Under a bending/stretching loading scenario, the strain energy may be expressed in terms of the square of the axial stress. Given Eq. (5.75), coupling occurs between the bending and stretching terms. However, unlike the classic Kirchhoff beam theory, this coupling remains after the infinitesimal energy dU is integrated over the domain. In other words, $U \propto [K_I^b + K_I^s]^2$. Because the energy released during crack propagation is simply a conversion of the strain energy, it should not be surprising that G is also proportional to the squared sum of the stress intensity factors. The particular expression for the energy release rate for plane strain problems is presented in Ref. [38] and takes the form

$$G(a/b) = \frac{1 - v^2}{E} (K_I^b + K_I^s)^2 \tag{5.76}$$

where the stress intensity factors are functions of the nondimensional crack depth a/b and are given in Ref. [39]. This expression for the energy release rate is used in the following section to obtain the change in compliance due to a crack for a tensioned beam undergoing transverse and axial vibrations.

5.7.2   The Effect of a Crack on the Local Stiffness

A schematic of the wire model is shown in Figure 5.30b. The system consists of two pin supports fixed in inertial space, which support the beam. The beam has length L, height b, cross-sectional area A, axial transport speed c, and is subjected to a constant tension force P. It also has a crack of fixed depth a that translates with the beam. The crack location changes constantly and is given by the normalized variable $\xi_o = x_o/L$. The presence of the crack reduces the stiffness of the beam both in axial stretching and in bending. In other words, a given load will produce more deformation in the cracked beam than in its uncracked counterpart. This additional deformation is highly localized near the crack and not distributed evenly over the structure. The asymmetric crack also causes the NA to dip down from the centerline in the vicinity of the crack. This is shown schematically with a dashed line in Figure 5.30b. This asymmetry produces an eccentric loading scenario which leads to mild coupling between the bending and stretching deformations.

To begin, consider the *net* effect of the crack and forget, for the moment, that the additional deformation is highly localized. In this case, the global force-deformation relations are

$$\left\{ \begin{matrix} d \\ \theta \end{matrix} \right\} = [C] \left\{ \begin{matrix} P_1 \\ P_2 \end{matrix} \right\} \tag{5.77}$$

where d and θ are measures of the overall (global) axial and rotational deformations, respectively, $P_1 = P$ is the axial tension, $P_2 = M$ is the bending moment, and $[C]$ is the compliance matrix. In the absence of a crack, the compliance matrix is diagonal with terms $c_{11} = (L/AE)_o$ and $c_{22} = (L/EI)_o$, where the subscript indicates that this refers to the uncracked structure. The addition of the crack increases the diagonal elements and introduces off-diagonal terms coupling the axial and bending deformation:

$$[C] = \begin{bmatrix} \left(\frac{L}{EA}\right)_o + \Delta c_{11} & \Delta c_{12} \\ \Delta c_{12} & \left(\frac{L}{EI}\right)_o + \Delta c_{22} \end{bmatrix} \tag{5.78}$$

where Δc_{ij} is the change in flexibility which leads to the net increase in the global deformation variables. The appropriate expression for the change in flexibility is

$$\Delta c_{ij} = \frac{\partial^2}{\partial P_i \, \partial P_j} \int_0^a G(\bar{a}) d\bar{a} \tag{5.79}$$

where a is the depth of the crack, G is the elastic energy release rate, and P_i are the generalized forces described previously. It should be noted that this description of the compliance matrix is only valid for situations where the crack remains open. If the crack should close, the compliance will experience a discontinuity (a strong nonlinearity) and a very different analysis would be required. The limitations and applicability of the present open-crack theory are discussed in the next section.

The global change in compliance may be computed using Eqs. (5.76) and (5.79), along with the expressions for the stress intensity factors for a beam with an edge crack in bending and stretching. These are found in Ref. [39] and take the form

$$K_I^s = \frac{P}{bt} \sqrt{\pi a} \; F_1(a/b) \tag{5.80}$$

$$K_I^b = \frac{6M}{b^2 t} \sqrt{\pi a} \, F_2(a/b) \tag{5.81}$$

where $F_1(a/b)$ and $F_2(a/b)$ are

$$F_1(a/b) = \sqrt{\frac{2b}{\pi a} \tan\left(\frac{\pi a}{2b}\right)} \; \frac{0.752 + 2.02\left(\frac{a}{b}\right) + 0.37\left(1 - \sin\left(\frac{\pi a}{2b}\right)\right)^3}{\cos\left(\frac{\pi a}{2b}\right)} \tag{5.82}$$

and

$$F_2(a/b) = \sqrt{\frac{2b}{\pi a} \tan\left(\frac{\pi a}{2b}\right)} \; \frac{0.923 + 0.1999\left(1 - \sin\left(\frac{\pi a}{2b}\right)\right)^4}{\cos\left(\frac{\pi a}{2b}\right)} \tag{5.83}$$

From this formulation, the global stiffness matrix may be computed by simply inverting the flexibility matrix, Eq. (5.78).

Thus far, a global approach has been taken despite the fact that the additional deformation is known to be localized near the crack. This approach was taken because the expression for the change in compliance, Eq. (5.79), was defined using the global deformations d and θ. To transform this global approach into a local one, two steps are taken. First, the deformation measures are changed to the local measures of stretching strain and curvature. In other words, d and θ are replaced by $\epsilon_s(x)$ and $\kappa(x)$, respectively. As a result, the stiffness is multiplied by L. To localize the additional deformation, the changes in the stiffness ΔK_{ij} are multiplied by L and the Dirac delta function centered at the crack, x_o. This produces the following stiffness matrix:

$$[K] = \begin{bmatrix} (EA)_o - \Delta K_{11} L \delta(x - x_o) & \Delta K_{12} L \delta(x - x_o) \\ \Delta K_{12} L \delta(x - x_o) & (EI)_o - \Delta K_{22} L \delta(x - x_o) \end{bmatrix} \tag{5.84}$$

where the elements ΔK_{ij} may be computed by inverting both the uncracked global compliance matrix and the cracked global compliance matrix (Eq. 5.78) and subtracting the former from the latter.

5.7.3 Limitations

To this point, the discussion has focused on the open-crack scenario. In other words, crack closure and the associated discontinuity in the stiffness have not been considered. But under what circumstances is this a reasonable assumption? When does this model break down? To build some physical insight, first consider the static case. If the beam is subjected to a tensile load and a positive bending moment, the crack will always remain open; see Figure 5.30b. Conversely, a compressive load and a negative bending moment will lead to a completely closed crack. If the sign of the axial load and the moment are opposite, the status of the crack will depend on the relative magnitudes of the loads and the result is not immediately obvious. This is the underlying difficulty in the dynamic problem because the beam vibration causes the internal bending moment to periodically change sign. The objective here is to determine what combinations of the constant axial tension and the bending moment lead to the initiation of crack closure.

The initiation of closure may begin either at the crack tip or mouth, as shown schematically in Figure 5.30c and d, respectively. For the moment, focus on the crack tip closing scenario shown in Figure 5.30c. From linear elasticity, it has been shown that the displacement field near the crack tip follows a $\sqrt{r}$ law [38]. Specifically, $u(r) = (K_I/E)(8r/\pi)^{1/2}$, where the stress intensity factor is the sum of the stress intensity factors for bending and stretching: $K_I = K_I^b + K_I^s$. Using this expression along with Eqs. (5.80) through (5.83), the crack tip opening is described by

$$u(r) = \left[\frac{6M}{b} F_b(a/b) + PF_s(a/b) \right] \frac{\sqrt{\pi a}}{bE} \left(\frac{8r}{\pi} \right)^{1/2}. \tag{5.85}$$

As the crack tip begins to close, the displacement field of the crack drops to zero even for nonzero r. As a result, the crack tip closure criterion is found by setting the coefficient in Eq. (5.85) equal to zero. This yields

$$\frac{M}{Eb^3} = \frac{P}{Eb^2} \left[\frac{0.923 + 0.199\left(1 - \sin\left(\frac{\pi}{2}\frac{a}{b}\right)\right)^4}{6\left(0.752 + 2.02\left(\frac{a}{b}\right) + 0.37\left(1 - \sin\left(\frac{\pi}{2}\frac{a}{b}\right)\right)^3\right)} \right] \tag{5.86}$$

which is a straight line in the (P, M) parameter plane.

Next, consider the crack mouth closing scenario, as shown in Figure 5.30d. Crack closure begins when the mouth displacement due to bending is equal and opposite to the mouth displacement due to stretching. In other words, $\delta_b + \delta_s = 0$. Using the expressions presented in Ref. [39] for the mouth deflections, the criterion for mouth closure is

$$\frac{M}{Eb^3} = \frac{P}{Eb^2} \left[\frac{1.46 + 3.42\left(1 - \cos\left(\frac{\pi}{2}\frac{a}{b}\right)\right)}{6\cos^2\left(\frac{\pi}{2}\frac{a}{b}\right)\left(0.8 - 1.7\left(\frac{a}{b}\right) + 2.4\left(\frac{a}{b}\right)^2 + \frac{0.66}{(1-a/b)^2}\right)} \right]. \tag{5.87}$$

This is also a straight line in the (P, M) parameter plane.

Equations (5.86) and (5.87) are shown in Figure 5.31 with a solid and dashed line, respectively. This is for a crack depth of $a/b = 0.3$. As indicated, the combination of a tensile load $(P > 0)$ and a positive bending moment $(M > 0)$ leads to an open crack. Similarly, a compressive load $(P < 0)$ and a negative bending moment $(M < 0)$ will ensure a closed crack. However, the second and fourth quadrant of this figure are more complicated because they contain the parameter combinations that initiate crack closure. Fortunately, because this particular damage problem deals with open cracks, only the transition from an open crack to the initiation of closure is critical. The *initiation of closure* lines have been darkened in Figure 3. It is also

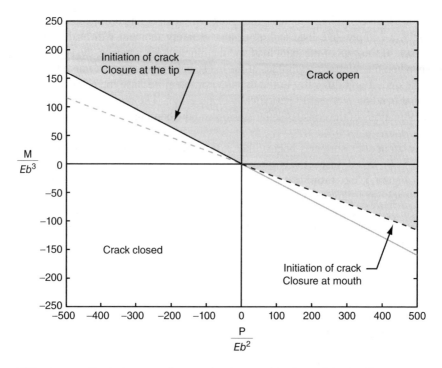

Figure 5.31 A normalized parameter diagram showing combinations of the bending moment M and the axial load P that leads to the initiation of crack closure

worth mentioning that under a tensile (compressive) load, the crack will always begin to close at the mouth (tip).

As discussed previously, the objective of this section is to focus on the dynamics of a translating beam containing an open crack. For the results of this section to remain valid, the amplitude of the bending moment must satisfy the following:

$$\left| \frac{M}{Eb^3} \right| < \left| \frac{P}{Eb^2} \left[\frac{1.46 + 3.42 \left(1 - \cos\left(\frac{\pi}{2} \frac{a}{b} \right) \right)}{6\cos^2\left(\frac{\pi}{2} \frac{a}{b} \right) \left(0.8 - 1.7\left(\frac{a}{b} \right) + 2.4\left(\frac{a}{b} \right)^2 + \frac{0.66}{(1 - a/b)^2} \right)} \right] \right| \quad (5.88)$$

where P is the specified axial load.

5.7.4 Equations of Motion

Hamilton's principle (see Section 4.3.5) is used to obtain the equations of motion for this system. To do so requires us to develop expressions for the internal strain energy, the kinetic energy, and the external work.

The strain energy of the beam may be written as

$$U = \int_0^L \frac{1}{2} \mathbf{q}^t [K] \mathbf{q} \, dx \quad (5.89)$$

where U is the strain energy, (K) is the stiffness matrix defined by Eq. (5.84), and $\mathbf{q}^t = \{\epsilon_s, \kappa\}$ is a generalized deformation vector. Under the typical small slope assumption, the curvature is simplified to $\kappa = v_{,xx}$. Also, ϵ_s is the stretching strain, which is expressed as $\epsilon_s = (u_{,x} + P/EA)$. The last term in this expression, P/EA, represents a static strain resulting from the constant axial load P.

Under the assumption that the crack is shallow and that the beam is long and thin, the coupling terms are extremely small and may be neglected [40]. Eliminating the off-diagonal terms leads to the following expression for the strain energy

$$U = \int_0^L \left[\frac{1}{2} EI v_{,xx}^2 + \frac{1}{2} EA \left(u_{,x} + \frac{P}{EA} \right)^2 \right] dx \tag{5.90}$$

where $EI = (EI)_o - \Delta K_{11} L \delta(x - x_o)$, $EA = (EA)_o - \Delta K_{22} L \delta(x - x_o)$, and the ΔK_{ij} are computed using the methods described previously.

The kinetic energy is given by

$$T = \frac{1}{2} \int_0^L \rho \left[(cv_{,x} + v_{,t})^2 + (c + cu_{,x} + u_{,t})^2 \right] dx \tag{5.91}$$

where $\rho = \rho_o - m\delta(x - x_o)$ is the mass per unit length of the beam, ρ_o is the mass per unit length of the uncracked beam, m is the reduction in the mass per unit length occurring at the crack location, and c is the axial transport speed of the beam.

Using the expressions for the potential energy, kinetic energy, and mass conservation $(c\partial\rho/\partial x + \partial\rho/\partial t = 0)$ with Hamilton's principle lead to the following uncoupled, linear partial differential equations which govern the unforced, undamped motion of a translating beam with an open crack:

$$\rho c^2 v_{,xx} + 2\rho c v_{,tx} + \rho v_{,tt} + (EI v_{,xx})_{,xx} - (P v_{,x})_{,x} = 0 \tag{5.92}$$

and

$$\rho c^2 u_{,xx} + 2\rho c u_{,xt} + \rho u_{,tt} - \left[EA \left(u_{,x} + \frac{P}{EA} \right) \right]_{,x} = 0. \tag{5.93}$$

These equations are recast using the full expressions for EI and EA, and the following dimensionless quantities: a new spatial coordinate $\xi = x/L$, two new deformation coordinates $V = v/L$ and $U = u/L$, and a dimensionless time $\tau = t\sqrt{EI_o/\rho_o L^4}$. The result is

$$(1 - \mu\delta)V_{,\tau\tau} + k_1(1 - \mu\delta)V_{,\xi\tau} + k_2(1 - \mu\delta)V_{,\xi\xi}$$
$$-k_3(\delta_{,\xi\xi}V_{,\xi\xi} + 2\delta_{,\xi}V_{,\xi\xi\xi} - \delta V_{,\xi\xi\xi\xi}) + V_{,\xi\xi\xi\xi} - k_4 V_{\xi\xi} = 0 \tag{5.94}$$

and

$$(1 - \mu\delta)U_{,\tau\tau} + k_1(1 - \mu\delta)U_{,\xi\tau} + k_2(1 - \mu\delta)U_{,\xi\xi}$$
$$+k_5[r\delta_{,\xi}U_{,\xi} - (1 - r\delta)U_{,\xi\xi}] = 0 \tag{5.95}$$

where $\mu = m/\rho_o$, $r = \Delta K_{11}L/EA_o$, $k_1 = 2c\sqrt{\rho_o L^2/EI_o}$, $k_2 = \rho_o L^2 c^2/EI_o$, $k_3 = \Delta K_{22}L/EI_o$, $k_4 = PL^2/EI_o$, $k_5 = A_o L^2/I_o$. Furthermore, it should be understood that $\delta = \delta(\xi - \xi_o)$. Spatial derivatives of $\delta(\xi - \xi_o)$ are evaluated using the technique described in Ref. [41].

5.7.5 Natural Frequencies and Stability

Now that we have a model for the cracked, translating beam, we just need to calculate the linear natural frequencies as the crack translates through the domain (between the supports) –so that the user can avoid driving the system at these frequencies.

To obtain the natural frequencies, the governing equations are discretized using a Galerkin procedure with the following expansions:

$$U = \sum_{i=0}^{n} \alpha_i(\tau)\Psi_i(\xi) \text{ and } V = \sum_{i=0}^{n} \beta_i(\tau)\Phi_i(\xi) \qquad (5.96)$$

where $\Psi_i(\xi) = \Phi_i(\xi) = \sin(i\pi\xi)$ for the simply supported boundary conditions under consideration here. These boundary conditions are justified by the fact that, in the wire EDM application, the wire wraps around a pulley-type mechanism at the boundary, which permits no displacement and applies no moment.

The discretized equations may be written in matrix form

$$[M]\ddot{\mathbf{x}} + [G]\dot{\mathbf{x}} + [K]\mathbf{x} = \mathbf{0} \qquad (5.97)$$

where $\mathbf{x} = \{\alpha_1, \alpha_2, ..., \beta_1, \beta_2, ...\}^t$, $[M]$ is the mass matrix, $[G]$ is the skew-symmetric gyroscopic matrix, and $[K]$ is the linear stiffness matrix. The natural frequencies for this system are found by rewriting Eq. (5.97) in first-order form and solving the associated eigenvalue problem numerically [42]. Note that the assumed solution will have the form: $\mathbf{x} = \mathbf{A}e^{\lambda t}$, where $\lambda = \text{Re}(\lambda) + i\text{Im}(\lambda)$. So the imaginary part is the natural frequency, while the real part gives an indication of the relative stability of the response ($\text{Re}(\lambda) < 0$ is stable).

5.7.6 Results

We first consider a beam with a thickness-to-length ratio of $b/L = 0.01$, a transport speed of half its critical speed $c/c_{cr} = 0.5$, and a crack depth of $a/b = 0.1$. This crack depth gives a mass loss ratio of $\mu = 0.1$. Using typical material properties and dimensions for an EDM wire, the nondimensional parameters become $k_1 = 3.164, k_2 = 2.503, k_3 = 2.57 \times 10^{-3}, k_4 = 0.151$, $k_5 = 1200$, $r = 1.27 \times 10^{-9}$, and the critical speed is $c_{cr} = 33$ m/s. Figure 5.32a shows the imaginary part of the first and second eigenvalues (natural frequencies) as a function of the crack location for this case. As the crack enters the domain, the natural frequencies are those of the uncracked, translating beam (given by the dashed lines). As the crack translates to the right with the beam, there is a change in the frequencies. Both the first and second mode frequencies begin to increase. The first mode frequencies continue to increase until a local maximum is achieved at the midspan and then they decrease monotonically until the crack exits the domain at $\xi_o = 1$. The second mode eigenvalue locus experiences two local maxima and a local minimum. Because the frequencies increase from the uncracked case, the mass loss is more significant than the stiffness loss ($\omega \approx \sqrt{k_{eff}/m_{eff}}$) in these shallow crack cases. Also, the darkened regions constitute *resonant zones*; a driving frequency in any of these regions would mean that a resonant condition would exist, albeit briefly, as the crack travels through the domain.

The stability portions of the eigenvalues are perhaps more interesting; Figure 5.32b shows how they vary. The first mode is fundamentally unstable ($\text{Re}(\lambda) > 0$) about the straight

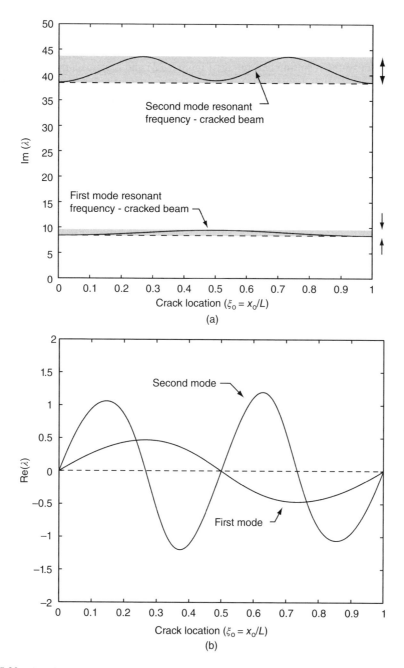

Figure 5.32 (a) The variation in the first and second natural frequency $(\mathrm{Im}(\lambda))$ as the crack translates through the domain at speed c, (b) the stability aspect of the complex eigenvalue $(\mathrm{Re}(\lambda))$ as the crack moves through the domain

configuration during the first half of the problem and stable in the second half. The second mode has two unstable portions.

These eigen results clearly tell a negative story about the wire EDM vibration problem – and the accuracy of the resulting cut. In fact, Figure 5.32a and b paints a pretty bleak picture. However, damping (not included thus far) would improve the vibration picture. This damping comes via the structure itself or from the flushing fluid, which is introduced into the kerf to eliminate removed material and to cool the system.

In this chapter we have introduced a number of different forms of structural damage and the associated modeling approaches; several of these we have illustrated through experiment. In subsequent chapters we will focus on the detection and identification of these types of damage.

References

[1] M. Williams, Stress singularities resulting from various boundary conditions, Journal of Applied Mechanics 19 (1952) 526–528.

[2] R. S. Barsoum, On the use of isoparametric finite elements in linear fracture mechanics, International Journal for Numerical Methods in Engineering 10 (1976) 25–37.

[3] R. D. Henshell, K. G. Shaw, Crack tip finite elements are unnecessary, International Journal for Numerical Methods in Engineering 9 (1975) 495–509.

[4] E. Z. Moore, J. M. Nichols, K. D. Murphy, Crack identification in a freely vibrating plate using Bayesian parameter estimation, Mechanical Systems and Signal Processing 25 (6) (2011) 2125–2134.

[5] L. E. Malvern, Introduction to the Mechanics of a Continuous Medium, Prentice-Hall, Englewood Cliffs, NJ, 1969.

[6] E. H. Dowell, Aeroelasticity of Plates and Shells, Noordhoff International Publishing, Leyden, The Netherlands, 1975.

[7] A. S. Vol'mir, Flexible Plates and Shells, Wright-Patterson AFB Technical Report: AFFDL-TR-66-216, Dayton, OH, 1967.

[8] K. D. Murphy, L. N. Virgin, S. A. Rizzi, The effect of thermal prestress on the free vibration characteristics of clamped rectangular plates: theory and experiment, Journal of Vibration and Acoustics 119 (1997) 243–149.

[9] R. V. Southwell, On the analysis of experimental observations in problems of elastic stability, Proceedings of the Royal Society of London 135A (1932) 601–616.

[10] S. A. Clevenson, E. F. Daniels, Capabilities of the Thermal Acoustic Fatigue Apparatus, NASA Technical Memorandum: 104106, Hampton, VA, 1992.

[11] Y. Zou, L. Tong, G. P. Steven, Vibration-based model-dependent damage (delamination) identification and health monitoring for composite structures – a review, Journal of Sound and Vibration 230 (2) (2000) 357–378.

[12] H. Schwarts-Givli, O. Rabinovitch, Y. Frostig, High-order nonlinear contact effects in cyclic loading of delaminated sandwich panels, Composites: Part B 38 (2007) 86–101.

[13] H. Luo, S. Hanagud, Dynamics of delaminated beams, International Journal of Solids and Structures 37 (2000) 1501–1519.

[14] G. W. Hunt, B. Hu, R. Butler, D. P. Almond, J. E. Wright, Nonlinear modeling of delaminated struts, AIAA Journal 42 (11) (2004) 2364–2372.

[15] K. D. Murphy, J. M. Nichols, A low-dimensional model for delamination in composite structures: theory and experiment, International Journal of Nonlinear Mechanics 44 (2009) 13–18.

[16] J. E. Wright, Compound bifurcations in the buckling of a delaminated composite strut, Nonlinear Dynamics 43 (1) (2004) 59–72.

[17] D. Young, R. Felgar, Tables of Characteristic Functions Representing Normal Modes of Vibration of a Beam, University of Texas Publications, 1949.

[18] J. M. Nichols, S. T. Trickey, M. Seaver, L. Moniz, Use of fiber optic strain sensors and holder exponents for detecting and localizing damage in an experimental plate structure, Journal of Intelligent Material Systems and Structures 18 (1) (2007) 51–67.

[19] K. D. Murphy, D. M. Ferriera, Thermal buckling of rectangular plates, International Journal of Solids and Structures 38 (2001) 3979–3994.

[20] D. P. Hess, Vibration- and shock-induced loosening, in: J. H. Bickford, S. Nasser (Eds.), Handbook of Bolts and Bolted Joints, Marcel Dekker, New York, 1998.

[21] N. G. Pai, D. P. Hess, Experimental study of loosening of threaded fasteners due to dynamic shear loads, Journal of Sound and Vibration 253 (3) (2002) 585–602.

[22] O. S. Bursi, J. P. Jaspart, Basic issues in the finite element simulation of extended end plate connections, Computers and Structures 69 (1998) 361–382.

[23] S. W. Doebling, C. R. Farrar, M. B. Prime, A summary review of vibration-based damage identification methods, Shock and Vibration Digest 205 (5) (1998) 631–645.

[24] M. Palacz, M. Krawczuk, Vibration parameters for damage detection in structures, Journal of Sound and Vibration 249 (5) (2002) 999–1010.

[25] M. D. Todd, J. M. Nichols, C. J. Nichols, and L. N. Virgin, An assessment of modal property effectiveness in detecting bolted joint degradation: theory and experiment, Journal of Sound and Vibration 275 (2004) 1113–1126.

[26] J.-N. Juang, Applied System Identification, Prentice-Hall, New York, 1994.

[27] A. Milanese, P. Marzocca, J. M. Nichols, M. Seaver, S. T. Trickey, Modeling and detection of joint loosening using output-only broad-band vibration data, Structural Health Monitoring 7 (4) 309–328.

[28] K. Worden, G. R. Tomlinson, Nonlinearity in Structural Dynamics, Institute of Physics Publishing, Bristol, Philadelphia, PA, 2001.

[29] E. Z. Moore, K. D. Murphy, E. G. Rey, J. M. Nichols, Modeling and identification of uniform corrosion damage on a thin plate using a Bayesian framework, Journal of Sound and Vibration 340 (2015) 112–125.

[30] D. Zhou, Vibrations of Mindlin rectangular plates with elastically restrained edges using static Timoshenko beam functions with the Rayleigh-Rritz method, International Journal of Solids and Structures 38 (32–33) (2001) 5565–5580.

[31] M. Hetenyi, Beams on Elastic Foundations, University of Michigan Press, Ann Arbor, MI, 1946.

[32] S. P. Timoshenko, Theory of Elastic Stability, McGraw-Hill, New York, 1961.

[33] Y. Weitsman, On foundations that react in compression only, Journal of Applied Mechanics 37 (1970) 1019–1030.

[34] Y. Weitsman, Onset of separation between a beam and a tensionless elastic foundation under a moving load, International Journal of Mechanical Sciences 13 (1971) 707–711.

[35] Y. Weitsman, A tensionless contact between a beam and an elastic half space, International Journal of Engineering Sciences 10 (1972) 73–81.

[36] L. Lin, G. G. Adams, Beam on tensionless elastic foundation, Journal of Engineering Mechanics 113 (1987) 542–553.

[37] W. H. Press, B. P. Flannery, S. A. Teukolsky, W. T. Vetterling, Numerical Recipes, Cambridge University Press, Cambridge, 1992.

[38] B. Lawn, Fracture of Brittle Solids, Cambridge University Press, Cambridge, 1993.

[39] H. Tada, P. Paris, G. Irwin, The Stress Analysis of Cracks Handbook, Del Research Corp., Hellertown, PA, 1972.

[40] K. D. Murphy, Y. Zhang, Vibration and stability of a cracked translating beam, Journal of Sound and Vibration 237 (2000) 319–335.

[41] D. J. Colwell, J. R. Gillett, A property of the Dirac delta function, International Journal of Mathematical Education in Science and Technology 18 (1987) 657–658.

[42] L. Meirovitch, Computational Methods in Structural Dynamics, Noordhoff Publishing, The Netherlands, 1980.

6

Estimating Statistical Properties of Structural Response Data

In Chapter 3, we defined a number of properties of stationary random processes. These properties are important in the analysis of structural response data, particularly in the detection and estimation of structural damage. However, up to this point, we have only considered these quantities in theory. In practice, we are handed sampled data and are required to *estimate* them. While we discuss estimation of damage parameters in Chapter (7), we also must think about how to approximate the expected value operator, correlation functions, and spectral properties from observations. Moreover, we need to consider how to quantify the accuracy of these estimates.

To set the stage, assume that we have placed a resistive strain gage on a concrete pile used in bridge construction. The pile is subject to dynamic wind, wave, and traffic loading and we record this vibrational response. Properly calibrated and conditioned, the sensor produces a signal in the ± 1 V range. Because we are unable to accurately predict the exact vibrational response due to the uncertainty in the loading, we model the response $y(t) \in [-1 \text{ V}, 1 \text{ V}]$ as a random process $Y(t)$. We have already shown that we can define a probability space on this voltage range and a function $P_Y(y(t))$ that assigns a probability to the event $Y(t) < y(t)$ at any time t (the cumulative distribution function, CDF). Correspondingly, we have shown that if we treat $Y(t)$ as a continuous random variable, we can also define $p_Y(y(t))$, i.e., a probability density function (PDF). The sensor records N discrete observations $\mathbf{y} \equiv (y(t_1), y(t_2), \cdots, y(t_N))$ from which we may also define the joint CDF and PDF, $P_\mathbf{Y}(\mathbf{y}), p_\mathbf{Y}(\mathbf{y})$. To begin, we must also assume that we have recorded K realizations of this random process $\mathbf{y}^{(k)}(t), k = 1 \cdots K$. That is to say, we have K samples recorded for each time $t_1, t_2, \cdots, t_N$ measured relative to the time at which recording begins.

Many of the quantities we would like to estimate involve the operator $E[\cdot]$, which we defined as an average, or expected value of a function of random variables with respect to their joint PDF (see Eq. 2.48). For example, assume we are interested in the expected value of the system response at time t_1 (relative to the start of the experiment) $E[Y(t_1)] \equiv \int p_Y(y(t_1)) y(t_1) dy(t_1)$. If all we have is a single realization (a single data value $y(t_1)$), we are obviously limited in trying to approximate this integral! However, with multiple realizations, a reasonable path forward

---

Modeling and Estimation of Structural Damage, First Edition. Jonathan M. Nichols and Kevin D. Murphy.
© 2016 John Wiley & Sons, Ltd. Published 2016 by John Wiley & Sons, Ltd.

is to approximate the integral with the Riemann sum

$$E[Y(t_1)] \approx \sum_{k=1}^{K} y^{(k)}(t_1) p_Y(y^{(k)}(t_1)) \Delta_{y(t_1)}. \tag{6.1}$$

To implement (6.1), we would take each of the K data values and plug into the (known) probability model, discretized to uniform intervals $\Delta_{y(t_1)}$, and then carry out the sum. This approach weights each of the data values $y^{(k)}(t_1)$ by its probability of occurrence $p_Y(y^{(k)}(t_1))\Delta_{y(t_1)}$. Of course, in some cases we may not know (or may be unwilling to specify) the underlying probability model.

Instead, consider the approximation to the PDF shown in Figure 6.1. In this case, the size of each interval in the Riemann approximation is allowed to vary such that it encompasses a single sample. Denote these intervals as $\Delta_{y(t_1)}^{(k)}, k = 1 \cdots K$. Now recall that the definition of a PDF is the probability of lying in a given interval normalized by the interval length. By construction, each of our samples appears in exactly one bin with probability $1/K$; thus, the PDF approximation is $p_Y(y^{(k)}(t_1)) \approx \frac{1}{K\Delta_{y(t_1)}^{(k)}}$ and we have (for arbitrary PDF)

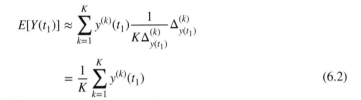

$$E[Y(t_1)] \approx \sum_{k=1}^{K} y^{(k)}(t_1) \frac{1}{K\Delta_{y(t_1)}^{(k)}} \Delta_{y(t_1)}^{(k)}$$

$$= \frac{1}{K} \sum_{k=1}^{K} y^{(k)}(t_1) \tag{6.2}$$

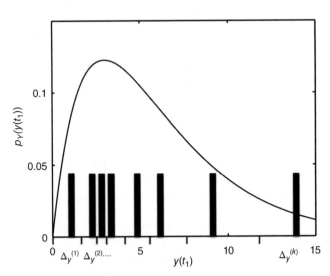

Figure 6.1 Illustration of one possible discrete approximation of a probability density function. In this example, each "bin" is adjusted to include exactly one sampled value. Thus, we have K variably sized bins. In this case, the approximation to the probability density is $\frac{1}{K\Delta_{y(t_1)}^{(k)}}$ which, when substituted into the Riemann approximation to the expected value operator (Eq. 6.1), yields the sample mean as an estimator.

We would further expect this approximation to improve with more data, because a more densely sampled PDF would tend to yield smaller intervals in the Riemann approximation. As we see shortly in Section 6.3, there is an important theorem that relates an expectation of any function of a random variable to a normalized sum of sample values.

Although this is a crude first example, what we have just illustrated is the basic problem of point estimation: given data $\mathbf{y}$ modeled with joint PDF $p_{\mathbf{Y}}(\mathbf{y})$ and containing unknown parameter(s) $\boldsymbol{\theta} \equiv (\theta_1, \theta_2, \cdots)$, compute a test statistic $s(\mathbf{y}) = \hat{\theta} \approx \theta$. In this example, the parameter was $\theta = E[Y(t_1)] \equiv \mu_Y$, and the test statistic was $s(\mathbf{y}) = 1/K \sum_k y^{(k)}(t_1)$.

A natural question to ask is: How close is our estimate to the true value? Does the error in the estimate improve with more data? These are well-studied questions, and a formal framework exists for supplying answers. Perhaps of more practical concern is whether we can make such an estimate in the absence of multiple realizations (a question answered in the affirmative in Section 6.3). After all, this example has focused on a particular instance in time "t_1" and on how to make use of multiple observations recorded at that time. Of course, this is *not* the type of data we are typically asked to work with in practice. It is far more common to be given a *single* realization of our random process (i.e., a time series) recorded at some sampling interval and to be tasked with estimating parameters from this signal alone. What we would like to be able to do is take a *temporal* average from a single realization, say $k = 1$, and write

$$E[Y(t_1)] \approx \frac{1}{N} \sum_{n=1}^{N} y^{(1)}(t_n). \tag{6.3}$$

We show in Section 6.3 that this more familiar estimator is, in fact, a good one provided certain assumptions about the nature of the signal are made. However, absent those assumptions, we cannot use (6.3). The reason, of course, is that the whole concept of an expectation is taken with regard to the random variable, not "time." Nonetheless, it is important to keep in mind that absent any assumptions about the nature of the time series (stationarity and the yet-to-be-defined ergodicity), we can still estimate meaningful quantities so long as we remember to treat each observation (each point in time) as a realization of a random variable following a given probability model. Although more thorough treatments of estimation are given in Refs [1]–[3], a formal estimation framework is important enough to our end goals to warrant discussion here. The next two sections provide such a framework.

Before proceeding, a few points to make regarding notation. When referring to an estimated quantity, it is common to use a carrot "$\hat{\cdot}$" over the quantity. Secondly, we note with the $\approx$ symbol that our test statistic (our estimate) will never be exactly equal to the unknown parameter because of the uncertainty inherent in a probabilistic model. Finally, we should mention that in estimation theory, the more general expression would be $s(\mathbf{y}) = \hat{\theta} = \hat{\Gamma}(\Theta)$, that is to say, the quantity we are estimating might be a *function* of a different, unknown parameter vector, denoted Θ [3]. This situation is quite common and will, in fact, be encountered in this book (see, e.g., estimation of information theoretics which are functions of the cross-correlation coefficient, Section 6.6). Perhaps the simplest example is using an estimate for the sample variance $\hat{\Theta} = \hat{\sigma}^2$ to estimate the standard deviation $\hat{\theta} = \hat{\sigma}$. In this case, $\Gamma(\cdot)$ is just the $\sqrt{\cdot}$ function. It is shown in the next section that if we can produce good estimates of our parameter, then functions of that parameter are also well estimated.

6.1   Estimator Bias and Variance

Before we can talk about methods for generating good estimators, we must state specifically what we want in a "good" estimator. There can be many different estimators of a parameter and we need some objective way of deciding among them. As is often the case, we define a good estimator as one that minimizes the mean square error (MSE)

$$
\begin{aligned}
MSE(\hat{\theta}) &= E[(\hat{\theta} - \theta)^2] \\
&= E[((\hat{\theta} - E[\hat{\theta}]) - (\theta - E[\hat{\theta}]))^2] \\
&= E[(\hat{\theta} - E[\hat{\theta}])^2] - 2(\theta - E[\hat{\theta}])E[\hat{\theta} - E[\hat{\theta}]] \\
&\quad + E[(\theta - E[\hat{\theta}])^2] \\
&= E[(\hat{\theta} - E[\hat{\theta}])^2] + E[(\theta - E[\hat{\theta}])]^2 \\
&= var(\hat{\theta}) + b(\hat{\theta})^2 \qquad\qquad\qquad\qquad\qquad (6.4)
\end{aligned}
$$

where, in the last line, we have introduced the *bias* of the estimator as $E[(\theta - E[\hat{\theta}])]$. The bias, b, is simply how far off, on average, our estimate is from the true parameter value. The first term, the estimator variance, is our familiar definition of variance for a random variable. Thus, we see that the MSE for an estimator is composed of two parts, bias and variance. We will therefore define a good estimator to be one that ideally has no bias and as small a variance as possible. There is a rich body of work devoted to finding minimum variance, unbiased estimators (which are not guaranteed to exist), both for finite sample sizes K, and for the large sample cases where $K \to \infty$. A thorough discussion can be found in Kay [2].

We first consider a simple example that illustrates both properties. Assume we have been given K realizations of a random variable Y, denoted as $y^{(k)}, k = 1 \cdots K$. Considering our approximation to the expected value operator from the previous section, we have from Chapter (2), Eqs. (2.45, 2.50) and our approximation to the expected value operator (6.2) that the mean and variance of a random variable Y may be estimated by

$$
\hat{\mu}_Y = \frac{1}{K} \sum_{k=1}^{K} y^{(k)}
$$

$$
\hat{\sigma}_Y^2 = \frac{1}{K} \sum_{k=1}^{K} (y^{(k)} - \hat{\mu}_Y)^2 \qquad\qquad\qquad\qquad (6.5)
$$

For these estimators, we can calculate the bias for each as

$$
b_\mu = E\left[\frac{1}{K} \sum_{k=1}^{K} Y^{(k)} \right] - \mu_Y = \frac{1}{K} \sum_{k=1}^{K} E[Y^{(k)}] - \mu_Y = \frac{K\mu_Y}{K} - \mu_Y = 0
$$

$$
b_{\sigma^2} = E\left[\frac{1}{K} \sum_{k=1}^{K} ((Y^{(k)} - \mu_Y) - \frac{1}{K} \sum_{k'=1}^{K} (Y^{(k')} - \mu_Y))^2 \right] - \sigma_Y^2
$$

$$
= \frac{1}{K} \sum_{k=1}^{K} E[(Y^{(k)} - \mu_Y)^2] - \frac{2}{K^2} \sum_{k=1}^{K} E[(Y^{(k)} - \mu_Y)^2]
$$

$$+ \frac{1}{K^3} \sum_{k=1}^{K} \sum_{k'=1}^{K} E[(Y^{(k')} - \mu_Y)^2] - \sigma_Y^2$$

$$= -2\frac{1}{K}\sigma_Y^2 + \frac{1}{K}\sigma_Y^2 = -\frac{\sigma_Y^2}{K}. \tag{6.6}$$

Thus, while our mean estimate is unbiased, our variance parameter estimate is not. However, it is *asymptotically* unbiased, that is to say,

$$\lim_{K \to \infty} E[\hat{\sigma}_Y^2] - \sigma_Y^2 = 0 \tag{6.7}$$

so that for large data sets we can expect good results. The variances of each estimator (6.5) can be similarly calculated and are found to be [1]

$$var_\mu = \frac{\sigma_Y^2}{K}$$

$$var_{\sigma^2} = \frac{2\sigma_Y^4}{K}. \tag{6.8}$$

Again, we see that asymptotically (for large K) the estimators will perform well, possessing a vanishingly small variance. An estimator for which this property holds is referred to as *consistent*. Although our estimator is consistent, in general, this does not mean that the estimator has the smallest possible variance for finite K. It turns out that, given the joint distribution of the data $p_Y(\mathbf{y}|\theta)$, we can derive a lower bound on the minimum possible error variance for any estimator of θ possessing either a constant or zero bias (see, e.g., [1, 2]). Specifically, given an unknown parameter vector $\theta = (\theta_1, \theta_2, \cdots, \theta_M)$, define the $M \times M$ Fisher information matrix

$$I_{ij}(\theta) = -E\left[\frac{\partial^2 \ln(p_Y(\mathbf{y}|\theta))}{\partial \theta_i \partial \theta_j}\right] \quad i, j = 1 \ldots M \tag{6.9}$$

in which case it can be shown that

$$E[(\hat{\theta}_i - \theta_i)^2] \geq I^{-1}(\theta)_{ii} \quad i = 1 \ldots M \tag{6.10}$$

provided that the condition $E[\partial \ln(p_Y(\mathbf{y}|\theta))/\partial\theta] = 0$ (the so-called regularity condition) is met. The limit implied by Eq. (6.10) is referred to as the Cramér–Rao lower bound (CRLB) and is the minimum conditional error variance possible for an unbiased estimator (note that the CRLB for a biased estimator can also be calculated and is given on page 400 of [1]). This bound represents the smallest theoretically attainable variance of any estimator and is therefore an excellent yardstick with which to measure estimator performance [1, 3]. An unbiased estimator may or may not achieve the CRLB but can never be better.

Again, we illustrate with a simple example. Assume now that our K realizations are collected from a random process we model as a sequence of independent, normally distributed random variables with variance σ_Y^2 and mean μ_Y. The joint PDF is

$$p_Y(\mathbf{y}) = \frac{1}{\left(2\pi\sigma_Y^2\right)^{K/2}} e^{-\frac{1}{2\sigma_Y^2} \sum_{k=1}^{K} (y^{(k)} - \mu_Y)^2} \tag{6.11}$$

in which case we have

$$
I(\theta) = -E \begin{bmatrix} -\dfrac{K}{\sigma_Y^2} & -\dfrac{1}{\sigma_Y^4}\sum_{k=1}^{K} Y^{(k)} - \mu_X \\[2ex] -\dfrac{1}{\sigma_Y^4}\sum_{k=1}^{K} Y^{(k)} - \mu_Y & \dfrac{N}{2\sigma_Y^4} - \dfrac{1}{\sigma_Y^6}\sum_{k=1}^{K}(Y^{(k)} - \mu_Y)^2 \end{bmatrix}
$$

$$
= \begin{bmatrix} \dfrac{K}{\sigma_Y^2} & 0 \\[2ex] 0 & \dfrac{K}{2\sigma_Y^4} \end{bmatrix} \tag{6.12}
$$

so that upon inversion, we have the CRLB for both parameters

$$
E[(\hat\mu_Y - \mu_Y)^2] \geq \frac{\sigma_Y^2}{K}
$$

$$
E[(\hat\sigma_Y - \sigma_Y)^2] \geq \frac{2\sigma_Y^4}{K}. \tag{6.13}
$$

This demonstrates that our estimators (6.5) achieve the CRLB and are therefore referred to as *efficient* estimators. In the sense of minimizing (6.4), an estimator that is both unbiased and efficient is the best we can hope to do.

Deriving estimator bias and variance is a bit more challenging for most structural estimation problems of interest. We discuss both of these quantities in the context of spectral estimation in Sections 6.4.1 and 6.4.2. In fact, bias and variance of spectral estimators relate directly to the fundamental limit with which we can accurately detect the presence of structural damage using the approaches described in Section (8.3.2). The case study in Section (8.3.3), in fact, shows how the probability of damage detection is influenced by the variance of a bispectral estimator.

In the next section, we also provide an approach to estimation that is guaranteed to reach the CRLB asymptotically as the number of data used in the estimation becomes large. These estimators are shown to have a number of other useful properties that we can leverage in estimating properties of structures and, ultimately, in helping us detect and identify those properties related to structural damage.

Before leaving this section, however, it is worth mentioning one particularly interesting fact about the CRLB. Obviously, the CRLB depends on both the signal model (captured by the parameters θ) and the probability model $p_Y(y|\theta)$. Interestingly, the CRLB associated with a jointly Gaussian probability model is the highest of all probability models that obey the aforementioned regularity condition [4]. This is a fascinating statement to make. It says that Gaussian noise is the *worst* possible contaminate from the perspective of estimation, as it guarantees that the smallest variance we could hope to achieve in the estimator is larger than for any other distribution! That perhaps the easiest distribution to handle from an analytical perspective is the worst possible case from an estimation point of view is certainly frustrating. However, it points to a very practical and useful point: namely, that *if* the corrupting noise source is non-Gaussian in nature, making the Gaussian assumption when constructing the estimator is a bad idea. We should, instead, take advantage of the nonnormality of the corrupting noise and design a more accurate estimator. We provide a specific example of this later in Section (7.1.5) in an ultrasonic damage detection application.

6.2 Method of Maximum Likelihood

While we have described simple estimators for basic statistical properties of a random variable (mean and variance) and properties of those estimators (bias and variance), we have said nothing about how to find such estimators. For these simple quantities, we relied on basic intuition to arrive at the sample mean as an approximation to an expected value. However, for more complicated models, we require a less heuristic means of estimating the quantities of interest. The approach we describe in this section provides a formal approach to generating reliable estimators that minimize Eq. (6.4).

To introduce this approach, we begin our discussion with the same simple estimation problem we considered at the beginning of the chapter. We are given a collection of K realizations of a random variable $Y(t_1)$, denoted $y^{(k)}(t_1), k = 1 \cdots K$. This time, we give a specific probability model for each of these observations, $p_Y(y^{(k)}(t_1)) = \dfrac{1}{\sqrt{2\pi\sigma_Y^2}} e^{-\frac{1}{2\sigma_Y^2}(y^{(k)}(t_1)-\mu_Y)^2}$. Now let us further assume that each realization $\mathbf{y}(t_1) \equiv y^{(k)}(t_1), k = 1 \cdots K$ is independent such that we have for the joint probability model

$$p_{\mathbf{Y}}(\mathbf{y}(t_1)|\boldsymbol{\theta}) = \prod_{k=1}^{K} \frac{1}{\sqrt{2\pi\sigma_Y^2}} e^{-\frac{1}{2\sigma_Y^2}(y^{(k)}(t_1)-\mu_Y)^2} = \frac{1}{(2\pi\sigma_Y^2)^{K/2}} e^{-\frac{1}{2\sigma_Y^2}\sum_{k=1}^{K}(y^{(k)}(t_1)-\mu_Y)^2}. \qquad (6.14)$$

This particular model depends on the parameter vector $\boldsymbol{\theta} \equiv (\mu_Y, \sigma_Y^2)$, and we have made this dependency explicit in the notation. One can read (6.14) as the probability of observing the data at time t_1 given the model parameters $\boldsymbol{\theta}$. However, from the perspective of estimation, where the acquired data set is fixed, it is often more appropriate to think of (6.14) as a function of the unknown parameters rather than the data in which case we write

$$p_Y(\boldsymbol{\theta}|\mathbf{y}(t_1)) = p_Y(\mathbf{y}(t_1)|\boldsymbol{\theta}) = \frac{1}{(2\pi\sigma_Y^2)^{K/2}} e^{-\frac{1}{2\sigma_Y^2}\sum_{k=1}^{K}(y^{(k)}(t_1)-\mu_Y)^2} \qquad (6.15)$$

and refer to the distribution as the *likelihood* function, read as the PDF of the model parameters given the data. In this example, the model parameters are the mean, θ_1, and variance θ_2. Note that functionally the probability of observing the data and the likelihood function are entirely equivalent. The only difference is which of the two quantities (data or model parameters) we consider to be unknown. This is consistent with our understanding of probability as a model of uncertainty. In estimation, we are almost always given the data and view the joint PDF (6.15) as a function of model parameters in which case it is formally the likelihood function, although we will sometimes refer to both notations as such.

Now, returning to the estimation problem, assume for the moment that we know σ_Y^2 but do not know the mean μ_Y. A reasonable approach to estimation is to find the value of μ_Y that maximizes the probability of having observed the data, i.e., maximize (6.14) [5]. This is, in fact, the method of maximum likelihood (ML), which is a powerful estimation approach that is revisited in the context of structural parameter estimation in Chapter (7). Parameter estimates obtained by maximizing the likelihood function are therefore referred to as maximum likelihood estimates (MLEs).

For many probability models, it makes more sense to maximize the logarithm of the distribution rather than the distribution itself. Because the logarithm is a monotonically increasing function, it does not change the parameter value at which the maximum occurs. With that in mind, we examine the log of our joint normal probability model

$$\log[p_{\mathbf{Y}}(\mathbf{y}(t_1))] = -\frac{K}{2}\log(2\pi\sigma^2) - \left[\frac{1}{2\sigma_Y^2}\left(\sum_{k=1}^{K} y^{(k)}(t_1) - \mu_Y\right)^2\right] \tag{6.16}$$

so that maximizing gives

$$\frac{\partial \log(p_{\mathbf{Y}}(\mathbf{y}(t_1)|\mu_Y))}{\partial \mu_Y} = -\frac{\partial}{\partial \mu_Y}\sum_{k=1}^{K}([y^{(k)}(t_1)]^2 - 2\mu_Y y^{(k)}(t_1) + \mu_Y^2) = 0$$

$$\rightarrow \sum_{k=1}^{K}(-2y^{(k)}(t_1) + 2\mu_Y) = 0$$

$$\rightarrow 2K\mu_Y = 2\sum_{k=1}^{K} y^{(k)}(t_1)$$

$$\rightarrow \hat{\mu}_Y = \frac{1}{K}\sum_{k=1}^{K} y^{(k)}(t_1) \tag{6.17}$$

which is the same estimator we derived heuristically before using a Riemann approximation to the expected value operator. Of course, we have assumed that (i) the derivative of the log-likelihood exists, (ii) that the point found is a maximum (not a minimum), and (iii) that it a unique maximizer of the likelihood. For the second requirement, it is sufficient to demonstrate that $\partial^2 \log[p_{\mathbf{Y}}(\mathbf{y}|\hat{\theta})]/\partial\hat{\theta}^2 < 0$. However, the first assumption suggests that the log-likelihood be a smooth, differentiable function. Indeed, this is the same regularity condition mentioned earlier in the context of the CRLB discussion and is required for the existence of an MLE. The last requirement speaks to the issue of *identifiability*. If the maximum is not unique, neither will be the identified parameters, hence a unique MLE does not exist in this case. In structural estimation problems, it is usually safe to assume that (i) and (ii) hold, while the issue of identifiability can be problematic. A specific example demonstrating the issue of identifiability is given in Section (7.1).

Now, unlike the estimator based on the Riemann approximation, there are certain important guarantees that are possessed by an MLE. It can be shown that asymptotically (as the number of realizations K becomes large) an MLE is normally distributed, unbiased, and efficient [3]. The fact that the MLE will asymptotically reach the CRLB is one of the prime reasons MLEs are so widely sought. Given all the other sources of uncertainty present in most estimation problems, it is very useful to at least claim that estimation errors have been minimized. We also rely on the fact that MLEs are approximately normally distributed when developing confidence intervals for our test statistics in Section 6.9.3.

Another outstanding property of MLEs states that if $\theta = \Gamma(\Theta)$ (the parameter of interest is a function of another parameter) then $\hat{\theta} = \Gamma(\hat{\Theta})$. That is to say, an ML estimator of Θ can be substituted into $\Gamma(\cdot)$ to yield an ML estimator of θ. This is referred to as the *invariance* property of the ML estimator [3]. We rely on this invariance property in developing estimators for spectral properties as well as information theoretics later in this chapter.

Now, the above-described properties may only be leveraged in the case where we have many observations. The good news is that for many engineering problems, data are plentiful as there is little cost to collecting many observations of the physical process of interest. The result is that we are nearly always in asymptopia[1] where the "large K" assumption is a good one. Hence, the method of ML offers perhaps the best available approach to generating point estimates of structural parameters.

Before concluding this section, we consider a more relevant example than those just discussed. After all, our aim is to eventually estimate more than just the mean or variance of a single temporal observation. Consider now the situation where we have a deterministic, physics-based model that predicts a structure's response at time t_1 and is itself a function of an unknown parameter vector θ (e.g., stiffness, damping, crack length, etc.). This basic situation was already introduced to the reader at the end of Chapter (3), Section (3.6). In this case, we would write for the kth realization for the response at time $t = t_1$ as

$$y^{(k)}(t_1) = x(t_1|\theta) + \eta^{(k)}(t_1). \qquad (6.18)$$

The $\eta^{(k)}(t_1)$ adds the uncertainty due to measurement and modeling error and is treated as a random variable H. As is commonly the case, we model the uncertainty with the normal PDF

$$p_H(\eta(t_1)) = \frac{1}{\sqrt{2\pi\sigma^2}} e^{-\frac{1}{2\sigma^2}(\eta(t_1))^2}. \qquad (6.19)$$

Now, if we prescribe that, for each realization we collect, the noise values are independently chosen from the *same* normal distribution, we can follow the procedure that led to Eq. (3.139) by noting that $\eta^{(k)}(t_1) = y^{(k)}(t_1) - x(t_1|\theta)$. Given that the joint density for the noise is the product of the marginal densities (by the assumption of independence), the likelihood function becomes

$$p_{\mathbf{H}}(\theta|\mathbf{y}) = \frac{1}{(2\pi\sigma^2)^{K/2}} e^{-\frac{1}{2\sigma^2}\sum_{k=1}^{K}(y^{(k)}(t_1)-x(t_1|\theta))^2} \qquad (6.20)$$

This is the likelihood function for the parameter vector θ given the data set (K realizations) and our model. Note that the model response, $x(t_1|\theta)$ plays the role of the "mean" of the random variable Y in (6.20). Note also that the PDF retains the subscript H, denoting that it is the additive noise PDF that gives rise to the likelihood function. This is always the case for the problems considered in this book. Absent any uncertainty, the concept of a PDF makes no sense, nor does the concept of estimation.

As we have already shown, maximizing this function is equivalent to minimizing the argument of the exponent (logarithm of the PDF), hence our MLE is given by

$$\hat{\theta} = \min_{\theta} \sum_{k=1}^{K} (y^{(k)}(t_1) - x(t_1|\theta))^2 \qquad (6.21)$$

---

[1] While this is not a technical term, it is one that I heard used frequently by my father throughout his career. It conjures up images of a magical place where asymptotic approximations are valid, and I have always found it useful in referring to such situations.

which says to find the parameters that minimize the MSE between data and model. This is common to most engineers as an approach for performing, among other things, linear regression analysis. However, it is worthwhile to remember the underlying assumptions: normally distributed, independent noise values chosen from the same PDF. Different models for the uncertainty $\eta(t)$ would have led to a different estimator. However, as we see in later chapters, finding an algorithm that can accurately solve for MLEs even in the Gaussian noise case is not a simple task.

As a non-Gaussian example, imagine we have the same physics-based model, but where the corrupting noise source is modeled using a Laplace distribution

$$p_H(\eta) = \frac{1}{2\sigma} e^{-|\eta|/\sigma}. \tag{6.22}$$

If we walk through the same procedure as for the Gaussian noise case (assuming independent noise values), we arrive at the Laplacian MLE

$$\hat{\theta} = \min_{\theta} \sum_{k=1}^{K} |y^{(k)}(t_1) - x(t_1|\theta)|. \tag{6.23}$$

In fact, this estimator is demonstrated in Section (7.1.5) to be the best estimator (closest to the CRLB) of ultrasonic pulse delay when the corrupting noise source is, in fact, Laplacian. Other noise models will lead to still other forms of the MLE. The main point here is that it is the form of the noise model that gives rise to the form of the MLE.

Now, of course, it is not very common to consider the system response at a single point in time, t_1 when estimating parameters, nor is it common to possess multiple realizations. Our reason for doing so up to this point was simply because we were interested in approximating expected values that are defined in terms of ensemble averages of a random variable. For the task of parameter estimation (as opposed to approximating expected values), we are not so limited.

Typically, we are asked to perform the estimation from a temporal sequence $y(t_n) = x(t_n|\theta) + \eta(t_n), n = 1 \cdots N$. We can imagine the example from the beginning of the chapter, where $y(t_n)$ are the measured data and $x(t_n|\theta)$ is our model of how we predict the concrete pile to respond. Again, considering each noise value to be independently chosen from the same Gaussian PDF, we arrive at the MLE

$$\hat{\theta} = \min_{\theta} \sum_{n=1}^{N} (y(t_n) - x(t_n|\theta))^2. \tag{6.24}$$

Note that we only arrive at this form by first assuming that each noise value is from the same PDF. Variability in the noise PDF with time would yield a different estimator. However, it is common in structural dynamics to assume the noise process is stationary, hence the MLE given by (6.24) is, in general, a good estimator of structural model parameters. We return to this estimator throughout the book, particularly in Sections (7.1), (9.1) and (9.3). Note that even though the random noise process may be stationary, the random process model for the measurements may not be. Assume, for example, that the model $x(t_n|\theta)$ is a free-decay response such that the signal possesses an amplitude A at $t_0 = 0$ and an amplitude of 0 at $t_n \rightarrow \infty$. Even if the noise is stationary, clearly the random process $Y(t)$ will not be. From the perspective of estimating model parameters, this is not a problem so long as we have a model.

Previous examples aside, much of our discussion in this introduction to estimation has focused on problems for which we have multiple realizations K of a random variable. Our reason for doing this is that, although uncommon in practice, using multiple realizations is the right way to think about probability and expectation, the key ingredients of an estimation problem. As a result, when we considered approximating expectations, we required no assumption about the temporal stationarity of the process being observed. In the next section, however, we demonstrate how additional assumptions on the nature of a temporal sequence of observations can free us from the need to collect multiple realizations. For signals that obey these assumptions, expectations can be approximated using a single realization, an extremely important and, in many cases, practically necessary property that we make heavy use of in estimating properties of structural response data.

6.3   Ergodicity

We now turn our attention to a concept that is extremely important in estimation. In the earlier discussion, we had to assume multiple realizations to estimate expected values of functions of random variables. Indeed, the whole notion of an expected value is defined with respect to an ensemble of observations of a random variable. In some applications, it may be feasible to acquire such a collection, however in many cases it will not. Given a single realization, $y(t)$, is there any way we can estimate, for example, $E[y(t_1)]$? It turns out that the answer is yes, provided the observed data obey certain properties.

We begin with some notation. Let $y(t)$ be the output of a dynamical system governed by a system of d differential equations $\dot{\mathbf{y}} = f(\mathbf{y}(0))$. These equations could describe the vibrational response of a multi-degree-of-freedom (MDOF) structure, for example, and is assumed to include both deterministic and stochastic components. In practice, we typically place a sensor on the structure and measure one of the outputs $y(n) \equiv y(n\Delta_t), n = 1 \cdots N$ at sample times $n\Delta_t$, where Δ_t is our usual sampling interval. If we treat the response as a random process, we can define $P_Y(y(t))$ as the CDF assigning a probability to the random variable $y(t)$ or, in discrete terms, $y(n)$.

Now, the particular realization we are dealing with is defined by the initial conditions, $\mathbf{y}(0) \in \mathbb{R}^d$ and the solution can be written, in general, as $\mathbf{y}(n) = \Gamma^n(\mathbf{y}(0))$. The function $\Gamma^n(\mathbf{y}(0))$ iterates the system dynamics forward n steps in time. For example, given that we know the state of this system at time $t = 0$, $\Gamma^3(\mathbf{y}(0))$ provides $\mathbf{y}(3\Delta_t)$. Applying Γ three consecutive times evolves our system forward three time steps. One of the keys to ergodic theory is to make the assumption that $\Gamma^n(\cdot)$ is *measure preserving*. A full discussion of measure theory is beyond the scope of this book. Suffice to say a "measure" assigns a number to a set Y and is usually denoted $\mu(Y)$ (not to be confused with the mean, μ_Y). In our case, the measure of interest will always be the probability measure $\mu(Y) \equiv P_Y(y(n))$ which assigns a probability to an event $y(n)$, defined on the space of possibilities Y. The "measure preserving" requirement can therefore be stated

$$P_Y(y(0)) = P_Y(\Gamma^{-n}(y(n))) \tag{6.25}$$

which states that the function that assigns a probability to our system at discrete time $n = 0$ is equally valid at discrete time n, i.e., the dynamics of the system do not change the CDF. If we recall, for most practical situations of interest, stationarity makes the same assumption, that is to say, $P_Y(y(1)) = P_Y(y(n))$ for any $n \in [1, N]$. Stationarity is therefore an assumption embedded in the concept of ergodicity. While ergodicity is more general (a stationary system is not always ergodic), we will always deal with random processes where assuming stationarity means we can apply the results of the forthcoming ergodic theorem.

Given the above notation, we first consider the early work of Poincaré [6]. The Poincaré recurrence theorem is stated concisely in Ref. [7]. Let Γ represent a measurable transformation $\Gamma : Y \to Y$ preserving a finite measure $\mu(Y)$ on metric space Y with distance D. Then, for almost every $\mathbf{y}(n) \in Y$ we have

$$\lim_{N \to \infty} \inf D(\Gamma^N \mathbf{y}(n), \mathbf{y}(n)) = 0$$

where

$$D(\mathbf{a}, \mathbf{b}) \equiv \|\mathbf{a} - \mathbf{b}\|_2^2 \tag{6.26}$$

is the Euclidean distance between the arguments. The term "metric space" simply refers to a set of points on which we can define a distance. In other words, Poincaré's theorem tells us that, given enough time, the orbit of points $\mathbf{y}$ which belong to subsets Y of finite measure returns arbitrarily close to the initial point. It is easy to imagine that if we monitored the output of a dynamical system for long enough, we could eventually find time indices n, m for which $D(y(n), y(m)) \approx 0$. If this were the case, we might think of treating each one-dimensional observation $y(t_n)$ as an independent realization of $y(t_m)$. In fact, if we denote N_ϵ the number of "nearby" points that satisfy $|y(t_n) - y(t_m)|_2^2 < \epsilon/2 \forall n \in [1, N]$, it would seem sensible to approximate the probability of the value $y(t_m)$ occurring as N_ϵ/N. Thus, for ϵ small, an estimate of the PDF is given by

$$p_Y(y(t_m)) \approx \frac{N_\epsilon}{\epsilon N} \tag{6.27}$$

In short, if we wait long enough (for large enough N), the observations will populate a region local to $y(t_m)$ with a relative frequency proportional to the probability of occurrence. Values of $y(t_m)$ which are not probable will have few N_ϵ and vice versa. Equation (6.27) describes the simple histogram approach that most of us have used throughout our careers and frees us from the need to collect multiple realizations as we required in (for example) Eq. (6.1). However, this result is not of tremendous practical value as it simply tells us that a stationary sequence will "recur" in a way that allows us to understand the basic probability distribution of our random variable. Moreover, the recurrence theorem provides no clear path toward estimating general statistical properties of a signal from a single temporal realization $y(n), n = 1 \cdots N$, our ultimate goal.

Fortunately, a more powerful result that underlies most of the field of signal processing exists. Using the same notation as earlier, the celebrated result of Birkhoff states [8, 9]:

Theorem 6.3.1 (*Birkhoff's ergodic theorem*) *Given a measurable transformation* $\Gamma : Y \to Y$ *that preserves the probability measure* $P_Y(y)$

$$\lim_{N \to \infty} \frac{1}{N} \sum_{n=0}^{N-1} f(\Gamma^n(y)) = \frac{1}{P_Y(Y)} \int_Y f(Y) dP_Y(y) \tag{6.28}$$

where the right-hand side is a Fourier–Stieljes integral of the kind already encountered in Section 3.3 (see Eq. 3.37). If we further assume that the CDF (our measure) is differentiable, we may alternatively write

$$\lim_{N\to\infty} \frac{1}{N} \sum_{n=0}^{N-1} f(\Gamma^n(y)) = \frac{1}{P_Y(Y)} \int_Y f(y) \frac{dP_Y(y)}{dy} dy$$

$$= \int_Y f(y) p_Y(y) dy \equiv E[f(Y)] \qquad (6.29)$$

which is the desired result. In forming this result, we have noticed that the normalization constant $1/P_Y(Y)$ is simply 1 if the measure is taken as the probability measure (the probability assigned to the entire space of possibilities is always 1 by definition, as we learned in Chapter 2).

Let's say, for example, that we have the observations $y(n), n = 0 \cdots N - 1$. Given the initial condition, $y(0)$, each subsequent observation at discrete time n is assumed to come from the nth application of our dynamical system map $y(n) = \Gamma^n(y(0))$. First, we consider $f(y) = y$. In this case, the ergodic theorem tells us

$$\lim_{N\to\infty} \frac{1}{N} \sum_{n=0}^{N-1} y(n) = \int_Y y p_Y(y) dy = E[Y] \qquad (6.30)$$

which simply states that *the average value of our temporal sequence of points is equal to the ensemble average of the random variable*. This is the statement we were hoping to make earlier in the chapter (see Eq. 6.3). We can similarly consider more complicated functions $f(\cdot)$. For example, if $f(\cdot)$ is the product $y(t)y(t + \tau)$, then by (6.29)

$$\lim_{N\to\infty} \frac{1}{N-T} \sum_{n=0}^{N-T} y(n)y(n+T) = \int_Y y(t)y(t+\tau) p_Y(y) dy = E[Y(t)Y(t+\tau)]. \qquad (6.31)$$

Thus, for long sequences of observations taken from an ergodic random process, averaging over time is a good approximation to ensemble averaging over different realizations. The importance of this result cannot be understated as it forms the basis of many well-established estimators. It is remarkable in that it relates all of the statistical properties of a random process discussed in Chapter (3) to averages over a single, discrete sequence of observations! Thus, we are on solid ground when we measure a single realization $y(n), n = 1 \cdots N$ and use it to estimate $E[f(Y)]$. With the exception of Section 6.8, where we discuss empirical tests for non-stationarity in our data, we only consider signals that are ergodic in deriving estimators in this chapter. In what follows, we provide estimators for a number of frequently used functions that implicitly make this assumption. For discussions of nonstationary signal processing, see for example, Bendat and Piersol [10].

Before leaving the topic, however, we note that we could just as easily have defined the measure preserving function Γ^t to be defined as a continuous time "flow." In this case, the discrete sum on the left-hand side of (6.29) becomes an integral, that is, [11]

$$\lim_{T\to\infty} \frac{1}{T} \int_0^T f(\Gamma^t(y)) dt = E[f(Y)] \qquad (6.32)$$

and the two relationships described in Eqs. (6.30) and (6.31) become

$$\lim_{T \to \infty} \frac{1}{T} \int_0^T y(t)dt = E[Y]$$

$$\lim_{T \to \infty} \frac{1}{T} \int_0^T y(t)y(t + \tau)dt = E[Y(t)Y(t + \tau)]. \tag{6.33}$$

The ergodic theorem also allows us to estimate other properties of interest. Consider again an observed sequence of data $x(n), n = 1 \cdots N$. Let's assume that we wish to estimate the PDF evaluated at a given location x^*, i.e., estimate $p_X(x^*)$. To derive an estimator $\hat{p}_X(x^*)$, we might follow the intuition discussed earlier in the context of the recurrence theorem and define $f(x) = \Theta(\epsilon/2 - |x - x^*|)$, where

$$\Theta(x) = \begin{cases} 1 & if \quad x \geq 0 \\ 0 & if \quad x < 0 \end{cases} \tag{6.34}$$

is the Heaviside step function. With this definition, the ergodic theorem tells us that

$$\lim_{N \to \infty} \frac{1}{N} \sum_{n=0}^{N-1} \Theta(\epsilon/2 - |x(n) - x^*|) = \int_X \Theta(\epsilon/2 - |x - x^*|)p_X(x)dx \tag{6.35}$$

First, we note that $\sum_{n=0}^{N-1} \Theta(\epsilon/2 - |x(n) - x^*|)$ is simply the number of points within an interval $[-\epsilon/2, \epsilon/2]$ around the fiducial location x^*. Denote this number of points $N_\epsilon(x^*)$ so that the left-hand side of (6.35) becomes simply $\frac{N_\epsilon(x^*)}{N}$. For the right-hand side, we note that the action of the Heaviside function is to change the limits of integration, thus

$$\int_X \Theta(\epsilon/2 - |x - x^*|)p_X(x)dx = \int_{x^*-\epsilon/2}^{x^*+\epsilon/2} p_X(x)dx$$

$$= P_X(x^* + \epsilon/2) - P_X(x^* + \epsilon/2) \tag{6.36}$$

Dividing both sides of (6.35) by ϵ yields

$$\frac{N_\epsilon(x^*)}{\epsilon N} = \frac{P_X(x^* + \epsilon/2) - P_X(x^* - \epsilon/2)}{\epsilon} \tag{6.37}$$

which, in the limit as $N \to \infty$ and $\epsilon \to 0$, yields

$$p_X(x^*) = \lim_{\substack{N \to \infty \\ \epsilon \to 0}} \frac{N_\epsilon(x^*)}{\epsilon N}. \tag{6.38}$$

Of course, when dealing with a finite number of observations, we are really dealing with the estimator

$$\hat{p}_X(x^*) = \frac{N_\epsilon(x^*)}{\epsilon N} \tag{6.39}$$

which is again the familiar histogram approach to estimating a PDF. Using the ergodic theorem we have placed our heuristic result, obtained using the recurrence theorem, on a more sound footing. We simply choose a "bin" width ϵ, count the fraction of total observations in $x(n), n = 1 \cdots N$ falling in the "bin" defined by the range $[x^* - \epsilon/2, x^* + \epsilon/2]$, and then normalize by ϵ.

There are many variations we could put on this basic theme. For example, instead of the Heaviside function, we could have used a more general form $f(x) \equiv K((x^* - x(n))/\epsilon)$, in which case we get the estimator

$$\hat{p}_X(x^*) = \frac{1}{N\epsilon} \sum_{n=0}^{N-1} K((x^* - x(n))/\epsilon). \qquad (6.40)$$

The function $K(\cdot)$ is often chosen to be a smooth, symmetric function (e.g., a Gaussian function) such that the resulting probability density estimate also has a smooth appearance. The kernel "bandwidth," ϵ, governs the width of the function just as it did for our initial Heaviside function example. However, we note that some kernel functions, like the Gaussian, do not have finite support, hence the limits on the integral in (6.36) do not change. Nonetheless, we still make the argument that if the kernel bandwidth is small enough, the associated weighting is negligible outside a narrow range about the fiducial point. In fact, regardless of the function used, the theory still requires a lot of data ($N \to \infty$) and an infinitesimal bandwidth ($\epsilon \to 0$). As with any probability density, we also require that the estimate integrate to unity. To make this true in our estimate (6.40), we therefore require that $\frac{1}{\epsilon} \int K((x^* - x(n))/\epsilon)dx^* = 1$. It is also worth noting that for symmetric kernels $K(u) = K(-u)$; hence, we have

$$E[x^*] \approx \int_{X^*} x^* \hat{p}_X(x^*)dx^* = \int_{X^*} x^* \frac{1}{N\epsilon} \sum_{n=0}^{N} K((x^* - x(n))/\epsilon)dx^*$$

$$= \frac{1}{N\epsilon} \sum_{n=0}^{N-1} \int_{X^*} x^* K((x^* - x(n))/\epsilon)dx^*$$

$$= \frac{1}{N\epsilon} \sum_{n=0}^{N-1} \int_U (\epsilon u + x(n))K(u)\epsilon du$$

$$= \frac{1}{N} \sum_{n=0}^{N-1} x(n) \int_U K(u)du + \epsilon \int_U uK(u)du$$

$$= \frac{1}{N} \sum_{n=1}^{N} x(n) \qquad (6.41)$$

which is, of course, the sample mean. The second integral over U is eliminated by the symmetry of the kernel, and it is at least partially for this reason that only symmetric kernels are usually considered in density estimation. The first integral over U evaluates to unity by definition.

We also have the flexibility to decide the number and locations of the kernel centers x^*, i.e., at what points do we seek the estimate $\hat{p}_X(x^*)$? It is common in standard histogramming to choose J nonoverlapping bins such that $\epsilon = \frac{\max(x) - \min(x)}{J-1}$ and the bin centers are defined by $x_j^* = j\epsilon/2, j = 1 \cdots J$. Another common approach is to use each datum as a fiducial point, that is to say, $x_j^* = x(j), j = 1 \cdots N$.

We have just used the ergodic theorem to derive a very large class of estimators for the PDF associated with a random variable given a time series of observed data. Indeed, entire books have been written on the subject of kernel density estimation and a number of procedures have been established for selecting the kernel bandwidth, number of "bin" centers, and so on, for a given problem (see, e.g., [12]). We use the Heaviside kernel later in the estimation of information theoretics from a single realization of a random process. As we show, this forms the core of a powerful approach to damage detection. We also rely on kernel density estimation techniques in taking samples from the Markov Chain Monte Carlo identification algorithm (see Section 7.2.3) and forming the PDFs of the parameters we seek to identify.

More generally, we have shown the ergodic theorem to be of great practical importance in estimation problems. Without this theorem, a large fraction of signal processing would not exist as it does at present. Rather than studying ensembles of signals, the ergodic theorem allows us to learn much about a random process based on a single realization. The caveats we have already given, namely, that (i) the data are produced by a system governed by a measure preserving function and (ii) the number of data are "large" ($N \to \infty$). With regard to (i), we are always talking about a probability preserving measure in this text. That is to say, the CDF governing the observations is invariant to where in the sequence we are sampling. Concerning (ii) it is a difficult question to answer: how many data are enough for a good estimate? The question does not have a general answer but depends, as we showed in the previous section, on the quantity being estimated and the quality required of the estimate (i.e., both bias and variance are usually dependent on N). We provide, where available, analytical results that speak specifically to what we mean by "quality" for different estimators and show the dependency on N.

6.4   Power Spectral Density and Correlation Functions for LTI Systems

In the detection and identification of structural damage, perhaps no property of a random process has seen more use than the power spectral density (PSD). At a heuristic level, this certainly makes sense. As we have already shown in Chapter 4, our models of structural dynamics are often solved by linear superpositions of sinusoids at various frequencies and amplitudes that depend on the structure's properties, for example, stiffness. A Fourier decomposition of the signal is therefore an appropriate way to highlight these properties and, more to the point of this work, detect changes to the structure properties arising because of damage.

Indeed, numerous papers and techniques have been devoted to this basic approach. Denote the ith estimated modal frequency when the structure is in the damaged state as $\widehat{\omega}_i^{(d)}$ and the corresponding undamaged frequency $\widehat{\omega}_i^{(u)}$. In some approaches, the changes in natural frequencies, $\widehat{\Delta}_i = \widehat{\omega}_i^{(d)} - \widehat{\omega}_i^{(u)}$, are used directly as a damage detection statistic (see, e.g., [13] and the references therein). In other model-based approaches that seek to identify stiffness loss (see Section 7.1.2), one of the key components to the associated estimation problem is $\widehat{\Delta}_i^2$ (see e.g., [14–16]).

In short, the PSD and autocorrelation functions are frequently used descriptors of structural response data that have found much use in damage identification applications. For this reason, we will provide estimators for these quantities along with analytical expressions for common cases.

6.4.1   Estimation of Power Spectral Density

Recall from Section (3.3.3) that we had two alternative ways to express the PSD

$$S_{XX}(f)df = E[dZ^*(f)dZ(f)]$$

$$S_{XX}(f) = \lim_{T \to \infty} E\left[\frac{X(f)^*X(f)}{2T}\right] \tag{6.42}$$

depending on the support of our signal $x(t)$. Given a random process extending infinitely in time, we showed that the Fourier–Stieljes representation was appropriate and defined the PSD according to the first definition. However, when it comes to estimating $S_{XX}(f)$, we are *always* dealing with data that exist over some finite temporal window $[0, T]$ so that we can always define the restricted signal $x_R(t) = x(t), t \in [0, T]$ and zero elsewhere. In this case, it makes sense to use the second form of (6.42) as the estimator. However, we are often further limited to a single realization $x(t)$ with which to approximate the expected value. Fortunately, under the assumption of ergodicity we have just shown (section 6.3) that we can circumvent this problem by replacing the expectation with a suitable temporal average.

Before giving the estimator, however, we mention that had we used the angular frequency variable ω, the estimator would instead be based on

$$S_{XX}(\omega) = \lim_{T \to \infty} \left(\frac{1}{2\pi}\right) E\left[\frac{X(\omega)^* X(\omega)}{2T}\right] \qquad (6.43)$$

an expression we derived in Eq. (3.58). Our reasons for using the frequency variable f in the estimation of spectral densities is that most software packages define the Fourier transform (FT) in terms of f, rather than ω.

To estimate (6.42) or (6.43), we require estimates of the FT $\hat{X}(f)$ (or $\hat{X}(\omega)$); we discuss these estimates first. Begin by denoting our finite sequences of sampled data $x_R(n), n = 1 \cdots N$, collected with fixed sampling interval Δ_t [2]. The sampling period of our signal is therefore $T \equiv (N-1)\Delta_t$ and we are implicitly defining the signal as zero outside this range, an assumption we already know guarantees the existence of the FT (see Section 3.3; see Eq. 3.41). An obvious approach to Fourier analysis considers the estimator

$$\hat{X}(f) \approx \sum_{n=0}^{N-1} x_R(n\Delta_t) e^{-2\pi i f n \Delta_t} \Delta_t$$

$$= \Delta_t \sum_{n=0}^{N-1} \left\{ x(n\Delta_t) \cos(2\pi f n \Delta_t) + i x_R(n\Delta_t) \sin(2\pi f n \Delta_t) \right\} \qquad (6.44)$$

which is just the familiar Riemann approximation for the Fourier integral. In principle, we can evaluate this expression for any frequency value we choose. We could choose to look at a few specific frequencies, or attempt to evaluate (6.44) over a very fine grid of frequencies. We return to this question later in the section.

First, however, it is important to note that

$$e^{2\pi i f n \Delta_t} = e^{2\pi i (f+s/\Delta_t) n \Delta_t}, \quad s = \cdots, -2, -1, 0, 1, 2, \cdots \qquad (6.45)$$

for integer values of s. Thus, the function that defines the discrete Fourier transform (DFT) repeats itself every $1/\Delta_t$ Hz and therefore the estimates given by (6.44) will also repeat as f is increased or decreased by integer multiples of this interval. More specifically, for a discrete sequence of observations, all possible negative and positive frequency values for which *unique* information is obtainable lie on the range $[-1/2\Delta_t \leq f < 1/(2\Delta_t)]$. A graphical depiction of this effect can be seen in Figure 6.2 for a particular set of discrete frequencies (discussed

---

[2] Certainly, nonuniform sampling intervals are possible, however in practice, one typically sets a digitizer to a fixed frequency $f_d = 1/\Delta_t$ and records N samples at this rate.

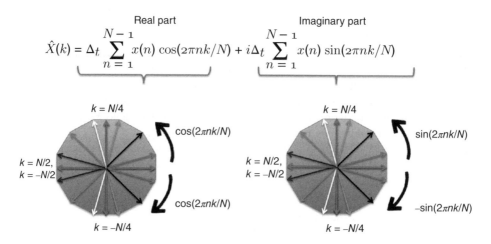

Figure 6.2 Graphical depiction of the relationship expressed by (6.45). For negative and positive discrete frequencies "k," unique frequency information is only available between $f_k = [-1/2\Delta_t, 1/2\Delta_t]$, which corresponds to $k = -N/2 \cdots N/2$

momentarily). In particular, we note that for a real-valued signal $x(n)$, the real Fourier coefficient for positive discrete frequencies is seen to be the same as that for negative frequencies on the aforementioned range. Similarly, we can see that for the imaginary components we have $\text{Im}[\hat{X}(f)] = -\text{Im}[\hat{X}(-f)]$. Hence, we correctly conclude that for a real-valued signal $x(n)$, $\hat{X}(f) = \hat{X}^*(-f)$.

The limiting frequency $f_{Ny} = 1/(2\Delta_t)$ is referred to as the Nyquist frequency in honor of Harry Nyquist [17]. It turns out that if the signal of interest is bandlimited such that it has zero frequency content outside of the Nyquist interval, then the sampled points $x_R(n\Delta_t)$ may be interpolated to yield the exact value of the signal at *any* point in time! This remarkable result is often referred to as the Shannon–Nyquist sampling theorem, as it is attributed to the work of Nyquist and Shannon [18]. Thus, it is often a goal in data acquisition to choose a small enough interval Δ_t such that most of the signal's frequency content lies within the Nyquist interval. If the signal has frequency content outside of this interval, however, those higher frequencies will be (falsely) interpreted by our estimator as lying in the Nyquist band. This phenomenon is commonly referred to as *aliasing* (a direct consequence of 6.45) and will result in a poor estimate of the true underlying frequency content.

For example, given $\Delta_t = 0.01\ s$, we have a Nyquist band of $[-50, 50]$ Hz. However, should the signal being sampled possess frequency content at 60 Hz, that information will appear in the estimated Fourier coefficients corresponding to a frequency of 40 Hz. Thus, the practitioner may come to believe signal energy is present at a frequency in which no signal energy exists. In practice, the situation is not so dire. Often, we know *a priori* the frequency content of the underlying phenomena and can therefore set our sampling interval (and hence the Nyquist band) accordingly. However, in situations where we are not so fortunate, we may place an analog antialiasing filter before sampling the data. In this manner, we may force the signal to have Nyquist limited frequency content before sampling. Regardless of the strategy used, it is important to be aware that discrete approximations to Fourier analysis may result in aliased information.

Regardless of how Δ_t is selected, it turns out that from both a practical and mathematical point of view it makes sense to divide our frequency space into N evenly spaced intervals of length $\Delta_f = 1/(N\Delta_t)$. Thus, we seek to evaluate (6.44) at discrete frequencies $f_k = k/(N\Delta_t), k = 0 \cdots N - 1$ (assuming N even). In this discrete setting, we may observe that $e^{-2\pi i k n/N} = e^{2\pi i (N-k)n/N}, k = 1 \cdots N/2$. Thus, it can be seen that the first $N/2$ coefficients in the discrete transform correspond to the positive frequencies $f_k = k/N\Delta_t, k = 0, \cdots, N/2$, while coefficients $N/2 + 1, \cdots, N - 1$ hold the corresponding negative frequency coefficients which, for a real-valued signal, we know to be the complex conjugate of the positive frequency coefficients. Note, however, that the negative frequency coefficients are stored in reverse order from the positive valued coefficients. Specifically, for N even, $\widehat{X}(f_1) = \widehat{X}^*(f_{N-1})$, $\widehat{X}(f_2) = \widehat{X}^*(f_{N-2})$, and so on until $\widehat{X}(f_{N/2-1}) = \widehat{X}(f_{N/2+1})$. The coefficient values $\widehat{X}(f_0)$ and $\widehat{X}(f_{N/2})$ and hold the real-valued, constant (DC) signal components which are zero for a zero-mean signal, that is, if $E[X(t)] = 0$.

With this discussion in mind, we may finally write for our DFT pair

$$\widehat{X}(f_k) \approx \sum_{n=0}^{N-1} x_R(n\Delta_t)e^{-2\pi i k n/N}\Delta_t, \quad k = 0 \cdots N$$

$$x_R(n\Delta t) \approx \sum_{k=0}^{N-1} \widehat{X}(f_k)e^{2\pi i k n/N}\Delta_f, \quad n = 1 \cdots N. \qquad (6.46)$$

Typically, in signal processing, these quantities are indexed simply by the integers k (frequency) and n (time) such that the more common representation is

$$\widehat{X}(k) \approx \sum_{n=0}^{N-1} x_R(n)e^{-2\pi i k n/N}\Delta_t, \quad k = 0 \cdots N$$

$$x_R(n) \approx \frac{1}{N\Delta_t}\sum_{k=0}^{N-1} \widehat{X}(k)e^{2\pi i k n/N}, \quad n = 1 \cdots N \qquad (6.47)$$

Note that we have left the sampling interval Δ_t in these expressions. In many books, this piece of information is omitted and one must implicitly assume its presence. Our reason for retaining the sampling interval is that it is important to retain the proper units when working with the spectral quantities (derived next). The units of an FT are therefore $[x] \cdot s$ or $[x]/\text{Hz}$, where $[x]$ denotes the units of the observations (e.g., strain or displacement).

As we mentioned, there are several benefits to defining the frequencies on such a grid. First, the problem of inversion, that is to say, solving for the $\widehat{X}(k), k = 0, \cdots, N - 1$ is a uniquely determined problem as we have N equations and N unknowns. Secondly, for this grid spacing, the resulting frequency vectors $\cos(2\pi n k/N), \sin(2\pi n k/N), k = 0, \cdots, N - 1$ form a *basis*. It is well known that a vector of length N can be described exactly as a linear combination of N basis vectors. The signal model given in Eq. (6.47) is therefore a very efficient signal model. Finally, it can be shown that the estimated coefficients are uncorrelated, i.e., $E[\widehat{X}(f)\widehat{X}(f')] = 0, f \neq f'$ [19]. Recall from Section (3.3.3) that for a stationary random process, the complex Fourier–Stieltjes coefficients are known to be independent. By selecting this particular frequency grid, we enforce this property among our estimated values as well.

6.4.1.1 Sampling Artifacts

Before we assess the quality of the estimator (6.47), there is one remaining implication of discrete data that is worth dealing with. We have clearly demonstrated that the estimated Fourier coefficients repeat outside the Nyquist bandwidth, that is to say, $\hat{X}(k) = \hat{X}(k + Np)$ for $p = 1, 2, \cdots$. If we were to inverse FT these data, we see that in the time domain we also have that the discrete signal is being effectively treated as if it too is T periodic, i.e., $x_R(n) = x(n + Np), p = 1, 2, \cdots$ The problem, of course, is that when we record a signal $x_R(n)$, it almost assuredly will *not* be a single period of a larger, repeating sequence. If this were true, the values $x_R(1)$ and $x_R(N)$ would be nearly identical as the signal prepared to repeat again. In general, our signal will be a random process, extending infinitely in time.

To illustrate the problem, let us first model our sampled signal as a *windowed* version of the true signal, that is,

$$x_R(t) = x(t)g(t) \tag{6.48}$$

where $g(t) \equiv \Theta(\frac{1}{2} - |t/T - \frac{1}{2}|)$, where Θ is again the Heaviside function, hence $g(t) = 1$, if $t \in [0, T]$ and $g(t) = 0$ otherwise. While this conveniently limits our random process, the windowing will introduce frequency-domain distortions referred to as spectral "leakage," which can be illustrated in the following manner. If we consider that multiplication in the time domain can be equivalently written as a convolution in the frequency domain, we have that

$$X_R(f) = \int_{-\infty}^{\infty} X(f')G(f - f')df' \tag{6.49}$$

where $G(f)$ is the FT of the window function and is given by

$$G(f) = T\frac{\sin(\pi fT)}{\pi fT}e^{-i\pi fT} \equiv T\text{sinc}[\pi fT]e^{-i\pi fT}. \tag{6.50}$$

Ideally, $G(f)$ would be the delta function, in which case the output of the convolution $X_R(f)$ would be equal to the true coefficient $X(f)$; however, this implies a time window of infinite extent.

In terms of our discrete data, assuming the rectangular window, the convolution (6.49) becomes

$$X_R(f_k) = \Delta_f \sum_{k'=1}^{N} X(f_{k'})G(f_k - f_{k'})$$

$$= \sum_{k=1}^{N} X(f_k')\text{sinc}[\pi(f_k - f_{k'})N\Delta_t] \tag{6.51}$$

where we have noted that $T \equiv N\Delta_t$ and recall that the discrete frequencies are given by $f_k = k/(N\Delta_t)$. Thus, the windowed response $X_R(f_k)$ on our standard Fourier grid is a linear combination of the true Fourier representation multiplied by our window function centered at f_k.

This magnitude of the rectangular window function in the frequency domain is illustrated in Figure 6.3 and is seen to possess zeros at integer multiples of the fundamental frequency increment $\Delta_f = 1/N/\Delta_t$. The implication is that if the true signal frequency $f_{k'}$ lies *exactly* on the discrete grid $k/N/\Delta, k = 0, 1, \cdots, N/2$, then each frequency sample $X_R(k)$ will be given unit

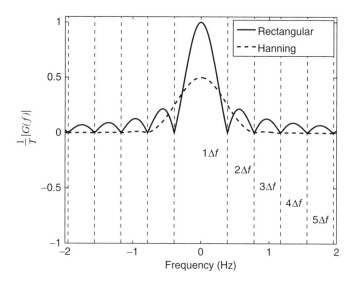

Figure 6.3   Frequency domain representation of the rectangular and Hanning windowing function. Using the rectangular window, if the signal frequency lies exactly on the discrete grid, then $X_R(f_k) = X(f_k)$ and each estimated Fourier amplitude will be correctly weighted unity, while all other frequencies will be weighted zero. Signal frequencies lying "off the grid" will be biased downward and will "leak" into neighboring frequency bins. For the Hanning window, tones that are "on the grid" will be estimated as a weighted sum of three samples: the true frequency (weighting 0.5) and the two neighboring frequencies (weighting 0.25 each). All other samples are essentially weighted zero, regardless of whether the tone is on or off the discrete frequency grid. Hence, if we cannot guarantee that the sampled data lie on the discrete grid, the Hanning window will generally yield a better estimate. In the rare event we know we are sampling a tone on the grid, the rectangular window is preferred

weighting, while all other discrete frequencies will be correctly ignored in (6.51). If the true frequency lies off the grid, i.e., $f_{k'} \neq f_k$, there are two problems. First, the sample at the true frequency will be weighted less than unity (biasing our estimate downward) and, secondly, other samples will factor into the convolution, thereby "leaking" the energy into adjacent frequency bins (and hence the term "leakage").

There are a few ways we can handle this problem. While we certainly cannot expect our signal to be periodic, we can enforce the condition $x(1) \approx x(N)$ with a more appropriate windowing function. This is typically an apodizing function, chosen such that $g(1) = g(N) = 0$. Examples include a sine function or a Hanning window given by

$$g(n) = \sin\left(\pi n/(N-1)\right) \quad \text{Sine window}$$

$$g(n) = 0.5\left(1 - \cos\left(\frac{2\pi n}{N-1}\right)\right) \quad \text{Hanning window} \tag{6.52}$$

respectively. In the frequency domain, the Hanning window appears as in Figure 6.3. The appeal is obviously that, unlike the rectangular window, the Hanning window quickly decays to zero without the "side lobes" so prevalent in the rectangular window. On the downside, even if a tone is located precisely on the grid, the Hanning window gives the output as $X_R(f_k) = 0.5X(f_k) + 0.25X(f_{k-1}) + 0.25X(f_{k+1})$, that is, as the weighted sum of three neighboring frequencies. This is a rare occurrence, however, and we will almost never know whether a given signal lies on or off the discrete frequency grid.

With windowing, the DFT becomes

$$\widehat{X}(k) \approx \sum_{n=0}^{N-1} g(n)x(n)e^{-2\pi ikn/N}\Delta_t, \; k = 0\cdots N. \tag{6.53}$$

While desirable in one sense, windowing does not come without a cost. Clearly, we are modifying the total signal power, which scales as $|X_R(f_k)|^2$ by a factor that should be accounted for if we are to accurately estimate the PSD (our eventual goal). In the case of the Hanning window, the factor is easy enough to compute as $C_g = \sqrt{0.5^2 + 0.25^2 + 0.25^2} = \sqrt{3/8}$. Hence, the windowed FT is more accurately written as

$$\widehat{X}(k) \approx \frac{1}{C_g} \sum_{n=0}^{N-1} g(n)x(n)e^{-2\pi ikn/N}\Delta_t \quad k = 0\cdots N. \tag{6.54}$$

As an example, consider a signal comprised of a single tone, i.e., $x(t) = 2N\Delta_t \cos(2\pi ft + \phi)$ sampled at intervals of $\Delta_t = 0.01$ s for $N = 256$ points. The random phase was taken as 4.027 radians. The results of applying either a rectangular or a Hanning window are shown in Figure 6.4.

Now, imagine that we wish to estimate the Fourier coefficients from these observations. We know analytically that the result should be a single complex coefficient with $\mathrm{Re}(X(f)) = \cos(\phi)$ and $\mathrm{Im}(X(f)) = -\sin(\phi)$. As we have discussed earlier, the FT is expected to be an accurate estimate for $f \in \{k/N/\Delta_t\}, k = 0, \cdots, N/2$. Indeed, if we choose $f = 9\Delta_f = 3.516$ Hz, we obtain the results on Figure 6.5a and b. If, on the other hand, we choose $f = 9.5\Delta_f = 3.711$, we obtain the results on Figure 6.5c and d. This example clearly highlights the trade-offs involved in windowing data. Of course, the case of a single tone located on the Fourier grid is a pathological one. While it is a useful case to look at in understanding the effects of windowing,

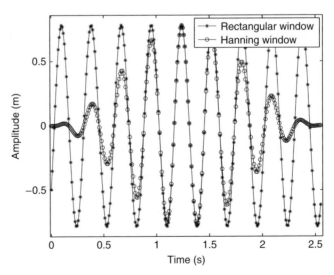

Figure 6.4 Time domain representation of a single tone with frequency $f = 3.52$ Hz and its windowed (Hanning) representation

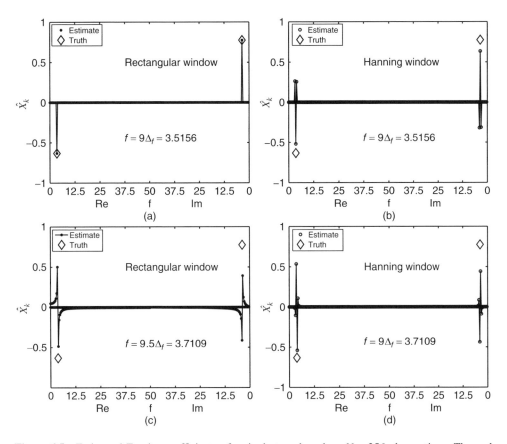

Figure 6.5 Estimated Fourier coefficients of a single tone based on $N = 256$ observations. The real parts of the coefficients are displayed for frequencies $0 \cdots 1/2/\Delta_t$, i.e., zero through Nyquist. The corresponding imaginary components are displayed for their corresponding frequencies on the second half of the plot. (a and b) Estimates obtained when the signal is located on the discrete Fourier grid. (c and d) Estimates obtained when the true signal is exactly at the midpoint of two Fourier grid points. In the former case, the rectangular window yields a better estimate, while in the latter it is clearly advantageous to use the Hanning window. In general, we recommend the use of a windowing function as the frequencies in a random process are always distributed across a wide band of frequencies. The use of the windowing function will therefore always result in more accurate estimates of the spectral properties. The pathological case illustrated in the top row here is useful in understanding the windowing process but is not of much practical utility

in practice we will never be given such data. Rather, we are asked to analyze the output of a random process which will have frequency content spread continuously over some number of frequencies. For this reason, the application of a windowing function is always recommended, as it will help a great deal in suppressing frequency leakage. It is worth keeping in mind, however, that the application of such a function will bias the spectral properties downward.

Finally, it is useful to observe how the window influences the magnitude squared coefficients, $|X_R(f_k)|^2$. This is the key term to be estimated in forming our PSD estimates and is shown in Figure 6.6, again for both the "on" and "off" grid cases for illustrative purposes.

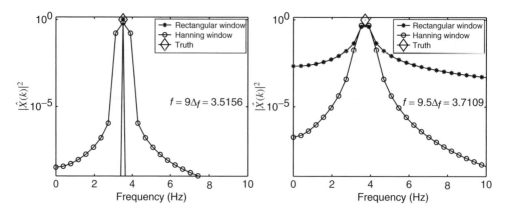

Figure 6.6 Estimated magnitude squared coefficients, $|\hat{X}_R(f_k)|^2$ for the same cases we explored in Figure 6.5. Results are shown on a log scale to better illustrate the side-lobe suppression afforded by a proper windowing function

6.4.1.2 Implementation

Now, recall that earlier we chose to evaluate the FT at the discrete frequencies $\Delta_f = 1/(N\Delta_t)$. To some extent, this was an arbitrary decision on our part, however it is one with some practical value. This can be seen when we consider algorithms for implementing (6.47). We first note that our DFT can also be expressed in matrix form

$$\mathbf{X} = \Delta_t \mathbf{\Psi} \mathbf{x} \tag{6.55}$$

where

$$\mathbf{\Psi} = \begin{bmatrix} 1 & 1 & 1 & \cdots & 1 \\ 1 & e^{-2\pi i/N} & e^{-4\pi i/N} & \cdots & e^{-2\pi i(N-1)/N} \\ 1 & e^{-4\pi i/N} & e^{-8\pi i/N} & \cdots & e^{-4\pi i(N-1)/N} \\ 1 & e^{-6\pi i/N} & e^{-12\pi i/N} & \cdots & e^{-6\pi i(N-1)/N} \\ \vdots & \vdots & \vdots & \vdots & \vdots \\ 1 & e^{-2\pi i(N-1)/N} & e^{-4\pi i(N-1)/N} & \cdots & e^{-2\pi(N-1)(N-1)/N} \end{bmatrix} \tag{6.56}$$

where $\mathbf{x} \equiv (x(0), x(1), \cdots, x(N-1))$ and $\mathbf{X} \equiv (X(0), X(1), \cdots, X(N-1))$ are the sampled random process and its Fourier coefficients, respectively. The matrix $\mathbf{\Psi}$ is an orthogonal matrix where each column has a norm of $\sqrt{N}$. Different references will therefore sometimes multiply the matrix by a scale factor of $1/\sqrt{N}$ to render the columns orthonormal. Here, however, we use no scale factor to keep our matrix representation consistent with the notation we have adopted in (6.47). The orthogonality allows us to write the inverse transform

$$\mathbf{x} = \frac{1}{N\Delta_t} \mathbf{\Psi}^T \mathbf{X} \tag{6.57}$$

where we note that by convention the transpose of a complex matrix also takes the complex conjugate of the entries. Thus, our matrix notation is again seen to be consistent with the definitions given in (6.47). We can now state that given a measured sequence of data values, a simple matrix multiplication is all that is required for both forward and inverse FTs. For small N, this is not a problem, however in signal and image processing, one frequently encounters values for N in the tens, if not hundreds of thousands, rendering storage of $\mathbf{\Psi}$ impossible. Alternatively, one could implement (6.47) in a simple "for" loop, which produces an $O(N^2)$ routine. Again, this is prohibitive for large N. In most cases, use is made of fast Fourier transform, or FFT algorithms, proposed by Tukey [20] and summarized in Ref. [10]. This algorithm is $O(N \log N)$ and is by far the preferred way to implement the transform in signal processing. Note that the FFT algorithm *only* works for our chosen frequency grid spacing $1/N\Delta_t$, giving yet another reason for making this choice.

As an aside, we point out that we could have just as easily divided our frequency space into some larger number of intervals $> N$. Our reason for restricting to N evenly spaced intervals is again a matter of practice, as it allows a simple, straightforward solution for the Fourier coefficients. However, we could, for example, have used a frequency spacing of $\Delta_f = 1/(3 \times N\Delta_t)$. While this $3\times$ higher frequency resolution may seem like a good idea, it comes at a significant computational cost as we are no longer guaranteed a unique solution for the Fourier coefficients (i.e., Eq. 6.57 is underdetermined). Nonetheless, recent work has shown that one can, in fact, solve this problem, provided that the signals being transformed have most of their energy localized in only a few Fourier coefficients [21]. The mathematics that make this possible have spawned the field of "Compressed Sensing," a topic that is beyond the scope of this book. We mention it here simply to alert the reader to the fact that traditional restrictions on frequency resolution in Fourier analysis no longer apply in certain situations.

6.4.1.3 Quality of the Estimator

Finally, while we have proposed an estimator for $\widehat{X}(f)$, we have said nothing about the quality of this estimator. To do so, as always, we require a noise model for our observations. Using the matrix notation of Eq. (6.57), we assume a jointly Gaussian distributed noise vector $\eta(n), n = 0, \cdots, N-1$ and write for the observed data

$$\mathbf{x} = \frac{1}{N\Delta} \mathbf{\Psi}^T \mathbf{X} + \mathbf{\eta}. \tag{6.58}$$

Because each of the noise values is independent, normally distributed we can easily see that the MLEs for the coefficients $\mathbf{X}$ are the values that minimize

$$\widehat{\mathbf{X}} = \min_{\mathbf{X}} \|\mathbf{x} - \frac{1}{N\Delta_t} \mathbf{\Psi}^T \mathbf{X}\|_2^2 \tag{6.59}$$

i.e., the least-squares solution. We can also recognize that the minimizer of this function is unchanged with multiplication by a scalar, and hence we choose to multiply by $\sqrt{N}\Delta_t$ for reasons that will shortly become apparent. Thus, we minimize

$$\widehat{\mathbf{X}} = \min_{\mathbf{X}} \| \sqrt{N}\Delta_t \mathbf{x} - \frac{1}{\sqrt{N}} \mathbf{\Psi}^T \mathbf{X}\|_2^2 \tag{6.60}$$

Solving this system of equations (see Section 7.1.1) results in

$$\widehat{\mathbf{X}} = \left[\left(\frac{1}{\sqrt{N}} \mathbf{\Psi} \right)^T \left(\frac{1}{\sqrt{N}} \mathbf{\Psi} \right) \right]^{-1} \left(\frac{1}{\sqrt{N}} \mathbf{\Psi} \right)^T (\sqrt{N}\Delta_t) \mathbf{x}$$

$$= \Delta_t \mathbf{I} \mathbf{\Psi}^T \mathbf{x} \tag{6.61}$$

so that

$$\widehat{X}(k) = \Delta_t \sum_{n=0}^{N-1} x(n)e^{-i2\pi kn/N}, \ k = 0, \cdots, N-1 \tag{6.62}$$

which is, of course, the discrete approximation to the Fourier integral that we have been using. Thus, in the event that the noise on our observations is jointly Gaussian distributed with independent values, the DFT is an MLE and therefore enjoys the properties of MLE (asymptotically minimum variance, unbiased). If other noise distributions are assumed, one *cannot* expect the quality of an MLE when using this estimator. A good discussion of Fourier coefficient estimation for other noise models (in particular, the Laplacian) can be found in Ref. [22].

We are now in a position to return to the task of estimating the PSD function. If we are willing to assume that the processes are ergodic, each record may be divided into a sequence of possibly overlapping segments $x_s(m) \equiv x(m + sM - L)$, $m = 0 \cdots M - 1, s = 0 \cdots S - 1$, where L denotes the degree of overlap ($0 \le L < M$). Each of these S segments may be viewed as a different realization of the same random process.

Now, we have just demonstrated that the FT of each segment may be approximated as the summation

$$\widehat{X}_s(f_k) \approx \Delta_t \sum_{m=1}^{M} x_s(m)e^{-2\pi i f_k m \Delta_t}$$

$$= \Delta_t \sum_{m=1}^{M} x_s(m)e^{-2\pi i km/M} \tag{6.63}$$

where the transform has been defined only at frequencies $f_k = k/M\Delta_t$. Given this discrete representation, we may form the expectation

$$E[X^*(f)X(f)] \approx \frac{1}{S} \sum_{s=1}^{S} \widehat{X}_s^*(f_k)\widehat{X}_s(f_k)$$

$$= \frac{\Delta_t^2}{S} \sum_{s=1}^{S} \sum_{m=1}^{M} x_s(m)e^{2\pi i km/M} \sum_{m=1}^{M} x_s(m)e^{-2\pi i km/M} \tag{6.64}$$

Referring back to the definition of the autospectral density function (6.42) and noting that the record length is now $M\Delta_t$, we may finally write

$$\widehat{S}_{XX}(f_k) = \frac{\Delta_t^2}{SM\Delta_t} \sum_{s=1}^{S} \sum_{m=1}^{M} x_s(m)e^{2\pi i km/M} \sum_{m=1}^{M} x_s(m)e^{-2\pi i km/M}$$

$$= \frac{\Delta_t}{SM} \sum_{s=1}^{S} \sum_{m=1}^{M} x_s(m)e^{2\pi i km/M} \sum_{m=1}^{M} x_s(m)e^{-2\pi i km/M} \tag{6.65}$$

which simply averages the product of windowed (possibly), Fourier transformed data segments of length M. Properties of this estimator are described in more detail in Ref. [10]. The two properties we are most interested in, of course, are estimator bias and variance. This estimator can be shown to be both unbiased as $M \to \infty$ and consistent as $S \to \infty$. With regard to the bias, the estimator bias can be derived by Taylor expansion as [10]

$$b \approx \frac{\Delta_f^2}{24} \frac{\partial^2 S_{XX}(f)}{\partial f^2}$$

$$= \frac{1}{24M^2\Delta_t^2} \frac{\partial^2 S_{XX}(f)}{\partial f^2}. \tag{6.66}$$

Not surprisingly, the coarser the frequency resolution (recall $\Delta_f = 1/M/\Delta_t$), the greater the bias. Also not surprising is the fact that the bias scales with the second derivative or curvature of the true PSD, thus we see that sharp peaks are most likely to be biased down in estimates. We saw this in the earlier example in Figure 6.6.

With regard to the estimator variance, it has been shown to scale with the number of averages. Specifically, denoting the variance of the PSD estimate σ_P^2, we have

$$\sigma_P^2 = S_{XX}^2(f)\frac{M}{N} = S_{XX}(f)^2\frac{1}{\#_{avgs}} \tag{6.67}$$

thus, we have the typical bias versus variance trade-off. Obviously, we would like M large to resolve sharp features in the frequency domain (reduce bias), however at the same time, we require the ratio M/N to be as small as possible to improve estimator variance. This trade-off is universal in estimation theory and is unavoidable. How the practitioner chooses to weight these two constraints in the final estimate will be task-dependent. We also note that this same basic approach to spectral estimation (block-averaging DFTs) will be used later when discussing the higher order spectra, where the same trade-offs will be evident.

It should also be pointed out that this approach to estimation will yield an MLE for the PSD function, provided (again) that the noise is jointly Gaussian distributed. As we have discussed earlier in this section, for such a noise model the DFT yields an MLE for the Fourier coefficients. We also know that the averaging approach used in approximating the expected value operator gives an MLE for the averaged quantity. Moreover, from the aforementioned invariance property of ML estimators, we know that an MLE for a function of some number of parameters will be obtained if one substitutes the MLEs for those parameters into the expression. Hence, our PSD estimate is an MLE under the joint normal noise assumption.

We also note that if the constituent DFTs that comprise the product (6.65) are premultiplied by a windowing function $g(n)$, the PSD estimate must be normalized accordingly. In the non-windowed case, we are effectively assuming $\hat{g}(n) = 1, n = 1 \cdots N$. A windowed PSD estimate is therefore multiplied by the factor $\sum_{n=1}^{N} g(n)^2$. Simply dividing the estimate (6.65) by this value provides the needed correction.

The same approach is also used to estimate the cross-spectral density function. If we are given two random processes $x(t), y(t)$ the cross-spectral density is defined as

$$S_{XY}(f) = \lim_{T \to \infty} E\left[\frac{X(f)^*Y(f)}{2T}\right] \tag{6.68}$$

Assuming the sample data $x(n), y(n), n = 1 \cdots N$, we use the same block-transform procedure for the data $y(n)$ as we did for the $x(n)$ earlier. Thus, our final estimate for the cross-spectral density function is

$$
\begin{aligned}
\hat{S}_{XY}(f_k) &= \frac{\Delta_t^2}{SM\Delta_t} \sum_{s=1}^{S} \sum_{m=1}^{M} x_s(m)e^{2\pi i k m/M} \sum_{m=1}^{M} y_s(m)e^{-2\pi i k m/M} \\
&= \frac{\Delta_t}{SM} \sum_{s=1}^{S} \sum_{m=1}^{M} x_s(m)e^{2\pi i k m/M} \sum_{m=1}^{M} y_s(m)e^{-2\pi i k m/M}
\end{aligned}
\tag{6.69}
$$

which possesses the same bias and variance properties as the autospectral density estimate.

6.4.1.4 Examples

Next, we demonstrate this estimator with an example. Consider a single degree of freedom (SDOF) spring-mass system with parameters $k = 1000$ N/m, $c = 3$N $\cdot$ s/m, and $m = 1.0$ kg. We have already shown (see Eq. 4.45) that the response of this system is given by

$$
Y(f) = H(f)X(f)
\tag{6.70}
$$

where

$$
H(f) = \frac{1}{k + ic(2\pi f) - m(2\pi f)^2}.
\tag{6.71}
$$

Note that if the response is given in terms of velocity or acceleration, we have $H_v(f) = i\omega H(f)$ and $H_a(f) = -\omega^2 H(f)$, respectively. Now consider the PSD associated with the acceleration response $\ddot{y}(t)$. From (6.68) we have

$$
\begin{aligned}
S_{\ddot{Y}\ddot{Y}}(f) &= \lim_{T\to\infty} E\left[\frac{\ddot{Y}(f)^*\ddot{Y}(f)}{2T}\right] \\
&= |H_a(f)|^2 \lim_{T\to\infty} E\left[\frac{X(f)^*X(f)}{2T}\right] \\
&= |H_a(f)|^2 S_{XX}(f)
\end{aligned}
\tag{6.72}
$$

As we have mentioned before, a signal with a constant PSD implies infinite signal energy (variance). Thus, even if the input is taken as an independent and identically distributed (i.i.d.) random process, we must restrict ourselves to a finite time range as in Eqs. (3.41) and (3.48). Again, this restriction is of not much consequence in practice as our estimates are always so restricted. Consider the input to be an i.i.d. jointly Gaussian random process with zero mean, variance σ^2 and duration T seconds.

We know from the ergodic theorem and Parseval's relationship (3.47) that

$$
\lim_{T\to\infty} \frac{1}{T} \int_0^T E[X^2(t)]dt = \sigma^2
$$

$$
= \lim_{T\to\infty} \int_{-\infty}^{\infty} \frac{1}{T} E[|X(f)|^2]df = \int_{-\infty}^{\infty} S_{XX}(f)df
\tag{6.73}
$$

We also note that any sampled signal is limited to the Nyquist band with frequency spacing $\Delta_f = \frac{1}{N\Delta_t}$, so that in discrete terms

$$\sigma^2 \approx \frac{1}{N\Delta_t} \sum_{k=0}^{N} S_{XX}(f_k). \tag{6.74}$$

Finally, it is noted that for an i.i.d. signal, $S_{XX}(f_k)$ is a constant across all frequencies, so that we write

$$\sigma^2 \Delta_t \approx \frac{1}{N} \times N S_{XX}(f_k). \tag{6.75}$$

Here we have an important relationship for sampled signals, namely, that for an i.i.d., jointly Gaussian random process, the PSD is a function of the signal variance and also the sampling interval Δ_t (the latter defining the Nyquist band). Thus, when simulating the response of a first-order system to i.i.d. Gaussian excitation with sampling interval Δ_t, the corresponding input PSD is $S_{XX}(f_k) \approx \sigma^2 \Delta_t$. Note, this is a result we derived earlier in Eq. (3.75).

Returning to our example, we assume $\Delta_t = 0.01$ s and $\sigma^2 = 1$, and so the input PSD is $S_{XX}(f) = 0.01 \text{N}^2/\text{Hz}$. Thus, we have completed the description of our predicted response spectral density. We generate a total of $N = 32768$ observations of this SDOF system and use the estimator defined by (6.65) with two different sets of parameters M, L. In the first case, we set $M = 2048$ and $L = 1024$, while in the second we use $M = 256, L = 128$. The result of both estimates is shown in Figure 6.7.

In this example, we would expect an estimator variance near the peak ($S_{xx}(f) \approx 1$) to be $\sigma_P^2 = 1/31 = 0.023$ and $\sigma_P^2 = 1/255 = 0.004$ for the left and right plots, respectively. Note that if we attempt to reduce the estimator variance, we incur the predicted bias penalty. This

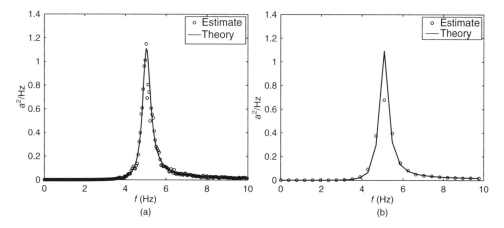

Figure 6.7 Comparison of theoretical and estimated power spectral density (PSD) associated with a first-order system. The input PSD was assumed to be $S_{XX}(f) = 0.01 \text{N}^2/\text{Hz}$ over all frequencies. For plot (a), $M = 2048$, $L = 1024$ resulting in $S = 31$ averages in approximating the expected value required of the estimate. In plot (b), we use $M = 256$, $L = 128$, resulting in $S = 255$ averages. While the estimate is quite smooth (low variance), we see the bias at the peak value resulting from the coarse frequency resolution

is seen in Figure 6.7b, where the coarser frequency spacing results in a poor estimate of the peak value. Recall from Eq. 6.66 that the bias scales with the curvature of the PSD, hence for structural response data (which often possess sharp peaks in the associated spectral response due to light damping), the effect can be pronounced.

The implications for structural damage detection will depend on the problem. If, for example, we seek a very small change in natural frequency, we may be willing to accept the larger variance in exchange for the higher frequency resolution which should allow better localization of the estimated natural frequency. On the other hand, reliably capturing a change in the amplitude of a particular frequency component may suggest we use more averages to minimize estimator variance. Although the amplitude itself will be biased, the amplitude change is probably better quantified with the low variance estimator. In Section 6.7.1 we, in fact, show that the change in Fourier amplitudes is, under certain circumstances, the optimal detection statistic when looking for changes in the stiffness of a structure (see Eq. 6.243).

As a second example we consider the response of a first-order system with a quadratic non-linearity as was done in Eq. (4.74). A recurring theme in this book is that structural damage results in nonlinear behavior. We therefore find it important to model and estimate the nonlinear response of structural systems. This was our motivation in introducing the Volterra series modeling approach in Section (4.6). Considering the excitation as a zero-mean, Gaussian random process we have the model

$$y(t) = y_1(t) + y_2(t)$$

$$= \int h_1(\tau_2)x(t - \tau_2)d\tau_2 + \int_{\mathbb{R}^2} h_2(\tau_3, \tau_4)x(t - \tau_3)x(t - \tau_4)d\tau_3 \, d\tau_4$$

$$\bar{y} = \int_{\mathbb{R}^2} h_2(\tau_3, \tau_4)E[x(t - \tau_3)x(t - \tau_4)]d\tau_3 \, d\tau_4. \tag{6.76}$$

By the Wiener–Khinchin theorem, we know that the PSD is the FT of the expectation

$$E[(Y(t) - \bar{Y})(Y(t + \tau_1) - \bar{Y})] = E\left[\int h_1(\tau_2)X(t - \tau_2)d\tau_2 \int h_1(\tau_3)X(t + \tau_1 - \tau_3)d\tau_3 \right.$$

$$+ \int_{\mathbb{R}^2} h_2(\tau_2, \tau_3)\{x(t - \tau_2)x(t - \tau_3) - E[X(t - \tau_2)X(t - \tau_3)]\}d\tau_2 \, d\tau_3$$

$$\left. \times \int_{\mathbb{R}^2} h_2(\tau_4, \tau_5) \left\{x(t + \tau_1 - \tau_4)x(t + \tau_1 - \tau_5) - E[X(t + \tau_1 - \tau_4)X(t + \tau_1 - \tau_5)]\right\} d\tau_4 \, d\tau_5\right]$$

$$\tag{6.77}$$

where we have noted that terms involving the product of three input variables will be zero, as per Isserlis' theorem (Eq. 3.11). Thus, the remaining two terms are simplified as

$$E\left[\int h_1(\tau_2)h_1(\tau_3)X(t - \tau_2)X(t + \tau_1 - \tau_3)d\tau_2 \, d\tau_3 \right.$$

$$+ \int_{\mathbb{R}^4} h_2(\tau_2, \tau_3)h_2(\tau_4, \tau_5) \left\{X(t - \tau_2)X(t - \tau_3)X(t + \tau_1 - \tau_4)X(t + \tau_1 - \tau_5)\right.$$

$$- X(t + \tau_1 - \tau_4)X(t + \tau_1 - \tau_5)R_{XX}(-\tau_3 + \tau_2)$$

$$-X(t-\tau_2)X(t-\tau_3)R_{XX}(-\tau_5+\tau_4)+R_{XX}(-\tau_3+\tau_2)R_{XX}(-\tau_5+\tau_4)\big\}\,d\tau_{2\to5}\bigg]$$

$$=\int_{\mathbb{R}^2}h_1(\tau_2)h_1(\tau_3)R_{XX}(\tau_1-\tau_3+\tau_2)d\tau_2d\tau_3$$

$$+\int_{\mathbb{R}^4}h_2(\tau_2,\tau_3)h_2(\tau_4,\tau_5)\big\{\cancel{R_{XX}(-\tau_3+\tau_2)R_{XX}(-\tau_5+\tau_4)}+$$

$$+R_{XX}(\tau_1-\tau_4+\tau_2)R_{XX}(\tau_1-\tau_5+\tau_3)+R_{XX}(\tau_1-\tau_5+\tau_2)R_{XX}(\tau_1-\tau_4+\tau_3)$$

$$-\cancel{R_{XX}(-\tau_3+\tau_2)R_{XX}(-\tau_5+\tau_4)}-\cancel{R_{XX}(-\tau_3+\tau_2)R_{XX}(-\tau_5+\tau_4)}$$

$$+\cancel{R_{XX}(-\tau_3+\tau_2)R_{XX}(-\tau_5+\tau_4)}\big\} \tag{6.78}$$

whereby noting that $h_2(a,b)=h_2(b,a)$, we have

$$E[(Y(t)-\bar{Y})(Y(t+\tau_1)-\bar{Y})]=\int_{\mathbb{R}^2}h_1(\tau_2)h_1(\tau_3)R_{XX}(\tau_1-\tau_3+\tau_2)d\tau_2d\,\tau_3$$

$$+2\int_{\mathbb{R}^4}h_2(\tau_2,\tau_3)h_2(\tau_4,\tau_5)R_{XX}(\tau_1-\tau_4+\tau_2)R_{XX}(\tau_1-\tau_5+\tau_3). \tag{6.79}$$

By the Wiener–Khinchin theorem (Eq. 3.69), we may make the substitution

$$R_{XX}(\tau_1-\tau_3-\tau_2)=\frac{1}{2\pi}\int S_{XX}(\omega)e^{i\omega(\tau_1-\tau_3+\tau_2)}d\omega$$

$$R_{XX}(\tau_1-\tau_4+\tau_2)=\frac{1}{2\pi}\int S_{XX}(\omega_1)e^{i\omega_1(\tau_1-\tau_4+\tau_2)}d\omega_1$$

$$R_{XX}(\tau_1-\tau_5+\tau_3)=\frac{1}{2\pi}\int S_{XX}(\omega_2)e^{i\omega_2(\tau_1-\tau_5+\tau_3)}d\omega_2 \tag{6.80}$$

where, upon changing variables $\omega\to\omega_1+\omega_2$ and simplifying yields

$$E[(Y(t)-\bar{Y})(Y(t+\tau_1)-\bar{Y})]=\frac{1}{2\pi}\int\big\{|H_1(\omega)|^2S_{XX}(\omega)$$

$$+\frac{1}{\pi}\int|H_2(\omega-\omega_2,\omega_2)|^2S_{XX}(\omega-\omega_2)S_{XX}(\omega_2)d\omega_2\big\}e^{i\omega\tau_1}d\omega. \tag{6.81}$$

By inspection we see that the right-hand side is the inverse FT of the linear PSD plus the inner integral over ω_2. Thus, taking the FT of both sides, we see that by definition

$$S_{XX}(\omega)=|H_1(\omega)|^2S_{XX}(\omega)+\frac{1}{\pi}\int|H_2(\omega-\omega_2,\omega_2)|^2S_{XX}(\omega-\omega_2)S_{XX}(\omega_2)d\omega_2. \tag{6.82}$$

The first term is the familiar linear PSD. The second term contains a product of the input spectral density and also involves a contour integral over the second Volterra kernel. This integral is carried out in Appendix (B) as an example. The result is an analytical expression for the PSD of this nonlinear system response. It is interesting to note that the nonlinearity shows up as a term by multiplying the product of two PSD functions. These functions are, for many structural dynamics applications, of order $S_{XX}(\omega) \sim O(10^{-2})$, hence the nonlinearity shows up as a higher order term. Later, in Section 6.5, we revisit this example in the context of higher order spectral analysis and see analytically that for the bispectrum, the nonlinear term is dominant, offering a means of identifying damage-induced nonlinearity in a structure's vibrational response.

To demonstrate, the system's acceleration response was simulated using a nonlinear stiffness of $k_2 = 100,000 \text{N/m}^2$ while holding the linear system parameters fixed to the same values as in the previous example. The estimate was performed using $N = 32768$ observations and the parameters $M = 1024$ and $L = 512$ with a Hanning window. The result is shown in Figure 6.8.

The addition of the nonlinearity results in a small peak at the second harmonic (≈ 10 Hz). This peak is captured in both the estimate and the theory. Analyses such as this can be used to explore the influence of damage-related parameters on the PSD of the response. Damage detection algorithms that focus on harmonic generation can use this analysis to predict detector performance (see, e.g., [23]). This simple problem also illustrates the challenge associated with detecting small damage-induced nonlinearities using spectral analysis, namely, that the influence of damage on the spectral density function is often small.

6.4.2 Estimation of Correlation Functions

There are two main approaches referenced in the estimation of correlation functions. As in the previous discussion, we assume that data $y_i(n\Delta_t), y_j(n\Delta_t), n = 0 \cdots N - 1$ have been recorded

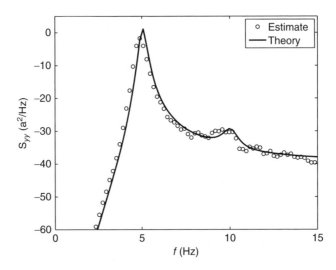

Figure 6.8 Comparison of theoretical and estimated power spectral density (PSD) associated with a first-order nonlinear system with $k_2 = 100,000 \text{N/m}^2$. The input PSD was assumed to be $S_{XX}(f) = 0.01\text{N}^2/Hz$ over all frequencies. The estimate used a Hanning smoothing window and the parameters $M = 1024, L = 512$, resulting in $S = 63$ averages in approximating the expected value required of the estimate

from the stationary random processes $\mathbf{Y}_i$, $\mathbf{Y}_j$, respectively. In the first, referred to as the "direct" approach, we make use of the ergodic theorem and write

$$\hat{R}_{Y_iY_j}(\tau) = \frac{1}{N-\tau}\sum_{n=0}^{N-\tau} y_i(n)y_j(n+\tau) \tag{6.83}$$

which, in the limit of $N \to \infty$, is equivalent to $E[Y_i(t)Y_j(t+\tau)]$. This is an unbiased estimator [10], however it is computationally costly to implement being of order $\tau_{max}N$. By far the more prevalent approach is to use the Weiner–Khinchin theorem discussed earlier (again, see Eq. 3.69), which states that the covariance function is the inverse FT of the cross-spectral density, for example, $R_{Y_iY_j}(\tau) = \int S_{Y_iY_j}(f)e^{i2\pi f\tau}df$. The problem is that we do not know $S_{Y_iY_j}(f)$, rather it too has to be estimated from a finite sequence of observations (see the previous section). It has been shown (see [10], page 420, 421) that for a finite data of length N and fixed sampling interval Δ_t

$$\hat{S}_{Y_iY_j}(k) = \frac{1}{N\Delta_t}\sum_{r=0}^{N-1}\left\{\frac{(N-r)}{N}\hat{R}_{Y_iY_j}(r\Delta_t) + \frac{r}{N}\hat{R}_{Y_iY_j}((N-r)\Delta_t)\right\}e^{-i2\pi kr/N}. \tag{6.84}$$

Thus, taking the inverse DFT of the spectral density estimate will *not* give the desired covariance estimate, but rather a mixture of the covariance estimate at times $r\Delta_t$ and $(N-r)\Delta_t$. This effect stems from the implicit assumption in Fourier analysis that a finite data record of length N is N-periodic, i.e., the sequence repeats indefinitely. This was discussed earlier in Section 6.4.1. If the correlations decay quickly such that $R_{Y_iY_j}(r) \approx 0$ for $r < N$, this effect does not hurt the estimate appreciably. However, the problem can be easily fixed by simply padding N zeros on the original data, for example, $y_i(N+1\cdots 2N) = 0$. By doing this we guarantee that $\hat{R}_{Y_iY_j}(r), r = 1\cdots N$ and $\hat{R}_{Y_iY_j}(2N-r), r = 1\cdots N$ do not overlap. Thus, a Fourier-based estimator proceeds by estimating the PDF function (see Section 6.4.1) using segments of length $2M$ (zero-padding the M point segments with M zeros) before FT and averaging. The inverse FT of this quantity yields the $2M$ point covariance function

$$\hat{R}_{Y_iY_j}(r) = \begin{cases} \frac{(N-r)}{N}\sum_{k=0}^{2M}\hat{S}_{Y_iY_j}(k)e^{i2\pi kr/(2M)} & : \quad r = 0,\cdots,M \\ \frac{r}{N}\sum_{k=0}^{2M}\hat{S}_{Y_iY_j}(k)e^{i2\pi kr/(2M)} & : \quad r = M,\cdots,2M-1 \end{cases}. \tag{6.85}$$

Next, we test the quality of this estimator against an analytical expression, derived for linear MDOF structures.

Deriving the cross-correlation function between signals collected from different points on a randomly forced structure is nontrivial. One situation where such a derivation is possible is the case of a linear structure. Consider the general M DOF system governed by the matrix equation

$$\mathbf{M}\ddot{\mathbf{y}}(t) + \mathbf{C}\dot{\mathbf{y}}(t) + \mathbf{K}\mathbf{y}(t) = \mathbf{x}(t). \tag{6.86}$$

The input and output are modeled as the random processes $\mathbf{X}(t)$ and $\mathbf{Y}(t)$, respectively. From Chapter (4), Eq. (4.53), we know that if we denote the system eigenvectors $\mathbf{u}_i, i = 1\cdots M$, the equations of motion may be uncoupled and written separately as

$$\ddot{\eta}_i + 2\zeta_i\omega_i\dot{\eta} + \omega_i^2\eta_i = \mathbf{u}_i^T\mathbf{x}(t) = q_i(t). \tag{6.87}$$

Note this form assumes that we may write $\mathbf{C} = \beta\mathbf{K}$, so that $2\zeta_i\omega_i = \beta\omega_i^2$. Solving for the dimensionless damping parameters yields $\zeta_i = \frac{1}{2}\beta\omega_i$. The parameter β is determined by considering the strength of the damping parameters c relative to the parameters governing the restoring force k. Typically, we have $c/k < 1$, i.e., a lightly damped system. We also know from Section 4.5, Eq. (4.46) that convolution may be used to write the solution as

$$\eta_i(t) = \int_0^\infty h_i(\theta)q_i(t-\theta)d\theta \qquad (6.88)$$

so that

$$y_i(t) = \sum_{l=1}^M u_{il}\eta_l(t)$$

$$= \int_0^\infty \sum_{l=1}^M u_{il}h_l(\theta)q_l(t-\theta)d\theta \qquad (6.89)$$

Because the excitation $\mathbf{x}(t)$ is modeled as a zero-mean random process, $\mathbf{X}(t)$, $q_l(t)$ will also be a random process, modeled as $Q_l(t)$. Using this model, we may construct the covariance

$$E[Y_i(t)Y_j(t+\tau)] =$$

$$E\left[\int_0^\infty \int_0^\infty \sum_{l=1}^M \sum_{m=1}^M u_{il}u_{jm}h_l(\theta_1)h_m(\theta_2)Q_l(t-\theta_1)Q_m(t+\tau-\theta_2)d\theta_1\,d\theta_2\right]$$

$$= \int_0^\infty \int_0^\infty \sum_{l=1}^M \sum_{m=1}^M u_{il}u_{jm}h_l(\theta_1)h_m(\theta_2)E[Q_l(t-\theta_1)Q_m(t+\tau-\theta_2)]d\theta_1\,d\theta_2 \quad (6.90)$$

which is a function of the mode shapes $\mathbf{u}_i$, the impulse response function $h(\cdot)$, and the covariance of the modal forcing matrix. Knowledge of this covariance matrix can be obtained from knowledge of the forcing covariance matrix $R_{X_lX_m}(\tau) \equiv E[X_l(t)X_m(t+\tau)]$. Recalling that

$$q_l(t) = \sum_{p=1}^M u_{lp}x_p(t) \qquad (6.91)$$

we write

$$E[Q_l(t-\theta_1)Q_m(t+\tau-\theta_2)] = \sum_{p=1}^M \sum_{q=1}^M u_{lq}u_{mp}E[X_q(t-\theta_1)X_p(t+\tau-\theta_2)]. \qquad (6.92)$$

It is assumed that the random vibration inputs are uncorrelated, i.e. $E[X_q(t)X_p(t)] = 0\,\forall q \neq p$. Thus, the above equation can be simplified as

$$E[Q_l(t-\theta_1)Q_m(t+\tau-\theta_2)] = \sum_{p=1}^M u_{lp}u_{mp}E[X_p(t-\theta_1)X_p(t+\tau-\theta_2)] \qquad (6.93)$$

The most common linear models assume that the input is applied at a single spatial location, P, i.e. $f_p(t)$ is nonzero only for $p = P$.

With these assumptions on the forcing, the expression for the covariance becomes

$$E[Y_i(t)Y_j(t + \tau)] = \int_0^\infty \int_0^\infty \sum_{l=1}^M \sum_{m=1}^M u_{il}u_{jm}u_{lP}u_{mP}h_l(\theta_1)h_m(\theta_2)$$

$$E[X_P(t - \theta_1)X_P(t + \tau - \theta_2)]d\theta_1 \, d\theta_2$$

$$= \sum_{l=1}^M \sum_{m=1}^M u_{il}u_{jm}u_{lP}u_{mP} \int_0^\infty h_l(\theta_1) \int_0^\infty h_m(\theta_2)$$

$$E[X_P(t - \theta_1)X_P(t + \tau - \theta_2)]d\theta_2 \, d\theta_1. \tag{6.94}$$

The inner integral can be further evaluated as

$$\int_0^\infty h_m(\theta_2)E[X_P(t - \theta_1)X_P(t + \tau - \theta_2)]d\theta_2 = \int_0^\infty h_m(\theta_2) \int_{-\infty}^\infty S_{XX}(\omega)e^{i\omega(\tau - \theta_2 + \theta_1)}d\omega \, d\theta_2. \tag{6.95}$$

If we consider the forcing to be comprised of statistically independent values (i.e., the process is i.i.d), we have that the forcing PSD $S_{XX}(\omega) = \text{const.} = S_{XX}(0)$. The FT of a constant is simply $\int_{-\infty}^\infty \text{const.} \times e^{i\omega t} dt = \text{const.} \times \delta(t)$, and hence our integral becomes

$$\int_0^\infty h_m(\theta_2)E[X_P(t - \theta_1)X_P(t + \tau - \theta_2)]d\theta_2 = \int_0^\infty h_m(\theta_2)S_{XX}(0)\delta(\tau - \theta_2 + \theta_1)d\theta_2$$

$$= h(\tau + \theta_1)S_{XX}(0). \tag{6.96}$$

Returning to Eq. (6.94) we then have

$$E[Y_i(t)Y_j(t + \tau)] = \int_0^t \sum_{l=1}^M \sum_{m=1}^M u_{il}u_{jm}h_l(\theta_1)h_m(\theta_1 + \tau)u_{lP}u_{mP}S_{XX}(0)d\theta_1. \tag{6.97}$$

At this point, we can simplify the expression by carrying out the integral. First, we note that for this system

$$h_i(\theta) = \frac{1}{\omega_{di}}e^{-\zeta_i\omega_i\theta}\sin(\omega_{di}\theta) \tag{6.98}$$

where we define $\omega_{di} \equiv \omega_i\sqrt{1 - \zeta_i^2}$. In this case, the expectation (6.97) becomes [24, 25]

$$R_{Y_iY_j}(\tau) = \frac{S_{XX}(0)}{4} \sum_{l=1}^M \sum_{m=1}^M u_{lP}u_{mP}u_{il}u_{jm} \left[A_{lm}e^{-\zeta_m\omega_m\tau}\cos(\omega_{dm}\tau) + B_{lm}e^{-\zeta_m\omega_m\tau}\sin(\omega_{dm}\tau) \right] \tag{6.99}$$

where

$$A_{lm} = \frac{8(\omega_l\zeta_l + \omega_m\zeta_m)}{\omega_l^4 + \omega_m^4 + 4\omega_l^3\omega_m\zeta_l\zeta_m + 4\omega_m^3\omega_l\zeta_l\zeta_m + 2\omega_m^2\omega_l^2(-1 + 2\zeta_l^2 + 2\zeta_m^2)} \tag{6.100}$$

$$B_{lm} = \frac{4(\omega_l^2 + 2\omega_l\omega_m\zeta_l\zeta_m + \omega_m^2(-1 + 2\zeta_m^2))}{\omega_{dm}(\omega_l^4 + \omega_m^4 + 4\omega_l^3\omega_m\zeta_l\zeta_m + 4\omega_m^3\omega_l\zeta_l\zeta_m + 2\omega_m^2\omega_l^2(-1 + 2\zeta_l^2 + 2\zeta_m^2))}.$$

If we further normalize this function in accordance with Eq. (2.52), we have

$$\rho_{Y_i Y_j}(\tau) = R_{Y_i Y_j}(\tau) / \sqrt{R_{Y_i Y_i}(0) R_{Y_j Y_j}(0)} \tag{6.101}$$

for the analytical cross-correlation function.

This analysis can also be used to define the covariance between forcing and response. Following the same steps as these, we define

$$R_{Y_i X_P}(\tau) = E[Y_i(t) X_P(t + \tau)] \tag{6.102}$$

as the covariance between the random process $Y_i(t)$ and the random process at a single forcing at location P, i.e. $X_P(t)$. As before, we use a modal solution for $y_i(t)$ and write

$$
\begin{aligned}
R_{Y_i X_P}(\tau) &= E\left[\int_0^\infty \sum_{m=1}^M u_{im} h_m(\theta) q_m(t - \theta) d\theta \times X_P(t + \tau) \right] \\
&= \int_0^\infty \sum_{m=1}^M \sum_{l=1}^M u_{im} u_{ml} h_m(\theta) E[X_l(t - \theta) X_P(t + \tau)] d\theta \\
&= \int_0^\infty \sum_{m=1}^M u_{im} u_{mP} h_m(\theta) E[X_P(t - \theta) X_P(t + \tau)] d\theta \\
&= \int_0^\infty \sum_{m=1}^M u_{im} u_{mP} h_m(\theta) S_{X_P X_P}(0) \delta(\tau + \theta) d\theta
\end{aligned}
\tag{6.103}
$$

which is only nonzero for negative delays, thus we have

$$R_{Y_i X_P}(-\tau) = \sum_{m=1}^M u_{im} u_{mP} h_m(\tau) S_{X_P X_P}(0) \, \forall \tau > 0. \tag{6.104}$$

We stress that the given analysis assumed stationary, i.i.d., zero-mean Gaussian excitation. Other assumptions on the input would require other analysis. For example, if the input covariance matrix is *not* diagonal (we do not assume independent increments on the forcing), a double integral over both θ_1, θ_2 would have been required. Finally, we note that because the impulse response function $h(\tau)$ is defined only for $\tau > 0$, negative time correlations are obtained by symmetry by recalling that $R_{Y_i Y_j}(-T) = R_{Y_j Y_i}(T)$. We also recall from Section (2.5) that the correlation coefficient lies in the range $-1 \leq \rho_{Y_i Y_j}(T) \leq 1$.

Now, consider the simple five degree of freedom (DOF) system with the constant coefficient mass, damping, and stiffness matrices given by

$$
\mathbf{M} = \begin{bmatrix} m_1 & 0 & 0 & 0 & 0 \\ 0 & m_2 & 0 & 0 & 0 \\ 0 & 0 & m_3 & 0 & 0 \\ 0 & 0 & 0 & m_4 & 0 \\ 0 & 0 & 0 & 0 & m_5 \end{bmatrix}
$$

$$\mathbf{C} = \begin{bmatrix} c_1 + c_2 & -c_2 & 0 & 0 & 0 \\ -c_2 & c_2 + c_3 & -c_3 & 0 & 0 \\ 0 & -c_3 & c_3 + c_4 & -c_4 & 0 \\ 0 & 0 & -c_4 & c_4 + c_5 & -c_5 \\ 0 & 0 & 0 & -c_5 & c_5 \end{bmatrix}$$

$$\mathbf{K} = \begin{bmatrix} k_1 + k_2 & -k_2 & 0 & 0 & 0 \\ -k_2 & k_2 + k_3 & -k_3 & 0 & 0 \\ 0 & -k_3 & k_3 + k_4 & -k_4 & 0 \\ 0 & 0 & -k_4 & k_4 + k_5 & -k_5 \\ 0 & 0 & 0 & -k_5 & k_5 \end{bmatrix} \quad (6.105)$$

respectively. Solving the undamped eigenvalue problem for this system $|\mathbf{M}^{-1}\mathbf{K} - \omega_i^2 \mathbf{I}|\hat{\mathbf{u}}_i = 0$ gives the eigenvalues ω_i and eigenvectors $\hat{\mathbf{u}}_i$, $i = 1 \cdots 5$. Normalizing with respect to the mass matrix by solving

$$\alpha_i^2 \hat{\mathbf{u}}_i^T \mathbf{M} \hat{\mathbf{u}}_i = 1$$

for α_i gives the mass-normalized eigenvectors $\mathbf{u}_i = \alpha_i \hat{\mathbf{u}}_i$, $i = 1 \cdots 5$ [26]. The normalized mode shapes will determine the relative strengths of the impulse response terms used in forming the correlation matrix. The form of the dissipative coupling is seen to be the same as that of the restoring force, thus the damping matrix may be written $\mathbf{C} = \beta \mathbf{K}$, where β is a small constant (defined momentarily). In this case, we may write

$$\zeta_i = \frac{1}{2}\beta\omega_i$$

which may solve for the unknown damping coefficients ζ_i. This provides yet another justification for the modal damping model (Eq. 4.25).

Now consider this system driven by a joint normally distributed random process applied at the end mass, i.e. $\mathbf{x}(t) = (0, 0, 0, 0, \mathcal{N}(0, 1))$ with constant spectral density $S_{X_5 X_5}(0) = 0.01 \mathrm{N}^2/\mathrm{Hz}$. In this example, we examine the correlations among response data collected from two different points on the structure. Specifically, allow $m_i = 0.01$ kg, $c_i = 0.1 \mathrm{N} \cdot \mathrm{s/m}$, and $k_i = 10$ N/m for each of the $i = 1 \cdots 5$ DOFs, thus $\beta = 0.01$ in our modal damping model. The system response to the stochastic forcing, $y_i(n\Delta_t)$, $n = 1 \cdots 2^{15}$, is then generated via numerical integration. For simulation purposes we used a time step of $\Delta_t = 0.01$ s, which is sufficient to capture all five of the system modes (the highest of which possesses a resonant frequency of 9.65 Hz). On the basis of these parameters, we first generate the analytical expressions $R_{Y_i Y_j}(T)$ and compare them to the estimates provided by Eqn. 6.85; these results are shown in Figure 6.9 for the case where $i = 2, j = 3$. Specifically, we show the autocovariances $R_{Y_2 Y_2}(T)$, $R_{Y_3 Y_3}(T)$, and the cross-covariance, $R_{Y_2 Y_3}(T)$. Estimate and theory are in good agreement, as we might have hoped. We also note that the variances of the random processes $Y_2(t)$ and $Y_3(t)$ are correctly given by $R_{Y_2 Y_2}(0), R_{Y_3 Y_3}(0)$, respectively. It is also instructive to compare theory and estimate of the cross-correlation between drive and response signals. Figure 6.10 shows the covariance between the drive signal and the response at mass 3. As predicted by Eq. (6.104), this quantity is *only* nonzero for negative delays. Clearly, future values of the drive are uncorrelated with the response. This type of behavior is typical of dynamical systems for which the coupling is unidirectional. In this case, the forcing is coupled to the response, but the response has no influence on the forcing.

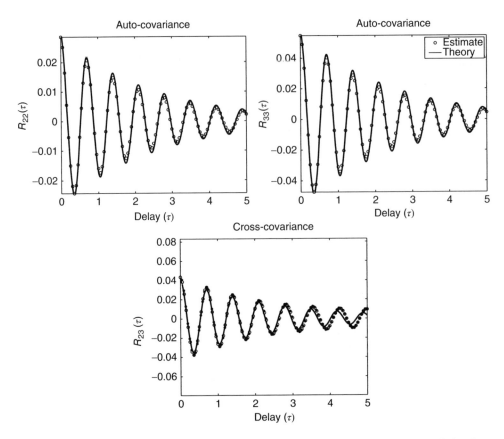

Figure 6.9 Comparison of theoretical and estimated (using Eqn. 6.85) auto- and cross-correlation functions for the five DOF systems driven at the end mass. For easier display, we have written $R_{Y_2 Y_3}(\tau)$ as simply $R_{23}(\tau)$, implicitly noting that the random variables of interest are those associated with the response, $Y_i(t)$

To complete the example, we can form an estimate for the cross-correlation function and compare to the theoretical prediction; this result is shown in Figure 6.11. The analytical covariance functions we have derived here are used later in this section to validate theoretical expressions for information-theoretic quantities. They may also be used to understand the influence of linear system parameters (mass, stiffness, and damping) on the auto- and cross-correlation functions frequently estimated in signal processing applications.

6.5 Estimating Higher Order Spectra

In the next few sections, we develop analytical expressions for the higher order spectra (HOS), described earlier in Section (3.3.5), and show how they may be estimated from observations. These expressions are essential to understanding how the HOS can be used to detect the

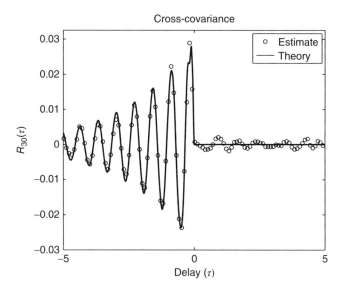

Figure 6.10 Comparison of theoretical and estimated cross-correlation functions between the drive and response at mass 3. The notation $R_{30}(\tau)$ is shorthand for $R_{Y_3X}(\tau)$

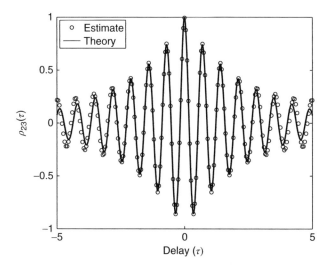

Figure 6.11 Comparison of theoretical and estimated cross-correlation functions for the five-DOF system driven at the end mass. The notation $\rho_{23}(\tau)$ is shorthand for $\rho_{Y_2,Y_3}(\tau)$

presence of damage-induced nonlinearity in structures. As has been our convention throughout, we use $X(t)$ as the random process model for the input signal $x(t)$ and similarly $Y_i(t)$ as the random process associated with the structural response at location i. Recall that the auto-bispectral density for the stationary response of an MDOF system is defined as the double

FT of the third cumulant (alternatively, third moment about the mean)

$$B_{Y_k Y_i Y_j}(\omega_1, \omega_2) =$$

$$\left(\frac{1}{2\pi}\right)^2 \int\limits_{-\infty}^{\infty} \int\limits_{-\infty}^{\infty} E[(Y_k(t) - \bar{Y}_k)(Y_i(t + \tau_1) - \bar{Y}_i)(Y_j(t + \tau_2) - \bar{Y}_j)] e^{-i(\omega_1 \tau_1 + \omega_2 \tau_2)} d\tau_1 d\tau_2$$

$$(6.106)$$

The distinction is sometimes not made between the terms "auto-bispectrum" and "auto-bispectral density." Equation (6.106) is, in fact, a density with units $[y]^3 / \mathrm{Hz}^2$ and is referred to as such throughout. The expected value under the integral will only be a function of the delays τ_1, τ_2 due to the assumption of stationarity.

Now, if our system is well approximated by a linear model, we have $y_i(t) = \int h_i^r(\theta) x_r(t - \theta) d\theta$, where $x_r(t)$ is the input signal applied at DOF r and $h_1^r(\tau)$ is the impulse response function relating input to output. Considering the expectation in (6.106), we have

$$E[(Y_k(t) - \bar{Y}_k)(Y_i(t + \tau_1) - \bar{Y}_i)(Y_j(t + \tau_2) - \bar{Y}_j)] =$$

$$E\left[\int h_k^r(\theta_1)(X_r(t - \theta_1) - \bar{X}_r) d\theta_1 \int h_i^r(\theta_2)(X_r(t + \tau_1 - \theta_2) - \bar{X}_r) d\theta_2\right.$$

$$\left. \times \int h_j^r(\theta_3)(X_r(t + \tau_2 - \theta_3) - \bar{X}_r) d\theta_3\right]$$

$$= \int \int \int h_k^r(\theta_1) h_i^r(\theta_2) h_j^r(\theta_3) E[(X_r(t - \theta_1) - \bar{X}_r)(X_r(t + \tau_1 - \theta_2) - \bar{X}_r)$$

$$\times (X_r(t + \tau_2 - \theta_3))] d\theta_1 \, d\theta_2 \, d\theta_3 \qquad (6.107)$$

in which case the third joint moment of the output is seen to be expressible as linear combinations of third joint moments of the input. Thus, if the input is modeled as a stationary, jointly Gaussian random process we have by Isserlis' theorem (3.1.1) that

$$E[(X_r(t - \theta_1) - \bar{X}_r)(X_r(t + \tau_1 - \theta_2) - \bar{X}_r)(X_r(t + \tau_2 - \theta_3) - \bar{X}_r)] = 0$$

and so $B_{kij}(\omega_1, \omega_2) = 0$ for all frequency pairs ω_1, ω_2. Now, there are two ways in which the expectation (6.107) can attain nonzero values. The first way this can happen is if the input possesses a nonzero joint third moment. Even if the system is linear, this will result in a nonzero bispectral density. Indeed, we will revisit the non-Gaussian excitation case shortly. The second mechanism for producing a nonzero bispectrum is the presence of a nonlinearity in the system, that is, the impulse response is no longer sufficient to model the input/output relationship. Even if the input *is* jointly Gaussian distributed, a quadratic nonlinearity (for example) forces $E[(Y_k(t) - \bar{Y}_k)(Y_i(t + \tau_1) - \bar{Y}_i)(Y_j(t + \tau_2) - \bar{Y}_j)] \neq 0$. The presence of a nonzero bispectral density can therefore be used to infer the presence of the nonlinearity.

Before we can derive the auto-bispectral density for a nonlinear system, we require a model capable of capturing nonlinearity. To this end, we make use of the Volterra series model introduced in Section (4.6) and write for the response [27, 28]

$$y_i(t) = y_{1,i}(t) + y_{2,i}(t) + \cdots$$

$$\approx \int h_{1,i}^r(\theta) x_r(t - \theta) d\theta + \int \int h_{2,i}^r(\theta_1, \theta_2) x_r(t - \theta_1) x_r(t - \theta_2) d\theta_1 \, d\theta_2 \qquad (6.108)$$

where $y_{1,i}$ and $y_{2,i}$ are the first- and second-order components of the Volterra expansion for the response at DOF i (it has been assumed, without loss of generality, that the static term $y_{0,i}(t)$ is zero). The quantities $h_{1,i}^r(\theta)$ and $h_{2,i}^r(\theta_1, \theta_2)$ are the Volterra kernels associated with the ith response DOF, with excitation located at point r. In what follows, we certainly could have chosen to use more terms in the Volterra series model, for example, $y_{3,i}(t), y_{4,i}(t)$. However, as with a Taylor series, each additional term provides higher order contributions to the signal model. These terms have a correspondingly minor influence in the resulting bispectrum, and hence we exclude them in what follows. Moreover, we are eventually interested in damage detection applications, where the goal is to capture low levels of nonlinearity. For these reasons, the two-term model is sufficient for our analysis. We see later in Section (6.5.4) that for cubic nonlinearities we indeed require a third-order Volterra model.

Consider the excitation, applied at DOF r, to be i.i.d. Gaussian noise with zero mean and unit standard deviation, i.e. $x_r(t) \sim N(0, 1)$. In this case, the mean of the system response is given by

$$\bar{y}_i = \bar{y}_{1,i} + \bar{y}_{2,i}$$

$$= 0 + \int \int h_{2,i}^r(\theta_1, \theta_2) E[X_r(t - \theta_1) X_r(t - \theta_2)] d\theta_1 \, d\theta_2. \qquad (6.109)$$

At this point, it is convenient to define the zero-mean signal

$$\tilde{y}_i \equiv (y_{1,i}(t) - \bar{y}_{1,i}) + (y_{2,i}(t) - \bar{y}_{2,i})$$

$$= \int h_{1,i}^r(\theta) x_r(t - \theta) d\theta$$

$$+ \int_{\mathbb{R}^2} h_{2,i}^r(\theta_1, \theta_2) \{ x_r(t - \theta_1) x_r(t - \theta_2) - E[X_r(t - \theta_1) X_r(t - \theta_2)] \} d\theta_1 \, d\theta_2 \qquad (6.110)$$

Substituting Eqs. (6.109) and (6.108) into Eq. (6.106) yields for the main expectation

$$E[\tilde{Y}_k(t) \tilde{Y}_i(t + \tau_1) \tilde{Y}_j(t + \tau_2)]$$

$$= E[\tilde{Y}_{1,k}(t)\tilde{Y}_{1,i}(t + \tau_1)\tilde{Y}_{1,j}(t + \tau_2)] + E[\tilde{Y}_{1,k}(t)\tilde{Y}_{2,i}(t + \tau_1)\tilde{Y}_{1,j}(t + \tau_2)]$$

$$+ E[\tilde{Y}_{2,k}(t)\tilde{Y}_{1,i}(t + \tau_1)\tilde{Y}_{1,j}(t + \tau_2) + E[\tilde{Y}_{2,k}(t)\tilde{Y}_{2,i}(t + \tau_1)\tilde{Y}_{1,j}(t + \tau_2)]$$

$$+ E[\tilde{Y}_{1,k}(t)\tilde{Y}_{1,i}(t + \tau_1)\tilde{Y}_{2,j}(t + \tau_2)] + E[\tilde{Y}_{1,k}(t)\tilde{Y}_{1,i}(t + \tau_1)\tilde{Y}_{2,j}(t + \tau_2)]$$

$$+ E[\tilde{Y}_{2,k}(t)\tilde{Y}_{1,i}(t + \tau_1)\tilde{Y}_{1,j}(t + \tau_2)] + E[\tilde{Y}_{2,k}(t)\tilde{Y}_{2,i}(t + \tau_1)\tilde{Y}_{2,j}(t + \tau_2)]$$

$$= E[\tilde{Y}_{2,k}(t)\tilde{Y}_{2,i}(t + \tau_1)\tilde{Y}_{2,j}(t + \tau_2)] + E[\tilde{Y}_{1,k}(t)\tilde{Y}_{1,i}(t + \tau_1)\tilde{Y}_{2,j}(t + \tau_2)]$$

$$+ E[\tilde{Y}_{2,k}(t)\tilde{Y}_{1,i}(t + \tau_1)\tilde{Y}_{1,j}(t + \tau_2)] + E[\tilde{Y}_{1,k}(t)\tilde{Y}_{2,i}(t + \tau_1)\tilde{Y}_{1,j}(t + \tau_2)] \qquad (6.111)$$

where the four terms can be eliminated because they involve the expectation of products of an odd number of Gaussian distributed random variables (again, see Isserlis' theorem 3.1.1).

Taking the double FT of the remaining four terms and simplifying therefore yields the bispectrum. The general procedure is illustrated in the second of these terms. First, we establish the following notation

$$E[X(t - \tau_a)X(t - \tau_b)] \equiv \phi_{XX}^{a,b} = \frac{1}{2\pi} \int_{\mathbb{R}} S_{XX}(\omega)e^{i\omega(-\tau_b + \tau_a)} d\omega$$

$$E[X(t + \tau_a - \tau_b)X(t - \tau_c)] \equiv \phi_{XX}^{a,b,c} = \frac{1}{2\pi} \int_{\mathbb{R}} S_{XX}(\omega)e^{i\omega(-\tau_c - \tau_a + \tau_b)} d\omega$$

$$E[X(t + \tau_a - \tau_b)X(t + \tau_c - \tau_d)] \equiv \phi_{XX}^{a,b,c,d} = \frac{1}{2\pi} \int_{\mathbb{R}} S_{XX}(\omega)e^{i\omega(\tau_c - \tau_d - \tau_a + \tau_b)} d\omega \qquad (6.112)$$

where we have made use of the Weiner–Khinchin theorem (3.69) in relating the correlation functions to the PSD. Substituting in the Volterra model for the third term in (6.111) gives

$$E[\tilde{Y}_{1,k}(t)\tilde{Y}_{1,i}(t + \tau_1)\tilde{Y}_{2,j}(t + \tau_2)] = E\left[\int h_{1,k}(\tau_3)X(t - \tau_3)d\tau_3 \int h_{1,i}(\tau_4)X(t + \tau_1 - \tau_4)d\tau_4 \right.$$

$$\left. \times \int_{\mathbb{R}^2} h_{2,j}(\tau_5, \tau_6) \left[X(t + \tau_2 - \tau_5)X(t + \tau_2 - \tau_6) - \phi_{XX}^{2,5,2,6} \right] d\tau_5 \, d\tau_6 \right] \qquad (6.113)$$

where we have noted that the mean of the Volterra model is $\int_{\mathbb{R}^2} h_{2,j}(\tau_5, \tau_6)E[X(t + \tau_2 - \tau_5)X(t + \tau_2 - \tau_6)]d\tau_5 \, d\tau_6$. Simplifying this expression further yields

$$E[\tilde{Y}_{1,k}(t)\tilde{Y}_{1,i}(t + \tau_1)\tilde{Y}_{2,j}(t + \tau_2)] =$$

$$\int_{\mathbb{R}^6} h_{1,k}(\tau_3)h_{1,i}(\tau_4)h_{2,j}(\tau_5, \tau_6)\{E[X(t - \tau_3)X(t + \tau_1 - \tau_4)X(t + \tau_2 - \tau_5)X(t + \tau_2 - \tau_6)]$$

$$- E[X(t - \tau_3)x(t + \tau_1 - \tau_4)\phi_{xx}^{5,6}]\}d\tau_1 \, d\tau_2 \, d\tau_3 \, d\tau_4 \, d\tau_5 \, d\tau_6$$

$$= \int_{\mathbb{R}^6} h_{1,k}(\tau_3)h_{1,i}(\tau_4)h_{2,j}(\tau_5, \tau_6) \left\{ -\phi_{XX}^{1,4,3}\phi_{XX}^{5,6} \right.$$

$$+ E[X(t - \tau_3)X(t + \tau_1 - \tau_4)]E[X(t + \tau_2 - \tau_5)X(t + \tau_2 - \tau_6)]$$

$$+ E[X(t - \tau_3)X(t + \tau_2 - \tau_5)]E[X(t + \tau_1 - \tau_4)X(t + \tau_2 - \tau_6)]$$

$$\left. + E[X(t - \tau_3)X(t + \tau_2 - \tau_6)]E[X(t + \tau_1 - \tau_4)X(t + \tau_2 - \tau_5)]\right\}$$

$$= \int_{\mathbb{R}^6} h_{1,k}(\tau_3)h_{1,i}(\tau_4)h_{2,j}(\tau_5, \tau_6) \left\{ -\cancel{\phi_{XX}^{1,4,3}}\cancel{\phi_{XX}^{5,6}} + \cancel{\phi_{XX}^{1,4,3}}\cancel{\phi_{XX}^{5,6}} + \phi_{XX}^{2,5,3}\phi_{XX}^{1,4,2,6} \right.$$

$$\left. + \phi_{XX}^{2,6,3}\phi_{XX}^{1,4,2,5} \right\} d\tau_1 \, d\tau_2 \, d\tau_3 \, d\tau_4 \, d\tau_5 \, d\tau_6. \qquad (6.114)$$

Note that in lines 5–7 of the given expression we have made use of Isserlis' theorem 3.1.1 in expanding the expectation of the product of four Gaussian distributed random variables $X(t)$. This allows us to write the product as the sum of three products of covariances. At this point, we also note the symmetry of the second-order Volterra kernel $h(\tau_5, \tau_6) = h(\tau_6, \tau_5)$, and hence we arrive at

$$E[\tilde{Y}_{1,k}(t)\tilde{Y}_{1,i}(t + \tau_1)\tilde{Y}_{2,j}(t + \tau_2)]$$

$$= 2 \int_{\mathbb{R}^6} h_{1,k}(\tau_3)h_{1,i}(\tau_4)h_{2,j}(\tau_5, \tau_6)\phi_{XX}^{2,5,3}\phi_{XX}^{1,4,2,6} d\tau_1 \, d\tau_2 \, d\tau_3 \, d\tau_4 \, d\tau_5 \, d\tau_6. \quad (6.115)$$

We once again make use of the Weiner–Khinchin theorem (3.69) and write each of the covariance functions ϕ_{XX} as the inverse FT of an associated PSD function (see Eq. 6.112). Specifically, letting $\phi_{XX}^{2,5,3} = \frac{1}{2\pi} \int S_{XX}(\xi)e^{i\xi(-\tau_3 - \tau_2 + \tau_5)}d\xi$ and $\phi_{XX}^{1,4,2,6} = \frac{1}{2\pi} \int S_{XX}(\xi)e^{i\xi(-\tau_1 + \tau_4 + \tau_2 - \tau_6)}d\xi$ so that upon substitution we have

$$E[\tilde{Y}_{1,k}(t)\tilde{Y}_{1,i}(t + \tau_1)\tilde{Y}_{2,j}(t + \tau_2)] = 2 \times \left(\frac{1}{4\pi^2}\right) \int_{\mathbb{R}^8} h_{1,k}(\tau_3)h_{1,i}(\tau_4)h_{2,j}(\tau_5, \tau_6)$$

$$\times S_{XX}(\xi_1)S_{XX}(\xi_2)e^{i\xi_1(-\tau_3 - \tau_2 + \tau_5)}S_{XX}(\xi_2)e^{i\xi_2(-\tau_1 + \tau_4 + \tau_2 - \tau_6)}d\xi_1 \, d\xi_2$$

$$= 2 \times \left(\frac{1}{4\pi^2}\right) \int_{\mathbb{R}^2}\int_{\mathbb{R}} h_{1,k}(\tau_3)e^{-i\xi_1\tau_3}d\tau_3 \int_{\mathbb{R}} h_{1,i}(\tau_4)e^{i\xi_2\tau_4}d\tau_4$$

$$\int_{\mathbb{R}^2} h_{2,j}(\tau_5, \tau_6)e^{i\xi_1\tau_5 - \xi_2\tau_6}d\tau_5 \, d\tau_6$$

$$\times S_{XX}(\xi_1)S_{XX}(\xi_2)e^{i(-\xi_1 + \xi_2)\tau_2}e^{-i\xi_2\tau_1}d\xi_1 \, d\xi_2 \quad (6.116)$$

Now, let $\omega_1 = -\xi_2$ and $\omega_2 = -\xi_1 + \xi_2$. Under this change of variables, the expression becomes ($\xi_2 \to -\omega_1$ and $\xi_1 \to -\omega_1 - \omega_2$)

$$E[\tilde{Y}_{1,k}(t)\tilde{Y}_{1,i}(t + \tau_1)\tilde{Y}_{2,j}(t + \tau_2)]$$

$$= 2 \times \left(\frac{1}{4\pi^2}\right) \int_{\mathbb{R}^2}\int_{\mathbb{R}} h_{1,k}(\tau_3)e^{i(\omega_1 + \omega_2)\tau_3}d\tau_3 \int_{\mathbb{R}} h_{1,i}(\tau_4)e^{-i\omega_1\tau_4}d\tau_4$$

$$\times \int_{\mathbb{R}^2} h_{2,j}(\tau_5, \tau_6)e^{-i(\omega_1 + \omega_2)\tau_5 + \omega_1\tau_6}d\tau_5 d\tau_6 S_{XX}(-\omega_1 - \omega_2)$$

$$S_{XX}(-\omega_1)e^{i(\omega_1\tau_1 + \omega_2\tau_2)}d\omega_1 \, d\omega_2$$

$$= 2 \times \left(\frac{1}{4\pi^2}\right) \int_{\mathbb{R}^2} H_{1,k}(-\omega_1 - \omega_2)H_{1,i}(\omega_1)H_{2,j}(\omega_1 + \omega_2, -\omega_1)$$

$$\times S_{XX}(-\omega_1 - \omega_2)S_{XX}(-\omega_1)e^{i(\omega_1\tau_1 + \omega_2\tau_2)}d\omega_1 \, d\omega_2. \quad (6.117)$$

At this point, we can recognize that the right-hand side is in the form of an inverse, double FT. Thus, by definition, the term inside the integrals is the double FT of the left-hand side and is therefore a portion of the bispectrum. In other words,

$$\int_{\mathbb{R}^2} E[\tilde{Y}_{1,k}(t)\tilde{Y}_{1,i}(t + \tau_1)\tilde{Y}_{2,j}(t + \tau_2)]e^{-i(\omega_1\tau_1 + \omega_2\tau_2)} =$$

$$2H_{1,k}(-\omega_1 - \omega_2)H_{1,i}(\omega_1)H_{2,j}(\omega_1 + \omega_2, -\omega_1)S_{XX}(\omega_1 + \omega_2)S_{XX}(\omega_1) \quad (6.118)$$

where we have also taken advantage of the symmetry of the PSD function, i.e., $S_{XX}(-\omega) = S_{XX}(\omega)$.

The procedure is exactly the same for each of the other terms in (6.111): substitute the Volterra model, combine terms under a multidimensional integral, take expectations and expand using Isserlis' theorem, use the Weiner–Khinchin theorem to provide a spectral representation of the covariances, and simplify taking into account the symmetry of the Volterra kernels. It is a tedious, but straightforward process that is more an excercise in book-keeping than mathematics. In the end, we have for the bispectral density [29]

$$
B_{Y_k Y_i Y_j}(\omega_1, \omega_2) =
$$

$$
\frac{8}{2\pi} \int_{-\infty}^{\infty} H_{2,k}(\omega_1 + \xi, \omega_2 - \xi) H_{2,i}(-\omega_1 - \xi, \xi) H_{2,j}(-\omega_2 + \xi, -\xi)
$$

$$
S_{xx}(\xi) S_{XX}(\omega_1 + \xi) S_{XX}(\omega_2 - \xi) d\xi
$$

$$
+ 2H_{2,i}^2(\omega_1 + \omega_2, -\omega_2) H_{1,k}(-\omega_1 - \omega_2) H_{1,j}(\omega_2) S_{XX}(\omega_1 + \omega_2) S_{XX}(\omega_2)
$$

$$
+ 2H_{2,j}(\omega_1 + \omega_2, -\omega_1) H_{1,k}(-\omega_1 - \omega_2) H_{1,i}(\omega_1) S_{XX}(\omega_1 + \omega_2) S_{XX}(\omega_1)
$$

$$
+ 2H_{2,k}(-\omega_1, -\omega_2) H_{1,i}(\omega_1) H_{1,j}(\omega_2) S_{XX}(\omega_1) S_{XX}(\omega_2) \tag{6.119}
$$

where $S_{XX}(\omega) = \frac{1}{2\pi} \int_{-\infty}^{\infty} E[X_r(t) X_r(t - \tau_1)] e^{-i\omega\tau_1} d\tau_1$ is, by definition the autospectral density function. Note that if we assume that the input values $x(t)$ are independent, identically Gaussian distributed that $S_{xx}(\omega) = $ const. $= P$. Note also that for the nonlinear auto-spectral density function (6.82) the nonlinearity showed up as a higher order term in "P", whereas here the terms involving the nonlinearity scale as the product of two autospectral density functions.

Expressions for the frequency domain kernels, $H_1 = \int_{-\infty}^{\infty} h_1(\tau_1) e^{-i\omega\tau_1} d\tau_1$ and $H_2 = \int_{-\infty}^{\infty} \int_{-\infty}^{\infty} h_2(\tau_1, \tau_2) e^{-i(\omega_1\tau_1 + \omega_2\tau_2)} d\tau_1 d\tau_2$, may be obtained via the harmonic probing technique described in Section (4.6). These kernels will be dependent on the specific system under study. However, it is not immediately clear how to deal with the integral term. First, consider a constant input PSD $S_{xx}(\omega) = P$ in units of x^2 per Hz. Now recall that for the Volterra model to hold, we are required to consider the weakly nonlinear response regime which requires that the excitation spectral density be small in magnitude. For most physically meaningful examples, we may therefore write $P < 1 [x]^2/\text{Hz}$. The result is that the integral term is typically a higher order term, i.e., $O(P^3) < O(P^2)$.

Nonetheless, the integral term can be evaluated analytically as a contour integral just as we did for the higher order term in the nonlinear autospectral density function. In Appendix B, we provide the details on how to solve for this higher order term. Specific examples of bispectral density functions are considered shortly, however we first introduce a related concept that will prove useful in damage detection applications where the input signal does not adhere to the jointly Gaussian model.

6.5.1   Coherence Functions

Isserlis' theorem is the cornerstone of many techniques in signal processing and was featured prominently in the last section in deriving the analytical bispectral density function for nonlinear systems. Many of the random processes we observe in the field are, in fact, appropriately modeled as jointly Gaussian distributed and so application of the theorem is usually reasonable. Nonetheless, there are many instances in which the input random process $\mathbf{X}$ is comprised of independent observations drawn from some *non-Gaussian* distribution, hence use of the theorem cannot be justified. For example, consider a linear system driven by a stationary, i.i.d. random process where the third cumulant is nonzero, i.e., $k_{XXX}(0,0) = E[(X(t) - \mu_X)^3] \neq 0 \forall t$. In this linear case, the first term in the output expectation (6.111) is the only term that remains, that is, $E[\tilde{Y}_i(t)\tilde{Y}_j(t + \tau_1)\tilde{Y}_k(t + \tau_2)] = E[\tilde{Y}_{1,k}(t)\tilde{Y}_{1,i}(t + \tau_1)\tilde{Y}_{1,j}(t + \tau_2)]$. Thus, following the same process we used to derive the expression for the nonlinear response bispectrum of the previous section, we can readily observe that

$$B_{Y_k Y_i Y_j}(\omega_1, \omega_2) = H_{1,i}(\omega_1)H_{1,j}(\omega_2)H_{1,k}(-\omega_1 - \omega_2)B_{XXX}(\omega_1, \omega_2) \qquad (6.120)$$

where the input bispectrum $B_{XXX}(\omega_1, \omega_2) = \text{const.} \neq 0$. This follows from the definition of the HOS (3.94), where we see that if the stationary cumulant $k_{XXX}(0,0) \neq 0$ (see Eq. 3.21), and $k_{XXX}(\tau_1, \tau_2) = 0$ for $\tau_1, \tau_2 > 0$ (due to the i.i.d. assumption), then the bispectral density is a constant. For this more general forcing, the output bispectral density for a linear system will no longer be zero, but will possess the general shape of the product of linear transfer functions as given in (6.120). This presents a difficulty for the nonlinearity detection schemes introduced in Chapter 8, as one cannot easily distinguish the nonlinear response of a Gaussian-driven structure with a linear structure driven by a stationary i.i.d. random process with non-Gaussian marginal distribution.

Thus, it was suggested in Ref. [30] that the bispectrum be normalized such that deviations from linearity in the driven system produce a readily identifiable signature when the system is nonlinear, regardless of the joint PDF of the input. This can be accomplished by first noting that for a linear system, the PSD of the output is related to the PSD of the input via the transfer function by $S_{YY}(\omega) = |H_1(\omega)|^2 S_{XX}(\omega)$. A convenient normalization for the bispectrum is therefore obtained by simply dividing the magnitude of Eq. (6.120) by an appropriate product of output spectra, giving the *bicoherence function*

$$b_{Y_k Y_i Y_j}(\omega_1, \omega_2) = \frac{B_{Y_k Y_i Y_j}(\omega_1, \omega_2)}{\sqrt{S_{Y_i Y_i}(\omega_1)S_{Y_j Y_j}(\omega_2)S_{Y_k Y_k}(\omega_1 + \omega_2)}} \qquad (6.121)$$

the magnitude of which, for linear systems, becomes

$$|b_{Y_k Y_i Y_j}(\omega_1, \omega_2)| = \frac{|B_{XXX}(\omega_1, \omega_2)||H_{1,i}(\omega_1)H_{1,j}(\omega_2)H_{1,k}(-\omega_1 - \omega_2)|}{|H_{1,i}(\omega_1)H_{1,j}(\omega_2)H_{1,k}(-\omega_1 - \omega_2)|\sqrt{S_{XX}(\omega_1)S_{XX}(\omega_2)S_{XX}(\omega_1 + \omega_2)}}$$

$$= \frac{|B_{XXX}(\omega_1, \omega_2)|}{\sqrt{S_{XX}(\omega_1)S_{XX}(\omega_2)S_{XX}(\omega_1 + \omega_2)}} \qquad (6.122)$$

For an i.i.d. input process **X**, the magnitude bicoherence will therefore be a constant because we know that for a sequence of independent observations spanning time T that $S_{XX}(\omega) = E[(X(t) - \mu_X)^2]\Delta/(2\pi) = \sigma_X^2 \Delta/(2\pi)$ (see eqn 3.75). Similarly, we can show that $B_{XXX}(\omega_1, \omega_2) = E[(X(t) - \mu_X)^3]\Delta^2/(4\pi^2) \equiv k_3\Delta^2/(4\pi^2)$ (see discussion surrounding Eq. 3.93). The expectation is, by definition, the third cumulant (see Eq. 2.81) and is therefore simply denoted k_3. Thus, we arrive at the magnitude bicoherence function for an i.i.d. random process as being the constant

$$|b_{Y_k Y_i Y_j}(\omega_1, \omega_2)| = \text{const.}$$

$$= \frac{k_3 \Delta^2}{(\sigma_X^2 \Delta)^{3/2}} \frac{(2\pi)^{-2}}{(2\pi)^{-3/2}}$$

$$= \frac{k_3 \Delta^{1/2}}{\sigma_X^3 (2\pi)^{1/2}} \tag{6.123}$$

and possessing units $(\text{rad/s})^{-1/2}$. The sampling interval is frequently left out of the derivation for this constant, however the bicoherence is *not* a dimensionless quantity and correct units are certainly required if one is to match theory to estimate. In later chapters (see Chapter 8), we see how the estimated bicoherence leads to a simple test for damage-induced nonlinearity in structures. As usual, we point out that had we used frequency variable f instead of ω, the result would be the same, but without the factor of $(2\pi)^{1/2}$ in the denominator.

A similar normalization can be used for the output trispectrum of Section (3.3.5), i.e., we may form the tricoherence

$$t_{Y_p Y_i Y_j Y_k}(\omega_1, \omega_2, \omega_3) = \frac{T_{Y_p Y_i Y_j Y_k}(\omega_1, \omega_2, \omega_3)}{\sqrt{S_{X_i X_i}(\omega_1) S_{X_j X_j}(\omega_2) S_{X_k X_k}(\omega_3) S_{X_p X_p}(\omega_1 + \omega_2 + \omega_3)}} \tag{6.124}$$

As with the bicoherence, for a linear system driven by an i.i.d. random process X, the magnitude tricoherence reduces to the constant

$$|t_{Y_p Y_i Y_j Y_k}(\omega_1, \omega_2, \omega_3)| = \frac{|T_{XXXX}(\omega_1, \omega_2, \omega_3)|}{\sqrt{S_{X_i X_i}(\omega_1) S_{X_j X_j}(\omega_2) S_{X_k X_k}(\omega_3) S_{X_p X_p}(\omega_1 + \omega_2 + \omega_3)}}$$

$$= \frac{k_4 \Delta^3 (2\pi)^{-3}}{\sigma_X^4 \Delta^2 (2\pi)^{-2}}$$

$$= \frac{k_4 \Delta}{2\pi \sigma_X^4} \tag{6.125}$$

where k_4 is the fourth cumulant associated with the input X and given in Eq. (2.81). However, before using the HOS or the associated coherence functions, we require a means of estimating these quantities, for example, (6.119) and (6.121) from observed data, the main thrust of this chapter.

6.5.2 Bispectral Density Estimation

The estimation of $B_{X_k X_i X_j}(\omega_1, \omega_2)$ is made with finite data and will therefore be subject to both bias and variance. This results in nonzero values of the bispectrum even when no nonlinearity is present. Ideally, the estimator used will be consistent, that is to say, both the bias and variance will go to zero as the number of data points become large. This is of some importance given that we plan to use the HOS as detectors of nonlinearity, where the estimated values are to be used as detection statistics.

There are two main approaches used in bispectral density estimation, and our choice depends on the type of data we have at our disposal. Consider K independent realizations of the three signals $y_i^{(\kappa)}(t), y_j^{(\kappa)}(t), y_k^{(\kappa)}(t), \kappa = 1 \cdots K$ and the general definition of the bispectral density function

$$B_{kij}(\omega_1, \omega_2) = \frac{1}{4\pi^2} \int_{\mathbb{R}^2} E[(Y_k(t) - \bar{Y}_k)(Y_i(t - \tau_1) - \bar{Y}_i)(Y_j(t - \tau_2) - \bar{Y}_j)] e^{-i(\omega_1 \tau_1 + \omega_2 \tau_2)} d\tau_1 \, d\tau_2.$$

(6.126)

The first approach, referred to as the *indirect method*, involves averaging over K separate data records (realizations) to approximate the expected value

$$E[(Y_k(t) - \bar{Y}_k)(Y_i(t - \tau_1) - \bar{Y}_i)(Y_j(t - \tau_2) - \bar{Y}_j)] \approx$$

$$\frac{1}{K} \sum_{\kappa=1}^{K} (y_k^{(\kappa)}(n) - \bar{y}_k)(y_i^{(\kappa)}(n + T_1) - \bar{y}_i)(y_j^{(\kappa)}(n + T_2) - \bar{y}_j) \qquad (6.127)$$

and then taking the double, discrete FT with regard to T_1, T_2. Details of this approach are discussed in Ref. [31].

The more commonly used *direct* method for estimating the bispectrum and bicoherence functions has been discussed at length in a number of references (see, e.g., Nikias and Raghuveer [31] and Huber *et al.* [32]) and proceeds in much the same way as for the PSD function. We again assume that the observed data are ergodic and possess a Fourier representation so that our estimate can be formed from a single realization. As we have already shown (see Eq. 3.90), given three signals $y_i(t), y_j(t), y_k(t)$, the bispectral density function can be written in the form

$$B_{kij}(\omega_1, \omega_2) = \lim_{T \to \infty} \left(\frac{1}{4\pi^2} \right) \frac{E[Y_i(\omega_1)Y_j(\omega_2)Y_k(-\omega_1 - \omega_2)]}{2T}$$

$$B_{kij}(f_1, f_2) = \lim_{T \to \infty} \frac{E[Y_i(f_1)Y_j(f_2)Y_k(-f_1 - f_2)]}{2T} \qquad (6.128)$$

that is, as an expectation over the FTs of the random processes of interest. We choose this estimator (as opposed to 3.87), because for our limited data records we can guarantee the existence of the FT. An estimator can be formed by averaging the discrete FT of the data over some number of segments, just as is commonly done for power spectrum estimation. Because most software packages implement the DFT with frequency f instead of ω, we opt for the second expression in (6.128) in forming the estimate.

To begin, assume the observed signals are sampled with frequency $f_s = 1/\Delta_t$ and may be denoted $y_i(n), y_j(n), y_k(n), n = 0 \cdots N - 1$. Dividing the observed data $y(n), n = 1 \cdots N$ into overlapping segments of length M gives $y_{s,i}(m) = y_i(m + sM - L), m = 0 \cdots M - 1$, $s = 0 \cdots S - 1$. The variable L denotes the degree of overlap ($0 \leq L < M$). The mean of the data segment is subtracted (in accordance with the bispectrum definition), windowed, and then Fourier transformed to give $Y_{s,i}(f_p) = \sum_m g(m) y_{s,i}(m) e^{-i2\pi pm/M}, p = 0, \cdots, M - 1$, where $g(m)$ is the windowing function (see page 223). The final estimator for Eq. (6.106) is then the average

$$\hat{B}_{kij}(f_p, f_q) = \frac{\Delta_t^2}{SM} \sum_{s=0}^{S-1} Y_{s,i}(f_p) Y_{s,j}(f_q) Y_{s,k}^*(f_p + f_q) \qquad (6.129)$$

for discrete frequencies f_p, f_q. Estimating the bicoherence (6.121) simply involves dividing by the needed product of estimated PSDs

$$\widehat{S}_{ii}(f_p) = \frac{\Delta_t}{SM} \sum_{s=0}^{S-1} Y_{s,i}(f_p) Y_{s,i}^*(f_p) \qquad (6.130)$$

to give

$$\widehat{b}_{kij}(f_p, f_q) = \frac{\widehat{B}_{kij}(f_p, f_q)}{\sqrt{\widehat{S}_{ii}(f_p) \widehat{S}_{jj}(f_q) \widehat{S}_{kk}(f_p + f_q)}} \qquad (6.131)$$

Both bispectrum and bicoherence are complex normally distributed [30, 32] with variance

$$\sigma_e^2 = \kappa \frac{M^2 \Delta_t}{N} S_{ii}(f_1) S_{jj}(f_2) S_{kk}(f_1 + f_2) \qquad (6.132)$$

and

$$\sigma_e^2 = \kappa \frac{M^2 \Delta_t}{N} \qquad (6.133)$$

respectively. The prefactor κ accounts for the possibility of overlapping segments ($L > 0$) and windowing and is discussed at length in Huber *et al.* [32]. It is also worth noting that the variance doubles along the line $f_1 = f_2$, as discussed in Ref. [33].

As the number of data increases toward infinity, the bias in the estimate approaches zero, thus Eq. (6.129) is an unbiased estimate [31]. In fact, we have shown earlier that the estimates for the Fourier coefficients used in the estimate are MLEs if the noise on the observed data is joint normally distributed. Thus, we would also expect the estimator (6.129) to enjoy the properties of an MLE. As a check, in what follows we derive the bias in the estimate in a manner analogous to the derivation used in PSD estimates [10].

The expected value for the estimate may be written as the value of the magnitude auto-bispectral density averaged over each frequency bin, that is,

$$E[|\widehat{B}_{YY}(f_1, f_2)|^2] \approx \frac{1}{(2\Delta_f)^2} \int_{f_2 - \Delta_f}^{f_2 + \Delta_f} \int_{f_1 - \Delta_f}^{f_1 + \Delta_f} |B_{YY}(\xi_1, \xi_2)|^2 d\xi_1 \, d\xi_2 \qquad (6.134)$$

where $\Delta_f = 1/(2M\Delta_t)$ is the half-frequency bin width. For small Δ_f, the magnitude auto-bispectral density may be expanded as a Taylor series up to the second order, giving

$$E[|\widehat{B}_{YY}(f_1, f_2)|^2] \approx \frac{1}{(2\Delta_f)^2} \int_{f_2 - \Delta_f}^{f_2 + \Delta_f} \int_{f_1 - \Delta_f}^{f_1 + \Delta_f} |B_{YY}(f_1, f_2)|^2 + (\xi_1 - f_1)\frac{\partial}{\partial f_1}|B_{YY}(f_1, f_2)|^2 +$$

$$(\xi_2 - f_2)\frac{\partial}{\partial f_2}|B_{YY}(f_1, f_2)|^2 + \frac{(\xi_1 - f_1)(\xi_2 - f_2)}{2}\frac{\partial^2}{\partial f_1 \partial f_2}|B_{YY}(f_1, f_2)|^2 +$$

$$\frac{(\xi_1 - f_1)^2}{2}\frac{\partial^2}{\partial f_1^2}|B_{YY}(f_1, f_2)|^2 + \frac{(\xi_1 - f_2)^2}{2}\frac{\partial^2}{\partial f_2^2}|B_{YY}(f_1, f_2)|^2 d\xi_1 d\xi_2$$

$$+ \text{H.O.T} \qquad (6.135)$$

The two terms involving the first derivative of the auto-bispectral density vanish as does the mixed partial derivative owing to the fact that $\int_{f-\Delta_f}^{f+\Delta_f} (\xi - f)d\xi = 0$. Carrying out the integration for the remaining terms results in the expression

$$E[|\hat{B}_{YY}(f_1,f_2)|^2] \approx |B_{YY}(f_1,f_2)|^2 + \frac{\Delta_f^2}{6}\frac{\partial^2}{\partial f_1^2}|B_{YY}(f_1,f_2)|^2 + \frac{\Delta_f^2}{6}\frac{\partial^2}{\partial f_2^2}|B_{YY}(f_1,f_2)|^2 \qquad (6.136)$$

thus, the bias is simply

$$b = \frac{\Delta_f^2}{6}\frac{\partial^2}{\partial f_1^2}|B_{YY}(f_1,f_2)|^2 + \frac{\Delta_f^2}{6}\frac{\partial^2}{\partial f_2^2}|B_{YY}(f_1,f_2)|^2. \qquad (6.137)$$

Using the closed-form expression for the auto-bispectral density (Eq. 6.119) in combination with Eq. (6.137), an expression for the bias is obtained. One point of interest regarding Eq. (6.137) concerns the scaling with the nonlinearity parameters c_2, k_2. The bias term grows proportional to the size of the nonlinearity squared, i.e., $b \sim k_2^2, c_2^2$. Thus, for linear increases in the nonlinearity strength, quadratic increases in the bias are expected. In addition, it is evident that the bias also scales as $1/M^2$, thus a linear increase in frequency resolution provides a large reduction in bias. However, as is shown next, for a consistent estimate of the auto-bispectral density, there exists a rather severe constraint on allowable M.

The bias in bispectral estimates is entirely analogous to that found in power spectrum estimation and scales as the second derivative of the auto-bispectral density with respect to the frequencies [34]. For lightly damped systems, such as the structural systems we are interested in, there will tend to be sharp peaks such as the one displayed in Figure 8.8, resulting in a potentially large bias. Accurate resolution of sharp peaks requires a high-frequency resolution. This means that the number of points in the estimation segments M must be large, but this adversely affects estimator variance.

Thus, we arrive at the familiar bias versus variance trade-off. We wish M large to improve the resolution of peaks in the bispectral density function for structural systems, however for a consistent estimator, we have from Eq. (6.132) that the number of data in each segment must fulfill $M < N^{\frac{1}{2}}$. Using windowed, overlapping segments can help ease this constraint but the scaling of M with N remains.

As a first example, define the structural model of interest to be an SDOF system governed by

$$m\ddot{y} + c_1\dot{y} + c_2\dot{y}^2 + k_1 y + k_2 y^2 = Ax(t) \qquad (6.138)$$

where m (kg) is the mass, c_1 (N $\cdot$ s/m) and k_1 (N/m) the linear damping and stiffness coefficients, $c_2[N \cdot s^2/m^2]$ and $k_2[N/m^2]$ the nonlinear damping and stiffness coefficients, and A (N) the amplitude of excitation. Using the acceleration $\ddot{y}(t)$ as the response of interest, harmonic probing (see Section 4.6) yields

$$H_1(\omega) = \frac{-\omega^2}{k_1 + ic_1\omega - m\omega^2}$$

$$H_2(\omega_1,\omega_2) = \frac{1}{\omega_1^2\omega_2^2}(-k_2 + c_2\omega_1\omega_2)H_1(\omega_1)H_1(\omega_2)H_1(\omega_1+\omega_2) \qquad (6.139)$$

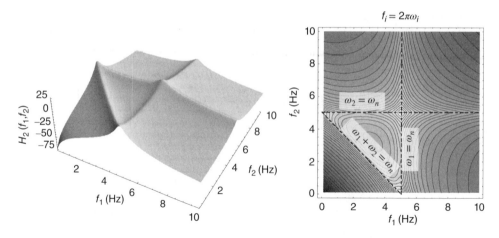

Figure 6.12   Magnitude of $H_2(\omega_1, \omega_2)$ (dB scale) showing the location of the poles and the ridge $\omega_1 + \omega_2 = \omega_n$. *Source*: Reproduced from [29], Figure 1, with permission of Elsevier

as the first- and second-order Volterra kernels. Consider the parameters $m = 1$ kg, $k_1 = 1000$ N/m, $c_1 = 3$ N$| \cdot$ s/m, $k_2 = 10^5$ N/m^2, $c_2 = 0$ N $\cdot$ s^2/m^2. Before looking at the bispectrum, it is useful to first consider the structure of the second-order Volterra kernel. The presence of nonlinearity in either the damping or stiffness terms will lead to nonzero values in $H_2(\omega_1, \omega_2)$, resulting in "peaks" at both the natural frequency $\omega_n = \sqrt{k_1/m}$ and the bifrequency $\omega_1 + \omega_2 = \omega_n$. This can be seen simply by considering the location of the poles associated with $H_2(\omega_1, \omega_2)$ (see Figure 6.12).

If we substitute these two kernels into Eq. (6.119), we obtain a closed-form solution for the auto-bispectral density in terms of the physical parameters m, k_1, k_2, c_1, c_2 and the excitation PSD. For the excitation, we consider $x(t)$ to be an i.i.d., jointly Gaussian random process with a constant PSD, $S_{xx}(\omega) = P$. In this example, we choose $P = 0.01$N^2/Hz. Given these parameters, the analytical and estimated bispectral density functions associated with the displacement, velocity, and acceleration system responses are shown in Figure 6.13. Note that in this, and other magnitude HOS plots we label the "z" axis according to the units of the quantity being estimated in order to clearly differentiate among them.

The estimates were performed using $N = 2^{17}$ observations with $M = 1024$ and $L = 512$ (50% overlap), giving a total of $S = 255$ averages. A Gaussian window was applied to each data segment before taking the FT. The nonlinearity manifests itself in different ways, depending on which response variable we are monitoring. For example, the nonlinearity shows up most strongly in the displacement bispectrum at the locations $(\omega_n, 0), (0, \omega_n)$, while the peak at ω_n, ω_n is not as strong. Conversely, the acceleration response yields a more pronounced peak at the bifrequency ω_n, ω_n. From a nonlinearity detection standpoint, this is precisely the information we are interested in. As we see in later chapters, knowing which bispectral peak to monitor is essential to using the bispectrum as a damage detection tool.

A number of other interesting features can be seen in the bispectral density, but these are more easily seen if the data are plotted on a log scale. Figure 6.14 shows both theory and estimate for the log (base 10) of the auto-bispectral density function. In addition to the pronounced peaks, there is a clear "ridge" along the line $\omega_1 + \omega_2 = \omega_n$, a feature easily predicted

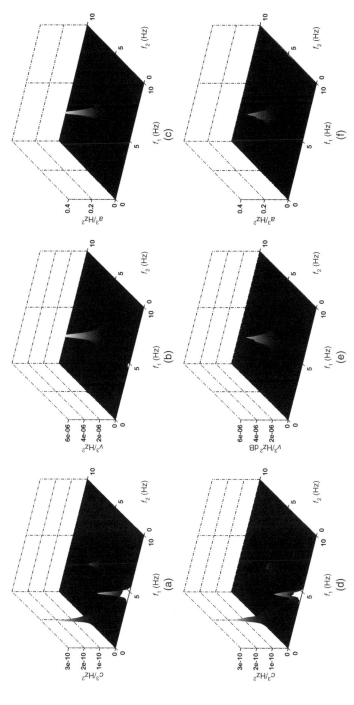

Figure 6.13 Magnitude Bispectral density functions as given by Eq. (6.119) without the integral term (top row) and the associated estimates (bottom row) for the displacement (a and d), velocity (b and e), and acceleration (c and f) responses. For the purpose of comparison the peak heights are labeled according to the bispectral density units, "d" (displacement), "v" (velocity), and "a" (acceleration)

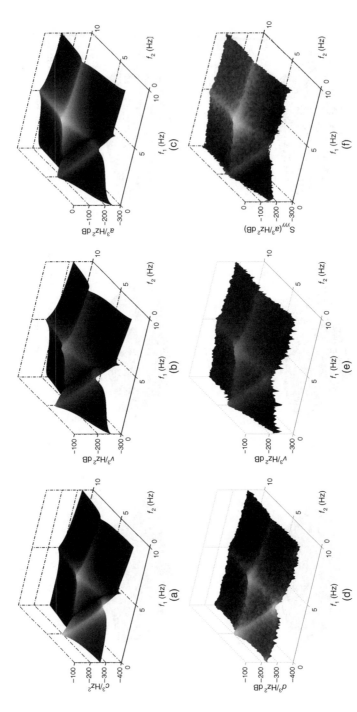

Figure 6.14 Log of the magnitude bispectral density functions as predicted by theory (top row) and the associated estimates (bottom row) for the displacement (a and d), velocity (b and e), and acceleration (c and f) responses. For the purpose of comparison the peak heights are labeled according to the bispectral density units, "d" (displacement), "v" (velocity), and "a" (acceleration)

by examining the structure of the second order Volterra kernel (Eqn. 6.139). Also of interest is the sensitivity of the bispectral peaks to the presence of a nonlinearity. Differentiating Eq. (6.119) with respect to k_2 or c_2 shows that the peak location $\omega_1 = \omega_2 = \omega_n$ is the most sensitive to the presence of both stiffness and damping nonlinearities. This is precisely the information required to understand the performance of bispectral detection schemes. Indeed, the analytical bispectrum will be needed later to derive the distributions for the magnitude auto-bispectral density peak heights.

We should also point out that these examples exclude the higher order term in the analytical expression (i.e., the integral term in Eqn. 6.119). To illustrate what is lost in this simplification, we again consider the acceleration bispectrum *with* the higher order term included (Figure 6.15). No difference can be observed on a linear scale, however on a log scale one can see some of the details found in the estimate that are not present in the analytical solution that excludes the higher order term. In particular, there is a small ridge along the line $\omega_1 + \omega_2 = 2\omega_n$ that is visible in both theory and estimate. In what follows, we ignore the higher order term, simply because it contributes very little to the bispectral peak heights that will ultimately be used as our statistic in damage detection applications.

From this simple one-dimensional example, we move on to more realistic, MDOF systems. Consider now a three-DOF system governed by the equations

$$[\mathbf{M}]\ddot{\mathbf{y}}(t) + [\mathbf{C}]\dot{\mathbf{y}}(t) + [\mathbf{K_L}]\mathbf{y}(t) + \mathbf{G}(\dot{\mathbf{y}}, \mathbf{y}) = \mathbf{x}(t) \qquad (6.140)$$

where

$$[\mathbf{M}] = \begin{bmatrix} m_1 & 0 & 0 \\ 0 & m_2 & 0 \\ 0 & 0 & m_3 \end{bmatrix} \quad [\mathbf{C}] = \begin{bmatrix} c_1 + c_2 & -c_2 & 0 \\ -c_2 & c_2 + c_3 & -c_3 \\ 0 & -c_3 & c_3 \end{bmatrix} \quad [\mathbf{K}] = \begin{bmatrix} k_1 + k_2 & -k_2 & 0 \\ -k_2 & k_2 + k_3 & -k_3 \\ 0 & -k_3 & k_3 \end{bmatrix}$$

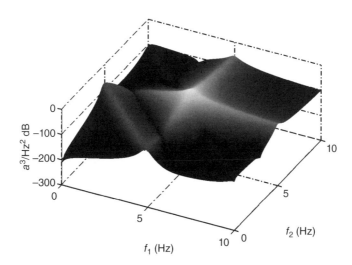

Figure 6.15 Log of the analytical bispectral density function associated with the acceleration response of the single DOF system. Unlike in Figure 6.14, this result includes the higher order "contour integral" term as part of the solution. Peak heights are displayed in terms of the bispectral density units

are constant coefficient mass, damping, and stiffness matrices. The nonlinear function $\mathbf{G}(\cdot)$ provides quadratic coupling between masses. Here, we consider both a quadratically nonlinear damping and restoring force between masses. For example, if the nonlinearity is located between masses 1 and 2, we have

$$\mathbf{G}(\dot{\mathbf{y}}, \mathbf{y}) = \begin{Bmatrix} -k_{non}(y_2(t) - y_1(t))^2 \\ k_{non}(y_2(t) - y_1(t))^2 \\ 0 \end{Bmatrix} + \begin{Bmatrix} -c_{non}(\dot{y}_2(t) - \dot{y}_1(t))^2 \\ c_{non}(\dot{y}_2(t) - \dot{y}_1(t))^2 \\ 0 \end{Bmatrix}$$

where k_{non}, c_{non} are the nonlinear stiffness and damping coefficients, respectively. The structure's linear parameters are given by $m_1 = m_2 = m_3 = 1\text{kg}$, $k_1 = k_2 = k_3 = 2000\text{N/m}$, $c_1 = c_2 = c_3 = 3.0\text{N} \cdot \text{s/m}$, and nonlinear parameters $k_{non} = 100,000\text{N/m}^2$ and $c_{non} = 20\text{N} \cdot \text{s}^2/\text{m}^2$. Again, we assume a constant input PSD of $P = 0.01\text{N}^2/\text{Hz}$, where the input is applied to the third DOF. To show the generality of the expression we have derived, we compute $B_{312}^3(\omega_1, \omega_2)$, that is to say, we choose the signals comprising the bispectrum to all be different (one from each DOF) and place the forcing at mass 3 (forcing location will henceforth be denoted with a superscript). As before, we use blocks of size $M = 1024$ to form the estimates with $L = 512$ (50%) overlap. Figure 6.16 compares the estimated and

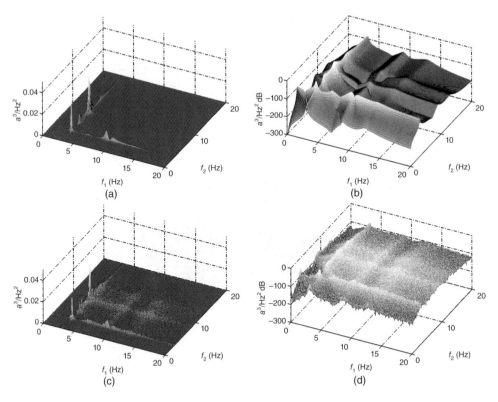

Figure 6.16   Magnitude bispectral density $|B_{312}^3(f_1, f_2)|$ plotted on both linear and log scales. Analytical predictions are shown on (a and b), while estimates are given on (c and d). Peak heights are displayed in terms of the bipectral density units

theoretical bispectral densities $B_{312}^3(\omega_1, \omega_2)$. The estimates and theory are in close agreement and show peaks at each of the resonant frequencies as well as combinations of the resonant frequencies. Again, this is expected as the bispectrum will always have poles at the same locations as the poles of the linear system and additional poles at some and differences of the resonant frequencies.

6.5.3 Analytical Bicoherence for Non-Gaussian Signals

As we stressed in Section 6.5.1, if the input to a linear system possesses a non-Gaussian marginal distribution, it becomes difficult to distinguish the system output from that of a non-linear system driven with a Gaussian input. Hence, it was suggested that through appropriate normalization we could instead consider the bicoherence function (6.121). Deriving an expression for the bicoherence of a nonlinear system driven with non-Gaussian noise is a rather challenging endeavor. Nonetheless, it is possible, again with the help of Isserlis' theorem.

As in previous sections, the first step is to develop an analytical expression for $y_i(t)$ capable of capturing second-order nonlinearities. Here, again, the subscript i is used to denote a different random process, usually taken as a signal acquired from a different point on a structure. To this end, we again use the Volterra functional series approach described in Chapter 4. As we have detailed in Section 4.6, the Volterra model is essentially a Taylor series expansion for functions with memory [35]. As in the Taylor series, the Volterra approximation is only appropriate for systems with smooth (differentiable) nonlinearities and will not converge for arbitrary system parameters. Solutions far from linear (i.e., strongly nonlinear behavior) cannot be captured using this approach. For detecting structural damage, however, the focus is typically on weak nonlinearities for which the Volterra model is appropriate.

As in the preceding section, we use the two-term Volterra series model given by Eqs. (6.108) and (6.109). While more terms could be used in the expansion, their contribution to the response is increasingly higher order and is therefore not considered. We could, for example, have added a cubic term to the model (6.108), which would better capture the quadratic nonlinearity and would also reflect the influence of smooth, third-order nonlinearities. The resulting solution for the bispectrum in this case will contain contributions from $h_{3,i}(\tau_1, \tau_2, \tau_3)$. However, several forms of structural damage produce weak, quadratic nonlinearities (see Chapter 8), in which case the additional terms turn out to be orders of magnitude smaller than those containing $h_{2,i}(\tau_1, \tau_2)$. Inclusion of the higher order kernels in this case results in a minimal gain in model accuracy but comes at a huge cost analytically, as will become apparent in the following derivation. We should also mention that if the goal is the detection of third-order nonlinearities, the trispectrum is probably a more appropriate tool. For the trispectrum, third-order nonlinearities are the leading order terms, as was demonstrated in Ref. [36].

Deriving the bispectrum involves substituting the signal model (6.108) into (6.106) and simplifying. However, as we have shown, the assumption of a Gaussian distributed input was essential as it allowed Isserlis' theorem [37] (see 3.11) to be used in the simplification. Unfortunately, no analogous theorem exists for non-Gaussian joint distributions of the input $x(t)$.

To circumvent this problem, we consider the input to be a static (memoryless), nonlinear transformation of a jointly Gaussian distributed random process $W(t) \sim \mathcal{N}(0, \sigma_W^2)$ as

$$x(t) = a_0 + a_1 w(t) + a_2 w(t)^2 + a_3 w(t)^3. \tag{6.141}$$

This particular transformation appears to have been first used by Fleishman [38] in creating non-Gaussian distributions. Although the transformation is simply a third-order polynomial, it can be used to generate a very large variety of probability models for $x(t)$, depending on the values of the parameters $a_{0\to3}$. Note that this model does not assume that the input is spectrally white (i.i.d.). The resulting random process $X(t)$ can be white or spectrally colored, depending on the process $W(t)$. This model is limited, however, to random processes with linear temporal correlations. For example, (6.141) could not be used to describe an input that is itself the output of a nonlinear system (although one could conceivably replace (6.141) with a second Volterra series model, for example, Eq. 6.108, and carry the derivation forward). Nonetheless, modeling the input as a spectrally colored random process governed by the broad class of PDFs specified by (6.141) is a very general approach.

A derivation for the bispectrum subject to the class of non-Gaussian inputs generated by Eq. (6.141) can now be carried out. Substituting Eq. (6.141) into Eq. (6.108) and then into Eq. (6.106) yields a very large number of integrals, each involving the products of an even number of Gaussian distributed random variables $w(t)$. Some of the products turn out to be quite large, in fact some terms involve the product of 16 $w(t)$'s. According to Isserlis' theorem, the number of covariance terms in the sum is $\frac{(2n)!}{n!\times 2^n}$ where $2n$ is the number of Gaussian variables involved. For $2n = 16$, this gives 2,027,025 terms, thus implementation of the theorem 3.1.1 was carried out using Mathematica software. Fortunately, a number of the resulting terms turn out to be higher order and can be discarded. Specifically, in the simplification, one encounters a number of integrals where the following inequality has been assumed to hold

$$\int H_2(\omega_1 - \omega, \omega) S_{ww}(\omega) d\omega \ll \int S_{WW}(\omega) d\omega \qquad (6.142)$$

for all choices of ω_1. For structural systems, one can get a sense of when (6.142) is expected to hold by considering a specific $H_2(\omega_1, \omega_2)$ kernel. For an SDOF, quadratically nonlinear structure (see Eq. 8.39), the second-order Volterra kernel can be shown to have the form:

$$H_2(\omega_1 - \omega, \omega) = -k_N H_1(\omega_1 - \omega) H_1(\omega) H_1(\omega_1) \qquad (6.143)$$

where $H_1(\omega) = 1/(-M\omega^2 + ic\omega + k_L)$ is the familiar linear transfer function. Many structural systems are modeled as lightly damped and the denominator of $H_1(\omega)$ is dominated by the linear stiffness, k_L. The second-order kernel contains a product of three linear transfer functions, thus the ratio k_N/k_L^3 is expected to be a good predictor of when (6.142) can be expected to hold. We have carried out the integral needed in (6.142) and have found that if $k_N/k_L^3 < 1$, the inequality (6.142) is indeed satisfied. We should point out that this is precisely what has already been assumed in modeling the system response, i.e., a weakly nonlinear system.

The end result is the bispectral density function for any weakly quadratic nonlinear system subject to an input that can be described by the transformation (6.141) as

$$B_{Y_k Y_i Y_j}(\omega_1, \omega_2) = C(\omega_1, \omega_2) H_{1,i}(\omega_1) H_{1,j}(\omega_2) H_{1,k}(-\omega_1 - \omega_2)$$

$$+ \left[C_2 S_0^{(a)} + C_3 S_1^{(a)} + C_4 S_2^{(a)} + 8a_2^4 S_3^{(a)} \right] H_{1,j}(\omega_2) H_{1,k}(-\omega_1 - \omega_2) H_{2,i}(-\omega_2, \omega_1 + \omega_2)$$

$$+ \left[C_2 S_0^{(b)} + C_3 S_1^{(b)} + C_4 S_2^{(b)} + 8a_2^4 S_3^{(b)} \right] H_{1,i}(\omega_1) H_{1,k}(-\omega_1 - \omega_2) H_{2,j}(-\omega_1, \omega_1 + \omega_2)$$

$$+ \left[C_2 S_0^{(c)} + C_3 S_1^{(c)} + C_4 S_2^{(c)} + 8a_2^4 S_3^{(c)} \right] H_{1,i}(\omega_1) H_{1,j}(\omega_2) H_{2,k}(-\omega_1, -\omega_2)$$

$$+ 2\bar{x}C(\omega_1,\omega_2)\left[H_{1,j}(\omega_2)H_{1,k}(-\omega_1-\omega_2)H_{2,i}(\omega_1,0)\right.$$

$$+ H_{1,i}(\omega_1)H_{1,k}(-\omega_1-\omega_2)H_{2,j}(\omega_2,0) + H_{1,i}(\omega_1)H_{1,j}(\omega_2)H_{2,k}(-\omega_1-\omega_2,0)\Big]$$

$$+ 2\bar{x}\left[C_2 S_0^{(a)} + C_3 S_1^{(a)} + C_4 S_2^{(a)} + 8a_2^4 S_3^{(a)}\right]H_{1,k}(-\omega_1-\omega_2)H_{2,i}(-\omega_2,\omega_1+\omega_2)H_{2,j}(\omega_2,0)$$

$$+ 2\bar{x}\left[C_2 S_0^{(b)} + C_3 S_1^{(b)} + C_4 S_2^{(b)} + 8a_2^4 S_3^{(b)}\right]H_{1,k}(-\omega_1-\omega_2)H_{2,i}(\omega_1,0)H_{2,j}(-\omega_1,\omega_1+\omega_2)$$

$$+ 2\bar{x}\left[C_2 S_0^{(c)} + C_3 S_1^{(c)} + C_4 S_2^{(c)} + 8a_2^4 S_3^{(c)}\right]H_{1,i}(\omega_1)H_{j,2}(\omega_2,0)H_{k,2}(-\omega_1,-\omega_2)$$

$$+ 2\bar{x}\left[C_2 S_0^{(a)} + C_3 S_1^{(a)} + C_4 S_2^{(a)} + 8a_2^4 S_3^{(a)}\right]H_{1,j}(\omega_2)H_{2,i}(-\omega_2,\omega_1+\omega_2)H_{2,k}(-\omega_1-\omega_2,0)$$

$$+ 2\bar{x}\left[C_2 S_0^{(b)} + C_3 S_1^{(b)} + C_4 S_2^{(b)} + 8a_2^4 S_3^{(b)}\right]H_{1,i}(\omega_1)H_{2,j}(-\omega_1,\omega_1+\omega_2)H_{2,k}(-\omega_1-\omega_2,0)$$

$$+ 2\bar{x}\left[C_2 S_0^{(c)} + C_3 S_1^{(c)} + C_4 S_2^{(c)} + 8a_2^4 S_3^{(c)}\right]H_{1,j}(\omega_2)H_{i,2}(\omega_1,0)H_{k,2}(-\omega_1,-\omega_2)$$

$$+ 4\bar{x}^2 C(\omega_1,\omega_2)\left[H_{1,k}(-\omega_1-\omega_2)H_{2,i}(\omega_1,0)H_{2,j}(\omega_2,0)\right.$$

$$+ H_{1,j}(\omega_2)H_{2,i}(\omega_1,0)H_{2,k}(-\omega_1-\omega_2,0) + H_{1,i}(\omega_1)H_{2,j}(\omega_2,0)H_{2,k}(-\omega_1-\omega_2,0)\Big]$$

$$(6.144)$$

where the functions $C(\omega_1,\omega_2)$, C_2, C_3, C_4, $\bar{x}$, and a_2 describe the probability distribution of the input and the functions $S_{0,1,2,3}^{(a,b,c)}$ describe the input spectral properties. These functions are given by

$$C(\omega_1,\omega_2) = 2a_1^2 a_2 S_0 + 6a_1 a_2 a_3 (S_1 + 2\sigma_W^2 S_0) + 8a_2^3 \tilde{S} + 18a_2 a_3^2 (2S_2 + \sigma_W^2 S_1 + \sigma_W^4 S_0)$$

$$C_2 = 2a_1^4 + 24\sigma_W^2 a_1^3 a_3 + 108\sigma_W^4 a_1^2 a_3^2 + 216\sigma_W^6 a_1 a_3^3 + 162\sigma_W^8 a_3^4$$

$$C_3 = 4a_1^2 a_2^2 + 24\sigma_W^2 a_1 a_2^2 a_3 + 36\sigma_W^4 a_2^2 a_3^2$$

$$C_4 = 12a_1^2 a_3^2 + 72\sigma_W^2 a_1 a_3^3 + 108\sigma_W^4 a_3^4$$

$$\bar{x} = E[X(t)] = a_0 + \sigma_W^2 a_2$$

$$(6.145)$$

$$S_0^{(a)} = S_{WW}(\omega_1+\omega_2)S_{WW}(\omega_2)$$

$$S_0^{(b)} = S_{WW}(\omega_1+\omega_2)S_{WW}(\omega_1)$$

$$S_0^{(c)} = S_{WW}(\omega_1)S_{WW}(\omega_2)$$

$$2\pi S_1^{(a)} = S_{WW}(\omega_2)\int S_{WW}(\omega_1+\omega_2+\omega_3)S_{WW}(\omega_3)d\omega_3 + S_{WW}(\omega_1+\omega_2)$$

$$\times \int S_{WW}(\omega_2+\omega_3)S_{WW}(\omega_3)d\omega_3$$

$$2\pi S_1^{(b)} = S_{WW}(\omega_1) \int S_{WW}(\omega_1 + \omega_2 + \omega_3)S_{WW}(\omega_3)d\omega_3 + S_{WW}(\omega_1 + \omega_2)$$

$$\times \int S_{WW}(\omega_1 + \omega_3)S_{WW}(\omega_3)d\omega_3$$

$$2\pi S_1^{(c)} = S_{WW}(\omega_1) \int S_{WW}(\omega_2 + \omega_3)S_{WW}(\omega_3)d\omega_3 + S_{WW}(\omega_2)$$

$$\times \int S_{WW}(\omega_1 + \omega_3)S_{WW}(\omega_3)d\omega_3$$

$$4\pi^2 S_2^{(a)} = S_{WW}(\omega_1 + \omega_2) \int\int S_{WW}(\omega_2 + \omega_3 + \omega_4)S_{WW}(\omega_3)S_{WW}(\omega_4)d\omega_3 d\omega_4$$

$$+ S_{WW}(\omega_2) \int\int S_{WW}(\omega_1 + \omega_2 + \omega_3 + \omega_4)S_{WW}(\omega_3)S_{WW}(\omega_4)d\omega_3 d\omega_4$$

$$4\pi^2 S_2^{(b)} = S_{WW}(\omega_1 + \omega_2) \int\int S_{WW}(\omega_1 + \omega_3 + \omega_4)S_{WW}(\omega_3)S_{WW}(\omega_4)d\omega_3 d\omega_4$$

$$+ S_{WW}(\omega_1) \int\int S_{WW}(\omega_1 + \omega_2 + \omega_3 + \omega_4)S_{WW}(\omega_3)S_{WW}(\omega_4)d\omega_3 d\omega_4$$

$$4\pi^2 S_2^{(c)} = S_{WW}(\omega_1) \int\int S_{WW}(\omega_2 + \omega_3 + \omega_4)S_{WW}(\omega_3)S_{WW}(\omega_4)d\omega_3 d\omega_4$$

$$+ S_{WW}(\omega_2) \int\int S_{WW}(\omega_1 + \omega_3 + \omega_4)S_{WW}(\omega_3)S_{WW}(\omega_4)d\omega_3 d\omega_4$$

$$4\pi^2 S_3^{(a)} = \int\int S_{WW}(\omega_1 + \omega_2 + \omega_4)S_{WW}(\omega_2 + \omega_3)S_{WW}(\omega_3)S_{WW}(\omega_4)d\omega_3 d\omega_4$$

$$4\pi^2 S_3^{(b)} = \int\int S_{WW}(\omega_1 + \omega_2 + \omega_4)S_{WW}(\omega_1 + \omega_3)S_{WW}(\omega_3)S_{WW}(\omega_4)d\omega_3 d\omega_4$$

$$4\pi^2 S_3^{(c)} = \int\int S_{WW}(\omega_1 + \omega_3)S_{WW}(\omega_2 + \omega_4)S_{WW}(\omega_3)S_{WW}(\omega_4)d\omega_3 d\omega_4$$

$$2\pi\tilde{S} = \int S_{WW}(\omega_1 + \omega_3)S_{WW}(\omega_2 + \omega_3)S_{WW}(\omega_3)d\omega_3 \qquad (6.146)$$

and $S_0 = S_0^{(a)} + S_0^{(b)} + S_0^{(c)}$, $S_1 = S_1^{(a)} + S_1^{(b)} + S_1^{(c)}$, $S_2 = S_2^{(a)} + S_2^{(b)} + S_2^{(c)}$.

Each of the integrals in Eq. (6.145) extends over the real number line, that is, $-\infty \cdots + \infty$ which we know causes problems when dealing with a random process that is both continuous and i.i.d. The reason for this, again, is that a truly i.i.d random process possesses a constant PSD, i.e., $S_{WW} =$ const. $\equiv P$. By Parseval's relationship, the integral of the PSD gives the variance, however this implies that an i.i.d process has infinite variance. However, we know from Section 3.3.4 that when dealing with sampled data the limits on the integrals are dictated by the sampling interval, Δ_t. For example, assuming an i.i.d random process (constant PSD, P), the last equality in Eq. (6.146) becomes

$$2\pi\tilde{S} = P^3 \int_{-\pi/\Delta_t}^{\pi/\Delta_t} d\omega$$

$$\tilde{S} = P^3/\Delta_t. \tag{6.147}$$

To derive the bicoherence, an expression is also required for the power spectrum in terms of the coefficients a_0, a_1, a_2, a_3, and the spectral properties of the input $S_{WW}(\omega)$. This can be accomplished using the same general procedure as was used for the bispectrum. The expression for $S_{Y_iY_i}(\omega_1)$ was obtained using the two-term Volterra model, Eq. (6.108). Using this approach, the power spectrum can be written as

$$S_{Y_iY_i}(\omega_1) = [(a_1^2 + 6\sigma_w^2 a_1 a_3 + 9\sigma_w^4 a_3^2)S_4^{(a)} + 2a_2^2 S_4^{(b)} + 6a_3^2 S_4^{(c)}]$$
$$\times \{H_{1,i}(\omega_1)H_{1,i}(-\omega_1) + 2\bar{x}(H_{2,i}(0,-\omega_1)H_{1,i}(\omega_1) + H_{2,i}(0,\omega_1)H_{1,i}(-\omega_1))\} \tag{6.148}$$

where

$$S_4^{(a)} = S_{WW}(\omega_1)$$

$$2\pi S_4^{(b)} = \int S_{WW}(\omega_1 + \omega_2)S_{WW}(\omega_2)d\omega_2$$

$$4\pi^2 S_4^{(c)} = \int\int S_{WW}(\omega_1 + \omega_2 + \omega_3)S_{WW}(\omega_2)S_{WW}(\omega_3)d\omega_2\,d\omega_3 \tag{6.149}$$

An analytical expression for the bicoherence function can therefore be obtained by simply dividing the magnitude of Eq. (6.144) by the product of PSDs required by Eq. (6.121).

Again, allow the structural parameters in Eq. (6.140) to be $m_1 = m_2 = m_3 = 1.0$ (kg), $k_1 = k_2 = k_3 = 2000$ (N/m), $c_1 = c_2 = c_3 = 3.0$ (N $\cdot$ s/m). The Volterra kernels for this system are already provided by Eq. (6.139). The linear parameters are fixed to these values for the remainder of this example. First, consider a restoring nonlinearity $k_N = 10^5$ (N/m)^2 located between masses 1 and 2. Figure 6.17 shows both the theoretical and estimated bispectrum $(\hat{B}_{\ddot{y}_2\ddot{y}_2\ddot{y}_1}(f_1,f_2)$, and $\hat{B}_{\ddot{y}_1\ddot{y}_2\ddot{y}_2}(f_1,f_2))$ associated with the acceleration response of this nonlinear system subject to Gaussian excitation ($a_0 = a_2 = a_3 = 0, a_1 = 1, \sigma_w = 1$) applied at the third mass. Estimation of the bispectrum and bicoherence functions has already been discussed in Section 6.5, Eqs. (6.129) and (6.131). Figure 6.17 shows two examples of the estimated magnitude bispectrum associated with acceleration measurements. For the estimation, time series of length $N = 2^{18}$ were generated by numerically integrating Eq. (6.140). The sampling interval was chosen to be $\Delta_t = 0.01$ (s), while the estimation parameters were chosen to be $M = 1024, L = 512$, and the windowing function $g(m)$ was set to a Hanning window (see again discussion starting on page 223). In this example, we have used time series from different points on the structure (e.g., DOFs 1 and 2) to illustrate the very different bispectra that can result.

As a second example, consider the input to be governed by a highly non-Gaussian distribution (but with the same nonlinearity). Figure 6.18 shows the distribution along with the predicted and estimated bispectrum. The same estimation parameters were used as for the previous example. The solution given by Eq. (6.144) clearly serves as a good approximation for the bispectrum for a very large number of input distributions and works for any combination of time series collected from the structure.

More to the point of this section, the solution can also be used to develop an expression for the analytical bicoherence function. Figure 6.19 provides a comparison between theory and estimate for the bicoherence function associated with a nonlinear system driven with an input

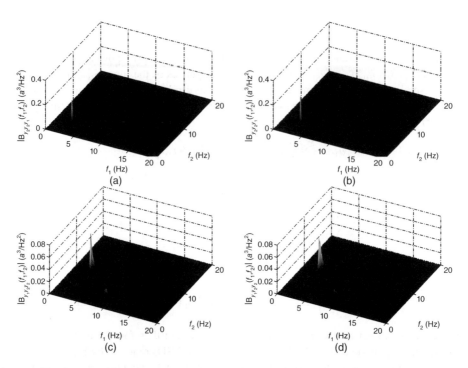

Figure 6.17 (a) Theoretical and (b) estimated bispectrum magnitude based on multivariate acceleration response, $|B_{\ddot{Y}_2\ddot{Y}_2\ddot{Y}_1}(f_1,f_2)|$. (c) Theoretical and (d) estimated bispectrum magnitude $|\hat{B}_{\ddot{Y}_1\ddot{Y}_2\ddot{Y}_2}(f_1,f_2)|$. The forcing was a Gaussian distributed random process, obtained by setting $a_0 = a_2 = a_3 = 0$ and $a_1 = 1$. The power spectral density of the forcing was taken to be 0.01 (N²/Hz]). *Source*: Reproduced from [39], Figure 1, with permission of Elsevier

conforming to the χ^2 distribution ($a_0 = a_1 = a_3 = 0, a_2 = 1$). For this example, both quadratic damping and stiffness terms were placed between masses 1 and 2 with values $c_N = 75$ (N · s²/m²) and $k_N = 10^5$ (N/m²). Had the system been linear, theory predicts that the bicoherence would have been perfectly flat across the entire f_1,f_2 Plane, with the value $|b_{\ddot{y}_k\ddot{y}_i\ddot{y}_j}(f_1,f_2)| = 0.283$ due to the highly skewed nature of the input distribution. The nonlinearity, however, gives rise to various peaks, located along the combination resonances for this system, rising above this constant value. As was discussed previously, the estimator associated with the bicoherence has a much higher variance. For this reason, the estimate was made with $M = 512$ point segments, sacrificing bias for variance. The estimate still clearly possesses a higher variance than the bispectrum estimate, yet it also clearly captures the features predicted by the theory.

This section has shown how we can predict the bispectral density and bicoherence functions for structural response data to a wide variety of excitations. As we show later in Section 8.4.2, this information can be used to predict which bicoherence function to estimate, where to excite the structure, and which probability distribution to use in constructing the input signal for the purpose of detecting damage-induced nonlinearity in structures.

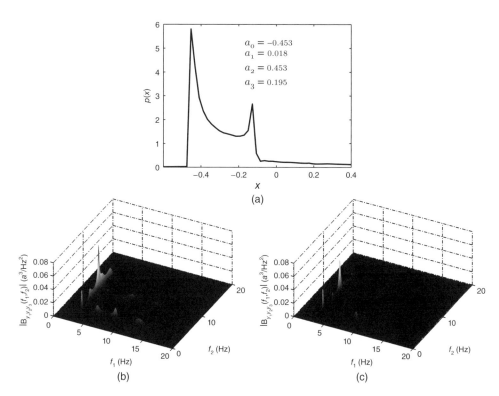

Figure 6.18 (a) Non-Gaussian input distribution, obtained by setting $a_0 = -0.453, a_1 = 0.018$, $a_2 = 0.453, a_3 = 0.195, \sigma_w = 1$ in Eq. (6.141). (b) The theoretical and (c) the estimated output accelerance bispectrum magnitude $|B_{\ddot{Y}_1 \ddot{Y}_2 \ddot{Y}_3}(f_1,f_2)|$ associated with this input applied at mass 3. A quadratic nonlinearity $k_N = 10^5$ (N/m^2) was placed between the first and second masses. *Source*: Reproduced from [39], Figure 2, with permission of Elsevier

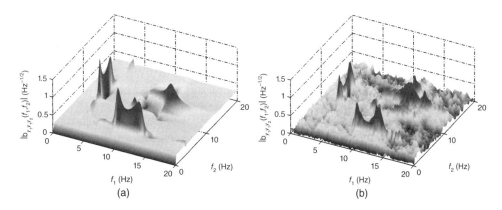

Figure 6.19 (a) The theoretical and (b) estimated output accelerance bicoherence magnitude $|b_{\ddot{Y}_1 \ddot{Y}_1 \ddot{Y}_2}(f_1,f_2)|$ associated with the input applied at mass 3. The input distribution was χ^2 ($a_0 = a_1 = a_3 = 0, a_2 = 1, \sigma_w = 1$). Quadratic nonlinearities $k_N = 10^4$(N/m^2) and $c_N = 75$ (N · s^2/m^2) were placed between the first and second masses. *Source*: Reproduced from [39], Figure 3, with permission of Elsevier

6.5.4   Trispectral Density Function

We have already derived the general expression for the trispectrum of an MDOF system in Eq. (3.104). As with the bispectral density, an analytical expression for the trispectral density first requires an analytical expression for $y_i(t), y_j(t), y_k(t), y_p(t)$ capable of capturing (up to) third-order nonlinearities. Again, we use the Volterra series approach [35] and again acknowledge that the approach is only effective for systems with smooth (differentiable) nonlinearities and may not converge for arbitrary system parameters. Solutions far from linear (i.e., strongly nonlinear behavior) may not be captured using this approach. However, given that our goal is the *early* detection of damage-induced nonlinearity, this assumption is a good one.

A three-term Volterra series model may be used to capture nonlinearities up to third order, thus the response is modeled as

$$y_i^r(t) = y_{i,1}^r(t) + y_{i,2}^r(t) + y_{i,3}^r(t)$$

$$= \int_{\mathbb{R}} h_{1,i}^r(\tau) x_r(t-\tau) d\tau + \int_{\mathbb{R}^2} h_{2,i}^r(\tau_1, \tau_2) x_r(t-\tau_1) x_r(t-\tau_2) d\tau_1 d\tau_2$$

$$+ \int_{\mathbb{R}^3} h_{3,i}^r(\tau_1, \tau_2, \tau_3) x_r(t-\tau_1) x_r(t-\tau_2) x_r(t-\tau_3) d\tau_1 \, d\tau_2 \, d\tau_3 \qquad (6.150)$$

where $h_{i,1}^r, h_{i,2}^r, h_{i,3}^r$ are the first-, second-, and third-order Volterra kernels relating excitation at location "r" to output at location "i". For notational convenience, the "r" is dropped from the Volterra kernel notation and is left instead as a subscript on the excitation $x_r(t)$. As we have done previously, assume that the input signal $x_r(t)$ is Gaussian distributed with zero mean and unit standard deviation, i.e., $x_r(t) \sim N(0, 1)$.

Recall again that this allows us to use the theorem (3.11) which states that all expectations of products of the input random variable can be factored as sums of products of covariances. For the case of four Gaussian random variables $\eta_1, \eta_2, \eta_3, \eta_4$, for example, application of the theorem 3.11 gives the relation $E[\eta_1 \eta_2 \eta_3 \eta_4] = E[\eta_1 \eta_2] E[\eta_3 \eta_4] + E[\eta_1 \eta_3] E[\eta_2 \eta_4] + E[\eta_1 \eta_4] E[\eta_2 \eta_3]$. Under the assumption of a joint normally distributed input, the mean of Eq. (6.150) can therefore be readily obtained by applying the theorem to give

$$\bar{y} = \int_{\mathbb{R}^2} h_{2,i}^r(\tau_1, \tau_2) E[X_r(t-\tau_1) X_r(t-\tau_2)] d\tau_1 \, d\tau_2. \qquad (6.151)$$

The signal model (Eqs. 6.150 and 6.151) can now be used to derive an expression for the trispectrum. Returning to Eq. (3.104), there are four terms that must be solved to obtain the analytical trispectrum. As we have done previously, instead of writing the constituent random processes in subscripts, we assume they are all associated with the system output and simply denote their location. In other words, the term $S_{Y_p Y_i Y_j Y_k}(\omega_1, \omega_2, \omega_3)$ can simply be denoted $S_{pijk}(\omega_1, \omega_2, \omega_3)$. With this notation, the first term in the trispectrum is $S_{pijk}(\omega_1, \omega_2, \omega_3)$, which is the triple FT of the expected value

$$C_{pijk}(\tau_1, \tau_2, \tau_3) = E[Y_p(t) Y_i(t+\tau_1) Y_j(t+\tau_2) Y_k(t+\tau_3)]. \qquad (6.152)$$

Substituting (6.150) into Eq. (6.152) for each of the arguments $y_p(t), y_i(t+\tau_1), y_j(t+\tau_2), y_k(t+\tau_3)$ and simplifying yields 81 separate terms. Only 41 of these terms are nonzero for Gaussian excitation, again owing to the fact that in expectation, the products of odd numbers of normally distributed random variables is zero (Theorem 3.11). Most of the remaining terms lead to higher order contributions in the final trispectrum expression. All terms in the trispectrum contain products of the PSDs of the input. For the types of signal amplitudes observed in experiment, the PSD is typically of order $P \sim O(10^{-2})$. It turns out that the leading order terms involve the triple product of input PSDs, while other terms involve higher powers of P. That is to say, if the input is i.i.d Gaussian noise with constant PSD of P (N^2/Hz), only terms to $O(P^3)$ are retained. After discarding higher order terms, the following 11 expected values remain:

$$C^I_{pijk}(\tau_1, \tau_2, \tau_3) = E[Y_{p,1}(t)Y_{i,1}(t+\tau_1)Y_{j,1}(t+\tau_2)Y_{k,1}(t+\tau_3)]$$

$$C^{II}_{pijk}(\tau_1, \tau_2, \tau_3) = E[Y_{p,2}(t)Y_{j,1}(t+\tau_2)Y_{k,1}(t+\tau_3)Y_{i,2}(t+\tau_1)]$$

$$C^{III}_{pijk}(\tau_1, \tau_2, \tau_3) = E[Y_{p,2}(t)Y_{i,1}(t+\tau_1)Y_{k,1}(t+\tau_3)Y_{j,2}(t+\tau_2)]$$

$$C^{IV}_{pijk}(\tau_1, \tau_2, \tau_3) = E[Y_{p,1}(t)Y_{k,1}(t+\tau_3)Y_{i,2}(t+\tau_1)Y_{j,2}(t+\tau_2)]$$

$$C^V_{pijk}(\tau_1, \tau_2, \tau_3) = E[Y_{p,2}(t)Y_{i,1}(t+\tau_1)Y_{j,1}(t+\tau_2)Y_{k,2}(t+\tau_3)]$$

$$C^{VI}_{pijk}(\tau_1, \tau_2, \tau_3) = E[Y_{p,1}(t)Y_{j,1}(t+\tau_2)Y_{i,2}(t+\tau_1)Y_{k,2}(t+\tau_3)]$$

$$C^{VII}_{pijk}(\tau_1, \tau_2, \tau_3) = E[Y_{p,1}(t)Y_{i,1}(t+\tau_1)Y_{j,2}(t+\tau_2)Y_{k,2}(t+\tau_3)]$$

$$C^{VIII}_{pijk}(\tau_1, \tau_2, \tau_3) = E[Y_{p,3}(t)Y_{i,1}(t+\tau_1)Y_{j,1}(t+\tau_2)Y_{k,1}(t+\tau_3)]$$

$$C^{IX}_{pijk}(\tau_1, \tau_2, \tau_3) = E[Y_{p,1}(t)Y_{i,3}(t+\tau_1)Y_{j,1}(t+\tau_2)Y_{k,1}(t+\tau_3)]$$

$$C^X_{pijk}(\tau_1, \tau_2, \tau_3) = E[Y_{p,1}(t)Y_{i,1}(t+\tau_1)Y_{j,3}(t+\tau_2)Y_{k,1}(t+\tau_3)]$$

$$C^{XI}_{pijk}(\tau_1, \tau_2, \tau_3) = E[Y_{p,1}(t)Y_{i,1}(t+\tau_1)Y_{j,1}(t+\tau_2)Y_{k,3}(t+\tau_3)] \tag{6.153}$$

so that Eq. (6.152) can be written as $C_{pijk}(\tau_1, \tau_2, \tau_3) = \sum_{r=I}^{XI} C^r_{pijk}$. The first term that arises in this expectation consists entirely of linear terms, an undesirable property because the trispectrum is frequently used as a nonlinearity detector. It can be easily shown, however, that the δ terms in Eq. (3.104) exactly cancel out this linear contribution, highlighting the importance of considering cumulants as opposed to moments in the derivation. Simplification of $S^I_{pijk}(\omega_1, \omega_2, \omega_3)$ is carried out in Appendix C.

Each of the expected values in Eq. (6.153) must be triple FT with regard to τ_1, τ_2, τ_3 to give the corresponding portion of the trispectrum, for example, $S^{I \to XI}_{pijk}(\omega_1, \omega_2, \omega_3)$. Simplification of these terms is tedious but straightforward and proceeds in the same manner as for the bispectrum.

Each of the calculations required of the trispectrum proceeds in the exact same manner and the process can be significantly aided by noting that certain terms involve the same combination of first-, second-, and third-order kernels. For example, by inspection one can see that the terms $C^{III}_{pijk}(\tau_1, \tau_2, \tau_3)$ and $C^V_{pijk}(\tau_1, \tau_2, \tau_3)$ have the exact same form as $C^{II}_{pijk}(\tau_1, \tau_2, \tau_3)$. Therefore, these three terms can be obtained by simply making the appropriate changes in subscripts i, j, k, p and frequency variables $\omega_1, \omega_2, \omega_3$. Other terms also follow the same form so that in the end, only components $C^{II,IV,VIII,IX}_{pijk}(\tau_1, \tau_2, \tau_3)$ need to be simplified and the rest obtained by simply interchanging subscripts and frequencies. A sample calculation for $C_{pijk}^{VIII}(\tau_1, \tau_2, \tau_3)$ can be found in Appendix C.

Each of the terms in Eq. (6.153) contributes to the third *moment* spectrum $S_{pijk}(\omega_1, \omega_2, \omega_3)$; However, there are also the additional three terms in the cumulant spectrum involving the δ function

$$\delta(\omega_1 + \omega_2)S_{ji}(\omega_1)S_{pk}(\omega_3)$$

$$\delta(\omega_1 + \omega_3)S_{ki}(\omega_1)S_{pj}(\omega_2)$$

$$\delta(\omega_2 + \omega_3)S_{pi}(\omega_1)S_{kj}(\omega_2). \tag{6.154}$$

The procedure for evaluating these terms is the same, that is, substitute in a Volterra series model for the cross-spectral density terms (e.g., $S_{ji}(\omega_1)$) and simplify. Sample calculations for these terms are also carried out in Appendix C. As we have mentioned these terms exactly cancel three nonzero terms that arise in the fourth moment spectrum of a *linear* structure. Without them, the trispectrum would therefore be nonzero (at least on the three submanifolds) even for the response of a Gaussian excited, linear structure.

Nonetheless, we are most interested in the terms that are valid everywhere on the primary manifold, i.e., the $\omega_1, \omega_2, \omega_3$ plane. Much of the interesting structure in the trispectrum can be found here and it is therefore this region that is the focus of this section and not the submanifolds, i.e., the $\omega_1 = -\omega_2$, $\omega_1 = -\omega_3$ and $\omega_2 = -\omega_3$ planes. The final expression for the trispectrum, valid over the primary manifold, is

$$T_{ijkp}(\omega_1, \omega_2, \omega_3) = 6H_{i,1}(\omega_1)H_{j,1}(\omega_2)H_{k,1}(\omega_3)H_{p,3}(-\omega_1, -\omega_2, -\omega_3)S_{XX}(\omega_1)S_{XX}(\omega_2)S_{XX}(\omega_3)$$

$$+ 6H_{p,1}(-\omega_1 - \omega_2 - \omega_3)S_{xx}(\omega_1 + \omega_2 + \omega_3)$$

$$\times \{H_{j,1}(\omega_2)H_{k,1}(\omega_3)H_{i,3}(\omega_1 + \omega_2 + \omega_3, -\omega_2, -\omega_3)S_{XX}(\omega_2)S_{XX}(\omega_3)$$

$$+ H_{i,1}(\omega_1)H_{k,1}(\omega_3)H_{j,3}(\omega_1 + \omega_2 + \omega_3, -\omega_1, -\omega_3)S_{XX}(\omega_1)S_{XX}(\omega_3)$$

$$+ H_{i,1}(\omega_1)H_{j,1}(\omega_2)H_{k,3}(\omega_1 + \omega_2 + \omega_3, -\omega_1, -\omega_2)S_{XX}(\omega_1)S_{XX}(\omega_2)\}$$

$$+ 4H_{j,1}(\omega_2)H_{k,1}(\omega_3)H_{p,2}(-\omega_2, -\omega_1 - \omega_3)H_{i,2}(-\omega_3, \omega_1 + \omega_3)S_{XX}(\omega_1 + \omega_3)S_{XX}(\omega_2)S_{XX}(\omega_3)$$

$$+ 4H_{j,1}(\omega_2)H_{k,1}(\omega_3)H_{p,2}(-\omega_3, -\omega_1 - \omega_2)H_{i,2}(-\omega_2, \omega_1 + \omega_2)S_{XX}(\omega_1 + \omega_2)S_{XX}(\omega_2)S_{XX}(\omega_3)$$

$$+ 4H_{i,1}(\omega_1)H_{k,1}(\omega_3)H_{p,2}(-\omega_1, -\omega_2 - \omega_3)H_{j,2}(-\omega_3, \omega_2 + \omega_3)S_{XX}(\omega_2 + \omega_3)S_{XX}(\omega_1)S_{XX}(\omega_3)$$

$$+ 4H_{i,1}(\omega_1)H_{k,1}(\omega_3)H_{p,2}(-\omega_3, -\omega_1 - \omega_2)H_{j,2}(-\omega_1, \omega_1 + \omega_2)S_{XX}(\omega_1 + \omega_2)S_{XX}(\omega_1)S_{XX}(\omega_3)$$

$$+ 4H_{i,1}(\omega_1)H_{j,1}(\omega_2)H_{p,2}(-\omega_1, -\omega_2 - \omega_3)H_{k,2}(-\omega_2, \omega_2 + \omega_3)S_{XX}(\omega_2 + \omega_3)S_{XX}(\omega_1)S_{XX}(\omega_2)$$

$$+ 4H_{i,1}(\omega_1)H_{j,1}(\omega_2)H_{p,2}(-\omega_2, -\omega_1 - \omega_3)H_{k,2}(-\omega_1, \omega_1 + \omega_3)S_{XX}(\omega_1 + \omega_3)S_{XX}(\omega_1)S_{XX}(\omega_2)$$

$$+ 4H_{p,1}(-\omega_1 - \omega_2 - \omega_3)S_{XX}(\omega_1 + \omega_2 + \omega_3)$$

$$\times \{H_{k,1}(\omega_3)H_{i,2}(\omega_1 + \omega_2 + \omega_3, -\omega_2 - \omega_3)H_{j,2}(-\omega_3, \omega_2 + \omega_3)S_{XX}(\omega_2 + \omega_3)S_{XX}(\omega_3)$$

$$+ H_{k,1}(\omega_3)H_{j,2}(\omega_1 + \omega_2 + \omega_3, -\omega_1 - \omega_3)H_{i,2}(-\omega_3, \omega_1 + \omega_3)S_{XX}(\omega_1 + \omega_3)S_{XX}(\omega_3)$$

$$+ H_{j,1}(\omega_2)H_{i,2}(\omega_1 + \omega_2 + \omega_3, -\omega_2 - \omega_3)H_{k,2}(-\omega_2, \omega_2 + \omega_3)S_{XX}(\omega_2 + \omega_3)S_{XX}(\omega_2)$$

$$+ H_{j,1}(\omega_2)H_{k,2}(\omega_1 + \omega_2 + \omega_3, -\omega_1 - \omega_2)H_{i,2}(-\omega_2, \omega_1 + \omega_2)S_{XX}(\omega_1 + \omega_2)S_{XX}(\omega_2)$$

$$+ H_{i,1}(\omega_1)H_{k,2}(\omega_1 + \omega_2 + \omega_3, -\omega_1 - \omega_2)H_{j,2}(-\omega_1, \omega_1 + \omega_2)S_{XX}(\omega_1 + \omega_2)S_{XX}(\omega_1)$$

$$+ H_{i,1}(\omega_1)H_{j,2}(\omega_1 + \omega_2 + \omega_3, -\omega_1 - \omega_3)H_{k,2}(-\omega_1, \omega_1 + \omega_3)S_{XX}(\omega_1 + \omega_3)S_{XX}(\omega_1)\}.$$

$$(6.155)$$

This function is defined everywhere in the $\omega_1, \omega_2, \omega_3$ plane. The expression is zero everywhere for linear systems (provided the terms 6.154 are subtracted) and for cubic or quadratic nonlinear systems is defined in terms of the linear and nonlinear Volterra kernels $H_{i,1}(\omega_1)$, $H_{i,2}(\omega_1, \omega_2)$ and $H_{i,3}(\omega_1, \omega_2, \omega_3)$ relating input at location "r" to output at location "i."

By inspecting Eq. (6.155) we see that the trispectrum is sensitive to both second- and third-order nonlinearities. A structure with quadratic nonlinearities only will still possess a positive trispectrum, showing peaks at the poles of the first- and second-order Volterra kernels. This is in contrast to the derived expression for the bispectrum [40], which includes *only* second-order Volterra kernels and thus is only non-zero for second-order nonlinearities. Had we used a third order Volterra model in the bispectrum derivation we would have seen nonzero third-order terms, but they would have been of a higher order in P. For the trispectrum, the contributions of both second- and third-order nonlinearities show up in the analysis at the same order.

As an example, consider the system defined by the second-order, ordinary differential equation

$$m\ddot{y}(t) + c_1\dot{y}(t) + c_2\dot{y}^2 + c_3\dot{y}^3 + k_1 y(t) + k_2 y(t)^2 + k_3 y(t)^3 = x(t) \qquad (6.156)$$

For this SDOF system, the linear and nonlinear transfer functions are given by

$$H_1(\omega) = \frac{1}{k_1 + ic_1\omega - m\omega^2}$$

$$H_2(\omega_1, \omega_2) = -(k_2 - c_2\omega_1\omega_2)H_1(\omega_1)H_1(\omega_2)H_1(\omega_1 + \omega_2)$$

$$H_3(\omega_1, \omega_2, \omega_3) = (1/3)H_1(\omega_1 + \omega_2 + \omega_3)\left[-3k_3 H_1(\omega_1)H_1(\omega_2)H_1(\omega_3)\right.$$

$$+ i3c_3\omega_1\omega_2\omega_3 H_1(\omega_1)H_1(\omega_2)H_1(\omega_3) - 2k_2 H_1(\omega_3)H_2(\omega_1, \omega_2) + 2c_2\omega_1\omega_3 H_1(\omega_3)H_2(\omega_1, \omega_2)$$

$$+ 2c_2\omega_2\omega_3 H_1(\omega_3)H_2(\omega_1, \omega_2) + 2k_2 H_1(\omega_2)H_2(\omega_1, \omega_3) + 2c_2\omega_1\omega_2 H_1(\omega_2)H_2(\omega_1, \omega_3)$$

$$\left. - 2k_2 H_1(\omega_1)H_2(\omega_2, \omega_3) + 2c_2\omega_1\omega_2 H_1(\omega_2, \omega_3) + 2c_2\omega_1\omega_3 H_1(\omega_1)H_2(\omega_2, \omega_3)\right] \qquad (6.157)$$

respectively. For this SDOF system, the subscript denoting "DOF" is dropped so that, for example, $H_{1,1}(\omega) \to H_1(\omega)$. The nonlinear kernels, $H_2(\omega_1, \omega_2)$, $H_3(\omega_1, \omega_2, \omega_3)$ were obtained via the well-established harmonic probing method outlined in Worden and Tomlinson [28]. In accordance with the theory, the excitation was chosen to be i.i.d Gaussian noise with unit variance and zero mean. The PSD of the input is therefore given by a constant $S_{XX}(\omega) = P\,(N^2/Hz)$. The system parameters were initially fixed to the values $P = 0.01\,(N^2/Hz)$, $m = 1$ (kg), $c_1 = 3.0\,(N \cdot s/m)$, $c_2 = 0.0\,(N \cdot s^2/m^2)$, $c_3 = 0.0\,(N \cdot s^3/m^3)$, $k_1 = 2000$ (N/m), $k_2 = 0.0\,(N/m^2)$, and $k_3 = 10^7\,(N/m^3)$. The linear natural frequency for this system is denoted $f_n = 7.11$ (Hz). Figure 6.20 provides some insight into how the trispectrum for this system is formed. Shown are the various components that comprise the diagonal of the trispectrum along with the final plot of $T(\omega_1, \omega_1, \omega_1)$ (again, because we only have a one-DOF system, each of the constituent time series are the same, thus the subscripts

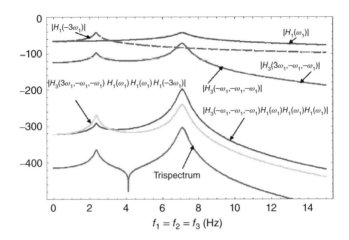

Figure 6.20 Various components of $T(\omega_1, \omega_1, \omega_1)$. *Source:* Reproduced from [36], Figure 1, with permission of Elsevier

on the trispectrum can be dropped, i.e., $T_{ijkp} \rightarrow T$). The main peaks for this trispectrum, with only a cubic nonlinearity, appear at both the natural frequency and $1/3$ of the natural frequency.

The full trispectrum exists in a four-dimensional space (three frequency variables + magnitude) and therefore cannot be plotted with conventional graphing techniques. Some researchers have elected to use "bubble plots," whereby the values in a few frequency bins correspond to the diameter of a sphere plotted in three-dimensional frequency space [41]. However, these plots tend to miss some of the structures of the trispectrum. Here, various three-dimensional "slices" of the trispectrum are compared to estimation. Figure 6.21 shows one particular slice of the trispectrum, $T(f_1, f_1, f_2)$, obtained by setting $f_2 = f_1$. Because there are no quadratic terms, the trispectrum reduces to the expression

$$T_{ijkp}(\omega_1, \omega_2, \omega_3) = 6H_{i,1}(\omega_1)H_{j,1}(\omega_2)H_{k,1}(\omega_3)H_{p,3}(-\omega_1, -\omega_2, -\omega_3)S_{XX}(\omega_1)S_{XX}(\omega_2)S_{XX}(\omega_3)$$

$$+ 6H_{p,1}(-\omega_1 - \omega_2 - \omega_3)S_{xx}(\omega_1 + \omega_2 + \omega_3)$$

$$\times \left\{ H_{j,1}(\omega_2)H_{k,1}(\omega_3)H_{i,3}(\omega_1 + \omega_2 + \omega_3, -\omega_2, -\omega_3)S_{XX}(\omega_2)S_{XX}(\omega_3) \right.$$

$$+ H_{i,1}(\omega_1)H_{k,1}(\omega_3)H_{j,3}(\omega_1 + \omega_2 + \omega_3, -\omega_1, -\omega_3)S_{XX}(\omega_1)S_{XX}(\omega_3)$$

$$\left. + H_{i,1}(\omega_1)H_{j,1}(\omega_2)H_{k,3}(\omega_1 + \omega_2 + \omega_3, -\omega_1, -\omega_2)S_{XX}(\omega_1)S_{XX}(\omega_2) \right\} \quad (6.158)$$

Displayed in Fig. (6.21) are the four components of Eq. (6.158), ordered clockwise from the upper left. The final plot shows the complete trispectrum associated with this particular slice. There is a great deal of structure in the trispectrum, and Eq. (6.155) provides a convenient expression for understanding how the various nonlinearities will influence the estimate.

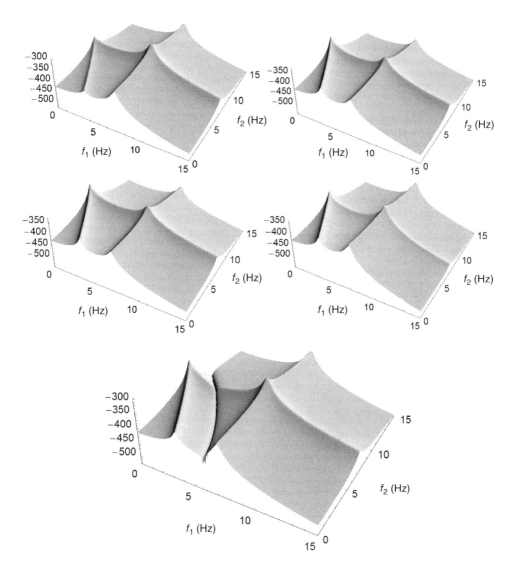

Figure 6.21 The four components of the "cubic only" trispectrum as given by Eq. (6.155) along with the complete trispectrum slice $T(f_1,f_1,f_2)$. *Source*: Reproduced from [36], Figure 2, with permission of Elsevier

6.5.4.1 Trispectral Density Estimation

The trispectrum estimate involves the various components in expression (3.104). The estimator for the main portion of the trispectrum is given by Eq. (3.106) (see also [42])

$$S_{Y_p Y_i Y_j Y_k}(f_1,f_2,f_3) = \lim_{T \to \infty} \frac{E[Y_i(f_1)Y_j(f_2)Y_k(f_3)Y_p(-f_1 - f_2 - f_3)]}{2T} \qquad (6.159)$$

where $Y_i(f)$ is the discrete FT of $y_i(t)$ (again, we are using f in the estimate instead of ω because of the implementation of the DFT in most software routines). Similarly, the terms involving the delta function must also be estimated, for example,

$$\delta(f_1+f_2)\hat{S}_{ji}(f_1)\hat{S}_{pk}(f_3) = \delta(f_1+f_2)\lim_{T\to\infty}\frac{E[\hat{Y}_i(f_1)\hat{Y}_j(-f_1)]}{2T}\frac{E[\hat{Y}_k(f_3)\hat{Y}_p(-f_3)]}{2T}. \quad (6.160)$$

Under the assumption that the acquired data are ergodic, the expected values required in Eqs. (6.159) and (6.160) may be realized by averaging over windowed segments of data, as is done for estimates of the PSD. The N point sampled data are divided into windows of length M, possibly overlapping by L points, and an appropriate windowing function $g(m)$ is applied. Each windowed, FT data segment is denoted

$$\hat{Y}_i^{(q)}(u) = \sum_{m=1}^{M} g(m)y_i(m+(q-1)M-L)e^{-i2\pi um/M}.$$

This gives a total of $N_w = (N-L)/(M-L)$ windows. Using this notation, the complete estimate becomes

$$\hat{T}_{ijkp}(u,v,z) = \frac{\Delta_t^3}{N_w\|W\|_4}\sum_{q=1}^{N_w}\hat{Y}_i^{(q)}(u)\hat{Y}_j^{(q)}(v)\hat{Y}_k^{(q)}(z)\hat{Y}_p^{(q)*}(u+v+z)$$

$$-\delta(u+v)\frac{\Delta_t^2}{N_w^2\|W\|_2^2}\sum_{q=1}^{N_w}\hat{Y}_i^{(q)}(u)\hat{Y}_j^{(q)*}(u)\sum_{q=1}^{N_w}\hat{Y}_k^{(q)}(z)\hat{Y}_p^{(q)*}(z)$$

$$-\delta(u+z)\frac{\Delta_t^2}{N_w^2\|W\|_2^2}\sum_{q=1}^{N_w}\hat{Y}_i^{(q)}(u)\hat{Y}_k^{(q)*}(u)\sum_{q=1}^{N_w}\hat{Y}_j^{(q)}(v)\hat{Y}_p^{(q)*}(v)$$

$$-\delta(v+z)\frac{\Delta_t^2}{N_w^2\|W\|_2^2}\sum_{q=1}^{N_w}\hat{Y}_i^{(q)}(u)\hat{Y}_p^{(q)*}(u)\sum_{q=1}^{N_w}\hat{Y}_j^{(q)}(v)\hat{Y}_k^{(q)*}(v) \quad (6.161)$$

where $\|W\|_4 = \sum_{m=1}^{M}|g(m)|^4$ and $\|W\|_2 = \sum_{m=1}^{M}|g(m)|^2$ are the norms of the window function. The discrete frequencies u,v,z are related to their continuous counterparts via, $\omega_1 = 2\pi\Delta_f(u-1-M/2)$ $u=1\cdots M$ where $\Delta_f = 1/(M\Delta_t)$ is the frequency bin width. Typically, however, we are only concerned with the positive frequency domain, and the portions of the estimate affected by the δ terms are ignored. This is not simply for convenience. Our theory tells us that if there is a structural nonlinearity, it will show up as a peak at a combination resonance among multiple structural modes, all of which exist only for $f_1,f_2,f_3 > 0$.

The biggest obstacle to trispectrum estimation is the data requirements. For each frequency bin, the sample variance is given by the expression

$$\sigma_e^2 = \frac{M^3\Delta_t^2}{N}\beta(f_1,f_2,f_3)S_{ii}(f_1)S_{jj}(f_2)S_{kk}(f_3)S_{pp}(f_1+f_2+f_3) \quad (6.162)$$

where $S_{ii}(f)$ is the auto PSD of the response at DOF "i" and $\beta(f_1,f_2,f_3)$ is a scale factor that depends on where in the tri-frequency domain the estimate is being examined [43]. Again, the primary concern in this work is the principle domain for which the scale factor is either

$\beta(f_1, f_2, f_3) = 1$ (all frequencies different) or $\beta(f_1, f_2, f_3) = 2$ (two of the three frequencies identical). The main problem is the prefactor in the variance expression, $\frac{M^3 \Delta_t^2}{N}$, which implies that for a consistent estimate (bias and variance go to zero as $N \to \infty$), the segment size must obey $M < N^{1/3}$. For a block size of $M = 128$ points (a very poor frequency resolution), this implies that the time series be roughly $N = 2^{21}$ points long! Increasing the block size for a fixed number of data improves the bias, but severely degrades the variance. We have discussed this trade-off at length for both the power spectrum and bispectral density functions; for the trispectrum, finding an appropriate balance is particularly problematic. Compounding the problem for structural systems is the fact that there tend to be sharp peaks in the transfer function, translating directly into sharp peaks in the HOS. Thus, the bias issue is even more pronounced and it becomes extremely difficult, we have found, to correctly estimate the trispectral peak heights. These issues are best illustrated via examples.

To generate the time-series data used in the trispectrum estimation, Eq. (6.156) was numerically integrated with a fixed time step of $\Delta_t = 0.01s$. for a total of $N = 2^{21}$ observations. The system parameters were fixed to the same values used in generating Figure 6.21, but with the addition of a nonlinear quadratic stiffness, $k_2 = 300,000$ (N/m^2). Estimates of the trispectrum were obtained using $N_w = 16384$ nonoverlapping segments, each $M = 128$ points in length. Figure 6.22 compares the analytical and estimated trispectrum on a logarithmic scale for different slices in the f_1, f_2, f_3 plane. The analytical trispectrum expression correctly predicts the height of the main peak as well as the "ridges" that arise in the trispectrum due to the frequency interaction. The ridge along the $f_1 + f_2 + f_3 = f_n$ can be seen clearly in both theory and estimate. The addition of the quadratic nonlinearity boosts the amplitude of the main peak and also introduces a new "ridge" along the sum frequency $f_1 + f_2 = f_n$. The aforementioned bias versus variance trade-off makes it difficult to capture the sharp features exhibited in spectral estimation for structural systems. Even for the large amount of data considered here, the variance is sufficiently high to obscure some of the details of the estimate while the bias makes it difficult to clearly resolve the peaks. Nonetheless, Eq. (6.155) clearly predicts the main features of the trispectrum.

As a final SDOF example, we consider what happens to the trispectrum estimated in Figure 6.22 if we increase the frequency resolution. For this example, the number of data were taken to be $N = 2^{23} = 8,388,608$ points and the quadratic stiffness nonlinearity was removed (i.e., $k_2 = 0$ (N/m^2)). An FT block size of $M = 512$ points was used in the estimate with an overlap of $L = 256$. Despite the large number of points, there are still too few data for generating a consistent estimate ($M^3 = 2^{27}$). Nonetheless, the estimate was computed, taking 5.5 h to complete on a PC with a single 2.66 GHz processor. The results are shown on both linear and log scales in Figure 6.23. The estimate clearly captures the predicted behavior; however, from the linear scale it can be seen that the predicted peak height is roughly 2.5 times what is being estimated due primarily to the estimator bias. Again, for structural systems, the trispectrum needs to be able to resolve very sharp peaks. These features are difficult to resolve even in bispectrum estimates, where the data requirements and bias are not nearly as severe. In short, it may be some time before the data required for accurate trispectrum estimates are able to be processed in a reasonable amount of time. Until then, the practitioner should understand that for structural systems, the estimated trispectrum may possess significant inaccuracies (primarily due to bias). This SDOF example is useful in studying the general features of the trispectrum; however, it is clearly more practical to look at MDOF nonlinear systems.

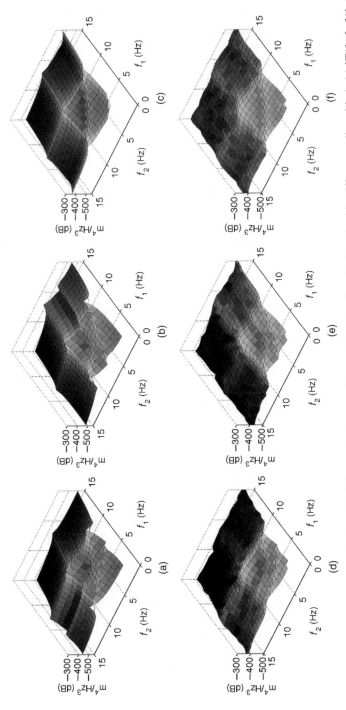

Figure 6.22 Theoretical (a–c) and estimated (d–f) magnitude auto-trispectral densities (quadratic and cubic stiffness nonlinearities): (a) $|T(f_1, f_2, f_2)|$, (b) $|T(f_1, f_1, f_2)|$, (c) $|T(f_n, f_2, f_3)|$. Peak heights are displayed in terms of the trispectral density units. *Source:* Reproduced from [36], Figure 4, with permission of Elsevier

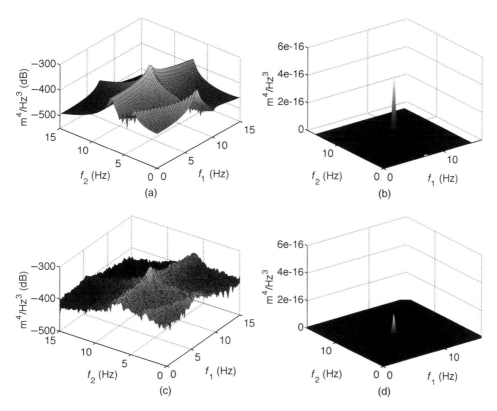

Figure 6.23 Theoretical (a and b) and estimated (c and d) auto-trispectrum $|T(f_1,f_2,f_2)|$ (same case as Figure 6.21b) on both linear and log scales. For this estimate, the number of data were increased to $N = 2^{23}$ points, while $M = 512$ points were used in the averaged FT segments. Although the estimate is improved, the dominant peak value is underestimated by a factor of 2.5. Peak heights are displayed in terms of the trispectral density units. *Source*: Reproduced from [36], Figure 5, with permission of Elsevier

The derivation of Eq. (6.155) was purposefully kept general to handle the MDOF case. Consider the three-DOF system governed by

$$[M]\ddot{\mathbf{y}}(t) + [C]\dot{\mathbf{y}}(t) + [K_L]\mathbf{y}(t) + \mathbf{g}(\dot{\mathbf{y}}, \mathbf{y}) = \mathbf{x}(t). \tag{6.163}$$

where $[M]$ is the diagonal mass matrix and $[C], [K]$ are the familiar tri-diagonal matrices containing linear damping and stiffness matrices. The function $\mathbf{g}(\dot{\mathbf{y}}, \mathbf{y})$ allows for the possibility of a quadratic or cubic stiffness or damping element between any of the three masses. The excitation, as before, is taken to be an i.i.d Gaussian distributed random process applied at any one of the three masses.

To illustrate, consider this MDOF system with $k_{11} = k_{12} = k_{13} = 6000$ N/m, $c_{1,1}, c_{1,2}, c_{1,3} = 3$ N $\cdot$ s/m, and $m_1 = m_2 = m_3 = 1$ kg. For this example, we consider both cubic and quadratic nonlinearities $k_2 = 100$ N/m^2, $c_2 = 100$ N $\cdot$ s^2/m^2, and $k_3 = 10^8$ N/m^3. The addition of the quadratic nonlinearity adds a second "ridge" in the trispectrum plot, where the sum of the two frequencies equals the dominant resonant frequency. A comparison between theory

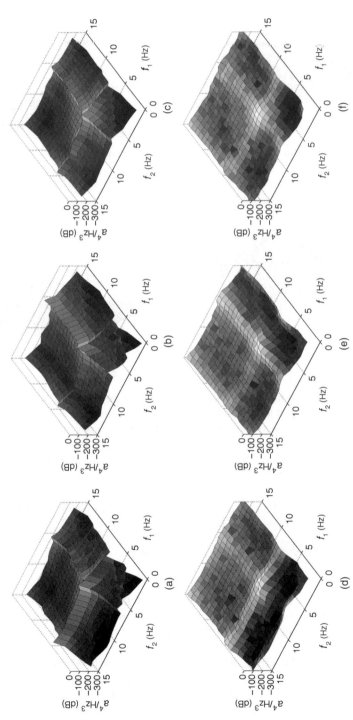

Figure 6.24 Theoretical (a–c) and estimated (d–f) acceleration magnitude trispectral densities for the MDOF system with quadratic damping and stiffness nonlinearities as well as a cubic stiffness nonlinearity between mass 1 and mass 2: (a) $|T^3_{1,2,3,3}(f_1,f_2,f_2)|$, (b) $|T^3_{1,2,3,3}(f_1,f_2,f_2)|$, (c) $|T^3_{1,2,3,3}(f_n,f_2,f_3)|$. Peak heights are displayed in terms of the trispectral density units *Source:* Reproduced from [36], Figure 7, with permission of Elsevier

and estimate is presented in Figure 6.24 for several different trispectrum slices. The theory correctly predicts the location and magnitude of the peaks and ridges of the estimated trispectrum. The poles of the trispectrum can be seen by Eq. (6.155) to be located at both the system's natural frequencies *and* along the ridges defined by $f_1 + f_2 + f_3 = f_n$ (for cubic nonlinearities) and, for example, $f_1 + f_2 = f_n$, for quadratic nonlinearities. These features can be observed in the estimate just as they were for the SDOF system. The estimate, of course, suffers from the aforementioned bias due to the severe data requirements (poor frequency resolution), but is still able to match the theoretical trispectrum fairly well.

It should also be mentioned that the trispectrum will show peaks even in the absence of cubic nonlinearities. This can be seen by simply examining the complete expression for the trispectrum. In short, the theory does a good job predicting the location and heights of the main trispectral peaks. While the theory does also correctly predict the locations of the ridges of the trispectrum, these are much more difficult to resolve than the main peaks, particularly for MDOF systems. As computing power improves, these features will perhaps be more clearly seen in the estimates as well.

The chief benefit of possessing an analytical expression is that the performance of a trispectral-based damage detection scheme can be quantified. The variance associated with the trispectrum estimator has already been given in this section and can be used in conjunction with the analytical expression for the trispectrum to predict our ability to detect damage-induced nonlinearity. This is done later in Section (8.5) in demonstrating the limits to detecting a delamination in a composite structure.

6.6   Estimation of Information Theoretics

In Chapter 3, we advocated the use of information theoretics, mutual information, and transfer entropy (TE) specifically, for studying coupling among components of nonlinear dynamical systems. By definition, they capture any order correlation among structural response data. These higher order correlations frequently are the hallmark of damage-induced nonlinearity, as is demonstrated in Chapter (8). In this section, we provide estimators for both the time-delayed mutual information and time-delayed transfer entropy (TDTE) functions.

In the following discussion, we require a slight change in our notation. To this point, we have discussed different random processes $\mathbf{X}$ and $\mathbf{Y}$. This notation becomes cumbersome, however, when more than two random processes are being considered. In our TE examples, we consider MDOF systems that produce a number of output signals, each of which is modeled as a different random process. In what follows, we therefore assume that we have observed the signals $x_i(t_n), i = 1 \cdots M$ as the output of an M-DOF dynamical system and that we have measured these signals at particular times $t_n, n = 0 \cdots N - 1$. As we have done throughout this book, we choose to model each sampled value $x_i(t_n)$ as a continuous random variable X_i with probability density $p_{X_i}(x(t_n))$. The vector of random variables $\mathbf{X}_i$ defines the ith random process and will be used to model the sequence of observations $\mathbf{x}_i \equiv x_i(t_n), n = 1 \cdots N$. Using this notation, we can also define the joint PDF $p_{\mathbf{X}_i}(\mathbf{x}_i)$, which specifies the probability of observing such a sequence.

Note that in discussing the estimates we are using the discrete time index n (as opposed to the continuous time $t_n = n\Delta_t$). From Eqs. (3.127) and (3.133) we have closed form expressions for information theoretics in terms of the linear cross-correlation coefficient, $\rho_{X_i X_j}(\tau)$. excitation. Using the methods of Section 6.4, we can estimate $\hat{\rho}_{X_i X_j}(T)$. This estimator is asymptotically

consistent and unbiased, and hence it must be the MLE. We can therefore substitute this estimate into the linearized expressions for both mutual information and TE to give

$$\widehat{I}_{X_iX_j}(T) = -\frac{1}{2}(1 - \widehat{\rho}^2_{X_iX_j}(T))$$

$$\widehat{TE}_{j \to i}(\tau)$$

$$= \frac{1}{2}\log_2\left[\frac{(1 - \widehat{\rho}^2_{X_iX_i}(\Delta_t))(1 - \widehat{\rho}^2_{X_iX_j}(\tau))}{1 - \widehat{\rho}^2_{X_iX_j}(\tau) - \widehat{\rho}^2_{X_iX_j}(\tau - \Delta_t) - \widehat{\rho}^2_{X_iX_i}(\Delta_t) + 2\widehat{\rho}_{X_iX_i}(\Delta_t)\widehat{\rho}_{X_iX_j}(\tau)\widehat{\rho}_{X_iX_j}(\tau - \Delta_t)}\right]$$

$$(6.164)$$

which, by the invariance property of the MLE (see Section 6.2), will also be an MLE.

However, if the system is nonlinear, it becomes necessary to use a different estimator. Computing the time-delayed mutual information and/or TDTE between two time series involves
the estimation of the single and joint entropies that comprise Eqs. (3.124) and (3.130). The
entropies, in turn, require estimates of the single and joint PDFs. To keep the discussion
general, assume that we record a d-dimensional data vector $\mathbf{x}(n) = (x_1(n), x_2(n), \cdots, x_d(n))$ at
discrete times $n = 1 \cdots N$. At each point in time, we may model the vector with a joint PDF
$p_\mathbf{X}(\mathbf{x}(n))$. We further assume the data are stationary so that we may drop the time index n in
what follows.

We begin by recalling (see Eqn. 3.113) that the expression for entropy is really that of the
expectation of the log of the probability density. That is to say, the joint entropy is

$$h_\mathbf{X} = -\int_{\mathbb{R}^d} p_\mathbf{X}(\mathbf{x})\log_2(p_\mathbf{X}(\mathbf{x}))\mathbf{dx} \equiv -E[\log_2(p_\mathbf{X}(\mathbf{X}))].\tag{6.165}$$

Previously in this chapter we have discussed the finite data approximation to the expected value
operator. Using this simple approach, we form repeated estimates of the probability density
$p_\mathbf{X}(\mathbf{x})$ from the observed data and average the result. One way to accomplish this is to estimate
the PDF at each point in the data vector $\mathbf{x}(n)$ and use the approximation

$$\int p_\mathbf{X}(\mathbf{x}(n))\log_2[p_\mathbf{X}(\mathbf{x}(n))] \approx \frac{1}{N}\sum_{n=0}^{N-1}\log_2[\widehat{p}_\mathbf{X}(\mathbf{x}(n))]\tag{6.166}$$

where we have explicitly acknowledged that we will still have to estimate the PDF in the
logarithm.

If the practitioner has access to an ensemble of measurements, the density estimates may
be obtained from the collection at any point in time (as we discussed at the beginning of this
chapter), thus allowing information theoretics to be estimated from nonstationary time series.
In most cases, however, the practitioner only has a single measurement (time series) with
which to work. For this situation, estimates of mutual information and/or TE are only possible
if the data can be safely assumed to be strictly stationary and ergodic. This latter assumption
is particularly important because, as we have shown, it allows us to replace ensemble averages
with averages over data observed in similar dynamical states. Given this assumption the
practitioner may use kernel methods of Section 6.3 to estimate the local densities. A thorough
discussion of the estimation of information theoretics using kernel-based methods has already
been given by Kaiser and Schreiber [44] and is briefly summarized here. In the limiting

case of Gaussian processes (e.g., linear structures driven with Gaussian noise) estimates of the linear auto- and cross-correlation functions are sufficient to estimate the information theoretics. This approach is also discussed.

Given a d-dimensional data vector, the goal is to estimate the density local to the point with time index n. Two of the more popular choices are the "fixed-bandwidth" kernel and the "fixed-mass" kernel. The former involves simply counting points within a specified radius (bandwidth) ϵ of the fiducial point and normalizing by the number of data. We have already used the ergodic theorem of Birkhoff (Section 6.3) to show how density estimates can be obtained using an appropriate kernel. Here, we use the aforementioned Heaviside function to give the estimate

$$\hat{p}_X(\mathbf{x}(n), \epsilon) = \frac{1}{N} \sum_{\substack{m=1 \\ m \neq n}}^{N} \Theta(\epsilon - \|\mathbf{x}(n) - \mathbf{x}(m)\|) \tag{6.167}$$

where

$$\Theta(\epsilon - \|\mathbf{x}(n) - \mathbf{x}(m)\|) = \begin{cases} 1 & : & \epsilon - \|\mathbf{x}(n) - \mathbf{x}(m)\| \geq 0 \\ 0 & : & \epsilon - \|\mathbf{x}(n) - \mathbf{x}(m)\| < 0 \end{cases}$$

and the operator $\|\cdot\|$ takes the vector norm (here, the Euclidean norm is used). Equation (6.167) takes the local density estimate about point n to be the number of points in a hypersphere of size ϵ about that point divided by the total number of points in the time series N. Equation (6.167) represents a simple form of kernel density estimation using the "step" kernel with fixed bandwidth ϵ.

Another approach to kernel density estimation is to use the so-called fixed-mass approach rather than a fixed-bandwidth one. In this approach, the local densities are estimated by the ratio of a fixed number of points to the volume of space occupied by those points. In this case, Eq. (6.167) is rewritten

$$\hat{p}_X(\mathbf{x}(n), M) = \frac{1}{N} \frac{M}{V(\mathbf{x}(n))} \tag{6.168}$$

where $V(\mathbf{x}(n))$ is the minimum volume containing these points. The volumes may be taken as hyperspheres, hyperrectangles, and so on. Here, the hypersphere is used, thus $V(\mathbf{x}(n)) = \alpha \epsilon_j^d$, where $\alpha = \pi^{d/2}/(\Gamma[d/2 + 1])$ and $\Gamma[\cdot]$ is the gamma function. The kernel is adaptive because it allows the volume to adjust depending on the local density of points. However, this approach is also more computationally costly. Finding the nearest M points to a given data vector is inherently more difficult than finding all points within a given radius ϵ. As an example, if an estimate of the joint density for $p_X(x_1(n), x_2(n))$ is required (i.e., a $d = 2$ dimensional vector $\mathbf{x} \equiv (x_1, x_2)$), one computes the distances $\sqrt{(x_1(n) - x_1(m))^2 + (x_2(n) - x_2(m))^2}$ for all $m = 1 \cdots N$ and then implements Eq. (6.167) or (6.168).

Returning to the quantities to be estimated, we have shown that for random processes X and Y, the mutual information can be written as a function of the individual and joint entropies. Using the above-described estimation procedure, these quantities become

$$\hat{I}_{X_i X_j}(T, \epsilon) = \frac{1}{N} \sum_{n=1}^{N} \left\{ \log_2(\hat{p}_{X_i X_j}(x_i(n), x_j(n + T), \epsilon)) - \log_2(\hat{p}_{X_i}(x_i(n), \epsilon)) \right.$$

$$\left. - \log_2(\hat{p}_{X_j}(x_j(n + T), \epsilon)) \right\} \tag{6.169}$$

where the dependence on the ϵ has been left explicitly in the equation. Given two time series x_i and x_j, one may estimate $\hat{I}_{X_iX_j}(T, \epsilon)$ directly from the data using either the fixed-bandwidth kernel 6.167 (as given by 6.169) or the fixed-mass kernel (6.168). The above-mentioned formulation is generic and may be extended to look at shared information in more than two variables or between groups of variables. This extension of mutual information is referred to in the literature as *redundancy* and is discussed thoroughly by Prichard and Theiler [45] and Paluš [46]. The algorithm for the multivariate case stays basically the same, the only difference being that the densities must be evaluated in a higher dimensional space. The main difficulty in implementing this scheme lies in finding near neighbors within a radius ϵ to a given point n. For time series of even modest size, using the naive $O(N^2)$ approach is prohibitive. However, there exist a number of fast near-neighbor search algorithms that can significantly reduce computation time. For example, the box-assisted approach of [47] or a variant of the K-D tree algorithm [48] are two frequently used options. We have found that for low-dimensional analysis ($d \leq 5$), the box-assisted approach yields faster results. The K-D tree algorithm is generally superior for larger dimensions.

As a test case, assume a linear two-DOF structure governed by the usual coupled, ordinary differential equations

$$\mathbf{M\ddot{x}}(t) + \mathbf{C\dot{x}}(t) + \mathbf{Kx}(t) = \mathbf{f}(t) \tag{6.170}$$

Further, assume that the forcing for the structure is applied at the end mass and is a sequence of i.i.d., jointly Gaussian distributed random variables. In this case, it was shown how an analytical expression for the linear cross-correlation coefficient $\rho_{X_iX_j}(T)$ could be obtained. Thus, it is a simple matter to substitute this function into Eq. (3.127) to get a closed form solution for $I_{X_iX_j}(T)$. Figure 6.25 shows the time-delayed mutual information $I_{X_iX_j}(T)$ based on theory and as estimated from the time series $x_i(n), x_j(n)$. Shown are the linearized estimate (based

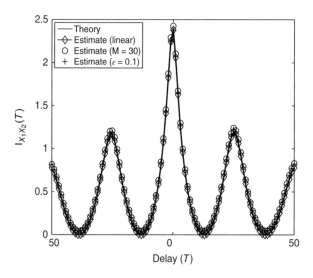

Figure 6.25 Mutual information $I_{X_1X_2}(T))$ based on theory (solid line) and estimated from linear time series data using the FT-based approach (diamond), fixed-bandwidth kernel (cross), and fixed-mass kernel (circle)

on the FFT) and the two different kernel-based approaches: fixed-bandwidth with $\epsilon = 0.1$ and fixed-mass with $M = 30$. The mutual information is shown in units of bits resulting from the use of the base 2 logarithm in Eq. (3.123). As $|T|$ increases, the two processes x_i, y_j become decorrelated with each other and the mutual information shows the expected decay. Also seen is the dominant damped period of oscillation in both forward and reverse time. Because the two processes are near in space (directly coupled), there is only a small lag before the mutual information curve reaches a peak. In this case, the joint probability density $p_{X_i X_j}(x_i(n), x_j(n + T))$ is a maximum at $T < 0$, while reversing the computation, that is, estimating $I_{X_i X_j}(T)$, the mutual information peaks for $T > 0$.

Each of the estimation techniques produces a result in agreement with the theory. There exists a slight difference, however, between the theory and the curve obtained using the fixed-bandwidth estimate (see right plot of Figure 6.25). Figure 6.26 illustrates the effect of varying ϵ on the resulting mutual information curve. This error appears to grow as the radius ϵ decreases and can be attributed to the lack of data at finer scales of resolution. As the radius is increased to $\epsilon = 0.1$, the results appear to match the theoretical result well. However, there are two difficulties associated with using large values for ϵ. As ϵ approaches the signal amplitude, all points are counted in Eq. (6.167) and the resulting mutual information curves

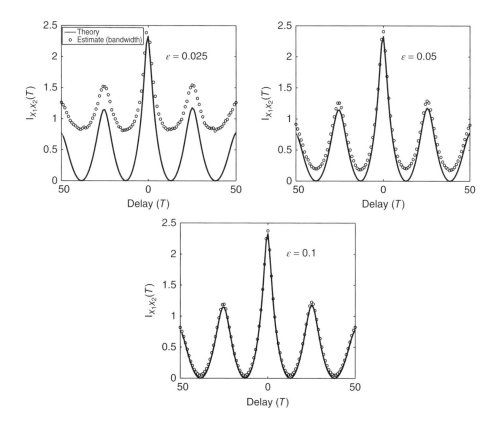

Figure 6.26 Analytical and estimated mutual information curves associated with the linear system response for varying ϵ

are flattened toward zero. The second issue is purely computational. The time it takes to find near neighbors within a radius ϵ scales as a power of ϵ. For large values, the computations take prohibitively long times. In this work, $\epsilon = 0.1\sigma$ is used where σ is the signal variance. This value was also utilized by Prichard and Theiler [45] in the analysis of chaotic data and has produced excellent agreement with theory for all applications the author has tried. The fixed-mass approach yields results in agreement with the theory for a number of choices of M, but is more computationally costly for reasons already mentioned.

TE also reduces to a simple function of the local density estimates given by Eq. (6.167).

$$TE_{j \to i}(T, \epsilon) = \frac{1}{N} \sum_{n=1}^{N} \left\{ \log_2 \left[\hat{p}_{\mathbf{X}_i \mathbf{X}_j}(x_i(n+1), x_i(n), x_j(n+T), \epsilon) \right] + \log_2 \left[\hat{p}_{X_i}(x_i(n), \epsilon) \right] \right.$$

$$\left. - \log_2 \left[\hat{p}_{\mathbf{X}_i}(x_i(n+1), x_i(n), \epsilon) \right] - \log_2 \left[\hat{p}_{X_i X_j}(x_i(n), x_j(n+T), \epsilon) \right] \right\} . \quad (6.171)$$

Again, dependence on the bandwidth ϵ has been left in the equation. For illustrative purposes, we examine the first term on the right-hand side of Eq. (6.171). Expanding Eq. (6.167) gives the needed probability density estimate as

$$\hat{p}_{\mathbf{X}_i \mathbf{X}_j}(x_i(n+1), x_i(n), x_j(n+T), \epsilon) = \frac{1}{N} \sum_{\substack{m=1 \\ |m-n|>t}}^{N} \Theta \left(\epsilon - \left\| \begin{matrix} x_i(n+1) - x_i(m+1) \\ x_i(n) - x_i(m) \\ x_j(n+T) - x_j(m+T) \end{matrix} \right\| \right) \quad (6.172)$$

where $\| \cdot \|$ takes the Euclidean norm of the three vector components. As with the mutual information, both "fixed-mass" and fixed-bandwidth approaches can be used, that is, we could have replaced (6.172) with a corresponding fixed-mass estimator (6.168).

As with the time-delayed mutual information, we may test the quality of the estimate against the known, linear solution of the previous example. Figure 6.27 shows the theoretical

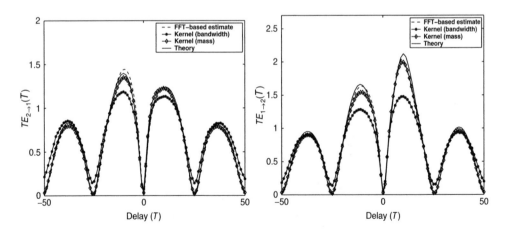

Figure 6.27 Different estimates of the time-delayed transfer entropy plotted against theoretical predictions (solid line) for both $TE_{2 \to 1}(T)$ and $TE_{1 \to 2}(T)$. Linearized estimates obtained via FFT (dashed), kernel estimates using a fixed bandwidth (star), and kernel estimates using a fixed mass (diamond). *Source*: Reproduced from [49], Figure 6, with permission of Elsevier

predictions for TE plotted against the linearized TE along with the two kernel-based methods. Again, because the TE is asymmetric in its arguments, both $TE_{2\rightarrow1}(T)$ and $TE_{1\rightarrow2}(T)$ are shown. As with the mutual information example, the fixed-mass approach yields results which are in good agreement with the theory. However, the fixed-bandwidth approach ($\epsilon = 0.075$) does not produce quantitative agreement. This disagreement was present for all choices of ϵ tried. The adaptive kernel is therefore recommended for estimating the TE. The fixed-bandwidth approach is simply not capable of resolving the subtle differences between the density $p_{X_iX_j}(x_i^{(1)},x_i,x_j^{(T)})$ and $p_{X_iX_j}(x_i,x_j^{(T)})$.

Looking at Figure 6.27, it is noted that the TE is zero for $T = 0$. In other words, neither $x_i(t)$ or $x_j(t)$ are providing additional information about the other, because of the high redundancy for zero lag. Local maxima are reached after $\sim \pm10$ time steps (0.2 s), implying that there is some delay before the information provided by one mass begins to add information about the transition probabilities (dynamics) of the other. In this regard, TDTE is essentially the opposite of time-delayed mutual information, showing peaks where the mutual information shows valleys. This is interesting as both quantities have been proposed as ways of quantifying coupling in dynamical systems. As with mutual information, there is again an associated periodicity corresponding to the dominant natural frequency. As the time series go in and out of phase with one another with T, the joint probability densities are changing periodically. However, unlike mutual information, where the joint pdf $p_{X_iX_j}(x_i,x_j^{(T)})$ controls the qualitative behavior, here it is the *ratio* of joint densities $p_{X_iX_j}(x_i^{(1)},x_i,x_j^{(T)})/p(x_i,x_j^{(T)})$ that is governing. The other densities required of Eq. (3.130), $p_{X_i}(x_i^{(1)},x_i)$ and $p_X(x_i)$, are not varying with delay.

Estimates of the TDTE based on the fixed-mass kernel are shown in Figure 6.28. As was demonstrated earlier, the fixed-bandwidth kernel produces poor estimates for this metric and was therefore not used for estimating the TE. The fixed-mass kernel also has difficulty matching theory over all delays. For this example, as M gets larger, the peaks of the TE are reduced from their theoretical values. The peaks of the TE occur for delays where the joint probability densities $p_{X_iX_j}(x_i,x_j^{(T)}),p_{X_iX_j}(x_i^{(1)},x_i,x_j^{(T)})$ are at their minimum (for the linearized metric, delays where the cross-correlation is zero), making it difficult to get good estimates (not many points falling in the joint "bins"). The valleys of the TDTE are preserved, however, for a wide variety of M values. Both the asymmetry in the peaks of the curves and the overall differences in magnitudes between $TE_{1\rightarrow2}$ and $TE_{2\rightarrow1}$ are also preserved, regardless of M.

As a final estimation example, we consider a five-DOF structural system defined by the constant coefficient matrices. Take, for instance, the five-DOF system governed by Eq. (3.135), where

$$\mathbf{M} = \begin{bmatrix} m_1 & 0 & 0 & 0 & 0 \\ 0 & m_2 & 0 & 0 & 0 \\ 0 & 0 & m_3 & 0 & 0 \\ 0 & 0 & 0 & m_4 & 0 \\ 0 & 0 & 0 & 0 & m_5 \end{bmatrix}$$

$$\mathbf{C} = \begin{bmatrix} c_1+c_2 & -c_2 & 0 & 0 & 0 \\ -c_2 & c_2+c_3 & -c_3 & 0 & 0 \\ 0 & -c_3 & c_3+c_4 & -c_4 & 0 \\ 0 & 0 & -c_4 & c_4+c_5 & -c_5 \\ 0 & 0 & 0 & -c_5 & c_5 \end{bmatrix}$$

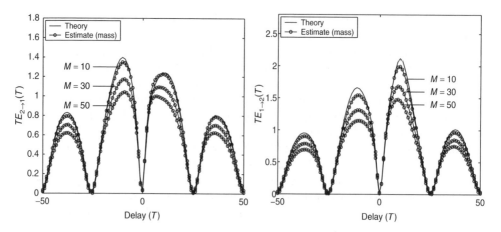

Figure 6.28   Transfer entropy estimates using the fixed-mass kernel with $M = 10, M = 30$, and $M = 50$.
Source: Reproduced from [49], Figure 7, with permission of Elsevier

$$\mathbf{K} = \begin{bmatrix} k_1 + k_2 & -k_2 & 0 & 0 & 0 \\ -k_2 & k_2 + k_3 & -k_3 & 0 & 0 \\ 0 & -k_3 & k_3 + k_4 & -k_4 & 0 \\ 0 & 0 & -k_4 & k_4 + k_5 & -k_5 \\ 0 & 0 & 0 & -k_5 & k_5 \end{bmatrix} \tag{6.173}$$

are constant coefficient matrices commonly used to describe structural systems. In this case, these particular matrices describe the motion of a cantilevered structure, where we assume a joint normally distributed random process applied at the end mass, i.e., $\mathbf{f}(t) = (0, 0, 0, 0, \mathcal{N}(0, 1))$. In this first example, we examine the TDTE between response data collected from two different points on the structure. We fix $m_i = 0.01$ kg, $c_i = 0.1$ N $\cdot$ s/m, and $k_i = 10$ N/m for each of the $i = 1 \cdots 5$ DOFs (thus, we are using $\alpha = 0.01$ in the modal damping model $\mathbf{C} = \alpha \mathbf{K}$). The system response data $x_i(n\Delta_t), n = 1 \cdots 2^{15}$ to the stochastic forcing is then generated via numerical integration. For simulation purposes, we used a time step of $\Delta_t = 0.01$ s, which is sufficient to capture all five of the system modes (the lowest of which resides at a frequency of $\omega_1 = 9.00$ rad/s). On the basis of these parameters, we generated the analytical expressions $TE_{3\rightarrow2}(\tau)$ and $TE_{2\rightarrow3}(\tau)$ and also $TE_{5\rightarrow1}(\tau)$ and $TE_{1\rightarrow5}(\tau)$ for illustrative purposes. These are shown in Figure 6.29 along with the estimates formed using the FT-based procedure. The solutions are plotted as a function of continuous (as opposed to discrete) time to highlight the dominant period of oscillation in the function.

With Figure 6.29 in mind, first consider negative delays only where $\tau < 0$. Clearly, the further the random variable $X_j(n + \tau)$ is from $X_i(n)$, the less information it carries about the probability of X_i transitioning to a new state Δ_t seconds into the future. This is to be expected from a stochastically driven system and accounts for the decay of the TE to zero for large $|\tau|$. However, we also see periodic returns to the point $TE_{j\rightarrow i}(\tau) = 0$ for even small temporal separation. Clearly, this is a reflection of the periodicity observed in second-order linear systems. In fact, for this system, the dominant period of oscillation is $2\pi/\omega_1 = 0.698$ seconds. It can be seen that the argument of the logarithm in (3.133) periodically reaches a minimum value of unity at precisely half this period; thus, we observe zeros of the TDTE

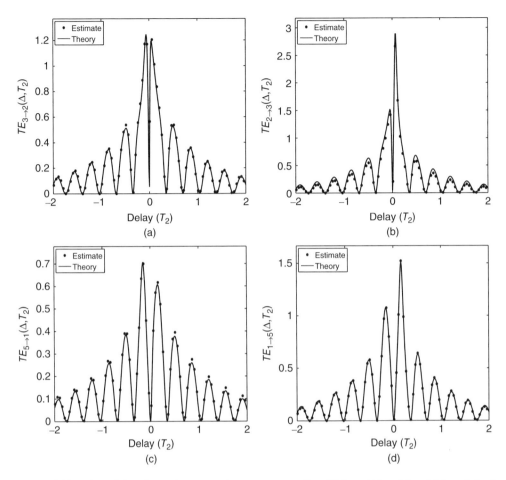

Figure 6.29   Time-delayed transfer entropy between masses 2 and 3 (a and b) and 1 and 5 (c and d) of a five-degree-of-freedom system driven at mass $P = 5$

at times $(i - 1) \times \pi / \omega_1, i = 1, 2, \cdots$ In this case, the TDTE is going to zero *not* because the random variables $X_j(n + \tau), X_i(n)$ are unrelated, but because knowledge of one allows us to exactly predict the position of the other (no additional information is present). We believe this is likely to be a feature of most systems possessing an underlying periodicity and is one reason why using the TE as a measure of coupling must be done with care.

We also point out that values of the TDTE are nonzero for positive delays as well. Again, so long as we interpret the TE as a measure of predictive power, this makes sense. That is to say, future values X_j can aid in predicting the current dynamics of X_i. Interestingly, the asymmetry in the TE peaks near $\tau = 0$ may provide the largest clue as to the location of the forcing signal. Consistently, we have found that the TE is larger for negative delays when the mass closest to the driven end plays the role of X_j; conversely, it is larger for positive delays when the mass furthest from the driven end plays this role. So long as the coupling is bidirectional, results such as those shown in Figure 6.29 can be expected in general.

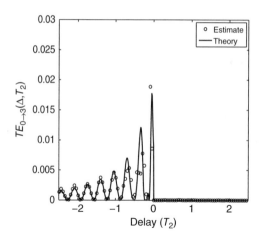

Figure 6.30 Time-delayed transfer entropy between the forcing and mass 3 for the same five-degree-of-freedom system driven at mass $P = 5$. The plot is consistent with the interpretation of information moving from the forcing to mass 3

However, the situation is quite different if we consider the case of unidirectional coupling. For example, we may consider $TE_{f \rightarrow i}(\tau)$, that is, the TDTE between the forcing signal and response variable i. This is a particularly interesting case as, unlike in previous examples, there is no feedback from DOF i to the driving signal. Figure 6.30 shows the TDTE between drive and response and clearly highlights the directional nature of the coupling. Past values of the forcing function clearly help in predicting the dynamics of the response. Conversely, future values of the forcing say nothing about transition probabilities for the mass response, simply because the mass has not "seen" that information yet. Thus, for unidirectional coupling, the TDTE can easily diagnose whether X_j is driving X_i or vice versa. It can also be noticed from these plots that the drive signal is not that much help in predicting the response as the TDTE is much smaller in magnitude than when computed between masses. We interpret this to mean that the response data are dominated by the physics of the structure (e.g., the structural modes), which is information not carried in the drive signal. Hence, the drive signal offers little in the way of additional predictive power. While the drive signal puts energy into the system, it is not very good at predicting the response. It should also be pointed out that the kernel density estimation techniques are not able to capture these small values of the TDTE. The error in such estimates is larger than these subtle fluctuations. Only the "linearized" estimator is able to capture the fluctuations in the TDTE for small ($O(10^{-2})$) values.

Our reasons for studying the TDTE will become apparent in subsequent chapters, in particular, Sections 8.7 and 8.8. The TDTE turns out to be quite sensitive to the presence of nonlinearity in a system. As such, estimates of the TDTE can be used with great effect to detect damage-induced nonlinearity in structures.

6.7 Generating Random Processes

Before leaving this chapter we discuss how to generate a realization of a random process with particular statistical properties. The importance of this topic in structural health monitoring

(SHM) stems from the fact that we are frequently trying to mimic (simulate) naturally occurring or even man-made loadings on structures in our models. Anytime we wish to simulate wave, wind, or some other band-limited loading on a structure, we require a reliable means of generating a corresponding sequence of data values. We have found that fast, efficient means of generating such sequences to be of some use in developing SHM techniques and have therefore chosen to provide a few options here. In fact, this is a well-studied but challenging problem in random vibrations that come into play in Chapters 8 and 10. The topic appears in this chapter because the realizations we generate are of finite length and the best we can hope for is to generate loading signals with statistical properties that are good estimates of the true, desired properties.

As in Chapter 3, define a stationary random process to be one that results in a sequence of observations $x(t_1), x(t_2), \cdots, x(t_N)$ for which the joint probability distribution $p_{\mathbf{X}}(x(t_1), x(t_2), \cdots, x(t_N))$ is invariant to temporal shifts, that is,

$$p_{\mathbf{X}}(x(t_1), x(t_2), \cdots, x(t_N)) = p_{\mathbf{X}}(x(t_1 + \tau), x(t_2 + \tau), \cdots, x(t_N + \tau)).$$

As we have already discussed, to completely define such a process we would have to specify the entire joint PDF. In practice, however, it is far more common to describe a random process in terms of (i) an unchanging marginal distribution associated with each observation, $p_X(x) \equiv p_X(x(t_1)) = p_X(x(t_2)) = \cdots p_X(x(t_N))$, and (ii) the stationary auto-covariance function $R_{XX}(\tau) = E[\tilde{X}(t)\tilde{X}(t + \tau)]$, where $\tilde{x}(t) = x(t) - E[X(t)]$. These two quantities are perhaps the most commonly used descriptors of a random process. Alternatively, we can use the Weiner–Khinchin theorem to instead specify the marginal distribution and autospectral density function. More specifically, assuming the FT of $R_{XX}(\tau)$ exists (a sufficient condition being that $\int_{-\infty}^{\infty} |R_{XX}(\tau)| d\tau < \infty$)) the Weiner–Khinchin theorem (and eqn. 3.67) state $S_{XX}(\omega) = \frac{1}{2\pi} \int_{-\infty}^{\infty} R_{XX}(\tau) e^{-i\omega\tau} d\tau$. Thus, specifying the auto-covariance function is the same as specifying the PSD. In engineering applications, it is more common for a random process to be described by $S_{XX}(\omega)$.

Thus, in this section we consider the problem of generating a sequence of observations with a specific $p_X(x)$ and $S_{XX}(\omega)$. Other quantities, for example, $R_{XXX}(\tau_1, \tau_2) = E[\tilde{X}(t)\tilde{X}(t + \tau_1)\tilde{X}(t + \tau_2)]$ (equivalently, the bispectrum) are typically not specified when describing random processes and are often implicitly assumed to be negligible. This may or may not be a good assumption. A number of researchers have detected the presence of significant third-order correlations in a variety of random processes. For example, Richardson and Hodgkiss found clear evidence of higher order correlations in underwater acoustic data. Kim and Powers [50] detected the presence of third-order correlations in plasma density fluctuations [50] as did Hajj *et al.* [51] in fluid flow. Gurley *et al.* found evidence of higher order correlations in wave height data and, in fact, proposed an approach for generating time series consistent with such correlations [52]. Nonetheless, we assume in this section that $p_X(x)$ and $S_{XX}(\omega)$ are sufficient descriptors of our random process.

6.7.1   Review of Basic Concepts

The problem of how to generate observations $x(n), n = 0 \cdots N - 1$ with a given PDF and auto-covariance (or PSD) has been tackled by a number of researchers. While the details vary considerably, all approaches tend to follow the same general prescription. This procedure is

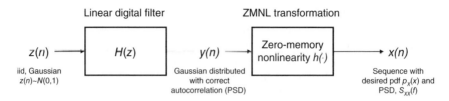

Figure 6.31 General process for generating signals with a user-defined probability density function and power spectral density function

illustrated schematically in Figure 6.31. First, a spectrally white, Gaussian distributed sequence is linearly filtered to produce a signal with the proper auto-covariance (equivalently PSD). This step makes use of the fact that if the input to a linear filter is Gaussian distributed, so too will be the output. We have already shown why this is so in Section (2.4). The job of the filter is therefore to impose the correct spectral coloring on the output signal.

After the filtering operation, the data are then subject to a zero memory, nonlinear (ZMNL) transformation to produce a signal with the appropriate non-Gaussian PDF. By "zero memory" we mean that the transformation does not impose any temporal correlations on the data, i.e., the transformation influences the signal value at t_n independently of t_m. The requirement of a memoryless transformation ensures that the covariance (and hence spectral properties) of the signal generated in the first step will not be significantly altered by the transformation step (see, e.g., Wise *et al.* [53] and Liu and Munson [54]). Depending on the method, one either stops at this point or iterates between these two steps until a convergence criteria is met.

In the context of structural dynamics, there are numerous papers on the topic of colored noise generation (see Bocchini and Deodatis [55] for a review). It is the second step, finding the appropriate transformation for altering the PDF, that is the main source of difficulty. Yamazaki and Shinozuka [58] proposed what appears to be the core algorithm for matching both spectral properties *and* the marginal distribution. The method iteratively adjusts the original Gaussian distributed signal (generated using the method of [57]) to have the correct PDF and then corrects for the changes this operation caused to the PSD. Again, the goal is to approximate the influence of the unknown $h(\cdot)$ without adversely influencing the PSD.

Initially, we might construct an ZMNL function, $h(\cdot)$, using concepts from Section (2.4). First, we note that if Y is a normally distributed random variable, then $P_Y(y)$ will be uniformly distributed [62] (this is true for *any* distribution $P_Y(y)$, not just the normal distribution). We also know from Section (2.4) that applying the inverse CDF of a random variable X to a uniform random variable gives us a random variable that is distributed according to $P_X(x)$. Hence, the function

$$h(Y) = P_X^{-1} P_Y(Y) \qquad (6.174)$$

takes the normally distributed variable Y and returns the variable $X = h(Y)$ with CDF $P_X(x)$ and therefore PDF $p_X(x)$ [62]. The ZMNL function $h(\cdot)$ can be expanded in terms of Hermite polynomials, which enables the autocorrelation of the ZMNL output to be written as a power series of the autocorrelation of $y(n)$. One then solves for the autocorrelation associated with $y(n)$, which makes the output of the ZMNL best approximate the desired autocorrelation, and then designs the linear filter to approximate this autocorrelation arbitrarily closely. The main problem with this method is that $P_X(\cdot)$ is sometimes not invertible analytically (e.g., the Gamma CDF), and finding $P_X(\cdot)^{-1}$ numerically reduces the simplicity and accuracy of the method. In response to this, Filho and Yacoub [63] proposed an approach that used a combination of the Hermite polynomials (for generating the proper autocorrelation) followed by a rank-reordering step for transforming the PDF to the desired non-Gaussian target. This rank-reordering or "shuffling" procedure attempts to approximate the influence of $h(\cdot)$ and has the distinct advantage of not requiring an analytical expression for the static, nonlinear transformation. A similar approach to approximating this function was tried earlier by Hunter and Kearney [64] and dubbed "stochastic minimization."

In what follows, we describe a few different approaches for generating the needed temporal sequences in order of increasing complexity. In the process we discuss covariance factorization and rejection sampling, two useful tools in statistical signal processing.

6.7.2 Data with a Known Covariance and Gaussian Marginal PDF

In the case where the marginal PDF for each observation is Gaussian, construction of a sequence $x(n), n = 0, \cdots, N-1$ with a specific covariance is straightforward. Following the schematic (6.31), we need only concern ourselves with developing the linear filter $\mathbf{H}$ such that $\mathbf{x} = \mathbf{Hz}$ (the nonlinear transformation is not required in the Gaussian case). The question is how to design $\mathbf{H}$ so that the resulting samples have a known covariance $\mathbf{C}_{XX}$?

To see how this can be done, we first note that the covariance matrix is, by definition, positive semi-definite. As such, it lends itself to a Cholesky decomposition [65], sometimes referred to as the "square root" of a matrix. Performing a Cholesky decomposition on the covariance yields

$$\mathbf{C}_{XX} = \mathbf{LL}^T \tag{6.175}$$

where $\mathbf{L}$ is a lower right triangular matrix. Now consider the N-dimensional sequence $\mathbf{x} = \mathbf{Lz}$, where $\mathbf{z}$ is a sequence of N i.i.d., normally distributed random variables with zero mean and unit variance. The mean of this new vector is zero and the associated covariance is

$$E[\mathbf{XX}^T] = E[(\mathbf{LZ})(\mathbf{LZ})^T]$$
$$= E[\mathbf{LZZ}^T\mathbf{L}^T]$$
$$= \mathbf{L}E[\mathbf{ZZ}^T]\mathbf{L}^T$$
$$= \mathbf{LIL}^T = \mathbf{LL}^T = \mathbf{C}_{XX} \tag{6.176}$$

where, in the last line, we have noted that the covariance matrix for a sequence of independent, unit variance random variables is the identity matrix. Thus, the $N \times N$ filter matrix is simply $\mathbf{H} = \mathbf{L}$ and the filtered sequence $\mathbf{x} = \mathbf{Hz}$ will have the desired covariance matrix. Moreover, as we have already demonstrated, a linear superposition of normally distributed random variables is itself a normally distributed random variable, and hence the sequence is joint normally distributed.

As an example, consider the covariance matrix

$$\mathbf{C}_{XX} = \begin{bmatrix} 1.0 & 0.85 & 0.7 & 0.55 & \cdots \\ 0.85 & 1.0 & 0.85 & 0.7 & \cdots \\ 0.7 & 0.85 & 1.0 & 0.85 & \cdots \\ \vdots & \ddots & \ddots & \ddots & \cdots \end{bmatrix} \tag{6.177}$$

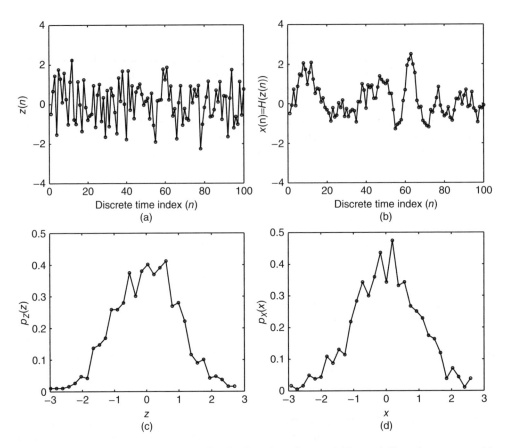

Figure 6.32 Sequence of i.i.d., normally distributed random variables and filtered sequence with desired covariance (a and b). Estimated marginal probability density functions (c and d) associated with each sequence. In both cases, the sequence is normally distributed as predicted

As is often the case, the covariance matrix possesses a banded-diagonal structure, in this case falling off to zero after eight samples (at a rate of 0.15 units for every step away from the diagonal). We generate a sequence of $N = 1000$ independent, normally distributed samples with mean zero and unit variance. We then perform a Cholesky decomposition of the 1000×1000 matrix $\mathbf{C}_{XX} = \mathbf{L}\mathbf{L}^T$ and form the filtered sequence $\mathbf{x} = \mathbf{H}\mathbf{z}$, where the filter is given simply by $\mathbf{H} = \mathbf{L}$. The original and filtered sequence of samples are shown in Figure 6.32 along with the estimated PDFs formed via histogram. Both sequences are clearly normally distributed, as predicted by theory. In addition, in the filtered sequence we can see the correlation structure imposed by the covariance matrix. The downside to this approach is that one is required to factor a potentially large matrix (in this case, 1000×1000). Spectral methods (discussed next) are preferred if long sequences are desired.

In fact, the more common situation is for the autospectral density to be specified (as opposed to the covariance matrix). The two are, of course, equivalent (by Wiener–Khitchine); however, the way we handle these cases in practice is slightly different. Although the next section handles a more general case, it can be easily used to generate a sequence with joint normal PDF and known autospectral density. One can simply proceed to Eq. (6.192) and simply ignore the iterative portion of the following algorithm.

Before leaving this section, however, we briefly describe an approach for generating normally distributed random processes of higher dimension. This is of some practical utility as we require realizations of a random process with known covariance in two temporal and one spatial dimension in Chapter (10). It turns out that the filtering approach described here can be extended in a straightforward way to the higher dimensional case. We consider the two-dimensional case first and examine the vector of random processes $\mathbf{X} = X_i(t_n), i = 0, \dots, M, n = 0, \dots, N - 1$. This situation arises when, for example, modeling time series data consisting of N observations each, recorded at M different spatial locations. Further, assume that these data are stationary (in both time and space) and possess correlations in both the spatial and temporal dimensions governed by the covariance matrices $C_{\mathbf{SS}}, C_{\mathbf{TT}}$, respectively. That is to say, at each point in time we have the $M \times M$ covariance matrix $C_{\mathbf{SS}} \equiv C_{S_i S_j}(n) = E[(X_i(n) - \bar{X}_i(n))(X_j(n) - \bar{X}_j(n))], i, j = 0, \dots, M - 1$. Spatial stationarity requires that the matrix is invariant to the absolute indices i, j but is rather a function of thes on the difference $i - j$ only. Similarly, for each spatial location, we have the $N \times N$ temporal covariance $C_{\mathbf{TT}} \equiv C_{TT}(n, m) = E[(X_i(n) - \bar{X}_i(n)(X_i(m) - \bar{X}_i(m))], n, m = 0, \dots, N - 1$ which, assuming stationarity, depends on the time difference $n - m$. Given $C_{\mathbf{SS}}$ and $C_{\mathbf{TT}}$, how might we generate realizations of $\mathbf{X}(t)$?

Consider the filters $\mathbf{L}_S$ and $\mathbf{L}_T$ formed by Cholesky decomposition of both spatial and temporal covariance matrices, for example, $C_{\mathbf{SS}} = \mathbf{L}_S \mathbf{L}_S^T$ and $C_{\mathbf{TT}} = \mathbf{L}_T \mathbf{L}_T^T$. Now consider an $M \times N$ matrix $\mathbf{Z}$ comprised of i.i.d. Gaussian entries with zero mean and unit standard deviation, i.e., $\mathbf{Z} \equiv z_{ij} \sim \mathcal{N}(0, 1), i = 0, \dots, M, j = 0, \cdots, N - 1$. Finally, consider the desired data matrix $\mathbf{X}$ to be formed as

$$\mathbf{X} = \mathbf{L}_S \mathbf{Z} \mathbf{L}_T^T. \tag{6.178}$$

To see that this matrix possesses the desired covariance in both space and time, consider the $MN \times 1$ vector formed by stacking each of the columns of $\mathbf{X}$ end to end. We will denote this vector as simply $vec(\mathbf{X})$. The entire $MN \times MN$ covariance matrix associated with $\mathbf{X}$ is therefore by definition

$$C_{\mathbf{XX}} = E[vec(\mathbf{X})vec(\mathbf{X})^T] \tag{6.179}$$

that is, the expectation of the product of any two entries in $\mathbf{X}$. Substituting (6.178) into (6.179) and simplifying yields

$$\begin{aligned} C_{\mathbf{XX}} &= E\left[vec(\mathbf{L}_S \mathbf{Z} \mathbf{L}_T^T)vec(\mathbf{L}_S \mathbf{Z} \mathbf{L}_T^T)^T\right] \\ &= E\left[(\mathbf{L}_T \otimes \mathbf{L}_S)vec(\mathbf{Z})\{(\mathbf{L}_T \otimes \mathbf{L}_S)vec(\mathbf{Z})\}^T\right] \\ &= E\left[(\mathbf{L}_T \otimes \mathbf{L}_S)vec(\mathbf{Z})vec(\mathbf{Z})^T(\mathbf{L}_T \otimes \mathbf{L}_S)^T\right] \\ &= (\mathbf{L}_T \otimes \mathbf{L}_S)E[vec(\mathbf{Z})vec(\mathbf{Z})^T](\mathbf{L}_T^T \otimes \mathbf{L}_S^T) \\ &= (\mathbf{L}_T \otimes \mathbf{L}_S)\mathbf{I}(\mathbf{L}_T^T \otimes \mathbf{L}_S^T) \\ &= \mathbf{L}_T \mathbf{L}_T^T \otimes \mathbf{L}_S \mathbf{L}_S^T \\ &= C_{\mathbf{TT}} \otimes C_{\mathbf{SS}} \end{aligned} \tag{6.180}$$

that is to say, the total space-time covariance structure is governed independently by C_{TT} and C_{SS} and can be expressed as the Kronecker product between those two matrices. Thus, our filtering procedure (6.178) produces data with *separable* covariance structure. The derivation given here makes use of several useful identities for Kronecker products [66]

$$vec(\mathbf{ABC}) = (\mathbf{C}^T \otimes \mathbf{A})vec(\mathbf{B})$$

$$(\mathbf{A} \otimes \mathbf{B})^T = \mathbf{A}^T \otimes \mathbf{B}^T$$

$$(\mathbf{A} \otimes \mathbf{B})(\mathbf{C} \otimes \mathbf{D}) = \mathbf{AC} \otimes \mathbf{BD} \tag{6.181}$$

as well as the fact that the covariance of $\mathbf{Z}$ is the identity matrix (following from the i.i.d., zero mean, normally distributed entries). This result can be generalized to still higher dimensions, that is, create a normal i.i.d. tensor $\mathbf{Z}$ and filter in an arbitrary number of dimensions (see [67]).

As an example, consider the generation of a spatiotemporal random process with auto-covariance given by the matrices

$$C_{S_i S_j} = \max(1.0 - |(i-j)\alpha_S|, 0) = \mathbf{L_S L_S^T}, i,j = 0, \dots, M-1$$

$$C_{TT}(n,m) = \max(1.0 - |(n-m)\alpha_T|, 0) = \mathbf{L_T L_T^T}, n,m = 0, \dots, N-1 \tag{6.182}$$

so that each covariance matrix is linearly decaying away from the diagonal (a common covariance structure found in naturally occurring processes). For this example, we select $\alpha_S = 0.625$ while $\alpha_T = 0.050$ so that the resulting random process is strongly correlated in time and very loosely correlated in space. These two covariance matrices are shown in Figure 6.33 for $M = 30$ and $N = 40$. According to the theory we have developed, the total space-time covariance should be the $MN \times MN$ matrix $C_{\mathbf{XX}} = C_{TT} \otimes C_{SS}$. To test this relationship, we formed $K = 4000$ realizations of the random process $\mathbf{X}^{(k)}, k = 1, 4000$ and estimated the full covariance via

$$\hat{C}_{\mathbf{XX}} = \frac{1}{K} \sum_{k=1}^{K} vec(\mathbf{X}^{(k)})vec(\mathbf{X}^{(k)})^T \tag{6.183}$$

Figure 6.33c and d shows both predicted and estimated full space-time covariance matrices.

A single realization of $\mathbf{X}$ is shown in Figure 6.34. By inspection we can see that this realization of the random process is much more strongly correlated in time than in space (the desired effect). Fast efficient means of generating high-dimensional random processes can be extremely useful in simulating the naturally occurring random processes that are frequent sources of structural vibration data. In fact, we use precisely this approach when simulating spatiotemporal loading on a ship hull in the main example of Chapter 10.

6.7.3 Data with a Known Covariance and Arbitrary Marginal PDF

Assume now that we wish to create a sequence of observations $x(n), n = 0, \cdots, N-1$ with PDF $p_X(x)$ and a two-sided PSD $S_{XX}(\omega)$. The first step is to sample the desired PSD function at N discrete frequencies $\omega_k = (k - N/2)\Delta_\omega$, $k = 0, \cdots, N-1$ giving $S_{XX}(\omega_k)$. The frequency bin width is dictated by the desired temporal sampling interval Δ_t. The two are related via

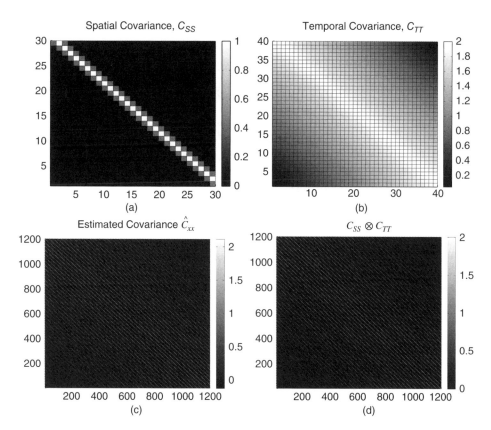

Figure 6.33 (a and b) Individual spatial (a) and temporal (b) covariance matrices used in generating the $M \times N$ observation matrix **X**. (c and d) Predicted (c) and estimated (d) spatiotemporal covariance matrix $C_{\mathbf{XX}}$. The estimated covariance was obtained using the estimator (6.183) with $K = 4000$

$\Delta_\omega = 2\pi/N\Delta_t$ and, as usual, Δ_t should be chosen in accordance with the Nyquist criterion for the maximum resolvable frequency. We also define the DFT as we did earlier

$$X(k) = \mathrm{FT}(x(n)) \equiv \sum_{n=0}^{N-1} x(n)e^{-i2\pi kn/N} \tag{6.184}$$

(i.e., Eq. 6.53 with a rectangular window) and inverse FT

$$x(n) = \mathrm{FT}^{-1}(X(k)) \equiv \frac{1}{N}\sum_{k=0}^{N-1} X(k)e^{i2\pi kn/N}. \tag{6.185}$$

Before proceeding, it should be mentioned that $p_X(x)$ and $S_{XX}(\omega)$ cannot be specified independently, as they are linked through the signal mean $\bar{x}$ and variance σ_x^2. For the signal mean, we have the relationship

$$\frac{X(0)}{N} = \frac{\Delta_t}{N}\sum_{n=0}^{N-1} x(n) = \bar{x} \tag{6.186}$$

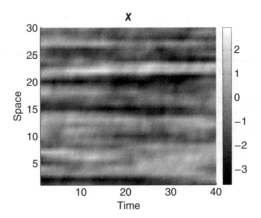

Figure 6.34 Single realization of a spatiotemporal random process **X** specified by the covariance matrices (6.182). By design, the random process is clearly more strongly correlated in time than in space

so that in terms of the PSD

$$2\pi S_{XX}(0) = N\Delta_t \bar{x}^2. \tag{6.187}$$

It is more natural to specify the mean as part of $p_X(x)$ and then simply adjust $S_{XX}(0)$ to conform. For this reason, we set $S_{XX}(0) = 0$ and $\bar{x} = 0$ before proceeding. The mean will be put back into the signal in the last step of the procedure. By Parseval's theorem, we also have the relationship

$$\sigma_X^2 = \frac{N}{N-1} \sum_{k=0}^{N-1} S_{XX}(\omega_k)\Delta_\omega. \tag{6.188}$$

In constraining the variance, we use Eq. (6.188) and adjust $E[X(n)^2]$ to conform (effectively forcing $p_X(x)$ to have the variance imposed by the PSD). For most engineering applications, we believe this to be the most common way of satisfying the constraints and simply note that the mean and variance can be specified either through the PDF or PSD functions.

The work of Schreiber and Schmitz [68] presented a simple, iterative solution for generating the static, nonlinear transformation needed to get the correct probability distribution. The only change that needs to be made in our version of the cited algorithm is that $p_X(x)$ and $S_{XX}(\omega)$ are specified seperately rather than coming from the same signal (as they would with surrogates). This necessitates an additional scaling step. Our algorithm works as follows:

The first step is to extract the target Fourier amplitudes $X^{(t)}(k)$ $k = 0 \cdots N - 1$ from the prescribed PSD function (superscript t denotes target). These are given by

$$X^{(t)}(k) = \sqrt{2\pi N S_{XX}(\omega_k)/\Delta_t} \quad k = 0 \cdots N - 1. \tag{6.189}$$

The sampling interval and factor of 2π need to be included to convert the density function with units $[x]^2$/rad/s to an amplitude distribution, units $[x]^2$. Had we included the sampling interval in our definition of the DFT (6.184), the Δ_t would instead have appeared in the numerator under the square root. Moreover, because PSD functions are typically given in terms of angular frequency variable ω, we require the 2π conversion. Finally, if it is the one-sided density that was specified *a priori*, divide by two before taking the square root.

The next step is to generate a sequence of N i.i.d random variables from the specified distribution $p_X(x)$. This sequence is denoted $s(n), n = 0, \cdots, N-1$ and can be produced with a random number generator (most packages, e.g., Matlab, provide samples from a wide variety of distributions) or from sampled data (e.g., experimental observation). In addition, if the CDF of the desired distribution is invertible, we have already shown at the end of Section 2.4 that one can simply generate a sequence of uniformly distributed random numbers U on the interval $[0, 1]$ and take $s \sim P_X^{-1}(U)$ (this is, in fact, the approach used in many canned routines). In the instance one cannot apply one of these standard approaches, the needed samples can be generated using rejection sampling, which is discussed in Section 6.7.4.

Next, compute the variance of the signal, $\sigma_s^2 = \frac{1}{N-1}\sum_n(s(n) - \bar{s})^2$, and generate the scaled sequence

$$x_o(n) = \frac{\sigma_s}{\sigma_x}s(n) \tag{6.190}$$

so that $x_o(n)$ possesses the variance dictated by Eq. (6.188). We then subtract and store the signal mean $\bar{x}_o = \frac{1}{N}\sum_n x_o(n)$ (this will be put back into the signal in the final step). The final step in the initialization process is to reorder this signal from the smallest to the largest value, giving $x_o(I_n), n = 0, \cdots, N-1$, where I_n gives the indexing such that $x_o(I_1) < x_o(I_2) < \cdots < x_o(I_N)$.

At this point, we have a sequence of values that are distributed according to $p_X(x)$ (the $x_o(n)$), but with the mean removed and with the variance consistent with $S_{XX}(\omega)$. We also have the Fourier amplitudes $X^{(t)}(k)$ corresponding to the specified PSD function $S_{XX}(\omega)$. The iterative process to follow attempts to shuffle the values of $x_o(n)$ such that the Fourier amplitudes are given by $X^{(t)}(k)$. The PDF associated with $x_o(n)$ will be invariant to temporal reordering so the shuffled signal will automatically preserve the PDF.

Before iterating, we create a duplicate time series $x(n) = x_o(n)$ so that the original values $x_o(n)$ will not be destroyed in the subsequent manipulations. Next, the phases of the values $x(n)$ are determined via FT, $X(k) = \text{FT}(x(n))$, giving

$$\phi(k) = \tan^{-1}\left(\frac{\text{Im}(X(k))}{\text{Re}(X(k))}\right) \tag{6.191}$$

A new signal is then created with the same phases, but with the target Fourier amplitudes. This is accomplished by substituting in the target Fourier amplitudes and inverse Fourier transforming

$$x(n) = \frac{1}{N}\sum_{k=0}^{N-1} e^{i\phi(k)}X^{(t)}(k)e^{i2\pi kn/N} = \text{FT}^{-1}(e^{i\phi(k)}X^{(t)}(k)) \tag{6.192}$$

resulting in a signal $x(n)$ with exactly the target PSD. However, the inverse FT is a summation of N random variables, therefore as a result of the central limit theorem (see Eq. 2.75), this step has the effect of "whitening" the signal, that is, altering the marginal density toward a Gaussian distribution. The next step is therefore to bring the PDF in line with the target PDF. As with most methods, this is done by assuming that there exists an ZMNL transformation that accomplishes this. However, instead of specifying a particular form to this transformation, we try to simulate its effect on the PDF. This is accomplished by a rank-reordering procedure, whereby the values of the signal with the correct distribution, $x_o(n)$, are shuffled to match the rank of $x(n)$. That is to say, the smallest value of $x_o(n)$ is given the same position in the signal as the smallest value of $x(n)$. The next smallest value of $s(n)$ is given the position of the next smallest value of $x(n)$, and so on. The sorted values of $x(n)$ are denoted $x(J_n) : x(J_0) < x(J_1) < \cdots < x(J_{N-1})$. We then take

$$x(J_n) = x_o(I_n) \quad n = 0 \cdots N - 1 \tag{6.193}$$

giving a signal $x(n)$ that will, by definition, possess the correct PDF *and* have a very similar PSD. The rank reordering attempts to preserve the auto-covariance by giving similar values of the time series $x_o(n)$ the same relationships among each other as the values in $x(n)$. We should point out that this reordering step attempts to simulate the influence of the ZMNL function of Figure 6.31. The procedure is taking a signal with an incorrect, whitened PDF ($x(n)$) and mapping it onto a signal $x_o(n)$ that consists of samples drawn directly from the target PDF. Thus, a major advantage of this approach is that a functional form of the ZMNL function need not be specified. This was also pointed out in both [69] and [63].

Algorithm I: Algorithm for generating a sequence of N observations x that are consistent with a user-specified power spectral density Sxx(ω) and user-specified probability distribution $p_X(x)$ (in this example a Rayleigh distribution is assumed, see line 4).

```
Sxx(N/2+1)=0;                          % zero out the DC component (remove mean)
Xf=sqrt(2*pi*N*Sxx/dt);                % Convert PSD to target Fourier amplitudes
Xf=ifftshift(Xf);                      % Put in Matlab format
x=raylrnd(N,1);                        % generate N iid samples conforming to p(x)
vs=(2*pi/N/dt)*sum(Sxx)*(N/(N-1));     % Get signal variance (as determined by PSD)
x=x*sqrt(vs/var(x));                   % guarantee new data match this variance
mx=mean(x); x=x-mx;                    % subtract the mean
[xo,indx]=sort(x);                     % store sorted signal xo with correct p(x)
k=1;                                   % initialize counter
while(k)
   R=fft(x);                           % Compute FT
   Rp=atan2(imag(R),real(R));          % Get phases
   x=real(ifft((exp(i.*Rp)).*abs(Xf)));  % Give signal correct PSD
   [zx,indx]=sort(x);                  % Get rank of signal with correct PSD
   x(indx)=xo;                         % rank re-order (simulate nonlinear transform)
   k=k+1;                              % increment counter
   if(indx==indxp) k=0; end            % if we converged, stop
   indxp=indx;                         % re-set ordering for next iter
end
x=x+mx;                                % Put back in the mean
```

The act of shuffling the $x(n)$ will, however, alter the Fourier amplitudes and hence the PSD. It was clearly shown by Schreiber [68] that one can simply repeat the described procedure, correcting for the PSD first (keep phases, but replace Fourier amplitudes), then the PDF (rank reorder). For univariate data, the procedure does converge to the point where no more reordering is possible. Thus, when the rank obtained in the final step matches the rank from the previous iteration, the algorithm terminates. Finally, after the last iteration, the signal mean is added back in, i.e., $x(n) = x(n) + \bar{x}_o$.

Because the final signal consists of samples drawn directly from the PDF of interest, there is an exact match between the target PDF and that obtained. The error in the procedure is therefore left in the signal PSD, which will be slightly different from the target PSD. The important questions, of course, are (i) how much error can one expect? and (ii) does it depend on the class of PDF? PSD? The algorithm is effectively solving a constrained optimization problem, where deviations from the target PDF and PSD are being minimized. The work of Filho and Yacoub [63] addressed the issue of algorithm convergence. They were able to prove that that the algorithm asymptotically converges to the correct PSD as the number of data N becomes large. Thus, we can expect the error to monotonically decrease with the number of data. As for the second issue, few restrictions need be placed on the type of PSD or PDF other than that they follow the compatibility conditions as put forth by Liu and Munson [54]. Practically, these restrictions state that very strongly auto-correlated data (sharp spectral peaks) coupled with strongly asymmetric PDFs (e.g., exponential) can sometimes be incompatible and cannot be realized. An example of such a process is provided shortly. Otherwise, there are no restrictions on the algorithm. We demonstrate, for example, that processes with multiple spectral peaks do not pose a problem for the algorithm, nor do processes with nonstandard PDFs. While undoubtedly a few counterexamples can be found, the algorithm works extremely well for most random processes found in engineering applications.

The algorithm is quite simple to implement as it involves only repeated applications of the FT and a sorting routine. Both of these can be found in most standard software libraries. Furthermore, there are no parameters to set or thresholds to specify. The practitioner simply provides the spectral density function, the form of $p_X(x)$, and a chosen discretization, that is, the number of points N and sampling interval Δ_t. The Matlab code for implementing this algorithm is given in the box (Algorithm I). The only line of the code that the practitioner will need to alter is the line where data from a specified distribution are generated with the function "raylrnd" (line 3). This assumes that we seek data that follow a Rayleigh distribution for $p_X(x)$. In Matlab, one might also choose the functions "chi2rnd" for chi-squared distributed variables or "exprnd" for exponentially distributed random variables. The generation of other distributions for which a simple algorithm is not available is discussed in one of the following examples. For very large signals ($N > O(10^5)$), the practitioner may wish to stop the iterations before strict convergence is achieved in the reordering process. This is easily accomplished by simply changing the "while" loop to a "for" loop over some prespecified number of iterations.

As a final point, we point out that solutions to the colored noise-generation problem bear close resemblance to the generation of the so-called surrogate sets, an approach we use heavily in Section (8.6) to determine whether an observed signal is consistent with a linear random process [70]. The goal of most surrogate algorithms is to produce a new (surrogate) signal with the same PSD and marginal PDF as the original data, precisely the task we have just finished discussing.

6.7.4   Examples

This section demonstrates how the algorithm of the previous section can be used to generate colored, non-Gaussian data. For the following three examples, the two-sided PSD function is

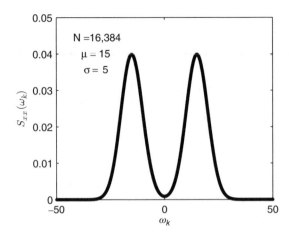

Figure 6.35 Two-sided spectral density function based on a mixed-Gaussian distribution with $\mu = 15, \sigma = 5$. The mixed-Gaussian PDF provides a convenient way to generate a range of two-sided spectral density functions. *Source*: Reproduced from [71], Figure 2, with permission of Elsevier

given by the mixed-Gaussian PDF

$$S_{XX}(\omega) = \frac{1}{2\sqrt{2\pi}\sigma}\left(e^{-\frac{(\omega-\mu)^2}{2\sigma^2}} + e^{-\frac{(\omega+\mu)^2}{2\sigma^2}}\right). \tag{6.194}$$

In general, PDFs are a good model for spectral density functions. A normalized spectral density function, in fact, shares all the mathematical properties of a probability density (nonnegative, integrates to unity, etc.) [19]. The mixed Gaussian has the added benefit of being symmetric, as is required of a two-sided spectral density. Because the function integrates to unity, we have $\sigma_X^2 = 1$. Figure 6.35 shows a sample two-sided spectral density function, obtained by setting $\mu = 15, \sigma = 5$ in Eq. (6.194). By adjusting μ, σ, this function can produce a PSD consistent with a low-pass, high-pass, or band-pass filtered output (Figure 6.35, for example, shows a two-sided "band-pass" PSD function with energy concentrated at roughly 15 Hz).

Next, we turn our attention to generating signals consistent with a particular spectral representation and PDF. The first example focuses on the simple case of generating a marginally Gaussian distributed signal with a low-pass spectral representation. Figure 6.36 shows the results of applying the algorithm. The PDF is, by definition, matched to the target realization. Spectrally, the algorithm is also able to accurately match the desired characteristics. For this example, $N = 4096$ points were used in generating the sample and the PSD was defined by $\mu = 25, \sigma = 25$.

As a second example, consider the task of simulating a Rayleigh-distributed random process with narrow bandwidth spectral properties. In this example, the PSD was sampled at $N = 8192$ points and defined by the parameters $\mu = 15, \sigma = 1$. Figure 6.37 shows the generated signal, the probability distribution, and the power spectrum. The specified PSD and PDF are again fairly easy for the algorithm to match. In general, any PDF that does not vary dramatically from Gaussian is easy to match, almost regardless of the spectral characteristics. Spectrally,

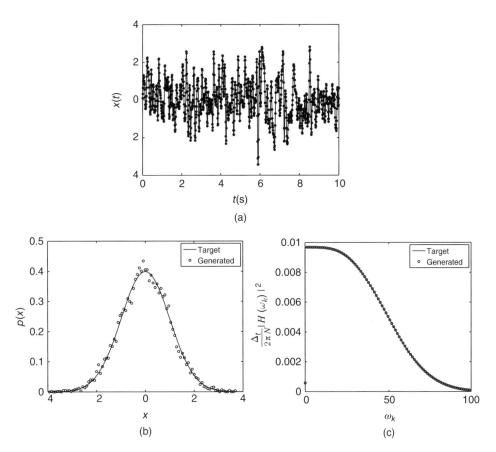

Figure 6.36 Simulation of a colored, Gaussian random process. (a) Time-series segment. (b) Probability density function. (c) Power spectral density function. *Source*: Reproduced from [71], Figure 3, with permission of Elsevier

very narrow peaks can be difficult to resolve, particularly if the PDF is far from Gaussian as is shown in the next example.

Consider the case of an exponentially distributed random process. The exponential distribution is highly non-Gaussian and is therefore particularly susceptible to the "whitening" effect described earlier. Figure 6.38 shows the results of trying to match both the narrowband PSD from the previous example and the exponential probability density. The PDF is matched exactly; however, the algorithm is unable to produce the correct spectral peak height. Nonetheless, the agreement is quite good for even this difficult case.

In the next few examples, we turn our attention to the simulation of a few naturally occurring stochastic processes. We require this capability in Chapter (10) when simulating the influence of ambient excitation on structural vibration. Assume we wish to generate a time series consistent with the PSD and PDF of a random process that describes ocean wave surface height $H(t)$ at a given point in space. In Reference [72], it was mentioned that the

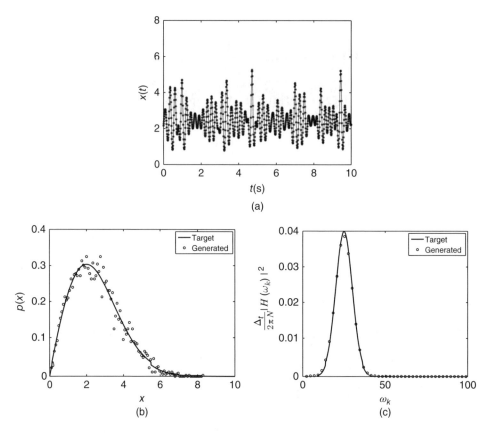

Figure 6.37 Simulation of a colored, Rayleigh-distributed random process. (a) Time-series segment. (b) Probability density function. (c) Power spectral density function. *Source:* Reproduced from [71], Figure 4, with permission of Elsevier

PDF of this random process is traditionally described by the Rayleigh distribution, while the PSD (in $m^2 \cdot s$) is described by the Pierson–Moskowitz distribution [72, 73] giving

$$S_{HH}(\omega) = \frac{A_0}{\omega^5} \exp\left[-\frac{4A_0}{H_s^2}\omega^{-4}\right]$$

$$p_H(H) = 2\left(\frac{H}{H_{rms}}\right) \exp\left[\left(\frac{H}{H_{rms}}\right)^2\right]. \qquad (6.195)$$

The results of applying the algorithm are shown in Figure 6.39 for the parameters $A_0 = 0.7796 \ (m^2/s^4)$ and $H_s = 5$ (m). The parameter H_{rms} in the probability distribution can be fixed arbitrarily as it represents the wave height variance specified via $S_{HH}(\omega)$. Thus, we generate a sequence of Rayleigh-distributed random variables and scale them according to the needed variance (line 4 of Algorithm I).

We should also point out that the algorithm is not limited to unimodal spectra. Consider again a Rayleigh-distributed random process, but let the process be described by a bimodal

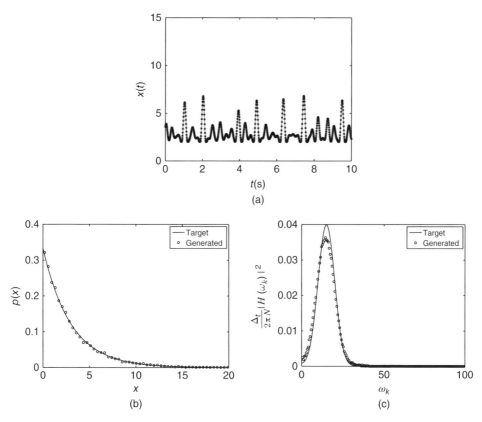

Figure 6.38 Simulation of a colored, exponentially distributed random process. (a) Time-series segment. (b) Probability density function. (c) Power spectral density (PSD) function. In this last case, the constraints are such that the PSD function is difficult to match; however, the approximation is still quite good. *Source*: Reproduced from [71], Figure 5, with permission of Elsevier

spectral distribution

$$S_{XX}(\omega) = \sum_{i=1}^{2} 0.11 H_i^2 T_i (T_i \omega/(2\pi))^{-5} \exp\left[-0.44(T_i\omega/(2\pi))^{-4}\right] \gamma^{\exp\left[-(1.296 T_i\omega/(2\pi)-1)^2/(2\sigma^2)\right]}$$

$$(6.196)$$

This spectrum was introduced by Soares [74] as a spectral model for wave height that incorporates both wind-induced and swell components. This model was shown to provide a good fit for a number of data sets from the North Atlantic and the North Sea and is used again in Chapter (10) in the generation of various wave loading conditions. Here, we use the model with $H_1 = 0.521$ (m), $H_2 = 0.253$ (m), $T_1 = 0.774$ (s), $T_2 = 1.625$ (s), $\gamma = 3$, and $\sigma = 0.1$. Figure 6.40 shows the algorithm has no difficulty in generating the desired Fourier amplitudes and probability distribution.

There is the possibility, however, that the random process being simulated will not admit a simple algorithm for generating samples of the specified distribution. One's ability to easily generate random variables from a distribution depends on whether the desired CDF

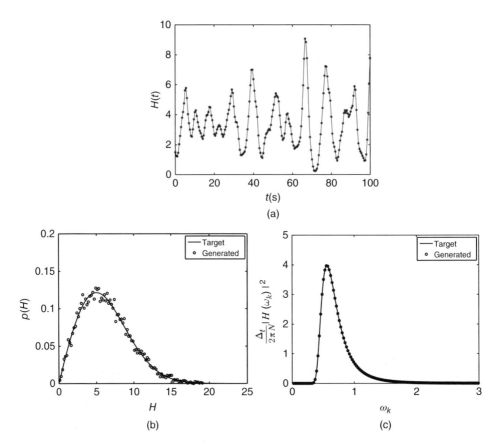

Figure 6.39 Simulation of a random process conforming to a Rayleigh probability distribution and the Pierson–Moskowitz frequency distribution. (a) Time-series segment. (b) Probability density function. (c) Power spectral density function. *Source*: Reproduced from [71], Figure 6, with permission of Elsevier

is invertible or whether there exists a known relationship between the desired variable and another, easy-to-generate distribution. For example, an exponentially distributed random variable has a CDF $P_X(x) = 1 - e^{-\lambda x}$, which can be generated by taking

$$x = -\ln(U)/\lambda \tag{6.197}$$

where U is a uniform random deviate taken from the interval $U \in [0, 1]$. Uniform random numbers are easily generated on a digital computer, thus knowledge of P_X^{-1} is all that is required to generate the samples x. Other distributions can be expressed as simple functions of a simple-to-generate distribution. The Rayleigh distribution, for example, can be obtained by summing two normally distributed random variables and then taking the square root.

There are cases, however, when one does not know P_X^{-1} and thus cannot employ the above-described approach to random number generation. In this case, one can still generate samples from the needed distribution using *rejection sampling*. The basic idea behind rejection sampling is to "carve" the needed target distribution $p_X(x)$ out of an existing candidate distribution $c_X(x)$ that is easily sampled. Rejection sampling requires choosing a constant

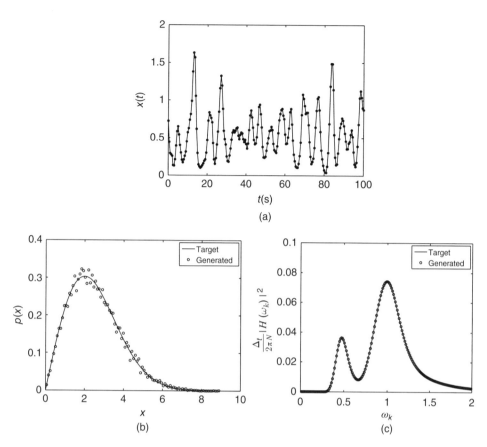

Figure 6.40 Simulation of a random process conforming to a Rayleigh probability distribution and the bimodal frequency distribution of Soares [74], used in describing structural loading due to deep-water ocean waves. (a) Time-series segment. (b) Probability density function. (c) Power spectral density function. *Source*: Reproduced from [71], Figure 7, with permission of Elsevier

M such that $p_X(x) \leq M_X(x)$ over the entire range of $p_X(x)$. The rejection sampling algorithm proceeds by repeating the following procedure until the required number of samples have been generated. First, draw a sample from $x \sim c_X(x)$ and calculate the ratio $r = p_X(x)/(Mc_X(x))$. Proper choice of M leads to $0 \leq r \leq 1$. Next, a Bernoulli trial is conducted with success probability r. That is to say, we draw a uniform random number $R \sim U(0,1)$ and keep the sample x as an observation from $p_X(x)$ if the trial was successful, i.e., if $r < R$, otherwise discard. The process is repeated until the required number of samples of $p_X(x)$ have been generated. The candidate distribution $c_X(x)$ can be thought of as the original block of material from which we chip away the rejected samples to get $p_X(x)$. Obviously, it is advantageous to start with a $c_X(x)$ that is close to, but always greater than, $p_X(x)$.

Consider the above-described wave height example with the Pierson–Moskowitz PSD, but with a modified PDF given by Mendez and Castanedo [75]

$$p_H(H) = \frac{2\phi^2(k)\xi}{(1 - \kappa\xi)^3} \exp\left[\phi^2(\kappa)\left(\frac{\xi}{1 - \kappa\xi}\right)^2\right] \tag{6.198}$$

where $\xi = H/H_{rms}$ and $0 \leq \kappa < 1$ is a dimensionless parameter controlling the shape of the wave height distribution. For $\kappa = 0$ we recover the Rayleigh distribution from the previous example, but for $\kappa > 0$ we require samples from a very different distribution.

Algorithm II (rejection sampling): Given a target PDF p(.) and a uniform candidate distribution with support $[a, b]$ and amplitude M, return samples of p(.) in the vector x

```
a=0.0;b=30.0;                         % Set constants for candidate distribution
x=zeros(N,1);                         % initialize vector for target distribution
M=1.48*(b-a);                         % fix inequality constraint M
k=1;                                  % initialize counter
while(k≤N)
    testpt=a+(b-a)*rand;              % generate a candidate point
    r=p(testpt)/(M/(b-a));            % assess ratio
    if(rand<r) x(k)=testpt; k=k+1;end % accept new point with uniform prob.
end
```

Here, the function $p(\cdot)$ in line (6) is simply the target distribution (in this case, Eq. 6.198). For this example, we use a uniform distribution $c_X(x) = U(a,b) = 1/(b-a)$ and set $a = 0$ and $b = 30$. Thus, the constant M must be chosen such that $M/(b-a) > p(H) \; \forall \; H$. Differentiating Eq. (6.198) and setting equal to zero yields the maximum value $p_{max} = 1.48$; thus, $M \geq 1.48 \times (b-a)$ (this value will be different for different applications, of course). Technically, we should have chosen $c_X(x)$ to have support $[0, \infty]$; however, by setting $b = 30$, we are effectively assuming $p(H) = 0$ for $H > 30$. The output of this algorithm, $x(n)$, $n = 1 \cdots N$, replaces line (4) of Algorithm I if a distribution is needed, for which a simple random number generator does not exist.

Figure 6.41 shows the resulting time series, probability distribution, and frequency distribution obtained by using Algorithm I in conjunction with the rejection sampling to generate the i.i.d sequence from $p(H)$. For the PDF we used the shape parameter $\kappa = 0.5$ in Eq. (6.198) and used Algorithm II to generate $N = 8192$ samples of that distribution. The resulting time series clearly matches both the specified PSD and probability distribution.

We have just described a rather simple algorithm for generating random processes with a certain marginal probability distribution and PSD function. The algorithm involves only repeated application of the FT and inverse FT and, as such, should be relatively easy to implement. Although there are certain combinations of probability distribution and spectral density that are difficult to replicate, the approach is capable of faithfully generating a wide variety of colored, non-Gaussian signals. The algorithm also relies on the practitioner's ability to generate samples from the target probability distribution. In the instance that standard methods are not applicable, we suggest using a simple Monte Carlo rejection sampling technique to provide the needed samples.

6.8   Stationarity Testing

All of the estimation techniques we have described thus far rely on the assumption of an ergodic (therefore stationary) random process. Indeed, these concepts underlie much of what we do in signal processing. However, to this point we have said nothing about how one might go about testing the validity of this assumption given a particular realization of a random process.

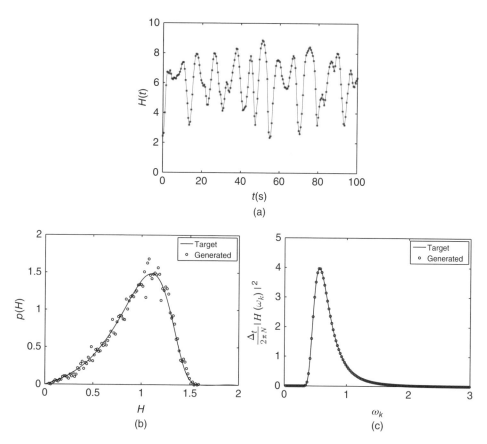

Figure 6.41 Simulation of a random process conforming to the probability distribution given by Eq. (6.198) and the Pierson–Moskowitz frequency distribution. (a) Time-series segment. (b) Probability density function. (c) Power spectral density function. *Source*: Reproduced from [71], Figure 8, with permission of Elsevier

We can either accept this assumption based on visual inspection of the signal (as is frequently done) or attempt to be a bit more rigorous and develop a formal test for the property. In what follows, we present two such tests, both of which can be applied to a single realization of a random process.

Before proceeding, it is worth recalling (Eq. 3.14) that the strict definition of stationarity implies that the entire joint PDF is invariant to shifts in the data record (e.g., time or space). It is quite difficult, as one might imagine, to verify strict stationarity as it requires a test for shift invariance of all joint moments. However, a test for weak, or "wide-sense" stationarity is much easier to conceive. Recall that a "weakly" stationary random process is one where both the mean and covariance are shift invariant. It is often the case that the practitioner tests for weak stationarity and then presumes strict stationarity based on the results. This may or may not be a good idea. Certainly, if the analysis being conducted relies on second-order statistics, such a procedure seems reasonable. However, in the case of other, higher order properties (e.g., the bispectrum), such an assumption may be unwarranted. That is to say, a random process could appear stationary in covariance and mean but be nonstationary with respect to third (or higher) moments in the sequence. We view such a situation as unlikely, however, and both of the tests described next test the assumption of a weakly stationary random process only.

6.8.1 Reverse Arrangement Test

The first test is described succinctly in Bendat and Piersol [10] and is referred to as a "reverse arrangement" test. We have found this test to be both simple and useful and therefore summarize its implementation here and provide an example.

Given a sequence of discrete observations of random variable $x(t_n), n = 0, \cdots, N - 1$, we wish to assess whether the underlying random process $\mathbf{X}$ is weakly stationary, that is, stationary in mean and covariance. An informal way to check for stationarity is to test for the presence of a trend in either of these properties. Begin by dividing the sequence into P nonoverlapping blocks of length $M = N/P$ (we assume here that N is an integer multiple of P). For each block, estimate the mean or covariance, that is, form

$$y_p = \frac{1}{M} \sum_{m=1}^{M} x(t_{pM+m}), p = 0, \cdots, P$$

or

$$y_p = \frac{1}{M-1} \sum_{m=1}^{M-d} (x(t_{pM+m}) - \bar{x}_p)(x(t_{pM+m+d}) - \bar{x}_p), p = 0, \cdots, P \tag{6.199}$$

where d is an integer measure of delay (e.g., set $d = 0$ if the goal is to test stationarity in variance). In either case (mean or covariance), we denote these sequences $y_p, p = 1 \cdots P$. In what follows, we assume that each of the y_p are independent random variables. In this case, we may follow [10] and define

$$a_{pq} = \begin{cases} 1 \text{ if } y_p > y_q \\ 0 \text{ otherwise} \end{cases} \tag{6.200}$$

as the number of times $y_p > y_q$ for every $p < q$. The total number of times this occurs for each block p is

$$A_p = \sum_{q=p+1}^{P-1} a_{pq}, p = 0, \cdots, P - 2 \tag{6.201}$$

and for all blocks combined

$$A = \sum_{p=0}^{P-2} A_p. \tag{6.202}$$

This new random variable A is normally distributed (by the central limit theorem) with mean and variance given by [10]

$$\mu_A = \frac{D(D-1)}{4}$$

$$\sigma_A^2 = \frac{D(2D+5)(D+1)}{72} \tag{6.203}$$

Thus, we may test for the presence of a trend in either the mean or variance at a prescribed level of significance $\alpha \in [0, 1]$ by checking the interval inequality (see Section 6.9.3)

$$\left[\mu_A - Z_{\alpha/2}\sigma_A < A < \mu_A + Z_{1-\alpha/2}\sigma_A\right] \tag{6.204}$$

where $Z_{\alpha/2}$ is the inverse standard normal distribution evaluated at $\alpha/2$. While checking for a trend in mean or variance is certainly not guaranteed to identify nonstationarity, it is certainly capable of detecting one particular class of nonstationary processes that does occur in practice.

As an example, consider a simple mechanical system governed by the second-order differential equation

$$m\ddot{x} + c\dot{x} + (k + \epsilon t^2)x = F(t) \tag{6.205}$$

where m, c, k are the usual mass, damping, and stiffness constants. The parameter ϵ has units $N/m/s^2$ and thus defines a nonstationarity in the system stiffness. The forcing $F(t)$ is taken as the output of a joint normally distributed random process with constant PSD $S_{FF}(f) = 0.01, N^2/Hz$. In this case, the variance of the stationary response ($\epsilon = 0$) is the integral of the response PSD function and is shown in Appendix (B) to be

$$E[X^2] = \frac{\pi S_{FF}(0)}{ck} \tag{6.206}$$

so that an increase in stiffness decreases the variance of the signal. Figure 6.42 shows the numerically generated signal $x(t_n), n = 0, \cdots, 2^{15} - 1$ for $\epsilon = 0.0001$ and $\epsilon = 0.01$, respectively. Other parameters were set to the values $m = 1.0\,kg$, $c = 3\,N \cdot s/m$, and $k = 100\,N/m$, while a sampling interval of $\Delta_t = 0.01\,s$ was assumed.

As time progresses, we see the expected decrease in variance. We applied the above-described trend test in both mean and variance using a block size of $P = 1310$ points ($D = 25$) and a confidence interval defined by $\alpha = 0.05$. The results are shown in Figure 6.43 for varying degrees of nonstationarity. For each level of nonstationarity, 10 separate realizations were analyzed and the associated A values computed. The mean of the signal is unchanged with time ($E[X(t)] = 0$); however, the variance clearly exhibits a

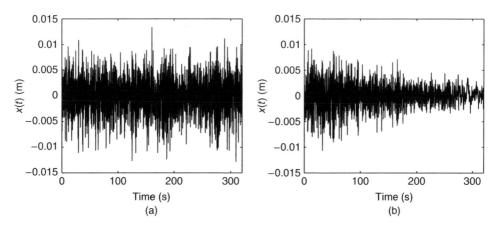

Figure 6.42 (a) Nonstationarity time series with $\epsilon = 0.0001$ and (b) nonstationary time series with $\epsilon = 0.01$

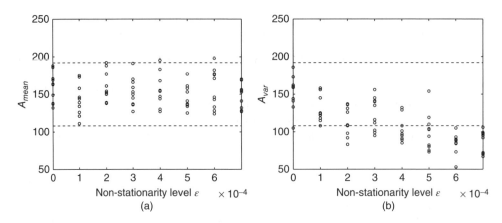

Figure 6.43 Progression of A with degree of nonstationarity for both mean (a) and variance (b) for the $\epsilon = 0.01$ signal (Figure 6.42b)

trend that is inconsistent with the stationary hypothesis for $\epsilon > 0.0005$. Smaller values of ϵ can sometimes produce too small a nonstationarity to detect. The approach is both simple and general in the sense that we could have tested for a trend in any statistical property.

6.8.2   Evolutionary Spectral Testing

The second test for stationarity is more formal in the sense that it looks for more than simply a trend in statistical properties. Instead, it focuses on the spectral properties of a nonstationary random process. We know from Chapter (3) that there exists an equivalence between spectral density functions and second-order statistical properties for a stationary random process. Thus, if spectral properties are determined to be stationary, we can at least claim a wide-sense stationary process.

6.8.2.1   Evolutionary Spectral Density Function

This particular test was developed by Priestly and Rao [76] and was based on Priestly's earlier work in "evolutionary spectra" [77]. The test begins with a model for a zero-mean, nonstationary random process

$$x(t) = \int_{-\infty}^{\infty} e^{i2\pi ft} A_t(f) dZ(f) \tag{6.207}$$

where $A_t(f)$ is a function of both time "t" and frequency, while $dZ(f)$ is a stationary, complex random process as discussed in Section 3.3, Eq. (3.43). Given this representation, we write for the *evolutionary PSD* function

$$S_{XX}(t,f)df = |A_t(f)|^2 E[|dZ(f)|^2] \tag{6.208}$$

Clearly, if the time dependence in $A_t(f)$ is eliminated, we have a stationary random process and $S_{XX}(t,f) \to S_{XX}(f)$. However, a continuously varying time dependence in the filter characteristics will inevitably result in a modulation of the spectral density. Moreover, a time-dependent PSD is an indication of a nonstationary auto-covariance function, hence a nonstationary random process. The question is, can we form accurate estimates of (6.208) and can we further use those estimates to infer the presence of the time dependency?

To explore this question, we first note that the frequency modulating function can be further represented by $A_t(f) = \int_{-\infty}^{\infty} e^{i2\pi u} dH_f(u)$. That is to say, for any frequency f the time-dependent characteristics of the modulation can be described by the Fourier amplitudes $dH_f(u)$. It turns out that this test for nonstationarity will hinge on the properties of the $dH_f(u)$.

First, consider the extreme case that we "freeze" the properties of $A_t(f)$ at time t^*. In this case, the frequency domain properties of the signal are not being modulated in time, that is to say, $dH_f(u) = \delta(u)$. Moreover, we could appropriately view the spectral function $A_{t^*}(f)$ as describing a linear, time-invariant filter acting on the stationary signal $s(t) = \int_{-\infty}^{\infty} e^{2\pi i f t} dZ(f)$. That is to say, we could write $A_{t^*}(f) = \int_{-\infty}^{\infty} h_{t^*}(\tau) e^{-i2\pi f \tau} d\tau$ and, finally, $x(t) = \int_{-\infty}^{\infty} s(t) h_{t^*}(t - \tau) d\tau$.

Now consider the case where the modulation $A_t(f)$ is *slowly varying*. In this case, we have that $dH_f(u) \neq \delta(u)$; however, most of the energy in the modulation is concentrated around $u = 0$, i.e., $dH_f(u)$ is a narrow-band random process centered at $u = 0$. In fact, if this modulation is "slow enough," we could envision the discussion of the previous paragraph holding over some time interval. This is, in fact, the key argument used to form estimates of (6.208) and then subsequently testing for consistency of those estimates in time.

First, define our finite length (sampled) signal $x(t), t = 0, \cdots, T$. Following Priestly [76], we proceed as follows. Define a localized (in time) filter $g(u)$ and write for any frequency

$$U(t,f) = \int_0^T g(t - \tau) x(\tau) e^{-i2\pi f \tau} d\tau \tag{6.209}$$

which is recognized as simply the windowed FT. Two important characteristics are required of the window function

1. the "width" of $g(u)$ as measured by $B_g = \int_{-\infty}^{\infty} |u| |g(u)| du$ be much less than the width of the modulating function,

$$B_H = \max_f \left[\int_{-\infty}^{\infty} |u| |dH_f(u)| du \right]^{-1}, \tag{6.210}$$

i.e., $B_g / B_H < 1$. Should this hold, we are essentially making the same argument as earlier, namely, that we can locally "freeze" the properties of the signal and treat them as stationary. Priestly refers to this as a "semi-stationary" random process [77].

2. the filter be chosen such that

$$2\pi \int_{-\infty}^{\infty} |g(u)|^2 du = \int_{-\infty}^{\infty} |\Gamma(f)|^2 df = 1 \tag{6.211}$$

where $\Gamma(f) = \int_{-\infty}^{\infty} g(u) e^{-i2\pi fu} du$. This normalization is chosen so as not to affect the magnitude of the PSD estimate.

The function $U(t,f)$ clearly reflects time-varying spectral properties, however it is not the needed PSD (6.208). This latter quantity is defined as an expectation over many realizations. However, we know from earlier in this chapter that we may estimate a PSD from a single record by averaging estimates over portions of that record. To this end, introduce the weighting function $w_{T'} \geq 0$ such that

1. $w_{T'}(t) \to 0$ as $|t| \to \infty$

2. $\int_{-\infty}^{\infty} w_{T'}(t)dt = 1$ for all T'

3. $\int_{-\infty}^{\infty} w_{T'}(t)^2 dt < \infty$ for all T'

where the spectral properties are given by

$$W_{T'}(f) = \int_{-\infty}^{\infty} w_{T'}(t)e^{-i2\pi ft} \, dt \qquad (6.212)$$

and where

$$\lim_{T' \to \infty} \left\{ T' \int_{-\infty}^{\infty} |W_{T'}(f)|^2 df \right\} = C \qquad (6.213)$$

In this case, Priestly [77] defines the estimate

$$\hat{S}_{XX}(f,t) = \int_0^T w_{T'}(t-\tau)|U(\tau,f)|^2 d\tau. \qquad (6.214)$$

More importantly, it is also shown that [76]

$$E[\hat{S}_{XX}(t,f)] \approx S_{XX}(t,f)$$

$$E[(\hat{S}_{XX)}(t,f) - E[\hat{S}_{XX}(t,f)])^2] \approx \frac{C}{T'} \left\{ \int_{-\infty}^{\infty} |\Gamma(\theta)|^4 d\theta \right\} S_{XX}^2(t,f) \qquad (6.215)$$

that is, the estimate is approximately unbiased in the neighborhood of time t with known variance. These properties form the basis of the nonstationarity test that follows.

As a quick example, we consider the familiar spring-mass system $m\ddot{x} + c\dot{x} + kx = F(t)$, where $F(t)$ is taken as a joint normally distributed random process with unit variance. Parameters used were $m = 1.0, kg$, $c = 3.0, \text{N} \cdot s/m$, and $k = 100, \text{N/m}$. A signal $x(t_n), n = 1 \cdots 1024$ was generated by numerically integrating these equations with a sampling interval of $\Delta_t = 0.01s$, hence the PSD of the input was $S_{FF}(f) = 0.01, \text{N}^2/Hz$. We applied the above-described estimation procedure, replacing all integrals with summations over all $N = 1024$ points and replacing the windowed Fourier integral (6.209) with the discrete FT as described by Eq. (6.54). For the windowing parameters, we chose $h = 32$ and $T' = 256$. The resulting estimate for the evolutionary PSD is shown in Figure 6.44.

For this stationary case, we see that the signal power resides in the same frequency over all sampled times. For nonstationary signals this will not be the case, as will be demonstrated in the following test.

6.8.2.2 Testing Procedure

The first step is to choose the windowing functions $g(u)$ and $w_{T'}(u)$ required of the estimate (6.214). Again, following the lead of Priestly [76], we choose

$$g(u) = \begin{cases} \frac{1}{2\sqrt{h\pi}}, & |u| \leq h \\ 0, & |u| > h \end{cases} \qquad (6.216)$$

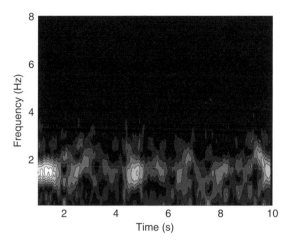

Figure 6.44 Evolutionary power spectral density estimate based on $N = 1024$ samples of a linear oscillator's response to a stationary random process input. The natural frequency of the oscillator is roughly 1.6 Hz and can be clearly seen to contain most of the signal power

and

$$w_{T'}(t) = \begin{cases} \frac{1}{T'}, & -\frac{1}{2}T' \le t \le \frac{1}{2}T' \\ 0, & \text{otherwise} \end{cases} \tag{6.217}$$

Both choices satisfy the above-stated requirements and further can be shown to yield $\int_{-\infty}^{\infty} |\Gamma(\theta)|^4 d\theta = \frac{2h}{3\pi}$ and $C = 2\pi$. The parameter h dictates the width of a rectangular windowing function, while T' specifies the width of a triangular weighting kernel of finite support. While other choices are certainly possible, we use these definitions in the subsequent testing procedure.

Given that the variance of the estimate is governed by (6.215), it can be demonstrated that the natural logarithm transformation

$$Y_{XX}(t,f) \equiv \log \hat{S}_{XX}(f,t) \tag{6.218}$$

produces a test statistic that has an approximate mean of $\bar{Y}_{XX}(t,f) = \log S_{XX}(t,f)$ and variance

$$var(Y_{XX}(t,f)) = \frac{C}{T'} \left\{ \int_{-\infty}^{\infty} |\Gamma(\theta)|^4 d\theta \right\}$$

$$= \frac{4h}{3T'} \equiv \sigma^2 \tag{6.219}$$

Moreover, this test statistic is approximately joint normally distributed [76]. Thus, by performing an ANalysis Of VAriance (ANOVA) we may check for the consistency of these estimates over both time and frequency. A complete description of the ANOVA test appears in many statistics references and so is not described in detail here. In short, the test seeks to determine the influence of different variables (termed "effects") on the variability in the observed data (in this case, our estimated test statistic $Y_{XX}(t,f)$. The "effects" in this case are "time" and "frequency" and we are particularly interested in the influence of the "time" parameter on our estimates, a significant influence being indicative of nonstationarity. Significant, in this case, means variability beyond that which we expect in our estimates.

More specifically, we perform a two-factor ANOVA on the data

$$Y_{ij} \equiv Y_{XX}(t_i, f_j) \tag{6.220}$$

where we have explicitly acknowledged the discrete nature of our estimate at times $t_i, i = 1 \cdots I$ and frequencies $f_j, j = 1 \cdots J$. The times/frequencies at which we perform the ANOVA may be selected from any point in the data record. However, it would seem sensible to define the time/frequency points to span the entire data record.

In a classical two-factor ANOVA test, one computes

$$S_T = J \sum_{i=1}^{I} (\bar{Y}_i - \bar{Y})^2$$

$$S_F = I \sum_{j=1}^{J} (\bar{Y}_j - \bar{Y})^2$$

$$S_I = \sum_{i=1}^{I} \sum_{j=1}^{J} (Y_{ij} - \bar{Y}_i - \bar{Y}_j + \bar{Y})^2 \tag{6.221}$$

where $\bar{Y} = \frac{1}{IJ} \sum_{i=1}^{I} \sum_{j=1}^{J} Y_{ij}$ and $\bar{Y}_i = \frac{1}{J} \sum_{j=1}^{J} Y_{ij}$ (similarly for $\bar{Y}_j$). Now, if each Y_{ij} is normally distributed with mean $\bar{Y}$, we may use the methods of Section 2.4 to show that $\bar{Y}_i - \bar{Y}$ will also be normally distributed with $\mu_Y = 0$ and $\sigma_Y^2 = \sigma^2/J$. Thus, the quantity $J(\bar{Y}_i - \bar{Y})/\sigma^2$ is standard normal with unit variance. Moreover, we can also use the methods of Section 2.4 to then show that $\frac{J}{\sigma^2} \sum_{i=1}^{I} (\bar{Y}_i - \bar{Y})^2$ is a chi-squared random variable with $I - 1$ DOFs, i.e., $S_T/\sigma^2 \sim \chi_{I-1}^2$. Given that we know this distribution we can ask whether the test statistic S_T/σ^2 is consistent with that distribution. We may similarly test to see that $S_F/\sigma^2 \sim \chi_{J-1}^2$ and $S_I/\sigma^2 \sim \chi_{(I-1)(J-1)}^2$.

Denote $P_\chi(x|n)$ as the χ_n^2 CDF which is conditional on the degrees-of-freedom parameter n. We are interested in values of our test statistic that are unlikely in the sense that they exceed the upper $\alpha \times 100$ percentile of expected values. For example, we may test $S_T/\sigma^2 > P_\chi^{-1}(\alpha|n)$ where P_χ^{-1} denotes the inverse chi-squared CDF. Following the classical ANOVA procedure, we first look at the "interaction" statistic and test

$$S_I/\sigma^2 > P_\chi^{-1}(\alpha|(I-1)(J-1)). \tag{6.222}$$

If this statement is true, we immediately conclude that the random process $x(t)$ is nonstationary. If it is not true, the structure of our test statistic implies that the random process is *uniformly modulated*, possessing the form $x(t) = a(t)x_o(t)$ where $x_o(t)$ is a stationary random process [76]. In this case, we proceed further and test

$$S_T/\sigma^2 > P_\chi^{-1}(\alpha|I-1) \tag{6.223}$$

exceedances indicating nonstationarity. Interestingly, should the interaction term be positive, this particular test can be used to see if particular frequencies are nonstationary in time. Denoting as $j_k, k = 1, \cdots, K$ as the discrete frequencies of interest, we test

$$\frac{1}{\sigma^2} \sum_{k=1}^{K} \sum_{i=1}^{I} (Y_{ij} - \bar{Y}_j)^2 > P_\chi^{-1}(\alpha | K(I-1)) \tag{6.224}$$

for this effect. Note that testing S_F/σ^2 is not usually of interest as we do not care whether there is variation in the frequency domain; in fact, for most systems of interest, we would expect it!

As an example, we reconsider the system (6.205) with $\epsilon = 0.1$, that is, a highly nonstationary signal. We form the evolutionary spectral density estimate as described in the previous section using $N = 1024$ points with $h = 32$ and $T' = 200$. The time-series and evolutionary spectral density estimate are shown in Figure 6.45.

We consider the set of discrete times $i = \{101, 251, 401, 551, 701, 851\}$ and frequencies $j = \{25, 100, 175, 250, 325, 400, 475\}$. Table (6.1) shows the test statistic Y_{ij} for each of the $I = 6$ and $J = 7$ time/frequency combinations. from which our test statistics are found to be $S_I/\sigma^2 = 163.9$ and $S_T/\sigma^2 = 20.74$. The respective χ^2 thresholds for $\alpha = 0.95$ are $P_\chi^{-1}(0.95|30) = 43$ and $P_\chi^{-1}(0.95|5) = 11.07$, respectively. Thus, we conclusively state that this random process is nonstationary.

As a final illustration of the procedure, we repeat the same numerical experiment that was used in evaluating the reverse arrangement test (see Figure 6.43). The time-series data consisted of $N = 1640$ observations and we chose the discrete frequencies $\{25, 100, 175, \cdots, 775\}$ and times $\{101, 251, 401, \cdots, 1451\}$ at which to form $Y_{ij}, i = 1, \cdots, 10, j = 1, \cdots, 11$. We also use the same parameters as before in our evolutionary spectral estimate, that is to say, $h = 32$ and $T' = 200$. The results are shown in Figure 6.46. As with the reverse arrangement test, this approach seems to pick up the presence of a nonstationarity at around the same level, $\epsilon = 0.0005$. This test, however, is on a more sound theoretical footing and is a true test for wide-sense stationary random processes (as opposed to a "trend" test). While the reverse arrangement test is easier to implement, the evolutionary spectral approach provides a

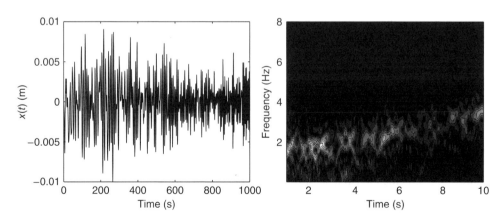

Figure 6.45 Evolutionary power spectral density estimate based on $N = 1024$ samples of a nonstationary oscillator response. The natural frequency is increasing with time, while the signal variance clearly drops

Table 6.1 Test statistic Y_{ij} estimated for various discrete time and frequency indices

Times\Frequencies	25	100	175	250	325	400	475
101	−12.88	−11.39	−14.76	−15.53	−16.50	−16.76	−17.12
251	−12.99	−11.12	−14.27	−15.30	−16.37	−16.51	−16.77
401	−13.03	−12.30	−13.60	−15.17	−16.80	−16.71	−17.24
551	−14.61	−12.67	−12 57	−15.09	−16.34	−16.79	−17.23
701	−14.76	−13.91	−12.70	−15.43	−16.57	−17.17	−17.37
851	−15.45	−14.15	−12.96	−14.47	−16.06	−17.17	−17.31

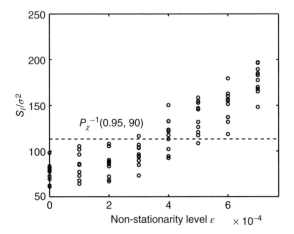

Figure 6.46 Progression of "interaction" statistic S_I/σ^2 as a function of nonstationarity level ϵ. For this experiment, we used $I = 10$ different time points and $J = 11$ different frequencies in the evaluation; hence, the relevant χ^2 test statistic for comparison is $P_\chi^{-1}(0.95|90)$, where we assume an $\alpha = 0.95$ level of significance

more powerful diagnostic for the kinds of time-series data observed in structural monitoring applications. We also note that beyond its usefulness as a means of testing for stationarity in time-series data, the evolutionary spectrum has seen use in SHM applications, perhaps most notably in diagnosing the presence of flaws in rotating machinery.

6.9 Hypothesis Testing and Intervals of Confidence

To this point, our discussion of random processes has focused largely on the quantities we use later to characterize structural response data. However, before leaving this chapter it is essential to discuss how these quantities can also be used to detect the presence of structural damage. Many of the aforementioned quantities, for example, Fourier coefficients, can be used as *test statistics* for comparing a "damaged" to a "healthy" structural response and deciding which of these two possible responses we observed. To understand this process, we need to understand how to formally test the hypothesis of structural damage and how to form intervals of confidence for our test statistic. In doing so, we introduce several important concepts required of the next chapter in structural parameter estimation.

6.9.1 Detection Strategies

Perhaps the most basic goal in identifying the state of a structure is to detect the presence of the damage. This is an easier problem than the one posed by identification (handled in Chapters 7 and 9) in that the solution is a binary decision, damaged or undamaged. In detection, we are given N observations $\mathbf{y} \equiv y(n)$, $n = 1 \cdots N$ from a structure in an unknown condition and asked to make the diagnosis. As we have done throughout this text, these data are modeled as a realization of a random process and are therefore described by a joint probability model, i.e., a joint PDF. In problems of detection, it is common to denote these models (likelihoods) as $p_{\mathbf{Y}_0}(\mathbf{y}|\mathcal{H}_0)$ and $p_{\mathbf{Y}_1}(\mathbf{y}|\mathcal{H}_1)$, which describe the data under the null (undamaged) hypothesis, $\mathcal{H}_0$, and alternative (damaged) hypothesis, $\mathcal{H}_1$, respectively. The quantity "model," denoted $\mathcal{H}$, is admittedly abstract. When we say (and write) "model," we could be referring to the specific functional form of a governing equation or simply a key parameter value that defines our hypothesis. In fact, for damage detection applications, $\mathcal{H}_0$ and $\mathcal{H}_1$ will often represent the same basic model (same functional form) and differ only in the value a particular parameter might take. For example, a beam will often be modeled to possess a stiffness given by the product of elastic modulus and mass moment of inertia $k \equiv EI$. We might consider $\mathcal{H}_0 \equiv k < 50$ MPa and $\mathcal{H}_1 \equiv k > 50$ MPa. Given available data, $\mathbf{y}$, the detection problem asks which of these statements is most likely true.

Of course, the uncertainty inherent in our observations means that this decision must be made probabilistically. We will never know precisely whether the structure is damaged, rather we speak in terms of the likelihood, or probability, that the structure is damaged. Such an assessment should consider the probability models $p_{H_0}(\mathcal{H}_0|\mathbf{y})$ and $p_{H_1}(\mathcal{H}_1|\mathbf{y})$, read as the probabilities that the structure is undamaged or damaged given the available data. Returning to the beam example, we would write $p_{K_u}(k_u|\mathbf{y})$ and $p_{K_d}(k|\mathbf{y})$, read as the PDFs associated with the "undamaged" and "damaged" stiffness values k_u, k_d. Figure 6.47 shows sample PDFs for the beam stiffness under both damaged and undamaged hypothesis. Obviously, if we *had* these distributions, the detection problem would be fairly simple and we could proceed in any number of ways. For example, we might plug in our observations and the values for k_u, k_d and simply ask whether

$$p_{K_u}(k_u|\mathbf{y}) > p_{K_d}(k_d|\mathbf{y}). \tag{6.225}$$

i.e., is the probability of having observed the undamaged stiffness greater than the damaged stiffness? Conceptually, this makes sense, however this approach is entirely impractical. First and foremost, the parameters of these distributions clearly will depend on the values k_u, k_d. The more damaged incurred, the further the latter parameter will be reduced, thereby shifting the damaged PDF to the left in Figure 6.47. In a typical damage detection problem, however, we will *not* know the values of the competing model parameters at the outset. We could always try to directly estimate the stiffness distribution, $\hat{p}_K(k|\mathbf{y})$, at which point we could easily make a judgment. For example, if we declared "damaged" to be a stiffness less than 30 MPa, we could compute the damage probability as $\int_0^{30} \hat{p}_K(k|\mathbf{y})dk$. In fact, we consider precisely this approach and methods for estimating the damage parameter distribution in Section (7.2.6). However, as we also see, this estimation problem can be extremely challenging and, for detection problems, not necessary.

Instead, we will proceed for the moment as if we did know the models that generated the data. The only uncertainty we are considering is therefore the uncertainty carried in the observations $\mathbf{y}$. Returning to the general case, the first thing to note is that the distribution $p_{H_0}(\mathcal{H}_0|\mathbf{y})$

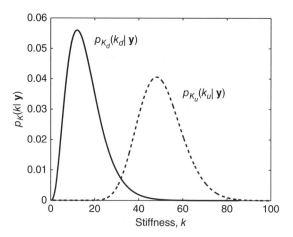

Figure 6.47 Probability density functions for both "damaged" and "undamaged" beam structures. In the former case, the stiffness has degraded and possesses a PDF of $p_{K_d}(k_d|\mathbf{y})$. The probability model for the healthy beam is $p_{K_u}(k_u|\mathbf{y})$. Both models are conditioned on the observed data and we would like to know which of these two PDFs is best supported by these observations. The problem is that these distributions depend on the underlying stiffness values, which we do not know *a priori*

is *not* the same as the likelihood $p_{\mathbf{Y}}(\mathbf{y}|\mathcal{H}_0)$. However, using the law of conditional probability (see Section 2.3), we can write

$$p_{H_0}(\mathcal{H}_0|\mathbf{y}) = \frac{p_{\mathbf{Y}}(\mathbf{y}|\mathcal{H}_0)p_{H_0}(\mathcal{H}_0)}{p_{\mathbf{Y}}(\mathbf{y})}$$

$$p_{H_1}(\mathcal{H}_1|\mathbf{y}) = \frac{p_{\mathbf{Y}}(\mathbf{y}|\mathcal{H}_1)p_{H_1}(\mathcal{H}_1)}{p_{\mathbf{Y}}(\mathbf{y})}. \tag{6.226}$$

These relationships form what is known as Bayes' rule in estimation theory, a subject we discuss at length in Section (7.2). The distributions $p_{H_0}(\mathcal{H}_0) \equiv P_{H_0}$ and $p_{H_1}(\mathcal{H}_1) \equiv P_{H_1}$ are referred to as *prior* PDFs (probabilities) for the two models and are specified by the practitioner at the outset. Our desired distributions on the left-hand side are referred to as the *posterior* probability distributions. In general, we can ask the question posed by (6.225) as

$$\frac{p_{H_1}(\mathcal{H}_1|\mathbf{y})}{p_{H_0}(\mathcal{H}_0|\mathbf{y})} > 1, \quad \text{or,}$$

$$\frac{p_{\mathbf{Y}}(\mathbf{y}|\mathcal{H}_1)P_{H_1}}{p_{\mathbf{Y}}(\mathbf{y}|\mathcal{H}_0)P_{H_0}} > 1 \tag{6.227}$$

where we have noticed that the joint PDF for the data disappears as it is common to both distributions. We have also noticed that the prior density ratio can equivalently be written as the ratio of prior probabilities P_{H_1}/P_{H_0} that our data were produced by the damaged and undamaged structures, respectively. In damage detection problems, the prior probabilities are usually assumed to be equal, hence the detection strategy is to compute the *likelihood ratio*

$$\lambda(\mathbf{y}) = \frac{p_{\mathbf{Y}}(\mathbf{y}|\mathcal{H}_1)}{p_{\mathbf{Y}}(\mathbf{y}|\mathcal{H}_0)} > 1. \tag{6.228}$$

Equivalently, taking the logarithm of both sides, we declare a detection if

$$\log[\lambda(\mathbf{y})] > 0. \tag{6.229}$$

The likelihood ratio (or log-likelihood) plays a very large role in detection and estimation theory as is evidenced by what follows in this and subsequent chapters (see, e.g., the Markov Chain Monte Carlo algorithm given in Section 7.2.3). The criteria (6.227) we have just derived is referred to in the detection literature as the *maximum a posteriori probability* (MAP) criterion.

However, perhaps a better detector for damage detection applications follows the "Neyman–Pearson" criterion. Rather than asking (6.225) as the relevant question, the Neyman–Pearson detector seeks to maximize the probability of detection (correctly detecting damage under $\mathcal{H}_1$) for a given, fixed probability of "false alarm" (declaring damage under $\mathcal{H}_0$). It can be shown [1] that for this detection strategy, one also compares the likelihood ratio to a threshold, that is,

$$\lambda(\mathbf{y}) = \frac{p_{\mathbf{Y}}(\mathbf{y}|\mathcal{H}_1)}{p_{\mathbf{Y}}(\mathbf{y}|\mathcal{H}_0)} > \lambda_0 \tag{6.230}$$

where the threshold λ_0 is set by the desired (acceptable) false-alarm probability. This detector is likely to be more useful for damage detection application where the practitioner is typically in a position to assign costs to false alarms and "missed" detections. In both detectors (MAP and Neyman–Pearson), the key is the likelihood ratio; the only real differences being the incorporation of prior information and the threshold for comparison. The former is, in fact, the same as the latter for (i) equal prior probabilities and (ii) $\lambda_0 = 1$. We proceed assuming the Neyman–Pearson detector as the one most relevant to the damage detection problem, noting that it is fairly straightforward to use the MAP criterion if desired. As with the MAP detector, one frequently compares the log-likelihood ratio to the threshold, $\lambda_0' = \log(\lambda_0)$

$$\log[\lambda(\mathbf{y})] > \lambda_0' \tag{6.231}$$

The log transformation is often helpful in such problems as the functional form of many common probability distributions includes an exponential term; the logarithm is a convenient way to eliminate this term without changing the detector performance or the max/min of the likelihood function.

If we again return to our stiffness example, our detector becomes

$$\log\left(\frac{p_{\mathbf{Y}}(\mathbf{y}|k_d)}{p_{\mathbf{Y}}(\mathbf{y}|k_u)}\right) > \lambda_0' \tag{6.232}$$

That is to say, we would plug in our data to the log-likelihood function, along with values for k_u, k_d and see if the result is greater than λ_0', deciding in favor of $\mathcal{H}_1$ if it is. Of course, we still have not addressed what values to use for k_u, k_d in making this assessment, moreover we now have the problem of how to set λ_0. With regard to the former issue, one approach is to simply average over the unknown parameters, that is (assuming equal priors), compute

$$\lambda(\mathbf{y}) = \frac{E_{K_d}\left[p_{\mathbf{Y}}(\mathbf{y}|k_d)\right]}{E_{K_u}\left[p_{\mathbf{Y}}(\mathbf{y}|k_u)\right]} > \lambda_0 \tag{6.233}$$

However, again we are taking the expectation with respect to a distribution that we do not know! A far more practical approach is to compute the *generalized likelihood ratio*

$$\lambda(\mathbf{y}) = \frac{p_{\mathbf{Y}}(\mathbf{y}|\widehat{k}_d)}{p_{\mathbf{Y}}(\mathbf{y}|\widehat{k}_u)} > \lambda_0 \tag{6.234}$$

where

$$\widehat{k}_d, \widehat{k}_u = \max_{k_d, k_u} p_{K_d K_u}(\mathbf{y}|k_d, k_u) \tag{6.235}$$

are the *maximum likelihood estimates* of the unknown parameters. This topic of ML estimation is covered at length in the next chapter. The comparison of (6.234) is referred to in the literature as the *generalized likelihood ratio test*, or GLRT.

At this point, it is useful to consider a specific example of the GLRT detector. Consider an undamped, spring-mass system driven with a sinusoidal input $F\cos(\omega t)$ where $|\omega - \omega_n| \gg 1$, that is to say, the system is being driven far from the resonant frequency. Let us also assume that the initial position is given by $x(0) = 0\,\mathrm{m}$. In this case, we have already seen that the steady-state solution is

$$x(t|k) = \frac{F}{k - m\omega^2}\cos(\omega t)$$
$$= A(k)\cos(\omega t) \tag{6.236}$$

so that the amplitude is a function of the stiffness, k. Using a laser vibrometer (or some other displacement measuring device), we record the position of the mass, $y(n)$, at discrete times $t = n\Delta_t, n = 0 \cdots N$ and assume the observation model

$$y(n) = x(n|k) + \eta(n) \tag{6.237}$$

where each $\eta(n)$ is an independent, normally distributed random variable with zero mean and variance σ^2. We have left the dependence of the solution on the unknown stiffness parameter k. The probability distribution for the entire observed sequence is therefore

$$p_{\mathbf{H}}(\eta) = \frac{1}{(2\pi\sigma^2)^{N/2}} \prod_{n=0}^{N-1} e^{-\frac{1}{2\sigma^2}\eta(n)^2}$$
$$= \frac{1}{(2\pi\sigma^2)^{N/2}} e^{-\frac{1}{2\sigma^2}\sum_{n=0}^{N-1}(y(n)-x(n|k))^2}$$
$$\equiv p_{\mathbf{H}}(\mathbf{y}|k) \tag{6.238}$$

where our notation now alludes to the fact that the noise distribution is what gives rise to the form of the joint data distribution. The dependence of our data on the parameter k is also left in the notation. Because the other model parameters are known, we may estimate the signal amplitude and note that this uniquely determines the needed stiffness. First we obtain the MLEs for the unknown parameters by maximizing with respect to $A(k)$

$$
\begin{aligned}
\widehat{A}(k) &= \max p_{\mathbf{H}}(\mathbf{y}|k) \\
&= \max \log(p_{\mathbf{H}}(\mathbf{y}|k)) \\
&= \max \left\{ \sum_{n=0}^{N-1} (y(n) - A(k)\cos(\omega n))^2 \right\} \\
&= \max \left\{ \sum_{n=0}^{N-1} y(n)^2 - 2A(k) \sum_{n=0}^{N-1} y(n)\cos(\omega n) + A(k)^2 \sum_{n=0}^{N-1} \cos^2(\omega n) \right\}
\end{aligned}
$$
(6.239)

which, taking the derivative with regard to $A(k)$ and setting equal to zero, yields

$$
\widehat{A}(k) \sum_{n=0}^{N-1} \cos^2(\omega n) = \sum_{n=0}^{N-1} y(n)\cos(\omega n)
$$

$$
\widehat{A}(k) = \frac{\sum_{n=0}^{N-1} y(n)\cos(\omega n)}{\sum_{n=0}^{N-1} \cos^2(\omega n)}
$$

$$
\widehat{A}(k) = \frac{2}{N} \sum_{n=0}^{N-1} y(n)\cos(\omega n).
$$
(6.240)

This is immediately recognized as a discrete estimator of the real Fourier coefficient (3.28). (Note that in the last step of (6.240), we have made the large-N approximation that $\sum_{n=0}^{N-1} \cos^2(\omega n) \approx N/2$.) In fact, what we have just derived is a particular instance of the so-called *matched filter* in detection theory. If, for example, our structural response was *not* a sinusoid, but instead any deterministic function $s(n)$ with unknown amplitude, we would have the amplitude estimate $\widehat{A} = \frac{\sum_{n=0}^{N-1} y(n)s(n)}{\sum_{n=0}^{N-1} s^2(n)}$. What Eq. (6.240) says is that to estimate the stiffness we can first estimate the response amplitude via Fourier analysis, and then simply set

$$
\widehat{k} = \frac{F}{\widehat{A}} + m\omega^2.
$$
(6.241)

If we denote the "undamaged" and "damaged" amplitudes as A_u, A_d, respectively, the log-likelihood ratio required of the GLRT test becomes

$$
\begin{aligned}
\log(\lambda(\mathbf{y})) &= \left(-\frac{1}{2\sigma^2} \sum_{n=0}^{N-1} (y(n) - \widehat{A}_d \cos(\omega n))^2 \right) - \left(-\frac{1}{2\sigma^2} \sum_{n=0}^{N-1} (y(n) - \widehat{A}_u \cos(\omega n))^2 \right) \\
&= -\frac{1}{2\sigma^2} \left[\sum_{n=0}^{N-1} y(n)^2 - N\widehat{A}_d \frac{2}{N} \sum_{n=0}^{N-1} y(n)\cos(\omega n) + \widehat{A}_d^2 \sum_{n=0}^{N-1} \cos^2(\omega n) \right.
\end{aligned}
$$

$$-\sum_{n=0}^{N-1} y(n)^2 + N\widehat{A}_u \frac{2}{N} \sum_{n=0}^{N-1} y(n)\cos(\omega n) - \widehat{A}_u^2 \sum_{n=0}^{N-1} \cos^2(\omega n) \Bigg]$$

$$= -\frac{1}{2\sigma^2} \left[-N\widehat{A}_d^2 + \frac{N}{2}\widehat{A}_d^2 + N\widehat{A}_u^2 - \frac{N}{2}\widehat{A}_u^2 \right] \qquad (6.242)$$

so that our GLRT detector becomes

$$\widehat{A}_d^2 - \widehat{A}_u^2 > \frac{2\sigma^2 \log(\lambda(\mathbf{y}))}{N/2} \qquad (6.243)$$

which says to compare the difference in estimated Fourier coefficient magnitudes to a threshold. Differences larger than this threshold indicate that the amplitude has exceeded a critical value which, in this case, indicates a loss of stiffness. The particular threshold we choose will depend on the degree to which we tolerate false alarms and also on the required POD. We discuss the performance of this detector at length in the next section.

This example illustrates a well-defined process for detection that has firm roots in probability theory. The GLRT detector is used frequently in radar and sonar problems and is certainly appropriate for use in structural dynamics problems as well. The difficulty of course, is that for all but the simplest structural dynamics problems it is challenging to analytically develop a test statistic based on the GLRT. The matched filter is certainly a useful detector in certain damage detection applications (see, e.g., Section (7.1.5) for estimating time of flight in ultrasonic applications), however for more complicated problems, derivations such as this one may simply not be possible.

Instead, in problems of detection it often makes sense to choose a quantity, s, that is easy to estimate but which is also clearly linked to the physics we are trying to capture. In keeping with the beam example, we might measure the beam's vibrational response to a broad-band input (broad spectral content), estimate the PSD using the approach described earlier in this chapter, and then set $s = \widehat{f}_1$ where $\widehat{f}_1$ is the estimated first natural frequency of the structure. We know from Chapter (4) that this frequency is directly related to the needed stiffness, and hence we might use natural frequency as a detection statistic. Note that the reason this did not make for a sensible test statistic in our earlier example is that we were driving the beam far from resonance, in which case only the response amplitude carried stiffness information. In short, our models should lead us toward a specific *detection statistic*, s, which we will use to make the final judgment, damaged or undamaged. Ideally, we will choose s so that

$$p_{\mathbf{Y}}(\mathbf{y}|\theta, s) = p_{\mathbf{Y}}(\mathbf{y}|s). \qquad (6.244)$$

This means that all of the information about our damage parameter of interest, θ, is contained in s. A statistic s chosen such that (6.244) holds is referred to as a *sufficient statistic*. A good goal from the perspective of damage detection should always be to choose the test statistic to be sufficient.

Often, the property (6.244) is not so straightforward to verify. However, it can be shown [1] that if we are able to factor the joint distribution

$$p_{\mathbf{Y}}(\mathbf{y}|\theta) = g(\widehat{\theta}(\mathbf{y})|\theta) f(\mathbf{y}) \qquad (6.245)$$

then $\widehat{\theta}$ is a sufficient statistic. Returning once more to our running example, we have that

$$p_{\mathbf{Y}}(\mathbf{y}|A) = \frac{1}{(2\pi\sigma^2)^{N/2}} e^{-\frac{1}{2\sigma^2}\sum_{n=0}^{N-1} y(n)^2} e^{\frac{AN}{2\sigma^2}\frac{2}{N}\sum_{n=0}^{N-1} y(n)\cos(\omega n) + A^2 \sum_{n=0}^{N-1} \cos^2(\omega n)}$$

$$= \underbrace{\frac{1}{(2\pi\sigma^2)^{N/2}}}_{f(\mathbf{y})} \underbrace{e^{-\frac{1}{2\sigma^2}\sum_{n=0}^{N-1} y(n)^2} e^{\frac{AN}{2\sigma^2}\hat{A}+A^2 \sum_{n=0}^{N-1} \cos^2(\omega n)}}_{g(\hat{\theta}(\mathbf{y})|\theta)} \qquad (6.246)$$

thus, the matched filter is a sufficient statistic for the amplitude A. Moreover, we have that the function of a sufficient estimator is also sufficient, hence the matched filter is a sufficient statistic for distinguishing among different stiffness values. Note that this analysis holds true for this particular physical model (spring-mass system) and this particular noise model (joint normal distribution). As will always be the case, our ability to detect and identify structural damage is predicated on both types of models.

In damage detection, myriad test statistics have been applied to the problem; they are frequently referred to in the SHM literature as "features." We continue to use "test statistic", however, as this terminology is more closely tied to the detection literature and the associated formalism. A variety of test statistics and their associated performance will be discussed in Chapter 8.

6.9.2 Detector Performance

To this point, we have discussed a formal detection approach that compares a detection statistic s to a threshold. In one case, the test statistic is selected on the basis of the GLRT and the other based on what we hope is an informed choice reflecting the physics of the problem. However, we have so far not said anything about how we will evaluate the performance of this detector.

As always, it makes sense to also treat the test statistic as a random variable with PDF $p_{S_0}(s|H_0)$ and $p_{S_1}(s|H_1)$ under the null and alternative hypotheses, respectively. Hypothetical distributions are shown schematically in Figure 6.48 under the "undamaged" and "damaged" hypotheses, respectively. The key question then becomes where to set the detection threshold

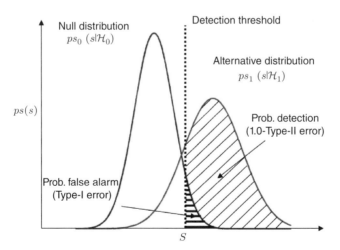

Figure 6.48 Illustration of Type I and Type II errors associated with a detection problem. Shown are two hypothetical distributions for a test statistic S under both the null (undamaged) and alternative (damaged) hypotheses

when deciding in favor of H_0 or H_1. This will depend on the distributions of the test statistic and how we assign costs to errors in our detector. The best we can hope to do is minimize these errors according to a predefined criteria for what we consider to be optimal.

The key pieces of information needed to characterize any detector are the Type I and Type II errors. Type I error is also known as the probability of false alarm (PFA), that is, how often do we declare a detection when the data were produced by the model $\mathcal{H}_0$? If we denote a "detection" as D_0, we can write the type I error as $P(D_1|\mathcal{H}_0)$. Type II error, on the other hand, is the probability of deciding in favor of $\mathcal{H}_0$, that is to say, declaring D_0, when the data were produced by the model $\mathcal{H}_1$. The POD is simply one minus the Type II error. A schematic depicting Type I and Type II errors is shown in Figure 6.48. These errors can only be obtained if we have both null *and* alternative hypotheses. One cannot properly quantify the performance of a detection scheme in SHM with only an "undamaged" set of test statistics. While an undamaged distribution is sufficient to estimate Type I error, the Type II error cannot be obtained. The POD, however, is critical in damage detection applications (the danger of ignoring a Type II error and subsequently misinterpreting a Type I error has been well documented in the statistics literature [78, 79]). If the practitioner truly wants to assess his/her detector, some knowledge of the distribution of damaged feature values is necessary, i.e., an alternative hypothesis is needed. Ideally, both null and alternative distributions would be known analytically. In cases where they are not, both must be estimated empirically (referred to in the damage detection literature sometimes as "supervised learning"). Once these distributions are known, the detector is characterized.

On the basis of the null and alternative distributions we can now set the detection threshold based on a chosen criterion for optimality. For example, we might simply choose a detection threshold based on which feature distribution of the test statistic is most likely to have come from, yielding our previously derived MAP criterion of optimality that we just discussed in the preceding section. Another possibility is to use a Bayes cost criterion. In damage detection, we are typically more interested in maximizing POD (minimizing Type II error) for a given (tolerable) Type I error. This leads to the Neyman–Pearson criteria for choosing a threshold (each of the above-described definitions of optimality are discussed in Ref. [1]). The point is that once the Type I and Type II errors are known, we can choose a criterion for optimality. In the end, this is the information that potential users of a damage detection system will need. The owner of a structure should be able to assign costs to both Type I and Type II errors and can decide whether or not to implement the detection system. Absent knowledge of these costs, the best we can do in the SHM community is provide Type I and Type II errors associated with our detection algorithms.

To illustrate, we revisit our example of the previous discussion. We have already shown that the Neyman–Pearson detector of stiffness loss for a spring-mass system driven far from resonance is

$$\widehat{A}_d^2 - \widehat{A}_u^2 > \lambda'' \tag{6.247}$$

where λ'' is a scalar threshold depending on the signal-to-noise ratio (SNR) as seen in Eqn. 6.246. To determine Type I and Type II errors for this detector, we require $p(A_d|\mathcal{H}_0)$ and $p(A_u|\mathcal{H}_1)$. To calculate these probabilities, we require the PDFs for both damaged and undamaged signal amplitudes. Our test statistic is a summation of a large number of normally distributed random variables (see Eq. 6.240) and, hence, (by the central limit theorem) will also be normally distributed. In particular, we have that

$$E[\hat{A}] = \frac{2}{N}E\left[\sum_{n=0}^{N-1}y(n)\cos(\omega n)\right]$$

$$= \frac{2}{N}E\left[\sum_{n=0}^{N-1}(A\cos(\omega n)+\eta(n))\cos(\omega n)\right]$$

$$= \frac{2}{N}E\left[\sum_{n=0}^{N-1}A\cos^2(\omega n)\right] + \sum_{n=0}^{N-1}E[\eta(n)]\cos(\omega n)$$

$$= \frac{2A}{N}\sum_{n=0}^{N-1}\cos^2(\omega n) + 0 = A \tag{6.248}$$

thus, our estimator is asymptotically unbiased (again, for large N we have that $\sum_{n=0}^{N-1}\cos^2(\omega n) = \frac{N}{2}$). The associated variance is

$$E[(\hat{A}-E[\hat{A}])^2] = E\left[\left(\frac{2}{N}\sum_{n=0}^{N-1}(A\cos(\omega n)+\eta(n))\cos(\omega n)-A\right)^2\right]$$

$$= E\left[\left(\left(\frac{2A}{N}\sum_{n=0}^{N-1}\cos^2(\omega n) - \frac{2}{N}\sum_{n=0}^{N-1}\eta(n)\cos(\omega n)\right)-A\right)^2\right]$$

$$= E\left[\left(A - \frac{2}{N}\sum_{n=0}^{N-1}\eta(n)\cos(\omega n) - A\right)^2\right]$$

$$+ E\left[\frac{4}{N^2}\sum_{n=0}^{N-1}\eta^2\cos^2(\omega n)\right]$$

$$= E[\eta^2]\frac{4}{N^2}\sum_{n=0}^{N-1}\cos^2(\omega n) = \frac{2}{N}\sigma^2 \equiv \sigma'^2 \tag{6.249}$$

which is simply the noise variance divided by the signal energy for unit amplitude, i.e., the signal energy for $A=1$. Note that in the second to last line we invoked the assumption that the noise was i.i.d., which allowed the expected value of the square of the summation to instead be written as the sum of the expected square of the argument. Reducing the noise σ^2 or increasing the number of data N will both improve the precision of the estimate as one might expect. Each $\hat{A} \sim \mathcal{N}(A, \sigma')$ is therefore a normally distributed random variable with mean A and variance $\sigma'^2 = \frac{2}{N}\sigma^2$. Applying the transformation rules from Chapter (2.4), it can be shown that the square of a single, normally distributed random variable with nonzero mean is given by the noncentral chi-squared distribution with one DOF (see Appendix A). In this example, we therefore have that

$$p_A(\hat{A}^2) = \frac{1}{(2\pi\sigma'^2\hat{A}^2)^{1/2}}e^{-(\hat{A}^2+A^2)/2\sigma'}\cosh\left(\sqrt{\hat{A}^2A/\sigma'^2}\right) \tag{6.250}$$

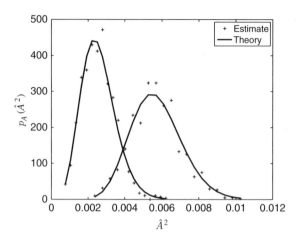

Figure 6.49 Theoretical and estimated probability models for the squares of the signal amplitudes. Shown are the PDFs for $A_u = 0.05$ and $A_d = 0.075$, which correspond to two different stiffness values. Given these probability models, we can begin to talk about the probabilities of successfully distinguishing A_d from A_u

for our PDF for the estimated square of the signal amplitude. This probability model is a special case of the *noncentral chi-squared* distribution for the sum of squares of n normally distributed random variables [1]. In Figure 6.49, we plot this distribution for $A_u = 0.05, A_d = 0.075$ and an SNR of 30. By definition, this means that the noise variance in (6.250) is $\sigma^2 = A^2 N/(2SNR)$ so that

$$\sigma'^2 = A^2/SNR. \tag{6.251}$$

The estimated probabilities were obtained by repeatedly generating $N = 1000$ length signals according to the signal model (6.236, 6.237) with different noise realizations, and then using the Fourier coefficient estimator given by the final line of (6.240). The basic histogram estimator was used in forming the estimated PDF. These two distributions are sufficient to characterize the performance of the detector described by (6.247). The aforementioned Type I and Type II errors can be computed via integration. Specifically, we can compute

$$
\begin{aligned}
P(D_d|A_u) &= \int_\lambda^\infty p_A(\hat{A}^2|A_u)d\hat{A} \\
&= \frac{1}{(2\pi\sigma'^2)^{1/2}} \int_\lambda^\infty \frac{1}{\hat{A}} e^{-(\hat{A}^2+A_u^2)/2\sigma'} \cosh\left(\sqrt{\hat{A}^2}A_u/\sigma'^2\right) d\hat{A} \\
&= \frac{1}{2}\left(2 + Erf\left[\frac{A_u - \sqrt{\lambda}}{\sqrt{2}\sigma'}\right] - Erf\left[\frac{A_u + \sqrt{\lambda}}{\sqrt{2}\sigma}\right]\right) \\
&\equiv \text{Type-I error} \tag{6.252}
\end{aligned}
$$

where $Erf(\cdot)$ is the error function given by

$$Erf(z) = \frac{2}{\pi} \int_0^z e^{-t^2} \, dt. \tag{6.253}$$

Similarly, we have that the Type II error is given by

$$
\begin{aligned}
P(D_u|A_d) &= \int_0^\lambda p_A(\hat{A}^2|A_d)\hat{A} \\
&= \frac{1}{(2\pi\sigma'^2)^{1/2}} \int_0^\lambda \frac{1}{\hat{A}} e^{-(\hat{A}^2 + A_d^2)/2\sigma'} \cosh\left(\sqrt{\hat{A}^2 A_d/\sigma'^2}\right) d\hat{A} \\
&= \frac{1}{2}\left(-Erf\left[\frac{A_d - \sqrt{\lambda}}{\sqrt{2}\sigma'}\right] + Erf\left[\frac{A_d + \sqrt{\lambda}}{\sqrt{2}\sigma}\right]\right)
\end{aligned}
$$

$$\equiv \text{Type-II error} \tag{6.254}$$

so that the POD is

$$
\begin{aligned}
P(D_d|A_d) &= 1 - P(D_u|A_d) \\
&= \frac{1}{2}\left(2 + Erf\left[\frac{A_d - \sqrt{\lambda}}{\sqrt{2}\sigma'}\right] - Erf\left[\frac{A_d + \sqrt{\lambda}}{\sqrt{2}\sigma}\right]\right). \tag{6.255}
\end{aligned}
$$

At this point, we have all of the information needed to characterize our detector. To display this information, we may use the so-called receiver operating characteristic (ROC) curve. The ROC curve is used extensively by the radar community in problems of detection and is wholly appropriate for use in the damage detection problem as well. Quite simply, the ROC curve plots the POD versus PFA as a function of the detection threshold, λ. Figure 6.50 plots (6.255) versus (6.252) as a function of the detection threshold, λ. In this example, we take the "healthy" structure's amplitude to be $A_u = 0.05$ m. We then define the "damaged" structure to be associated with amplitudes of $A_d = 0.55, 0.65, 0.75, 0.85$ m, respectively. As expected, if we hope to detect smaller levels of damage, we see a reduced probability of detection for a constant level of false alarms.

It should also be noted that we can, in general, write these expressions more compactly in terms of the CDFs of the probability models. Specifically, we note that

$$P(D_d|\mathcal{H}_u) = 1 - P_{H_u}(\lambda)$$

$$P(D_d|\mathcal{H}_d) = 1 - P_{H_d}(\lambda). \tag{6.256}$$

These expressions are valid for *any* probability models that we use for our test statistics. In fact, we may take the problem one step further and rewrite the threshold in terms of a range of desired false alarm probabilities. This allows us to specify the probability of detection directly in terms of the probability of false alarm as

$$\lambda = P_{H_u}^{-1}(1 - P_{fa})$$

$$P_d(P_{fa}) = 1 - P_{H_d}(P_{H_u}^{-1}(1 - P_{fa})). \tag{6.257}$$

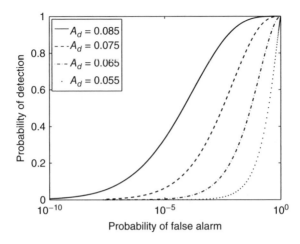

Figure 6.50 Theoretical detection performance for varying "damaged" amplitude values. For each curve, $A_u = 0.05$ m was assumed

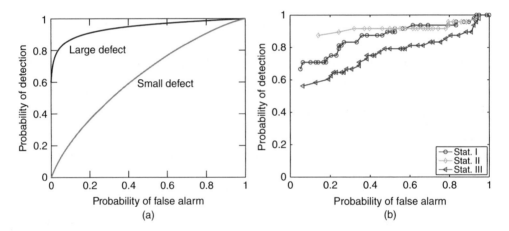

Figure 6.51 (a) Illustration of two ROC curves. A small defect may give a ROC curve that is not much better than chance. Large defects will likely improve detection performance. (b) Experimentally obtained ROC curves for detection of bolt loosening in strongly varying ambient conditions. Three different test statistics are being compared in terms of their Type I and Type II errors (we elaborate on the specifics of these curves in Chapter 8)

The result of Figure 6.50 is typical of those found in damage detection problems in the sense that larger damage is usually more easily detected for a fixed level of false alarms. In general, damage detection ROC curves will be qualitatively similar to those shown in the left plot of Figure 6.51, where the trade-off between Type I and Type II errors is evident. If the practitioner is willing to accept a large PFA, a high POD is usually obtainable. Conversely, a low Type I error is typically accompanied by a reduced POD.

As an experimental example, we consider a study (discussed in detail in Section 8.8.2) to compare the performance of three different detectors of bolt loosening in a composite-to-metal bolted joint. The goal was to perform these detections in the presence of strongly varying ambient fluctuations. Data were repeatedly acquired as the temperature of the structure was varied from 23 to 50 °C. The experiment was repeated many times over the span of several weeks in an effort to capture any other sources of variability. In the end, 78 undamaged and 48 damaged values were obtained from each of three different test statistics. Each test statistic was designed to capture nonlinearity in the structural response, a process that is discussed in detail in Chapter (8). For each group of test statistics, both POD and PFA were estimated by simply counting the fraction of correct diagnosis. Figure 6.51 shows the associated ROC curves. Which test statistic is optimal will depend, of course, on the costs associated with the different detection errors. For most applications, we would imagine test statistic II would be preferred as it offers a high power of detection with relatively few false alarms.

In the instance that the distributions of the test statistics are unknown and cannot be easily derived, we may use an empirical approach to estimate the needed probabilities as we vary the detection threshold. This approach is outlined here. Assume we are able to collect data from the structure in both a healthy and a damaged state. We collect M such signals in each instance and estimate the corresponding test statistics $s_u(m), s_d(m), m = 1 \cdots M$. Each test statistic is itself a random variable S_u, S_d with its own governing PDF. Taking both the maximum and minimum over both sets of statistics, we set a range of threshold values as $\lambda \in [S_{min}, S_{max}]$. For each threshold value we choose, set:

$$\widehat{P}_{fa}(\lambda) = \frac{\sum_{m=1}^{M} \Theta(s_u(m) - \lambda)}{M}$$

$$\widehat{P}_d(\lambda) = \frac{\sum_{m=1}^{M} \Theta(s_d(m) - \lambda)}{M} \tag{6.258}$$

that is to say, we simply look at the fraction of test statistics that exceed the threshold; these are then plotted as a function of the threshold value.

As a practical matter, it is often safe to assume that the test statistic under question is approximately normally distributed. This is a consequence of the central limit theorem (see discussion surrounding Eq. 2.75) which essentially says that any test statistic that is formed as a summation of a large number of data values should obey a normal probability model. In fact, the noncentral chi-squared distribution just mentioned can be well approximated by a normal distribution for large values of the noncentrality parameter. This is expected as the test statistic $\widehat{A}^2$ is formed as a linear combination (summation) of the data multiplying a sinusoid.

This suggests a simple way to construct a ROC curve given both undamaged and damaged test statistics, $s_u(m), s_d(m), m = 1 \cdots M$. For each group of test statistics, estimate the mean and variance $\widehat{\mu}_u, \widehat{\mu}_d$ and $\widehat{\sigma}_u, \widehat{\sigma}_d$, respectively. The probability of false alarms and the POD then become, respectively

$$P(D_d|S_u) = \frac{1}{\sqrt{2\pi\widehat{\sigma}_u^2}} \int_\lambda^\infty e^{-\frac{1}{2\widehat{\sigma}_u^2}(s_u - \widehat{\mu}_u)^2} ds_u$$

$$= \frac{1}{2}\left(1 - Erf\left(\frac{\lambda - \widehat{\mu}_u}{\sqrt{2}\widehat{\sigma}_u}\right)\right)$$

$$P(D_d|S_d) = 1 - P(D_u|S_d) = \frac{1}{\sqrt{2\pi\hat{\sigma}_d^2}} \int_\lambda^\infty e^{-\frac{1}{2\hat{\sigma}_d^2}(s_d - \hat{\mu}_d)^2} \, ds_d$$

$$= \frac{1}{2}\left(1 - Erf\left(\frac{\lambda - \hat{\mu}_u}{\sqrt{2}\hat{\sigma}_u}\right)\right) \tag{6.259}$$

We may go one step further and solve the first of these equations for the value of the threshold that yields a certain probability of false alarm. This is easily accomplished using the inverse normal CDF. A simple algorithm for taking two test statistic vectors (undamaged and damaged), assuming they are normally distributed, and estimating the ROC performance is given below.

Algorithm III (ROC curves from normally distributed test statistic)

Given $su(m), sd(m), m = 1 \cdots M$

Pfa=[0:0.01:1];	% False alarms to evaluate
mu1=mean(su); mu2=mean(sd);	% get population means
stdv1=std(su); stdv2=std(sd);	% get population variances
Pd=1-normcdf(norminv(1-Pfa,mu1,stdv1),mu2,stdv2);	% Pd vs. Pfa

---

One can then simply plot *Pd* versus *Pfa* to see the ROC performance. The same approach can be used for non-Gaussian PDFs. For example, assume that we are able to show the two test statistic distributions (undamaged and damaged) each follow a Gamma PDF. Estimates of the Gamma parameters may be first obtained via the Matlab pseudo-code: pars1 = gamfit(su), pars2 = gamfit(sd) for each vector of test statistics. The final line of the algorithm (6.9.2) is then replaced by $Pd = 1 - $ gamcdf(gaminv(1 − Pfa, pars1(1), pars1(2)), pars2(1), pars2(2)).

In short, the utility of the ROC curve is that it shows exactly the information that is of importance in problems of detection and only that information. It offers an unambiguous means of comparing different detectors of structural damage. Regardless of the detection strategy, e.g., the hardware used, the damage-sensitive test statistic chosen, and so on, the ROC curve is an appropriate forum for comparison as it reduces all results to the relevant probabilities. Results can't be "tweaked" to make a better looking ROC curve. Adjusting the threshold to make Type I or Type II error look better will inevitably be to the detriment of the other (as a colleague of mine is fond of saying, "you can't hide from a ROC curve"). The downside to ROC curves is that they require both healthy *and* damaged "training" data (or known probability models). Note that with a "healthy" test statistic distribution, one can *only* evaluate Type I error, i.e., P_{fa}. Counting outliers from a healthy distribution says nothing about the probability that the data we declare "damaged" were truly generated by a damaged structure.

In addition, when generated in a laboratory setting, the practitioner can control most sources of uncertainty and make his/her ROC curves look better than would be obtained *in situ*. Thus, in addition to presenting the ROC curve, it would be useful to report which sources of variability were considered. Still, it remains an industry standard in the radar and sonar communities and is an appropriate tool for assessing damage detection algorithms as well. Many of the results provided in subsequent chapters are given in terms of ROC curves for varying levels of structural damage.

6.9.3   Intervals of Confidence

A question that naturally arises in damage detection applications is how to quantify detection performance when only data collected under $\mathcal{H}_0$ is observed, that is, we only have healthy data. Now, if we have a well-formed alternative $\mathcal{H}_1$, we can form "surrogate" data that conform to this hypothesis and therefore still get estimates of both Type I and Type II errors (this approach forms the basis for much of chapter 8). However, absent a well-defined alternative, the best we can do is form the PDF for the "healthy" test statistic and declare values that are inconsistent with this probability model to be damaged. This is, in fact, the hallmark of many detection approaches based on statistical pattern recognition.

Consider a test statistic s that is known to be normally distributed with $\mu = 0$ and $\sigma^2 = 1$ when the structure is in the undamaged condition. The probability model for this test statistic is shown in Figure 6.52. Also shown is the 90% *confidence interval* (shaded region), taken as the open interval $[s_L, s_H]$ for which

$$\int_{s_L}^{s_H} p_S(s|\mathcal{H}_0)ds = 0.9 \tag{6.260}$$

In this example, we would say that future values of s that lie outside the range $[s_L, s_H]$ are unlikely to have occurred under the hypothesis that the structure is healthy (roughly 10% of the time we would expect exceedances of this range). Note that an exceedance may say little, or nothing at all, about the likelihood that the structure is truly damaged. This will be particularly true if the unknown alternative distribution closely overlaps the null distribution. Nonetheless, it may be useful to form confidence limits in this manner and declare "damage" accordingly. So long as the practitioner understands that this approach may or may not yield good detection properties when the damage is present (i.e., the data are produced under the model $\mathcal{H}_1$).

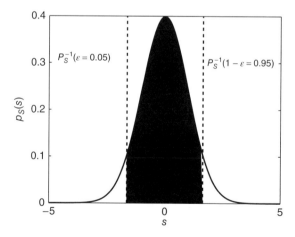

Figure 6.52   Process by which an interval of confidence is formed from a known PDF, in this example, a standard normal distribution. The upper and lower $100 \times (1 - 2\epsilon)$% confidence interval is formed by evaluating the associated inverse CDF at the points ϵ and $1 - \epsilon$. In this example, the shaded region was generated by setting $\epsilon = 0.05$ and therefore encompassing 90% of the probability predicted by the standard normal distribution

In general, we may determine a $100 \times (1 - 2\epsilon)$ confidence interval by inverting Eq. (6.260), yielding

$$s_L = P_S^{-1}(\epsilon)$$

$$s_H = P_S^{-1}(1 - \epsilon) \qquad (6.261)$$

where $P_S^{-1}(\cdot)$ is the inverse CDF associated with the test statistic probability model. For the example shown in Figure 6.52, we assume a standard normal random variable for which $s_L = -1.645$ and $s_H = 1.645$. In many practical cases, the test statistic is, in fact, well approximated by a normal distribution. This stems ultimately from the central limit theorem (Section 2.6, page 45) which states that the sum of a large number of independent random variables, regardless of their distribution, is approximately normal. Many test statistics are formed via summation of various functions of the values in the signal (e.g., Fourier coefficients) and are therefore modeled with a Gaussian PDF. Even in the non-Gaussian case, we can often use the methods of Section (2.4) to derive the PDF for the test statistic. For example, in Section (8.3.2) we use a test statistic which is formed as the sum of two normal distributions with nonzero mean. We will show the resulting distribution for the test statistic to be noncentral chi-squared distributed and form confidence intervals and ROC curves accordingly.

Of course, in some instances we may not have any information about the functional form (model) of the test statistic distribution with which to define the confidence limits as in (6.260). In such cases, we have little recourse but to take a frequentist approach and develop statements about probability through replication. For example, we might simply have a number of samples of the test statistic, $s_i, i = 1 \cdots M$. Ideally, these would be MLEs due to the minimal variance associated with such estimates. On the basis of these replicates, we could estimate the probability distribution by forming a histogram, possibly using the kernel density estimation techniques described earlier in Section 6.3. If the density is well approximated by a PDF with known functional form, we might even fit this PDF to the data and develop confidence intervals for S using (6.261).

However, if we are not willing to make such an assumption, a simple approach is to rank order the test statistic values $s_{k_i} : s_{k_1} < s_{k_2} <, \cdots, < s_{k_M}$ and discard the highest and lowest $100\epsilon\%$ of the samples. The first and last elements of the remaining vector then define an approximate interval of $100 \times (1 - 2\epsilon)\%$ confidence. For example, let's say we have $M = 100$ estimates of the first natural frequency. If we were to rank order the estimates and exclude the highest and lowest 5 estimated frequencies, the remaining 90 frequencies span a confidence interval of roughly 90% confidence. It is precisely this approach to interval estimation that was used to develop the intervals shown in the introductory example, page 12, Figure (1.8). The quality of this approximation will improve with the number of test statistic realizations, M.

One interesting aspect of the data-driven approach to confidence intervals is that it quantifies variability in the *estimation procedure*. The hope is that the variance among the values s_i is the same as the variance of the underlying probability model $p_S(s)$. This may or may not be the case and brings us to an important point about intervals formed through replication. Empirically formed intervals reflect statements about the variability in the estimation procedure, and not necessarily the parameter of interest. This is a subtle, but sometimes important, difference that will be discussed in greater detail in Section (7.2).

References

[1] R. N. McDonough, A. D. Whalen, Detection of Signals in Noise, 2nd ed., Academic Press, San Diego, CA, 1995.

[2] S. M. Kay, Fundamentals of Statistical Signal Processing: Estimation Theory, Vol. I, Prentice Hall, New Jersey, 1993.

[3] A. M. Mood, F. A. Graybill, D. C. Boes, Introduction to the Theory of Statistics, 3rd ed., McGraw-Hill, New York, 1974.

[4] P. Stoica, P. Babu, The Gaussian data assumption leads to the largest Cramèr-Rao bound, IEEE Signal Processing Magazine 28 (2011) 132–133.

[5] R. A. Fisher, On the mathematical foundations of theoretical statistics, Philosophical Transactions of the Royal Society of London, Series A 222 (1922) 309–368.

[6] H. Poincaré, Sur le problème des trois corps et les équations de la dynamique, Acta Mathematica 13 (1890) 1–270.

[7] L. Barreira, Poincaré recurrence: old and new, in: XIV International Congress on Mathematical Physics (Lisbon, 2003), World Scientific, 2005, pp. 415–422.

[8] G. D. Birkhoff, Proof of the ergodic theorem, Proceedings of the National Academy of Sciences of the United States of America 17 (1931) 656–660.

[9] G. D. Birkhoff, What is the ergodic theorem? American Mathematical Monthly 49 (4) (1942) 222–226.

[10] J. S. Bendat, A. G. Piersol, Random Data Analysis and Measurement Procedures, 3rd ed., John Wiley & Sons, Inc., New York, 2000.

[11] J.-P. Eckmann, D. Ruelle, Ergodic theory of chaos and strange attractors, Reviews of Modern Physics 57 (3) (1985) 617–656.

[12] B. W. Silverman, Density Estimation for Statistics and Data Analysis, Chapman and Hall, 1986.

[13] S. W. Doebling, C. R. Farrar, M. B. Prime, A summary review of vibration-based damage identification methods, Shock and Vibration Digest 205 (5) (1998) 631–645.

[14] Y. Xia, H. Hao, Statistical damage identification of structures with frequency changes, Journal of Sound and Vibration 263 (2003) 853–870.

[15] J.-T. Kim, Y.-S. Ryu, H.-M. Cho, N. Stubbs, Damage identification in beam-type structures: frequency-based method vs. mode-shape-based method, Engineering Structures 25 (2003) 57–67.

[16] G. Y. Xu, W. D. Zhu, B. H. Emory, Experimental and numerical investigation of structural damage detection using changes in natural frequencies, ASME Journal of Vibration and Acoustics 129 (2007) 686–700.

[17] H. Nyquist, Certain topics in telegraph transmission theory, Transactions of the AIEE 47 (1928) 617–644.

[18] C. E. Shannon, Communication in the presence of noise, Proceedings of the IEEE (reprinted from "Proceedings of the IRE 37(1), 10–21, 1949") 86 (2).

[19] M. B. Priestly, Spectral Analysis and Time Series, Probability and Mathematical Statistics, Elsevier Academic Press, London, 1981.

[20] J. W. Cooley, J. W. Tukey, An algorithm for the machine calculation of complex Fourier series, Mathematics of Computation 19 (1965) 297–301.

[21] E. J. Candes, T. Tao, Decoding by linear programming, IEEE Transactions on Information Theory 51 (12) (2005) 4203–4215.

[22] A. V. D. Bos, Estimation of Fourier coefficients, IEEE Transactions on Instrumentation and Measurement 38 (5) (1989) 1005–1007.

[23] M. Haroon, D. E. Adams, Time and frequency domain nonlinear system characterization for mechanical fault identification, Nonlinear Dynamics 50 (2007) 387–408.

[24] S. H. Crandall, W. D. Mark, Random Vibration in Mechanical Systems, Academic Press, New York, 1963.

[25] H. Benaroya, Mechanical Vibration: Analysis, Uncertainties, and Control, Prentice Hall, New Jersey, 1998.

[26] L. Meirovitch, Introduction to Dynamics and Control, John Wiley & Sons, Inc., 1985.

[27] M. Schetzen, The Volterra and Wiener Theories of Nonlinear Systems, John Wiley & Sons, Inc., New York, 1980.

[28] K. Worden, G. R. Tomlinson, Nonlinearity in Structural Dynamics, Institute of Physics Publishing, Bristol, Philadelphia, PA, 2001.

[29] J. M. Nichols, P. Marzocca, A. Milanese, On the use of the auto-bispectral density for detecting quadratic non-linearity in structural systems, Journal of Sound and Vibration 312 (4-5) (2008) 726–735.

[30] D. R. Brillinger, An introduction to polyspectra, Annals of Mathematical Statistics 36 (5) (1965) 1351–1374.

[31] C. L. Nikias, M. R. Raghuveer, Bispectrum estimation: a digital signal processing framework, Proceedings of the IEEE 75 (7) (1987) 869–891.

[32] P. J. Huber, B. Kleiner, T. Gasser, G. Dumermuth, Statistical methods for investigating phase relations in stationary stochastic processes, IEEE Transactions on Audio and Electroacoustics AU-19 (1) (1971) 78–86.

[33] A. M. Richardson, W. S. Hodgkiss, Bispectral analysis of underwater acoustic data, Journal of the Acoustical Society of America 96 (2) (1994) 828–837.

[34] K. Sasaki, T. Sato, Y. Yamashita, Minimum bias windows for bispectral estimation, Journal of Sound and Vibration 40 (1) (1975) 139–148.

[35] M. Schetzen, Nonlinear system modeling based on the Wiener theory, Proceedings of the IEEE 69 (12) (1981) 1557–1573.

[36] J. M. Nichols, P. Marzocca, A. Milanese, The trispectrum for Gaussian driven, multiple degree-of-freedom, non-linear structures, International Journal of Non-Linear Mechanics 44 (2009) 404–416.

[37] L. Isserlis, On a formula for the product-moment coefficient of any order of a normal frequency distribution in any number of variables, Biometrika 12 (1/2) (1918) 134–139.

[38] A. I. Fleishman, A method for simulating non-normal distributions, Psychometrika 43 (4) (1978) 521–532.

[39] J. M. Nichols, C. C. Olson, Optimal bispectral detection of weak, quadratic nonlinearities in structural systems, Journal of Sound and Vibration 329 (8) (2010) 1165–1176.

[40] P. Marzocca, J. M. Nichols, A. Milanese, M. Seaver, S. T. Trickey, Second-order spectra for quadratic nonlinear systems by Volterra functional series: analytical description and numerical simulation, Mechanical Systems and Signal Processing 22 (2008) 1882–1895.

[41] D. Hickey, K. Worden, M. F. Platten, J. R. Wright, J. E. Cooper, Higher-order spectra for identification of nonlinear modal coupling, Mechanical Systems and Signal Processing 23 (4) (2009) 1037–1061.

[42] M. J. Hinich, Higher order cumulants and cumulant spectra, Circuits, Systems, and Signal Processing 13 (4) (1994) 391–402.

[43] J. W. D. Molle, M. J. Hinich, Trispectral analysis of stationary random time series, Journal of the Acoustical Society of America 97 (5) (1995) 2963–2978.

[44] A. Kaiser, T. Schreiber, Information transfer in continuous processes, Physica D 166 (2002) 43–62.

[45] D. Prichard, J. Theiler, Generalized redundancies for time series analysis, Physica D 84 (1995) 476–493.

[46] M. Paluš, Testing for nonlinearity using redundancies: quantitative and qualitative aspects, Physica D 80 (1995) 186–205.

[47] P. Grassberger, An optimized box-assisted algorithm for fractal dimensions, Physics Letters A 148 (1990) 63–68.

[48] J. L. Bentley, Multidimensional binary search trees in database applications, IEEE Transactions on Software Engineering SE-5 (4) (1979) 333–340.

[49] J. M. Nichols, Examining structural dynamics using information flow, Probabilistic Engineering Mechanics 21 (2006) 420–433.

[50] Y. C. Kim, E. J. Powers, Digital bispectral analysis and its applications to nonlinear wave interactions, IEEE Transactions on Plasma Science PS-7 (2) (1979) 120–131.

[51] M. R. Hajj, R. W. Miksad, E. J. Powers, Perspective: measurements and analyses of nonlinear wave interactions with higher-order spectral moments, Journal of Fluids Engineering 119 (1997) 3–13.

[52] K. R. Gurley, A. Kareem, M. A. Tognarelli, Simulation of a class of non-normal random processes, Probabilistic Engineering Mechanics 31 (5) (1996) 601–617.

[53] G. L. Wise, A. P. Traganitis, J. B. Thomas, The effect of a memoryless nonlinearity on the spectrum of a random process, IEEE Transactions on Information Theory IT-23 (1977) 84–89.

[54] B. Liu, D. C. Munson Jr., Generation of a random sequence having a jointly specified marginal distribution and autocovariance, IEEE Transactions on Acoustics, Speech, and Signal Processing ASSP-30 (6) (1982) 973–983.

[55] P. Bocchini, G. Deodatis, Critical review and latest developments of a class of simulation algorithms for strongly non-Gaussian random fields, Probabilistic Engineering Mechanics 23 (2008) 393–407.

[56] U. G. Gujar, R. J. Kavanagh, Generation of random signals with specified probability density functions and power density spectra, IEEE Transactions on Automatic Control AC-13 (1968) 716–719.

[57] M. Shinozuka, C. M. Jan, Digital simulation of random processes and its applications, Journal of Sound and Vibration 25 (1) (1972) 111–128.

[58] F. Yamazaki, M. Shinozuka, Digital generation of non-Gaussian stochastic fields, Journal of Engineering Mechanics 114 (7) (1988) 1183–1197.

[59] G. Deodatis, R. C. Micaletti, Simulation of highly skewed non-Gaussian stochastic processes, Journal of Engineering Mechanics 127 (12) (2001) 1284–1295.

[60] Y. Shi, G. Deodatis, A novel approach for simulation of non-Gaussian field: application in estimating wire strenghts from experimental data, in: S. Srinivasan, B. Bhattacharya (Eds.), Proceedings of the 9th ASCE Speciality Conference on Probabilistic Mechanics and Structural Reliability, 2004.

[61] M. Grigoriu, Simulation of stationary non-Gaussian translation processes, Journal of Engineering Mechanics 124 (2) (1998) 121–126.

[62] L. Devroye, Non-Uniform Random Variate Generation, Springer-Verlag, New York, 1986.

[63] J. C. S. S. Filho, M. D. Yacoub, Coloring non-Gaussian sequences, IEEE Transactions on Signal Processing 56 (12) (2008) 5817–5822.

[64] I. W. Hunter, R. E. Kearney, Generation of random sequences with jointly specified probability density and autocorrelation functions, Biological Cybernetics 47 (1983) 141–146.

[65] G. H. Golub, C. F. V. Loan, Matrix Computations, Johns Hopkins Press, Baltimore, MD, 1996.

[66] P. D. Hoff, Separable covariance arrays via the tucker product, with applications to multivariate relational data, Bayesian Analysis 6 (2) (2011) 179–196.

[67] T. G. Kolda, Multilinear operators for higher-order decompositions, Tech. Rep. SAND2006-2081, Sandia National Laboratories, Albequerque, NM (2006).

[68] T. Schreiber, A. Schmitz, Improved surrogate data for nonlinearity tests, Physical Review Letters 77 (4) (1996) 635–638.

[69] T. Schreiber, A. Schmitz, Surrogate time series, Physica D 142 (2000) 346–382.

[70] J. Theiler, S. Eubank, A. Longtin, B. Galdrikian, J. D. Farmer, Testing for nonlinearity in time series: the method of surrogate data, Physica D 58 (1992) 77–94.

[71] J. M. Nichols, C. C. Olson, J. V. Michalowicz, F. Bucholtz, A simple algorithm for generating spectrally colored, non-Gaussian signals, Probabilistic Engineering Mechanics 25 (3) (2010) 315–322.

[72] D. J. Whitford, J. K. Waters, M. E. C. Vieira, Teaching time-series analysis. II. Wave height and water surface elevation probability distributions, American Journal of Physics 69 (4) (2001) 497–504.

[73] S. M. Han, H. Benaroya, Nonlinear and Stochastic Dynamics of Compliant Offshore Structures, Kluwer Academic Publishers, Dordrecht, Netherlands, 2002.

[74] C. G. Soares, Representation of double-peaked sea wave spectra, Ocean Engineering 11 (2) (1984) 185–207.

[75] F. J. Mendez, S. Castanedo, A probability distribution for depth-limited extreme wave heights in a sea state, Coastal Engineering 54 (2007) 878–882.

[76] M. B. Priestly, T. S. Rao, A test for non-stationarity of time-series, Journal of the Royal Statistical Society, Series B (Methodological) 31 (1) (1969) 140–149.

[77] M. B. Priestly, Evolutionary spectra and non-stationary processes, Journal of the Royal Statistical Society, Series B (Methodological) 27 (2) (1965) 204–237.

[78] D. H. Johnson, The insignificance of statistical significance testing, Journal of Wildlife Management 63 (3) (1999) 763–772.

[79] G. Gigerenzer, Mindless statistics, Journal of Socio-Economics 33 (2004) 587–606.

7

Parameter Estimation for Structural Systems

As was mentioned in the introduction, understanding the process of estimating model parameters from observed data is critical for a practicing scientist. Models are how we understand and predict observed phenomena, which is the primary objective of all scientific disciplines. Many models have parameters and the values of those parameters are often unknown *a priori*. Even if the values are known to within some interval (e.g., ultimate strength of a material is somewhere between 235 and 300 MPa), those values carry with them some uncertainty as to which value to use in the model predictions. As we will see, the parameter value we eventually use (our estimate) will be different depending on how we define "best" value. This chapter is therefore devoted to setting up the type of estimation problems frequently encountered in damage identification applications and providing solutions. The technical challenges to estimating structural damage parameters have been documented in a number of outlets (see, e.g., Friswell [1]); several of the larger obstacles are described in what follows.

At this point in the book, we have developed all the tools necessary to estimate the parameters of a given model from observations. We begin with the random process model for our data. In structural dynamics, our observations typically consist of a sequence of N temporally ordered samples (i.e., time series) measured at multiple locations $j = 1 \cdots M$ on the structure. These could be strain, acceleration, or displacement measurements. We will denote this data $y_j(n), n = 1 \cdots N, j = 1 \cdots M$. Next, we require a model for predicting what we observed. This model will be comprised of a physics-based component, and a component that captures the uncertainty in our measurement process (e.g., noise). Thus, we arrive at a commonly used descriptor of our observations

$$y_j(n) = x_j(n, \boldsymbol{\theta}) + \eta_j(n) \tag{7.1}$$

where $x_j(n, \boldsymbol{\theta})$ is our model response at discrete time n, governed by parameters $\boldsymbol{\theta}$. The quantity $\eta_j(n)$ is taken as additive noise for sensor j and sample n. This model makes the assumption that our observed signal is governed by a physical model $x_j(n, \boldsymbol{\theta})$ and a probabilistic model governing the additive noise. While other noise models exist (e.g., multiplicative noise), Eq. (7.1) accurately describes the vast majority of the structural vibrations one encounters in practice.

Modeling and Estimation of Structural Damage, First Edition. Jonathan M. Nichols and Kevin D. Murphy.
© 2016 John Wiley & Sons, Ltd. Published 2016 by John Wiley & Sons, Ltd.

For this reason, we use Eq. (7.1) as the basic building block for all of the estimation approaches described in this book.

At this point, we restate the common objection: "what if the model isn't the 'right' model?" This question makes no sense, of course, as a model is by definition wrong! We never have the "right" model, rather we have different descriptions of reality, some better than others. It's fair to ask if we might find a *better* model, but never fair to criticize the approach for not having the "right" one. Again, the goals of the model are (i) good predictive power and (ii) parsimony; to these ends, Eq. (7.1) is often a very good representation for structural response data, provided we keep the number of unknown parameters to a minimum.

Returning to (7.1), there are a number of different noise models to choose from. For most of the measurements we take, electronic or thermal noise results in a sequence of noise values that is normally distributed. To see that this is a good probability model, and not just a matter of mathematical convenience, consider strain data collected from one of the deck plates of a CG class Navy vessel in 2007 (deck plate shown in Figure 1.2). An estimated probability density function (PDF) of that data is shown in Figure 7.1 along with the estimated auto-correlation function. Clearly the data are well approximated by the normal probability model. Moreover, the sequence of values are very nearly independent as $\hat{R}_{XX}(\tau) \approx 0$ for $\tau > 0$.

Given the normal probability model, consider the PDF associated with a single noise value

$$p_H(\eta_j(n)) = \frac{1}{\sqrt{2\pi\sigma^2}}e^{-\frac{1}{2\sigma^2}\eta_j(n)^2} \tag{7.2}$$

where σ^2 is the variance of the noise. This does not have to be known *a priori* and can also be estimated from the data. Remember that our goal is to form a probabilistic model for the observations $y_j(n)$ in terms of the model parameters $\boldsymbol{\theta}$. We can do this by simply noting from

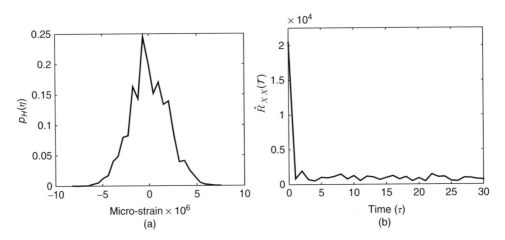

Figure 7.1 (a) Estimated probability density function (PDF) associated with strain data collected from the CG class vessel before transit. The sensors in this case are recording the ambient vibrations with no loading applied, that is, noise. Clearly, the normal probability model is a good one for describing the observed data. Moreover, (b) shows the estimated auto-correlation function for a few hundred delays. The function is approximately zero for $\hat{R}_{XX}(\tau) > 0$, indicating the observations are appropriately modeled as independent

Eq. (7.1) that $\eta_j(n) = y_j(n) - x_j(n, \boldsymbol{\theta})$ and writing

$$p_H(\eta_j(n)) = p_H(y_j(n) - x_j(n, \boldsymbol{\theta}))$$

$$\equiv p_H(y_j(n)|x_j(n, \boldsymbol{\theta})) = \frac{1}{\sqrt{2\pi\sigma^2}} e^{-\frac{1}{2\sigma^2}(y_j(n) - x_j(n,\boldsymbol{\theta}))^2} \qquad (7.3)$$

where the notation $p_H(y_j(n)|x_j(n, \boldsymbol{\theta}))$ captures the fact that the data $y_j(n)$ depends (is conditional) on the model $x_j(n, \theta)$. Now this is just a model describing the "nth" time sample from the "jth" sensor. What we really want is a model describing the *entire* acquired data set $y_j(n), n = 1 \cdots N, j = 1 \cdots M$. This means we need the *joint* probability density $p_H(\mathbf{y}|\mathbf{x}(\boldsymbol{\theta}))$, where $\boldsymbol{H} \equiv (\eta_1(1), \eta_2(1), \ldots, \eta_1(2), \eta_2(2), \ldots, \eta_M(N))$ and where we use bold-face type to denote a vector.

We will further state that the noise value associated with each sensor location and time step is independent of the noise values at other locations and time steps. Figure 7.1 provides strong evidence for this assumption in the form of the estimated auto-correlation function. This has important consequences in forming our model. This is because for independent random variables we know that, the joint distribution factors as a product of the individual (marginal) distributions. Mathematically speaking, if the random variables are independent, we know we can write

$$p_H(\eta_1(1), \eta_2(1), \ldots, \eta_M(N)) = p_H(\eta_1(1))p_H(\eta_2(1)) \cdots p_H(\eta_M(N)) = \prod_{n=1}^{N} \prod_{j=1}^{M} p_H(\eta_j(n)) \quad (7.4)$$

so that a probabilistic model for our entire observed data set becomes.

$$p_H(\boldsymbol{\eta}) = \prod_{n=1}^{N} \prod_{j=1}^{M} p_H(\eta_j(n))$$

$$= \prod_{n=1}^{N} \prod_{j=1}^{M} p_H(y_j(n) - x_j(n, \boldsymbol{\theta}))$$

$$= \prod_{n=1}^{N} \prod_{j=1}^{M} \frac{1}{\sqrt{2\pi\sigma^2}} e^{-\frac{1}{2\sigma^2}(y_j(n) - x_j(n,\boldsymbol{\theta}))^2}$$

$$= \frac{1}{(2\pi\sigma^2)^{\frac{N \times M}{2}}} e^{-\frac{1}{2\sigma^2} \sum_{n=1}^{N} \sum_{j=1}^{M} (y_j(n) - x_j(n,\boldsymbol{\theta}))^2}.$$

$$\equiv p_H(\mathbf{y}|\boldsymbol{\theta}). \qquad (7.5)$$

Equation (7.5) is similar to the one we arrived at in Chapter 3 (eqn. 3.139), that is, a probabilistic model describing how our entire sequence of observations $\mathbf{y}$ are distributed. This description is conditional on our model response $x_j(n, \boldsymbol{\theta})$. Because the model response is completely specified by the parameters $\boldsymbol{\theta}$, we often write Eq. (7.5) as $p_H(\mathbf{y}|\boldsymbol{\theta})$ for shorthand. Our data model is also conditional on σ^2, of course, which becomes another parameter we must estimate (unless we happen to know our noise variance *a priori*).

This equation is of central importance in estimation theory and is the likelihood function referred to in Chapter (6). It is so named because it describes the probability of having observed our data $y_j(n)$ given our chosen description of reality (our model) $x_j(n, \boldsymbol{\theta})$. This is the point at which the two main approaches to estimating $\boldsymbol{\theta}$ diverge.

7.1   Method of Maximum Likelihood

The method of maximum likelihood is one of the most powerful estimation tools available for the reasons mentioned in Section (6.2). In that section, we covered the basics of maximum likelihood estimates (MLEs), focusing on why they are desirable as point estimators of parameters. In this section, we begin to tailor the discussion toward structural dynamics applications, as our likelihood usually has the specific form given by (7.5).

The basic concept behind the MLE is straightforward: find the parameters $\boldsymbol{\theta}$ that maximize our probability of having observed our structural response $\mathbf{y}$, that is, find

$$\hat{\boldsymbol{\theta}} = \max_{\boldsymbol{\theta}} p_H(\mathbf{y}|\boldsymbol{\theta}). \tag{7.6}$$

As in Section (6.2), we note that neither the constant in front of the likelihood nor taking the logarithm influence the maximum, hence the maximization problem (7.6) is equivalent to the minimizer

$$\hat{\boldsymbol{\theta}} = \min_{\boldsymbol{\theta}} \sum_{n=1}^{N} \sum_{j=1}^{M} (y_j(n) - x_j(n, \boldsymbol{\theta}))^2. \tag{7.7}$$

We therefore see that in the instance that the noise is jointly Gaussian distributed, with independent values, the MLE is the familiar least-squares estimate. It should be stressed that it is *only* for the jointly Gaussian noise model that the estimator (7.7) is an MLE; other noise models will result in different functional forms of the MLE. Nonetheless, the Gaussian assumption is a common one and we therefore focus largely on (7.7).

While conceptually simple, this is often not a straightforward optimization problem. First of all, most structural response data do not depend on the structure's parameters in a linear way. Assume our spring-mass system from Eq. (3.135) and use the proportional damping model $c = 2\zeta m \sqrt{k/m}$, in which case the equations of motion for free vibration become

$$\ddot{x}(t) + 2\zeta \omega_n \dot{x}(t) + \omega_n^2 x(t) = 0 \tag{7.8}$$

where $\omega_n^2 = k/m$. The parameter vector that completely specifies the response is $\boldsymbol{\theta} \equiv (m, \zeta, \omega_n, x(0), \dot{x}(0))$. The solution to this initial value problem was given in Chapter (4) (see Eq. 4.33) and is

$$x(t, \boldsymbol{\theta}) = e^{-\zeta \omega_n t} \left[x(0) \cos(\omega_d t) + \frac{\dot{x}(0) + \zeta \omega_n x(0)}{\omega_d} \sin(\omega_d t) \right] \tag{7.9}$$

Using the additive noise model for our observations, $y(t) = x(t, \boldsymbol{\theta}) + \eta(t)$, and substituting into the likelihood (7.5) clearly renders (7.6) a nonlinear optimization problem even for this simple model. In general, minimizing (7.7) is a nonlinear least-squares problem for which we will require rather sophisticated minimization algorithms (see Section 7.1.3). However, for some structural dynamics problems, where a lumped parameter model is used, (7.7) can be recast as a linear least-squares problem, the mechanics of which are discussed in Section 7.1.1.

Before we get to specific MLEs, it is worth recalling a few basic properties. We have just shown that under the assumption of independent, Gaussian noise, the familiar "least-squares" solution is an MLE in the sense that it maximizes Eq. (7.5). As we showed in Section (6.2), the MLE is asymptotically (large N) the minimum variance, unbiased estimator of our parameter. Moreover, the estimator is normally distributed with known variance given by Eq. (6.10), thus allowing us to form asymptotic expressions for a confidence interval. However, rather than rely on asymptotics to form the interval, it is common to simply repeat the estimation procedure some number of times and estimate the variance of the resulting spread of values. We discussed this approach in section 6.9.3 and will, in fact, use this approach in later examples.

Another important issue to keep in mind is that of *identifiability*. Particularly in structural dynamics, we have found that there exist multiple combinations of parameters that can yield the same (or very nearly the same) maximum in the likelihood function. However, recall from Section (6.2) that for an MLE to exist, we require that the maximum be unique.

Perhaps the simplest example that illustrates this problem is the following. Assume our system (7.8) and that we know ζ (the dissipative constant) but do not know k or m. By inspection we can see the problem: only the *ratio* of unknown parameters (k/m) governs the response in Eqn. 7.9. Now assume we have observed $N = 5000$ values of the response $y(n), n = 1 \cdots N$ to an initial displacement $x(0) = 1.0 \ m$, each corrupted with an independently chosen, normally distributed, zero-mean random variable with variance $\sigma_Y^2 = 0.1$. The negative log of the likelihood is

$$-\log(p_H(\mathbf{y}|k, m, \zeta)) = \sum_{n=1}^{N} (y(n) - x(n, k, m, \zeta))^2 \qquad (7.10)$$

which we would like to minimize to find k, m. The true parameter values used in this numerical experiment are $m = 0.01$ kg and $k = 1$ N/m. The dimensionless damping ratio was fixed to $\zeta = 0.05$. Figure 7.2 shows the log likelihood as a function of the parameters m, k. The true parameter values clearly maximize this function, but so too do *any* parameter combinations that yield the ratio $k/m = 100$. Thus, these parameters cannot be identified individually by the MLE procedure. In the second plot, the likelihood function $p_Y(\mathbf{y}|k, \zeta)$ is shown (i.e., mass is known but damping and stiffness unknown). The true parameters are $k = 1$ N/m and $\zeta = 0.05$. For this estimation problem, the parameters *are* identifiable and there exists a unique MLE. A fair question might be: how do we know *a priori* whether our model is identifiable? We have found the most straightforward (if not elegant) approach is to simply take the likelihood and plot it for various parameter combinations. Constant ridges in parameter space are indications that any optimization routine, no matter how sophisticated, will have difficulty in maximizing the likelihood function.

In this case, the way forward is rather clear. Either one modifies the physics, for example, changing the sensor locations, choosing a different driving signal, and so on, or one makes an additional modeling assumption. We would imagine the latter solution to be preferable in most situations. For example, in a situation such as the above-mentioned, we might assume the mass to be known and fix it in the model to a value m^*. The resulting estimator for stiffness will be well defined in this case, but should be noted as a conditional estimator. That is to say, our MLE for stiffness would be $\hat{k} = \max_{k} p_Y(\mathbf{y}|k, m = m^*)$. In a later experimental example, we make exactly such an assumption. In what follows, we describe the mechanics of a few specific MLE approaches suitable for structural dynamics problems.

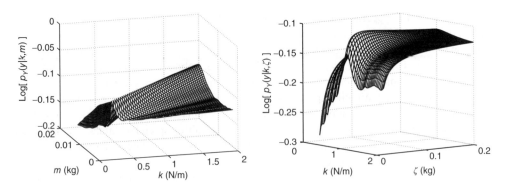

Figure 7.2 Log-likelihood (7.10) as a function of the unknown parameters. In the first example, the unknown parameters are stiffness k and mass m. This particular estimation problem does not yield a unique solution as evidenced by the constant-valued "ridge" for $k/m = 100$. In the second example, the unknown parameters are the damping ratio ζ and stiffness k. In this case, there is a well-defined maximum at the true parameter values, $\zeta = 0.05$, $k = 1$ N/m

7.1.1 Linear Least Squares

Consider a single measurement of the displacement x of the system defined by Eq. (7.8) but with a harmonic forcing term. Also assume that the measurement $y(t) = x(t) + \eta(t)$ is corrupted by an additive Gaussian random process, $\eta(t)$. Further, assume that we may numerically differentiate the observations (see, e.g., [2]) to yield $\dot{y}$ and $\ddot{y}$ or that we can measure these signals directly. Using this model, we may write

$$m\ddot{y} + c\dot{y} + ky = A\cos(\omega t) + \eta_t \tag{7.11}$$

where we have replaced the model displacement, velocity, and accelerations $x, \dot{x}, \ddot{x}$ with the measured versions (e.g., $x = y - \eta$). At the same time, we have "lumped" all of the measurement and modeling errors into a single noise term, η_t. This is a common "sleight-of-hand," but one with strong theoretical underpinnings. The noise term on the right-hand side is actually $\eta_t = m\eta^{(1)} + c\eta^{(2)} + k\eta^{(3)}$, which acknowledges that the noise associated with each of the three measurements is (potentially) different. However, we know from prior chapters that a linear combination of normally distributed random variables is also normally distributed, hence η_t is indeed normally distributed with a variance that depends on the noise level and the system parameters. It is also assumed in this analysis that the forcing function is known, however this is often the case (after all, we often can control the forcing). Now, because of the Gaussian noise assumption, the negative log-likelihood becomes

$$\sum_{n=1}^{N} (\ddot{y}(n) + c\dot{y}(n) + ky(n) - A\cos(\omega n))^2 \tag{7.12}$$

and the minimization problem (7.7) can be written in matrix form as

$$\hat{\theta} = \min_{\theta} (\mathbf{F} - \mathbf{Y}\theta)^T (\mathbf{F} - \mathbf{Y}\theta) \tag{7.13}$$

where

$$
\mathbf{Y} = \begin{bmatrix} \dot{y}(1) & y(1) \\ \dot{y}(2) & y(2) \\ \vdots & \vdots \\ \dot{y}(N) & y(N) \end{bmatrix}
\tag{7.14}
$$

$$
\mathbf{F} = \begin{Bmatrix} A\cos(\omega\Delta_t) - m\ddot{y}(1) \\ A\cos(\omega 2\Delta_t) - m\ddot{y}(2) \\ \vdots \\ A\cos(\omega N\Delta_t) - m\ddot{y}(N) \end{Bmatrix}
\tag{7.15}
$$

and

$$
\boldsymbol{\theta} = \begin{Bmatrix} c \\ k \end{Bmatrix}
\tag{7.16}
$$

Notice that owing to the aforementioned issue of identifiability, we have assumed that the mass is known, and hence the product of mass and observed acceleration can be subtracted off both sides of the equation. Our reasons for working toward this form are that the solution to (7.13) is the linear least-squares estimator referred to in numerous references. This solution is found by taking the derivative of (7.13) with regard to $\boldsymbol{\theta}$ and setting equal to zero, yielding

$$
0 = \frac{d}{d\theta}\left(\mathbf{F}^T\mathbf{F} - 2(\mathbf{Y}\theta)^T\mathbf{F} + (\mathbf{Y}\theta)^T(\mathbf{Y}\theta)\right)
$$

$$
0 = 0 - 2\mathbf{Y}^T\mathbf{F} + 2(\mathbf{Y}^T\mathbf{Y})\theta
$$

$$
\theta = (\mathbf{Y}^T\mathbf{Y})^{-1}\mathbf{Y}^T\mathbf{F}
\tag{7.17}
$$

The key to the least-squares solution is the inversion of the data matrix $\mathbf{Y}^T\mathbf{Y}$.

Typically, the number of parameters is much less than the number of data N, hence this is an *overdetermined* least-squares problem. In this case, the matrix $\mathbf{Y}^T\mathbf{Y}$ is not invertible, and hence we cannot simply form the estimator (7.17) without invoking some type of constraint. To this end, the Moore–Penrose pseudoinverse $\mathbf{Y}^+$ can be used to minimize (7.13) via

$$
\hat{\boldsymbol{\theta}} = (\mathbf{Y}^T\mathbf{Y})^{-1}\mathbf{Y}^T\mathbf{F} = \mathbf{Y}^+\mathbf{F}
\tag{7.18}
$$

Most software packages have a pseudoinverse function built in. The implicit constraint in the solution (7.18) is that it is the solution with the smallest mean square error between true and estimated parameter values. One of the simplest ways to compute the pseudoinverse is via singular value decomposition

$$
\mathbf{Y} = \mathbf{U}\boldsymbol{\Sigma}\mathbf{V}^*
\tag{7.19}
$$

from which we can form

$$
\mathbf{Y}^+ = \mathbf{V}\boldsymbol{\Sigma}^{-1}\mathbf{U}^*
\tag{7.20}
$$

where $*$ denotes complex conjugate (as usual) and the diagonal matrix $\boldsymbol{\Sigma}$ is inverted by simply taking the reciprocal of its values. This estimator will be used in an upcoming example in Section (7.3.1); we give two brief examples here.

For a numerical example, we consider (7.11) with $k = 1000$ N/m, $c = 10\text{N} \cdot s/m$, and forcing with $A = 1N$ amplitude and $\omega = 10$ rad/s. We collected $N = 1000$ observations of the displacement $y(n) = x(n) + \eta(n)$, where each $\eta(n) \sim N(0, 4.6e - 4)$. Numerically differentiating twice to get velocity and displacement, we use the estimator (7.18) which yielded $\hat{c} = 9.98\text{N} \cdot s/m$ and $\hat{k} = 997.60$ N/m. Both values are quite close to the true parameter values.

As a second example, we can demonstrate this estimator in an experiment. Figure 7.3 shows an experimental implementation of the nonlinear Duffing's equation [3]

$$\ddot{x} + c\dot{x} - k_1 x + k_3 x^3 = A\cos(\omega t). \qquad (7.21)$$

This equation has proved useful as a model for post-buckled vibration in structural systems [4] and therefore models a common form of structural damage. As shown in the figure, a cart moving on a track shaped as a quartic potential traces out the solution $x(t)$ and therefore mimics a buckled structure's response. Although this is a nonlinear system, it can be cast in the desired linear least-squares form. Following the same procedure as before, we form the matrices

$$\mathbf{Y} = \begin{bmatrix} \dot{y}(1) & y(1) & y(1)^3 \\ \dot{y}(2) & y(2) & y(2)^3 \\ \vdots & \vdots & \vdots \\ \dot{y}(N) & y(N) & y(N)^3 \end{bmatrix} \quad \mathbf{F} = \begin{Bmatrix} A\cos(\omega\Delta_t) - \ddot{y}(1) \\ A\cos(\omega 2\Delta_t) - \ddot{y}(2) \\ \vdots \\ A\cos(\omega N\Delta_t) - \ddot{y}(N) \end{Bmatrix} \qquad (7.22)$$

and

$$\boldsymbol{\theta} = \begin{Bmatrix} c \\ k_1 \\ k_3 \end{Bmatrix} \qquad (7.23)$$

The estimator $\hat{\boldsymbol{\theta}} = \mathbf{Y}^+\mathbf{F}$ yielded values of all three parameters that were within 5% of their expected (designed) values. Substituting these estimates into the model produced the estimated response curve shown in Figure 7.3, which is in excellent agreement with the experimentally obtained data.

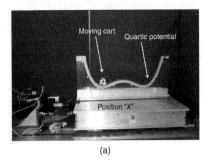

(a)

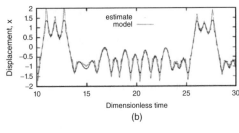

(b)

Figure 7.3 (a) Experimental cart-track system that mimics the structural response of a post-buckled beam. (b) Linear least-squares fit to the equations of motion using the estimator (7.18)

These examples show that with knowledge of the input and a willingness to deal with an overdetermined problem, we can obtain meaningful estimates of parameter values in structural system. This basic approach also underlies a popular approach in damage detection, referred to as finite element (FE) model updating. This approach to damage detection is discussed next.

7.1.2   Finite Element Model Updating

There exists a broad class of problems in structural damage detection that can loosely be described as some form of FE model updating. Just as it sounds, one forms an FE model of the structure and parameterizes the elements with a damage-related parameter. As data are acquired, a cost function is minimized to yield the parameter vector for all elements. Those elements for which the damage parameter is large are taken as damaged.

To describe the approach, we return to the FE description of a structural system given in Section (4.3.4), Eq. (4.26). An FE model for an unforced, N-degree-of-freedom (NDOF) structure was shown to be modeled by the second-order system of equations

$$\mathbf{M}\ddot{\mathbf{x}} + \mathbf{C}\dot{\mathbf{x}} + \mathbf{K}\mathbf{x} = 0. \tag{7.24}$$

As we have also shown in Section 4.5.1 (see Eq. 4.51), the solution can be given in terms of the eigenvalues and eigenvectors which are obtained by solving the eigen-value problem

$$(-\omega_i^2 \mathbf{M} + \mathbf{K})\mathbf{u}_i = 0, \ i = 1 \cdots N \tag{7.25}$$

where $\omega_i^2, \mathbf{u}_i$ are, by definition, the system natural frequencies and mode shapes, respectively. As we have alluded to many times, it is appropriate to model damage as a change in the structure's stiffness properties while the mass properties remain constant. Hence, we write for the damage system stiffness

$$\tilde{\mathbf{K}} = \mathbf{K} + \Delta\mathbf{K} \tag{7.26}$$

which gives rise to the altered eigenvalues and eigenvectors

$$\tilde{\omega}_i^2 = \omega_i^2 + \Delta\omega_i^2$$
$$\tilde{\mathbf{u}}_i = \mathbf{u}_i + \Delta\mathbf{u}_i \tag{7.27}$$

where we have denoted the *change* in measured eigenvalues as $\Delta\omega_i^2 = \tilde{\omega}_i^2 - \omega_i^2$. Now, if we replace the stiffness and eigenvalues in (7.25) with their damaged counterparts, we have the modified eigenvalue problem

$$-(\omega_i^2 + \Delta\omega_i^2)\mathbf{M}(\mathbf{u}_i + \Delta\mathbf{u}_i) + (\mathbf{K} + \Delta\mathbf{K})(\mathbf{u}_i + \Delta\mathbf{u}_i) \tag{7.28}$$

If we expand this expression and neglect all higher order terms (products of two Δ quantities), we are left with

$$-\omega_i^2\mathbf{M}\mathbf{u}_i - \Delta\omega_i^2\mathbf{M}\mathbf{u}_i - \omega_i^2\mathbf{M}\Delta\mathbf{u}_i + \mathbf{K}\mathbf{u}_i + \Delta\mathbf{K}\mathbf{u}_i + \mathbf{K}\Delta\mathbf{u}_i \tag{7.29}$$

Premultiplying both sides by $\mathbf{u}_i^T$, and noting that $-\omega_i^2\mathbf{u}_i^T\mathbf{M}\mathbf{u}_i + \mathbf{u}_i^T\mathbf{K}\mathbf{u}_i = 0$ and $-\omega\mathbf{u}_i^T\mathbf{M}\Delta\mathbf{u}_i + \mathbf{u}_i^T\mathbf{K}\Delta\mathbf{u}_i = 0$, we arrive at

$$-\Delta\omega_i^2(\mathbf{u}_i^T\mathbf{M}\mathbf{u}_i) + \mathbf{u}_i^T\Delta\mathbf{K}\mathbf{u}_i = 0 \tag{7.30}$$

or, if we have mass-normalized the mode shapes (recall the discussion surrounding Eq. (4.52)

$$\mathbf{u}_i^T \Delta \mathbf{K} \mathbf{u}_i = \Delta \omega_i^2 \tag{7.31}$$

which relates the change in stiffness properties to the change in natural frequencies. We know the $\mathbf{u}_i$ from our model and can measure $\Delta \omega_i^2$ using the frequency-domain estimation methods described in Chapter (6). Our ultimate goal is to use this information to estimate the change in stiffness due to the damage, however the problem is not yet in a form that easily lends itself to our estimation methods.

To get it into such a form, consider the parameter vector $\boldsymbol{\alpha}$ which denotes the stiffness value of each element (EI for an Euler–Bernoulli beam). Using this formulation, the change in stiffness can be denoted

$$\begin{aligned}
\Delta \mathbf{K} &= \frac{\partial K}{\partial \alpha_1} \Delta \alpha_1 + \frac{\partial K}{\partial \alpha_2} \Delta \alpha_2 + \cdots + \frac{\partial K}{\partial \alpha_M} \Delta \alpha_M \\
&= \sum_{j=1}^{M} \frac{\partial K}{\partial \alpha_j} \Delta \alpha_j \\
&= \sum_{j=1}^{M} \mathbf{K}_j^{(e)} \Delta \alpha_j
\end{aligned} \tag{7.32}$$

where $\mathbf{K}^{(e)}$ is the elemental stiffness matrix for the type of FE element being used. Pre- and post-multiplying by the eigenvectors (i.e., substituting into 7.31) yields

$$\left(\mathbf{u}_i^T \mathbf{K}_j^{(e)} \mathbf{u}_i \right) \Delta \alpha_j = \Delta \omega_i^2 \tag{7.33}$$

or, in matrix form

$$\mathbf{S} \Delta \alpha = \Delta \omega^2 \tag{7.34}$$

where the dimensions of $\mathbf{S}$ are $N \times M$ and where N are the number of degrees of freedom and M are the number of elements used in the model. The dimension of $\Delta \omega^2$ is therefore $N \times 1$, while the unknown damage parameter vector $\boldsymbol{\alpha}$ is $M \times 1$. As written, this is an overdetermined system of equations as $N > M$ for a typical FE model. However, in practice, this system of equations must be solved using estimates of the changes in the first N' natural frequencies, that is, $\Delta \omega_i^2$, where $N' \ll N$. This is because we can only reliably excite, and subsequently estimate, the first few modes of a structure. Hence, in practice, Eq. (7.34) is actually an *underdetermined* system for which a unique solution does not exist unless a constraint is imposed on the problem.

For these types of problems, it is common to use the estimator

$$\Delta \hat{\alpha} = \mathbf{S}^+ \Delta \omega^2 \tag{7.35}$$

where $\mathbf{S}^+$ is again the Moore–Penrose pseudoinverse discussed in the previous section. Now, in this approach, the connection to maximum likelihood is tenuous. This estimator is an MLE, provided that system noise and estimation error can be lumped into a Gaussian noise term corrupting the right-hand side of (7.33). Nowhere in the description of the method was there an explicit acknowledgment of the noise model, and hence there is no way to formally assess

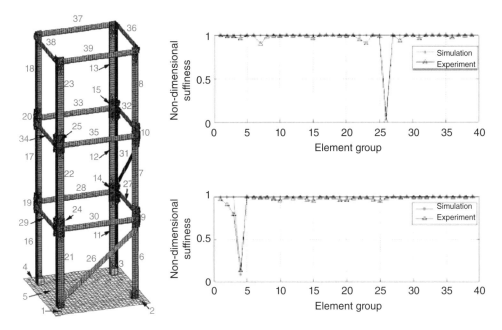

Figure 7.4 Subset of results presented in Ref. [11]. Shown are the experimental truss structure used in the experiment, with numbers corresponding to different truss member groups. Also shown are the experimental results obtained by loosening (respectively) groups 26 and 4. On the basis of the estimated natural frequencies, the modified version of the algorithm described in this section is correctly able to identify the near total loss of stiffness that occurs when the connections in these members are loosened to a hand-tight configuration (*Source:* Adapted from [11] with permission from ASME)

properties of the estimator. Suffice to say, anytime we obtain a least-squares solution, we are implicitly assuming a jointly normal probability model for the errors. In most cases, this is probably a very good assumption. However, it should always be kept in mind that this is what is going on "behind the scenes" of the estimator.

A number of researchers have applied variations to this basic approach in both numerical and experimental examples (see, e.g., [5–10]). Given the availability, fidelity, and speed of today's FE codes, this approach represents a powerful means of using the physics of the problem to draw inferences about the state of the structure. As an example of model updating, consider the experimental work described in Zhu and He [11]. While the specifics of the identification algorithm differ from those outlined earlier, the basic approach is the same. In this example, the goal was to locate damage (taken as a stiffness degradation) to the truss structure shown in Figure 7.4. The different truss elements are numbered in the figure. Also shown are the results of a damage detection scheme, whereby the bolted joint connection in groups 4 and 26 are loosened to a "hand-tight" configuration. In both cases, the algorithm correctly identifies the condition as a near total loss of stiffness in the element.

The key assumptions to this approach are that (i) the elemental mass is not changing because of the damage, (ii) the nature of the damage is a linear reduction in stiffness, (iii) the deviation from the pristine condition is so small that the higher order terms in the model can be safely neglected. For more general problems in structural system identification, particularly damage detection, (7.7) is a complicated function of the system parameters and is therefore nontrivial to minimize. The likelihood associated with damage parameters is often characterized by multiple maxima and therefore simple maximization algorithms such as gradient descent can fail to find the global maximum. In Section 7.1.3 we present an algorithm designed specifically to provide MLEs for structural system parameters.

7.1.3 Modified Differential Evolution for Obtaining MLEs

Without restricting to small parameter changes (as in model updating), the function defining our minimization problem (7.7) becomes extremely difficult to search, possessing a number of local minima. This property renders the more classical optimization algorithms (e.g., gradient descent) ineffective. We can therefore turn to a different class of optimization routines referred to as evolutionary algorithms (EAs).

The idea behind EAs is to use the principles of natural selection and biological evolution for the purpose of exploring multimodal functions for which the global minimum is difficult to obtain. The EA works by proposing a population of solutions (parameter vectors), and then iteratively modifies that population toward areas of higher likelihood. The algorithm terminates when some convergence criteria is reached. In problems of structural damage identification, EAs have proved particularly useful as they are known to be adept at exploring complex solution spaces. The early work of Friswell *et al.* [12] used an EA to search for damage in an experimental cantilevered plate. Additional work by Horibe and Watanabe [13] also used an EA to identify cracks in plate structures, while Stull *et al.* used EAs to identify initial geometric imperfections in shell structures [14].

While these works emphasized the utility of EAs in solving structural inverse problems, the associated estimates incur a considerable computational cost. Some EA-based optimizations can require $O(10^3)$ evaluations of the forward model, which, for many structural problems, can be prohibitive, taking days or weeks to perform a single optimization. This is particularly true if one is considering a detailed FE model (as is done next). Thus, for structural dynamics applications, the goal is to develop an algorithm that retains the exploratory power of an EA, but also converges quickly for structural model parameter estimation problems.

In this section, we describe one particular variant of an EA, referred to as differential evolution (DE). DE is characterized by mutation and crossover operations (to be described) of the potential solution vectors to form trial solutions. The original DE algorithm was proposed by Storn and Price [15], and has since enjoyed widespread popularity among a number of different scientific communities, including structural system parameter identification (see, e.g., Das and Suganthan [16]). Structural system parameters identified via DE have included the stiffness matrix coefficients of a short, cantilevered beam [17]; the mass and stiffness properties of a three-floor frame and a steel-concrete bridge [18]; as well as the mass, damping, and stiffness properties of a two-dimensional shear frame style structure [19]. The authors in Ref. [20] utilize DE to identify the components of the excitation forces of an experimental dynamic system. In what follows, we describe the basic DE algorithm and then later show how it can be modified to speed convergence of the solution (see Section 7.1.3).

Recall that our goal is to find the minimizer of (7.7). To do so, we define the cost function (our negative log-likelihood)

$$C(\mathbf{\theta}) = \sum_{n=1}^{N} \sum_{j=1}^{M} (y_j(n) - x_j(n, \mathbf{\theta}))^2 \qquad (7.36)$$

and seek the values of $\mathbf{\theta}$ for which this function is minimized. An initial population $\theta^{(q)}, q = 1 \cdots Q$ of P-dimensional parameter vectors is created, with values assigned at random from the parameter support. Sequentially, for each of the q parameter vectors within the population, one selects with uniform probability, the discrete indices $u, v, w \sim U(1, Q)$ such that $u \neq v \neq w \neq q$. A new parameter vector is then formed as

$$\theta^{u'} = \theta^{(u)} + \gamma(\theta^{(v)} - \theta^{(w)}) \qquad (7.37)$$

where the real constant, γ, controls the amplitude of the differential perturbation. In Reference [15] it was required that $\gamma \in [0, 2]$. The operation described by Eq. (7.37) is referred to as the *mutation* operation, and, accordingly, the resulting new vector is called the *mutant vector*. This mutant vector is then subjected to a so-called crossover operation to create a "trial vector", whereby each of the P parameters in $\theta_p^{(u')}$ is assigned to θ_p^{trial} with "crossover" probability Pr_{cr}. That is to say, for each of the p vector elements we set

$$\theta_p^{trial} = \begin{cases} \theta_p^{(u')} & : \text{ if } U(p) \sim [0, 1] < Pr_{cr} \\ \theta_p^{(q)} & : \text{ otherwise} \end{cases} \qquad (7.38)$$

where initially $Pr_{cr} = 0.5$ as in the standard DE algorithm from the literature. The final step of the algorithm is to retain

$$\theta^{(q)} = \begin{cases} \theta^{(q)} & : \text{ if } C(\theta^{(q)}) < C(\theta^{trial}) \\ \theta^{trial} & : \text{ otherwise} \end{cases} \qquad (7.39)$$

as the parameter vector carried forward to the next iteration. This defines one iteration k of the algorithm and the process is continued until a user-specified number of iterations K have been completed (or some other stopping criteria is used). Each iteration, therefore, produces the parameter vector $\theta^{(q)}(k)$ and we can take the best of the Q parameter vectors at the final iteration as our MLE. In other words, denote the DE-based MLE as

$$\hat{\theta} = \min_q C(\theta^{(q)}(K)) \qquad (7.40)$$

The schematic in Figure 7.5 illustrates one iteration of the DE algorithm.

While the DE algorithm is effective, it is time consuming to implement because of the repeated execution of the forward model. However, we can use what we know about typical parameter spaces for structural systems to generate improvement. Specifically, we note that although structural parameter spaces are often multimodal, they are typically smoothly varying and often possess well-spaced maxima (see, e.g., Figure 9.6). For this type of space, we can dynamically (as the algorithm progresses) alter the probability of selecting a perturbation term in (7.37) that results in a direction of lower cost, as in Figure 7.6. The idea is to

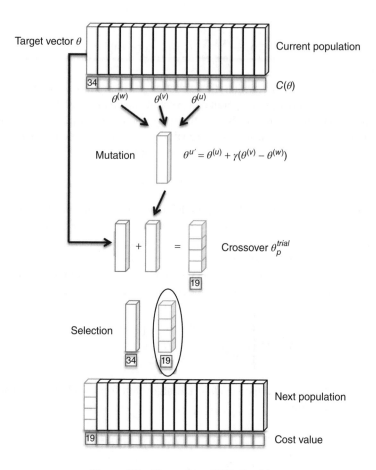

Figure 7.5 Illustration of DE algorithm

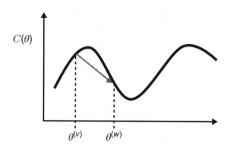

Figure 7.6 Parameter vector selection that results in decreasing cost. (*Source:* Reproduced from [21], Figure 5, with permission of Elsevier)

gradually tune the algorithm away from perturbations that lead in a direction of lower like-lihood, so that at the last iteration ($k = K$) only negative cost differentials are chosen. We also modify the crossover parameter dynamically, and show that this allows further gains in convergence.

Let $d_{qq'}$ denote the scalar difference in cost between parameter vector q and parameter vector q'. That is to say

$$d_{qq'} = C(\theta^{(q)}) - C(\theta^{(q')}) \qquad (7.41)$$

The resulting collection $d_{qq'}, q, q' = 1 \cdots Q$ is guaranteed to be symmetric about zero because $d_{qq'} = -d_{qq'}$. However, instead of choosing θ^v, θ^w with uniform probability, choose

$$\theta^{(v)}, \theta^{(w)} : \quad \mathrm{Pr}(d_{vw} < 0) \equiv Pr_d(k) \qquad (7.42)$$

That is to say, define $Pr_d(k)$ as the probability that the parameter vectors have a difference pointing in a direction of lower cost, and where $Pr_d(k)$ increases with iteration $k = 1, \ldots, K$. For a constant $Pr_d(k) = 0.5$ (for all k), the traditional DE algorithm is recovered. In this case, we are just as likely to generate a candidate vector using a perturbation toward a higher cost as toward a lower cost. Figure 7.7 illustrates how the probability of choosing a parameter vector differential in the direction of lower cost (i.e., $-d_{qq'}$) increases as the iterations k increase, so that for $k = K$ only difference vectors pointing toward lower cost are chosen.

By allowing $Pr_d(k)$ to increase with iterations k, we may slowly bias these perturbations in directions of lower cost (i.e., higher likelihood). Early in the algorithm, when broad explo-ration of the parameter space is beneficial, $Pr_d(0) = 0.5$ is appropriate as this choice allows for moves in all different directions within the parameter space. However, once the solutions begin to converge toward the global minimum, using $Pr_d(k) > 0.5$ offers an increased rate of convergence, as will be shown. In short, we dynamically change the DE algorithm from a true EA to a basic gradient descent.

At the same time, we have found that further gains can be made by allowing $Pr_{cr}(k)$ to increase with successive iterations. The idea is that, as we become more confident that our mutated vectors are pointing toward the true minimum, we would like to increase the prob-ability of selecting parameters from the mutated vectors (versus retaining the original target vectors). In the next section, we use this algorithm on a simple numerical example, while in Chapter (9) we use it in a much more complicated dent identification problem for shell structures.

7.1.4 Structural Damage MLE Example

As we have just stressed, identifying parameters in structural systems is a challenging propo-sition because of the lack of uniqueness associated with different parameter combinations. Moreover, even in cases where an MLE does exist, the parameter space frequently exhibits multiple "ridges" of near equal likelihood. This challenge is further compounded in the dam-age identification problem. One of our primary goals is to find the damage early, before it becomes large. Thus, we are often looking for the proverbial needle in a haystack, trying to determine the values of parameters that have very little effect on the response data. It was sug-gested in the previous section that a modified differential evolution approach is well suited to finding MLEs of structural parameters even in this challenging case.

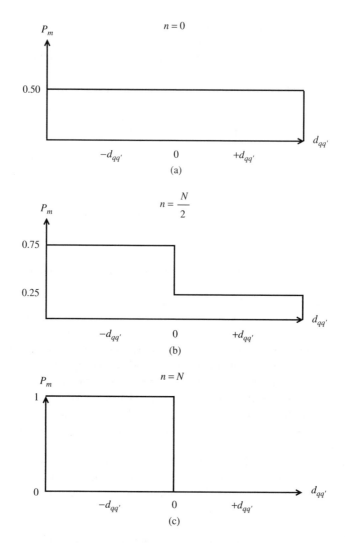

Figure 7.7 Distribution of cost differentials $d_{qq'}$ for increasing iterations k. (a) $k = 0$; (b) $k = \frac{K}{2}$; (c) $k = K$. (*Source:* Reproduced from [21], Figure 6, with permission of Elsevier)

In this section, we give a simple example depicted in Figure 7.8. A one-dimensional FE model of a cantilevered beam is modeled using D Euler–Bernoulli elements (as discussed in Chapter 4). Each element has two degrees of freedom (rotation and deflection) so that there are $2D$ DOFs in this system. Each element is governed by a mass and stiffness matrix, parameterized by the mass per unit length $m(x)$ and a stiffness $k(x) = EI(x)$. It is assumed at the outset that the beam dimensions and weight can be measured so that we know $m(x)$. The stiffness function, on the other hand, is assumed unknown at the outset. In addition, we consider damage as a stiffness reduction in one or more collection of elements that is, a "notch." This simple damage model is shown schematically in Figure 7.8. The job of a damage identification

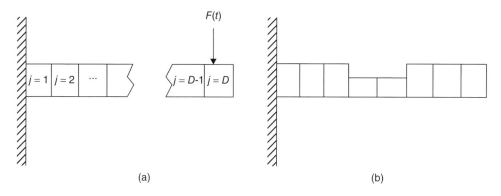

Figure 7.8   (a) Finite element model of a cantilevered beam structure. (b) Simple damage model, whereby one or more stiffness elements are degraded by a certain amount

procedure is to determine the presence, location, depth, and extent of this stiffness reduction based on the structural vibration data.

To accomplish this, an impact load is assumed to be applied to a particular element and we observe the displacement response at the jth element and model it with the (now) familiar expression

$$y_j(n) = x_j(n, \boldsymbol{\theta}) + \eta_j(n), \quad n = 1 \cdots N, \quad j = 1 \cdots M \tag{7.43}$$

using a sampling rate of Δ_t for a duration $T = N\Delta_t$ seconds (i.e., discrete time n corresponds to time $t = n\Delta_t$). As in our previous example, the jth model response signal is given by $x_j(n, \boldsymbol{\theta})$, where we denote the dependence on the model parameters explicitly. The observed data $y_j(n)$ are assumed corrupted with additive Gaussian noise $\eta_j(n\Delta_t)$ such that the signal-to-noise ratio (SNR) for the jth response is given by

$$SNR_j = 10 \log_{10} \left(\frac{\frac{1}{N} \sum_{n=1}^{N} x_j^2(n, \boldsymbol{\theta})}{\sigma_{\eta_j}^2} \right) \tag{7.44}$$

where $\frac{1}{N} \sum_{n=1}^{N} x_j^2(n, \boldsymbol{\theta})$ is the signal energy. While it is tempting, this quantity should not be confused with the signal variance. To be a variance estimate, $x_j(n)$ would have to be a stationary random process, which it is not. In fact, the observed signals $y_j(n) = x_j(n) + \eta_j(n)$ are technically each being modeled as a *nonstationary* random process. The reason for this is that each observation is a random variable with probability density $p_{Y_j}(y_j(n)) = \mathcal{N}(x_j(n), \sigma_\eta^2)$, that is, the mean is a function of the discrete time, n.

While this is certainly a problem from the standpoint of estimating statistics from a single realization, it does not affect our ability to estimate the model parameters. Our likelihood function does not presuppose stationarity (see Section 6.2), however we will assume the noise is independent and identically distributed (i.i.d.), and hence the joint PDF for our entire collection of observations (i.e., the likelihood) is

$$p_{\mathbf{Y}}(\mathbf{y}|\boldsymbol{\theta}) = \frac{1}{(2\pi\sigma_\eta^2)^{MN/2}} e^{-\frac{1}{\sigma_\eta^2} \sum_{j=1}^{M} \sum_{n=1}^{N} (y_j(n) - x_j(n|\boldsymbol{\theta}))^2}. \tag{7.45}$$

As we have already mentioned, maximizing this function is equivalent to maximizing the log-likelihood, hence our MLEs should satisfy

$$\hat{\theta} = \min_{\theta} \sum_{j=1}^{M} \sum_{n=1}^{N} (y_j(n) - x_j(n|\theta))^2. \tag{7.46}$$

To this point in the problem, we have said nothing about how we are going to model the damage. One approach is to treat each element's stiffness as a random variable and attempt to estimate all D stiffness values. In this case, we would hope that our estimator would successfully indicate which stiffness elements were degraded and by how much. The problem is that this is a high-dimensional optimization problem as for many FE models the number of elements D is in the hundreds, if not hundreds of thousands!

Instead, we use a modeling approach that will appear in subsequent, more complex examples. We allow the damage to be governed by a particular shape function $d(x, \theta_S)$ with parameters θ_S. In this example, we are trying to identify a stiffness reduction at a specific location. We, therefore, define the damage function

$$d(x|A_S, \mu_S, \sigma_S) = A_S e^{-\frac{1}{\sigma_S^2}(x-\mu_S)^2} \tag{7.47}$$

which is nothing more than a Gaussian-like "bump," with amplitude A_S and width σ_S^2 at location μ_S on the beam. Rather than solving for each individual stiffness value, the stiffness at *any* point on the beam is assumed constant aside from the defect, that is,

$$k(x) = EI - d(x|\theta_S). \tag{7.48}$$

If there is no defect, we would expect $A_S = 0$ and the beam stiffness is just the constant EI. A reduction in stiffness at a particular location will produce a nonzero "bump" $d(x|\theta_S)$. Note that we could have assumed a rectangular shape function that would have exactly matched the damage profile that we have introduced (again, see Figure 7.8). However, this is not information we will likely know about the problem *a priori*. It is far more likely that we will know to look for some form of stiffness reduction, but not know the precise geometry of that reduction. The Gaussian "bump" is just one possible shape function, introduced here because it is used again in later chapters. In fact, part of the point of this example is to demonstrate that one does not need to know the precise physics of the damage.

Finally, for this problem, we assume a proportional damping model for which the damping constant ζ is also unknown. Thus, our model response is a function of the five parameters: $\theta \equiv (EI, \zeta, A_S, \mu_S, \sigma_S)$ where the true values are given by (44, 0.05, 11, 0.163, -). The final parameter (damage standard deviation) does not have a "true" value as the "true" damage was a notch spanning precisely 3 elements, each of (non-dimensional) length 0.065. Our job is to estimate these parameters from the observed data $y_j(n)$, $n = 1 \cdots N$. In this numerical example, we use $D = 20$ elements and consider a 25% stiffness reduction spanning elements $j = 2, 3, 4$ as the true damage state. We further assume the measured data consist of only $M = 3$ response time series collected from elements $j = 3, 5, 7$, each consisting of $N = 1024$ observations with an SNR of 20 dB.

To solve the problem posed by (7.46), we can use the modified DE algorithm of the previous section with a population size of $Q = 20$ and an initial crossover rate of $Pr_{cr} = 0.5$. The algorithm was run for 500 iterations and the resulting MLE for each of the parameters are provided in Table 7.1.

Table 7.1 Maximum likelihood estimates for the beam parameters associated with the example depicted in Figure 7.8

MLEs for beam model parameters				
$\hat{EI}$	$\hat{\zeta}$	$\hat{A}_S$	$\hat{\mu}_S$	$\hat{\sigma}_S$
43.738	0.048	12.222	0.173	0.057

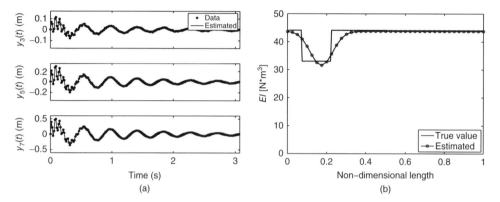

Figure 7.9 (a) Predicted response based on MLEs of the model parameters and observed data. (b) Estimated stiffness profile plotted against true stiffness profile used in generating the data

The predicted time-series responses are shown in Figure 7.9 along with the identified damage profile. The first thing to note is that the estimates for both stiffness and damping parameters are close to their true values. However, in the case of the damage parameters, the true damage (a "notch") is functionally different from our model of the damage (a "bump"). Although our damage model captures the basic physics, it is incapable of matching the exact stiffness profile. This illustrates one of the key points in this book, namely, that the model only need capture the physics required of the problem at hand. In this particular case, we are interested in identifying the location, depth, and width of a stiffness degradation. Although our damage model does not match the true profile, it is more than capable of providing the needed information. Again, we could have easily changed the functional form of $d(x)$ to more closely match a rectangular profile.

There are a host of issues associated with this example that we did not discuss. For example, if there is more than one area of stiffness reduction, can we simply add a second "bump" to our shape function? What can we say about the convergence of the algorithm used to find the MLE? Might other damage models have proved easier to identify? Do the forcing/response locations help/hurt identifiability? How might we form an interval of confidence for our result? Each of these questions is addressed in later chapters in the context of more realistic damage identification examples. The goal of this simple example was simply to provide the reader with a basic understanding of the process by which MLEs are obtained for a typical damage identification problem.

7.1.5   Estimating Time of Flight for Ultrasonic Applications

We conclude this section with a relevant example for which the familiar "mean-squared-error minimizers" do not yield MLEs. There are a class of damage detection strategies for which the practitioner relies on ultrasonic pulses to determine the presence and location of structural damage. Using this approach, one places a transducer at one point on the structure, and then uses that transducer to excite traveling waves that propagate across the structure (see [22] for a discussion of the physics behind the wave propagation). Also located on the structure are some number of receivers that measure this pulse, the so-called pulse-echo architecture shown in Figure 7.10. The pulses possess very high frequency content ($O(100 \text{ kHz})$) such that the defect will be "seen" by the pulse as a reflecting surface. Using this architecture, one estimates the time of flight between the initial pulse and one or more returns. On the basis of the structural properties and geometry, one can then use the time delay to infer the presence and location of the reflecting surface, which is assumed to coincide with structural damage (see, e.g., [23, 24]. A discussion of the physics of this problem is omitted here as this approach represents an entire field of nondestructive evaluation in structures. A detailed description of pulse echo and other acoustic damage detection methods can be found in various references including [25] and [26].

Instead, we focus on the problem of time-delay estimation (TDE) as this lies at the heart of the ultrasonic approach. As we will show, this is a nontrivial estimation problem, particularly when the corrupting noise source is non-Gaussian in nature. As we will also show, however, we can use the time-delayed mutual information function of Section (3.5.1) to provide an empirical MLE of the pulse delay, even in the non-Gaussian noise case. As such, this estimator

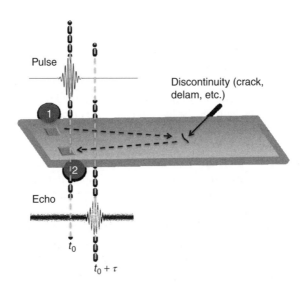

Figure 7.10   Basic architecture of a "pulse-echo" defect detection system. A clean, windowed, high-frequency pulse is sent by an actuator along the length of the structure under test. A discontinuity in the material (e.g., crack) yields a reflecting surface. By monitoring returns at the receiver, one can estimate the delay, τ, between sent and received pulse. On the basis of the material and pulse properties, an estimate of the time delay can be transformed into an estimate of the defect location (should any exist)

enjoys the properties of an MLE and, therefore, so too will the resulting estimate of the damage location.

7.1.5.1   Problem Statement

Denote the time-delayed "sent" pulse as the parametric signal $\mathbf{s}_\tau = [s_\tau(t_0), s_\tau(t_1), \ldots,$ $s_\tau(t_{N-1})]^T$, with $t_n = n\Delta_t$, Δ_t representing the sampling period and $s_\tau(t) \equiv s(t - \tau)$. The parameter τ will be used to represent the time of flight we seek to estimate. The noise-corrupted version, that is, the received signal, will be denoted $\mathbf{r} = [r(t_0), r(t_1), \ldots, r(t_{N-1})]^T$. We assume the usual additive noise model so that the measured digital signal of length N is given by

$$r(t_n) = s_\tau(t_n) + \eta(t_n), n = 0 \cdots N - 1. \tag{7.49}$$

In band-limited systems, typically we have that the sampling period should be at least as small as $\Delta_t = 1/(2B)$, with B (in Hertz) being the bandwidth of the digitization system [27, 28]. We note that in adopting the discrete model, the time delay τ can only be estimated for the discrete set $\tau \in \{0, \Delta_t, \ldots, (N - 1)\Delta_t\}$. If higher "resolution" in τ is required, the set to which τ belongs could be made larger (smaller shifts), or, alternatively, interpolation or approximation techniques can often be applied to that end [28–30]. Finally, when only N samples are available for the signal $s_\tau(t_n)$, it is also necessary to extend the signal by applying some kind of boundary condition so that $s_\tau(t_n)$ can be evaluated for all samples of interest. Here, we apply the periodic boundary conditions. That is, for $\tau = m\Delta_t$, we have that $s_\tau(t_n) = s(\{n - m \bmod N\}\Delta_t)$, where the 'mod' operation refers to the standard remainder after division by integer N. In what follows, we assume that the sequence $\mathbf{r}$ can be modeled as a stationary, ergodic random process.

The goal again is to use the discretely sampled model, $\mathbf{s}_\tau$ and N data samples $\mathbf{r}$ to estimate τ. To accomplish this, we turn to the maximum likelihood estimation framework of Section (6.2) and seek

$$\hat{\tau} = \arg\ \max_\tau p_{\mathbf{R}}(\mathbf{r}|\mathbf{s}_\tau) \tag{7.50}$$

or the equivalent estimator

$$\hat{\tau} = \arg\ \max_\tau \log(p_{\mathbf{R}}(\mathbf{r}|\mathbf{s}_\tau)) \tag{7.51}$$

where $p_{\mathbf{R}}(\cdot)$ is the joint likelihood function associated with the received signal. The most common noise model in (7.49) is the joint normal distribution such that the likelihood function becomes the familiar

$$p_G(\mathbf{r}|\mathbf{s}_\tau) = \prod_{n=0}^{N-1} \frac{1}{\sqrt{2\pi}\sigma} e^{-(r(t_n)-s_\tau(t_n))^2/(2\sigma^2)} \tag{7.52}$$

where σ^2 is the noise variance, and the MLE, $\max_\tau \log(p_G(\mathbf{r}|\mathbf{s}_\tau))$ is given by the linear cross-correlation between data and model as

$$\hat{\tau} = \arg\max_\tau \sum_{n=0}^{N-1} r(t_n)s(t_n - \tau) \tag{7.53}$$

assuming the signal energy $\sum_{n=0}^{N-1} [s_\tau(t_n)]^2$ is constant as a function of τ (see the "matched filter" described in Section 6.9.1). However, as we have stressed throughout, the form of the estimator depends largely on the model for the noise distribution. When the noise process is non-Gaussian, the linear correlation function (7.53) is no longer optimal. For example, if the noise is joint Laplacian distributed, we have the so-called average magnitude difference function (AMDF) described in Refs [28, 31, 32]. As the name implies, the estimator consists of minimizing

$$\hat{\tau} = \arg \min_\tau \sum_{n=0}^{N-1} |r(t_n) - s_\tau(t_n)|. \tag{7.54}$$

We have already seen in Chapter (6) (see Eq. 6.23) how maximizing the log of a Laplacian distribution naturally leads to such an estimator. This estimator has the advantage of low computational complexity, however as will be shown, it performs poorly for many noise models. The chief exception, of course, is the Laplace-distributed noise for which this estimator was proved in Ref. [31] to achieve the Cramer–Rao lower bound (CRLB).

For now, we keep the discussion general and free from any specific assumptions about the likelihood function other than the assumption of an i.i.d. noise process. This, of course, allows us to write the likelihood function as the product

$$p(\mathbf{r}|\mathbf{s}_\tau) = \prod_{n=0}^{N-1} p(r(t_n)|s_\tau(t_n)) \tag{7.55}$$

so that

$$\log(p(\mathbf{r}|\mathbf{s}_\tau)) = \sum_{n=0}^{N-1} \log(p(r(t_n)|s_\tau(t_n)). \tag{7.56}$$

To compute this quantity, it is necessary to be able to evaluate the probabilities $p(r(t_n)|s_\tau(t_n))$, for $n = 0, \ldots, N-1$. Under the assumption that τ is uniformly distributed, and that $s(t_n - \tau \bmod N)$ and $r(t_n)$ are stationary and ergodic, $p(r(t_n)|s_\tau(t_n))$ can be approximated 'empirically' from the sampled data pairs $\{r(t_n), s_\tau(t_n)\}$ via density estimation techniques [33]. The density-based estimation procedure we use here is the basic approach described in Chapter (6), which assigns a $1/N$ (with N the number of samples in our time series) weight to each sample. Specifically, given an equally spaced mesh $a_{j+1} - a_j = h, b_{q+1} - b_q = h$, the histogram estimate for the relative number of counts at $a_j < r < a_{j+1}, b_q < s_\tau < b_{q+1}$ is

$$\tilde{p}(r, s_\tau) = \frac{v_{j,q}}{h^2 N} \tag{7.57}$$

where $v_{j,q}$ is the number of data points $\{r(t_n), s_\tau(t_n)\}$ falling in the bin specified by j, q. Clearly, $v_{j,q}$ is a binomial random variable and its expectation is given by

$$E[\tilde{p}(r, s_\tau)] = \frac{E[v_{j,q}]}{h^2 N} = \int_{a_j}^{a_{j+1}} \int_{b_q}^{b_{q+1}} p(r, s_\tau) dr ds_\tau. \tag{7.58}$$

In what follows, we omit the h^2 term in the denominator (the two-dimensional histogram "bin" area) as it does not influence the result of the optimization problem shown subsequently. Given

M independent realizations of the signals (denoted by r^k, s_τ^k), one could compute an estimate for the expectation of the histogram defined above. We denote $\tilde{p}_M(r, s_\tau) = 1/M \sum_{k=1}^{M} \tilde{p}(r, s_\tau)$ and $p_d(r, s_\tau) = \lim_{M \to \infty} \tilde{p}_M(r, s_\tau) = E[\tilde{p}(r, s_\tau)]$.

We note that in choosing the parameters h, a_j, b_j, care should be taken to ensure that the estimated histogram $\tilde{p}(r, s_\tau)$ depends on the signal model, that is, it must be sensitive to shifts in the signal. Mathematically, the requirement can be stated $\tilde{p}(s_\lambda, s_\tau) \neq p(s_\lambda)$ for at least one value of the mesh. That is, the joint probability of a signal and its shifted version should differ from the marginal probability of the signal itself. This assumption also excludes certain types of signals such as constant signals, or signals that are periodic with period equal to the sampling period Δ_t, for example, from our analysis.

In practice, for signals such as the waveform we use in the simulations subsequently, this assumption is not restrictive and very likely to hold unless the number of bins used in the histogram is too few. For example, we test the performance of the maximum empirical likelihood (MEL) method using histograms consisting of 2^B, $B = 4 \cdots 10$ bins. For all of these histogram specifications, this assumption was found to hold.

In light of Eqs. (7.55) and (7.56) earlier, given M random replicates of a measured signal $r^k(t_n) = s_\lambda(t_n) + \eta^k(t_n)$, $n = 0, \ldots, N - 1$, with λ fixed, we seek to find the correct time delay by finding τ that minimizes

$$\hat{\tau} = \min_\tau \Psi_M(\tau)$$

where

$$\Psi_M(\tau) = -\frac{1}{MN} \sum_{k=1}^{M} \sum_{n=0}^{N-1} \log\left(\tilde{p}_M(r^k(t_n) | s_\tau^k(t_n))\right) \tag{7.59}$$

It turns out (see Ref. [34] for a proof of these propositions) that when there is no noise present, maximizing Eq. (7.59) guarantees that the correct result will be obtained (i.e., correct delay will be estimated). Under the same assumptions, it can also be shown that in the limit of infinite data, Eq. (7.59) can be used to recover the exact time delay.

7.1.5.2 Relationship to Mutual Information

We note that because $r^k(t_n)$ is not a function of τ, minimizing Eq. (7.59) is equivalent to maximizing

$$\frac{1}{MN} \sum_{k=1}^{M} \sum_{n=0}^{N-1} \log\left(\frac{\tilde{p}_M(r^k(t_n), s_\tau^k(t_n))}{\tilde{p}_M(r^k(t_n))\tilde{p}_M(s_\tau^k(t_n))}\right). \tag{7.60}$$

Interpreting this sum as an expectation (in the limit $M \to \infty$)), it can be written as

$$E\left\{\log\left(\frac{p_d(r, s_\tau)}{p_d(r)p_d(s_\tau)}\right)\right\} = I_{RS}(r, s_\tau) \tag{7.61}$$

where $I_{RS}(r, s_\tau)$ is the mutual information of Section (3.5.1, Eq. 3.120) between random variables R and S_τ (computed through the histogram definition in (7.57)). Even for $M = 1$ the estimator proposed in (7.59) is still equivalent to the maximum mutual information approach if we consider the estimator given by Eq. (6.169) (see also Mars and Arragon [35]). Here we have shown, however, that such estimators, in i.i.d. noise, are equivalent to the empirical ML approach. *Thus, we have demonstrated that maximizing the mutual information function, defined in Section 3.5.1, provides an ML estimator of time delay regardless of the noise model.* That is to say, for general i.i.d. noise

$$\hat{\tau} = \max_{\tau} I_{RS}(r, s_\tau). \tag{7.62}$$

By extension, this analysis has shown that the mutual information is effectively a general measure of cross-correlation between two signals that is appropriate for Gaussian and non-Gaussian cases. We already showed in Section (3.5.1) that the mutual information between two normally distributed random variables reduced to a simple function of the linear cross-correlation. The result we have just described is certainly more general and motivates our frequent use of the mutual information in applications where the signals being analyzed are being produced by a nonlinear system and therefore are described by non-Gaussian probability models.

Equation (7.59) is minimized by computing τ for all integers $0, \ldots, N - 1$ and choosing the value of τ that produces the minimum value. The computational results given later were calculated using $\sqrt{N}$ for the number of bins for the histogram computation defined in (7.57). We have found that this frequently used rule of thumb produced the best results in this application. This was determined by performing the estimate using a signal of length $N = 2^{14}$ observations and a histogram consisting of 2^B, $B = 4 \cdots 10$ bins. For each bin size, the degree to which the error in the estimated delay approached the CRLB (to be described next) was checked. The value $B = 7$ yielded the estimator error closest to the theoretical bound.

We note that the MEL approach, as just described, is more computationally expensive than the other methods we compare here. For a digital signal of length N, the computational complexity of the cross-correlation-based method computed with the aid of the fast Fourier transform algorithm is on the order of $O(N\log_2(N))$. The computational complexity of the MEL approach for the same problem, on the other hand, is $O(N^2K^2)$, with K being the number of bins used in each dimension of the joint histogram necessary for computing the probabilities involved.

7.1.5.3   Cramer–Rao Lower Bounds for Different Noise Models

As we mentioned in Section (6.1), a good way to assess the performance of the estimator (7.59) is to compare the estimator variance to the CRLB. The CRLB gives the minimum conditional error variance of an estimator and thus provides a convenient standard for judging estimator performance. If the estimator is unbiased, or the bias does not depend on the parameter value, in this case τ, the CRLB is given by the expression (see also Eq. 6.10)

$$var(\hat{\tau}) \geq \frac{1}{F} = \frac{1}{E\left[\left(\frac{\partial \log (p(\mathbf{r}|s_\tau))}{\partial \tau}\right)^2\right]} \tag{7.63}$$

where the denominator in (7.63) is the Fisher information (FI) given in Eq. (6.9).

The likelihood function for the TDE problem in Gaussian noise is given in Eq. (7.52). Taking the log and differentiating with respect to τ yields [36]

$$\frac{1}{\sigma^2} \sum_{n=0}^{N-1} [r(t_n) - s_\tau(t_n)] \frac{ds_\tau(t_n)}{d\tau}. \tag{7.64}$$

Therefore, the FI is simply

$$F_G = \left(\frac{1}{\sigma^2}\right)^2 \sum_{n=0}^{N-1} \sum_{m=0}^{N-1} E[\eta(t_n)\eta(t_m)] \frac{ds_\tau(t_n)}{d\tau} \frac{ds_\tau(t_m)}{d\tau}$$

$$= \frac{1}{\sigma^2} \sum_{n=0}^{N-1} \left(\frac{ds_\tau(t_n)}{d\tau}\right)^2 \tag{7.65}$$

where the i.i.d. assumption on the noise eliminates the expected value except for where $n = m$, that is, $E[\eta(t_n)\eta(t_n)] = \sigma^2$. The CRLB for an unconditionally biased (any bias is assumed constant for all τ) estimator of the pulse delay in Gaussian noise is thus given by

$$\text{var}(\hat{\tau}) \geq 1/F_G = \frac{\sigma^2}{\sum_{n=0}^{N-1} \left(\frac{ds_\tau(t_n)}{d\tau}\right)^2}.$$

The SNR, a necessary concept for the simulations demonstrated here, is defined as the energy in the pulse, normalized by the pulse duration $T = N\Delta_t$, divided by the noise variance. However, in what follows, the variance of the noise is considered fixed to unity so that a given SNR is achieved by varying the amplitude of the signal only. Thus, we have for the SNR of a digital signal

$$SNR = \frac{\frac{1}{N} \sum_{n=0}^{N-1} s_\tau(t_n)^2}{E[\eta(t)^2]} = \frac{1}{N} \sum_{n=0}^{N-1} s_\tau(t_n)^2.$$

The job of the estimator is to take the received data $\mathbf{r}$ and produce a value $\hat{\tau}$ that is as close as possible to the true delay time despite the (possibly large) uncertainty implied by the SNR. While most of this book is concerned with the Gaussian noise model, in what follows we illustrate the TDE problem on two very different, but very flexible, noise models that can be used to describe more complicated types of uncertainty.

7.1.5.4 Mixed-Gaussian Distribution

The mixed-Gaussian distribution provides a noise model where the deviation from normality can be easily controlled. This particular model appears in situations where the clutter contains one or more harmonic components (see, e.g., [37]). A sequence of i.i.d. jointly mixed Gaussian values follow

$$p_{MG}(\eta(t_n)) = \prod_{n=0}^{N-1} \frac{1}{2\sqrt{2\pi}\sigma} (e^{-(\eta(t_n)-\mu)^2/2\sigma^2} + e^{-(\eta(t_n)+\mu)^2/2\sigma^2}) \tag{7.66}$$

As $\mu \to 0$, the distribution given becomes normal with variance σ^2. A nonzero μ results in two separate noise "peaks"; the distance between the peaks increases with μ. This distribution provides a convenient way of continuously varying the degree of nonnormality and testing the quality of a proposed estimator. For this noise model, the conditional density of interest is

$$p_{MG}(\mathbf{r}|\mathbf{s}_\tau) = \prod_{n=0}^{N-1} \frac{1}{2\sqrt{2\pi}\sigma} \left(e^{-(r(t_n)-s_\tau(t_n)-\mu)^2/2\sigma^2} + e^{-(r(t_n)-s_\tau(t_n)+\mu)^2/2\sigma^2} \right). \tag{7.67}$$

Taking the log and differentiating yields leads to the expression

$$\frac{1}{\sigma^2} \sum_{n=0}^{N-1} \left[\eta(t_n) - \mu \tanh\left(\frac{\mu\eta(t_n)}{\sigma^2} \right) \right] \frac{ds_\tau(t_n)}{d\tau} \tag{7.68}$$

The FI for the mixed Gaussian distribution is therefore

$$\begin{aligned}
F_{MG} &= \left(\frac{1}{\sigma^2}\right)^2 \sum_{n=0}^{N-1}\sum_{m=0}^{N-1} \left\{ E\left[\eta(t_n)\eta(t_m)\right] - E\left[\mu\eta(t_n)\tanh\left(\frac{\mu\eta(t_m)}{\sigma^2}\right)\right] \right. \\
&\quad \left. -E\left[\mu\eta(t_m)\tanh\left(\frac{\mu\eta(t_n)}{\sigma^2}\right)\right] + E\left[\mu^2\tanh\left(\frac{\mu\eta(t_n)}{\sigma^2}\right)\tanh\left(\frac{\mu\eta(t_m)}{\sigma^2}\right)\right] \right\} \\
&\quad \times \frac{ds_\tau(t_n)}{d\tau}\frac{ds_\tau(t_m)}{d\tau} \\
&= \frac{1}{\sigma^4}\sum_{n=0}^{N-1} \left\{ E\left[\eta(t_n)\eta(t_n)\right] - 2E\left[\mu\eta(t_n)\tanh\left(\frac{\mu\eta(t_n)}{\sigma^2}\right)\right] \right. \\
&\quad \left. +E\left[\mu^2\tanh\left(\frac{\mu\eta(t_n)}{\sigma^2}\right)^2\right] \right\} \frac{ds_\tau(t_n)}{d\tau}^2
\end{aligned} \tag{7.69}$$

Where, again, the assumption of i.i.d. noise eliminates one of the sums. The first expected value yields the variance of the noise, $E[\eta(t_n)\eta(t_n)] = \mu^2 + \sigma^2$, while the second expectation becomes $2E[\mu\eta(t_n)\tanh[\frac{\mu\eta(t_n)}{\sigma^2}]] = 2\mu^2$. The final sum can be rearranged to give

$$\begin{aligned}
F_{MG} &= \frac{\sigma^2 + \mu^2(c(\mu)-1)}{(\sigma^2+\mu^2)^2} \sum_{n=0}^{N-1} \frac{ds_\tau(t_n)}{d\tau}^2 \\
&= \frac{1 + (c(\mu)-2)\mu^2}{(1-\mu^2)^2} \sum_{n=0}^{N-1} \frac{ds_\tau(t_n)}{d\tau}^2
\end{aligned} \tag{7.70}$$

where in the last step we have substituted $\sigma^2 = 1 - \mu^2$ for the scale parameter to enforce unit variance on the noise vector. The function $c(\mu)$ is the result of numerically taking the expectation (integrating) $\{\tanh[\frac{\mu\eta(t_n)}{1-\mu^2}]^2\}$ with respect to the mixed Gaussian density. This function, shown in Figure 7.11, depends on μ only and determines the degree to which F_{MG} differs from F_G. The FI for the Gaussian case is recovered for $\mu = 0$ as expected, while for $\mu \to 1$ the CRLB$\to 0$. As the peaks separate ($\mu \to 1$), the variance of the individual peaks must tend to zero, again due to the constraint that $E[\eta(t_n)\eta(t_n)] = 1$. In this limit, the noise PDF becomes two delta functions located at $\mu = \pm 1$, thus the uncertainty in the observations tends to zero as does the CRLB.

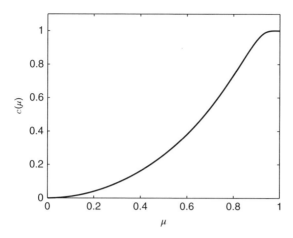

Figure 7.11 Function $c(\mu)$ in the mixed-Gaussian FI plotted as a function of the shape parameter μ. (*Source:* Reproduced from [34], Figure 1, with permission of Institution of Engineering and Technology)

7.1.5.5   Generalized Normal Distribution

As a second case, we also consider the generalized normal distribution as the contaminating noise source. The joint noise distribution in this case is given by

$$p_{GN}(\eta(t_n)) = \prod_{n=0}^{N-1} \frac{\beta}{2\sigma\Gamma\left(\frac{1}{\beta}\right)} e^{-(|\eta(t_n)-\mu|/\sigma)^\beta} \tag{7.71}$$

where $\sigma, \beta > 0$ are the distribution scale and shape parameters, respectively. As before, the constraint that the noise process variance be unity means we may write

$$\sigma = \sqrt{\frac{\Gamma(1/\beta)}{\Gamma(3/\beta)}}. \tag{7.72}$$

The generalized normal (7.71), therefore, allows a smooth transition from Gaussian to highly non-Gaussian distributions by varying a single parameter, β. For $\beta = 1$, one recovers the Laplace distribution, while $\beta = 2$ yields the familiar Gaussian. In the other direction, as $\beta \to \infty$ this distribution approaches the shape of the standard Uniform distribution. Thus, the generalized normal allows for both sharply peaked and very broadly peaked noise distributions. The FI in this case is derived as

$$F_{GN} = \frac{(1 + e^{i2\pi\beta})\beta\Gamma\left[2 - \frac{1}{\beta}\right]}{2\sigma^2\Gamma[1 + \frac{1}{\beta}]} \sum_{n=0}^{N-1} \frac{ds_\tau(t_n)}{d\tau}^2 \tag{7.73}$$

which can be seen to recover the correct CRLB for the Gaussian and Laplace distributions (the CR bound for the Laplace distribution is equal to one half the Gaussian CR bound). Figure 7.12 shows the CRLBs for both the mixed Gaussian and generalized normal distributions using the same parameters as in Figure 7.13. These results are consistent with what one observes for other parameters, namely, the larger the deviation from normality, the lower the CRLB. This

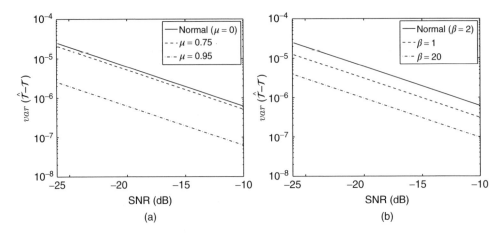

Figure 7.12   CRLB corresponding to the (a) mixed Gaussian and (b) generalized normal distributions for the same parameters as used in Figure 7.13. (*Source:* Reproduced from [34], Figure 2, with permission of Institution of Engineering and Technology)

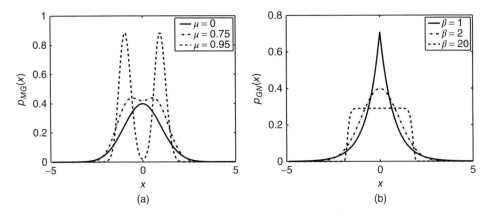

Figure 7.13   (a) Mixed Gaussian distribution and (b) generalized normal distribution for different values of their shape parameter. (*Source:* Reproduced from [34], Figure 3, with permission of Institution of Engineering and Technology)

is good news from an estimation standpoint as a lower CRLB implies that our estimates will be closer to truth than for the Gaussian noise model. In what follows, we will explore the performance of the mutual information-based estimator when the signal is corrupted by both mixed-Gaussian and generalized normal noise sources and demonstrate that we can, in fact, get better estimates (for a given SNR) when the noise probability model is non-Gaussian.

7.1.5.6   Results and Discussion

In this section, we evaluate the performance of the estimator stated in (7.59). As a baseline for better understanding the performance of this estimator, in each evaluation, we also plot the

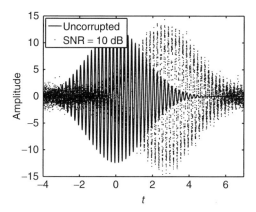

Figure 7.14   Clean and (Gaussian) noise corrupted pulse for $N = 16,384, \Delta_t = 0.0013, A = 12.4$, $f = 5, \alpha = 1.58$. (*Source:* Reproduced from [34], Figure 4, with permission of Institution of Engineering and Technology)

corresponding CRLB derived earlier, as well as the performance of the cross-correlation-based estimator (7.53), and the AMDF estimator (7.54). We use the following real-valued sinusoidal "pulse" as the signal model:

$$s_\tau(t_n) = Ae^{-(t_n-\tau)^2/(2\alpha^2)} \sin(2\pi f(t_n - \tau)).$$

In all experiments, the signals consist of 2^{14} points sampled at $\Delta_t = 0.0013$, while the frequency and scale parameters were taken to be $f = 5, \alpha = 1.58$. The numerical experiment proceeds by selecting τ from a uniform distribution in the interval of $[0, 2^{14} - 1]\Delta_t$ and computing $r(t_n)$ as in Eq. (7.49), for the corresponding noise model. In these simulations, the number of signal replicates M used to produce each delay estimate was $M = 1$ (while $M > 1$ would likely produce better estimates, experimentally we typically send one pulse and must estimate the delay from a single return). The SNR is controlled entirely by the signal amplitude A (recall the noise variance is always unity). It can be shown that, given these parameters, the signal amplitude is related to *SNR* by

$$A = 3.921 \times 10^{SNR/20}$$

assuming *SNR* has been specified on a dB scale. Figure 7.14 shows two pulses, one clean and one with additive Gaussian noise where $SNR = +10$ dB. For each type of noise, 1000 trials were conducted whereby two pulses were simulated, noise added to one, and the delay time estimated using the cross-correlation, AMDF, and MEL estimators. The average error (over the 1000 trials) was computed. The procedure was repeated for a variety of SNR ratios.

Results showing the mean-square-error as a function of SNR are given in Figure 7.15 for various distributions. As shown in Figure 7.15a, the MEL, AMDF, and the cross-correlation estimator perform similarly, with the advantage to the linear cross-correlation estimator, as it is an exact MLE in the Gaussian noise case. However, the MEL estimator (7.59) outperforms the estimator based on linear cross-correlation for most other noise models. As the parameter governing the degree of nonnormality is increased, so too are the improvements over the linear cross-correlation. In fact, the estimator based on (7.59) leads to estimates

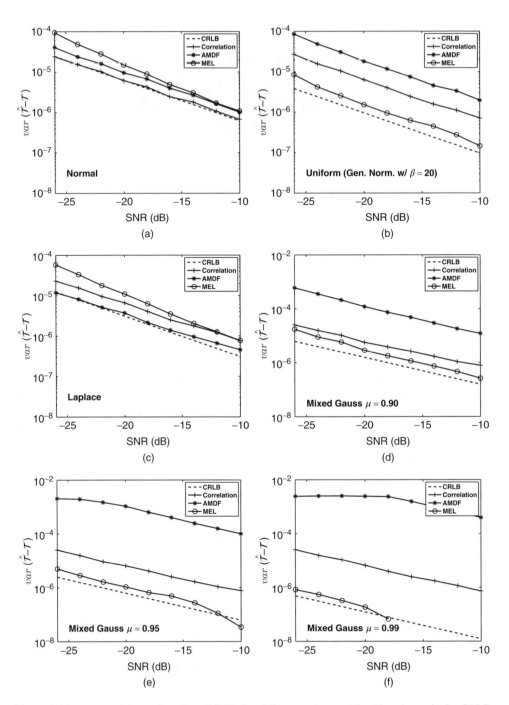

Figure 7.15 m.s.e. of $\hat{\tau}$ as a function of SNR for different noise models. Also shown is the CRLB associated with this estimation problem. (*Source:* Reproduced from [34], Figure 5, with permission of Institution of Engineering and Technology)

that, depending on the SNR level, can be an order of magnitude closer to the CRLB than the cross-correlation-based approach. Particularly in the mixed-Gaussian noise case, one sees very large improvements over other estimators. The AMDF estimator typically underperforms relative to (7.59), the exception being the case of Laplace noise.

It should be mentioned that errors less than a single time step cannot be resolved in this discrete experiment as all estimates are rounded to the nearest integer multiple of Δ. The result is a biasing of the estimate downward in high SNR regions (low error variance) and is the reason why the value of the MEL error on Figure 7.15f goes to zero beyond $SNR = -18$ dB. To see why this is so, we first note that MLEs asymptotically follow a normal distribution (see section 6.2). Now consider an SNR such that $2 \times E[|\hat{\tau} - \tau|] \ll 1/2\Delta_t$. In this case, we would expect 95% of the trials to yield estimates that are rounded down to zero integer delay difference, resulting in $\hat{\tau} = \tau$ (2 standard deviations encompasses roughly 95% of the area under the Gaussian distribution). Thus, as the CRLB drops below $var(\hat{\tau} - \tau) = (\frac{1}{4}\Delta_t)^2$ a large fraction of the trials will produce a zero error estimate; as a result, the estimated error variance becomes biased toward zero. Following this logic, the error variance at which the estimator should suffer is predicted to be $(\frac{1}{4}\Delta_t)^2 = 1.06 \times 10^{-7}$, which is consistent with the observed results (Figure 7.15e,f).

This section has shown that an empirical maximum likelihood method can be used for estimating the time delay of a signal with respect to a known model, in the regime where noise samples can be considered i.i.d. This is a common problem in "pulse-echo" damage detection methods, used frequently by the damage detection community to detect the presence of small subsurface defects in a structural component. We have also shown that the resulting estimation method is directly related to the mutual information of Section (3.5.1) and can be expected to perform well, regardless of the corrupting noise model. Indeed, the results provided show that in the event of highly non-Gaussian noise, the MEL approach produces results closer to the CRLB than other methods. The exceptions are Laplacian and Gaussian noise for which estimators tailored specifically to those noise models (ADMF, correlation) perform better because of the difficulty in accurately implementing the MEL (i.e., estimating the mutual information).

Another interesting outcome of this section is further evidence that the Gaussian PDF is the worst possible corrupting noise model from an estimation standpoint (produces highest CRLB). Specifically, we saw that if the noise source for a given structural estimation problem can reasonably be modeled as non-Gaussian, we can improve the quality of our estimates over those obtained by making the Gaussian noise assumption. The challenge, as we saw in this example, is that such estimators can be difficult to implement so that practical performance gains may be realizable only in the highly non-Gaussian case. Nonetheless, it is worth keeping in mind that estimators that minimize mean-square-error are MLEs *only* under the Gaussian assumption and that other, better estimators may be found for other noise models.

7.2 Bayesian Estimation

At several points in this book we have touched on the issue of forming confidence intervals for the estimated parameters. As we have already discussed, the most straightforward approach is to repeat the MLE some number of times (for some number of realizations) and simply look at the resulting distribution of estimates. An interval containing 95% of these estimates is said to be a 95% confidence interval. An alternative would be to evaluate the Hessian of the log-likelihood at the MLE. As we pointed out in Section (6.2), the Hessian is the key ingredient in the Fisher information matrix which provides an asymptotic expression for the variance of the parameter estimates. Because the MLEs are asymptotically normally distributed, this variance can be used to generate confidence limits. We must keep in mind, however, that these confidence intervals speak to the estimator, that is, if we repeat the *same estimation process* some number of times, what is the likelihood of finding the parameter value in a particular interval. While useful, such statements may not be what we would like. Often, we would like to state "there is an $x\%$ chance that my parameter lies in a given interval." To make this statement, we turn to a Bayesian view of estimation.

As we mentioned in Section 7.1, the implicit viewpoint of the MLE is that there is no uncertainty in the parameter itself, rather any parameters θ are fixed quantities with one true (best) value—the MLE. An alternative is to view the parameters as random variables and try to estimate the joint PDF associated with those variables. Given an estimate of the entire PDF, it is a simple matter to develop intervals that bound the desired fraction of probability. This approach is referred to as Bayesian estimation and is described here.

The Bayesian approach is fundamentally different in that it treats the unknown parameters as random variables with joint distribution $p_\Theta(\theta)$. For us, this distribution will always be conditional on what we observed, thus $p_\Theta(\theta) \equiv p_\Theta(\theta|\mathbf{y})$. Obviously, if we can estimate the entire distribution, the formation of confidence intervals becomes straightforward.

To this end, we turn to Bayes' rule, named for the Reverend Thomas Bayes. The rule itself is little more than a restatement of the law of conditional probability and reads

$$p_\Theta(\theta|\mathbf{y}) = p_H(\mathbf{y}|\theta)p_\pi(\theta)/p_\mathbf{Y}(\mathbf{y}). \qquad (7.74)$$

This says that the distribution of our parameters, conditional on what we observed, is equal to our likelihood multiplied by what is called a *prior distribution*, denoted $p_\pi(\theta)$. The prior incorporates any *a priori* information we might have about our parameter value. Although simple mathematically, Eq. 7.74 has had a large impact on estimation theory both for its treatment of the unknown parameter as a random variable, and for the explicit manner in which prior problem knowledge is built in to the estimate. In fact, the ability to assign a prior is a key strength of the approach as it provides a formal way to incorporate information one might have going into the estimation problem. If we are trying to estimate crack length, for example, and the component is brand new, we might make our prior crack length distribution an exponential distribution, with a small variance. We can make this choice because (i) crack length cannot be negative (the exponential distribution has only positive support) and (ii) if the component is new, the crack length distribution is likely near zero (hence, the small variance assumption).

Notice also that the key ingredient relating the prior to posterior distributions is again the likelihood function that we have already derived. The term in the denominator, $p_\mathbf{Y}(\mathbf{y}) = \int p_H(\mathbf{y}|\theta)p_\pi(\theta)d\theta$, is simply a normalizing constant guaranteeing that Eq. (7.74) is a true probability density that integrates to 1.

Conceptually, Eq. (7.74) provides a great deal of information about the parameter value(s) of interest. However, the expression is often difficult to evaluate analytically even for simple problems. Compound this by the fact that we are typically not interested in the entire joint parameter distribution $p_\Theta(\theta|\mathbf{y})$ but in the marginal distributions (distribution of the individual parameters, discussed in section 2.3). For example, say we want to estimate the "kth" parameter distribution and that there are a total of "P" parameters being estimated. The desired marginal distribution is found by integrating

$$p_{\Theta_k}(\theta_k|\mathbf{y}) = \int_{\mathbb{R}^{P-1}} p_{\Theta}(\boldsymbol{\theta}|\mathbf{y})d\boldsymbol{\theta}_{-k} = \frac{1}{p_{\mathbf{Y}}(\mathbf{y})}\int_{\mathbb{R}^{P-1}} p_H(\mathbf{y}|\boldsymbol{\theta})p_{\pi}(\boldsymbol{\theta})d\boldsymbol{\theta}_{-k} \qquad (7.75)$$

where the notation $\int_{\mathbb{R}^{P-1}} d\boldsymbol{\theta}_{-k}$ denotes the multidimensional integral over all parameters other than θ_k. Thus, we require a means of performing a potentially high-dimensional integral involving a likelihood function, for which we will have a very complicated expression. In some examples, for specific choice of prior distribution, analytical solutions to (7.75) are possible and rely on the notion of *conjugacy* among the prior and the likelihood. This concept is explained next.

7.2.1   Conjugacy

To emphasize the difficulty of solving for a posterior PDF analytically, consider a very simple estimation problem. Assume our job is to estimate the value of a DC signal using N observations from a single sensor, $y(n), n = 1 \cdots N$. Our signal model is $x(n) = c = $ const. for all n. Our likelihood is taken as Eq. (7.5) (all the usual assumptions on the noise). Let's also assume that our prior on c follows a normal distribution with mean M and variance τ^2. In other words, we have reason to believe that the DC value is M and our uncertainty in this belief is reflected in τ^2. Substituting into Bayes theorem we have

$$p_C(c|\mathbf{y}) = K_1^{-1}\frac{1}{(2\pi\sigma^2)^{N/2}}e^{-\frac{1}{2\sigma^2}\sum_{n=1}^{N}(y(n)-c)^2} \times \frac{1}{\sqrt{2\pi\tau^2}}e^{-\frac{1}{2\tau^2}(c-M)^2} \qquad (7.76)$$

where K_1 is the normalizing integral

$$K_1 = \int_{-\infty}^{\infty} \frac{1}{(2\pi\sigma^2)^{N/2}}e^{-\frac{1}{2\sigma^2}\sum_{n=1}^{N}(y(n)-c)^2} \times \frac{1}{\sqrt{2\pi\tau^2}}e^{-\frac{1}{2\tau^2}(c-M)^2}dc. \qquad (7.77)$$

Carrying out the integration shows that the posterior is also a Gaussian PDF and is given by

$$p_C(c|\mathbf{y}) = \frac{\sqrt{\frac{1}{\tau^2}+\frac{N}{\sigma^2}}}{\sqrt{2\pi}}\text{Exp}\left[\frac{-\left(\frac{1}{\tau^2}+\frac{N}{\sigma^2}\right)\left(c-\frac{\frac{M}{\tau^2}+\frac{N\bar{y}}{\sigma^2}}{\frac{1}{\tau^2}+\frac{N}{\sigma^2}}\right)^2}{2}\right] \qquad (7.78)$$

where $\bar{y} = \frac{1}{N}\sum_{n=1}^{N} y(n)$ is the sample mean. The mean of the posterior distribution is

$$\bar{c} = \frac{\frac{M}{\tau^2}+\frac{N\bar{y}}{\sigma^2}}{\frac{1}{\tau^2}+\frac{N}{\sigma^2}} \qquad (7.79)$$

and the variance

$$E[C^2] = \frac{1}{\frac{1}{\tau^2}+\frac{N}{\sigma^2}}. \qquad (7.80)$$

A reasonable final estimate for c would be to take $\bar{c}$ (the mean of the posterior). We can see that as the number of data gets larger, the influence of the prior is diminished as we weight the sample mean higher. If we hardly have any data, the prior factor is large. If we have a lot of uncertainty in our prior (τ^2 large), our prior "guess" M is given little weight. The converse is also true. This is a sensible result, but quite a bit work, especially considering the only unknown was c! For each additional parameter, one has to integrate the posterior distribution over all other parameters, in other words, form the marginal distribution given in Eq. (7.75). Clearly, this is intractable for a time-varying structural response with multiple unknowns. Fortunately, there is a numerical procedure that enables fast, efficient sampling of the posterior in even the most complicated situations. This procedure is described in some detail in Section 7.2.3.

The given example illustrates an important property referred to in Bayesian analysis as *conjugacy*, in this case between a Gaussian likelihood (known variance and unknown mean) and a Gaussian prior. This relationship is sometimes summarized with the following notation

$$\mathcal{L}(\mathbf{y}|\mu) \sim N(\mu, \sigma^2)$$

$$\Pi(\mu) \sim N(M, \tau^2)$$

$$p(\mu|\mathbf{y}) \sim N\left(\frac{\left(\frac{M}{\tau^2} + \frac{N\bar{y}}{\sigma^2}\right)}{\left(\frac{1}{\tau^2} + \frac{N}{\sigma^2}\right)^{-1}}, \left(\frac{1}{\tau^2} + \frac{N}{\sigma^2}\right)^{-1} \right) \tag{7.81}$$

where we have used the generic functions $\mathcal{L}(\cdot)$ for "likelihood" and $\Pi(\cdot)$ for the prior. A list of different likelihoods with specified unknown parameters and the associated conjugate priors can be found in Ref. [38].

7.2.2 Using Conjugacy to Assess Algorithm Performance

Consider the situation where we are trying to evaluate the efficacy of an automated damage identification system such as the one shown in Figure 7.16. The algorithm is fed with time-series strain response data. For example, let's assume that we have instrumented the deck plates of 20 different naval vessels with fiber-optic strain gages. Installations such as the one depicted in Figure 7.16 have, in fact, been performed with the expressed goal of assessing cracking and corrosion damage in navy ship deck plates (see [39]). Using a damage identification algorithm (such as those described in this book), we use the acquired data to assess both damage presence and damage type.

In addition, we assume the ability to visually inspect the plates, and that the visual inspection method is 100% accurate. That is to say, we can easily tell whether the plate was actually damaged and also what type of damage. We therefore have on-hand 20 different independent trials of our algorithm to compare against "truth." This is a common training scenario for many damage detection algorithms. The goal of this section is to show how to use these data to quantify the algorithm performance. Specifically, we show how to use these training data to estimate:

- probability of detection
- probability of false alarm
- probability of identifying a particular damage type (cracking or corrosion)

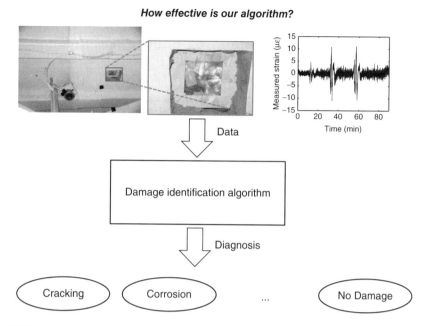

Figure 7.16 Hypothesized architecture for an automated damage classification system. Sensor data are fed into an algorithm that decides which damage class most likely produced those data. This basic scheme lies at the heart of numerous papers in the damage detection literature. The question addressed by this section is: given a number of sample diagnosis in cases where the true damage class is known (i.e., training data), how might we quantify the efficacy of the algorithm?

Algorithm performance estimates are necessary if the owner/operator of a structure is to decide whether to implement such a system.

The results of our hypothetical experiment (algorithm + visual inspection) are listed in Table 7.2. We are interested in predicting the performance of our algorithm in correctly identifying a particular type of damage. Owing to the aforementioned uncertainty in the identification process, there is no way to know precisely if the algorithm will correctly identify the damage. Rather, the algorithm's performance is better described by a probabilistic model that predicts the probability that a particular damage class is identified. As we show, this model will be governed by real-value parameters $p_{is} \in [0, 1]$ $i = 1 \cdots S$, which are the probabilities of identifying damage of class i given that the true damage class was s. We are most often interested in p_{ss}, that is, the conditional probability of correctly identifying damage class s given that the true damage class is also s. However, the model we develop permits more general inference as well, for example, probability that the algorithm incorrectly misclassifies class s as class i. The p_{is} are considered unknown and must be estimated from (possibly) limited experimental data. These probabilities will, therefore, carry with them a potentially large amount of uncertainty. Thus, our goals will be to estimate the probability parameters and quantify the uncertainty in those estimates.

The first step is to develop a probabilistic model for algorithm performance. We will refer to a "trial" as an instance of damage class $s \in [1, S]$. There is also the possibility that no damage is present, for which we will denote $s = 0$. The algorithm attempts to identify the damage and

Table 7.2 Outcomes associated with 20 trials of a hypothetical damage detection experiment

Trial	Visual			Algorithm		
	ID crack	ID corrosion	No damage	ID crack	ID corrosion	No damage
1	1	0	0	1	0	0
2	1	0	0	1	0	0
3	1	0	0	0	1	0
4	1	0	0	1	0	0
5	1	0	0	0	1	0
6	1	0	0	0	0	1
7	1	0	0	1	0	0
8	1	0	0	1	0	0
9	1	0	0	1	0	0
10	1	0	0	0	0	1
11	0	1	0	0	1	0
12	0	1	0	0	1	0
13	0	1	0	0	0	1
14	0	1	0	1	0	0
15	0	1	0	0	1	0
16	0	0	1	1	0	0
17	0	0	1	0	0	1
18	0	0	1	0	0	1
19	0	0	1	0	0	1
20	0	0	1	0	1	0

* We know truth, via visual inspection, and we also have the diagnosis (cracking, corrosion, or no damage) as recorded by our automated damage detection algorithm

declares class $j \in [1, S]$ to be present or $j = 0$ if no damage is observed. The binary outcome of this trial is therefore $k_{js} = 1$ and $k_{is} = 0 \, \forall \, i \neq j$. A probabilistic model for the outcome of this single trial is the probability mass function (PMF, see section 2.2)

$$f_{\mathbf{K}}(\mathbf{k}|\mathbf{p}_s) = \left(1 - \sum_{i=1}^{S} p_{is}\right)^{1 - \sum_{i=1}^{S} k_{is}} \prod_{i=1}^{S} p_{is}^{k_{is}} \tag{7.82}$$

where we have used bold font to denote a vector, for example, $\mathbf{p}_s \equiv (p_{1s}, p_{2s}, \ldots, p_{Ss})$. This model simply assigns the probability p_{is} to the outcome of the trial. The first term in the product arises from the constraint that for any PMF, $\sum f_{\mathbf{K}}(\mathbf{k}_s, \mathbf{p}_s) = 1$ (total probability is unity). For damage identification experiments, this term can be used to model the probability that no damage was identified. Thus, in the event that the algorithm declared "no damage present" ($i = 0$) the model assigns $f_{\mathbf{K}}(\mathbf{k}_s|\mathbf{p}_s) = p_{0s} = (1 - \sum_{i=1}^{S} p_{is})$. What this means in practice is that there are only S independent probabilities that need to be estimated and that from these estimates we can uniquely determine the p_{0s} parameter.

Of course, a single trial is not sufficient to draw the needed inference. Rather, it is assumed that a number of *independent* trials N_s are performed for damage class s. This assumption allows us to write the PMF for all trials as the product of the individual trials. Consider trial t of our experiment, where the true damage class is again s and the algorithm identified category j. Define

$$k_{is}^{(t)} = \begin{cases} 1 : i = j \\ 0 : \text{otherwise} \end{cases} \tag{7.83}$$

to be the collection of vectors consisting of all zeros and with a single 1 value placed in position j. The visual inspection results are considered to be "truth." Of the $N = 20$ total trials, we have $N_1 = 10$ cases where there was a crack, $N_2 = 5$ cases where we had corrosion, and $N_0 = 5$ cases where there was no damage.

Now, if we assume each of the N_s trials to be independent events, we can create a model for the entire collection of trials. For truth class s, the PMF for the entire table of outcomes is

$f_{\mathbf{K}}(\mathbf{k}_s | \mathbf{p}_s)$

$$= \frac{N_s!}{\left(N_s - \sum_{t=1}^{N_s} \sum_{i=1}^{S} k_{is}^{(t)}\right)! \prod_{i=1}^{S} \left(\sum_{t=1}^{N_s} k_{is}^{(t)}\right)!} \prod_{t=1}^{N_s} \left(1 - \sum_{i=1}^{S} p_{is}\right)^{1 - \sum_{i=1}^{S} k_{is}^{(t)}} \prod_{i=1}^{S} p_{is}^{k_{is}^{(t)}}$$

$$= \frac{N_s!}{\left(N_s - \sum_{t=1}^{N_s} \sum_{i=1}^{S} k_{is}^{(t)}\right)! \prod_{i=1}^{S} \left(\sum_{t=1}^{N_s} k_{is}^{(t)}\right)!} \left(1 - \sum_{i=1}^{S} p_{is}\right)^{N_s - \sum_{t=1}^{N_s} \sum_{i=1}^{S} k_{is}^{(t)}} \prod_{i=1}^{S} p_{is}^{\sum_{t=1}^{N_s} k_{is}^{(t)}}$$

$$= \frac{N_s!}{\prod_{i=0}^{S} \tilde{k}_{is}!} \left(1 - \sum_{i=1}^{S} p_{is}\right)^{\tilde{k}_{0s}} \prod_{i=1}^{S} p_{is}^{\tilde{k}_{is}} \tag{7.84}$$

where we have defined $\tilde{k}_{is} \equiv \sum_{t=1}^{N_s} k_{is}^{(t)}$ as the total number of trials in which state i was identified. The last line in Eq. (7.84) is the *multinomial distribution*; it serves as a good model for any classification experiment, where the data consist of the number of selections of a particular class given some number of independent trials. Equation (7.84) is our probabilistic model for algorithm performance, cast in terms of the unknown identification probabilities $\mathbf{p} \equiv (p_{1s}, p_{2s}, \ldots, p_{Ss})$. Again, the probability associated with the "no damage" class is automatically determined to be $1 - \sum_{i=1}^{S} p_{is}$ and the number of trials identified as such is $k_{0s} \equiv N_s - \sum_{i=1}^{S} \tilde{k}_{is}$ due to the aforementioned constraint.

Typically, these probabilities are unknown and we seek to estimate them from the data. We could, for example, obtain the MLE by finding the p_{is} that maximize (7.84). Proceeding along these lines, we take the log of (7.84)

$$\log\left(\frac{N_s!}{\prod_{i=0}^{S} \tilde{k}_{is}!}\right) + \left(N_s - \sum_{i=1}^{S} \tilde{k}_{is}\right) \log\left(1 - \sum_{i=1}^{S} p_{is}\right) + \sum_{i=1}^{S} \tilde{k}_{is} \log(p_{is}) \tag{7.85}$$

and then set the derivative to zero, that is,

$$0 = \frac{\partial}{\partial p_{is}} \left\{ \log \left(\frac{N_s!}{\prod_{i=1}^{S} k_{is}!} \right) \right\} - \frac{\left(N_s - \sum_{i=1}^{S} \tilde{k}_{is} \right)}{\left(1 - \sum_{i=1}^{S} p_{is} \right)} + \left(\frac{\tilde{k}_{is}}{p_{is}} \right)$$

$$\rightarrow \left(\frac{\tilde{k}_{is}}{p_{is}} \right) = \frac{\left(N_s - \sum_{i=1}^{S} \tilde{k}_{is} \right)}{\left(1 - \sum_{i=1}^{S} p_{is} \right)}. \tag{7.86}$$

We can rewrite this equality in terms of the fraction of observations of category i

$$\bar{k}_{is} \equiv \tilde{k}_{is}/N_s \tag{7.87}$$

giving

$$\left(\frac{\bar{\mathbf{k}}_{is}}{p_{is}} \right) = \frac{\left(1 - \sum_{i=1}^{S} \bar{\mathbf{k}}_{is} \right)}{\left(1 - \sum_{i=1}^{S} p_{is} \right)}. \tag{7.88}$$

It can be seen by inspection that the only way for this equality to be true is for

$$\hat{p}_{is} = \bar{\mathbf{k}}_{is}. \tag{7.89}$$

The MLE for each of these probabilities is therefore simply the fraction of the total observations made of class i. Thus, for the data in Table 7.2, we have the MLEs shown in Table 7.3. We now repeat the analysis using the Bayesian approach to estimation. We again start with the multinomial PMF as the likelihood

$$f_{\mathbf{K}_s}(\mathbf{k}_s | \mathbf{p}_s) = \frac{N!}{\prod_{i=0}^{S} \tilde{k}_{is}!} \left(1 - \sum_{i=1}^{S} p_{is} \right)^{\tilde{k}_{0s}} \prod_{i=1}^{S} p_{is}^{\tilde{k}_{is}}. \tag{7.90}$$

It turns out that the conjugate prior distribution to the multinomial likelihood is the Dirichlet distribution, given by

$$\Pi_{P_s}(\mathbf{p}_s) = \frac{1}{B(\boldsymbol{\alpha})} \prod_{i=1}^{S} p_{is}^{\alpha_i - 1} \tag{7.91}$$

Table 7.3 Maximum likelihood estimates of the damage classification probabilities

True Class	MLE		
	Crack	Corrosion	No damage
Crack	$\hat{p}_{11} = 6/10 = 0.6$	$\hat{p}_{21} = 2/10 = 0.2$	$\hat{p}_{01} = 2/10 = 0.2$
Corrosion	$\hat{p}_{12} = 1/5 = 0.2$	$\hat{p}_{22} = 3/5 = 0.6$	$\hat{p}_{02} = 1/5 = 0.2$
No Damage	$\hat{p}_{10} = 1/5 = 0.2$	$\hat{p}_{20} = 1/5 = 0.2$	$\hat{p}_{00} = 3/5 = 0.6$

where the normalizing constant is given by the (multivariate) beta function

$$B(\boldsymbol{\alpha}) \equiv \frac{\prod_{i=1}^{K} \Gamma(\alpha_i)}{\Gamma\left(\sum_{i=1}^{K} \alpha_i\right)} \tag{7.92}$$

and $\Gamma(\cdot)$ is the Gamma function (see Appendix A). As with the multinomial, the Dirichlet distribution must obey the constraint $\sum p_{is} = 1$.

Just as the product of two normally distributed random variables with unknown mean led to a normally distributed posterior (again, see Eq. 7.78), the product of a Dirichlet prior and multinomial likelihood results in a Dirichlet posterior distribution for the joint parameter vector $\mathbf{p}$

$$\mathcal{L}(\mathbf{k}_s | \mathbf{p}_s) \sim Multi(\mathbf{k}_s; N, \mathbf{p}_s)$$

$$\Pi_{P_s}(\mathbf{p}_s) \sim Dir(\mathbf{p}_s; \boldsymbol{\alpha})$$

$$f_{\mathbf{P}_s}(\mathbf{p}_s | \mathbf{k}_s) \sim Dir(\mathbf{p}_s; \boldsymbol{\alpha} + \mathbf{k}_s). \tag{7.93}$$

The parameters for the posterior distribution are simply updated by the counts for each class given in the vector $\mathbf{k}_s$ The noninformative prior sets $\alpha_i = 1$ for each of the S classes being considered. An example of the uniform Dirichlet prior is shown in Figure 7.17 for an $S = 2$ class problem. We can take the analysis further by noting that if the posterior distribution for our *joint* probability vector $\mathbf{p}_s$ is a Dirichlet distribution with parameter vector $\mathbf{a}$, the desired *marginal* distributions are also Beta distributions (see Appendix A) with parameters $a_i, \sum_{j \neq i} a_j$. In the given example, this means the marginal Beta distribution will have parameters $k_i + \alpha_i, \alpha^* - \tilde{k}_i - \alpha_i$ where $\alpha^* \equiv \sum_{i=1}^{S}(\tilde{k}_i + \alpha_i)$. Thus, we have for the estimated posterior probability distributions for each of our probability parameters

$$f_{P_{is}}(p_{is}) = \frac{1}{B(\tilde{k}_{is} + \alpha_i, \alpha^* - \tilde{k}_{is} - \alpha_i)} p_{is}^{\tilde{k}_{is}+\alpha_i-1} (1 - p_{is})^{\alpha^*-\tilde{k}_{is}-\alpha_i-1} \tag{7.94}$$

where $B(\cdot, \cdot)$ is the Beta function as defined in Appendix A. Figure 7.18 shows the posterior probability distributions associated with each damage class (no damage, crack, corrosion). A noninformative Dirichlet prior was assumed ($\alpha_1 = \alpha_2 = \alpha_3 = 1$) and the associated numbers of correct classification were $\mathbf{k}_{11} = 6, \mathbf{k}_{22} = 3, \mathbf{k}_{33} = 3$. The benefit to this analysis (as opposed to the MLE) is that we get the entire posterior distribution. From this distribution, we can readily quantify a *credible interval*, the Bayesian analog to the confidence interval. For example, defining the CDF for the Beta distribution as

$$I_x(a, b) \equiv \int_0^x Beta(a, b) \tag{7.95}$$

the 95% credible interval for $\hat{p}_{11}$ is defined by the upper and lower limits

$$CL = x: \quad I_x(\mathbf{k}_1 + \alpha_1, \alpha^* - \mathbf{k}_1) = 0.025$$

$$CU = x: \quad I_x(\mathbf{k}_1 + \alpha_1, \alpha^* - \mathbf{k}_1) = 0.975 \tag{7.96}$$

Note the function $I_x(a, b)$ is sometimes referred to as the regularized incomplete Beta function. In our example, this interval is calculated to be

$$0.277 \leq \hat{p}_{11} \leq 0.789$$

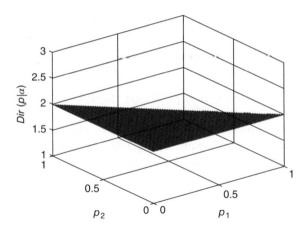

Figure 7.17   Noninformative Dirichlet prior ($\alpha_1 = \alpha_2 = \alpha_3 = 1$) for a two-class problem

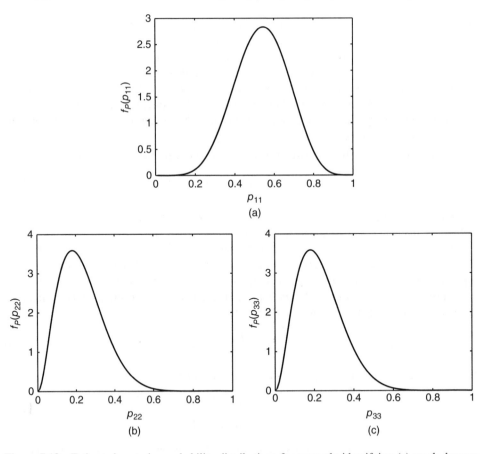

Figure 7.18   Estimated posterior probability distributions for correctly identifying (a) crack damage, (b) corrosion damage, and (c) no damage, respectively. Estimates are based on the data in Table 7.2 and a noninformative Dirichlet prior

Despite a rather limited data set, we are able to clearly draw inference about identification probabilities. Moreover, the form of the obtained distributions makes sense from a mathematical point of view. The support of the Beta distribution is the interval $[0, 1]$ which, of course, should be the support of any estimated probability.

We can further use the same basic framework to estimate probability of detection and probability of false alarm. In both cases, the multinomial distribution reduces to the binomial distribution as there are only two possibilities for each. Furthermore, the conjugate distribution reduces from the Dirichlet to the Beta distribution, which we have already defined (the binomial and Beta distributions are just special cases of the multinomial and Dirichlet distributions for the $S = 1$ problem).

When estimating probability of detection, p_D, damage of either type is detected or not detected. The probability of a "miss" in any detection problem is our Type II error, that is, $1 - p_D$ (see section 6.9.2). On the basis of the earlier table, of the 15 cases where damage was indeed present, the algorithm correctly detected damage $\mathbf{k}_{11} + \mathbf{k}_{12} + \mathbf{k}_{22} + \mathbf{k}_{21} = 12$ of those times (with $15 - 12 = 3$ misses). Thus, the posterior distribution for p_D (again, assuming noninformative Dirichlet prior) is given by

$$f_{P_D}(p_D) = \frac{1}{B(13, 4)} p_D^{12}(1 - p_D)^3. \tag{7.97}$$

Similarly, the two possibilities in the "no damage present case" are to incorrectly declare it present, our Type I error, or correctly declare the damage absent. The probability of false alarm is therefore the same as Type I error and is the quantity whose distribution we seek. Looking at the Table 7.2 data, we see that damage was absent in five cases and the algorithm falsely declared it present two times (correct three times). Assuming a Dirichlet prior with $\alpha_1 = \alpha_2 = 1$ we have for the posterior

$$f_{P_{fa}}(p_{fa}) = \frac{1}{B(3, 4)} p_{fa}^2(1 - p_{fa})^3. \tag{7.98}$$

These distributions are shown in Figure 7.19. As these examples show, if one can take advantage of conjugate priors, a full Bayesian analysis is possible without having to resort to numerical methods. When estimating detection and identification probabilities from discrete data such as those given in Table 7.2, the above-mentioned analysis provides a statistically sound approach. This, or related analyses, would ideally be performed whenever a new damage identification algorithm is developed. Without understanding the performance, and the uncertainty associated with that performance, it is very difficult to make a decision as to whether the approach should be implemented in practice.

This particular problem is significantly different from most of the estimation problems encountered in this book. For most of our applications, estimates of physics-based model parameters are of most interest, while in this example we sought probability (multinomial) model parameters. However, the framework we have developed is agnostic to the particular model or application, and hence its power. Whether or not we seek parameters of a physics-based or purely probabilistic model, we are armed with the appropriate estimation tools.

Unfortunately, for the applications we are often interested in, the parameters we seek to estimate do not have analytically tractable posterior distributions (we cannot rely on conjugacy). The good news is that there is a powerful approach to evaluating Bayes' theorem numerically. This algorithm is derived next.

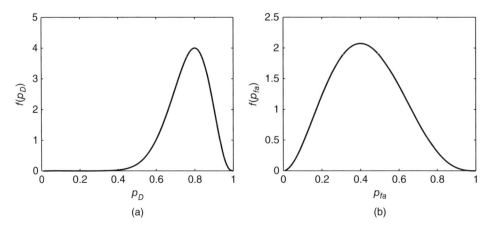

Figure 7.19   Estimated posterior probability distributions for probability of detection and probability of false alarm (a and b, respectively). Estimates are based on the hypothetical system performance given in Table 7.2 and a noninformative Dirichlet prior

7.2.3   Markov Chain Monte Carlo (MCMC) Methods

MCMC methods are designed to draw samples directly from the posterior distribution without having to simplify it analytically. In the past 40 years, the MCMC algorithm has made Bayesian analysis tractable for a wide variety of problems as witnessed by the explosion of papers that use MCMC as a means to a solution. Owing to the importance of these methods, we present a derivation of one popular variant of MCMC referred to as the Metropolis–Hastings algorithm.

Consider, for now, a single parameter θ (the multivariate case is considered shortly). The idea behind MCMC is to build a first-order Markov chain that has $p_\Theta(\theta|\mathbf{y})$ as its stationary PDF. In what follows, we simply denote this posterior distribution $p_\Theta(\theta)$ with the understanding that the function depends on our data $\mathbf{y}$.

Recall that a Markov chain defines a sequence $\theta_1, \theta_2, \ldots$, where the values in the sequence are determined in a probabilistic manner via a probability density, $f(\cdot)$ (see Section 3.4). The order of the Markov chain O specifies the dependence on past values allowing us to write the Markov model as

$$f(\theta_k|\theta_{k-1}, \theta_{k-2}, \ldots, \theta_{k-O}) = f(\theta_k|\theta_{k-1}, \theta_{k-2}, \ldots, \theta_{k-O}, \theta_{k-O-1}, \ldots). \qquad (7.99)$$

That is to say, the probability of attaining value θ_k is conditional on the past O values *only* (and not on any values that came earlier). Thus, a Markov chain is similar in spirit to an autoregressive (AR) model in that the new value θ_k is a function of some number of past values (however, the Markov chain is clearly more general). Here, we consider a first-order Markov chain $f(\theta_k|\theta_{k-1})$ which, for notational reasons, we will simply denote as $f(\theta^*|\theta)$.

Now, the MH algorithm attempts to create a sequence θ_k, $k = 1 \ldots K$ such that the values in the sequence *are samples from* $p_\Theta(\theta)$. Obviously, if we could construct such a chain, we could generate as many samples as we desire, and then estimate $\hat{p}_\Theta(\theta)$ using a simple histogram or one of the other kernel-based approaches to probability density estimation (discussed in Section 6.3). However, to do this, we require that the Markov chain be ergodic. A nonstationary Markov chain is of little value as the samples generated would be from different distributions, not necessarily $p_\Theta(\theta)$.

So how does one construct such a chain $f(\theta^*|\theta)$? This is precisely the problem addressed by Hastings in his 1970 paper [40]. The first step is to establish certain properties that the chain must possess, one being the so-called reversibility condition

$$p_\Theta(\theta)f(\theta^*|\theta) = p_\Theta(\theta^*)f(\theta|\theta^*). \tag{7.100}$$

Simplifying, we get

$$p_\Theta(\theta)\frac{f(\theta^*,\theta)}{f_M(\theta)} = p_\Theta(\theta^*)\frac{f(\theta,\theta^*)}{f_M(\theta^*)} \tag{7.101}$$

where $f_M(\theta) = \int f(\theta^*, \theta)d\theta^*$ is the marginal distribution. The only way this holds, in general, is if $p_\Theta(\theta) = f_M(\theta)$. In other words, our Markov chain will have as its marginal distribution $p_\Theta(\theta)$.

This condition, therefore, also guarantees that

$$p_\Theta(\theta^*) = \int p_\Theta(\theta)f(\theta^*|\theta)d\theta$$

$$= \int \frac{p_\Theta(\theta)}{f_M(\theta)}f(\theta^*,\theta)d\theta$$

$$= \int f(\theta^*,\theta)d\theta = f_M(\theta^*) \tag{7.102}$$

so that the values in the chain, θ^*, are values from the desired parameter posterior distribution. The question therefore becomes, what type of Markov chains guarantee (7.100)? The first requirement is that there be a nonzero probability of reaching state θ^* from state θ in a finite number of steps for all possible pairs of states θ, θ^*, that is, we have to be able to fully explore the solution space. This property is referred to as the "irreducibility" of $f(\theta^*|\theta)$.

To meet these criteria, Hastings suggests factoring the chain as the product [40]

$$f(\theta^*|\theta) = \alpha(\theta^*|\theta)q(\theta^*|\theta) \tag{7.103}$$

where $q(\theta^*|\theta)$ is an arbitrarily chosen first-order Markov chain and $\alpha(\theta^*|\theta)$ is a second first- order Markov chain to be solved for. The chain $q(\theta^*|\theta)$ is sometimes referred to as a "proposal" distribution or a "candidate-generating" distribution (for reasons that will become clear shortly) and is specified by the practitioner at the outset. Given this formulation, we substitute (7.103) into (7.100) to get

$$p_\Theta(\theta)\alpha(\theta^*|\theta)q(\theta^*|\theta) = p_\Theta(\theta^*)\alpha(\theta|\theta^*)q(\theta|\theta^*) \tag{7.104}$$

which can be solved for $\alpha(\theta^*|\theta)$. Note that this distribution must obey the restriction that $\alpha(\theta^*|\theta)d\alpha \leq 1$, that is, the probability of going from $\theta \to \theta^*$ is, in fact, a probability. It is also worth noting that the inclusion of the probability element $d\alpha$ in the notation. Hastings' work [40] dealt with discrete probabilities, however in the continuous case (using probability densities), we are dealing with probability elements $d\alpha$. For the remainder of this discussion, we leave out the $d\alpha, dq, df$ on the understanding that we are working with continuous probability densities, not discrete probabilities.

To meet the restriction $\alpha(\theta^*|\theta) \le 1$, we include a symmetric scaling function $s(\theta^*, \theta) = s(\theta, \theta^*)$. Because this function is symmetric, it does not influence Eq. (7.104). In this case, we may write

$$\hat{\alpha}(\theta^*|\theta) = s(\theta^*, \theta)\alpha(\theta^*|\theta) \tag{7.105}$$

and from Eq. (7.104) we have

$$\hat{\alpha}(\theta^*|\theta) = \hat{\alpha}(\theta|\theta^*) \left[\frac{p_\Theta(\theta^*)q(\theta|\theta^*)}{p_\Theta(\theta)q(\theta^*|\theta)} \right]. \tag{7.106}$$

Letting

$$r(\theta^*, \theta) = \frac{p_\Theta(\theta^*)q(\theta|\theta^*)}{p_\Theta(\theta)q(\theta^*|\theta)} \tag{7.107}$$

we can write the two equations

$$\hat{\alpha}(\theta^*|\theta) = \hat{\alpha}(\theta|\theta^*)r(\theta^*|\theta)$$

$$\hat{\alpha}(\theta|\theta^*) = \hat{\alpha}(\theta^*|\theta)r(\theta|\theta^*).$$

Subtracting one from the other yields

$$\hat{\alpha}(\theta^*|\theta)[1 + r(\theta, \theta^*)] = \hat{\alpha}(\theta|\theta^*)[1 + r(\theta^*|\theta)]. \tag{7.108}$$

In general, the only way for this equality to hold is if

$$\hat{\alpha}(\theta^*|\theta) = \frac{1}{1 + r(\theta, \theta^*)}$$

or equivalently

$$\hat{\alpha}(\theta|\theta^*) = \frac{1}{1 + r(\theta^*, \theta)}$$

This means we can write our expression for α as

$$\alpha(\theta^*|\theta) = \frac{s(\theta^*, \theta)}{1 + r(\theta, \theta^*)} \tag{7.109}$$

The final step in the derivation is to choose $s(\theta^*, \theta)$ so that $\alpha(\theta^*|\theta) \le 1$ and represents a true probability.

There are numerous choices one could make to guarantee that this inequality holds. However, some choices will clearly lead to higher probability of moving from one state to the next in our Markov chain. A poor choice, for example, would be to set $s(\theta^*, \theta) = \epsilon \ll 1$. This ensures that $\alpha(\theta^*, \theta) \le 1$, however it would result in a chain $f(\theta^*|\theta)$ that would take a very long time to explore the parameter space (i.e., converge). The probability of going from $\theta \to \theta^*$ would always be nonzero (as required), but would be extremely small (recall that we need to guarantee that there is some probability of getting from θ to θ^* in some finite number of steps). Convergence in finite time is guaranteed by many functions $s(\theta^*, \theta)$, however efficiency is clearly not.

A better solution is to make the choice dependent on the ratio $r(\theta, \theta^*)$ that is, a state-dependent function. Consider the case where $r(\theta, \theta^*) \geq 1$. In this case,

$$\alpha(\theta^*|\theta) = \frac{s(\theta^*, \theta)}{1 + r(\theta, \theta^*)} \leq 1 \tag{7.110}$$

so long as $s(\theta^*, \theta) \leq 1 + r(\theta, \theta^*)$. This is automatically guaranteed if $s(\theta^*, \theta) = 1 + r(\theta^*, \theta)$ since by definition (Eqn. 7.107)

$$r(\theta^*, \theta) = \frac{1}{r(\theta, \theta^*)} \tag{7.111}$$

so

$$\frac{1 + r(\theta^*, \theta)}{1 + r(\theta, \theta^*)} \leq 1 \tag{7.112}$$

if $r(\theta, \theta^*) > 1$. However, we still must consider the case where $r(\theta, \theta^*) < 1$. Clearly, $s(\theta^*, \theta)$ must be altered, otherwise $\alpha(\theta^*|\theta) \geq 1$. The solution is to simply switch the arguments, that is, if $r(\theta, \theta^*) < 1$ let $s(\theta^*, \theta) = 1 + r(\theta, \theta^*)$. This yields

$$\alpha(\theta^*|\theta) = \frac{1 + r(\theta, \theta^*)}{1 + r(\theta, \theta^*)} = 1. \tag{7.113}$$

So, for our scaling function we can choose

$$s(\theta^*, \theta) = \begin{cases} 1 + r(\theta^*, \theta) & : \quad \text{if } r(\theta, \theta^*) \geq 1 \\ 1 + r(\theta, \theta^*) & : \quad \text{if } r(\theta, \theta^*) < 1 \end{cases} \tag{7.114}$$

This leads to

$$\alpha(\theta^*, \theta) = \begin{cases} r(\theta^*, \theta) & : \quad \text{if } r(\theta, \theta^*) \geq 1 \\ 1 & : \quad \text{if } r(\theta, \theta^*) < 1 \end{cases}. \tag{7.115}$$

Noting that $r(\theta^*, \theta) = 1/r(\theta, \theta^*)$, we could have also written

$$\alpha(\theta^*, \theta) = \begin{cases} r(\theta^*, \theta) & : \quad \text{if } r(\theta^*, \theta) < 1 \\ 1 & : \quad \text{if } r(\theta^*, \theta) \geq 1 \end{cases} \tag{7.116}$$

which is exactly the main result arrived at in Ref. [40].

We now have everything needed to construct our Markov chain $f(\theta^*|\theta) = \alpha(\theta^*|\theta)q(\theta^*|\theta)$ and can write

$$f(\theta^*|\theta) = \begin{cases} q(\theta^*|\theta) & : \text{if } r(\theta^*, \theta) \geq 1 \\ r(\theta^*, \theta)q(\theta^*|\theta) & : \text{if } r(\theta^*, \theta) < 1 \end{cases} \tag{7.117}$$

Equation (7.117) gives a prescription for building a Markov chain in terms of (i) a proposal distribution $q(\theta^*|\theta)$ and (ii) the distribution we wish to draw samples from, $p_\Theta(\theta)$. The samples in that Markov chain will, by construction, be samples from $p_\Theta(\theta)$. The algorithm, therefore, works as follows: First, choose an initial value of θ from the parameter prior distributions, that is, $\theta \sim p_\pi(\theta)$. Given this value, use $q(\theta^*|\theta)$ to generate a value θ^*. Next, evaluate the ratio (7.107). If $r(\theta^*, \theta) \geq 1$, we keep θ^* with probability 1. If the ratio is < 1, we keep the value

θ^* with probability $r(\theta^*, \theta)$. More concisely, we move to θ^* with probability $\min\{r(\theta^*, \theta), 1\}$. The problem can be made even simpler by a clever choice of $q(\theta^*|\theta)$. Specifically, we allow $q(\theta^*|\theta) = q(\theta|\theta^*)$. In this case, our "proposal distribution" is symmetric and our probability of going from parameter θ to θ^* is the same as our probability of going from θ^* to θ. This means that we only have to consider the ratio of probability distributions when computing our ratio, i.e., $r(\theta^*, \theta) = p_\Theta(\theta^*)/p_\Theta(\theta)$.

To summarize, we wish to sample from a distribution

$$p_\Theta(\theta|\mathbf{y}) = p_H(\mathbf{y}|\theta)p_\pi(\theta)/p_\mathbf{Y}(\mathbf{y}). \tag{7.118}$$

For problems in structural dynamics, the evaluation of the posterior is usually analytically intractable, and hence we can use the above-described MCMC algorithm to numerically generate samples from $p_\Theta(\theta|\mathbf{y})$. An initial value θ is sampled from the parameter prior distribution $p_\pi(\theta)$. We then apply $q(\theta^*|\theta)$ to provide the candidate chain value θ^*. In principle, this can be an arbitrary distribution, however we would like to make it symmetric for simplicity. A popular choice is the uniform distribution, centered about the current parameter value

$$q(\theta^*|\theta) = \begin{cases} \frac{1}{2A} & : \text{if } |\theta^* - \theta| < A \\ 0 & : \text{otherwise} \end{cases}. \tag{7.119}$$

Choice of A is discussed momentarily. The function $q(\theta^*|\theta)$ moves us from θ to θ^* with probability $1/(2A)$. We then form the ratio of posteriors which, assuming a symmetric $q(\theta^*|\theta)$, becomes

$$r(\theta^*|\theta) = \frac{p(\theta^*|\mathbf{y})}{p(\theta|\mathbf{y})} = \frac{p_H(\mathbf{y}|\theta^*)p_\pi(\theta^*)/p_\mathbf{Y}(\mathbf{y})}{p_H(\mathbf{y}|\theta)p_\pi(\theta)/p_\mathbf{Y}(\mathbf{y})} \tag{7.120}$$

Here we see that a large benefit of casting the problem in terms of this ratio is that the scaling constant $p_\mathbf{Y}(\mathbf{y})$ does not appear! In addition, if we choose a uniform prior, we have $p_\pi(\theta^*) = p_\pi(\theta)$ in which case

$$r(\theta^*|\theta) = \frac{p_H(\mathbf{y}|\theta^*)}{p_H(\mathbf{y}|\theta)} \tag{7.121}$$

which is just the ratio of likelihoods. We then keep θ^* as the next value in the chain with probability

$$\min\{r(\theta^*|\theta), 1\} \tag{7.122}$$

In practice, this can be accomplished by drawing a uniform random number from the interval $u \sim U(0, 1)$ and comparing to r. If $u < r$, retain θ^*; otherwise, retain the value θ (see Algorithm I). In heuristic terms, we are continually perturbing the parameter θ according to $q(\theta^*|\theta)$, testing the likelihood ratio, and tending to retain values that are more consistent with the likelihood.

The above-described process repeats (generate candidate, test ratio, accept reject) for some number of iterations $k = 1, \ldots, K$ until we have the desired number of samples from $p_\Theta(\theta|\mathbf{y})$. For a final parameter estimate, we may take the average or median value of the K samples as $\hat{\theta}$. We can also look at the interval that contains say 95% of the values as the confidence interval (referred to as "credible intervals" in the literature). The pseudocode for the process will be given shortly in Algorithm I.

First, however, we conclude this section with an important example. Given our dependence on the independent, joint Gaussian noise model, we can substitute (7.45) into (7.121) in which case the ratio becomes

$$
r(\theta^*|\theta) = \frac{e^{-\frac{1}{\sigma_\eta^2} \sum_{j=1}^{M} \sum_{n=1}^{N} (y_i(n)-x_j(n|\theta^*))^2}}{e^{-\frac{1}{\sigma_\eta^2} \sum_{j=1}^{M} \sum_{n=1}^{N} (y_j(n)-x_j(n|\theta))^2}}
$$

$$
= e^{\frac{1}{\sigma_\eta^2}\left(\sum_{j=1}^{M} \sum_{n=1}^{N} (y_j(n)-x_j(n|\theta))^2 - \sum_{j=1}^{M} \sum_{n=1}^{N} (y_j(n)-x_j(n|\theta^*))^2 \right)}
$$

$$
= e^{\frac{1}{\sigma_\eta^2}(C(\mathbf{y},\theta)-C(\mathbf{y},\theta^*))} \tag{7.123}
$$

where $C(\mathbf{y},\theta)$ is the negative log-likelihood of the independent, joint Gaussian model which we defined previously with Eq. (7.36). The ubiquity of this cost function in estimation problems warrants defining it separately; it will recur throughout the remainder of the book.

7.2.4 Gibbs Sampling

An immediate question that arises is how to handle the multivariate parameter case. It turns out there's a very simple solution to this problem. We may simply run the Metropolis–Hastings algorithm in turn for each parameter in our vector $\theta \in \mathbb{R}^P$ holding all the other parameters fixed at their current values in their own respective chains, a process referred to as "Gibbs sampling."

Following this process, we are technically not sampling from the true marginal distribution, but are rather drawing samples from the conditional distribution $p_{\Theta_i}(\theta_i|\theta_{-i})$ where $\theta_{-i} = (\theta_1, \theta_2, \ldots, \theta_{i-1}, \theta_{i+1}, \ldots, \theta_P) \in \mathbb{R}^{P-1}$ is the vector containing the rest of the system parameters. It is typically assumed that this conditional distribution will be a good approximation to the desired marginal distribution. In fact, it can be shown that for a large number of samples (large K) we are indeed sampling from the desired marginal distribution [41]. We can think of this distribution as the marginal distribution for parameter i if each of the other parameters are known and fixed to their current values in their respective chains. Implementing Gibbs sampling is therefore a straightforward process. For each parameter, give a "kick" with the proposal distribution $q(\theta_i^*|\theta_i)$ while holding all the other parameters fixed. Next, the ratio of posteriors (7.120) is computed and we either keep θ_i^* or retain the previous value θ_i as the next value in the chain, depending on the outcome (with probability r). We then move to the next parameter, hold all others fixed, and repeat the process. After all parameters have been accounted for, we move to the next value in our Markov chain and start the process over.

In the end, there will be P Markov chains, one for each of the P parameters being estimated. The remaining challenge is to choose the "tuning parameters" A in the proposal distribution? Currently, there is no universally accepted method to accomplish this. Large values for these parameters decrease the likelihood of accepting a new value, thus the samples in the chain tend to be highly correlated. Too small a step, however, and the algorithm can take a prohibitively long time to converge. The trade-off between excessive computation time (small steps) and not generating independent samples (big steps) is well known. One simple approach is to adjust the tuning parameter on-the-fly in order to achieve an appropriate acceptance rate of between 30% and 50%. This range of acceptance probabilities has been demonstrated to produce Markov chains with low auto-correlation (good mixing) [42]. These adjustments are made during a "burn-in" period, where the Markov chains are settling down to (hopefully) a stationary process (a key assumption of the MCMC method). We will denote this number of burn-in iterations B. As with most dynamical processes, we have to wait for transients to die out before drawing valid samples that meet this assumption. In many cases, it may take hundreds to tens of thousands of iterations before the chain settles down to a stationary distribution. During the burn-in period, we simply divide A by a constant value $c_1 = 1.007$ after each rejection and multiply by $c_2 = 1.01$ after each acceptance. Thus, an acceptance causes us to expand our parameter search, while rejection results in smaller "kicks" to the previous value in the chain. The asymmetry in the constants causes a slight bias in favor of rejection. This simple approach is quite effective in producing acceptance rates of 40%, a good target for the MCMC algorithm. Following the approach, we simply "shut off" the adjustments once the burn-in period is over.

A slightly more complicated approach decreases the degree of adjustment to A as the number of iterations approach the end of the burn-in period. The idea is to create a smooth transition between burn-in and sampling periods (as opposed to simply shutting off the adjustments when the burn-in period is reached). To this end, we use the approach described in Ref. [43]. After acceptance of θ^*, we apply

$$A = A \times \left(c_2 - (c_2 - 1)\frac{k}{B} \right) \tag{7.124}$$

and after rejection

$$A = A / \left(\frac{1 - p_{accept}}{1 - p_{accept} \left(c_2 - (c_2 - 1)\frac{k}{B} \right)} \right) \tag{7.125}$$

where p_{accept} is a target acceptance probability (usually in the aforementioned 30–50% range), $c_2 = 1.01$ is again a small constant, and the index k denotes the current iteration of the algorithm. Many other tuning strategies exist; however, we have found these to be simple to implement and perform quite well in practice.

While we have shown how to estimate our parameter posterior distributions, we have not yet dealt with the noise variance σ^2 in our likelihood function. This is an unknown parameter and needs to be estimated as well. Fortunately, this *can* be done analytically without having to resort to Metropolis–Hastings.

7.2.5 Conditional Conjugacy: Sampling the Noise Variance

We have already discussed the concept of conjugate distributions and what they mean for a Bayesian analysis in Section 7.2.1. Here, we discuss the closely related topic of conditional conjugacy in the context of sampling from the type of posterior distribution we typically encounter in structural identification problems. In particular, we are interested in sampling the unknown noise variance parameter σ^2 from our jointly normal noise process.

To see how this is accomplished, we again define the "cost" as the log-likelihood (see again Eq. 7.36)

$$C(\mathbf{y}, \boldsymbol{\theta}) = \sum_{n=1}^{N} \sum_{j=1}^{M} (y_j(n) - x_j(n, \boldsymbol{\theta}))^2 \tag{7.126}$$

and let $N_T = M \times N$ be the total number of observations. The posterior for the noise variance parameter is written

$$p_H(\sigma^2, \boldsymbol{\theta}|\mathbf{y}) = \frac{\frac{1}{(2\pi\sigma^2)^{N_T/2}} \mathrm{Exp}\left[-\frac{1}{2\sigma^2} C(\mathbf{y}, \boldsymbol{\theta})\right] p_\pi(\sigma^2) p_\pi(\boldsymbol{\theta})}{\int_0^\infty \frac{1}{(2\pi\sigma^2)^{N_T/2}} \mathrm{Exp}\left[-\frac{1}{2\sigma^2} C(\mathbf{y}, \boldsymbol{\theta})\right] p_\pi(\sigma^2) p_\pi(\boldsymbol{\theta}) d\sigma^2 d\boldsymbol{\theta}}. \tag{7.127}$$

Now, in this equation we have assumed independent priors so that we can split out the prior for σ^2 as separate from the other priors. Also note that the scaling constant in the denominator integrates from 0, not $-\infty$. The variance can never be negative and is only supported on the semi-infinite interval $[0, \infty)$. Let $\tau = 1/\sigma^2$ and write

$$p_T(\tau, \boldsymbol{\theta}|\mathbf{y}) = \frac{\tau^{N_T/2} \mathrm{Exp}\left[-\frac{\tau}{2} C(\mathbf{y}, \boldsymbol{\theta})\right] p_\pi(\tau) p(\boldsymbol{\theta})}{\int_0^\infty \tau^{N_T/2} \mathrm{Exp}\left[-\frac{\tau}{2} C(\mathbf{y}, \boldsymbol{\theta})\right] p_\pi(\tau) p(\boldsymbol{\theta}) d\tau \, d\boldsymbol{\theta}} \tag{7.128}$$

By the law of conditional probability $p_T(\tau|\boldsymbol{\theta}, \mathbf{y}) = \frac{p_T(\tau, \boldsymbol{\theta}|\mathbf{y})}{p_\Theta(\boldsymbol{\theta})}$ so that the full conditional distribution for τ becomes

$$p_T(\tau|\boldsymbol{\theta}, \mathbf{y}) = \frac{\tau^{N_T/2} \mathrm{Exp}\left[-\frac{\tau}{2} C(\mathbf{y}, \boldsymbol{\theta})\right] p_\pi(\tau)}{\int_0^\infty \tau^{N_T/2} \mathrm{Exp}\left[-\frac{\tau}{2} C(\mathbf{y}, \boldsymbol{\theta})\right] p_\pi(\tau) d\tau} \tag{7.129}$$

Now, we could evaluate this numerically, or try and find a prior distribution such that the posterior distribution for this parameter has a closed-form solution. To this end, we let $p_\pi(\tau)$ be the Gamma distribution, that is,

$$p_\pi(\tau) = \tau^{\alpha-1} e^{-\beta\tau} \frac{\beta^\alpha}{\Gamma(\alpha)} \qquad \tau > 0$$

$$\equiv Gamma(\alpha, \beta) \tag{7.130}$$

where $\Gamma(\alpha)$ is the Gamma function (see Appendix A), defined as $\Gamma(\alpha) = \int_0^\infty \tau^{\alpha-1} e^{-\tau} d\tau$ In this case, the posterior becomes

$$p_T(\tau|\boldsymbol{\theta}, \mathbf{y}) = \frac{\frac{\beta^\alpha}{\Gamma(\alpha)} \tau^{\alpha-1+N_T/2} \mathrm{Exp}\left[-(\beta + \frac{1}{2} C(\mathbf{y}, \boldsymbol{\theta}))\tau\right]}{\frac{\beta^\alpha}{\Gamma(\alpha)} \int_0^\infty \tau^{\alpha-1+N_T/2} \mathrm{Exp}\left[-(\beta + \frac{1}{2} C(\mathbf{y}, \boldsymbol{\theta}))\tau\right] d\tau} \tag{7.131}$$

which, after performing the integral in the denominator, can also be recognized as a Gamma distribution. Thus, the Gamma distribution is conjugate to the likelihood for a jointly normal distribution. Letting $\alpha' = \alpha + N_T/2$ and $\beta' = \beta + C(\mathbf{y}/2, \boldsymbol{\theta})$ and integrating the denominator to get the normalizing constant yields

$$p_T(\tau|\boldsymbol{\theta}, \mathbf{y}) = \frac{\beta'^{\alpha'}}{\Gamma(\alpha')} \tau^{\alpha'-1} e^{-\beta'\tau}. \tag{7.132}$$

This means that we can sample $p(\tau|\boldsymbol{\theta}, \mathbf{y})$ directly from a Gamma distribution with parameters α', β'.

However, recall that we are actually trying to sample the noise variance $\sigma^2 = \tau^{-1}$. It can be shown that replacing τ by $1/\sigma^2$ in the Gamma distribution, and applying the rules for transforming a random variable, one obtains the so-called inverse Gamma distribution with parameters $\alpha', 1/\beta'$ (by conventional definition of the inverse Gamma, the second parameter is inverted). In other words we want to sample

$$\sigma^2 \sim IG(\alpha', 1/\beta') = \frac{(\beta')^{\alpha'}}{\Gamma(\alpha)} (\sigma^2)^{-\alpha'-1} e^{-\beta'/\sigma^2} \tag{7.133}$$

where $IG(\cdot, \cdot)$ denotes the inverse Gamma distribution. Note, however, that most software packages allow us to numerically draw from a Gamma, but not an inverse Gamma. Thus in practice one can sample a value $X \sim Gamma(\alpha', 1/\beta')$ and then invert the result, thereby giving $1/X \sim IG(\alpha', 1/\beta')$.

The remaining question becomes how to choose the specific form of the prior, i.e. what to choose for α, β? Often we assume the *diffuse* prior, $\alpha = 1, \beta = \epsilon \ll 1$ resulting in a (roughly) Uniform distribution for $p_\pi(\tau) \approx \epsilon$. Thus, at each step in the Markov chain we can draw a sample from the posterior distribution for σ^2 via

$$X \sim Gamma\left(N_T/2 + 1, \frac{1}{C(\boldsymbol{\theta}, \mathbf{y})/2 + \epsilon}\right)$$

$$\approx Gamma\left(N_T/2 + 1, 2/C(\boldsymbol{\theta}, \mathbf{y})\right)$$

$$\sigma^2 = 1/X. \tag{7.134}$$

Thus, there is no need to perturb parameters and check the ratio (7.120), and so on. The variance just becomes another parameter (the $P + 1$th parameter) in the Gibbs sampling strategy, but does not require application of the Metropolis–Hastings algorithm.

It is worth pointing out, however, that if the ratio of likelihoods provides little information about the parameter of interest *and* we use a diffuse prior, the resulting Markov chain can become unstable and "drift" off to an infinitely large value. This might be expected as neither the data nor our prior information is providing any information with which to solve the problem. In fact, one could argue that if neither the likelihood nor the prior can at least bound the parameter value of interest, the practitioner needs to work harder on the model before implementing MCMC!

The pseudocode for implementing the entire MCMC algorithm, assuming a jointly Gaussian distributed likelihood function, is provided in Algorithm 1.

Task: Generate the posterior parameter distributions $p_{\theta_i}(\theta_i)$ for $i = 1 \cdots P$ given the model and the observed data $\mathbf{y} \equiv y_j(n)$, $n = 1 \cdots N$, $j = 1 \cdots M$.
Also estimate noise variance $\theta_{P+1} = \sigma^2$
Initialization: Initialize chain $k = 0$
    Initial guesses for parameter values (priors) $\boldsymbol{\theta}(0) \equiv \theta_i(0)$ $i = 1 \cdots P$
    Initial values for tuning parameters A_i $i = 1 \cdots P$
    Sample initial variance from inverse Gamma distribution
       $\theta_{P+1}(0) = IG(N/2 + 1, 0.5C(\mathbf{y}, \boldsymbol{\theta}(0))$
    Set number of burn-in iterations B
Main iteration:increment k by 1 and apply
    For each parameter i
      Generate candidate $\theta_i^* = \theta_i(k-1) + 2A_i \times U(-1,1)$
        where $U(a,b)$ is a uniformly distributed number on $[a,b]$
      Compute $r = (Pr_i(\theta_i^*)/Pr_i(\theta_i(k-1)) \times \exp((-0.5/\theta_{P+1}(k-1))$
      $\times (C(\mathbf{y}, \boldsymbol{\theta}^*) - C(\mathbf{y}, \boldsymbol{\theta}(k-1))))$
        where $\boldsymbol{\theta}^* \equiv (\theta_1(k), \ldots, \theta_{i-1}(k), \theta_i^*, \theta_{i+1}(k-1), \ldots, \theta_P(k-1))$
        and $\boldsymbol{\theta}(k-1) \equiv (\theta_1(k), \ldots, \theta_{i-1}(k), \theta_i(k-1), \theta_{i+1}(k-1), \ldots, \theta_P(k-1))$
      If $U(0,1) < r$ keep the new value, i.e. set $\theta_i(k) = \theta_i^*$, and adjust tuning
        parameter $A_i = A_i \times 1.01$
      Else reject the new value, keeping $\theta_i(k) = \theta_i(k-1)$ and adjust tuning
        parameter $A_i = A_i/1.007$.
    Directly sample the variance posterior $\theta_{P+1}(k) = IG(N/2 + 1, 0.5C(\mathbf{y}, \boldsymbol{\theta}(k))$
    After $k > B$ iterations, cease adjusting the A_i and record subsequent
    values $\theta_i(k)$ as members of $p_{\theta_i}(\theta_i)$.
    Stop when desired number of iterations K have been recorded.

Algorithm 1. The MCMC algorithm using Metropolis–Hastings with Gibbs sampling assuming Gaussian likelihood, a Uniform "proposal" distribution $q(\theta_i^*, \theta_i(k-1)) = 1/(2A_i)$, and parameter prior distributions $Pr_i(\theta_i)$.

7.2.6 Beam Example Revisited

In Section 7.1, we used the method of maximum likelihood to estimate the stiffness and damping characteristics of a finite element beam with Euler–Bernoulli elements. Here, we return to that same problem, only using the MCMC algorithm to evaluate the posterior probability distribution associated with each of the parameters.

The main difference is that in this application we are required to specify a prior PDF for each of the variables we wish to identify. There are certain constraints that we know the problem must obey. First of all, stiffness cannot be negative. Hence, for the parameter *EI*, for example, we require a nonnegative prior. We have found a Gamma distribution to be an ideal prior in this situation. The Gamma distribution (given in Table A.2) has support on the interval

$[0, \infty]$. The parameters of this distribution can be set so that the prior is uninformative (uniform distribution) or weights certain values more heavily than others. If the desired mean and variance for the Gamma distribution are μ, σ, respectively, the Gamma parameters take the values

$$\alpha = \mu^2/\sigma^2$$

$$\beta = \sigma^2/\mu. \tag{7.135}$$

In what follows, we use Gamma priors for each of the nonzero parameters $EI, \zeta, A_S, \sigma_S^2$. For the stiffness parameter EI (actually stiffness is given by EI/L^3; however, we assume the length of the beam is known so that the random variable is EI) and we choose a Gamma prior centered roughly on the true value with a fairly large variance (see Figure 7.20). The elastic modulus is usually known for a given material, while the moment of inertia is based on measured geometry. We therefore would imagine that a prior with a tighter variance may be more realistic. We assume a similar degree of knowledge for the damping parameter. A simple logarithmic decrement calculation (see, e.g., [3]), for example, would provide a very good initial estimate from which the expected variation would be small.

The damage parameters represent a far more challenging problem. As we have stated multiple times, these parameters have very little influence on the observed response, hence the likelihood has very broad, shallow peaks with respect to these parameters, making them difficult to identify. We also have the problem of not having a great deal of *a priori* information. For example, we are looking for small damage, hence we might choose a broad distribution for $p_A(A_s)$, with most of the probability weighted toward zero (no damage). Our prior for this parameter is also shown in Figure 7.20. We choose a similar prior for damage width for the same reasons.

For the location parameter, μ_S, a different prior is used. We know the damage location must be physically on the beam, hence a uniform prior over the range $[0, L]$ where L is the length of the beam and is an appropriate prior distribution. Thus, we are claiming ignorance as to where the damage is located, stating only that we know it is somewhere on the beam. Setting priors is therefore a formal way of bounding solutions to the values that are physically possible. Any Markov chain value that lies outside the bounds dictated by the prior will be assigned a zero weight, hence that value will produce a ratio of $r = 0$ in the algorithm and *never* be accepted. In the results that follow we plot both prior, and the posterior value to which the solution converged.

For this example, we consider again a rectangular damage cut in the same location as our previous example. Markov chains are initialized for each parameter with a draw from the parameter prior. We then iterate the chains using a uniform candidate distribution with tuning parameter set using the approach outlined in Section 7.2.4. The posterior parameter distributions are shown, along with the priors, in Figure 7.20. Notice that as part of the identification process we easily obtain the posterior for the noise variance as drawn from the inverse Gamma prior.

At this point, we could choose as our final estimate the posterior means, modes, or some other function of the distribution. Taking the posterior mode yields the stiffness profile shown in Figure 7.21. As with the MLE approach, we are able to obtain good estimates of the damage parameters with the possible exception of an overestimated damage amplitude. However, unlike the MLE approach, we have the entire posterior PDF at our disposal, which readily

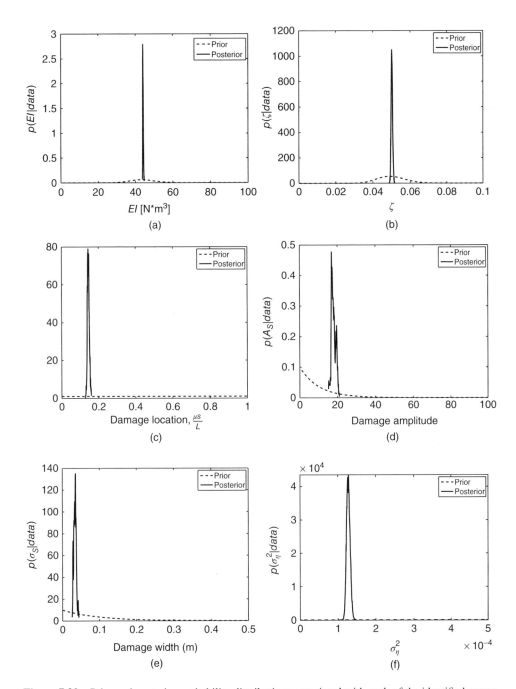

Figure 7.20 Prior and posterior probability distributions associated with each of the identified parameters: (a) EI, (b) ζ, (c) μ_S, (d) A_S, (e) σ_S^2, and (f) σ_η^2

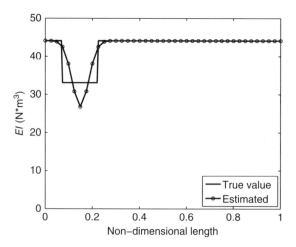

Figure 7.21   Identified stiffness profile associated with the mean values of the damage parameter posteriors. The identified profile closely matches the true profile denoted by the rectangular cut in the beam

provides a measure of uncertainty in the parameter estimates. From these distributions it is straightforward to construct credible intervals of confidence, as will be done frequently in Chapter (9).

7.2.7   Population-Based MCMC

In the MCMC method, the distribution $q(\theta_i^*|\theta_i)$ can be thought of as a rule for perturbing θ_i. The magnitude of this perturbation will be decided by the parameters associated with $q(\cdot)$. For example, it is common to choose the proposal distribution to be the uniform distribution centered at θ_i, that is,

$$q(\theta_i^*|\theta_i) = U(\theta_i - A, \theta_i + A) \tag{7.136}$$

where A is a real constant that plays the role of the perturbation size. Typically, A is tuned dynamically so that on average $r = 0.3 - 0.5$ (see [44] for a more thorough discussion and for sample code). Heuristically, it is easy to see how the approach works. The algorithm continually perturbs each parameter, checks whether a better fit (consistent with the prior) is achieved by computing r, and tends to keep parameter values that do well in this regard. The fact that in the end the chain of values are samples from $p_{\Theta_i}(\theta_i)$ is actually quite remarkable. However, one can immediately see where problems can arise.

Consider a bimodal posterior distribution of a single parameter θ,

$$p_{\Theta}(\theta) = \frac{1}{2\sqrt{2\pi f \sigma^2}} e^{-(\theta-\mu)^2/(2f\sigma^2)} + \frac{1}{2\sqrt{2\pi(1-f)\sigma^2}} e^{-(\theta+\mu)^2/(2(1-f)\sigma^2)} \tag{7.137}$$

for known, positive constants f, σ, μ. The constant $f < 1$ specifies the fraction of the distribution variance associated with the distribution peaks at $+\mu, -\mu$. For this example, the values $\sigma = 1, f = 0.3$ were used along with two different values for the peak locations, $\mu = 2, 4$. The resulting distribution presents difficulties for the conventional MCMC algorithm for the $\mu = 4$

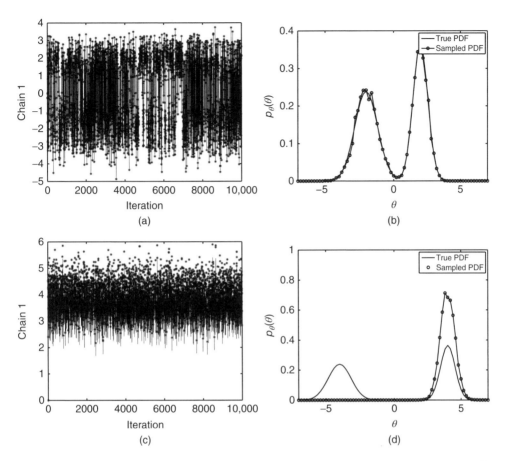

Figure 7.22 Markov chains and estimated bimodal posterior distribution given by Eq. (7.137) with $\mu = 2$ (a and b) and $\mu = 4$ (c and d), respectively. (*Source:* Reproduced from [45], Figure 1, with permission of Elsevier)

case, as shown in Figure 7.22. For $\mu = 2$, the peaks of the distributions are close enough that the algorithm can easily move back and forth between the two regions of high probability. However, for $\mu = 4$, the Markov chain quickly becomes "trapped" in a single portion of the posterior distribution. This is simply because the proposal distribution is not capable of moving the chain easily from one peak to the other. The solution, however, is not as simple as changing the proposal distribution to give us larger perturbations to our chain (i.e., increase A). If the proposal values are very far from the existing values, they will almost always be rejected, thus the Markov chains will take an extremely long time to converge. An efficient sampler is one that would allow us to *locally* explore a high probability region of the posterior while simultaneously providing a mechanism for covering large distances in parameter space to reach other high-probability regions. The population-based MCMC approach, described in this section, was designed specifically for this reason.

As with standard MCMC, the goal of the pop-MCMC algorithm is to draw samples from some desired posterior distribution $p_\Theta(\theta)$. A very nice introduction to the topic of population-based MCMC (pop-MCMC) is given by Jasra *et al.* [46]. The basic idea is to first create a new, composite posterior density

$$p_Q(\theta^{(1:Q)}) = \prod_{q=1}^{Q} p_q(\theta^{(q)}) \tag{7.138}$$

which is a function of the composite parameter vector $\theta^{(1:Q)} = (\theta^{(1)}, \dots, \theta^{(Q)})$. Here, we deviate from our usual notation and use Q to denote the composite density, formed as a product of densities denoted by the lower case "q," that is, the $p_q(\theta^{(q)})$ are the PDFs governing parameter vector $\theta^{(q)}$. It is required that $p_q(\theta^{(q)}) = p_\Theta(\theta)$ for at least one q, that is, one of the posteriors comprising the composite density is the true posterior distribution. In what follows, the authors use $q = 1$ to denote the true posterior distribution from which the desired samples will be drawn. If an irreducible, a-periodic Markov chain that has $p_Q(\theta^{(1:Q)})$ as its invariant distribution can be constructed, then samples from the marginal distribution

$$p_\Theta(\theta) = \int_{\mathbb{R}^{Q-1}} p_Q(\theta^{(1:Q)}) d\theta^{(2:Q)} \tag{7.139}$$

can be drawn where the notation $\int_{\mathbb{R}^{Q-1}} d\theta^{(2:Q)}$ denotes the multidimensional integral over all parameters vectors other than $\theta \equiv \theta^{(1)}$. This is accomplished numerically by running "Q" Markov chains concurrently, each exploring its own posterior distribution $p_q(\theta^{(q)})$. Each chain can be considered in turn, holding the others fixed (Gibbs sampling), and samples from $p_1(\theta^{(1)})$ are retained as samples from the desired joint posterior.

There exists a good deal of freedom in choosing the $p_q(\cdot)$ and in selecting different types of proposals for exploring the parameter space. With regard to the former, one would like to wisely choose the $p_q(\cdot)$ so as to facilitate easy exploration of the parameter space. While other approaches can be used (see Jasra *et al.* [46]), we have tended to use the so-called tempered sequence of distributions

$$p_q(\theta^{(q)}) \equiv p_\Theta^{\zeta_q}(\theta^{(q)}) \tag{7.140}$$

for $\zeta_q \in (0, 1]$. For $\zeta_1 = 1$, of course, one has the true posterior. For smaller values of ζ_q, one obtain successively smoother versions of the original posterior. The idea is that the smoothed distributions are easier for their corresponding Markov chains to explore, yet they are still related to the true posterior and therefore still carry information about high-probability regions of the parameter space. As an example, consider $Q = 4$ separate instances of the bimodal distribution (7.137) raised to the $\zeta_q = 1, 0.5, 0.1, 0.0125$ powers, respectively. These distributions are shown in Figure 7.23. The standard MCMC algorithm will have an easier time exploring these smoother parameter spaces. Of course, only samples from the distribution with $\zeta_1 = 1$ are of interest, however the additional distributions can clearly facilitate efficient sampling if one has a means of passing information between chains. The chain associated with the true posterior therefore needs to be informed by the chains exploring the smoother distributions. Perhaps the simplest type of move to accomplish this is the so-called swap move. This is similar to a Metropolis–Hastings move, where the proposal is to consider swapping the parameter values in chains u, v. Assuming an equal probability of selecting chains u, v from the N possibilities,

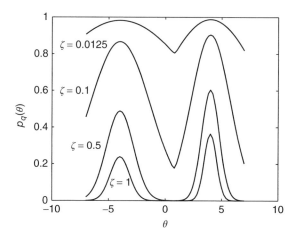

Figure 7.23   Successive $p_q(\theta) = p_\Theta^{\zeta_q}(\theta)$ corresponding to the bimodal posterior distribution Eq. (7.137). The true posterior distribution from which samples are sought is given by $\zeta_1 = 1$ *Source:* Reproduced from [45], Figure 2, with permission of Elsevier

this swap is accepted with probability

$$r_{swap} = \min\left\{\frac{p_u(\theta^{(v)})p_v(\theta^{(u)})}{p_u(\theta^{(u)})p_v(\theta^{(v)})}, 1\right\} \tag{7.141}$$

Typically, swap moves are not proposed after every iteration in the Markov chains, but are performed with some probability.

Returning to the bimodal example, consider $Q = 4$ chains exploring the composite target

$$p_Q(\theta^{(1:4)}) = \prod_{q=1}^{4} p_\Theta^{\zeta_q}(\theta^{(q)}) \tag{7.142}$$

We, therefore, have four Markov chains running concurrently and set $\zeta_q = 1, 0.75, 0.5, 0.25$, respectively. After each iteration in the Markov chain, a swap move is performed with 50% probability. This move consists of uniformly selecting two of the chains and evaluating Eq. (7.141). If the proposal is accepted, the values in the chains are exchanged and the algorithm continues to the next iteration in the Markov chains. Figure 7.24 shows the results of this sampler. All four Markov chains are informing each other as to the presence of multiple peaks in the distribution. The end result is that the chain associated with $\zeta_1 = 1$ contains samples from the desired posterior distribution; this distribution is shown in Figure 7.24b and compares favorably to the true posterior density.

While the swap move is an effective means of communicating among chains, a more sophisticated type of move is sometimes required. Particularly for multivariate parameter estimation, it is useful to design a move that can hold multiple parameters fixed while allowing other groups of parameters to move simultaneously. Such a move is not allowed with the standard Metropolis–within–Gibbs sampling (one parameter moved at a time) or by the swap move which moves all parameters simultaneously. A particularly effective move is to use the differential evolution algorithm of Section 7.1.3 to generate a trial vector. As we have already

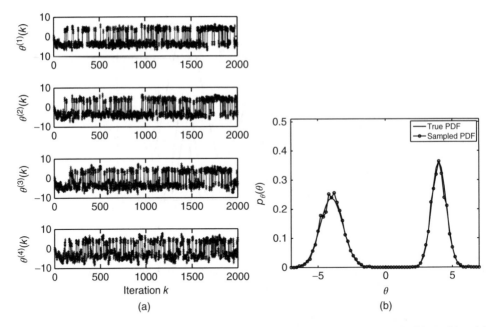

Figure 7.24 (a) Markov chains and (b) estimated posterior distribution associated with the bimodal distribution given by Eq. (7.137). (*Source:* Reproduced from [45], Figure 3, with permission of Elsevier)

discussed, DE is the engine of a popular genetic algorithm (GA) used in searching complex parameter space [15]. This particular move, therefore, draws at random three members of the population, $\theta^{(u)}, \theta^{(v)}, \theta^{(w)}$ and generates the trial vector

$$\theta^{(u')} = \theta^{(u)} + \gamma(\theta^{(v)} - \theta^{(w)}) \tag{7.143}$$

where γ is the same user-defined constant in Eq. (7.37). Each of the P elements in this trial vector replaces the elements of the original vector $\theta^{(u)}$ with 50% probability. This final step (keep new element or retain old) emulates the "crossover" step common to most GAs, and was given explicitly in our earlier discussion of DE, Eq. (7.38). Once the trial vector has been generated, it is accepted/rejected using r (Eq. 7.107). However, to differentiate among the types of moves, the acceptance ratio for the DE move will be denoted r_{DE}.

It should be mentioned that others have proposed a *crossover* move, whereby the parameter vector is split at a point $m \ll P$ for two "parent" vectors, generating the trial vectors:

$$\theta^{(u')} = \left[\theta_1^{(u)}, \ldots, \theta_m^{(u)}, \theta_1^{(v)}, \ldots, \theta_P^{(v)} \right]$$

$$\theta^{(v')} = \left[\theta_1^{(v)}, \ldots, \theta_m^{(v)}, \theta_1^{(u)}, \ldots, \theta_P^{(u)} \right] \tag{7.144}$$

and accepting with probability r_{swap} (Eq. 7.141) [46]. This is indeed a useful way of exchanging information between chains, however we have found that this mechanism is already provided

Normalized Dim.	Value
Width (L_x/L_y)	1.25
Length (L_y/L_y)	1
Thickness (h/L_y)	0.0026

Figure 7.25 Setup and physical dimensions of an experiment for identifying the properties of an aluminum plate

for in the differential evolution move, thus this type of move is not used in our implementation of pop-MCMC.

The above-described moves clearly borrow from the GA approach to optimization problems. In fact, pop-MCMC effectively combines the efficient search capabilities of GAs with the power of the Bayesian MCMC approach to sampling. Both types of moves, swap and differential evolution, will be used in the structural dynamics example presented in subsequent sections.

A simple experimental example that illustrates the utility of the pop-MCMC approach is depicted in Figure 7.25. The model that describes this particular experiment (along with more in-depth results) is provided in Section (9.3). In short, the plate response to an impulse excitation is recorded by three single-axis Vishay Micro-Measurements (model EA-13-125AD-120) strain sensors. Using this response, and assuming a jointly Gaussian noise model, we used the pop-MCMC approach to determine the posterior probability distribution of the plate thickness (again, the details of the estimator are provided in Section 9.3).

Our reason for gravitating toward this particular algorithm can be seen simply by inspecting the resulting distribution, shown in Figure 7.26. The posterior is clearly bimodal, with one peak located on the true value (as measured with a micrometer) and the other located at less than half the true value. This behavior is not uncommon in structural parameter estimation problems. Frequently, we have found that two very different values of a parameter can, when combined with similarly different values of a second parameter, predict very similar dynamical responses. In this case, a parameter governing compliance at the plate boundary (described in Section 9.3) had a similarly bimodal posterior distribution, thus two different pairs of these parameters produced nearly the same likelihood. Nonetheless, the algorithm clearly selected the correct parameter value as the more probable. Use of conventional optimization routines (tied to Bayesian or MLE estimation procedures) will have difficulty with these types of likelihoods and the pop-MCMC algorithm is one suitable alternative.

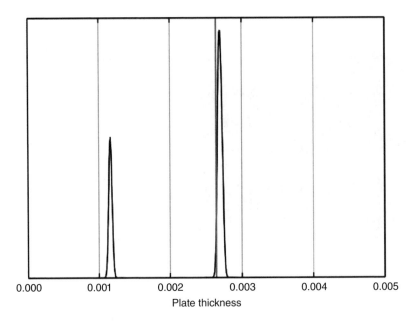

0.000   0.001   0.002   0.003   0.004   0.005

Plate thickness

Figure 7.26   Posterior probability distribution for the thickness of the plate shown in Figure 7.25. The
mode of the posterior distribution is a good estimate of the true parameter value given by the solid vertical
line

7.3   Multimodel Inference

A potential criticism of the Bayesian or any other damage identification procedure is that we
have so far considered only one damage model at a time. In reality, of course, there may be
multiple damages in a structure, the number of which we will not know *a priori*. It would
therefore seem useful to have a means of identifying both the number and type of damages *and*
the parameters associated with each of the damage models. The field of multimodel inference
or *model selection* is quite challenging as it forces the practitioner to ask what it is they really
want in a model. It also forces us to confront the fact that we will never have *the* model for
describing a given phenomenon, but can always search for a *better* model.

   While model selection has been used for many years in other fields (e.g., ecology [41, 47]),
its use in damage identification has been limited (see, e.g., [48] and [49]). In what follows,
we describe two approaches to multimodel inference and show how they can be applied to
problems in structural dynamics.

7.3.1   Model Comparison via AIC

In Section (3.5.1) we discussed the average mutual information function and mentioned that
one interpretation of the expression (3.119) was a "distance" between the hypothesis that the
random variables in a joint probability distribution were independent. This same framework
can be used in thinking about model comparisons. In what follows, we present a mathematical
tool for model comparison inspired by the presentation of Burnham and Anderson [47].

Consider two joint probability distributions for describing structural response data, $f_{\mathbf{X}}(\mathbf{x})$ and $g_{\mathbf{X}}(\mathbf{x}|\boldsymbol{\theta})$. Conceptually, consider the first distribution as "truth," that is, a model that predicts exactly the sequence $\mathbf{x}$. As will be shown, we will never have to know what this function looks like or what its parameters are (if it even has parameters, hence the lack of parameterization in the notation). It is merely a mathematical construct, playing the role of the most accurate possible model.

The second model $g_{\mathbf{X}}(\mathbf{x}|\boldsymbol{\theta})$ *is* a model that we postulate to forecast the observations. In the same way that the mutual information measured a "distance from independence," we could write

$$I_{FG} = \int_{\mathbf{X}} f_{\mathbf{X}}(\mathbf{x}) \log \left(\frac{f_{\mathbf{X}}(\mathbf{x})}{g_{\mathbf{X}}(\mathbf{x}|\boldsymbol{\theta})} \right) d\mathbf{x} \tag{7.145}$$

as a "distance" between our model and "truth." The quantity I_{FG} is therefore a reasonable way to assess how close a postulated model g is to the true model f. One could simply compute (7.145) for all possible g and select the one with the smallest value as the true model. Of course, this is not at all a practical approach as we know neither truth or the parameter values $\boldsymbol{\theta}$ governing our candidate models. Instead, we seek an approach by which we estimate (7.145) and use the estimate as our means for model comparison.

Expanding (7.145) as the sum of entropies (see Section 3.5.1), we have

$$I_{FG} = \int_{\mathbf{X}} f_{\mathbf{X}}(\mathbf{x}) \log (f_{\mathbf{X}}(\mathbf{x})) d\mathbf{x} - \int_{\mathbf{X}} f_{\mathbf{X}}(\mathbf{x}) \log (g_{\mathbf{X}}(\mathbf{x}|\boldsymbol{\theta})) d\mathbf{x}$$

$$= C - \int_{\mathbf{X}} f_{\mathbf{X}}(\mathbf{x}) \log (g_{\mathbf{X}}(\mathbf{x}|\boldsymbol{\theta})) d\mathbf{x}$$

$$= C - E_X[\log (g_{\mathbf{X}}(\mathbf{x}|\boldsymbol{\theta}))] \tag{7.146}$$

so that "truth" becomes a constant C and the second term is recognized as a joint expectation over the random process $\mathbf{X}$. However, as we mentioned in the previous paragraph, we also will not know *a priori* what the parameter values associated with our models are. Rather, these are values we need to estimate on the basis of available data. As usual, we assume our observations are contaminated by noise such that the observed data may be written $\mathbf{y} = \mathbf{x} + \boldsymbol{\eta}$. We, therefore, need to replace the true parameter values by their estimates $\boldsymbol{\theta} = \hat{\boldsymbol{\theta}}(\mathbf{y})$. Even with this substitution, the estimate

$$\hat{I}_{FG} = C - E_X[\log (g_{\mathbf{X}}(\mathbf{x}|\hat{\boldsymbol{\theta}}))] \tag{7.147}$$

cannot be obtained because there is no way to take this expectation in practice. Instead, we consider the expected estimated value over the unknown parameter vector and write

$$E_{\hat{\boldsymbol{\theta}}}[\hat{I}_{FG}] = C - E_Y E_X[\log (g_{\mathbf{X}}(\mathbf{x}|\hat{\boldsymbol{\theta}}))] \tag{7.148}$$

where it has been noted that expectation over a parameter vector that depends exclusively on $\mathbf{y}$ (our data) is equivalent to an expectation over the data.

Even then, it is a challenging matter to approximate the double expectation given by the right-hand side of (7.148). It turns out that expanding the log of $g_X(\cdot)$ in a Taylor series about the true parameter value, a good approximation for large samples is

$$E_Y E_X[\log (g_{\mathbf{X}}(\mathbf{x}|\hat{\boldsymbol{\theta}}))] \approx \log (g_{\mathbf{X}}(\hat{\boldsymbol{\theta}}|\mathbf{x}) - K \tag{7.149}$$

where K is the number of parameters in the vector θ [47]. Hence, we can write the estimated expected distance between a given model $g_{\mathbf{X}}(\cdot)$ and truth as

$$E_{\hat{\theta}}[\hat{I}_{FG}] = C - \log(g_{\mathbf{X}}(\hat{\theta}|\mathbf{x}) + K \qquad (7.150)$$

Given a collection of different models, and defining "best" model as the one with the smallest expected distance to truth as defined by (7.145), we choose the model such that

$$g_X^*(\mathbf{x}|\hat{\theta}) = \min_g\{K - \log(g_{\mathbf{X}}(\hat{\theta}|\mathbf{x})\} \qquad (7.151)$$

which some readers may recognize as the Akaike information criteria (AIC) for model selection. Sometimes the right-hand side appears in the literature premultiplied by the constant "2" as [50]

$$AIC \equiv 2K - 2\log(g_{\mathbf{X}}(\hat{\theta}|\mathbf{x}). \qquad (7.152)$$

The AIC is therefore shown to have a strong foundation in information theory, although it does involve several approximations in its development as an estimator. In addition to having a strong mathematical basis, the AIC has an intuitive appeal. It simply states that the best model will be one that fits the data well (small log-likelihood) while guarding against overparameterization (lack of parsimony) with the penalty term $2K$.

   Variants of the AIC exist, however these are usually designed to "fix" the approximation to the AIC in the instance of small sample size. In structural dynamics applications, the ratio of the number of data N to the number of parameters K is usually quite large, and hence these corrected formula are not presented here.

   Before making use of the AIC, it is worth noting that this quantity has no units. As a result, AIC values can range from large negative numbers to large positive numbers, that is, there is no natural scale for what constitutes a "big" or "small" AIC. As we have shown in the derivation, the AIC is a relative measure of model performance, hence the absence of a natural scale is not critical. Nonetheless, in model selection applications, it makes more sense to consider the differences in AIC values among models. Given M different models for comparison, and denoting the smallest AIC value among them as $AIC_{\min}$, we compute the relative values

$$\Delta_i = AIC_i - AIC_{\min} \qquad (7.153)$$

as recommended by Burnham and Anderson [47]. It is further noted in Ref. [47] that the likelihood of the model "i," given the data $\mathbf{x}$, is given by

$$L_I(i|\mathbf{x}) \propto \exp\left(-\frac{1}{2}\Delta_i\right). \qquad (7.154)$$

If we design the constant of proportionality such that this PMF integrates (sums) to unity, we arrive at the "Akaike weights" [46, 51]

$$w_i = \frac{\exp\left(-\frac{1}{2}\Delta_i\right)}{\sum_{m=1}^{M}\exp\left(-\frac{1}{2}\Delta_m\right)} \qquad (7.155)$$

By construction, the values for w_i lie on the interval [0, 1] and can be viewed as evidence in favor of model i given the data. It is also known from our discussion of Bayes' relationship, that premultiplying the likelihood by a prior probability yields a posterior PMF for the discrete variable "model" as

$$w_i = \frac{\exp\,(-\frac{1}{2}\Delta_i)\pi_i}{\sum_{m=1}^{M} \exp\,(-\frac{1}{2}\Delta_m)\pi_m} \tag{7.156}$$

where π_i is the ith model prior probability. Whether one uses (7.155) or (7.156) depends, as it should, on whether we have significant *a priori* information about the system dynamics to favor one model over another. Equal priors in (7.156) obviously lead us back to (7.155).

So how might we use (7.152) in damage detection applications? To date, we are not aware of situations where the AIC has been used to select among competing physics-based models of structural damage. However, several applications of the AIC have been proposed in selecting among time-series models of structural response data. In particular, the AR models of Section (3.4) are sometimes constructed to predict an "undamaged" response. As the dynamics of the structure change (hopefully due to damage), the prediction error should rise and will make a good damage-sensitive test-statistic. The AIC has been used by some in the field to select the order of the AR model used in this type of diagnostic procedure (see, e.g., [52–54]).

However, we view the potential use of AIC for selecting among competing physics-based models as a promising (and, to our knowledge, unexplored) avenue of research. Consider two models for a structural system

$$\begin{aligned}
\mathcal{M}_1 : \quad & m\ddot{x} + c\dot{x} + kx = A\cos(2\pi ft)) \\
\mathcal{M}_2 : \quad & m\ddot{x} + c\dot{x} - kx + \epsilon x^3 = A\cos(2\pi ft)
\end{aligned} \tag{7.157}$$

We do not know *a prior* which of these two models governs the structures dynamics, however on the basis of what we know about the physics, we believe the response $x(t)$ is accurately described by one of them.

For the observed data, we measure the model 2 acceleration $\ddot{y}(n\Delta_t) = \ddot{x}(n\Delta_t) + \eta_{n\Delta_t}$, $n = 1 \cdots 800$ using a sampling interval of $\Delta_t = 0.005$ s. This interval was based on the excitation frequency of $f = 10$ Hz. For the system parameters we assume: $m = 1$ (kg), $k = 10^3$ (N/m), $c = 10$ (N $\cdot$ s/m), and $\epsilon = 10^5$ (N/m^3). The amplitude of excitation A (N) will remain variable for illustrative purposes. Although the data were produced by $\mathcal{M}_2$, we do not know that *a priori*. Instead, we use the MLEs for the model parameters, along with the Akaike weights, to decide between $\mathcal{M}_1, \mathcal{M}_2$. In this example, we know enough about the problem (noise is normally distributed) to form the parameter MLEs via linear least squares. That is to say, if we let

$$\theta_1 \equiv (c, k, A)$$
$$\theta_2 \equiv (c, k, \epsilon, A) \tag{7.158}$$

we can form the estimates as discussed in Section 7.1 via

$$\hat{\theta}_1 = \mathbf{Y}_1^+\mathbf{y}$$
$$\hat{\theta}_2 = \mathbf{Y}_2^+\mathbf{y} \tag{7.159}$$

where, as before, the columns of the matrices $\mathbf{Y}_1, \mathbf{Y}_2$ are formed from the observed position and velocity data and $\mathbf{y}$ is the vector of observed acceleration.

Note that we again had to assume that the mass parameter was known for the reasons of identifiability mentioned in Section 7.1. In experiments, this is the easiest parameter to identify given that the geometry and density of the structure under study is typically well-defined. The MLE parameter vectors are then substituted into the likelihood functions

$$g(\hat{\theta}|\mathbf{y}) = \frac{1}{(2\pi\sigma)^{N/2}} \exp\left(-\frac{1}{2\sigma^2} \sum_{n=1}^{800} \left(\ddot{y}(n) - \ddot{x}(n|\hat{\theta})\right)^2\right) \qquad (7.160)$$

to give the AIC values

$$AIC_1 = -2\log(g(\hat{\theta}_1)) + 8$$
$$AIC_2 = -2\log(g(\hat{\theta}_2)) + 10 \qquad (7.161)$$

and, ultimately, the AIC weights given by Eq. (7.155). For this example, we assumed that both models were equiprobable ($\pi_1 = \pi_2 = 0.5$) initially.

In Figure 7.27a, we can observe the variable loading levels used in this study. Initially, for low amplitude excitation, the dynamics of the system are best (in the sense of AIC) described by a linear model. For small-amplitude motion, both $\mathcal{M}_1, \mathcal{M}_2$ describe the response equally well, however the addition of the extra model parameter required of $\mathcal{M}_2$ acts as a penalty, and hence $\mathcal{M}_1$ is favored.

As the excitation increases in strength, the characteristics of the nonlinear system begins to outweigh the addition of the ϵ parameter in the model selection criteria. This illustrates what we view as an important point. *If the measured dynamics of the system do not reflect the presence of the damage parameter we seek to estimate or exhibit the characteristics of the model we seek to identify, we have little hope of success.* This is yet another reason why the use of a physics-based approach is so important. Given a model of the type, extent, location,

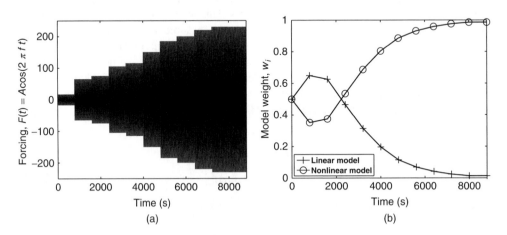

Figure 7.27 Progression in AIC model weights as a function of loading levels. For low loading levels, the system operates in a linear regime; hence, the linear model is selected over the nonlinear one as it is a more parsimonious description of the data. However, for large load levels, the linear likelihood diminishes and the AIC weights correctly select the nonlinear model. For all loading levels, truth is the nonlinear model

and so on, of the damage, we can directly test our ability to detect and identify it. Such a study lets us know immediately if the problem physics supports the design of an experiment or if the damage is simply too small and has too little influence on the dynamics to be identified. We can cite several examples where researchers have spent much time and effort searching for damage-related signatures in data but where the physics of the problem suggested such signatures were far too weak to detect.

Although this is a simple example, it illustrates the potential usefulness of AIC weights as a damage identification scheme. One could track the AIC weights as a function of time, giving a good estimate of the most likely underlying model. Moreover, in generating the AIC weights, one must develop MLEs for the model parameters, at least one of which presumably says something about the structural damage.

While we can view the AIC weights as model probabilities, this is not entirely correct. In generating the weights w_i, we rely on the principle of maximum likelihood in estimating the model parameters. There is certainly nothing wrong with this approach, however if one is to truly estimate the PMF for "model," a more sophisticated approach is required.

7.3.2 Reversible Jump MCMC

In this section, we describe how the MCMC algorithm of Section 7.2.3 can be generalized to generate not only the posterior parameter PDFs but also the PMF governing the candidate models. In other words, we consider "model" as a random variable $\mathcal{M}$ just as we typically do with parameters. The PMF for $\mathcal{M}$ is really what the model weights we derived using AIC were designed to approximate.

The idea of RJMCMC is to extend the Bayesian estimation philosophy we have used for individual model parameters to a collection of models and, ultimately, to estimate the model PMF. Estimating the posterior model PMF, $p_{\mathcal{M}}(m|\mathbf{y})$, is a more complicated endeavor than obtaining posterior parameter distributions. For example, different models will often have different numbers of parameters with (frequently) different physical meanings. The original algorithm was suggested by Green [55] as a modification to the standard MCMC approach. It has since seen only limited use in damage detection applications, the results in Section (9.1.3) serving as one of the few such studies.

Just as traditional MCMC prescribes "moves" from one parameter to the next in building the Markov chain, RJMCMC requires valid moves *between* models. Thus, at the end of the algorithm, there will be a Markov chain associated with the variable "model" as well as for the parameters associated with the various models. To begin, denote the d_m-dimensional parameter vector associated with model m as $\boldsymbol{\theta}^{(m)} \in \mathbb{R}^{d_m}$. The first conceptual difficulty is how to handle models of different dimension, for example, where $d_1 \neq d_2$. Recalling the traditional MCMC algorithm to the variable "model" would require a proposal distribution that can take us from model m to any other model in a finite number of steps. Moreover, the reverse of any proposed move must also exist such that Eq. (7.100) holds. Green [55] demonstrated that proposal distributions that accomplish this for models of different dimension can indeed be designed; we give one such construction here.

First, we rank the models such that $d_1 < d_2 < \cdots < d_M$ for the class of M models being considered. The increasing dimensionality can be accounted for in the analysis by considering an auxiliary random variable $\mathbf{u}^{(m)} \in \mathbb{R}^{d_{m+1}-d_m}$ with dimension equal to that of the difference between the more complex model parameter vector and the parameter vector associated with model m. That is to say, we have that $\theta^{(m+1)}$ and $(\theta^{(m)}, \mathbf{u}^{(m)})$ are of equal dimension. Conceptually, this allows us to create a one-to-one mapping between these parameter vectors; we will denote this mapping

$$\theta^{(m+1)} = g^{(m)}(\theta^{(m)}, \mathbf{u}^{(m)}). \tag{7.162}$$

The relationship is specific to model m and will be used henceforth to relate the parameter vector associated with model m to the next highest dimension model. Note that a more general derivation would create a mapping between any two models, not just between models $m, m + 1$. In practice, however, it is common to only consider proposals that increase or decrease the dimensionality by $d_{m+1} - d_m$. Now further assume that our auxiliary variables are independent from our model parameters such that the prior distribution for the joint parameter vector factors as $p_{\Theta,U}(\theta, \mathbf{u}) = p_{\Theta}(\theta)p_U(\mathbf{u})$.

Given this construction, it can be shown [41, 54] that we can evaluate the probability of moving from model m to a more complex model $m^* = m + 1$ using the same type of acceptance ratio used in traditional MCMC. Specifically, such a move is accepted with probability $\min(1, r)$ where r in this case is

$$r = \frac{p_Y(\mathbf{y}|\theta^*, m^*)p_{\Theta}(\theta^*|m^*)p_{\mathcal{M}}(m^*)j(m^*|m)}{p_Y(\theta|\mathbf{y,m})p_{\Theta}(\theta|m)p_U(\mathbf{u}|m)p_{\mathcal{M}}(m)j(m|m^*)}|J(g(\theta))| \tag{7.163}$$

and where $|J(g^{(m)}(\theta))| \equiv \frac{\partial(g^{(m)}(\theta^*))}{\partial(\theta^{(m)}, \mathbf{u}^{(m)})}$ is the Jacobian of the mapping (7.162). Because we assumed that θ^* is of greater dimension, there are no auxiliary variables in model m^*, and hence the term $p_U(\mathbf{u}|m^*)$ does not appear in the numerator (more general implementations may require such a term). Moreover, if we assume we're equally likely to consider model m^* as we are model m (i.e., we are just as likely to move to the model of next lower complexity as we are to the next higher complexity), the probabilities $j(m^*|m)$ and $j(m|m^*)$ are equal. In this slightly less general situation, the above-given ratio reduces to

$$r = \frac{p_Y(\mathbf{y}|\theta^*, m^*)p_{\Theta}(\theta^*|m^*)p_{\mathcal{M}}(m^*)}{p_Y(\mathbf{y}|\theta, m)p_{\Theta}(\theta|m)p_U(\mathbf{u}|m)p_{\mathcal{M}}(m)}|J(g(\theta))| \tag{7.164}$$

while for the reverse move, that is, moving from a model of higher to lower dimensionality, we evaluate

$$r = \frac{p_Y(\mathbf{y}|\theta, m^*)p_{\Theta}(\theta|m)p_U(\mathbf{u}|m)p_{\mathcal{M}}(m^*)}{p_Y(\mathbf{y}|\theta^*, m)p_{\Theta}(\theta^*|m^*)p_{\mathcal{M}}(m)} \frac{1}{|J(g(\theta))|} \tag{7.165}$$

and again accept with probability $\min(1, r)$.

Using this construction we now have a means of generating Markov chains for both the model parameters *and* the variable m itself. The algorithm proceeds by using the Gibbs sampling procedure of Section 7.2.4 to cycle through each of the parameters in θ while holding all others (including the current model state) fixed. This is accomplished in practice in the following manner. We first draw from the model prior $m(0) \sim p_{\mathcal{M}}(m)$, the associated parameter priors $\mathbf{u}^{(m(0))}(0) \sim p_U(\mathbf{u}(0)|m(0))$ and $\theta^{(m(0))}(0) \sim p_{\Theta}(\theta^{(m(0))})$. This initializes the chains for both the "model" variable m and each of the parameters. Next, we apply the ratio (7.163) to decide whether to accept a move to model m^* or remain at model m. Recall that by construction we are only considering "nearby" model moves, that is to say, from model m

we only consider either model $m^* = m + 1$ or model $m^* = m - 1$. We now have $m(1)$ and so we follow the standard MCMC algorithm with Gibbs sampling (see page (383), Algorithm 1) to get the updated parameter vector $\theta^{(m)}(1)$. Any auxiliary variables that exist for this model order are drawn from their prior, that is, $\mathbf{u}^{(m(1))}(1) \sim p_U(\mathbf{u}(1)|m(1))$. The process continues for $k = 2, 3, \dots, K$, where K is the number of iterations in the Markov chain as before (again neglecting the first B iterations for "burn-in").

Now, we have so far said nothing about how to choose $g(\cdot)$ or the auxiliary variables $\mathbf{u}$. This is simply because these choices are usually specific to the problem at hand and will be different for different collections of damage models. However, we introduce a few possibilities here and in Section (9.1.3) that may be common to certain classes of damage. For example, in many damage detection scenarios, the different damage models will correspond to the same *type* of damage, but will differ in the number of damages that are present. This means the RJMCMC approach will need to introduce a new damage (increasing model order) or remove a damage that the estimator has found to be inconsistent with prior and likelihood. This is consistent with our earlier derivation of the MCMC algorithm where we showed that any move must have a countermove to ensure the reversibility condition in creating the chains, that is, the function $g(\cdot)$ must have an inverse.

Denote the model order (number of damages) as $n_k \sim \mathcal{M}$, that is to say, our random variable for "model" is the model order. Increasing the model order by one means $\theta^{(m+1)} = (\theta^m, \mathbf{u}^{(m)})$, that is, $g(\cdot)$ is the identity. This type of move is sometimes referred to as a birth/death pair (see, e.g., [56]). These two operations are defined as

- *Death:* A particular damage (and its associated parameters) is uniformly selected on the range $[1, n_k]$ and is simply removed, hence $n_k \rightarrow n_k - 1$.
- *Birth:* A new instance of the damage is added, that is, $n_k \rightarrow n_k + 1$ so that the auxiliary parameter vector, sampled from its prior, is $\mathbf{u}^{(n_k)} \sim p_\Theta(\theta)$.

The vector $\mathbf{u}$ used in evaluating the ratio (7.163), therefore, simply contains the added damage parameters and the Jacobian of this move is unity. It should also be mentioned that the probability of a birth move must be zero when some maximum number of damages have been proposed. Similarly, for $n_k = 0$ the probability of a death move must be zero (negative damage is not physically meaningful). This is simply a way of imposing the physical constraint that we can't have negative damage, nor can we consider an infinite damage model, that is, we have to restrict the maximum number of damages in the model class. This particular pair of moves is but one possibility that we have found useful in RJMCMC for a particular class of damage identification problems. An example of this and other types of moves (e.g., split/merge) are provided in Section (9.1.3).

Multimodel inference is a challenging subject and we have given it only scant treatment here. We feel it is worth at least sketching some of the possible techniques, however, as we view multimodel inference as a logical next step in damage identification problems. The approach provides a clear path toward using observed data to draw an inference about the presence, type, severity, and extent of structural damage in a single identification procedure. This capability requires the practitioner to develop both deterministic and probabilistic modeling components. Not only must one be able to posit multiple physics-based damage models but one must also develop probabilistic models (priors) for the model type and the parameter vectors associated with each model. We view this not as a burden but as a strength, as it provides a formal mechanism for building in prior knowledge about every aspect of the damage identification problem.

As we have learned in earlier chapters, the more prior information that can be brought to bear on a problem, the better the quality of the resulting estimates.

References

[1] M. I. Friswell, Damage identification using inverse methods, Philosophical Transactions of the Royal Society of London, Series A 365 (2007) 393–410.

[2] W. H. Press, S. A. Teukolsky, W. T. Vetterling, B. P. Flannery, Numerical Recipes in C: The Art of Scientific Computing, 2nd ed., Cambridge University Press, 1992.

[3] J. M. Nichols, L. N. Virgin, H. P. Gavin, Damping estimates from nonlinear time-series, Journal of Sound and Vibration 246 (5) (2001) 815–827.

[4] L. N. Virgin, Introduction to Experimental Nonlinear Dynamics: A Case Study in Mechanical Vibration, Cambridge University Press, 2000.

[5] C. P. Fritzen, K. Bohle, Global damage identification of the "Steelquake" structure using modal data, Mechanical Systems and Signal Processing 17 (2003) 111–117.

[6] Y. Xia, H. Hao, Statistical damage identification of structures with frequency changes, Journal of Sound and Vibration 263 (2003) 853–870.

[7] E. Görl, M. Link, Damage identification using changes of eigenfrequencies and mode shapes, Mechanical Systems and Signal Processing 17 (1) (2003) 219–226.

[8] D. Balageas, C. P. Fritzen, A. Güemes, Structural Health Monitoring, John Wiley & Sons, Ltd, London, 2006.

[9] G. Y. Xu, W. D. Zhu, B. H. Emory, Experimental and numerical investigation of structural damage detection using changes in natural frequencies, ASME Journal of Vibration and Acoustics 129 (2007) 686–700.

[10] E. Reynders, A. Teughels, G. D. Roeck, Finite element model updating and structural damage identification using OMAX data, Mechanical Systems and Signal Processing 24 (2010) 1306–1323.

[11] W. D. Zhu, K. He, Detection of damage in space frame structures with l-shaped beams and bolted joints using changes in natural frequencies, ASME Journal of Vibration and Acoustics 135 (2013) 051001.

[12] M. I. Friswell, J. E. T. Penny, S. D. Garvey, A combined genetic and eigensensitivity algorithm for the location of damage in structures, Computers and Structures 69 (1998) 547–556.

[13] T. Horibe, K. Watanabe, Crack identification of plates using genetic algorithm, JSME International Journal 49 (2006) 403–410.

[14] C. J. Stull, C. J. Earls, W. Aquino, A *posteriori* initial imperfection identification in shell buckling problems, Computer Methods in Applied Mechanics and Engineering 198 (2008) 260–268.

[15] R. Storn, R. Price, Differential evolution - a simple and efficient heuristic for global optimization over continuous spaces, Journal of Global Optimization 11 (1997) 341–359.

[16] S. Das, P. N. Suganthan, Differential evolution: a survey of the state-of-the-art, IEEE Transactions on Evolutionary Computation 15 (1) (2011) 4–31.

[17] S. Casciati, Stiffness identification and damage localization via differential evolution algorithms, Structural Control and Health Monitoring 15 (2008) 436–449.

[18] M. Savoia, L. Vincenzi, Differential evolution algorithm for dynamic structural identification, Journal of Earthquake Engineering 12 (5) (2008) 800–821.

[19] H. Tang, S. Xue, C. Fan, Differential evolution strategy for structural system identification, Computer & Structures 86 (21–22) (2008) 2004–2012.

[20] L. A. Purcina, S. F. P. Saramago, M. A. Duarte, Differential evolution applied to solve problems of indirect identification of external forces, IV European Conference on Computational Mechanics Palais des Congress, Paris, Vol. 44 (May 16–21, 2010) 697–710.

[21] H. M. Reed, J. M. Nichols, C. J. Earls, A modified differential evolution algorithm for damage identification in submerged shell structures, Mechanical Systems and Signal Processing 39 (2013) 396–408.

[22] V. Giurgiutiu, Tuned lamb wave excitation and detection with piezoelectric wafer active sensors for structural health monitoring, Journal of Intelligent Material Systems and Structures 16 (2005) 291–305.

[23] L. Yu, V. Giurgiutiu, Advanced signal processing for enhanced damage detection with piezoelectric wafer active sensors, Smart Structures and Systems 1 (2) (2005) 185–215.

[24] M. R. Hoseini, W. Xiaodong, M. J. Zuo, Estimating ultrasonic time of flight using envelope and quasi-maximum likelihood method for damage detection and assessment, Measurement 45 (2012) 2072–2080.

[25] V. Giurgiutiu, Structural Health Monitoring: With Piezoelectric Wafer Active Sensors, Academic Press, 2008.

[26] R. A. Kline, Nondestructive Characterization Of Composite Media, Technomic Publishing Co., Lancaster, PA, 1992.

[27] S. M. Kay, Fundamentals of Statistical Signal Processing: Estimation Theory, Vol. I, Prentice-Hall, New Jersey, 1993.

[28] G. Jacovitti, G. Scarano, Discrete time techniques for time delay estimation, IEEE Transactions on Signal Processing 41 (2) (1993) 525–533.

[29] H. C. So, P. C. Ching, Y. T. Chan, A new algorithm for explicit adaptation of time delay, IEEE Transactions on Signal Processing 42 (7) (1994) 1816–1820.

[30] G. Rohde, A. Aldroubi, D. M. H. Jr., Sampling and reconstruction for biomedical image registration, in: Proceedings of the 41st IEEE Asilomar Conference on Signals, Systems and Computers, 2007, pp. 220–223.

[31] T.-H. Li, K.-S. Song, Estimation of the parameters of sinusoidal signals in non-Gaussian noise, IEEE Transactions on Signal Processing 57 (1) (2009) 62–72.

[32] J. Chen, J. Benesty, Y. Huang, Performance of GCC- and AMDF-based time-delay estimation in practical reverberant environments, EURASIP Journal on Applied Signal Processing 1 (2005) 25–36.

[33] B. W. Silverman, Density Estimation for Statistics and Data Analysis, Chapman and Hall, 1986.

[34] G. K. Rohde, F. Bucholtz, J. M. Nichols, Maximum empirical likelihood estimation of time delay in I.I.D. noise, IET Signal Processing 8 (7) (2014) 720–728.

[35] N. Mars, G. V. Arragon, Time delay estimation in nonlinear systems, IEEE Transactions on Acoustics, Speech, and Signal Processing 29 (3) (1981) 619–621.

[36] R. N. McDonough, A. D. Whalen, Detection of Signals in Noise, 2nd ed., Academic Press, San Diego, CA, 1995.

[37] Y. Tan, S. L. Tantum, L. M. Collins, Cramer-Rao lower bound for estimating quadrupole resonance signals in non-Gaussian noise, IEEE Signal Processing Letters 11 (5) (2004) 490–493.

[38] D. Fink, A compendium of conjugate priors (1997).

[39] J. M. Nichols, M. Seaver, S. T. Trickey, K. Scandell, L. W. Salvino, E. Aktaş, Real-time strain monitoring of a navy vessel during open water transit, Journal of Ship Research 54 (4) (2010) 225–230.

[40] W. K. Hastings, Monte Carlo sampling methods using Markov chains and their applications, Biometrika 57 (1) (1970) 97–109.

[41] G. Casella, E. I. George, Explaining the Gibbs sampler, The American Statistician 46 (3) (1992) 167–174.

[42] W. A. Link, R. J. Barker, Bayesian Inference with Ecological Examples, Academic Press, San Diego, CA, 2010.

[43] C. J. Stull, J. M. Nichols, C. J. Earls, Stochastic inverse identification of geometric imperfections in shell structures, Computer Methods in Applied Mechanics and Engineering 200 (1-4) (2011) 2256–2267.

[44] J. M. Nichols, W. A. Link, K. D. Murphy, C. C. Olson, A Bayesian approach to identifying structural nonlinearity using free-decay response: application to damage detection in composites, Journal of Sound and Vibration 329 (15) (2010) 2995–3007.

[45] J. M. Nichols, E. Z. Moore, K. D. Murphy, Bayesian identification of a cracked plate using a population-based Markov chain Monte Carlo method, Computers and Structures 89 (13–14) (2011) 1323–1332.

[46] A. Jasra, D. A. Stephens, C. C. Holmes, On population-based simulation for static inference, Statistics and Computing 17 (2007) 263–279.

[47] K. P. Burnham, D. R. Anderson, Model Selection and Inference: A Practical Information-Theoretic Approach, Springer-Verlag, New York, 1998.

[48] X. Guan, R. Jha, Y. Liu, Model selection, updating, and averaging for probabilistic fatigue damage prognosis, Structural Safety 33 (2011) 242–249.

[49] H. M. Reed, C. J. Earls, J. M. Nichols, Stochastic identification of imperfections in submerged shell structures, Computer Methods in Applied Mechanics and Engineering 272 (2014) 58–82.

[50] H. Akaike, Information theory as an extension of the maximum likelihood principle, in: B. N. Petrov, F. Csaki (Eds.), 2nd International Symposium on Information Theory, Akademiai Kiado, Budapest, 1973, pp. 267–281.

[51] H. Akaike, Information measures and model selection, International Statistical Institute 44 (1983) 277–291.

[52] R. Brincker, P. H. Kirkegaard, P. andersen, M. E. Martinez, Damage detection in an offshore structrue, in: Proceedings of the 13th International Modal Analysis Conference, vols 1 and 2, Vol. 2460 of Proceedings of the Society of Photo-Optical Instrumentation Engineers (SPIE), Society for Experimental Mechanics, 1995, pp. 661–667.

[53] P. S. Rao, C. Ratnam, Damage identification of welded structures using time series models and exponentially weighted moving average control, Jordan Journal of Mechanical and Industrial Engineering 4 (6) (2010) 701–210.

[54] E. Figueiredo, J. Figueiras, G. Park, C. R. Farrar, K. Worden, Influence of autoregressive model order on damage detection, Computer-Aided Civil and Infrastructure Engineering 26 (2011) 225–238.

[55] P. J. Green, Reversible jump Markov chain Monte Carlo computation and Bayesian model determination, Biometrika 82 (4) (1995) 711–732.

[56] P. Koutsourelakis, A multi-resolution, non-parametric, Bayesian framework for identification of spatially-varying model parameters, Journal of Computational Physics. 228 (2009) 6184–6211.

8

Detecting Damage-Induced Nonlinearity

At this point, we have covered sufficient background material to accomplish the goal of this book: the detection and identification of structural damage. The next several chapters are devoted to this topic and include both numerical and experimental examples. As promised in the introduction, we begin with very basic assumptions about the physics of structural damage and move toward the more specific. As we will see, the more general assumptions limit us in what we can infer about the damage.

We begin by noting that in many cases the vibrations of an undamaged, pristine structure is accurately described by a linear model. Engineers typically design structures in such a way as to avoid large-scale deflections. This makes sense when we think about a typical structure's intended functionality. Bridges, buildings, piers, aircraft, automobiles, and so on, represent structures for which large-amplitude vibrations and/or structural instabilities would undoubtedly be considered a problem. In terms of modeling, this means that the assumptions underlying a linear structural model will hold, namely, small-amplitude vibration. Moreover, if there is no nonlinearity present, there is no feedback mechanism that could cause the vibrations to grow, thus the structure's vibrations remain stable over time.

Perhaps, not surprisingly, it has been noted by a number of researchers that structural damage will sometimes result in nonlinear structural behavior. For example, a report by Sohn *et al.* [1] also notes that damage will often manifest itself as a nonlinearity (see also [2]). We have described how to model many of these forms of damage already in Chapter 5. These damage mechanisms, and other works that deal with the associated modeling and detection, include

1. **Cracking** A cracked structure will often result in nonlinear behavior. For example, Brandon [3] considers both crack and clearance nonlinearities as a nonlinear damage mechanism. Friswell and Penny consider the modeling of cracks using a bilinear stiffness term [4]. Zhang and Testa [5] also explored the nonlinearity due to the opening/closing of a crack in an experimental beam structure.
2. **Impacts and/or Rattling** Impacts, or clearance nonlinearities, also present an obvious situation where damage equates with nonlinearity. Brown and Adams [6], Rutherford *et al.*

Modeling and Estimation of Structural Damage, First Edition. Jonathan M. Nichols and Kevin D. Murphy.
© 2016 John Wiley & Sons, Ltd. Published 2016 by John Wiley & Sons, Ltd.

[7], and Nichols *et al.* [8] have all explored loosening connections as a nonlinear damage mechanism. This type of nonlinearity is also referred to in many works as backlash (see, e.g., [9, 10]).

3. **Delamination** Delamination has also been modeled as a damage-induced nonlinearity in composite structures. The work of Schwarts–Givli *et al.*, for example, considers both contact and geometric nonlinearity in modeling damaged sandwich panels. Earlier work by Hunt *et al.* [11] considered a low-dimensional model of a delaminated strut. The delamination was modeled as a local buckling, which resulted in a nonlinear response. A similar model was proposed by Murphy and Nichols and was shown to agree well with the experiment [12]; some of these results appear in Chapter 5. A separate nonlinear model for delaminated beams was also given in [13].

4. **Stick/slip, Rub** In the analysis of rotating machinery, rotor–stator "rub" is viewed as a damage mechanism. An experimental study illustrating rub as a source of nonlinearity can be found in [14]. Detection of this type of damage was studied in [15, 16], and will be further explored in the experimental example of section 8.8.1. In addition, jointed structures will often exhibit stick-slip behavior. Both modeling and simulation of this type of behavior in bolted joints was studied by Song *et al.* [17], among others. "Fretting" wear is another damage mechanism arising from stick-slip behavior and has been studied by a number of authors (see, e.g., [18]).

In these cases, rather than assuming that a specific model exists, we may assume that the *form* of the undamaged structural model is linear and that damage causes a nonlinearity. This assumption will certainly not be true for all structures/damages, however it is indeed valid for a large class of problems, several of which are described here.

Recall from Chapter 4 that a linear dynamical system is one in which the principle of superposition holds. That is to say, the output of a linear dynamical system is accurately modeled as the convolution

$$y(t) = \int h(t - \tau) x(\tau) d\tau \tag{8.1}$$

where $x(t)$ is the system input and the impulse response function $h(\tau)$ describes the influence of the structure yielding the output $y(t)$. Therefore, what is required is a formal way to discriminate between the hypothesis:

$$\mathcal{H}_0 : y(t) = \int h(t - \tau) x(\tau) d\tau$$

$$\mathcal{H}_1 : y(t) \neq \int h(t - \tau) x(\tau) d\tau. \tag{8.2}$$

However, we have to test this damage hypothesis without knowing $h(\tau)$. In other words, given only structural response data $\mathbf{y}(t) \equiv (y(t_1), y(t_2), \ldots, y(t_N))$, we wish to develop a formal statistical test that determines if the data are better described by a linear or a nonlinear model.

To accomplish this, we can use what we know about the statistical properties of random processes, established in Chapters 2 and 3. Consider a stationary input signal $x(t)$ and a function $P_\mathbf{X}(\mathbf{x})$ that assigns a probability to the sequence $x(t_1), x(t_2), \ldots$ (i.e., a joint cumulative distribution function (CDF). As we have pointed out in Chapter 2, for continuous random variables we can correspondingly define a joint probability density function (PDF) $p_\mathbf{X}(\mathbf{x})$.

We further showed that for stationary data, the joint statistical properties of this distribution become

$$C_X = E[X(t)] \equiv \mu_X$$

$$C_{XX}(\tau) = E[(X(t) - \mu_X)(X(t + \tau) - \mu_X)]$$

$$C_{XXX}(\tau_1, \tau_2) = E[(X(t) - \mu_X)(X(t + \tau_1) - \mu_X)(X(t + \tau_2) - \mu_X)]$$

$$\vdots \qquad (8.3)$$

Given a probabilistic description of the input signal, for example, (8.3), we can use Eq. (8.1) to define the corresponding descriptors of the output sequence $y(t)$

$$E[Y(t)] = E\left[\int_0^t h(t - \theta_1)x(\theta_1)d\theta_1\right]$$

$$= \int_0^t h(t - \theta_1)E[X(\theta_1)]d\theta_1$$

$$E[(Y(t) - \mu_Y)(Y(t') - \mu_Y)] = E\left[\int_0^t h(t - \theta_1)(X(\theta_1) - \mu_X)d\theta_1 \int_0^{t'} h(t' - \theta_2)(X(\theta_2) - \mu_X)d\theta_2\right]$$

$$= \int_0^t \int_0^{t'} h(t - \theta_1)h(t' - \theta_2)E[(X(\theta_1) - \mu_X)(X(\theta_2) - \mu_X)]d\theta_1 \, d\theta_2$$

$$E[(Y(t) - \bar{Y})(Y(t') - \mu_Y)(Y(t'') - \mu_Y)] = \int_0^t \int_0^{t'} \int_0^{t''} h(t - \theta_1)h(t' - \theta_2)h(t'' - \theta_3)$$

$$\times E[(X(\theta_1) - \mu_X)(X(\theta_2) - \mu_X)(X(\theta_3) - \mu_X]d\theta_1 \, d\theta_2 \, d\theta_3$$

$$\vdots$$

$$E[(Y(t) - \bar{Y})(Y(t') - \mu_Y)(Y(t'') - \mu_Y)\cdots(Y(t''\cdots') - \mu_Y)] =$$

$$\int_0^t \int_0^{t'} \int_0^{t''} \cdots \int_0^{t''\cdots'} h(t - \theta_1)h(t' - \theta_2)h(t'' - \theta_3)\cdots h(t''\cdots')$$

$$\times E[(X(\theta_1) - \mu_X)(X(\theta_2) - \mu_X)(X(\theta_3) - \mu_X)\cdots(X(\theta_n) - \mu_X)]$$

$$\times d\theta_1 d\theta_2 d\theta_3 \cdots d\theta_n \qquad (8.4)$$

In other words, for a linear system, the output correlation functions can be written as a linear combination of the input correlation functions. As we have already discussed in Section 6.7.1, if the input $x(t)$ is stationary, jointly Gaussian distributed (i.e., if each $x(t) \sim \mathcal{N}(\mu_X, \sigma_X^2)$), so too will be the output observations, that is, each $y(t) \sim \mathcal{N}(\mu_Y, \sigma_Y^2)$ (specifically, recall that in Chapter 2 we showed that linear combinations of Gaussians remain Gaussian, Section 2.4, page 29). Moreover, by Isserlis' theorem 3.11, the expectations of odd-ordered products of the input vanish, causing the corresponding expected output products to also vanish, for example, $E[(Y(\theta_1) - \mu_Y)(Y(\theta_2) - \mu_Y)(Y(\theta_3) - \mu_Y)] = 0$. Isserlis' theorem further tells us that the expectation of even-ordered products reduce to linear combinations of products of covariances. If we therefore look at the joint *cumulants* in the output (see Eq. 2.86), as opposed to joint moments, even these product terms disappear. It can therefore be stated that *for a linear, stationary Gaussian excited structure, all joint cumulants in the output greater than order n = 2 vanish.*

For nonlinear (damaged) systems, this will almost certainly not be true. Thus, a valid null hypothesis for a linear, Gaussian excited structure is that the auto-covariance function $C_{YY}(\tau) = E[(Y(t) - \mu_Y)(Y(t + \tau) - \mu_Y)]$, obtained by setting $t' = t + \tau$ in Eq. (8.4), is sufficient to describe all of the joint statistical properties of the response. More generally, given multivariate response data from a linear, Gaussian excited structure, $y_i(t), i = 1 \dots K$ where K might be the number of sensors attached to the structure, the correlations among any two response variables are dictated solely by $C_{Y_1 Y_2}(\tau) = E[(Y_1(t) - \mu_{Y_1})(Y_2(t + \tau) - \mu_{Y_2})]$ because higher order cumulants will not exist.

Thus, we have a clear path toward developing a practical test for (8.2). Recalling the definition of stationary joint cumulants given by Eqs. (3.21), the test

$$\mathcal{H}_0 : k_{Y_1 Y_2 \dots Y_n}(\tau_1, \tau_2, \dots, \tau_{n-1}) = 0 \; \forall \tau \geq 0, \quad n > 2$$

$$\mathcal{H}_1 : k_{Y_1 Y_2 \dots Y_n}(\tau_1, \tau_2, \dots, \tau_{n-1}) \neq 0 \; \forall \tau \geq 0, \quad n > 2 \tag{8.5}$$

assumes for the null hypothesis that *we have observed the response of a linear structure driven with a stationary, jointly Gaussian distributed random process.* The alternative (assuming a Gaussian input) is a nonlinear, hence damaged, structure. One promising approach to testing this hypothesis is described in the next section.

However, this is not nearly as general a hypothesis as we might like. Oftentimes the (stationary) input signal values are *not* accurately modeled by a Gaussian distribution. Naturally occurring phenomena such as wind or wave excitation are often modeled as stationary random processes, where the *joint* probability structure is still governed entirely by $k_{XX}(\tau) = C_{XX}(\tau) = E[(X(t) - \mu_X)(X(t + \tau) - \mu_X)] \neq 0 \forall \tau$, but where the *marginal* distribution

$$p_X(x(t)) = \int_{\mathbb{R}^{N-1}} p_{\mathbf{X}}(\mathbf{x}) dx_1 dx_2 \dots dx_{N-1} \tag{8.6}$$

is non-Gaussian. In this more general case, higher order, zero-lag cumulants associated with the input can be nonzero, that is,

$$E[(X(t) - \mu_X)^3] \neq 0$$

$$E[(X(t) - \mu_X)^4] - 3\sigma_X^4 \neq 0$$

$$\vdots \tag{8.7}$$

If this is the case, we can see from Eq. (8.4) that the expectations $E[(Y(t) - \mu_Y)^3] \neq 0$, $E[(Y(t) - \mu_Y)^4] - 3\sigma_Y^4 \neq 0$, that is, the higher order ($n > 2$) joint cumulants for $\tau_1 = \tau_2 = \dots, = \tau_{n-1} = 0$ no longer disappear from the output, even when there is no damage. Thus, the test (8.5) is incapable of distinguishing a damaged structure from an undamaged one subject to an input with non-Gaussian marginal distribution. A more general null hypothesis can therefore be written

$$\mathcal{H}_0 : k_{Y_1 Y_2, \dots, Y_n}(\tau_1, \tau_2, \dots, \tau_{n-1}) = 0 \; \forall \tau > 0, \quad n > 2$$

$$\mathcal{H}_1 : k_{Y_1 Y_2 \dots Y_n}(\tau_1, \tau_2, \dots, \tau_{n-1}) \neq 0 \; \forall \tau > 0, \quad n > 2 \tag{8.8}$$

which assumes for the null hypothesis that *we have observed the response of a linear structure driven by a stationary random process with (possibly) non-Gaussian marginal distribution but for which all joint cumulants beyond order $n = 2$ are zero for $\tau > 0$.* While the difference between (8.8) and (8.5) may appear small (the latter simply excluding zero-lag cumulants from the test), the difference is practically very important. Of course, there certainly exist input random processes $X(t)$ which will also possess nonzero higher order cumulants for positive time lags $\tau_1, \tau_2, \ldots > 0$. However, that would imply that the input is itself the output of some underlying nonlinear process, an unlikely situation in most structural dynamics applications. We have found the general hypothesis test (8.8) is sufficient to handle most cases of practical interest.

So how do we test either (8.5) or (8.8) in the sense defined in Section 6.9? In fact, how do we even know which of these two hypothesis tests to employ? To test the Gaussian assumption on the input, we might first use the histogram estimator of Section 6.3 and apply that estimator to the measured input sequence $x(n)$. Of course, we may not be able to directly observe the input, hence the more general hypothesis test (8.8) would be a safer choice.

In addition, the tests we have just described focus on a specific joint cumulant that may or may not possess a nonzero value for a particular type of nonlinearity (the reason why analytical expressions are so valuable!). However, we could also consider a more general test that considers the entire joint probability distribution of the response data. In the next section, we elaborate on both specific and general tests for the presence of higher order cumulants (hence, damage-induced nonlinearity) in structural response data. It is further demonstrated that we can test the hypothesis of damage-induced nonlinearity by looking at the joint statistical properties of the structures' response in either the time or frequency domains.

8.1 Capturing Nonlinearity

As we have just shown, to detect the presence of damage-induced nonlinearity, we require a test statistic s that will reflect the presence of higher order correlations in the data. We would expect estimates of "s" to possess a certain range of values when applied to a healthy structural response, but to attain a different value when applied to data that violate the null hypothesis (damage). [1] A number of test statistics exist for this purpose and depend, as one might imagine, on the particular damage physics (type of nonlinearity). One can estimate higher order joint moments directly or their Fourier transforms (FTs) as alluded to in the previous section. The bispectrum or bicoherence functions are a popular choice for capturing nonlinearity and are elaborated on in several case studies in this chapter. The theory behind the use of these measures has already been described in Section 6.5. Another popular choice are the information theoretics described in Sections 3.5 and 6.6. These have been used in several damage detection applications (see, e.g., [19]) and are featured in the surrogate data approach of Section 8.6. Because these quantities are functions of the entire joint probability distribution, they automatically capture the presence of any-order cumulant. Finally, one can apply a simple nonlinear prediction scheme to the data and use the prediction error as our nonlinearity detector. The idea is that a nonlinear prediction scheme will do a better job forecasting a nonlinear response

---

[1] When we say "different" value, we, of course, mean statistically significant in the sense of Section 6.9, minimizing type I and type II errors.

than it will a linear one. This particular test statistic has been used with the method of surrogate data (see Section 8.8.2) to form a powerful approach to the detection of structural damage (see, e.g., [20]). In what follows, we describe several of these test statistics.

8.1.1 Higher Order Cumulants

Assume that we have confirmed that the input is indeed jointly Gaussian distributed. In this case, we may use the test (8.5) to test for the presence of a nonlinearity. Perhaps the simplest approach is to estimate the zero-lag, higher order cumulants. In other words, we can estimate

$$k_{YY\ldots}(\tau_1 = 0, \tau_2 = 0, \ldots)\tag{8.9}$$

and compare to a detection threshold in the manner described in Section 6.9. This simple approach can yield good detection performance, as will be shown in Section 8.2.2.

Begin with the measured sequence $y(n) \equiv y(n\Delta_t), n = 1 \ldots N$. As an example, for the zero-lag third cumulant, we have the estimator

$$\hat{k}_{YYY} = \frac{1}{N}\sum_{n=1}^{N}(y(n) - \hat{\mu}_Y)^3\tag{8.10}$$

where

$$\hat{\mu}_Y = \frac{1}{N}\sum_{n=1}^{N}y(n)\tag{8.11}$$

is the estimated mean. By further normalizing by the standard deviation, we can create the dimensionless measure

$$\hat{\gamma}_{YYY} = \frac{\hat{k}_{YYY}}{\hat{k}_{YY}^{3/2}}\tag{8.12}$$

where k_{YY} is the second cumulant, that is, the variance, and is estimated by

$$\hat{k}_{YY} \equiv \hat{\sigma}_Y^2 = \frac{1}{N-1}\sum_{n=1}^{N}(y(n) - \hat{\mu}_Y)^2.\tag{8.13}$$

This normalized third cumulant is referred to as the "skew" and quantifies asymmetry in a distribution. As for all higher order cumulants, a jointly Gaussian distributed signal should have an estimated skew of near zero, while a non-Gaussian response may have a nonzero skew. It is important to note again that focusing on a particular cumulant can "miss" certain types of nonlinearity. For example, we know from our earlier work on higher order spectra (HOS) that an even-order nonlinearity *will* produce (to leading order in the Volterra expansion) a nonzero skew (see Section 6.5). However, this particular measure does not always capture a cubic nonlinearity as the response of a cubic-only nonlinear system will often be largely symmetric in its PDF.

For an odd-order nonlinearity, it may make more sense to estimate the fourth zero-lag cumulant, given by

$$\hat{k}_{YYYY} = \frac{1}{N}\sum_{n=1}^{N}(y(n) - \hat{\mu}_y)^4 - 3\hat{\sigma}_Y^4.\tag{8.14}$$

Just as with the skew of a distribution, it is common to normalize the cumulant to dimensionless form. As we have already seen on page 178, Section 5.4, the *kurtosis* is simply the normalized fourth, zero-lag, joint cumulant and is estimated using the direct definition

$$\hat{\gamma}_{YYYY} = \frac{\frac{1}{N}\sum_{n=1}^{N}(y(n)-\hat{\mu}_y)^4}{\hat{\sigma}_Y^4} - 3 \tag{8.15}$$

For a structure excited with a jointly Gaussian input, potential cumulant-based test statistics are denoted

$$s^{(k)} = \hat{\gamma}_{YYY}$$

or

$$s^{(k)} = \hat{\gamma}_{YYYY} \tag{8.16}$$

(depending on the type of nonlinearity) for use in testing (8.5).

Note that while these test statistics are appropriate for testing (8.5), they assume a Gaussian-distributed input. However, we argued that a more general hypothesis test would allow for the marginal distribution $p_Y(y)$ to be non-Gaussian (assuming a non-Gaussian input to a linear structure) but would still postulate that all higher order joint cumulants disappear, that is, the test (8.8). Toward this latter goal, it is more appropriate to consider the estimated cumulants for discrete time lags $T > 0$. In other words, given the measured sequence $y(n), n = 1 \cdots N$, form the test statistic

$$s^{(k)} \equiv \hat{k}_{YYY}(T_1, T_2) = \hat{C}_{YYY}(T_1, T_2)$$

$$= \frac{1}{N}\sum_{n=1}^{N}(y(n) - \hat{\mu}_Y)(y(n+T_1) - \hat{\mu}_Y)(y(n+T_2) - \hat{\mu}_Y)) \tag{8.17}$$

and declare

$$\begin{cases} \mathcal{H}_0 : s^{(k)} \geq \lambda \\ \mathcal{H}_1 : s^{(k)} < \lambda \end{cases} \tag{8.18}$$

for any $T_1, T_2 > 0$. Ideally, the threshold value would be $\lambda = 0$, however in practice, we will have neither a perfect Gaussian input or a perfect estimator, thus λ will be some small nonzero value. This "direct" estimation approach was used by Barnett and Wolff [21] in a nonlinearity detection application and further details can be found therein. In short, we can choose our detection statistic, $s^{(k)}$, from among a variety of cumulant functions and apply the methods described in Section 6.9 to obtain the Type I and Type II errors associated with this class of detector, something we illustrate in Section 8.2.

Alternatively, we could look at the spectral representation of the cumulants. This is, in fact, our reason for introducing the HOS in Section 3.3.5 and detailing methods for their estimation in Section 6.5. We know from these sections that we expect nonlinearities to show up as peaks at combinations of the structure's modes. We have also shown analytically near the end of Section 6.4 that even the power spectral density (PSD) of a nonlinear response can yield information about the nonlinearity, in particular an increase in spectral content at higher frequencies (harmonics), a fact we demonstrated in the modeling of bolted joints described in Section 5.4. The higher order spectral coefficients are therefore well suited for use as a test statistic in assessing (8.8).

8.1.2   Higher Order Spectral Coefficients

Recall from Section 3.3.5 that the HOS for a random process are formally the n-dimensional FT of the nth-order joint lagged cumulants,

$$HOS_{Y_1 Y_2 \ldots Y_n}(\omega_1, \omega_2, \ldots, \omega_n)$$

$$= \left(\frac{1}{2\pi}\right)^n \int_{\mathbb{R}^n} k_{Y_1, Y_2 \ldots Y_n}(t_1, t_2, \ldots, t_n) e^{-i(\omega_1 t_1 + \omega_2 t_2 + \cdots + \omega_n t_n)} dt_1 dt_2 \ldots dt_n. \qquad (8.19)$$

Under the assumption of stationarity, the HOS reduce to sums of FTs of joint moments (see again Section 3.3.5), which may be estimated using the methods in Section 6.5 as the block-averaged products of discrete FTs of the data (e.g., the estimators 6.129, 6.161 for the bispectral and trispectral densities). Moreover, as a direct consequence of Isserlis' theorem 3.11, we know that the HOS vanish for order $n > 2$ provided the response random process $\mathbf{Y}$ is joint normally distributed. Thus, we may evaluate (8.5) using the estimated bispectral density, for example, as the test statistic. That is to say, we estimate

$$s_{ij}^{(B)} = |\hat{B}_{YYY}(f_i, f_j)| \qquad (8.20)$$

for different discrete frequency values i, j and compare to a detection threshold. Of course, we know from our analytical work that certain frequency bins will produce better detection performance than others, namely, those at the combination resonances. For example, if the structure has a resonant frequency at f_n, we have showed that good choices for the monitored frequency are (f_n, f_n) although from the analytical expressions (see, e.g., Figures 6.16 and 6.19) we also know that other peaks may be similarly sensitive to the nonlinearity.

However, as we have just pointed out, if the input observations are not normally distributed, but possess some other marginal distribution, we have that $k_{Y_1, Y_2, \ldots, Y_n}(0, 0, \ldots, 0) \neq 0$ and so $HOS_{12 \ldots n}(\omega_1, \omega_2, \ldots, \omega_n) \neq 0$ even when the structure is linear (undamaged). In this more challenging case, we have shown that higher order *coherence* functions (see Section 6.5.1) can provide a nice alternative, provided that the input sequence is still independent, identically distributed (i.i.d.). In this case, we suggest for the family of test statistics

$$s_{ij}^{(b)} = |\hat{b}(f_i, f_j)| \qquad (8.21)$$

that is, the magnitude of the complex bicoherence function at a particular estimated frequency bin f_i, f_j.

Thus, for the Gaussian excitation case, the hypothesis test (8.5) becomes

$$s_{ij}^{(B)} \neq 0 \; \forall \; i, j \qquad (8.22)$$

again with the acknowledgment that some frequency pairs i, j are better than others. Alternatively, for non-Gaussian inputs, the test (8.8) becomes

$$s_{ij}^{(b)} \neq \text{const.} \; \forall \; i, j \qquad (8.23)$$

where we have already derived the constant in Section 6.5.1, Eq. 6.123. A similar pair of test statistics can be estimated from the trispectrum or tricoherence function, for example,

$$s_{ijk}^{(T)} = |\hat{T}(f_i, f_j, f_k)|$$

$$s_{ijk}^{(t)} = |\hat{t}(f_i, f_j, f_k)|. \qquad (8.24)$$

In cases where there is uncertainty about the location of the peak, one may simply sum the estimated spectral or coherence function over some number of spectral bins, defined by the discrete index arrays $i_0, i_1, \ldots, i_M, j_0, j_1, \ldots, j_M$, and $k_0, k_1, \ldots, k_M$, e.g.,

$$s^{(B)} = \sum_{i=i_0}^{i_M} \sum_{j=j_0}^{j_M} |\hat{B}(f_i, f_j)|$$

or

$$s^{(b)} = \sum_{i=i_0}^{i_M} \sum_{j=j_0}^{j_M} |\hat{b}(f_i, f_j)|$$

or

$$s^{(T)} = \sum_{i=i_0}^{i_M} \sum_{j=j_0}^{j_M} \sum_{k=k_0}^{k_M} |\hat{T}(f_i, f_j, f_k)|$$

or

$$s^{(t)} = \sum_{i=i_0}^{i_M} \sum_{j=j_0}^{j_M} \sum_{k=k_0}^{k_M} |\hat{t}(f_i, f_j, f_k)|. \tag{8.25}$$

The arrays may, or may not, describe a contiguous range of bins. For example, if we know the first several structural resonances, we know from our analytical expressions in Section 6.5 the locations of multiple peaks. In this case, we might specify the index arrays to include these multiple peak values. The performance of these metrics in damage detection applications appear throughout the remainder of this chapter. In some cases, we use the square of the detection statistic, for example, and define $s^{(B)} = \sum_{i=i_0}^{i_M} \sum_{j=j_0}^{j_M} |\hat{B}(f_i, f_j)|^2$ (see Section 8.3.2).

It is also worth noting that the presence of a nonlinearity can also be deduced from estimates of the standard PSD function described in Section 6.4. In that section, we showed using a Volterra series approach that a nonlinearity will produce spectral content at frequencies above the structural resonances. However, that approach also showed that the magnitude of this frequency "boost" is higher order in terms of the input PSD magnitude. Hence, for mild levels of excitation, the presence of the nonlinearity may be barely visible in the estimated PSD (see, e.g., Figure 6.8). Nonetheless, for strong levels of nonlinearity and/or for large-amplitude excitation, quantifying the appearance or rise of high-frequency content from a PSD estimate can be an effective approach to detection.

For example, consider Figure 5.21 where we see the amplification of high-frequency content due to the presence of a loose joint. In that example, the input response was filtered to remove signal power above a certain cutoff frequency, f_c. Thus, one possible test statistic is to estimate the integrated power in the response above this frequency, for example,

$$s^{(P)} = \int_{f_1}^{f_2} S_{YY}(f) df$$

$$\approx \sum_{k=k_1}^{k_2} \hat{S}_{YY}(f_k) \tag{8.26}$$

where $\hat{S}_{YY}(f_k)$ is estimated using the methods of Section 6.4 and k_1 and k_2 are the discrete indices associated with frequencies f_1, f_2, respectively. We, in fact, use a normalized version of this measure in Section 8.2 to detect joint loosening in an experimental system.

In the sections that follow, we use test statistics based on the estimated HOS and PSD to detect various types of nonlinearity, including the nonlinearity associated with the physics of delamination (see Section 8.5).

One of the potential pitfalls of relying on the analysis of particular cumulants is that we run the risk of "missing" the nonlinearity by monitoring a cumulant that is insensitive to the particular form of nonlinearity we are trying to capture. In practice, we have found that estimating properties related to both the third and fourth cumulants is sufficient to guard against this possibility. However, there are still other test statistics that are based on the entire joint probability distribution of the response and will therefore reflect changes to *any*-order cumulant in the data. These are described next.

8.1.3   Nonlinear Prediction Error

Assume again the observed structural response $y(n) \equiv y(n\Delta)$ with sampling interval Δ. Given this data, we would like to test the hypothesis (8.5) or (8.8). However, in some instances we may not have much *a priori* information about the type of damage-induced nonlinearity. Rather than risk a missed detection by focusing on a specific cumulant that is insensitive to the damage, it is worth considering test statistics that capture the entire joint probability structure of the response data. These test statistics lack the simple interpretation of the cumulants or cumulant spectra where we could test for nonzero or nonconstant values. Nonetheless, they can be used to effectively discriminate between a linear and nonlinear structural response, particularly when used with the method of surrogate data described in Section 8.6.

One such test statistic is obtained by forming a simple nonlinear model of the data and then using this model to forecast future data values. The resulting prediction error is then used as a damage-sensitive test statistic. To this end, we take a cue from the field of nonlinear time-series analysis [23]. Many such models in this field are based on the assumption that the observed data are ergodic. As we have already shown, ergodic data are particularly useful in estimation as they allow us to replace ensemble averages with averages collected from a single data set, but where the observations used in the average are collected from "similar" dynamical states. We have already used this property to show how we might estimate PDFs. Here, we use this same property to develop probabilistic models of our data.

The goal of any predictive model is to, given the current state, forecast future data. A general, probabilistic approach is to use a Markov model of order d (see Section 3.4) to predict the probability of taking the value $y(n + T)$ given the past history $(y(n), y(n - \Delta), \ldots, y(n - (d - 1)\Delta))$.

$$p_\mathbf{Y}(y(n + T)|\, y(n), y(n - \Delta), \ldots, y(n - (d - 1)\Delta))$$

$$\equiv p_\mathbf{Y}(y(n + T)|\mathbf{y}(n)). \tag{8.27}$$

where T is the prediction horizon. Now, we have already shown by the Poincare recurrence theorem that if we take a long enough sequence of observations, $y(n), n = 1 \cdots N$ where N is large, we may form the estimate

$$\hat{p}_\mathbf{Y}(\mathbf{y}(n)) = \frac{M}{NV(\mathbf{y}(n))} \tag{8.28}$$

where $V(\mathbf{y}(n))$ is the minimum volume required to encompass M points (see Section 6.6, Eq. 6.27). The points within this volume are denoted $\mathbf{y}(m_j)$, $j = 1 \cdots M$

As we showed in Section 6.6, if each of these near neighbors reside within some small radius ϵ, Eq. (8.28) is a reasonable density estimate for each point in the neighborhood, assuming the probability of observing a point in a local ϵ-ball is constant. Consider now what happens if we allow these M points to evolve forward in time by T time steps. Given their close proximity to $\mathbf{y}(n)$, it is reasonable to assume

$$p_Y(y(n+T)|\mathbf{y}(n)) \approx p_Y(y(m_j+T)|\mathbf{y}(m_j)), \forall j. \tag{8.29}$$

This expression tells us that we might forecast the probability of observing the future value $y(n+T)$ given its past history by using past knowledge of "nearby" values $\mathbf{y}(m_j)$ and their future states $y(m_j+T)$. As a final estimate of our future value, we therefore take the mean value of $p_Y(y(n+T)|\mathbf{y}(n))$ which, from Eq. (8.29), is well approximated by the mean of $p_Y(y(m_j+T)|\mathbf{y}(m_j))$, or

$$\hat{y}(n+T) = \frac{1}{M} \sum_{j}^{M} y(m_j+T). \tag{8.30}$$

Incidentally, it can be shown [24] that the estimator $E_Y[y(n+T)|\mathbf{y}(n)]$ is the optimal predictor in the sense that it minimizes the mean-square error $\|\hat{\mathbf{y}}(n+T) - \mathbf{y}(n+T)\|$ (see [24], Eq. (2.12.36)).

We are now in a position to form our detection statistic. Given a time-lag Δ, we choose the order of our Markov process d and form the collection of vectors $\mathbf{y}(n) = (y(n), y(n-\Delta), \dots, y(n-(d-1)\Delta)$ $n = 1 \cdots N$. For each point n in this collection, we find the M nearest neighbors $\mathbf{y}(m_j)$, $j = 1 \cdots M$

The T-step prediction $\hat{\mathbf{y}}(n+T) = (1/M) \sum_{j=1}^{M} \mathbf{y}(m_j+T)$ is then used to form the prediction error

$$\epsilon(n) = \|\hat{\mathbf{y}}(n+T) - \mathbf{y}(n+T)\|. \tag{8.31}$$

Repeating this procedure over each point n (or at least some representative subset), we compute the mean prediction error

$$s^{(PE)} \equiv \bar{\epsilon} = \sum_{n} \epsilon(n) \tag{8.32}$$

as our detection statistic. This prediction scheme is general, and appropriate for both linear and nonlinear models. Hence, the mean prediction error should be lower if there exists a nonlinearity. We show this to be precisely the case when using the prediction error feature with surrogate data method in the examples of Sections 8.7.1–8.8.2. The downside to using more general test statistics, such as $s^{(PE)}$, is that it lacks the simple interpretation of the cumulant-based approach. When testing cumulants (in either time or frequency domains), we are looking for something very specific, either nonzero or nonconstant values of s. For prediction errors, there exist no natural scale for setting the detection threshold λ. This clearly implies that we require both "healthy" and "damaged" distributions for $s^{(PE)}$, which implies that ample training data be acquired before we can assess Type I and Type II errors. As can be seen in Section 8.6, however, the real power in using $s^{(PE)}$ can be realized when combining this test statistic with surrogate data, which, by construction, will satisfy the linear system hypothesis. This will allow us to use $s^{(PE)}$ effectively in the absence of training data.

8.1.4 Information Theoretics

As we have already discussed in Chapter 2, the cross-correlation coefficient $\rho_{Y_i Y_j}(T)$ is the normalized covariance and therefore captures second-order relationships among the stationary random processes $Y_i(t)$ and $Y_j(t)$. However, $\mathcal{H}_1$ in our nonlinearity detection schemes require a test statistic that is sensitive to the presence of *any*-order joint statistical moment. As we have just shown with the prediction error statistic, one way to accomplish this is to design a measure that is a function of the entire joint PDF of the structural response signals. It turns out that the time-delayed information-theoretic quantities, introduced in Section 3.5, provide another useful test statistic for assessing (8.8).

It was shown in Chapter 3 that the time-delayed mutual information function among time-series data could play the role of a generalized correlation function

$$I_{XY}(T) = \int_X \int_Y p_{XY}(x(n), y(n+T)) \log_2 \left[\frac{p_{XY}(x(n), y(n+T))}{p_X(x(n)) p_Y(y(n+T))} \right] dx(n) dy(n+T) \tag{8.33}$$

Thus, we imagine using as our detection statistic the value of the estimated mutual information at various delays

$$s^{MI}(T) = \hat{I}_{XY}(T). \tag{8.34}$$

It was also shown that for stationary, zero-mean, jointly Gaussian random processes, the "linearized" time-delayed mutual information becomes

$$\hat{I}_{XY}(T) = -\frac{1}{2} \log_2 [1 - \hat{\rho}_{XY}^2(T)] \tag{8.35}$$

where $\hat{\rho}_{XY}(T)$ is an estimate of the linear cross-correlation function discussed in Section 6.4.2, Eq. (6.85). The resulting detection statistic can be denoted $s^{LMI}(T)$. Given a nonlinear input–output relationship, we might therefore expect these two different estimators to yield very different results. Indeed, in the first example of Section 8.7.1, we show the ability to detect structural nonlinearity (hence damage) using precisely this approach. We go on to show in that same section that a test statistic based on (8.33) is, by itself, appropriate for testing for damage-induced nonlinearities.

A slightly more sophisticated correlation function, the time-delayed transfer entropy, was also introduced in Section 3.5. For a stationary sequence of observations, this quantity is given by Eq. (3.130). Moreover, it was shown that for Gaussian processes the "linearized" transfer entropy (TE) estimate is written (see development leading to 6.164)

$$\hat{TE}_{j \to i}(T)$$

$$= \frac{1}{2} \log_2 \left[\frac{\left(1 - \hat{\rho}_{XX}^2(\Delta_t)\right) \left(1 - \hat{\rho}_{XY}^2(T)\right)}{1 - \hat{\rho}_{XY}^2(T) - \hat{\rho}_{XY}^2(T - \Delta_t) - \hat{\rho}_{XX}^2(\Delta_t) + 2\hat{\rho}_{XX}(\Delta_t)\hat{\rho}_{XY}(T)\hat{\rho}_{XY}(T - \Delta_t)} \right]. \tag{8.36}$$

As with mutual information, Eq. (8.36) requires only the computation of the dynamic linear cross-correlation coefficient and the autocorrelation function for the process **X**. More generally, however, the full estimate of the TE will be used to generate the detection statistic. That is to say, we will equate

$$s^{TE}(T) = \hat{TE}(T) \tag{8.37}$$

where the estimate $\hat{TE}(T)$ is obtained using the kernel density approach (see, e.g., 6.172) along with the full TE expression (6.171).

In short, we can take the mutual information function or the TE at one or more time delays as the damage-sensitive test statistic. Estimators for both quantities were given in Section 6.4.2 and can be compared to detect the presence of structural damage. As with the prediction error metric, information-theoretic values lack the simple interpretation of the cumulant-based tests. However, when used with surrogates, only the full estimator (e.g., Eq. 6.171) need be considered and can yield a good test for nonlinearity without healthy structural response data. This is demonstrated in Section 8.7.1.

8.2   Bolted Joint Revisited

In Section 5.4, we provided a model for the bolted joint structure shown schematically in Figure 8.1. The structure is clamped at one end and loose at the other, allowing for "rattling" between the beam and the bolt head. We further suggested that one model this physics as an impacting nonlinearity, in which case we would expect the presence of higher order statistical properties in the response as discussed in the previous section. This section therefore considers the strain response of this structure to a stationary, jointly Gaussian input signal and uses the hypothesis (8.5) to form a damage detection procedure. Both the fourth cumulant (kurtosis) and integrated PSD (8.26) will be used as test statistics.

8.2.1   Composite Joint Experiment

The experimental system studied is a composite-to-metal bolted joint. The experimental setup is similar to the one used in [20], and a schematic is provided in Figure 8.1. It consists of a composite beam, bolted at both ends to steel support plates, and subjected to a random excitation at the midspan. The beam is 1.219 m long, 17.15×10^{-2} m wide, and 19.05×10^{-3} m thick. The composite material is comprised of a quasi-isotropic layup consisting of (0/90) and ($\pm$45) 24 oz. knit EGlass fabric. The specific layup is [(+−45), (0/90)]6S, meaning there are six sets of (+−45), (0/90) plies stacked on top of each other in the first half of the laminate. The "S" is used to denote a symmetric laminate, meaning that the other half of the laminate is six sets of (90/0), (−+45) plies stacked on top of each other. Four instrumented 4×1.905 cm bolts measuring 8.89 cm in length are used, 2 at each extremity, for recording preload.

The strain data are collected by means of four fiber Bragg grating (FBG) strain sensors [25], corresponding to the measures labeled str_1, str_2, str_3, and str_4, at the locations shown in Figure 8.1. The vibrational response of the structure is measured using a fiber-optic strain gauge system that is described in detail elsewhere [25]. For these measurements, the fiber-optic system interrogated five FBGs simultaneously at a data rate of 1500 Hz with $\pm$1 Hz resolution. The FBGs were attached to the structure using M-Bond 200 adhesive (Vishay, Inc.) following the recommended application procedure. Four of the FBGs were attached to the composite beam, with the fifth sensor attached to a steel plate. All sensors were placed on the same side of the structure.

The excitation was applied at the midspan of the beam and consists of a broad-band Gaussian input with zero-mean and standard deviation set to 53.4 N and the frequency content of the input signal was filtered to a cutoff of 112 Hz. The force is physically applied by an MB

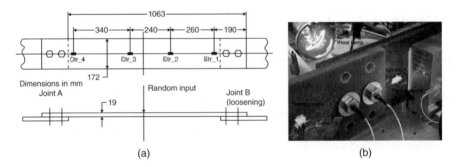

(a)   (b)

Figure 8.1 Schematic of the bolted joint used in the case study detailed in this section. The loosened joint is the one on the right side of (a). (b) Picture of the experimental structure. Shown are the composite beam, steel mounting plate, heat lamp, instrumented bolts, strain sensors, and thermocouples

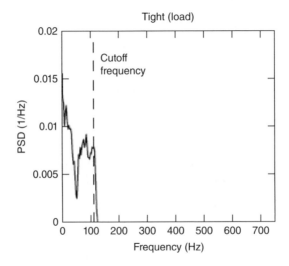

Figure 8.2 Estimated power spectral density function using the signal collected by the load cell

Dynamics PM50a electrodynamic shaker, instrumented with a Sensotec Model 31 load cell on the stinger. The estimated PSD of this load signal is shown in Figure 8.2. The recorded input information is not used in the damage detection procedures, which are based exclusively on the output measurements.

The damage is introduced as a progressive loosening of the two bolts at one end of the beam, joint B in Figure 8.1. Although the bolts themselves are instrumented so that a measure of the preload is also possible (see Section 8.8.2), in this initial study the joints are considered to be either in a tight or loose configuration. Two kinds of loose configurations are investigated. In the first, the bolts are tightened so that no finite gap between the surfaces is present, but no preload is acting on them: such a condition can be attained by simply tightening the bolts by hand, and hence the name "finger-tight" in the following. The joint is classified as loose also when a finite gap (smaller or larger) is included intentionally.

To test this approach in the presence of ambient variability, temperatures at the surface of the beam were adjusted using infrared heat lamps to warm only the beam and plate surrounding the joint under test. The rest of the structure remained at room temperature (23 °C). Local temperatures were monitored using four thermocouple gauges held onto the surface with thermally conductive putty. The thermocouples were placed on both sides of the plate and the beam. Lamp intensities were adjusted until the maximum temperature difference between the four thermocouples was 2 °C. For these measurements, four temperature settings were used, 23 (room temperature), 30, 40, and 50° C.

At each damage level we applied 30 s of dynamic loading and recorded the structural response from all five FBGs as well as the excitation (using a load cell). Figure 8.1 shows the end of the beam that was subjected to the bolt loosening and temperature fluctuations. Also shown are the locations of the two FBGs used to collect the data. All measurements were repeated three times and on different days. A total number of 190 acquisitions are recorded, yielding about one-and-a-half hours of vibration data.

Each of the recorded strain signals was 10 s long, which, given the 1500 Hz sampling rate, yields long (150 k samples) observations of the response. Each raw signal is denoted $y(n)$ and is first normalized to zero mean and unit standard deviation. We then numerically take the time derivatives of the acquired data. Our reason for doing so is that numerical differentiation is, in fact, a high-pass filtering operation that emphasizes the higher frequency content we seek to monitor in the analysis. Here we use a differencing scheme for both first and second derivatives, namely, $\dot{z}(n) \approx (z(n+1) - z(n))/\Delta_t$ and $\ddot{z}(n) \approx (z(n+1) - 2z(n) + z(n-1))/\Delta_t^2$. As shown in the results, the high-pass filtering appeared to cause nonnegligible changes in probability of detection (POD) for large levels of false alarm. At low levels of false alarms (the region of most interest), the high-pass filtering appeared to have little effect.

8.2.2 Kurtosis Results

The first statistic we focus on is the estimated kurtosis given by Eq. (8.15). As motivation for this choice, we look at the estimated PDFs associated with the normalized response. The first PDF in Figure 8.3 shows the estimated marginal PDF associated with the loading, which appears to be well described by a Gaussian distribution. The second PDF, however, shows the strain response in the "loose" joint configuration. This PDF is clearly more "peaked" than a Gaussian distribution, which is precisely what the kurtosis is designed to measure. This behavior was also predicted by the numerical model of Section 5.4 and was, in fact, why we chose to plot this measure in Figure 5.22. It should also be pointed out that had we used the third cumulant, or its normalized version "skew," we would likely not have seen any changes due to the damage. The skew measures asymmetry in a probability distribution which, at least for this type of damage, there appears to be none.

A summary of the kurtosis results are shown in Figure 8.4. The figure depicts the estimated kurtosis for a number of different signals (over multiple days and the four different temperature conditions) where the joint is in different stages of degradation. While plots such as (8.4) certainly suggest the approach to detection is solid, only by reporting Type I and Type II errors (see Section 6.9.2) can we make quantitative statements about this detector performance.

The receiver operating characteristic (ROC) curves associated with each of the four strain sensors (see Section 6.9.2) are shown in Figure 8.5. At very low levels of false alarm (the region of most practical interest), all signals (high-pass filtered or not) yield similar detection

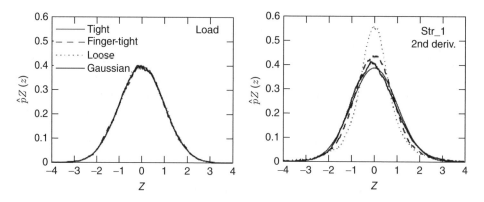

Figure 8.3 Estimated probability density functions (PDFs) associated with the tight and "loose" joint cases. A characteristic of the damage is to force the PDF of the response to become more "peaked." The estimated kurtosis of Section 8.1.1, Eq. (8.15) provides a measure of this particular deviation from normality. This also implies that had we used the skew, also described in Section 8.1.1, we would have significant difficulty detecting the damage because skew measures asymmetry in the distribution

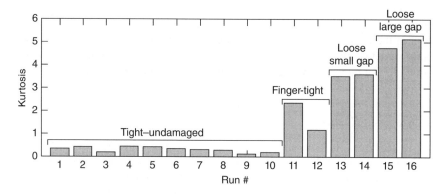

Figure 8.4 Estimated kurtosis, $\hat{k}_{YYY}$ as a function of damage. There is a clear increase in kurtosis as the joint becomes progressively more loose *Source:* Reproduced from [26], Figure 10, with permission of SAGE

probabilities. However, for large numbers of false alarms, the high-pass filtering (differentiation) appears to make a large difference in detection probabilities. It can also be noticed that the sensor closest to the damage yields the best detection performance, while the sensor furthest from the damage site has the most difficulty.

Although simple, the estimated kurtosis makes a reasonable damage detection statistic. However, keep in mind that this test statistic is based on the estimated marginal distribution associated with any single observation. While we use observations at different times to estimate the marginal PDF (and associated statistics such as the kurtosis) through the assumption of ergodicity, this is not the same as focusing on the joint distribution among observations at different times. As we now know, the spectral density functions (PSD or HOS) are the FTs

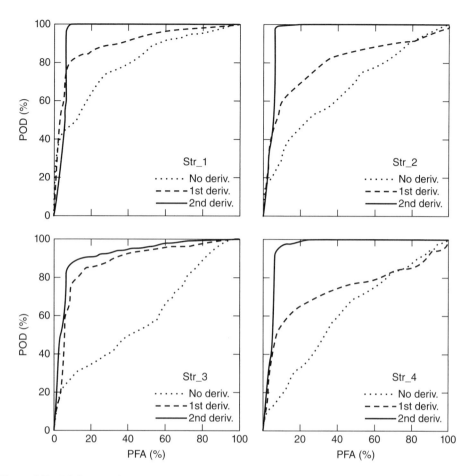

Figure 8.5   ROC curves for the kurtosis detection statistic. Performance varies with location, the closest sensor to the damage providing the best detection results

of these joint distributions and therefore carry potentially more information about the random process. In the next section, we use one such spectral approach to detecting the joint loosening.

8.2.3   Spectral Results

As we pointed out in Section 8.1.2, we may also form a test statistic based on the estimated power spectral density, that is, estimate $s^{(P)}$ (see Eq. 8.26). We know from our model (see Section 5.4) that this particular nonlinearity should amplify higher frequency content in the signal relative to the undamaged response. In particular, if the input PSD is limited to a cutoff frequency $f_c = 112$ Hz, from Figure 5.21 we see that damage will cause frequencies $f > f_c$ to be amplified. Figure 8.6 shows the estimated PSD functions associated with normalized strain signal and also the numerically differentiated and normalized strain signal. The "high-pass"

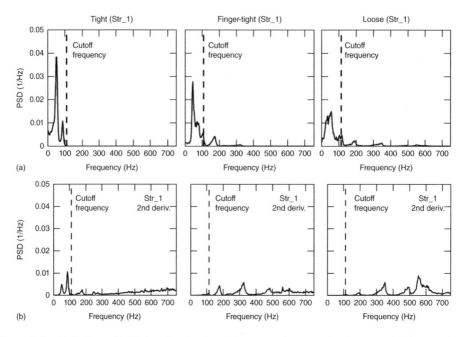

Figure 8.6   (a) Estimated PSD function for the strain sensor closest to the damage, with the joint in the three different configurations. (b) Estimated PSD function for the twice numerically differentiated strain sensor data

effect of the numerical differentiation can be clearly observed. Regardless of the filtering, however, the loosening joint clearly results in the increase in the higher frequency content, as predicted by the model.

The next question then becomes how to choose a test statistic that reflects this increase. To this end, the practitioner has a number of choices, each of which will lead to a slightly different detector performance. Here, we use the test statistic described by Eq. (8.26) with $f_1 = 112$ and $f_2 = 750$ Hz. Thus, we are looking at the portion of signal power at higher frequencies.

Figure 8.7 shows the ROC curve associated with this detection statistic for each of the strain sensors, just as was done for the kurtosis in the previous section. Again, we see that the sensor closest to the damage gives the best ROC performance. We also see that numerical differentiation helps emphasize the higher frequencies and improves performance for higher false-alarm probabilities (for low probability of false alarm (PFA), the results are unchanged with differentiation). Moreover, these ROC curves demonstrate improved detection probabilities (for a given PFA) over the performance realized by the kurtosis detector. This is perhaps not surprising as the estimated PSD carries information about the joint probability structure of the data (over many delays), while the kurtosis is a zero-lag (marginal probability) measure.

This first study demonstrates two basic approaches to capturing the physics of a common type of damage (nonlinearity). In the studies that follow, we explore our ability to detect the presence of different forms of damage physics. Performance will always be quantified by a ROC curve when possible.

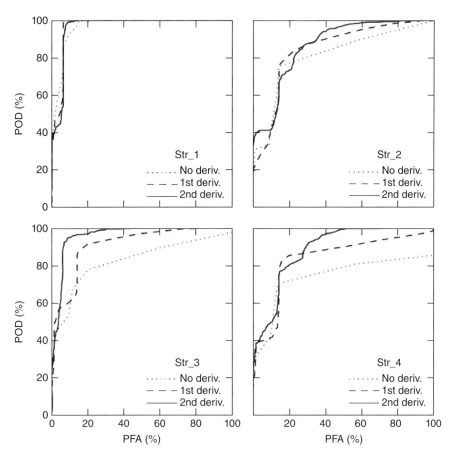

Figure 8.7 ROC curves for each of the strain sensors on the bolted joint. Performance for the test statistic $s^{(P)}$ (8.26) is quantitatively better than that for the kurtosis. This is perhaps expected as the PSD carries more information about the joint probability structure of the signal *Source:* Reproduced from [26], Figure 11, with permission of SAGE

8.3   Bispectral Detection: The Single Degree-of-Freedom (SDOF), Gaussian Case

We have already shown at the beginning of the chapter that if the input random process $X(t)$ is a jointly Gaussian-distributed random process, all of the stationary cumulants of order $n > 2$ will vanish for all lags $\tau > 0$. This led to the hypothesis test (8.5), which we mentioned could be accomplished in one of several different ways. As we have already mentioned, we can estimate the higher ordered cumulants directly from observed data using the estimator (8.17) and compare to a threshold λ.

However, we also mentioned a more general, second approach whereby one examines the FTs of the cumulants, that is, the HOS. Using this approach, one estimates (3.94) and compares the obtained values to a threshold, exceedances indicating nonlinearity. In the this section, we therefore provide a detailed description of how one can use the HOS to detect the presence of damage-induced nonlinearity in structures.

A great deal of attention has been given to the HOS in structural system identification. Worden and Tomlinson [27] used estimates of the auto-bispectrum to detect different types of nonlinearity in a spring-mass system. In the context of damage detection, Rivola and White [28] used the normalized auto-bispectrum to detect cracks in an experimental beam, while Zhang *et al.* [29] focused on detecting gear faults, also using the auto-bispectrum. Both the auto-bispectrum and auto-trispectrum were used by Teng and Brandon in detecting the deterioration of jointed structures [30].

As with any detection problem, we must also quantify the Type I and Type II errors associated with this type of detector. Ultimately, we would like ROC curves (see Section 6.9) associated with detecting different types and degrees of damage. This information is essential in understanding the efficacy of bispectrum-based detectors in applications such as damage detection mentioned earlier. However, before proceeding directly to estimating the HOS for damaged structural response data, it is worthwhile to understand exactly how damage will manifest itself in the estimates. A great deal of work has been performed in this area and was presented already in Section 6.5. In what follows, we review this section and show how we might use the information to detect damage-induced nonlinearity.

8.3.1 Bispectral Detection Statistic

The first question we should always ask in developing an approach to damage detection is: how do we expect damage to influence the test statistic? For this we require a specific structural model. We begin with perhaps the simplest structural nonlinearity of all, a quadratic dependency on either response position or velocity, that is,

$$m\ddot{y} + c_1\dot{y} + c_2\dot{y}^2 + k_1 y + k_2 y^2 = Ax(t) \tag{8.38}$$

where m(kg) is the mass, c_1(N $\cdot$ s/m) and k_1(N/m) the linear damping and stiffness coefficients, c_2(N $\cdot$ s^2/m^2) and k_2(N/m^2) the nonlinear damping and stiffness coefficients, and A(N) the amplitude of excitation. This particular form of nonlinearity shows up in various types of structural damage, among them loosened joints, delamination, and defects in rotating machinery. Recall from Section 6.5 that the auto-bispectral density for this single degree-of-freedom (SDOF) system is given by

$$B(\omega_1, \omega_2)$$

$$= \frac{8}{2\pi} \int_{-\infty}^{\infty} H_2(\omega_1 + \xi, \omega_2 - \xi) H_2(-\omega_1 - \xi, \xi) H_2(-\omega_2 + \xi, -\xi) S_{xx}(\xi) S_{xx}(\omega_1 + \xi) S_{xx}(\omega_2 - \xi) d\xi$$

$$\quad - 2H_2(-\omega_1 - \omega_2, \omega_2) H_1(\omega_1 + \omega_2) H_1(-\omega_2) S_{xx}(\omega_1 + \omega_2) S_{xx}(\omega_2)$$

$$\quad - 2H_2(-\omega_1 - \omega_2, \omega_1) H_1(\omega_1 + \omega_2) H_1(-\omega_1) S_{xx}(\omega_1 + \omega_2) S_{xx}(\omega_1)$$

$$\quad + 2H_2(\omega_1, \omega_2) H_1(-\omega_1) H_1(-\omega_2) S_{xx}(\omega_1) S_{xx}(\omega_2) \tag{8.39}$$

where $S_{xx}(\omega)$ is the PSD of the input and the Volterra kernels for the acceleration response, $\ddot{y}(t)$, were previously determined to be

$$H_1(\omega) = \frac{-\omega^2}{k_1 + ic_1\omega - m\omega^2}$$

$$H_2(\omega_1,\omega_2) = \frac{1}{\omega_1^2\omega_2^2}(-k_2 + c_2\omega_1\omega_2)H_1(\omega_1)H_1(\omega_2)H_1(\omega_1 + \omega_2). \tag{8.40}$$

Assume that the input process $X(t)$ is comprised of i.i.d. Gaussian values, and hence the power spectral density is the constant $S_{XX}(\omega) = P$. These expressions may be substituted into Eq. (8.39) to give a closed-form solution for the auto-bispectral density in terms of the physical parameters m, k_1, k_2, c_1, c_2, and excitation level P. For the system studied here, the presence of a nonlinearity in either the damping or stiffness terms will lead to nonzero values, resulting in "peaks" at both the natural frequency $\omega_n = \sqrt{k_1/m}$ and the bi-frequency $\omega_1 + \omega_2 = \omega_n$. Thus, for the detection statistic, we simply choose the first one mentioned in Section 8.1.2 and estimate

$$s_{nn}^{(B)} = |\hat{B}(\omega_n,\omega_n)|^2. \tag{8.41}$$

Detecting the presence of these nonlinearities therefore requires the practitioner to monitor this point in the ω_1,ω_2 plane and look for significant nonzero values. As we demonstrated in Section 6.9.3, characterizing Type I and Type II errors requires us to know the test statistic distribution under the null and alternative hypothesis. Fortunately, the analytical expression (8.39) can be used to derive the distributions for the estimated magnitude auto-bispectral density peak heights. In the following sections, we discuss the estimation problem and the problem of determining the significance of the magnitude auto-bispectral density peak heights as a nonlinearity detection measure.

8.3.2 Test Statistic Distribution

We have already discussed the estimation of the bispectral density and bicoherence functions in Section 6.5.2. Here, specific expressions are derived for both the bias and variance given the structural model. These expressions are essential for understanding the limitations of a bispectral-based nonlinearity detector. The real question for detection is "what is the smallest nonlinearity that can be detected with a given Type I or Type II error given the uncertainty in the estimate?". Of course, other noise sources will be present in practice and will degrade the detection performance further. By considering only the uncertainty in the estimate, the results that follow show the fundamental limitations of this detection scheme.

The direct method of estimation was described in Section 6.5 and rewritten here as

$$\hat{B}(f,g) = \frac{\Delta_t^2}{KM}\sum_{k=0}^{K-1} Y_k(f)Y_k(g)Y_k^*(f+g) \tag{8.42}$$

for discrete frequencies f,g. Here, as before, the signals are divided into K, possibly overlapping segments of length M, Fourier transformed, and then averaged. We note that obtaining the estimate at angular frequencies (ω_1,ω_2) simply requires the scaling $\omega = 2\pi\frac{2f-M}{2M\Delta_t}$, $f = 0\ldots M-1$ (and remembering to divide by $4\pi^2$) resulting in the estimate $\hat{B}(\omega_1,\omega_2)$. As we mentioned in Section 6.5, the bias in structural systems can be quite large owing to the sharpness of the spectral peaks (light damping). The variance associated with this estimate has already been derived [31, 32] and was discussed in Section 6.5. For nonoverlapping data segments and no windowing, the cited works have shown that both the real and imaginary parts of $\hat{B}(\omega_1, \omega_2)$ are asymptotically Gaussian distributed with common variance given by (6.132) for all nondiagonal frequency pairs ($\omega_1 \neq \omega_2$), while the variance is doubled ($\sigma^2 \to 2\sigma^2$) for frequencies on the diagonal ($\omega_1 = \omega_2$) [32]. In the absence of any bias, the means of these distributions are given by $E[\text{Re}\{\hat{B}(\omega_1, \omega_2)\}] = \text{Re}\{B(\omega_1, \omega_2)\}$ and $E[\text{Im}\{\hat{B}(\omega_1, \omega_2)\}] = \text{Im}\{B(\omega_1, \omega_2)\}$ respectively, for which we have analytical expressions.

From Eq. (6.132) it is seen that for a consistent estimate, the number of data points used in each segment must fulfill $M < N^{\frac{1}{2}}$. Using windowed, overlapping segments can help ease this constraint, but the scaling of M with N remains. As with the estimation of power spectra, there is clearly a trade-off between bias and variance. We also have by (6.132) that the variance depends on the autospectral density function associated with the response. From our work in Section 6.4.1, we know that the autospectral density $S_{yy}(\omega)$ used in Eq. (6.132) contains an additional term due to the nonlinearity. The full expression, assuming a second-order Volterra model with an input spectrum $S_{XX}(\omega) = P$, was found to be

$$S_{YY}(\omega) = P|H_1(\omega)|^2 + 2P^2 \int_{-\infty}^{\infty} |H_2(\omega_1, \omega - \omega_1)|^2 d\omega_1. \tag{8.43}$$

This expression was simplified via Cauchy integration (see Appendix B) and is given by

$$S_{YY}(\omega) = P|H_1(\omega)|^2$$

$$+ \frac{2\pi P^2 \left\{ \left[\omega_n^2 c_2^2 \omega^4 + \left(-3c_2^2 \omega_n^4 - 6c_2 k_2 \omega_n^2 + k_2^2 \right) \omega^2 + 4\omega_n^2 \left(c_2 \omega_n^2 + k_2 \right)^2 \right] m^2 + c_1^2 \left(\omega_n^2 \omega^2 c_2^2 + 4k_2^2 \right) \right\}}{m \omega_n^2 c_1 \left(m^2 \omega^2 + c_1^2 \right) \left[m^2 \left(\omega^2 - 4\omega_n^2 \right)^2 + 4\omega^2 c_1^2 \right]}$$

$$\times |H_1(\omega)|^2. \tag{8.44}$$

The second term of Eq. (8.44) is higher order $O(P^2)$ and therefore does not contribute significantly to the auto-bispectral density variance. It is included here for completeness. As we have mentioned, the detection statistic of interest is $s_{ij}^{(B)}$, that is, the magnitude of the auto-bispectral density at bifrequency i, j squared, $|B(\omega_i, \omega_j)|^2 = \text{Re}[B(\omega_i, \omega_j)]^2 + \text{Im}[B(\omega_i, \omega_j)]^2$. The distribution for the sum of the squares of two Gaussian random variables with the same variance, but different means, can be shown using the methods of Section 2.4 to be noncentral, χ^2 with two statistical DOFs (see also [33]) and is written

$$p_S(s) = \frac{1}{2} e^{-(s+\lambda)/2} I_0(\sqrt{s\lambda}) \tag{8.45}$$

where the normalized detection statistic is denoted

$$s = |\hat{B}(\omega_i, \omega_j)|^2 / \sigma^2, \tag{8.46}$$

$\lambda = (|B(\omega_i, \omega_j)|^2 - b)/\sigma^2$ is the noncentrality parameter, the estimator variance (σ^2) and bias (b) are given by Eqns. 6.132 and 6.137 respectively, and I_0 is the Bessel function of the first kind.

In the case of no nonlinearity ($\lambda = 0$) and an unbiased estimator, the Bessel function evaluates to unity and Eq. (8.45) reduces to the central χ^2 distribution with two DOFs, that is, the null distribution. The alternative is that there is a nonlinearity resulting in a non-zero $B(\omega_1, \omega_2)$ for which an analytical solution is known from the previous section. With both the null and the alternative distributions, a complete accounting of the Type I and Type II errors is possible. In the next section, some results are presented showing the performance of the auto-bispectral-density-based detector using the direct method of estimation. The detection of nonlinearity for both the damping ($c_2 \neq 0$) and stiffness ($k_2 \neq 0$) terms is considered.

8.3.3   Detector Performance

As we have consistently advocated, detector performance is quantified using the ROC curves of Section 6.9. In this section, we present theoretical ROC curves associated with this detection problem and compare them to those obtained via numerical simulation.

The system given by Eq. (8.38) was integrated using the Euler–Maruyama scheme. This approach is appropriate for stochastic differential equations and will preserve the correct output variance [34]. The parameters for the linear system were fixed to the values $m = 1.0$ (kg), $k_1 = 10^3$ (N/m), $c_1 = 3.0$ (N $\cdot$ s/m) with a sampling interval of $\Delta_t = 0.01$ s. The nonlinear system parameters, c_2 (N $\cdot$ s^2/m^2) and k_2 (N/m^2) were varied to understand their effects on the system response. For the auto-bispectral density estimates, the time series length was fixed to $N = 32,768$ points, while the window size was chosen to be $M = 128$.

Figure 8.8 shows both theoretical and estimated magnitude bispectral densities obtained for both stiffness (Figure 8.8a and b) and damping (Figure 8.8c and d) nonlinearities. The presence of the nonlinearity results in the peak observed at the system natural frequency. In the absence of the nonlinearity, this peak disappears.

The detection problem involves discerning the Type I and Type II errors associated with declaring a peak present. The magnitude auto-bispectral density peaks of Figure 8.8 are clearly visible, thus a POD of nearly unity would be expected for nearly zero PFA. However, as the degree of nonlinearity decreases, it becomes more difficult to distinguish these peaks from the fluctuations present in the estimate. As was stated earlier, in this study the uncertainty is simply given by the error in the integration routine (not accounted for in the theory but assumed negligible) and the estimation error (which is accounted for in the theory).

Next, we explore the limitations in detecting both stiffness and damping nonlinearities. To this end, the expected distributions for the magnitude auto-bispectral density peak height as a function of both c_2, k_2 was analytically obtained via Eqn. 8.46. The distributions were also obtained numerically by simulating 40 realizations of the random process described by Eq. (8.38). A bispectral density estimate was obtained for each realization and both the real and imaginary components were recorded for the peak frequency along with the magnitude bispectral density peak height. The variance for both real and imaginary components was then estimated and the resulting values were averaged to give $\hat{\sigma}$. The mean estimated peak value $|\hat{B}(\omega_1, \omega_2)|^2$ (which, of course, includes the bias term b), along with $\hat{\sigma}^2$ was then used to estimate the distributions given by Eq. (8.45). Results were obtained under the null hypothesis ($k_2 = c_2 = 0$) and for varying values of the two nonlinearity parameters.

Sample results are shown in Figure 8.9 for the SDOF system with a stiffness nonlinearity. Plotted are the distributions of the estimated magnitude auto-bispectral density (normalized)

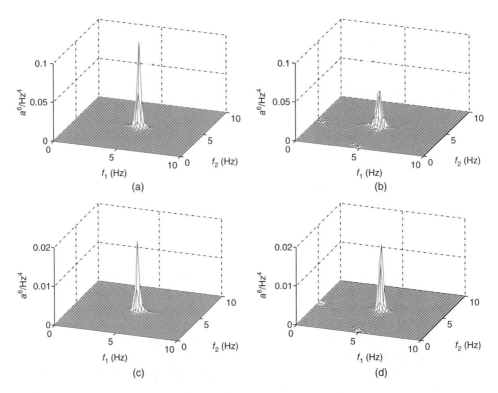

Figure 8.8 Magnitude auto-bispectral density obtained from theory (a and c) and from the direct estimation procedure (b and d). Results are plotted for a stiffness nonlinearity of $k_2 = 10^5$ N/m^2, with $c_2 = 0$ N $\cdot$ s^2/m^2 (a and b) and damping nonlinearity of $c_2 = 40$ N $\cdot$ s^2/m^2 with $k_2 = 0$ N/m^2 (c and d) Peak heights are displayed in terms of their units. *Source:* Reproduced from [34], Figure 2, with permission of Elsevier

at the peak frequency under the null hypothesis (no nonlinearity) and the alternative (nonlinearity present, $k_2 = 10,000$ N/m^2). These results are given for both theory (Figure 8.9a) and estimate (Figure 8.9b). Results show good agreement between observed and predicted values. Figure 8.9c compares both the theoretical and observed detector performance in the form of ROC curves. The agreement in the estimated and predicted distributions translates directly into close agreement in the estimated and predicted ROC curves.

The results associated with detecting nonlinear damping are shown in Figure 8.10. Again, the theory is still able to make good predictions regarding what levels of nonlinearity one can expect to detect and with what probability. Both Type I and Type II errors are correctly identified.

Finally, the complete set of detection results for both stiffness and damping nonlinearities is provided. Figures 8.11 and 8.12 show the family of ROC curves obtained for varying values of the nonlinearity parameters. For Figure 8.11, the nonlinear damping term was set equal to zero ($c_2 = 0$), while the nonlinear stiffness was allowed to vary from $0 \rightarrow 20,000$ N/m^2 in increments of 1000 N/m^2. Conversely, in Figure 8.12, the nonlinear stiffness was zero ($k_2 = 0$), while the nonlinear damping coefficient was varied from $0 \rightarrow 20$ N $\cdot$ s^2/m^2 in increments of 1 N $\cdot$ s^2/m^2.

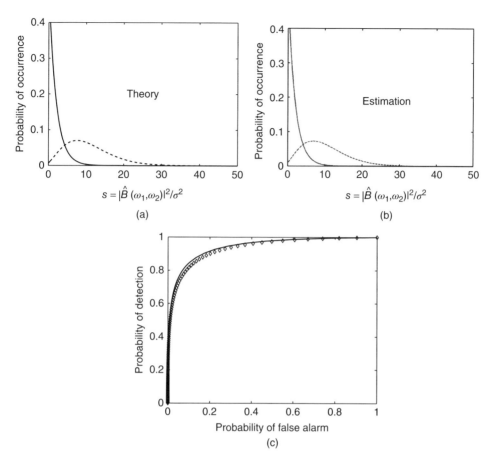

Figure 8.9 (a) Peak distributions obtained from theory and (b) those estimated from simulation. The solid line in both (a) and (b) represents the situation where no nonlinearity is present, while the dashed line corresponds to a nonlinear stiffness value of $k_2 = 10,000$ N/m^2 and no nonlinear damping term ($c_2 = 0$ N · s^2/m^2). The ROC curves associated with theory (solid line) and estimate (diamonds) are shown in (c) *Source:* Reproduced from [34], Figure 3, with permission of Elsevier

In both cases, good agreement between estimated and theoretically predicted ROC curves is observed. In the case of varying k_2, both theory and simulation predict that the smallest nonlinearity for which one can obtain 95% POD for 5% PFA is $\sim 13,000$ N/m^2 (specifically 13,470 N/m^2). Similarly, it can be seen from theory that a damping value of $c_2 = 13.17$ N · s^2/m^2 allows 95% POD for 5% PFA. Nearly the exact same Type I and Type II errors were obtained in simulation for this level of damping. It is also important to keep in mind that these results are dependent on the excitation level, P. These results were obtained for $P = 0.01$ (N^2/Hz). Larger P values will result in being able to detect smaller values of both k_2, c_2. However, it should be mentioned that if both the nonlinearity and excitation levels become large, the two-term Volterra model given here may break down and higher order terms may need to be considered. Again, for detection applications, the interest is in small levels of nonlinearity (damage) for which the two-term model is appropriate. The model presented here,

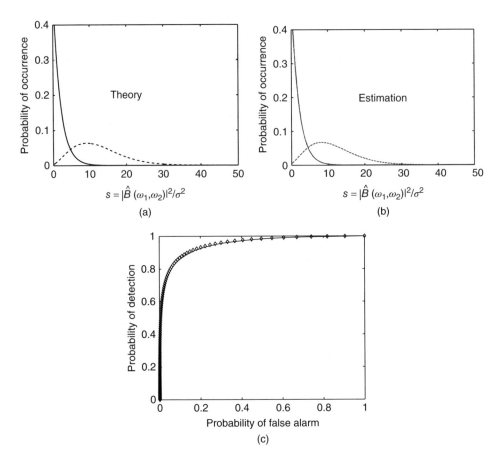

Figure 8.10 (a) Peak distributions obtained from theory and (b) those estimated from simulation. The solid line in both (a) and (b) represents the situation where no nonlinearity is present, while the dashed line corresponds to a nonlinear damping value of $c_2 = 10 \text{ N} \cdot \text{s}^2/\text{m}^2$ and no nonlinear stiffness term ($k_2 = 0 \text{ N/m}^2$). The ROC curves associated with theory (solid line) and estimate (diamonds) are shown in (c) *Source:* Reproduced from [34], Figure 4, with permission of Elsevier

combined with the statistical properties of the estimates, makes it easy to explore the influence of excitation.

It should also be pointed out that for very low levels of nonlinearity, the ROC curves associated with the simulation appear to give higher detection probabilities than those obtained from theory. The reason for this is an inability to generate a truly Gaussian time series in practice. In reality, there is always some skewness to the driving time series. This skewness results in peaks that, although small, can be detected, even under the null hypothesis of a linear structure. Thus, for small levels of nonlinearity, the estimated ROC curves do not exhibit the smooth transition toward the diagonal line predicted by theory for $c_2 = k_2 = 0$.

This simple study illustrated the process by which one may analytically obtain detection performance (Type I and Type II errors) for bispectral detection schemes. Even for this simple, one-DOF problem, a fair amount of work is required to provide such a solution. Moreover,

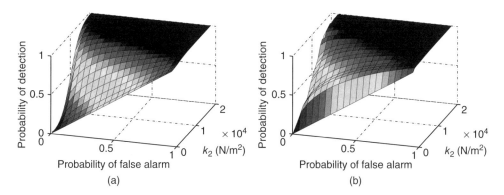

Figure 8.11 Comparison of ROC curves generated from theory (a) and simulation (b) as a function of the nonlinearity parameter k_2 *Source:* Reproduced from [34], Figure 5, with permission of Elsevier

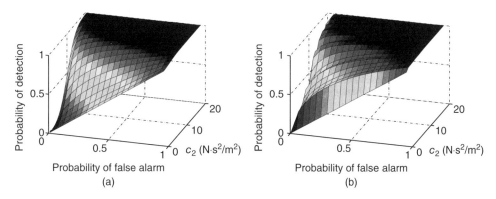

Figure 8.12 Comparison of ROC curves generated from theory (a) and simulation (b) as a function of the nonlinearity parameter c_2 ($k_2 = 0$) *Source:* Reproduced from [34], Figure 6, with permission of Elsevier

these results focus only on the uncertainty due to errors in the estimation. In this sense, the ROC curves presented show the fundamental limits of auto-bispectral-density-based detector performance for SDOF systems with quadratic nonlinearities using the direct method of estimation. In application, the measured and modeling uncertainties will increase, thus further diminishing our ability to detect. In subsequent Sections 8.4 and 8.5, we explore detection limits for more realistic structural models.

8.4 Bispectral Detection: the General Multi-Degree-of-Freedom (MDOF) Case

In this section, we consider the more realistic scenario of an multi-degree-of-freedom (MDOF) structure subject to (possibly) a non-Gaussian input. The specific structural model is the three-DOF system given by Eq. (6.140), characterized by linear mass, stiffness, and

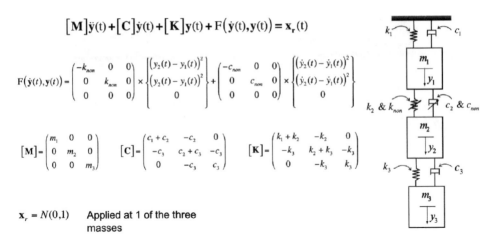

$$\left[\mathbf{M}\right]\ddot{\mathbf{y}}(t)+\left[\mathbf{C}\right]\dot{\mathbf{y}}(t)+\left[\mathbf{K}\right]\mathbf{y}(t)+F\left(\dot{\mathbf{y}}(t),\mathbf{y}(t)\right)=\mathbf{x}_r(t)$$

$$F(\dot{\mathbf{y}}(t),\mathbf{y}(t))=\begin{pmatrix}-k_{non}&0&0\\0&k_{non}&0\\0&0&0\end{pmatrix}\times\begin{Bmatrix}(y_2(t)-y_1(t))^2\\(y_2(t)-y_1(t))^2\\0\end{Bmatrix}+\begin{pmatrix}-c_{non}&0&0\\0&c_{non}&0\\0&0&0\end{pmatrix}\times\begin{Bmatrix}(\dot{y}_2(t)-\dot{y}_1(t))^2\\(\dot{y}_2(t)-\dot{y}_1(t))^2\\0\end{Bmatrix}$$

$$[\mathbf{M}]=\begin{pmatrix}m_1&0&0\\0&m_2&0\\0&0&m_3\end{pmatrix}\quad[\mathbf{C}]=\begin{pmatrix}c_1+c_2&-c_2&0\\-c_3&c_2+c_3&-c_3\\0&-c_3&c_3\end{pmatrix}\quad[\mathbf{K}]=\begin{pmatrix}k_1+k_2&-k_2&0\\-k_3&k_2+k_3&-k_3\\0&-k_3&k_3\end{pmatrix}$$

$\mathbf{x}_r=N(0,1)$ Applied at 1 of the three masses

Figure 8.13 Schematic of a three-degree-of-freedom system with quadratic nonlinearities in both restoring and dissipative terms inserted between masses 1 and 2. The structure is assumed to be driven by an i.i.d. Gaussian random process, applied as a force at one of the three masses

damping coefficients as well as a nonlinear stiffness (for this example, we assume no quadratic damping nonlinearity). Hence, our response data are assumed to come from multiple points on the structure, that is, we measure $y_i(n), n = 0,\dots,N-1$ for different structural locations $i = 1 \dots 3$. A pictorial representation of this system is shown in Figure 8.13.

As a first example, we may consider the case of Gaussian excitation just as we did in the SDOF example of the previous section. The difference is that now there are multiple peaks in the response that we could potentially use for the purpose of detection. For example, Figure 6.17 illustrates for this same three-DOF system a change in location of the dominant bispectral peak depending simply on which combination of response accelerations were used in the estimate. This is a common feature of bispectral analysis for MDOF structural systems that can yield drastic differences in detection performance.

Consider, for example, the system of Figure 8.13 with $k_1 = k_2 = k_3 = 2000$ N/m, $c_1 = c_2 = c_3 = 3.0$ N · s/m, and $m_1 = m_2 = m_3 = 1.0$ kg. The forcing PSD was the constant $S_{XX}(f) = 0.02$ N^2/Hz and we assumed a nonlinear stiffness of $k_N = 3000$ N/m^2 and no nonlinear damping ($c_N = 0$ N · s^2/m^2). Figure 8.14 shows a comparison between the performance of two detectors based on the estimated bispectral response obtained by driving the system at the end mass and using only the first mass acceleration response. The first detector is based on the dominant bispectral peak (labeled "peak 1" in the figure) and the second on the smaller peaks (labeled "peak 2" in the figure). The ROC curves in Figure 8.14 clearly show the difference in detection performance associated with our choice in test statistic. This discrepancy was accurately predicted by the theory we have previously developed.

In fact, results can change drastically depending not only on which peak we monitor but also which bispectrum we compute. Figure 8.15 compares two different bispectral densities and two different choices of detectors (as determined by peak location). Depending on which response one chooses to measure, and which frequencies are used in forming $s_{ij}^{(B)}$, very different performance is realized. This holds strong implications for the damage detection problem and is elaborated on further in the next section. Interestingly, similar comparisons for detecting

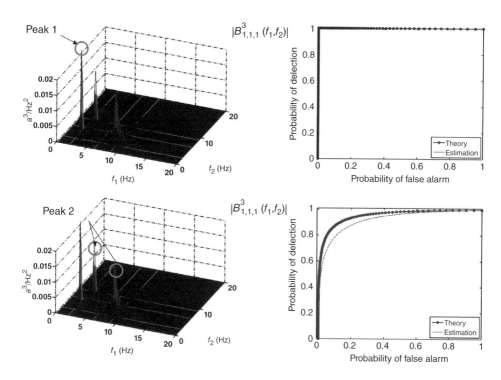

Figure 8.14   Difference in ROC curves associated with different detection statistics $s_{ij}^{(B)}$. In the first, the dominant peak located at the first natural frequency is used. For the second, the smaller peak located at the combination of the first two resonances is used. As predicted, the ROC performance associated with the larger peak indicates much lower Type I and Type II errors associated with that particular detector. Both sets of results are associated with the bispectral density $B_{1,1,1}^{3}(f_1, f_2)$

nonlinear damping (not shown) indicate that the resulting detector performance is much less sensitive to which bispectrum or even which peak is monitored.

What allowed us to focus on the bispectrum peaks for detection was the assumption of a driving signal with Gaussian marginal distribution. As we have already discussed, a more general null hypothesis assumes an i.i.d. input with a non-Gaussian marginal distribution, thus temporal correlations are still relegated to second-order statistics $C_{XX}(\tau) = E[(X(t) - \mu_X)(X(t + \tau) - \mu_X)] = 0 \ \forall \tau > 0$, but higher order, zero-lag central moments, for example,

$$E[(X(t) - \mu_X)^3] \neq 0 \tag{8.47}$$

still exist. If this is the case, we can see from Eq. (8.4) that the zero-lag statistical properties of the response are nonzero, for example, $E[(Y_i(t) - \mu_{Y_i})^3] \neq 0$, and we will get a nonzero bispectral density, even though the structure itself is linear. Thus, if $\hat{B}_{Y_k Y_i Y_j}(\omega_1, \omega_2) \neq 0$, there are two possibilities. One is that the structure obeys a nonlinear model such that a symmetric input distribution, similar to the Gaussian, leads to an asymmetric distribution on the output. The other possibility is that the structure is linear, but that the input distribution possesses a nonzero third (or higher) moment. In the latter case, we already showed that substituting the

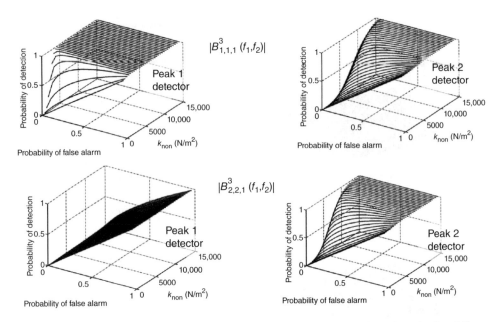

Figure 8.15 Difference in ROC performance as a function of the level of nonlinearity for different choices of bispectral density and detection statistic (peak 1 vs. peak 2). The choice of which bispectral density to estimate and which frequencies to use in the test statistic can make a very large difference in detection performance.

linear model into (6.106) yielded (see 6.120)

$$B_{Y_k Y_i Y_j}(\omega_1, \omega_2) = H_{1,i}(\omega_1)H_{1,j}(\omega_2)H_{1,k}(-\omega_1 - \omega_2)B_{XXX}(\omega_1, \omega_2) \qquad (8.48)$$

where $B_{XXX}(\omega_1, \omega_2)$ is the bispectrum of the input and $H_{1,i}$ is the linear system transfer function relating the input to the output at the "i"th DOF. In Section 6.5.1 we showed how to distinguish between these cases using the bicoherence (6.121), repeated here for convenience,

$$b_{Y_k Y_i Y_j}(\omega_1, \omega_2) = \frac{B_{Y_k Y_i Y_j}(\omega_1, \omega_2)}{\sqrt{S_{Y_i Y_i}(\omega_1)S_{Y_j Y_j}(\omega_2)S_{Y_k Y_k}(\omega_1 + \omega_2)}}. \qquad (8.49)$$

Analytical expressions and estimation of (8.49) for nonlinear structural systems was discussed at length in Section 6.5.3. For example, as we showed in Section 6.5.1, if the input consists of i.i.d. values chosen from a non-Gaussian distribution, and the structure is linear, the magnitude bicoherence will be the constant

$$|b_{Y_k Y_i Y_j}(\omega_1, \omega_2)| = \frac{|B_{XXX}(\omega_1, \omega_2)|}{\sqrt{S_{XX}(\omega_1)S_{XX}(\omega_2)S_{XX}(\omega_1 + \omega_2)}}$$

$$= \frac{k_3 \Delta_t^{1/2}}{\sigma^3 (2\pi)^{1/2}} \qquad (8.50)$$

Here, Δ_t is the sampling interval, the inclusion of which was already discussed in Section 6.5.1, and k_3 is the third cumulant (Eqn. 2.81) associated with the input random process X.

A nonlinear structural response will, however, result in nonconstant values of the bicoherence, regardless of the distribution of the input. Thus, a test for nonlinearity may be realized by simply testing the constancy of the estimated bicoherence function (see, e.g. Priestly [24]).

8.4.1   Bicoherence Detection Statistic Distribution

In this section, we follow the work of Garth and Bresler [35, 36], Richardson and Hodgkiss [37], and Hinich and Wilson [22] and suggest the detection statistic given by Eq. (8.25),

$$s^{(b)} = \frac{2}{\sigma_e^2} \sum_{p,q} |\hat{b}_{kij}^r(f_p, f_q)|^2 \tag{8.51}$$

where the sum is taken over all $D = M^2/16$ positive frequency bins in the estimated bicoherence function that satisfy $p, q \le M/4$. This constraint ensures that the sum frequency is below the Nyquist frequency, that is, $p + q \le M/2$. It was demonstrated in [35] that this test statistic was a sufficient statistic for the generalized likelihood ratio test (GLRT) discussed in Section 6.9.1. The shorthand notation $\hat{b}_{kij}^r(f_p, f_q)$ has been adopted in place of $\hat{b}_{\ddot{Y}_k \ddot{Y}_i \ddot{Y}_j}(f_p, f_q)$ as all results that follow are obtained using the acceleration output of Eq. (6.140). The superscript "r" will be used to denote the forcing location, that is, which DOF is subject to $x(t)$.

As was mentioned earlier, the bicoherence estimates in each frequency bin are complex Gaussian distributed with common variance σ_e^2, thus we know from Chapter 2 that the statistic S will follow a noncentral chi-squared distribution with $2D$ DOF (each frequency bin contributes a real and imaginary part, hence the factor of 2). For large D, however, the noncentral chi-squared distribution is well approximated by a Gaussian distribution with mean and variance

$$\mu = 2D + \Lambda$$

and

$$\sigma^2 = 2(2D + 2\Lambda)$$

where $\Lambda = \frac{2}{\sigma_e^2} \sum_{p,q} |b_{kij}^r(f_p, f_q)|^2$ is the noncentrality parameter.

The nonlinearity detection problem (8.8) can then be recast as discriminating between the hypotheses

$$\mathcal{H}_0 : s^{(b)} \sim \mathcal{N}(\mu_L, \sigma_L^2)$$
$$\mathcal{H}_1 : s^{(b)} \sim \mathcal{N}(\mu_N, \sigma_N^2) \tag{8.52}$$

where μ_L, σ_L and μ_N, σ_N are the linear and nonlinear mean and variance and are obtained by using the linear and nonlinear values of the non-centrality parameter for Λ are known analytically via Eqs. (6.144) and (6.148) for any combination of signals i, j, k and input locations r. We are therefore in a position to find the optimal bicoherence to estimate for the purpose of nonlinearity detection.

For this detection problem, we chose the Neyman–Pearson criteria for optimality, that is, to minimize the Type II error (maximize POD) for a given Type I error (PFA). For the problem stated in (8.52), assuming similar values for the variance, this may be accomplished by maximizing the deflection coefficient [38]

$$d^2 = \frac{|\mu_N - \mu_L|^2}{\sigma_N^2} \tag{8.53}$$

(even if the assumption of similar variance values is not made, the results of the following optimizations were not changed). In what follows, the goal is to maximize d with respect to the bicoherence test statistic parameters i, j, k, and r.

8.4.2 Which Bicoherence to Compute?

The goal of this section is to determine the optimal (in the sense described in the previous section) bicoherence to estimate. As we have already seen from the previous section, for an MDOF structural system, the nonlinearity can be located at different spatial points on a structure. For example, if the nonlinearity is located between masses 1 and 2, should we base Eq. (8.51) on $\hat{b}_{123}^1$ or $\hat{b}_{211}^3$? Specifically, we would like to find

$$\{i',j',k',r'\} = \max_{i,j,k,r} d$$

that is to say, the combination of input locations and output monitoring points such that we maximize our ability to successfully detect the nonlinearity. For this optimization, we considered the input random process $X(t)$ to be Gaussian, i.i.d., with a constant PSD of $S_{WW}(\omega) = $ const. $= 0.01$ N^2/Hz. For this particular choice of forcing $b_{kij}^r(f_p,f_q) = 0$ under the linear hypothesis, thus maximizing d strictly involves finding the largest value of μ_N. Fortunately, for this problem we can exhaustively search the possibilities and find the test statistic that maximizes Eq. (8.53). There are three possible locations for both excitation and response, giving a total of $3^4 = 81$ possible test statistics to estimate for a given nonlinearity location.

First, we consider the case where a quadratic stiffness nonlinearity, $k_N = 10^5$ (N/m^2), is placed between masses 1 and 2. Figure 8.16a plots the key component of μ_N (large μ_N equals large d) for all 81 possible bicoherence functions. A few of these are labeled in the figure and clearly show that some are more advantageous to use than others for detection purposes. It can be seen that the best bicoherence functions favor forcing at mass 3 and using the response at mass 1 (the mass furthest from the driven end) as the last argument in the estimate.

To test these predictions, the following numerical experiment was performed. Ten separate realizations, each consisting of $N = 32,768$ observations, were obtained from the system (6.140) using a Runge–Kutta numerical integration scheme. For each realization, the test statistic was estimated using Eqn. 8.52. The parameters used were $M = 256, L = 128$ and a Hanning window was applied for smoothing. Using these parameters, the adjustment factor in Eq. (6.133) becomes $\kappa = 0.273$. Figure 8.16b shows the estimated ROC curves associated with a few different choices of bicoherence function used in forming the test statistics. Large values of b_{kij}^r result in larger values μ_N, which give rise to better PODs for a given PFA. These results clearly indicate that choice of signals used in computing the bispectrum and the choice of excitation point are of paramount importance. The ROC performance is exactly in line with that predicted. Using the test statistic based on b_{221}^1, for example, gives very poor detection performance.

Next, we consider changing the location of the nonlinearity to lie between masses 2 and 3 and repeat the earlier analysis. These results are shown in Figure 8.17. Again, test statistics

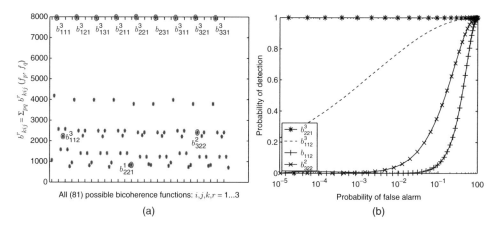

Figure 8.16 (a) All 81 possible values of b^r_{kij} under $\mathcal{H}_1$ when the nonlinearity is located between masses 1 and 2 and (b) the ROC curves corresponding to different choices of test statistic $s^{(b)}$ based on the associated bicoherence estimates *Source:* Reproduced from [39], Figure 4, with permission of Elsevier

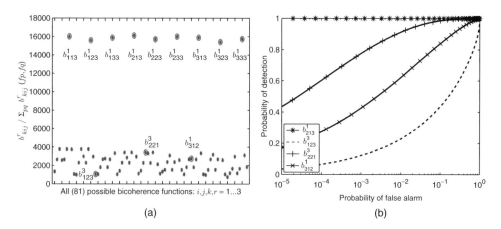

Figure 8.17 (a) All 81 possible values of b^r_{kij} under $\mathcal{H}_1$ with the nonlinearity located between masses 2 and 3 and (b) the ROC curves corresponding to different choices of test statistic s based on the associated bicoherence estimates *Source:* Reproduced from [39], Figure 5, with permission of Elsevier

based on the predicted optimal choice of b^r_{kij} clearly yield the best detection performance. Once again, the best choice of forcing location is away from the damage, in this case, mass 1. As before, the last argument in the estimated bicoherence should be the response collected furthest from the drive (mass 3). Regardless of where the damage is located, one can obtain drastically different detection performance depending on which bicoherence function is used in computing $s^{(b)}$.

In addition, these results give a straightforward way to locate the nonlinearity. An input can be applied at both masses 1 and 3 and the time series recorded from each. We can then estimate b^1_{113} and b^3_{311}. If $b^1_{113} \gg b^3_{311}$, the nonlinearity is between masses 2 and 3, while if $b^3_{311} \gg b^1_{113}$, the nonlinearity is between masses 1 and 2. In short, if we wish to locate a damage-induced, quadratic nonlinearity, we can do so by making only two sets of measurements, each obtained using a different excitation location. This scheme therefore provides one possible means of both detecting and locating a damage-induced nonlinearity in a structural system.

8.4.3 Optimal Input Probability Distribution for Detection

The problem formulation we have adopted in the previous section can also be used to ask an interesting question about the nature of the excitation that might be used in detecting the damage. Rather than using a Gaussian input, we could ask which input probability distribution yields the best detection performance. Recall from Section 6.5.3 that we could use a polynomial function of a Gaussian distribution to create a broad class of probability distribution functions, depending on the polynomial parameters $\mathbf{a} \equiv (a_0, a_1, a_2, a_3)$ (see 6.141). If we had control over the nature of the input distribution, which set of coefficients would yield the best detection performance? Specifically, for a fixed input power, we seek the parameter vector $\mathbf{a}$ that minimizes our Type I and Type II errors in making a detection. This optimization problem can also be written in terms of the deflection coefficient as

$$\mathbf{a} = \max_{a_1, a_2, a_3} d$$

An exhaustive search is not possible in this case as each of the parameters in $\mathbf{a}$ is a continuous variable. For this reason, we turn to the differential evolution algorithm already discussed at length in Section 7.1.3 to search the parameter space. It should also be pointed out that only zero-mean distributions are considered, that is, $a_0 = -a_2\sigma_W^2$, where, as in Section 6.5.3, σ_W^2 is the variance of the underlying white noise process being transformed by Eq. (6.141). This assumption is for practical reasons because the equipment used in vibration testing is often displacement limited. A large nonzero mean may help detection, but may not be feasible experimentally. The optimization was also limited to i.i.d. inputs, that is to say, there is no spectral coloring on the input signal. The goal here was to focus strictly on the form of the input probability density and not on where the signal energy should be concentrated in the frequency domain.

For this optimization, the nonlinearity location was chosen to be between masses 1 and 2 such that the bicoherence feature b_{221}^3 gives the best detection performance and was therefore taken to be the detection statistic. The results of the optimization are perhaps not entirely surprising. It turns out that for the broad class of input distributions considered, a Gaussian distribution is optimal. In general, symmetric distributions performed well. This is because, for a symmetric input distribution, the bicoherence is near zero if there is no nonlinearity, and it rises to some positive value with nonlinearity present. Because the variance of the test statistic increases with Λ, any input distribution that minimizes the value of the noncentrality parameter while maximizing the *difference* between linear and nonlinear noncentrality parameters is preferred. Highly asymmetric distributions performed poorly because even in the linear case they result in high values of Λ.

Figure 8.18 shows several different input distributions along with the generated ROC curves. The ROC curves were based on estimates of b_{221}^3, obtained from 20 independent realizations of Eq. (6.140). As predicted by theory, the best performing ROC curve is clearly associated with a Gaussian input. Thus, we conclude that for detecting a quadratic nonlinearity in a structural system, a Gaussian-distributed input is optimal. Yet one more way in which the Gaussian distribution holds a special place in the study of dynamical systems!

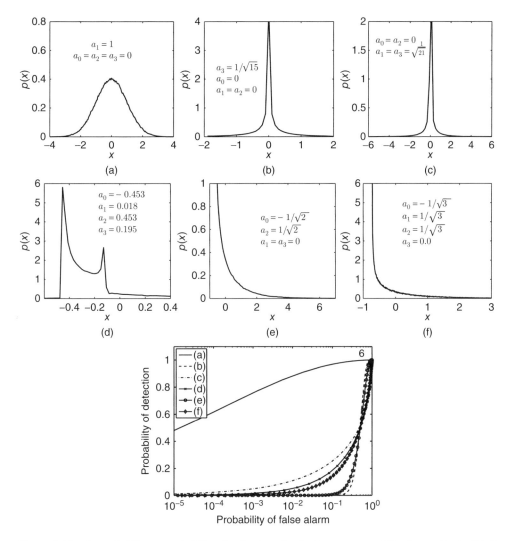

Figure 8.18 Several different probability distributions for the input process $X(t)$ along with the ROC curves associated with detecting a quadratic stiffness nonlinearity of $k_N = 30,000$ (N/m²). The Gaussian distribution was determined to be the optimal input probability distribution for detecting the nonlinearity. This is clearly seen to be the case in the estimated ROC curves *Source:* Reproduced from [39], Figure 6, with permission of Elsevier

Understanding this detection performance required the closed-form expressions for both the bispectral density and bicoherence functions derived earlier in this book. The main assumption underlying those derivations is that the system should be weakly nonlinear. This assumption is necessary both for the Volterra series model to hold and for the simplification of the full bispectrum expression. In damage detection applications, this is not much of a restriction as early detection (where damage-induced nonlinearity is small) is our primary goal.

This section has also demonstrated quite a bit about the performance of bispectral detectors. First, we saw that *which* bispectral estimate we use (i.e., which response DOFs are measured) can have a very large impact on detection performance. The results from our toy system suggest that the forcing should be applied away from the nonlinearity and that the signal in the last argument of the estimated bicoherence be taken close to the nonlinearity. While we will not know *a priori* where the damage is, this certainly suggests that multiple bicoherence estimates (from multiple DOFs) be used in practice.

This simple analysis also suggests that there are certain specific force/response combinations that lead to large values of the test statistic which can provide a convenient way to locate the nonlinearity. In addition, we recommend that, if the input signal can be controlled, one use a Gaussian-distributed input to obtain the best possible detection performance when using bispectral detection statistics (although it may be the case that deterministic inputs, e.g., sinusoid, or highly band-limited random inputs can produce still better detection performance; these were not considered in our class of input signals).

While we have focused thus far on very simple nonlinearities, we have found that these results are very much indicative of those seen in more sophisticated numerical models and in experimental systems as well. In other words, the lessons gleaned from these simple systems (which test statistic to use, how to drive the system, etc.) are very much applicable to more complex systems and experiments. The next section, in fact, employs some of these lessons in forming detection strategies for delamination damage.

8.5   Application of the HOS to Delamination Detection

The reason for deriving expressions for the HOS is so that we can better understand their potential use as a detection statistic with regard to different types of structural damage. While section 5.2 dealt with the static beam deflection problem (and associated localized buckling), section 9.2 will derive the dynamic equations of motion, appropriate for use in vibration-based analysis. As will be shown in Section 9.2, the resulting model contains both quadratic and cubic nonlinearities in the structure's response. Given that we know the form of the damage-induced nonlinearity, we may target a detector accordingly; the HOS are very well suited to this task.

The governing equations for the delaminated beam system of Figure 9.19 are given in terms of the time-dependent coefficients $\mathbf{q}(t)$

$$[M]\ddot{\mathbf{q}}(t) + [C]\dot{\mathbf{q}}(t) + [K_L]\mathbf{q}(t) + [K_M]q_1(t)\mathbf{q}(t) + [K_Q]\mathbf{q}^2(t) + [K_C]\mathbf{q}^3(t) = \mathbf{F}(t). \quad (8.54)$$

In examining this model (see the matrix Equations 9.45–9.50), we see that delamination is truly a local phenomenon. The only coupling between the global coordinate $q_1(t)$ and the delamination coordinates ($q_2(t)$ and $q_3(t)$) is inertial and occurs via the mass matrix. It will be shown that the local nature of the nonlinearity has significant implications for vibration-based detection. This section focuses on the detection of both the quadratic and cubic nonlinearities that arise due to the delamination using a polyspectral analysis. It is assumed throughout this section that the input vector acts at a location away from the delamination site, that is to say, $\mathbf{F}(t) \equiv (f(t), 0, 0)^T$.

As an example, consider the bispectrum and trispectrum associated with the delaminated beam response to Gaussian excitation. These expressions are formed analytically by using the harmonic probing method of Section (4.6) in conjunction with Eqn. 8.55, and then substituting into Eqns. (6.119) and (6.155). The beam parameters were fixed to the values obtained via

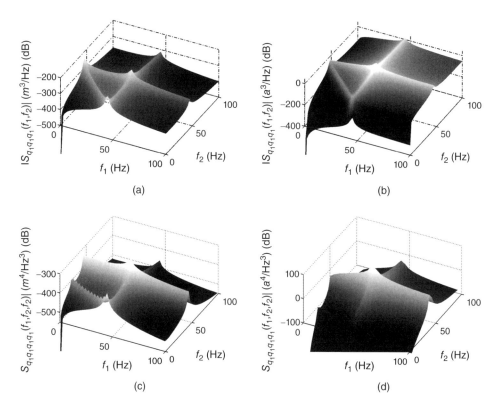

Figure 8.19 (a) Displacement and (b) acceleration bispectrum for delamination parameters $x_a = 0.05$ (m), $x_b = 0.11$ (m), and $a = h/6$ (m). (c) Displacement and (d) acceleration trispectrum for the same damage parameters. Shown is The-three dimensional "slice" obtained by plotting $|S_{q_1q_1q_1q_1}(f_1,f_2,f_2)|$
Source: Reproduced from [40], Figure 2, with permission of Elsevier

experiment in a prior work [12] giving $\rho = 1234.0$ (kg/m), $E = 75,889,600,000$ (GPa), $L = 0.23$ (m), $b = 0.06$ (m), $h = 0.0025$ (m). The delamination size and depth are governed by the parameters x_a, x_b and a. As an example, the value for delamination depth was fixed to $a = h/8$ (m) and the delamination was assumed to originate near the clamped end of the beam, $x_a = 0.05$ (m) and terminate at $x_b = 0.11$ (m). For these parameters, the natural frequency of the beam is $\omega_n = 278.57$ (rad/s) or $f_n = 44.34$ (Hz). The Gaussian excitation was applied at the beam tip with a PSD of $S_{FF}(\omega) = $ const. $= 0.04$ (N^2/Hz). Figure 8.19 shows sample bispectra and the trispectrum for both displacement and acceleration response data. The presence of the delamination results directly in the peaks and ridges observed in both the bispectrum and trispectrum. For the bispectrum, it is easy to see the dominant peak at $\omega_1 = \omega_2 = \omega_n$ and also the side peaks at $(\omega_n, 0), (0, \omega_n)$. In addition, the "ridges" along the sum frequency $\omega_1 + \omega_2 = \omega_n$ are also visible. Again, for no nonlinearity, the bispectrum and trispectrum are zero everywhere, thus the peaks and ridges provided a convenient way to detect the delamination presence. For the trispectrum, the full four-dimensional object cannot be plotted. Some researchers have opted to plot only the peaks in the form of "bubble plots" in three-dimensional frequency space [41]; however, here a three-dimensional "slice" is taken along the line $\omega_3 = \omega_2$. The dominant peak and the sum frequency ridge $\omega_1 + \omega_2 + \omega_3 = \omega_n$ can be clearly seen in the plot as well, particularly for the displacement response. For acceleration measurements, the linear kernel (from which all other kernels are constructed) is premultiplied by ω^2, thus both the bispectrum and trispectrum quickly drop to zero for $\omega_1, \omega_2, \omega_3 \to 0$. The detection problem is to therefore determine when the peaks reach a significant height, that is, rise above the noise floor. At a minimum, the noise floor is dictated by the uncertainty in the estimator of the polyspectra (see, e.g., Eqs. 6.132 and 6.162).

For the remainder of this section, the PSD of the jointly Gaussian input was fixed to the value $S_{FF}(\omega) = 0.04$ (N^2/Hz). The delamination size and depth are governed by the parameters x_a, x_b, and a. In what follows, the delamination was assumed to originate near the clamped end of the beam, $x_a = 0.05$ (m).

Figure 8.20a shows the theoretical PSD function along with the estimated PSD for no delamination and for a large delamination of $x_b - x_a = 3L/4$. For the estimate, time series of length $N = 16,384$ points were used with an assumed sampling interval of $\Delta_t = 0.001$ (s). Results for other delamination lengths and depths can be similarly obtained. In this particular case, a delamination running over three-quarters of the structure produces only a 3% change in frequency. Obviously, using higher modes (not available in this model) would provide better resolution, however Figure 8.20a captures analytically what is largely known to the experimentalist: natural frequencies cannot capture the influence of local damage unless that damage is very large. This is further supported by Figure 8.20b, which shows a plot of the dominant natural frequency as a function of the delamination length, $(x_b - x_a)$ (m) for two different delamination depths. For the shallow delamination ($a = h/12$), it is not until the delamination reaches nearly half of the beam length that even a 1 Hz change in the natural frequency is observed. The best probability of detecting the damage is obtained when the delamination is exactly in the center of the beam, where $a = h/2$. All other depths, for which $a < h/2$, produce smaller changes in natural frequency per unit change in delamination length and hence produce curves that lie above $a = h/2$ in Figure 8.20b. However, even at this depth, detection would be

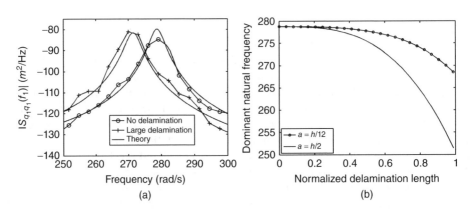

Figure 8.20 (a) Theoretical and estimated power spectral density function associated with delaminations of length $x_b - x_a = 0$ and $x_b - x_a = 3/4L$ respectively. The delamination depth was $a = h/12$ in both cases. (b) Progression of dominant natural frequency with normalized delamination length for two different delamination depths. The delamination start point was taken to be $x_a = 0$ *Source:* Reproduced from [40], Figure 3, with permission of Elsevier

difficult for all but very large delaminations. The other issue with this detection scheme is that it is sensitive to both damage-induced changes to the stiffness matrix and changes introduced via environmental fluctuations (e.g., temperature). The higher order statistics will also be influenced by covariates, but will *only* be present if there is damage. This provides the higher order detection schemes a degree of immunity to some of the confounding factors in second-order (spectral-based) damage detection schemes.

Next, the detection properties of the higher order detection schemes are explored. Using the beam displacement as the signal of interest, Figure 8.21 plots the analytical bispectral and trispectral peak heights as a function of the delamination length for various depths. Figure 8.21a shows this progression for a signal measured near the beam tip, while Figure 8.21b shows the relative displacement response at the center, top of the delaminated portion of the beam. Just as with the frequency-based detector, the bispectral peak heights grow exponentially with delamination length, however the interplay between delamination

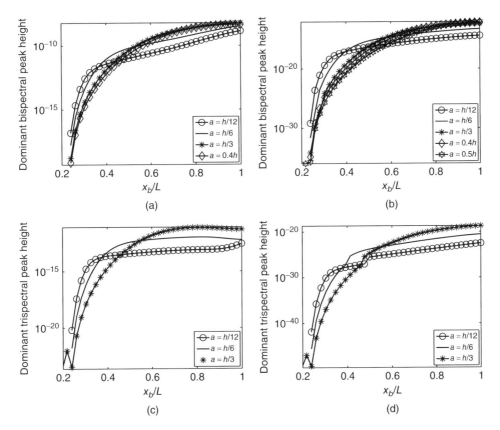

Figure 8.21 Progression of the main displacement bispectrum peak height as a function of both the normalized delamination endpoint x_b/L and depth assuming the response was measured at (a) the beam tip and (b) the center of the delamination. Progression of displacement trispectrum peak height as a function of the normalized delamination length and depth assuming the response was measured at (c) the beam tip and (d) at the center of the delamination. In all cases, the delamination originated at $x_a/L = 0.2$
Source: Reproduced from [40], Figure 4, with permission of Elsevier

depth and length is slightly more complicated. These results predict that for very long delaminations (x_b large), deep delaminations should be easier to detect. Conversely, for smaller delaminations (x_b small), delaminations closer to the beam surface should be easier to detect. The same holds true for the trispectrum, that is to say, a smaller delamination is easier to detect if it is shallow, whereas a more deeply located delamination is easier to detect if it is greater in size. These results can be explained physically by considering the motion of the top laminate (q_2) and recalling the nonlinearity results essentially from a localized buckling. For very small delamination length, only a very thin beam (delamination) can buckle. Thicker beams (deeper delaminations) are not able to buckle if they are short in length. However, provided that the delamination is long enough to cause buckling, the thicker beams (deeper delaminations) will have a greater influence on the response because of their larger inertia.

The real question from a detection standpoint is whether the observed peak heights are significant in the sense that they exceed (at a minimum) the values resulting from errors in the estimate. For a given frequency bin, if both the real and imaginary parts are Gaussian distributed, the magnitude will follow a Rician distribution. That is to say, the test statistic

$$s_{nn}^{(B)} \equiv |\hat{B}(\omega_n, \omega_n)| = \sqrt{Re\{\hat{B}(\omega_n, \omega_n)\}^2 + Im\{\hat{B}(\omega_n, \omega_n)\}^2}$$

will be distributed according to

$$p_S(s) = \frac{s}{\sigma^2} e^{\left(\frac{-(s^2+\lambda)}{2\sigma^2}\right)} I_0\left(\frac{s\lambda}{\sigma^2}\right) \tag{8.55}$$

where, assuming an unbiased estimator, the value of the noncentrality parameter is known via Eq. (6.119) as

$$\lambda^2 = |B(\omega_n, \omega_n)|^2$$

and I_0 is the modified Bessel function of the first kind with order zero. The variance parameter σ was already discussed and is given by Eq(s). (6.132). Therefore, given only the uncertainty present in the estimate, both Type I and Type II errors associated with a bispectral peak detector can be determined analytically. It turns out the results are entirely dependent on where along the beam the time series is measured. If the time series is recorded off the delamination site, the variance in the estimate completely obscures the bispectral peaks. Additive noise, such as that typically found in the experiment, would only strengthen this conclusion. However, if data are available from the delaminated region, the Type I and Type II errors are essentially zero. These results are summarized in the ROC curves shown in Figure 8.22. For these results, it was assumed that the time-series length was $N = 2^{17}$ points, sampled every $\Delta_t = 0.001$ s with estimation parameters $M = 512, L = 256$. The analysis also assumes that a Hanning window is applied in computing the FT of the data ($\kappa = 0.27$). The results shown are typical of those found for other values of a, x_b. For the beam response off the damage site, the strength of the nonlinearity simply cannot overcome the estimator variance. While more points could be considered, $N = 2^{17}$ already represents a large data set. In addition, for a more coarse frequency resolution $M < 512$, the estimator can no longer be assumed to be unbiased. For the beam response on the damage site, the detector performance is predicted to be excellent.

For trispectrum-based detection, the detection results for sensors located away from the delamination site are equally poor. The variance in the estimate (Eqn. 6.162) dominates so that for any reasonable frequency resolution, one cannot distinguish estimation error from damage. This problem will only be compounded by measurement noise. However, for trispectral

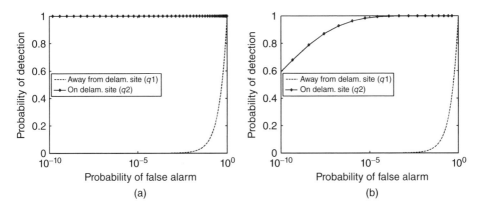

Figure 8.22 ROC curves showing the predicted detection and false-alarm probabilities associated with (a) bispectral and (b) trispectral peak detectors for $a = h/6$ and $(x_b - x_a)/L = 0.25$ (25% delamination). Data obtained away from the delamination site will have a very poor ability to detect the delamination presence. Data obtained from the delamination site can expect to have excellent detection performance *Source:* Reproduced from [40], Figure 5, with permission of Elsevier

estimates recorded from the delamination site, there will be very good detection performance. Again, however, these results are contingent on the damage causing a localized buckling. If the buckling does not occur, the damage will likely not be detected using global methods, regardless of damage size and location.

The results to this point have considered the beam's response to a jointly Gaussian input as the driving signal. However, as we have already pointed out, if the beam is instead driven by a non-Gaussian, i.i.d. random process, even an undamaged (linear) beam will yield the bispectral density (8.48), in which case there is ambiguity as to whether the observed peaks in the response bispectrum are due to a non-Gaussian input (nonzero $B_{XXX}(\omega_1, \omega_2)$) or a damage-induced nonlinearity.

We further showed that using the estimated bicoherence function (8.50) eliminated this ambiguity. Thus, for a non-Gaussian input to a linear system, we have $b_{YYY}(\omega_1, \omega_2) = \text{const.} = \beta \Delta_t^{1/2}/(\sigma_Y^3(2\pi)^{1/2})$, where σ_Y^2 is the variance of the system response and β the third moment about the mean. A nonlinear system, however, will cause deviations from this constant.

The estimator for the bicoherence was given by Eq. (6.131) and involves simply dividing estimates of the bispectrum by the product of estimates of the PSD function. The variance of the bicoherence estimator was given in Section 6.5.2 as $\sigma^2 = \frac{\kappa}{2}(\delta(\omega_1 - \omega_2) + 1)\frac{M^2 \Delta_t}{N}$. Similarly, the trispectrum can be normalized by the square root of the PSD product found in Eq. (2.50), giving the tricoherence function. The peak estimates will still follow a Rician distribution, as given by Eq. (8.55). The ROC curves predicted using the coherence functions, as opposed to the polyspectra, are qualitatively similar to those already shown in Figure 8.22, that is, the placement of the strain sensor is key to detectability.

To verify the predicted results, the following numerical experiment was conducted. The beam response was simulated for $N = 131{,}072$ points using a sampling interval of $\Delta_t = 0.001$ (s). The linear system parameters were fixed to the values used in the previous results, while the parameter governing the start of the delamination was fixed to the value $x_a = 0.05$ (m). The free parameters for this study were assumed to be the delamination length, governed by x_b and the delamination depth, governed by a. Displacement data were recorded both from locations near the beam tip and on the center-top of the delamination.

Figure 8.23 shows the peak height distribution for the estimated bicoherence collected from near the beam tip (a) and delamination site (b). Figure 8.23c shows the ROC curves associated with both the local and global bicoherence-based detectors. As predicted, for very small delaminations (5%), only the most shallow delamination is able to be detected. Successful detection depends entirely on the practitioner's ability to acquire data from the damage location. Longer delaminations are better detected if they are more deeply located. Again, the key factor in detecting delamination using a vibration-based approach is whether or not the delaminated portion buckles. If it does, *and* a measurement can be recorded from the delamination site, vibration-based detection is possible. However, nonlocal measurements are relying on the inertial influence of the delaminated portion, which is relatively small for most delamination lengths & depths.

As a final check, the model was modified to incorporate the possibility of the laminates coming into contact inside the domain of the delamination, that is, $x_a < x < x_b$. In the model just derived, no such physical constraint was imposed, creating the unrealistic possibility that the laminates may pass through one another. This assumption made the problem analytically tractable, and, indeed, a good deal of physical insight was gained from leaving out this constraint. However, in reality, these laminates may touch; so an impact model was developed and the simulations were repeated. Specifically, at each time step, the positions of the laminates were checked for an intersection in position. If contact occurred, a coefficient of restitution impact model was used and the simulation continued. We found essentially no quantifiable difference in the detection characteristics, regardless of whether or not impacts were included in the model.

In short, we conclude that vibration-based detection of delamination using polyspectral estimates is feasible, but only if the data can be collected from the damage site. There is simply not enough inertia in the delaminated portion to significantly influence measurements acquired from the nondelaminated portions. However, the polyspectra did an excellent job of capturing the damage-induced nonlinearity when the data were collected from the delaminated portion of the beam. These results held for a variety of damage parameters (delamination depth and length).

A more sensible approach might be to use ultrasonic-based approaches such as that described in Section 7.1.5. These approaches can be used to capture the physics of small-scale defects (see, e.g., [42]). The downside is that, in practice, it may be quite difficult to instrument and interrogate a structure by propagating ultrasonic waves. For this type of damage, the practitioner either accepts an experimentally simple, but challenging estimation problem, or an experimentally challenging yet simple estimation problem. The decision regarding which approach to use can be made quantitatively (see Chapter 10) provided that the practitioner can attach a cost to these two competing constraints.

8.6   Method of Surrogate Data

For some test statistics, e.g., information theoretics, we will likely not know the "undamaged" (linear) values a priori. Absent knowledge of the test statistic distribution under the null hypothesis, we would require training data to establish baseline values. This is a luxury we may also

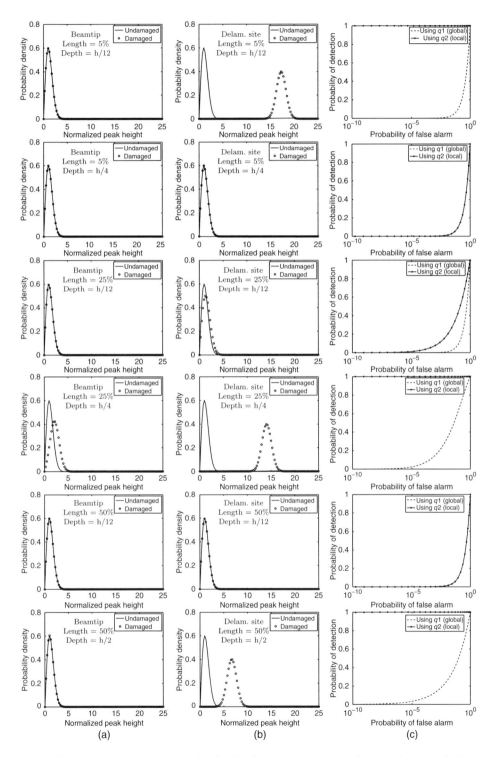

Figure 8.23 Bicoherence peak height distribution for measurement recorded near the beam tip (a) and at the delamination site (b). ROC curve comparing global and local detectors (c) *Source:* Reproduced from [40], Figure 6, with permission of Elsevier

not have in practice. In fact, in most cases we do not have access to a pristine structure with which to acquire such a data set.

Consider a different approach to testing (8.5) or more generally (8.8). What if we could *create* data that matched both the auto-covariance and marginal PDF of the observed data, but *forced* all higher joint cumulants to be zero? That is to say, manufacture a signal consistent with H_0. If we then estimate, for example, the bicoherence from this signal (see Section 8.4), we expect a constant function in the bifrequency plane, albeit with some variance depending on the noise levels. If the original data possessed a nonzero third cumulant, we would see statistically significant differences between its estimated bicoherence and that of the manufactured signal. We therefore might require that our manufactured data preserve the auto- and cross-correlation functions but destroy the higher order statistics associated with nonlinearity.

This may be accomplished by forming linear *surrogate* data that preserve both second-order statistical relationships among the response data (auto- and cross-correlations) and the marginal probability distribution of the data in accordance with 8.8. Appropriate comparisons between the original data and surrogates will then yield differences when damage (nonlinearity) is present. This approach is also insensitive to many forms of ambient variation. Regardless of how much is known *a priori* about the structure, the issue of ambient variability in temperature, humidity, or some other influence is a problem in structural health monitoring (SHM). Such variation adds a degree of uncertainty to the problem by causing the structural response to change even though no damage may have occurred. Again, by formulating the problem as one of nonlinearity detection (undamaged=linear, damage=nonlinear), any ambient variation that does not affect the form of the underlying model will not register as a damage-induced change. This is demonstrated experimentally in Section 8.8.2. The overall surrogate-based approach is shown schematically in Figure 8.24.

Surrogate data methods are by no means new, having been used for at least 15 years in the field of nonlinear time-series analysis. Early approaches such as the Fourier transform surrogate (FTS) method of Theiler *et al.* [43] were designed to test whether the data arose from a linear, jointly Gaussian process (i.e., test 8.5). In that same work, Theiler *et al.* extended the null to include non-Gaussian data, resulting in a procedure referred to as the amplitude adjusted Fourier transform (AAFT) method (testing 8.8). This more general null was further improved upon by Schreiber and Schmitz [44], resulting in the iterative amplitude adjusted Fourier transform IAAFT) method. Subsequent works by Kugiumtzis [45] have also addressed one of the key assumptions underlying the AAFT approach (resulting in the corrected amplitude adjusted Fourier transform or CAAFT algorithm). Finally, the digitally filtered surrogates (DFS) of Dolan and Spano [46], and simulated annealing [47] are yet additional ways of generating surrogates. A good overview of surrogate data methods is given in [48].

In what follows, we compare both IAAFT and DFS approaches. The general surrogate algorithm was described briefly by the authors in [19], but is given more thorough treatment here as there exist several caveats to beware of in generating appropriate surrogates for damage detection applications.

8.6.1 Fourier Transform-Based Surrogates

Recall that the goal of the surrogate is to preserve certain properties/relationships among the data (the null hypothesis H_0) but destroy others (the alternative H_1). One can then compare values of an appropriately chosen metric (more on this later) computed from the data to values

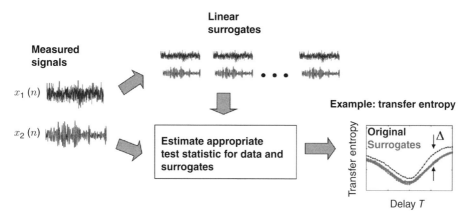

Figure 8.24   Schematic of the surrogate data approach to damage detection. If one equates damage with nonlinearity, a suitable null hypothesis is that the observed data were generated by a linear time-invariant system. Using the methods of this section, one can generate data that essentially conform to this hypothesis; these are the surrogate sets. One then estimates a test statistic on both the data and surrogates and compares the results. If the test statistic from the data lies outside the confidence interval obtained from the surrogate test statistics, it is concluded that the data are produced by a nonlinear system, that is, a damaged structure

obtained from the surrogates. Should the null be unlikely, there will be statistically significant differences in these values. In the case of detecting damage-induced nonlinearities, we want the null to correspond to a linear structural response. We have shown that if the random processes $X_i(t)$, $i = 1 \ldots K$ are stationary, zero mean, jointly Gaussian distributed, all correlations in the response data are dictated solely by $R_{X_iX_j}(\tau) = E[X_i(t)X_j(t + \tau)]$ because higher order correlations are not expected to exist. We therefore might require that our surrogate data preserve the auto- and cross-correlation functions but destroy the higher order statistics associated with nonlinearity.

A surrogate algorithm that preserves second-order relationships while destroying higher order correlations was described by Prichard and Theiler [49]. The approach makes use of the Weiner–Khinchin theorem developed earlier in Chapter 3

$$S_{X_iX_j}(f) = \int_{-\infty}^{\infty} R_{X_iX_j}(\tau)e^{-j2\pi f\tau}d\tau \qquad (8.56)$$

where $S_{X_iX_j}(f)$ is the two-sided cross-spectral density function. Preserving $S_{X_iX_j}(f)$ will therefore also preserve $R_{X_iX_j}(\tau)$. Estimates for $S_{X_iX_j}(f)$ can be obtained via the discrete FT over the data length N as we showed in Section 6.4.1. Let $x_i(n), n = 0 \cdots N - 1$ be the observed data at discrete time n with sampling interval Δt (i.e., $t = n\Delta t$). Denote the discrete FT of the data at frequencies $f_k = k/N\Delta t$ as $X_i(f_k) = \sum_{n=0}^{N-1} x_i(n)e^{-j2\pi kn/N}, k = 0 \cdots N - 1$. In this case, the sample two-sided auto- and cross-spectral densities may be estimated with no averaging ($S = 1$ in Eq. 6.69) as

$$S_{X_j X_i}(f_k) = \frac{\Delta t}{N} X_j^*(f_k) X_i(f_k) = \frac{\Delta t}{N} |X_j(f_k)||X_i(f_k)| e^{j(\phi_j(f_k) - \phi_i(f_k))}. \qquad (8.57)$$

Both complex quantities have been written in terms of magnitude and phases, $\phi_i(f_k), \phi_j(f_k)$. We can see from Eq. (8.57) that adding the *same* random phase to both $\phi_j(f_k)$ and $\phi_i(f_k)$ will not change $S_{X_j X_i}(f_k)$ (i.e., the phase difference will be preserved). The autospectral density estimate is simply given by $S_{X_i X_i} = \frac{\Delta t}{N} |X_i(f_k)|^2$ so that adding a random phase has no influence on this estimate either. Phase-randomized surrogates are therefore obtained by

$$\hat{x}_i(n) = \frac{1}{N} \sum_{k=0}^{N-1} X_i(f_k) e^{j\psi(f_k)} e^{j2\pi kn/N} \qquad (8.58)$$

where $\psi(f_k) = [0, 2\pi)$ is a random phase added to frequency component f_k for each of the $i = 1 \cdots K$ time series comprising the structure's response. Because the discrete FT is unique only for the first $N/2$ components, the phases should obey $\psi(f_{N-k}) = -\psi(f_k), k = 1 \cdots N/2$ ($\psi(f_0) = 0$). Note this is essentially the same technique we used in generating random processes in Section 6.7.3, specifically see Eq. (6.192).

The surrogate time series $\hat{x}_i(n)$ will possess the same PSD (and hence covariance structure) as the original data, however no other quantities (e.g., higher cumulants) will be preserved. In fact, should any other higher order properties have existed, they will be destroyed by the phase-randomization process. Thus, the $\hat{x}_i(n)$ provides a good estimate of what the structural response would have looked like *if it had been linear*. Comparing $x_i(n)$ to $\hat{x}_i(n)$ can be made using any of the nonlinearity detection statistics described in Section 8.1.2. Note that this approach is only valid for testing the null hypothesis 8.5. Surrogates generated with this approach will, by the central limit theorem, be normally distributed. Surrogates for testing the more general 8.8 are described next.

8.6.2 AAFT Surrogates

For a linear, Gaussian-excited structure, the above-described algorithm is sufficient for both univariate and multivariate data. However, we have already mentioned the more general case where the null too corresponds to data recorded from a linear structure subject to excitation, by a random process with an arbitrary marginal probability density. It is more common to describe this null in slightly different terms. The new null assumes a Gaussian-excited, linear system that is observed through an invertible (monotonic), instantaneous (does not depend on n) measurement function $x(n) = h(g(n))$, where $g(n)$ is the hypothesized underlying Gaussian time history and $x(n)$ the true (observed) time history with arbitrary amplitude (probability) distribution. The idea behind the AAFT method is to simulate h^{-1}, create an FT surrogate base on the resulting Gaussian series, and then simulate $h(\cdot)$ again to reclaim the correct distribution. If $h(\cdot)$ does not introduce any higher order correlations (hence the restriction on $h(\cdot)$ not depending on time), then time series that obey this null will share the same probability distribution as well as the auto- and cross-spectra. Surrogates produced using this approach are randomly shuffled versions of the original time series, and hence the amplitude distribution is automatically preserved. However, the shuffling is performed in such a way as to preserve the power spectrum of the response, that is, a *constrained* randomization. A description of this approach may be found in [43]. This algorithm can suffer from a bias resulting from the difficulty in approximating $h^{-1}(\cdot)$ and $h(\cdot)$. Specifically, the approach assumes that $h(\cdot)$ is monotonic, which may not be the case in practice. In response to these difficulties, a different approach was developed, the IAAFT algorithm.

8.6.3 IAFFT Surrogates

This algorithm was first described in [44] and later presented (with modifications for multi-variate data) in [48]. As with the AAFT surrogates, the goal is to find a temporal reordering of the data that preserves the auto- and cross-spectral densities. Again, because the surrogate is simply a shuffled version of the original time series, the amplitude distribution is automatically preserved. The spectral amplitudes and phases of the multivariate response data are given by their discrete FTs $FT(x_i(n)) = |X_i(f_k)|e^{j\psi_i(f_k)}$. According to the null hypothesis, the reordering needs to preserve both the magnitudes $|X_i(f_k)|$ and the phase relationships $\psi_i(f_k) - \psi_j(f_k)$ between each pair of time series i, j.

The algorithm begins by generating either an AAFT surrogate or simply a random shuffling of the original data, denoted $g_i(n)$. These data are Fourier transformed giving $G_i(f_k) = |G_i(f_k)|e^{j\hat{\psi}_i(f_k)}$. The next step is to replace the Fourier amplitudes with the desired ones, $|X_i(f_k)|$, and adjust the phases. The phases are adjusted such that the difference between the new phases, denoted $\phi_i(f_k)$, and the ones being replaced, $\hat{\psi}_i(f_k)$, are minimized over all time series $i = 1 \ldots K$ (i.e., make the adjustment minimal). At the same time, the procedure should maintain the "target" phase differences $\psi_i(f_k) - \psi_j(f_k)$ between the original time series. Schreiber and Schmitz [48] shows how this is accomplished by solving a minimization problem resulting in the adjusted phases $\phi_i(f_k) = \psi_i(f_k) + \alpha(f_k)$ where

$$\tan(\alpha(f_k)) = \frac{\sum_{m=1}^{K} \sin(\hat{\psi}_m(f_k) - \psi_m(f_k))}{\sum_{m=1}^{K} \cos(\hat{\psi}_m(f_k) - \psi_m(f_k))} \tag{8.59}$$

Each $\alpha(f_k)$ is solved for and adjusted to be in the correct quadrant by the following procedure. For each f_k, the function $\sum_{i=1}^{K} \cos(\alpha(f_k) - (\hat{\psi}_i(f_k) - \psi_i(f_k))$ is monitored. If the sum is less than zero, $\alpha(f_k) = \alpha(f_k) + \pi$; otherwise, it remains unchanged. Note that for univariate data, no phase adjustment is necessary and $\phi(f_k) = \hat{\psi}(f_k)$. Now that we have the amplitude and phase adjustment, the surrogate is given by

$$\hat{g}_i(n) = \frac{1}{N} \sum_{k=0}^{N-1} e^{j\phi_i(f_k)} |X_i(f_k)| e^{j2\pi kn/N} \tag{8.60}$$

This procedure results in the correct spectral properties, however the probability distribution will be altered. The data are therefore rank reordered according to the sorted values of the $\hat{g}_i(n)$. In other words, the smallest data value in $g_i(m)$ is shuffled to the same position as the smallest value in $\hat{g}_i(n)$. This step exactly preserves the probability distribution (we are shuffling the original data values). However, the reordering will corrupt the desired spectral properties. Correcting for the discrepancy involves repeating the above-mentioned procedure: computing $G_i(f_k)$; correcting the phase, inverse FT to give $\hat{g}_i(n)$; and then rank reordering. The error between desired and actual cross-spectra decreases with each iteration until convergence is achieved. The surrogate may then be given by either $\hat{g}_i(n)$ or the rank reordered series $g_i(n)$ depending on whether it is more important to match spectral properties or probability distribution. The convergence for univariate data is monotonic and the algorithm will eventually reach a point where the data are no longer being reordered. For multivariate data, the convergence properties are not monotonic, nor do they converge. Excellent agreement between desired and obtained correlations may, however, be achieved. Unlike the AAFT algorithm, in the aforementioned scheme, the hypothesized measurement function h need not be monotonic and can therefore accommodate a slightly more general null hypothesis.

One final point regarding implementation of the above-mentioned scheme concerns numerical artifacts arising from data that do not exactly repeat over the length of the record. Computing the spectral properties via FT implicitly assumes that the entire record repeats over the finite signal interval $[-T, T]$ as we discussed at length in Section 3. Differences in the first and last time-series values and the local slopes of the waveform at these points can result in well-known errors that are usually corrected for by windowing as we discussed in Section 6.4.1. In this case, however, windowing will destroy the invertibility of the transformation and cannot be used.

For this reason, the practitioner may wish to search for subsets of the time series for which the beginning and end values and slopes are similar, thus minimizing these artifacts. To account for this, we follow the procedure described by Schreiber and Schmitz [48]. Quite simply, this involves passing a window of size $M < N$ through the data and monitoring endpoint and slope mismatch. The M-point subset which minimizes these two metrics is chosen as the segment of time series used in the analysis.

8.6.4 DFT Surrogates

While the above-described approaches are simple and practically effective, there remains a problem that the astute reader may have noticed. The null hypothesis satisfied by these approaches does not quite correspond to what is needed for statistical hypothesis testing. The observed data should constitute a *statistically* independent realization of a linear stochastic process. The IAAFT algorithm attempts to match the *same* power spectrum for each surrogate, that is, a *single* realization. Thus, if the algorithm worked according to design, it would produce the exact same power spectrum for *each* surrogate, which is clearly incorrect. For example, if we were to repeatedly measure the response of a linear stochastic process and estimate the PSD of each realization, there would be some amount of variance that we know scales as the true PSD squared, that is, $\text{var}(\hat{S}_{XX}(f)) \propto S_{XX}^2(f)$ (see Eq. 6.67). The end result is that the IAAFT approach will sometimes produce surrogates with a smaller variance than desired. This fact has been pointed out by Kugiumtzis [50] as well as by Dolan and Spano [46].

The DFS algorithm of Dolan and Spano [46] (see also Dolan and Neiman [51]) attempts to correct this shortcoming by creating each surrogate from statistically independent permutations of the original data. First, the power spectrum of the data is estimated and a response function computed by taking the square root of the estimate and applying the inverse FT. A random permutation of the data (spectrally white) is then convolved (filtered) with the response function, resulting in matched spectral properties. Finally, the original data can be rank reordered according to the filtered series to match the probability distribution (this step is identical to that described in [48]). This algorithm has been shown to produce a spectral variance in the surrogate population that is consistent with independent realizations [46].

We now have a means of creating constrained randomizations of our response data. Our null is that the data were generated by a *linear, stochastic process observed through an instantaneous, possibly nonlinear, measurement function*. This null hypothesis is consistent with that given by 8.8 and may be used in a number of situations where nonlinearity detection is required. We can use any of the test statistics described thus far (see, e.g., Section 8.1) in conjunction with the surrogates to apply this approach to damage detection.

8.7 Numerical Surrogate Examples

Before testing the surrogate data approach on experimental systems, we first explore some of the basic properties and performance metrics using numerically generated data. In this section, we consider two such examples: detection of a bilinear stiffness and detection of a cubic stiffness. The former, as was discussed in Section 4.4.2, is sometimes used as the simplest model of a breathing crack and therefore represents a good initial test case for the method. The second form of nonlinearity captures the physics of post-buckled structures (see [52]) and provides a second relevant example. At the end of the section, we conclude with an example that highlights one of the main reasons we pursue the surrogate methods in the first place: invariance to certain forms of ambient variability that can confuse methods that rely on "change detection" in structural response data.

8.7.1 Detection of Bilinear Stiffness

As a simple example, assume that the healthy (undamaged) system of interest is the two- DOF, linear spring-mass-damper system depicted in Figure 8.25. The equations of motion for this system are

$$\mathbf{M} \left\{ \begin{matrix} \ddot{x} \\ \ddot{y} \end{matrix} \right\} + \mathbf{C} \left\{ \begin{matrix} \dot{x} \\ \dot{y} \end{matrix} \right\} + \mathbf{K} \left\{ \begin{matrix} x \\ y \end{matrix} \right\} = \mathbf{f}(t) \tag{8.61}$$

where the system parameter matrices are simply

$$\mathbf{M} = \begin{bmatrix} m_1 & 0 \\ 0 & m_2 \end{bmatrix}, \mathbf{C} = \begin{bmatrix} c_1 + c_2 & -c_2 \\ -c_2 & c_2 \end{bmatrix}, \mathbf{K} = \begin{bmatrix} k_1 + k_2 & -k_2 \\ -k_2 & k_2 \end{bmatrix}. \tag{8.62}$$

The driving signal is taken as a random process $\mathbf{F}(t) = \{0, f(t)\}^T$, where each value in the random process $\mathbf{F}$ is an independent, normally distributed Gaussian random variable, that is, each $f(t) \sim \mathcal{N}(0, 1)$. Let the two response random processes of interest be the positions of the first and second mass, respectively, denoted X and Y. Following the analytical approach outlined in Chapter 8.3.2, we work in modal coordinates by allowing $x = u_{11}\eta_1 + u_{21}\eta_2$ and $y = u_{12}\eta_1 + u_{22}\eta_2$ where $\mathbf{u}_1 \equiv \left\{ \begin{matrix} u_{11} \\ u_{21} \end{matrix} \right\}$ and $\mathbf{u}_2 \equiv \left\{ \begin{matrix} u_{12} \\ u_{22} \end{matrix} \right\}$ are the mass-normalized mode shapes. The resulting stationary correlation function was already shown to be given by

$$R_{XY}(\tau) = \frac{S_{FF}(0)}{4} \sum_{l=1}^{M} \sum_{m=1}^{M} u_{lP} u_{mP} u_{il} u_{jm} [A_{lm} e^{-\zeta_m \omega_m \tau} \cos(\omega_{dm}\tau) + B_{lm} e^{-\zeta_m \omega_m \tau} \sin(\omega_{dm}\tau)] \tag{8.63}$$

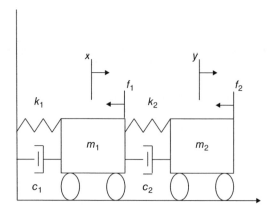

Figure 8.25 Schematic of linear spring-mass-damper system

where A_{lm}, B_{lm}, are already defined (see Eq. 6.100) and $S_{FF}(0)$ is the constant autospectral density associated with the driving signal. As before, $\omega_{di} \equiv \sqrt{1 - \zeta_i^2}\omega_i, i = 1, 2$ are the damped natural frequencies for the system. Proper normalization yields the cross-correlation coefficient $\rho_{XY}(T) = R_{XY}(T)/\sqrt{R_{XX}(0)R_{YY}(0)}$, which can be substituted into Eqs. (3.127) or (3.133) to give a closed form solution to either the time-delayed mutual information or TE between the stationary, jointly Gaussian random processes X, Y. This approach was already presented in Section 6.6 where we derived the linearized estimators (6.164).

As an example, consider the case where $k_1 = k_2 = 1.0 \, \text{N/m}, c_1 = c_2 = 0.01 \, \text{N·s/m}$, and $m_1 = m_2 = 0.01 \, \text{kg}$. In this case, the natural frequencies for the system are $\omega_1 = 6.18, \omega_2 = 16.18$ rad/s. Assuming a proportional damping model, the dimensionless damping ratios become $\zeta_1 = 0.0275, \zeta_2 = 0.0728$. Using these parameters, the system described by Eq(s). (8.61) was simulated using a fifth-order Runge–Kutta scheme with a time step of $\Delta t = 0.02s$. giving the time series for the two displacements $x(n), y(m), n, m = 1 \cdots N$, where N was chosen to be 50,000 points. Before analysis, each time series was normalized to zero mean and unit variance.

In an effort to test the surrogate approach with nonlinear coupling, the linear stiffness k_2 is replaced by a bilinear stiffness term so that the restoring force assumes the values

$$k_2(y - x) = \begin{cases} k_2(y - x) & : \quad (y - x) \geq 0 \\ \delta k_2(y - x) & : \quad (y - x) < 0 \end{cases}$$

where the parameter $0 < \delta \leq 1$ controls the degree to which the stiffness is decreased for negative displacements. The discontinuity was implemented numerically by utilizing the Henon integration scheme described in Section 4.4.1. Using this approach, the change in stiffness could be implemented at exactly $y - x = 0$ (to the error in the Runge–Kutta algorithm).

Before even moving to the surrogate-based approach, it is instructive to compare the full estimate of the mutual information (Eq. 6.169) to the linearized estimate (6.164). Figure 8.26 provides such a comparison. This full estimate was formed using a kernel bandwidth of $\epsilon = 0.1\sigma_X^2$ (the variance of the random process X) over a variety of values of δ. Linearized mutual information was computed by using Eq. (3.127) where $\hat{\rho}_{XY}(T)$ was obtained using the FT-based estimator alluded to in Section 6.4.2.

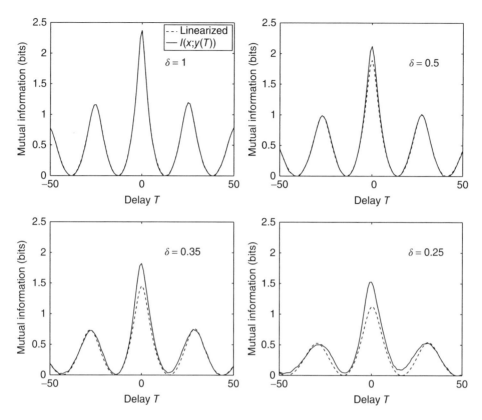

Figure 8.26 "Linearized" and nonlinear time-delayed mutual information for $\delta = 1.0$, $\delta = 0.5$, $\delta = 0.35$, and $\delta = 0.25$ *Source:* Reproduced from [53], Figure 4, with permission of Elsevier

For $\delta = 1.0$ (linear system), the curves show very close agreement. All of the information in this case is captured by $\rho_{XY}(T)$, that is, second-order correlations. As δ is reduced, the cross-correlation-based analysis shows the expected decrease in frequency as the peaks become wider. However, although the peaks scale differently (get smaller) as the degree of nonlinearity increases, they tend to retain the same shape as for the purely linear case.

By contrast, the various peaks associated with $\hat{I}_{XY}(T)$ tend to become wider and more skewed in their distribution in the presence of nonlinear coupling. For $\delta = 0.35$, the distortions can be clearly seen in the mutual information curves obtained using the kernel-based method. The magnitude of the distortion appears to be directly related to the degree of nonlinearity. For $\delta = 0.25$, the linearized results show large differences from those obtained using the kernel density approach. By utilizing the entire probability distribution function as part of a more general definition of coupling, higher order correlations introduced by the nonlinearity are easily quantified. Thus, the average mutual information function is a good candidate for capturing damage-induced nonlinearities.

The problem, of course, is that there are fundamental differences in the two estimators (linear vs. full) that could lead to erroneous declarations of damage where none exists. That is to say, even if there is no damage, a difference in estimator bias could easily lead one to falsely

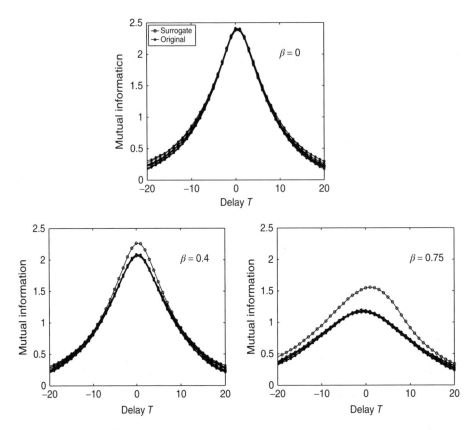

Figure 8.27 Comparison between the estimated mutual information function for both original signal and 10 different surrogate data sets *Source:* Reproduced from [19], Figure 1, with permission of IOP

conclude that the structure is damaged. For the mutual information, this is not much of an issue as the estimator is asymptotically unbiased, hence the two "undamaged" curves in Figure 8.26 line up almost exactly.

Nonetheless, it makes sense to always consider the full estimator, but compare the estimated values for both original signals and a family of surrogates. As we have already described, the practitioner takes the original time series $x(n), y(m), m, n = 1 \cdots N$ and constructs time reordered versions of the series $x(n'), y(m'), n', m' = 1 \cdots N$ using the algorithms described in Section (8.6). The temporal shuffling is done in such a way as to preserve the linear auto- and cross-correlation functions of the original signals. Thus, the kernel-based estimator of mutual information and/or TE may be computed for the original data and for some number of surrogates. Differences are then attributed to nonlinearity. Because the same estimator is used for both linearized and (possibly) nonlinear data, only relative differences in the results are important. A comparison of the mutual information estimated for a collection of surrogates and for the original signal is given in Figure 8.27 for three different values of the damage parameter δ.

In addition, the effects of the bilinear stiffness nonlinearity on the TE are also examined. These results are shown in Figure 8.28. The estimator in this case was taken as (6.171) using

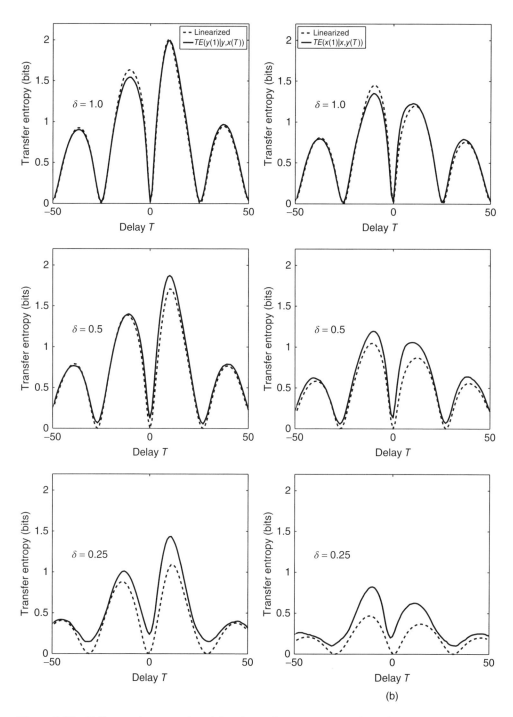

Figure 8.28 Difference between time-delayed transfer entropy as estimated using the fixed-mass approach ($M = 10$) and the linearized transfer entropy for increasing levels of nonlinearity. The left column shows $TE_{X \to Y}(T)$, while the right column displays $TE_{Y \to X}(T)$ *Source:* Reproduced from [53], Figure 8, with permission of Elsevier

a fixed-mass kernel (Eqn. 6.168) with spherical volume elements and $M = 10$ near neighbors. As with the mutual information, the TE peaks begin to take on a distorted shape as the degree of nonlinearity is increased. A common characteristic of both information theoretics appears to be wider, more skewed, and, in most cases, larger peaks. Furthermore, as the nonlinearity increases, the TE shows increased values relative to that predicted by the linearized TE. For example, the maximum peak of the linearized TE decreases from a value of 2.0 bits in the linear case ($\delta = 1$) and takes a value of 1.1 bits in the $\delta = 0.25$ case. By contrast, the TE obtained through the kernel-based procedure of Section 6.6 shows only a slight decrease from 2.0 bits to around 1.4 bits. The full estimator is capturing information associated with higher order correlations, whereas by definition, the linearized measure cannot. As with mutual information, the TE can be used to detect the presence of nonlinearity in time-series data. In fact, these results suggest that TE may be even more sensitive to nonlinearity as the difference from the linearized version is more pronounced than in the case of mutual information.

However, the difficulty in estimating the TE leads to small differences between the two algorithms (linearized and kernel-based) even in the case of a purely linear structure (top two plots). This is the problem alluded to earlier, whereby differences in estimator bias could easily be mistaken for nonlinearity. Again, we can make use of surrogate data to develop a nonlinearity index that responds only to the nonlinearity and not simply to differences in the estimator. Figure 8.29 shows $TE_{Y \rightarrow X}(T)$ computed between the time series and for 40 linearized surrogate data sets. By counting the number of outliers, it can be seen that the nonlinearity begins to cause significant deviations in the results for $\delta = 0.3$. This general approach to damage (nonlinearity) detection is shown extremely effectively in the experimental examples given later in this chapter.

The goal of this simple numerical study was to demonstrate the use of information theoretics as a means of quantifying damage-induced nonlinearity in structures. Furthermore, the method of surrogate data provides a convenient way to formally test this hypothesis in a structural system. If we are willing to equate nonlinearity with damage, we have a reference-free approach to detecting damage presence and extent. However, we obviously have many different ways of performing a surrogate-based hypothesis test. We have freedom in both how we generate the surrogates and which test statistic to use for comparison. This simple study has shown how information-theoretic quantities, in conjunction with IAAFT surrogates, can be used to this end. The subsequent two case studies illustrate how this basic approach can be used to detect different types of nonlinearity.

8.7.2   Detecting Cubic Stiffness

As we have just shown, the surrogate data approach provides one possible means of constructing a test for damage-induced nonlinearity by means of looking at the system response data. One of our motivations in choosing the particular test statistic (information theoretics) was that they should be sensitive to *any* higher order correlations and not just the bilinear stiffness model studied. We demonstrate this property of the information theoretics next by considering a five-DOF system with a cubic nonlinearity.

As we already learned in Section 5.2, cubic nonliniearities arise naturally in buckled structures (see also [52]). In our previous experimental example, we saw that the delaminated portion of a composite structure can buckle when a load is applied. We therefore turn our attention to detecting the presence of this form of damage using vibrational response data.

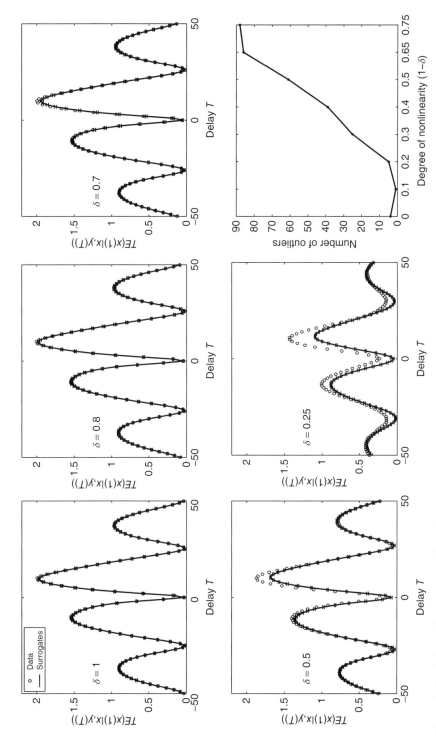

Figure 8.29 $TE_{Y \to X}(T)$ as computed from the original time series (circles) and from 40 linearized surrogate data sets (solid). Confidence limits computed from the surrogates represent an interval of 95%. Final plot shows a nonlinearity index obtained by counting the number of outliers *Source:* Reproduced from [53], Figure 9, with permission of Elsevier

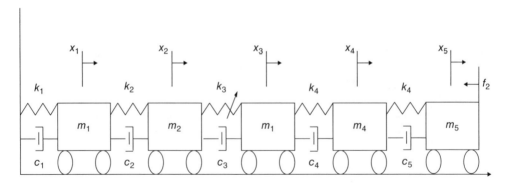

Figure 8.30 Schematic of linear spring-mass-damper system *Source:* Reproduced from [8], Figure 1, with permission of Elsevier

Assume that the undamaged system of interest is the five-DOF spring-mass-damper system depicted in Figure 8.30. The equations of motion for this system are (again)

$$\mathbf{M}\ddot{\mathbf{x}}(t) + \mathbf{C}\dot{\mathbf{x}}(t) + \mathbf{K}\mathbf{x}(t) = \mathbf{f}(t) \tag{8.64}$$

where $\mathbf{M}$, $\mathbf{C}$, and $\mathbf{K}$ are the mass, damping, and stiffness matrices. Let the dynamical variables of interest be the positions of the masses $x_i(t), i = 1 \cdots M$.

In an effort to test the algorithms, we consider a linear five-DOF structure ($M = 5$) with parameters $m_i = 0.01, c_i = 0.05, k_i = 10.0, i = 1 \ldots 5$. Further assume that the excitation is Gaussian (unit standard deviation) and is applied only at the end mass so that $\mathbf{f} = \{0, 0, 0, 0, \mathcal{N}(0, 1)\}^T$. The natural frequencies and damping ratios for this system are summarized in Table 8.1. Assuming a proportional damping model ($\mathbf{C} = \beta\mathbf{K}$), we have as an approximation $\zeta_i = \frac{1}{2}c_i\omega_i$. Using these parameters, the system described by Eq(s). (8.64) was simulated using a fifth-order Runge–Kutta scheme with a time step of $\Delta t = 0.01$ s giving time series for the displacements $x_i(n), n = 1 \cdots N$, where N was chosen to be 50,000 points. Before analysis, each time series was normalized to zero mean and unit variance.

For this analysis, we focus on the coupling between masses $i = 2$ and $j = 3$, that is, use $x_2(n), x_3(n)$ as the time series of interest. In an effort to demonstrate the utility of the approach in diagnosing nonlinearity, a cubic spring replaces k_3 such that the linear stiffness matrix is

Table 8.1 Modal parameters for five-DOF structure

Mode	ζ_i	ω_i (rad/s)	$\omega_i^{(A)}$ (rad/s)
1	0.02	9.0	8.8
2	0.07	26.3	25.4
3	0.10	41.4	40.0
4	0.13	53.2	51.4
5	0.15	60.7	58.7

now given by

$$
\mathbf{K} = \begin{bmatrix}
(k_1 + k_2) & -k_2 & 0 & 0 & 0 \\
-k_2 & (k_2 - k_3) & k_3 & 0 & 0 \\
0 & k_3 & (-k_3 + k_4) & -k_4 & 0 \\
0 & 0 & -k_4 & (k_4 + k_5) & -k_5 \\
0 & 0 & 0 & -k_5 & k_5
\end{bmatrix}
$$

$(-k_3$ replaces $k_3)$ and a nonlinear restoring force is added to the right side of Eq. (8.64)

$$
\mathbf{f}^{(N)} = \begin{bmatrix}
0 \\
-\mu k_3 (x_3 - x_2)^3 \\
\mu k_3 (x_3 - x_2)^3 \\
0 \\
0
\end{bmatrix}.
$$

The equilibrium point $x_3 - x_2 = 0$ is replaced by the two stable points $x_3 - x_2 = \pm\sqrt{1/\mu}$. As μ is increased, the asymmetry in restoring force associated with the nonlinearity also increases. For a large enough value $\mu = \mu^*$, this system will begin to oscillate between the two equilibria.

Figure 8.31 shows the results of computing the average mutual information function for increasing levels of nonlinearity. The curve with open circles represents the algorithm applied to the original data, while the solid lines show the results of applying the algorithm to each of 10 linear surrogates, generated using the approach described in Section 8.6. As μ increases, the curves begin to separate, particularly near the dominant peak. The average mutual information shows an increase in the amount of coupling relative to the linear surrogates. Our interpretation is that the higher order coupling is still present in the original data but absent in the surrogates. To quantify this difference, we form the confidence intervals at each delay

$$
CL(T) = \mu(T) - Z_{\alpha/2}\sigma(T)
$$

$$
CU(T) = \mu(T) + Z_{\alpha/2}\sigma(T)
$$

where $\mu(T), \sigma(T)$ are the mean and standard deviation of the surrogates at delay T. The values for $Z_{\alpha/2}$ are chosen by the practitioner and indicate the desired level of confidence associated with the null hypothesis that the dynamics of the structure are linear (see section 6.9.3 on the formation of confidence intervals). Consequently, values for the time-delayed mutual information that fall outside these bounds indicate nonlinearity (violation of the null). Here, we take $Z_{0.025} = 1.96$ giving confidence intervals of 95%. Because the information theoretics show increased values over their linearized counterparts, we need only focus on the upper bound. A convenient nonlinearity index for the mutual information can therefore be defined on the basis of the distance from this bound as

$$
Z_M = \sum_T \begin{cases}
0 & : \quad I_{X_i X_j}(T) \leq CU(T) \\
(I_{X_i X_j}(T) - CU(T))/CU(T) & : \quad I_{X_i X_j}(T) > CU(T)
\end{cases}
$$

where we are summing over all values of the mutual information that exceed the confidence interval. An alternative is to take the maximum distance (rather than the sum). We find few differences in the results regardless of which index is used. A plot of Z_M as a function of nonlinearity is shown in Figure 8.32. The resulting index is monotonic with μ and quantifies the separation between the surrogates and $I_{X_2 X_3}(T)$. For $\mu = 0$, we have the linear system and,

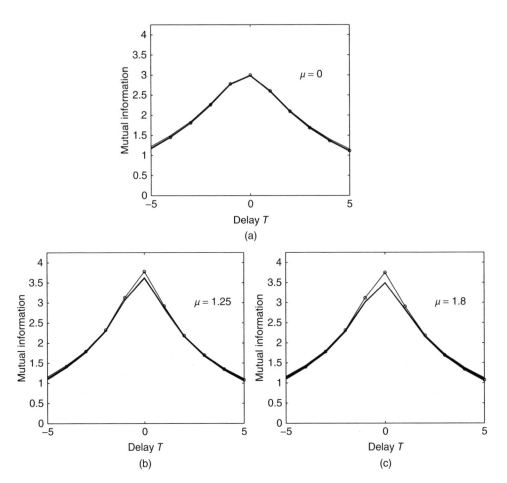

Figure 8.31 Dominant peak of $I_{X_2X_3}(T)$ showing increasing discrepancy between surrogates (solid line) and data (open circles) for $\mu = 0$ (linear system), $\mu = 1.5$, and $\mu = 1.75$ *Source:* Reproduced from [8], Figure 3, with permission of Elsevier

as expected, no difference is observed between the data and the surrogates. As the degree of nonlinearity is increased, so too does the amount of separation as quantified by Z_M. For values of $\mu > \mu^* = 1.75$, the dynamics begin to oscillate between the two equilibria and the nonlinearity is too trivial to detect. Values for Z_M for this case are an order of magnitude larger than those shown in Figure 8.32 and are therefore not presented. The subtle nonlinearity introduced for $\mu < \mu^*$ is much more difficult to detect and hence is the focus of this study.

The time-delayed transfer entropy was also used in analyzing the relationship between masses 2 and 3. As an example, consider the case where we examine the coupling between $x_3(t)$ to $x_2(t)$, that is, compute $TE_{3 \to 2}$. As with mutual information, results can be compared to those obtained from surrogate data as nonlinearity is introduced into the system. Figure 8.33 illustrates these results for several different values of the nonlinearity parameter μ. The TE appears more sensitive to increasing nonlinearity then does the mutual information.

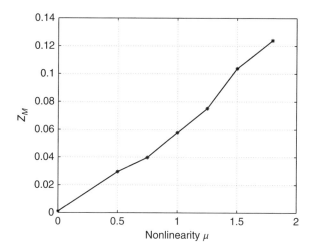

Figure 8.32   Index Z_M as a function of % nonlinearity (damage) in the system

Furthermore, the differences between surrogates and data are noticeable over a wider range of delay T. As with the mutual information, we may define a nonlinearity index

$$Z_T = \sum_T \begin{cases} 0 & : \quad TE_{j\rightarrow i}(T) \leq CU(T) \\ (TE_{j\rightarrow i}(T) - CU(T))/CU(T) & : \quad TE_{j\rightarrow i}(T) > CU(T) \end{cases}$$

where, again, the upper confidence limit $CU(T)$ is chosen on the basis of the mean and standard deviation of surrogate values for delay T. The values of this index are larger than are those for the mutual information as can be seen in Figure 8.34. This is consistent with other model systems we have examined. We conclude that for this simple system, TE is a more sensitive indicator of nonlinear coupling and is therefore a good candidate for detecting damage-induced nonlinearities.

8.7.3   Surrogate Invariance to Ambient Variation

Perhaps the greatest strength of the proposed approach is that it is insensitive to certain types of ambient variability such as those caused by temperature, humidity, and so on. Unless these changes affect the *form* of the underlying dynamics (linear/nonlinear), they will not affect the proposed indicators Z_M, Z_T. As a final numerical example, we therefore alter the stiffness values of the five-DOF system of the previous section so as to simulate a temperature gradient across the structure. This was accomplished by decreasing the first stiffness value by 2%, the next by 4%, and so on. The last stiffness, k_5, was therefore altered by 10%. Again, we explore the relationship between $x_2(t), x_3(t)$ using both the original simulated time series and 10 linear surrogates. Sample results are displayed in Figure 8.35 for both mutual information and TE in the case of $\mu = 1.45$. Again, we can see a clear difference between the surrogates and the data, indicating that there is a statistically significant degree of nonlinearity in the structure. Although the global stiffness properties have changed, the nonlinearity remains. The natural frequencies for the altered structure, denoted $\omega_i^{(A)}$, are summarized in Table 8.1.

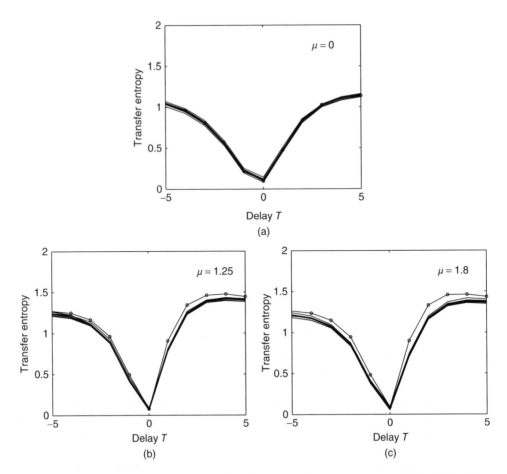

Figure 8.33 Plot of $TE_{3\rightarrow2}$ showing increasing discrepancy between surrogates (solid line) and data (open circles) for $\mu = 0$ (linear system), $\mu = 1.5$, and $\mu = 1.75$ *Source:* Reproduced from [8], Figure 6, with permission of Elsevier

Damage indices were computed in the same manner as in the previous example and the results are displayed in Figure 8.36. As the level of nonlinearity is increased, both indices rise in a monotonic manner. The main difference is that the values of both indices are slightly higher than in the previous case. For example, in the original structure $Z_T = 0.8$ for $\mu = 1.75$, while here $Z_T = 0.8$ for $\mu = 1.45$. This can be explained by the fact that the perturbed stiffness values have changed the point at which the system begins to oscillate between the two equilibria. Given our fixed level of excitation (Gaussian noise with unit standard deviation), these oscillations now occur for values $\mu > \mu^* = 1.45$ (as opposed to $\mu^* = 1.75$) and the nonlinearity becomes too trivial to detect, as before.

The important feature to note in Figure 8.36 is that the damage index for the linear structure remains zero even though the natural frequencies of the structure have shifted by between 2.2% and 5% (depending on the mode). Mode shapes for the structure are similarly altered. Health monitoring schemes based on modal properties would, in this example, produce false

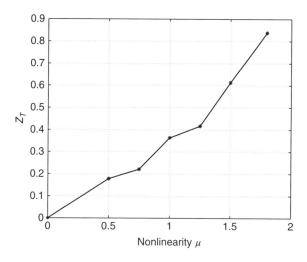

Figure 8.34 Index Z_T as a function of % nonlinearity (damage) in the system *Source:* Reproduced from [8], Figure 7, with permission of Elsevier

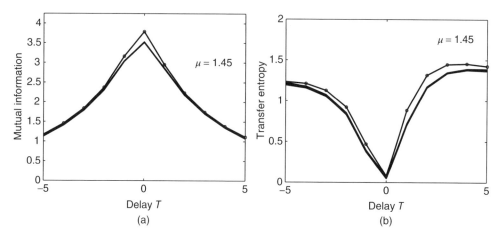

Figure 8.35 Mutual information (a) and transfer entropy (b) for the case where $\mu = 1.45$ *Source:* Reproduced from [8], Figure 8, with permission of Elsevier

positives (declare damage when none exists) or suffer a reduced sensitivity to the damage. We stress that this approach yields an absolute measure of nonlinearity as opposed to relative. This approach therefore obviates the need for baseline data sets for feature comparisons. We still require a baseline *assumption* equating "healthy" with linear, but no baseline time series. The surrogates are, in effect, a baseline (null hypothesis) against which the hypothesis of nonlinearity may be tested. A structure that has been retro-fitted with a health monitoring system (sensors/algorithms) and for which no baseline exists could still be monitored for damage using this approach.

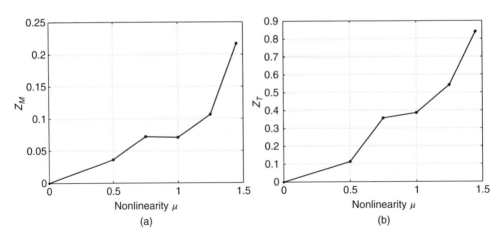

Figure 8.36 Damage indices Z_M, Z_T as a function of % nonlinearity (damage) in the system *Source:*
Reproduced from [8], Figure 9, with permission of Elsevier

The goal of these simple numerical studies was to demonstrate one possible means of
using surrogate data to estimate the degree of nonlinearity in a structural system. If we are
willing to equate nonlinearity with damage, we have a reference-free approach to detecting
damage presence and extent. However, we obviously have many different ways of performing a surrogate-based hypothesis test. We have freedom in both how we generate the
surrogates and which test statistic to use for comparison. This simple study has shown how
information-theoretic quantities, in conjunction with IAAFT surrogates, can be used to
this end.

Both the time-delayed mutual information and time-delayed transfer entropy were presented
as two alternative, probabilistic definitions of coupling. In contrast to standard signal processing techniques which focus on second-order correlations, these two quantities capture
coupling in all moments of the signal's underlying PDFs. Because both quantities capture
general dependencies among time series, they may be effectively used to diagnose when the
coupling becomes nonlinear.

The surrogate-based approach is demonstrated next on two different experimental systems.

8.8   Surrogate Experiments

Perhaps our chief reason for gravitating toward surrogate methods was demonstrated in the
final numerical example just provided. Namely, that the surrogate method is very good at
capturing changes in damage-induced nonlinearity while effectively ignoring any change in
linear system properties which we take to mean the covariance structure of the observed data.
Many naturally occurring sources of variability that have nothing to do with damage tend to
influence the covariance of the response, for example, the simulated temperature effects we just
examined. For this reason, the surrogate method is quite good at lowering the false positives
that often occur using "change detection" methods that focus on any source of change in the
statistical properties of a signal.

In what follows we provide two experimental examples where we have used the surrogate method to good effect. Still other experimental examples we have looked at can be found in [19]. We chose these two examples as they illustrate the surrogate method in two very different systems with different forms of structural damage. In the first we look at rotor–stator rub in a shaft-bearing assembly, and in the second we examine the strength of a bolted connection as it is slowly loosened.

8.8.1 Detection of Rotor–Stator Rub

In rotary systems, the owner/operator is often interested in diagnosing the presence of a rub between the shaft and the housing (stator). The undamaged rotor system is often modeled as a linear system with constant coefficient mass, stiffness, and damping matrices [54]. The presence of a rotor–stator rub is expected to produce a nonlinearity (stick-slip/impacting) in an otherwise linear system. Thus, we can use the surrogate data method along with one of the detection statistics of Section 8.1 to tell whether this form of damage is present. In the experimental example that follows, the TE is used as the detection statistic.

The system studied in this experimental example, depicted in Figure 8.37, consists of an active magnetic bearing (AMB) (Revolve Magnetic Bearing Inc.), inboard and outboard bearings (located at either end of the shaft), and a balance disk. The system also includes a PID controller for minimizing shaft position errors. In addition, two proximity probes (Bently-Nevada®) for monitoring the shaft vibration were positioned at approximately the 1/4 and 3/4 points along the length of the shaft. In this particular example, the system is "offline," that is to say, in a nonrotating state. The vibration data used in making the diagnosis were obtained by applying a small amount of jointly Gaussian input excitation using the AMB located near the midspan of the shaft. AMBs can be used to excite the shaft with arbitrary waveforms and thus provide a convenient mechanism for performing damage detection. Data were collected from the pristine structure (no rub) and for two different rub scenarios. A total of 40 surrogates were generated from the original data, each consisting of $N = 32,768$ points. The time-delayed transfer entropy was estimated using data recorded from the inboard

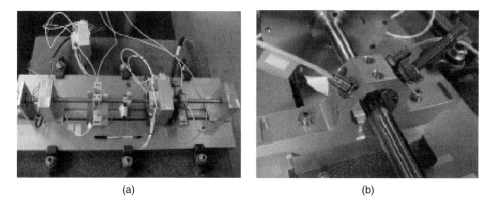

(a)   (b)

Figure 8.37 Overview of rotor experiment (a) and close-up of "rub" screw (b)

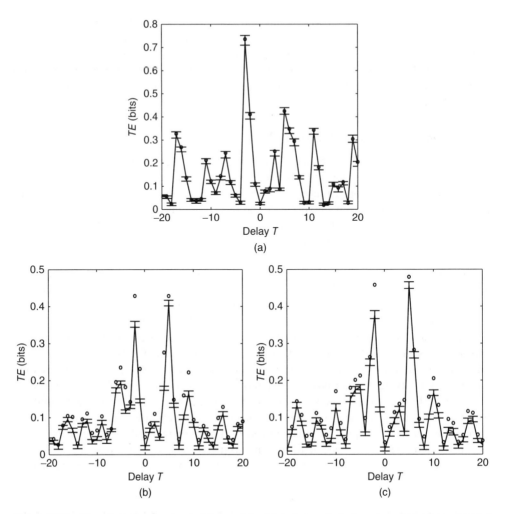

Figure 8.38   Transfer entropy estimated from original data (open circles) compared to values obtained from surrogates for undamaged (a) and damaged (b and c) rotor

and outboard horizontal probes using a fixed-mass kernel with $M = 40$ points. Confidence intervals, based on the surrogates, were again formed by discarding the low and high values of the surrogate data to give an interval of 95% confidence (as discussed in Section 6.9.3). These results are shown in Figure 8.38. As the damage is introduced, the estimated TE values begin to fall outside the confidence intervals formed from the surrogates. As the severity of the rub is increased (Figure 8.38c), the separation between surrogates and original data increases as expected. Damage in this case clearly manifests itself as the presence of a nonlinearity. The TE values across nearly all delays exceed the confidence limits for both rub scenarios. The strength of this general approach is that each of the plots in Figure 8.38 can be viewed independently to assess the level of damage. Each is providing an absolute measure of nonlinearity (damage), hence data from the healthy system is not needed to detect the presence of the rub.

8.8.2 Bolted Joint Degradation with Ocean Wave Excitation

As a second experimental test of the surrogate approach, we again consider the composite bolted joint already detailed in Section 8.2.1 and shown in Figure 8.1. This particular material is under consideration from the Navy for use in ship decks and/or superstructures, in part due to its resistance to the type of stress corrosion cracking mentioned in Section 1.3 and shown in Figure 1.1. The vibrations this component is expected to see *in situ* arise due to the natural wave loading the ship will experience in transit. One of the main goals of this section is to therefore demonstrate the utility of the surrogate data method in an example where the driving signal is not i.i.d. Gaussian noise, but is rather a signal conforming to a naturally occurring random process. The obvious advantage of using natural excitation is that the practitioner can perform *in situ* monitoring without needing the hardware or power used in interrogating the structure.

To this end, we take the driving signal to be a random process consistent with the Pierson–Moskowitz PSD function for wave height (we have already used this distribution in Section 6.7.4 and it occurs once more in Chapter 10). This distribution is frequently used in the ocean engineering community to describe dynamic loading on maritime structures (e.g., offshore platforms [55]). Using the method mentioned in Section 6.7.2, we generated an excitation signal possessing a Gaussian marginal distribution and the Pierson–Moskowitz spectral density. This signal was then used as the driving voltage for a shaker, mounted at the midspan of the beam. Figure 8.39 shows the estimated PSD of the forcing actually felt by the structure as recorded by the load cell. This spectrum accurately matches the intended spectral density.

The experimental approach is therefore to apply a realization of the loading signal, measure the dynamic strain response, and then use the method of surrogate data to decide whether the response came from a linear (healthy) or nonlinear (damaged) structure.

Complicating the detection problem is the fact that this component is likely to experience a wide range of temperatures during operation. As we have stressed, covariates provide

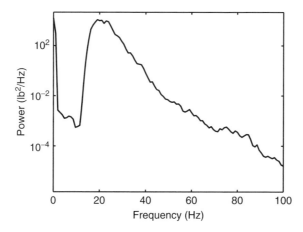

Figure 8.39 Estimated power spectral density (PSD) function for the signal recorded by the load cell. The PSD is that of the Pierson–Moskowitz frequency distribution for wave height *Source:* Reproduced from [56], Figure 2, with permission of ASME

unwanted sources of "change" in a signal that can significantly complicate the detection problem, particularly if we rely on a data-driven model developed in the absence of information about the influence of temperature on the vibration signals. By relying on the physics of the damage (linear vs. nonlinear), the surrogate approach should be largely immune to such variability. In an effort to test this hypothesis, for each preload level we also considered temperature as a covariate.

As discussed already in Section 8.2.1, a series of heat lamps were placed near the loosened end of the beam (see Figure 8.1) and were used to heat the structure while a thermocouple recorded the local beam temperature. The strain response data used in this study were acquired for thermocouple readings of $23, 30, 40,$ and $50°$ C just as we did in compiling the data set used in the analysis of Section 8.2.1. These temperature variations are consistent with those observed by such a structure on a day-to-day basis during operation (see, e.g., [57]).

Damage was introduced into the structure as before by reducing the preloads on both bolts connecting one end of the composite to the steel. The undamaged structure was represented by preloads near 10,000 lbs and an ambient temperature of $23°$ C (room temperature), with potential damage represented using preloads of roughly 5000, 2500, 1250, 750, 500, 250, and 0 lbs of preload. These are in addition to the damage states used in the prior study, Section 8.2.1. At 0 lb preload, the bolt conditions can be finger tight (the nut is touching the back of the steel), include a small gap (the nut is loose but the bolt is tightly held in the bolt holes), or include large gap (the nut is loose and the bolts were loosened in the holes).

The amplitude of excitation was such that the maximum beam deflection was $\ll 2.54$ cm for each case. Two time series were then recorded, each consisting of 60,000 observations (30 s of data) of the beam's dynamic strain response to the wave excitation. Transients were discarded and a moving window of $M = 2^{15}$ points was passed through the data and checked for end-point matches as described in Section 8.6. Forty surrogates were then created for each response time series using the IAFFT algorithm. Figure 8.40 shows the original time-series data (units of microstrain) and an associated surrogate for the data recorded from both the steel (sensor 1 of Figure 8.1) and composite (sensor 2 of Figure 8.1) for the undamaged structure. The bottom row shows both the cross-spectrum and cross-correlation functions associated with these two time series. Despite the fact that we have reordered the original time series, clearly both the cross-spectrum and cross-correlation are preserved. The variability in the loads recorded by the instrumented bolts was rather large owing to the resolution of the bolts (± 100 lbs) and the practitioner's ability to obtain the target loads manually (using a large wrench).

8.8.2.1 Nonlinear Prediction Error Results

As the detection statistic we focused initially on the nonlinear prediction error (NPE) (8.32) as the damage-sensitive measure. Recall that this approach forms a probabilistic, nonlinear model of the structural response data and then quantifies the error in the resulting predictions. Data produced by a nonlinear system should be better described by this model than data produced by a linear system. We therefore expect a lower prediction error for a damaged structural response than for the surrogate data. In this particular implementation, we fixed the prediction horizon at $T = 1$ time steps. We also chose to use a simple $d = 2$ dimensional Markov model with a delay of $\Delta = 10$ time steps, but note that the results are not sensitive to these choices.

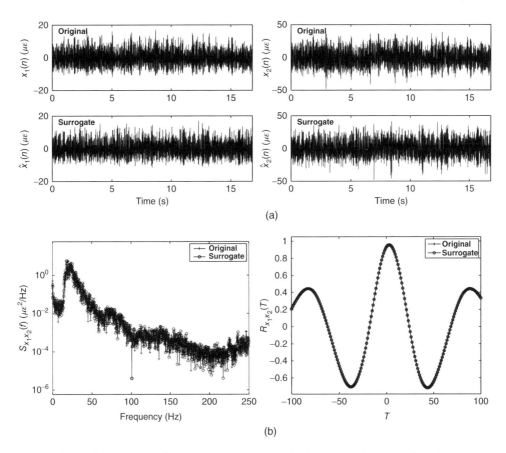

Figure 8.40 Original and sample surrogate time series for the data recorded at sensor (1) and sensor (2). (b) Sample cross-spectrum and cross-correlation functions associated with the original data and sample surrogates *Source:* Reproduced from [56], Figure 3, with permission of ASME

Beginning with the undamaged configuration, the bolts at one end of the beam were then loosened in fixed increments and the above-described procedure repeated. At each preload level, the data were tested for the presence of a damage-induced nonlinearity using the mean NPE as the feature of interest. Figure 8.41 shows the mean prediction error as a function of clamping force for both the data (crosses) and surrogates (dots). Because we expect the prediction error to decrease with nonlinearity, we may perform a one-sided test. We discard the minimum surrogate value for $\bar{e}$ so that the remaining 39 values span an interval of $39/40 = 97.5\%$ confidence [58] (see again Section 6.9.3). Clearly, the structure is well approximated by a linear model until the bolt begins to come loose (finger tight). Only one of the $\bar{e}$ values exceed the confidence limits spanned by the surrogates before the finger-tight stage. As the bolt begins to loosen, there is macro-slip as well as some subtle impact discontinuities and the dynamics are determined to be nonlinear at a high level of confidence. This can be seen as a large separation between the data and surrogates. Damage in this experiment is a loosening of the bolt, which does indeed cause an otherwise linear structure to behave nonlinearly.

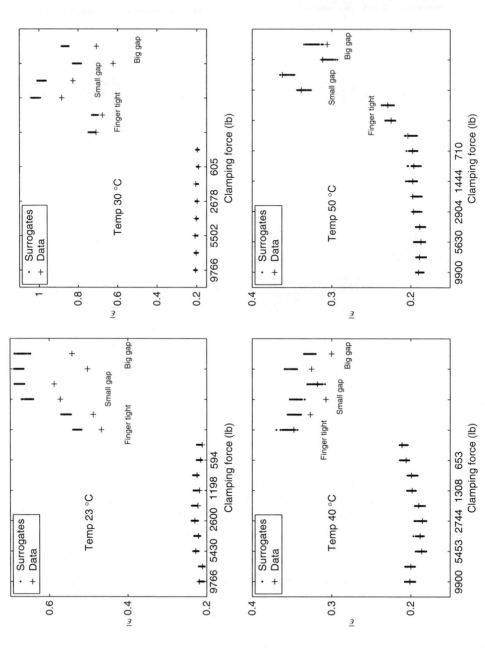

Figure 8.41   Mean prediction error as a function of damage for varying temperature for surrogates ·, and original data + *Source:* Reproduced from [56], Figure 4, with permission of ASME

To illustrate this property, the experiment was repeated as the temperature at the damaged end of the beam was fixed at the three additional temperatures of 30, 40, and 50° C. These results are also shown in Figure 8.41. While temperature certainly influences the measured response, the difference in $\bar{e}$ values between the data and surrogates remains nearly the same as for the 23° C case across all preload levels. The beam appears well described by a linear model until the bolts reach the "finger-tight" stage, at which point the dynamics become nonlinear. The main exception were the data collected from the beam at 50° C. The response appeared linear until the final damage level. One possible explanation is that there was less clearance between the composite and the bolt at this temperature. The composite and bolt expanded and filled in what little clearance existed at room temperature. The result was a bolt that despite not carrying a preload occasionally would become wedged so tightly that the structure remained well approximated by a linear model. However, it can be seen from the next set of results that this explanation is not sufficient.

We repeated the experiment again varying both the temperature and preload level as before. These results are displayed in Figure 8.42 and are very similar to those obtained on the first run. A third repeat of the experiment (results not shown) agreed well with both displayed sets of results. As before, the beam appears linear until the finger-tight stage, at which point the dynamics become nonlinear and damage is clearly indicated with confidence. In this second experiment, however, the data collected from the beam at 50° C also showed significant separation between data and surrogates for each of the last three damage levels (this was also the case for the third repeat of the experiment). We suspect that the raised temperature simply increases the likelihood of the bolt becoming stuck because of diminished bolt clearance (as was the case with the first set of results). For this elevated temperature, some runs may appear undamaged, while others appear damaged because of the large physical change to the system. There is also some evidence that the degree of separation between data and surrogates relates to the magnitude of the damage. For example, in both Figures 8.41 and 8.42, the separation tends to be smaller for the finger-tight case than for the fully loose conditions.

We also considered what happens to the natural frequencies of the undamaged beam following the temperature changes. Owing to the limited bandwidth of the excitation, only the first two natural frequencies were excited with significant power. The first natural frequency remained unchanged (51.7 Hz), however the second frequency dropped from 89.8 to 86.5 Hz as the temperature was raised. This shift is well outside the variance in the frequency estimate. If frequency had been used as a feature, the beam would have falsely been declared damaged at higher temperatures. For comparison purposes, we also looked at the second natural frequency of the beam in the finger-tight condition. Here, the frequency shifted even lower to 83 Hz. We conclude that using frequency as a feature would likely give a high probability of detection (low Type II error) but would also result in high Type I error (large numbers of false positives) as a result of the baseline information shifting.

We should also point out that only twice was one of the time series collected from the clamped configuration declared to have been generated by a nonlinear model. There are several mechanisms for generating these types of false-positive diagnosis. First, and most obvious, the confidence limits are constructed such that we expect 2.5% of the response data to generate a false-positive result. For the number of cases examined in this study, two false positives are expected. Secondly, there are errors in the surrogate generation process that could lead to such a result. While we have made every effort to match the spectral properties of the data/surrogates, there will always be some discrepancy. In addition, as we mentioned in Section 8.6.4 the

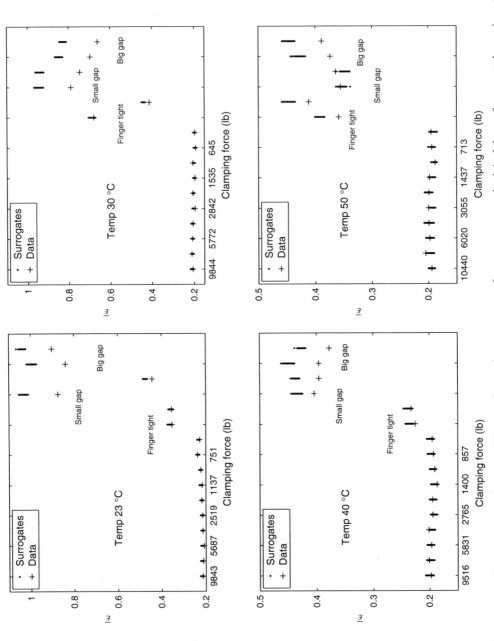

Figure 8.42 Mean prediction error as a function of damage for varying temperature for surrogates ·, and original data + for repeated experiment *Source:* Reproduced from [56] Figure 5, with permission of ASME.

IAAFT algorithm is known to result in a feature distribution with too small a variance (hence more false positives). We test this possible effect in the next section.

8.8.2.2 Comparison of Surrogate Methods and Detection Statistics

A correction to the IAAFT algorithm, the DFS method of Dolan and Spano, was already mentioned as one possibility of providing the correct surrogate variance [46, 51]. In this section, we compare the performance of several test statistics with the three prime surrogate generation algorithms in their respective ROC performance. Figure 8.41a and b shows several ROC curves associated with both the mutual information detector and the NPE detector. As expected, for large numbers of false positives, we also have a high probability of detection. Conversely, low numbers of false positives accompany lower detectability. The end-user in SHM applications will most likely be interested in the low range of false positives. We can see from Figure 8.43a that the mean prediction error registers roughly a 70% probability of detection for 5% false positives, regardless of the type of surrogate used. The DFS surrogates result in a slight improvement in the number of false positives (3%) for 70% POD, however the effect is not great. Although the IAAFT surrogates were not designed to possess the correct population variance, it appears they do in this case. We can only conclude, as did the authors of [46], that the random error in the IAAFT surrogate generation procedure roughly corresponds to that exhibited by independent realizations of the underlying process. It is also evident from the ROC curves that regardless of whether the error in IAAFT surrogates is left in the spectral properties (IAAFT-I), or probability distribution (IAAFT-II), the detection properties are the same. We should mention here that the point $(0,0)$ will exist on all ROC curves (threshold is set so that nothing is declared a detection) and so is included in each of the ROC plots presented here.

The mutual information detector (Figure 8.43b) based on IAAFT-I surrogates does not perform as well for low numbers of false positives giving $\ll 60\%$ detection probability for roughly

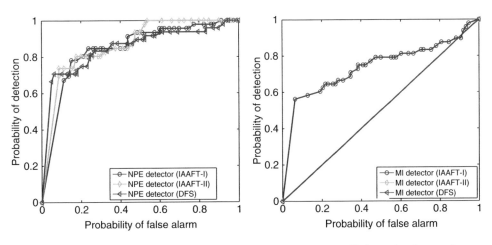

Figure 8.43 ROC curves associated with the NPE detector and the mutual information detector *Source:* Reproduced from [59], Figure 3, with permission of Elsevier

5% PFA. The mutual information detectors based on the IAAFT-II and DFS surrogates were not effective at all, producing >95% false positives for all thresholds. For these two detectors, all points cluster near (1, 1) on the ROC curves except for the point (0, 0) which, again, will always be present. Thus, both the IAAFT-II and DFS detectors produce diagonal lines connecting (1, 1) and (0, 0) in Figure 8.43b. The reason for this behavior, alluded to earlier in Section 6.9, is that the size of the test (as determined by the number of surrogates) does not match the PFA obtained. In this case the difference is so great as to make the test essentially useless. Because the mutual information is based on estimates of probability densities, the errors in probability distribution present in the IAAFT-II and DFS surrogates become critical. These errors cause all values of mutual information computed from the original data to appear significantly different than the surrogates, even when the original data conform to the null hypothesis (thus, nearly 100% false positives). For detectors that involve density estimates, the IAAFT-I surrogates are likely to be the better choice as they exactly match the probability distribution of the original data.

Figure 8.44 shows the ROC curves associated with the bicoherence-based detector. The bicoherence detector appears to offer better performance over a wide range of Type I errors. When using IAAFT-I surrogates, the bicoherence detector registers 90% or better POD for ≈ 10% PFA. Similar results are found when the bicoherence is combined with the other two surrogate generation techniques. While we cannot accurately estimate the performance for very low Type I errors, the shape of the curves would suggest that the performance for low levels of false alarms is likely quite good.

We should mention that the bicoherence detector will not work for all nonlinearities, only those that result in frequency coupling in the response. This may, in fact, be the reason

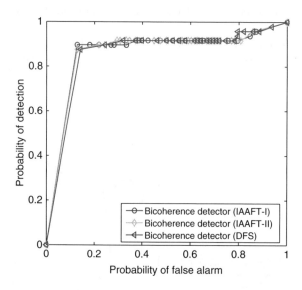

Figure 8.44 ROC curves associated with the bicoherence detector, $s^{(b)}$. For this bolted joint experiment, the bicoherence detector yielded the best probability of detection for a fixed, small number of false alarms. Because this measure is targeting a specific cumulant in the response (as opposed to the more general methods of Figure 8.43), and because the loosening joint does produce an increase in this cumulant, the detection performance is excellent *Source:* Reproduced from [59], Figure 4, with permission of Elsevier

for the improved ROC performance. In problems of detection, typically the more targeted approach yields higher PODs. For example, when detecting a sinusoid buried in noise, Fourier analysis (sinusoidal basis) will always outperform the very general power detector (looking for increased power due to the presence of an arbitrary signal). It is therefore not surprising that the approach that searches for a specific type of nonlinearity performs better than the two that are looking for arbitrary nonlinearities.

At present, we are assuming no knowledge of the costs associated with making a Type I or Type II error. If these costs are available, the practitioner can use the ROC curves to select among the different available detectors. For example, if 50% false positives are tolerable but a >99% POD is needed, the NPE detector using the IAAFT-II surrogates might be used. However, in cases where both false positives and false negatives are costly, the bicoherence detector would probably be considered optimal.

8.9   Surrogates for Nonstationary Data

In each of the above-mentioned examples, the null hypothesis has been consistent with the response of a linear, time-invariant system. However, these tests will also reject the null hypothesis for *nonstationary, linear* processes. The nonstationarity may come as a result of the properties of the input changing in time ($X(t)$ is nonstationary) or because the system parameters are changing in time, that is, $h(\tau) \rightarrow h(t, \tau)$. The latter situation is referred to as a linear time-varying system. In short, any test that is based on the linear, time-invariant null hypothesis may decide in favor of the alternative if the alternative is (i) a nonlinear process, (ii) a linear process with nonstationary inputs, or (iii) a time-varying linear process. We have already provided in Section 6.8.2 two ways of testing for the presence of wide-sense nonstationarity. Such a test should be viewed as a prerequisite for performing a surrogate-based analysis. Our reasons for stating this are clearly illustrated in the following example.

Figure 8.45 shows the results of applying the method of surrogate data to several different types of system output. The discriminating feature used in this case was the mean nonlinear prediction error described in Section 8.1.3, normalized by the standard deviation of the signal. The equations of motion for the system are given by

$$m\ddot{y} + c\dot{y} + (k + \epsilon t^2)y + ky^2 = x(t) \tag{8.65}$$

where $m = 1.0$ kg, $c = 3.0$ N $\cdot$ s/m and $k = 1000$ N/m were fixed parameters and ϵ N/m/s^2 and k_N N/m^2 were the nonstationarity and nonlinearity parameters, respectively. The excitation was taken as a stationary Gaussian random process with constant PSD $S_{XX}(f) = 0.01$ N^2/Hz. This system was numerically integrated, giving time series of $N = 32,768$ observations sampled at intervals of 0.01 s. For $\epsilon = k_N = 0$, the system is linear and the data and surrogates produce mean NPE values in the same range. As expected, when $k_N = 10,000$ N/m^2, $\epsilon = 0$ the system is nonlinear, stationary and the data fall outside the linear surrogate bound. However, when $k = 100$ N/m, $\epsilon = 0.025$ N/m/s^2, $k_N = 0$ N/m^2, the system is linear with a time-varying stiffness and, hence, is nonstationary.

On the basis of the surrogate analysis, the response still appears nonlinear despite the fact that it was produced by a linear system. Other tests for nonlinearity will similarly have difficulty in making the distinction. The main reason is that estimation, as we have already discussed, almost always involves the assumption of ergodicity (hence stationarity). The

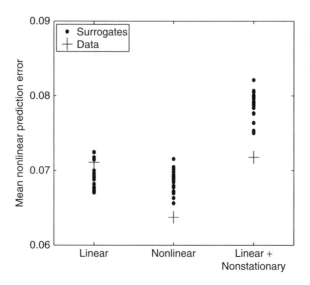

Figure 8.45 A surrogate test applied to the response of a single-degree-of-freedom system driven with stationary Gaussian excitation. The discriminating feature used was the mean nonlinear prediction error, normalized by the standard deviation of the signal. Dots represent the surrogate NPE values, while crosses represent the NPE values obtained from the original data. As expected, the nonlinear model better describes the nonlinear system response. However, the nonlinear model is also better able to describe the nonstationary data, producing results that can easily be confused with nonlinearity

surrogate algorithm, for example, requires estimates of the PSD function. Coherence-based tests for nonlinearity similarly involve estimates of the coherence function. In both cases, the estimates we use start with assumed ergodicity, a property that is violated for all nonstationary data of practical interest.

8.10   Chapter Summary

Many forms of damage manifest as a nonlinearity in a structure that is otherwise (when healthy) appropriately modeled as a linear dynamical system. By making this basic assumption, detecting the damage presence reduces to the problem distinguishing whether the observed data were produced by a linear or nonlinear dynamical system. To this end, a host of methods are available to the practitioner, several of which we have described in detail. We have described a number of test statistics in Section 8.1 that can be used to distinguish a damaged from an undamaged response. Specifically we have described and used (i) nonlinear modeling error, (ii) information theoretics, and (iii) higher order spectral properties. We have shown how each of these may be estimated from time-series data and applied to the problem of nonlinearity detection. Depending on the type of nonlinearity one is trying to capture, some may perform better than others. For example, if we know we are looking specifically for a quadratic nonlinearity, the estimated bispectral density function is an excellent candidate for a test statistic. If we know nothing about the form of the damage, perhaps one of the more general measures, for example, TE, would yield better results as it will, in principle, be sensitive to all higher order joint statistical properties in the data.

A particular point of focus was the HOS which we have found to be quite powerful in developing detectors that minimize both Type I and Type II errors. In fact, we showed how a detailed knowledge of the damage physics can tell us quite a bit about how these types of detectors are likely to perform. For example, we used a low-dimensional, dynamic model of a delaminated beam to describe the delamination in terms of its location, size, and depth. Expressions for the polyspectra were derived up to fourth order (power spectrum, bispectrum, and trispectrum) and these expressions linked the damage-related parameters directly to features in the polyspectra. When combined with previously derived results for the variance of polyspectral estimates, the analysis predicted both the Type I and Type II errors associated with polyspectral detection schemes.

One of the more powerful approaches to nonlinear damage detection is the method of surrogate data. Using this approach, one takes the measured data from the structure and "linearizes" it. Specifically, one shuffles the observed data in such a way that *all joint cumulants above second order are destroyed while preserving the covariance among observations*. The resulting data will be consistent with our null (undamaged) hypothesis that the observations were generated by a linear structure responding to a stationary input that is appropriately modeled as a memoryless transformation of a jointly normal random process. It is worth recapping what the latter portion of this hypothesis really means. All it is saying is that the input signal itself does not contain significant joint cumulants beyond second order. The marginal distributions associated with the input distribution can certainly be non-Gaussian under this hypothesis as this frequently occurs in practice (e.g., wind or wave loading). We just need to make sure that the test for nonlinearity is picking up higher order joint properties in the *response* and *not* in the input. Certainly if a linear structure was driven with the output of a nonlinear system, we would run the risk of declaring nonlinearity (damage) even if the structure of interest was perfectly linear (healthy). The null hypothesis we use in this chapter therefore requires that the practitioner has verified that the input to the structure is not itself the output of a nonlinear system. By comparing one of the aforementioned test statistics to the surrogates, we have shown how one can generate an absolute measure of nonlinearity that should be largely insensitive to changes in other covariates, so long as the changes caused by those covariates do not themselves result in nonlinear structural behavior. In many cases (e.g., temperature fluctuations), this would seem a safe assumption to make.

For example, we have shown that it is possible to detect bolt loosening in an ambiently excited structure in the presence of strongly varying temperatures. Specifically, we have demonstrated the surrogate approach in assessing whether the connection has come loose on a composite-to-metal bolted joint. As expected, a loose joint results in a dynamical response that is more appropriately described in a nonlinear model than a linear one. The assessment was made using ambient vibration conforming to the Pierson–Moskowitz frequency distribution and in the presence of variations in the local temperature from 23 to 50° C. The result is a robust detector of damage-induced nonlinearities that works for arbitrary excitation. The approach also worked well experimentally in diagnosing rotor–stator rub in rotating machinery as well as other experimental systems we have looked at (e.g., an airfoil and a composite plate [19]). The key question to ask from a health monitoring perspective is whether the nonlinearity shows up before it is too late to apply corrective action. This decision will be application specific.

References

[1] H. Sohn, C. R. Farrar, F. M. Hemez, J. J. Czarnecki, D. D. Shunk, D. W. Stinemates, B. R. Nadler, A review of structural health monitoring literature: 1996-2001, Tech. Rep. LA-13976-MS, Los Alamos National Laboratory (2003).

[2] H. Soon, H. W. Park, K. H. Laws, C. R. Farrar, Combination of a time reversal process and a consecutive outlier analysis for baseline-free damage diagnosis, Journal of Intelligent Material Systems and Structures 18 (4) (2007) 335–346.

[3] J. A. Brandon, Some insights into the dynamics of defective structures, Proceedings of the Institution of Mechanical Engineers Part C: Journal of Mechanical Engineering Science 212 (1998) 441–454.

[4] M. I. Friswell, J. E. T. Penny, Crack modeling for structural health monitoring, Structural Health Monitoring 1 (2) (2002) 139–148.

[5] W. Z. Zhang, R. B. Testa, Closure effects on fatigue crack detection, Journal of Engineering Mechanics - ASCE 125 (10) (1999) 1125–1132.

[6] R. L. Brown, D. E. Adams, Equilibrium point damage prognosis models for structural health monitoring, Journal of Sound and Vibration 262 (2003) 591–611.

[7] A. C. Rutherford, G. Park, C. R. Farrar, Non-linear feature identifications based on self-sensing impedance measurements for structural health assessment, Mechanical Systems and Signal Processing 21 (2007) 322–333.

[8] J. M. Nichols, S. T. Trickey, M. Seaver, Detecting damage-induced nonlinearities in structures using information theory, Journal of Sound and Vibration 297 (2006) 1–16.

[9] T. Tjahjowidodo, F. Al-Bender, H. V. Brussel, Experimental dynamic identification of backlash using skeleton methods, Mechanical Systems and Signal Processing 21 (2007) 959–972.

[10] I. Trendafilova, H. V. Brussel, Non-linear dynamics tools for the motion analysis and condition monitoring of robot joints, Mechanical Systems and Signal Processing 15 (2001) 1141–1164.

[11] G. W. Hunt, B. Hu, R. Butler, D. P. Almond, J. E. Wright, Nonlinear modeling of delaminated struts, AIAA Journal 42 (11) (2004) 2364–2372.

[12] K. D. Murphy, J. M. Nichols, A low-dimensional model for delamination in composite structures: theory and experiment, International Journal of Non-Linear Mechanics 44 (2008) 13–18.

[13] H. Luo, S. Hanagud, Dynamics of delaminated beams, International Journal of Solids and Structures 37 (2000) 1501–1519.

[14] F. L. Chu, W. X. Lu, Experimental observation of nonlinear vibrations in a rub-impact rotor system, Journal of Sound and Vibration 283 (2005) 621–643.

[15] J. M. Nichols, M. Seaver, S. T. Trickey, T. Bash, M. Kasarda, Use of information theory in structural monitoring applications, in: F.-K. Chang (Ed.), Proceedings of the 5th International Workshop on Structural Health Monitoring, DEStech Publications, Lancaster, PA, 2005.

[16] Z. K. Peng, F. L. Chu, P. W. Tse, Detection of the rubbing-caused impacts for rotor-stator fault diagnosis using reassigned scalogram, Mechanical Systems and Signal Processing 19 (2) (2005) 391–409.

[17] Y. Song, C. J. Hartwigsen, D. M. McFarland, A. F. Vakakis, L. A. Bergman, Simulation of dynamics of beam structures with bolted joints using adjusted Iwan beam elements, Journal of Sound and Vibration 273 (2004) 249–276.

[18] S. H. Jeong, J. M. Park, Y. Z. Lee, Transition of friction and wear by stick-slip phenomenon in various environments under fretting conditions, Key Engineering Materials 321-323 (2006) 1344–1347.

[19] J. M. Nichols, M. Seaver, S. T. Trickey, L. W. Salvino, D. L. Pecora, Detecting impact damage in experimental composite structures: an information-theoretic approach, Smart Materials and Structures 15 (2006) 424–434.

[20] J. M. Nichols, C. J. Nichols, M. D. Todd, M. Seaver, S. T. Trickey, L. N. Virgin, Use of data-driven phase space models in assessing the strength of a bolted connection in a composite beam, Smart Materials and Structures 13 (2004) 241–250.

[21] A. G. Barnett, R. C. Wolff, A time-domain test for some types of nonlinearity, IEEE Transactions on Signal Processing 53 (1), 26–33.

[22] M. J. Hinich, G. R. Wilson, Detection of non-Gaussian signals in non-Gaussian noise using the bispectrum, IEEE Transactions on Acoustics, Speech, and Signal Processing 38 (7) (1990) 1126–1131.

[23] M. Ragwitz, H. Kantz, Markov models from data by simple nonlinear time series predictors in delay embedding spaces, Physical Review E 65 (2002) 056201.

[24] M. B. Priestly, Spectral Analysis and Time Series, Probability and Mathematical Statistics, Elsevier Academic Press, London, 1981.

[25] M. D. Todd, G. A. Johnson, B. L. Althouse, A novel Bragg grating sensor interrogation system utilizing a scanning filter, a Mach-Zehnder interferometer, and a 3x3 coupler, Measurement Science and Technology 12 (2001) 771–777.

[26] A. Milanese, P. Marzocca, J. M. Nichols, M. Seaver, S. T. Trickey, Modeling and detection of joint loosening using output-only broad-band vibration data, Structural Health Monitoring 7 (4), 309–328.

[27] K. Worden, G. R. Tomlinson, Nonlinearity in experimental modal analysis, Philosophical Transactions of the Royal Society of London, Series A 359 (2001) 113–130.

[28] A. Rivola, P. R. White, Bispectral analysis of the bilinear oscillator with application to the detection of cracks, Journal of Sound and Vibration 216 (5) (1998) 889–910.

[29] G. C. Zhang, J. Chen, F. C. Li, W. H. Li, Extracting gear fault features using maximal bispectrum, Key Engineering Materials 293-294 (2005) 167–174.

[30] K. K. Teng, J. A. Brandon, Diagnostics of a system with an interface nonlinearity using higher order spectral estimators, Key Engineering Materials 204-205 (2001) 271–285.

[31] D. R. Brillinger, M. Rosenblatt, Asymptotic theory of estimates of k-th order spectra, in: B. Harris (Ed.), Advanced Seminar on Spectral Analysis of Time Series, John Wiley & Sons, Inc., New York, 1967, pp. 153–188.

[32] P. J. Huber, B. Kleiner, T. Gasser, G. Dumermuth, Statistical methods for investigating phase relations in stationary stochastic processes, IEEE Transactions on Audio and Electroacoustics AU-19 (1) (1971) 78–86.

[33] R. N. McDonough, A. D. Whalen, Detection of Signals in Noise, 2nd ed., Academic Press, San Diego, CA, 1995.

[34] J. M. Nichols, P. Marzocca, A. Milanese, On the use of the auto-bispectral density for detecting quadratic nonlinearity in structural systems, Journal of Sound and Vibration 312 (4-5) (2008) 726–735.

[35] L. M. Garth, Y. Bresler, A comparison of optimized higher order spectral detection techniques for non-Gaussian signals, IEEE Transactions on Signal Processing 44 (5) (1996) 1198–1213.

[36] L. M. Garth, Y. Bresler, The degradation of higher order spectral detection using narrowband processing, IEEE Transactions on Signal Processing 45 (7) (1997) 1770–1784.

[37] A. M. Richardson, W. S. Hodgkiss, Bispectral analysis of underwater acoustic data, Journal of the Acoustical Society of America 96 (2) (1994) 828–837.

[38] S. M. Kay, Fundamentals of Statistical Signal Processing: Detection Theory, Vol. II, Prentice-Hall, New Jersey, 1998.

[39] J. M. Nichols, C. C. Olson, Optimal bispectral detection of weak, quadratic nonlinearities in structural systems, Journal of Sound and Vibration 329 (8) (2010) 1165–1176.

[40] J. M. Nichols and K. D. Murphy, Modeling and detection of delamination in a composite beam: A polyspectral approach, Mechanical Systems and Signal Processing 24 (2010) 365–378.

[41] D. Hickey, K. Worden, M. F. Platten, J. R. Wright, J. E. Cooper, Higher-order spectra for identification of nonlinear modal coupling, Mechanical Systems and Signal Processing 23 (4) (2009) 1037–1061.

[42] V. Giurgiutiu, Structural Health Monitoring: With Piezoelectric Wafer Active Sensors, Elsevier Academic Press, New York, 2008.

[43] J. Theiler, S. Eubank, A. Longtin, B. Galdrikian, J. D. Farmer, Testing for nonlinearity in time series: the method of surrogate data, Physica D 58 (1992) 77–94.

[44] T. Schreiber, A. Schmitz, Improved surrogate data for nonlinearity tests, Physical Review Letters 77 (4) (1996) 635–638.

[45] D. Kugiumtzis, Surrogate data test for nonlinearity including nonmonotonic transforms, Physical Review E 62 (1) (2000) R25–R28.

[46] K. T. Dolan, M. L. Spano, Surrogate for nonlinear time series analysis, Physical Review E 64 (2001) Art. no. 046128.

[47] T. Schreiber, Constrained randomization of time series data, Physical Review Letters 80 (10) (1998) 2105–2108.

[48] T. Schreiber, A. Schmitz, Surrogate time series, Physica D 142 (2000) 346–382.

[49] D. Prichard, J. Theiler, Generating surrogate data for time series with several simultaneously measured variables, Physical Review Letters 73 (7) (1994) 951–954.

[50] D. Kugiumtzis, Test your surrogate data before you test for nonlinearity, Physical Review E 60 (3) (1999) 2808–2816.

[51] K. T. Dolan, A. Neiman, Surrogate analysis of coherent multichannel data, Physical Review E 65 (2), 026108.

[52] L. N. Virgin, Introduction to Experimental Nonlinear Dynamics: A Case Study in Mechanical Vibration, Cambridge University Press, 2000.

[53] J. M. Nichols, Examining structural dynamics using information flow, Probabilistic Engineering Mechanics 21 (2006) 420–433.

[54] G. Mani, D. D. Quinn, M. Kasarda, Active health monitoring in a rotating cracked shaft using active magnetic bearings as force actuators, Journal of Sound and Vibration 294 (3) (2006) 454–465.

[55] S. M. Han, H. Benaroya, Nonlinear and Stochastic Dynamics of Compliant Offshore Structures, Kluwer Academic Publishers, Dordrecht, Netherlands, 2002.

[56] J. M. Nichols, S. T. Trickey, M. Seaver, S. R. Motley, E. D. Eisner, Using ambient vibrations to detect loosening of a composite-to-metal bolted joint in the presence of strong temperature fluctuations, Journal of Vibration and Acoustics 129 (2007) 710–717.

[57] J. M. Nichols, M. Seaver, S. T. Trickey, K. Scandell, L. W. Salvino, E. Aktaş, Real-time strain monitoring of a navy vessel during open water transit, Journal of Ship Research 54 (4) (2010) 225–230.

[58] B. F. Manly, Randomization and Monte Carlo Methods in Biology, Chapman and Hall, 1991.

[59] J. M. Nichols, S. T. Trickey, M. Seaver, S. R. Motley, Using ROC curves to assess the efficacy of several detectors of damage-induced nonlinearities in a bolted composite structure, Mechanical Systems and Signal Processing 22 (7) (2008) 1610–1622.

9

Damage Identification

In this chapter, we combine material from the preceding chapters to form a more complete picture of model-based damage identification. The approach we follow in each of the examples will have the same basic construct: formulate a damage model, estimate the parameters of that model, and use those estimates to detect and identify the damage. In formulating our models, we make use of both the deterministic and stochastic modeling tools we have developed in earlier chapters. With regard to the former, the goal is parsimony: find the simplest model that accurately captures the phenomena of interest, in this case structural damage. Our probabilistic modeling, on the other hand, will focus on the joint distribution of the noise. As we have discussed at length in Chapter 7, if we have a model of our uncertainty we can leverage a powerful framework for minimizing the influence of this uncertainty on the quality of our damage estimates. Several different damage mechanisms will be considered including imperfections in shell structures, cracks in plates, and corrosion damage.

9.1   Modeling and Identification of Imperfections in Shell Structures

As we mentioned in the Introduction, the U.S. Navy has a vested interest in the advance of damage identification technologies due to the incredibly large number of assets it must maintain. For this reason, methods for identifying damage in ship hulls and/or other maritime assets are desired. In this section, we discuss the modeling and identification of imperfections in these types of structures.

Since the publication of Koiter's seminal dissertation in 1945, it has become well known that initial geometric imperfections in shell structures may lead to dramatic erosions in ultimate strength [1, 2, 3]. Such imperfections may arise because of manufacturing, fabrication, construction, or service conditions and are the primary cause of discrepancies between the theoretical and observed shell buckling strengths. A recent and comprehensive survey of research developments in this area during the period 1996–2006 [4] highlights the importance of understanding the nature and effects of imperfections in shell structures. The goal of this section is to take the work one step further and propose a specific approach to using shell models to identify the presence, size, and location of the imperfections.

Modeling and Estimation of Structural Damage, First Edition. Jonathan M. Nichols and Kevin D. Murphy.
© 2016 John Wiley & Sons, Ltd. Published 2016 by John Wiley & Sons, Ltd.

In keeping with the spirit of this book, we seek to infer the imperfection field from observed data. As always, the key ingredients are a deterministic model governing the structure's response, parameterized in such a way as to describe both the "healthy" and "damaged" response data, and a probabilistic model, describing the uncertainty. In what follows, we provide these models and then demonstrate a few of the estimators we proposed in Chapter 7.

9.1.1 Modeling of Submerged Shell Structures

The goal of this particular modeling effort is to describe the vibrations of a submerged shell that has been dented during assembly or *in situ*. The model therefore describes both the hull vibrations and the influence of those vibrations on the surrounding fluid. This way we can either attempt to infer the damage based on direct hull vibrations or from noncontact acoustic measurements made in the fluid (e.g., from hydrophones). In this latter measurement scenario, the goal is to provide a completely unintrusive approach to identifying flaws in ship structures.

A mathematical description of this fluid-structure interaction (FSI) problem is defined on the domain $\bar{\Omega} = \Omega \cup \partial\Omega$, $\Omega \subset \mathbb{R}^3$, closed and bounded with boundary $\partial\Omega$. The domain Ω is the union of the solid and fluid domains $\Omega = \Omega_s \cup \Omega_f$ and the boundary $\partial\Omega$ is the union of the solid and fluid boundaries, including the "fluid-structure interface" they share, subject to $\partial\Omega_{FSI} \cap \partial\Omega_s \cap \partial\Omega_f = 0$, as illustrated in Figure 9.1. On this problem domain, we require a coupled dynamic equilibrium between the solid and the fluid. Assuming linear elastic material behavior, the solid structural response is governed by (neglecting a body force term)

$$\nabla \cdot \bar{\sigma} = \rho \ddot{\mathbf{u}}(\mathbf{x}, t), \qquad \bar{\sigma} = C^{IV} : \bar{\epsilon}(\mathbf{u}(\mathbf{x}, t)), \qquad \mathbf{x} \in \Omega_s \subset \mathbb{R}^3 \qquad (9.1)$$

where $\mathbf{u}(\mathbf{x}, t)$ is the displacement field satisfying all prescribed boundary conditions on the solid boundary; C^{IV} is the fourth-order tensor relating the infinitesimal strain tensor, $\bar{\epsilon}$, and the Cauchy stress tensor, $\bar{\sigma}$; and ρ is the solid material density. The vector $\mathbf{x}$ represents the three-dimensional Cartesian coordinates (x, y, z) in Figure 9.1. The first portion of (9.1) is simply an expression of force balance. The second equation is a statement of Hooke's law, relating material stress to material strain. The third expression in (9.1) states that the entire problem is defined in three-dimensional space.

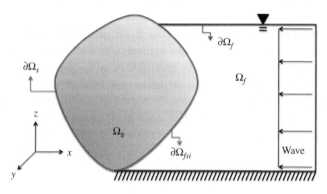

Figure 9.1 Schematic of an FSI problem. *Source*: Reproduced from [5], Figure 1, with permission of Elsevier

In the fluid domain, the requirement for the dynamic equilibrium of the response is governed by the homogeneous acoustic wave equation

$$\nabla^2 \mathbf{p}(\mathbf{x}, t) = \frac{1}{c^2} \ddot{\mathbf{p}}(\mathbf{x}, t) \qquad \mathbf{x} \in \Omega_f \subset \mathbb{R}^3 \tag{9.2}$$

where $\mathbf{p}$ denotes the scalar pressure in the fluid satisfying all prescribed boundary conditions on the fluid boundary; and c is the speed of sound in the fluid.

To derive the finite element (FE) model for the coupled fluid-structure system, we may multiply the partial differential equations (PDE) by kinematically admissible test functions and then apply Green's lemma [6] to the two weighted PDEs. The two systems of equilibrium are then coupled via continuity of linear momentum at the fluid-structure interface:

$$\rho \ddot{\mathbf{u}}_n = \frac{\partial \mathbf{p}}{\partial n} \tag{9.3}$$

that is, the force due to the nodal accelerations at the interface are offset by the pressure gradient in the fluid, normal to the interface. The normal derivative is interpreted in the same manner as in [7], where the pressure gradient is taken as proportional to the force of the fluid on the structure divided by the discrete nodal tributary area. The result is the coupled system of equations (9.4)

$$\begin{bmatrix} \bar{M} & \bar{0} \\ \rho(\bar{G}\bar{A})^T & \bar{Q} \end{bmatrix} \begin{Bmatrix} \ddot{\mathbf{u}} \\ \ddot{\mathbf{p}} \end{Bmatrix} + \begin{bmatrix} \bar{K} & \bar{G}\bar{A} \\ \bar{0} & \bar{H} \end{bmatrix} \begin{Bmatrix} \mathbf{u} \\ \mathbf{p} \end{Bmatrix} = \begin{Bmatrix} \mathbf{F} - \bar{G}\bar{A}p_i \\ \rho\bar{A}\ddot{\mathbf{u}}_{ni} \end{Bmatrix} \tag{9.4}$$

where $\bar{M}$ and $\bar{Q}$ are the solid and fluid mass matrices, respectively; $\bar{K}$ and $\bar{H}$ are the solid and fluid stiffness matrices, respectively; $\bar{A}$ is the matrix of tributary areas along the fluid-structure interface, and $\bar{G}$ is a matrix of direction cosines, that converts a vector of outward normals at the interface to its components in the global space. On the right-hand side of the equation, $\mathbf{F}$ is the vector of applied mechanical forces on the structure; p_i are the incident pressures in the fluid, and $\ddot{\mathbf{u}}_{ni}$ are the incident nodal accelerations on the interface [8].

To solve (9.4), we require both boundary (spatial) and initial (temporal) conditions for the displacements and pressures. The shell is assumed to be loaded along the top edge with boundary conditions as depicted in Figure 9.2. The structure is excited sinusoidally at an acoustic frequency (~ 28 kHz) as depicted. The idea behind such a loading scenario is to explore the possibility of nondestructively "pinging" the structure using readily available equipment (e.g., sonar) and then "listening" to the response. The question is then whether there is enough information in such a response to reliably identify damage-related parameters. The initial pressure profile is assumed zero across the surface of the plate, while the initial displacement profile provides us an opportunity to model dents or any other initial structural imperfections.

As was done in the FE example of Section 7.1, we require a way to describe the shell surface in a way that accounts for the presence of the damage, but leaves us with only a few unknown parameters to identify. Specifically, we parametrize the geometric imperfection in the shell using Gaussian radial basis functions (RBFs) [9, 10] to model the "dented" condition. Figure 9.3 shows a sample (magnified) dent at the center of the structure. Specifically, we model the imperfections

$$\mathbf{u}(\mathbf{x}, \boldsymbol{\theta}) = \sum_{i=1}^{n_k} A_i e^{-\left(\frac{\| \bar{c}_i - \mathbf{x} \|_2^2}{\sqrt{2}\bar{\sigma}_i} \right)^2} \tag{9.5}$$

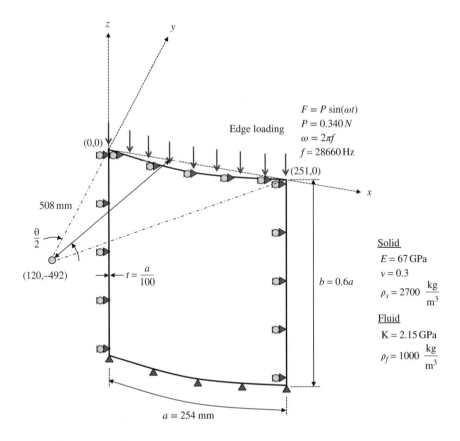

-Denotes x,y translation restraint

-Denotes x,y,z translation restraint

Figure 9.2 Depiction of an edge-loaded barrel vault shell structure. *Source*: Reproduced from [5], Figure 2, with permission of Elsevier

where $\vec{c}_i$ is the ith radial basis center given by two parameters, $\vec{c}_i \equiv (c_{i1}, c_{i2})$ (i.e., we need both "x" and "y" locations) ; $\tilde{\sigma}_i$ is the standard deviation or "width" of the ith dent, and A_i are the associated amplitudes. Thus, for "n_k" RBFs we may construct the unknown parameter vector consisting of $P = 4n_k + 1$ entries

$$\theta = \left\{ \begin{array}{c} \vec{c}_1 \\ A_1 \\ \tilde{\sigma}_1 \\ \vdots \\ \vec{c}_{n_k} \\ A_{n_k} \\ \tilde{\sigma}_{n_k} \\ \sigma^2 \end{array} \right\} \tag{9.6}$$

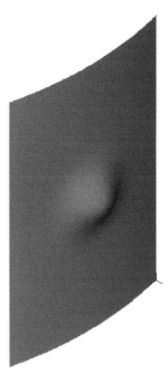

Figure 9.3 Ground truth dent imperfection $u(\mathbf{x}, \theta)$ (magnified 500 ×). The Gaussian RBF serves as a reasonable model for this type of structural damage. *Source*: Reproduced from [11], Figure 2, with permission of Elsevier

where we have also included the observational noise variance term σ^2 in the unknowns. It is possible that one may know this parameter *a priori* and not need to solve for it, however as we showed in Section 7.2.5, adding the noise variance as a parameter incurs little computational cost when it is estimated using the Bayesian approach (a problem we tackle in Section 9.1.3). Substituting Eq. (9.5) into Eq. (9.4) therefore defines the model system response in terms of the unknown imperfection parameters. All other parameters (density, stiffness, etc.) are assumed fixed and known in what follows. Our goal is to estimate the parameters of the imperfection on a portion of a cylindrical shell (barrel vault) that is being loaded acoustically via the fluid/structure interface.

The structural model was implemented using the nonlinear FE method (geometric nonlinearity only). Specifically, the four-node, shell FE given by Bathe and coworkers (i.e., MITC4 Shell) [12, 13, 14] is employed in conjunction with the classical Newton–Raphson solution approach [15]. A detailed mesh convergence study (see Figure 9.4 for a summary of these results) reveals that a MITC4 mesh with approximately 200,000 degrees of freedom represents an accurate mesh; suitable for use as the *ground truth* case within the context of the current inverse problem solution. A less dense mesh (having approximately 50,000 degrees of freedom) is thought to represent a reasonable compromise between modeling accuracy and computational expedience and is used in generating the forward model results required of our estimation techniques. The large number of analyses that are required for the solution

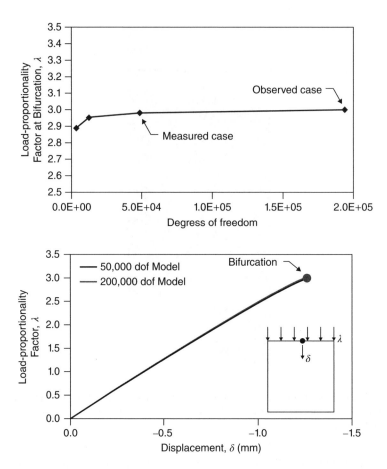

Figure 9.4 Summary of mesh convergence study results. The "load proportionality factor" is an output of many FE packages and is often used to assess FE mesh convergence. Here the quantity is plotted as a function of both number of elements and displacement and illustrates the sufficiency of the 50,000 DOF mesh. *Source*: Reproduced from [11], Figure 3, with permission of Elsevier

of this type of inverse problem (i.e., potentially thousands of runs of the forward model) justifies the trade-off in accuracy for solution speed. It is pointed out that while the standard Newton–Raphson nonlinear solution algorithm is used in the solution of the inverse problems, a hybrid solution algorithm, combining a modified spherical arc length method and the constant increment of external work method [16], is used in the collapse analyses associated with the mesh convergence study. It is further noted that imperfect meshes were used in the above-mentioned convergence study, as per [17]. An approximation to the first global buckling mode, as obtained from a linearized eigenvalue buckling analysis, was employed as a means for perturbing the mesh geometry used during the nonlinear collapse analyses.

Our goal is to use the sensor observations to infer the "true" imperfection state as defined by the parameter vector θ. Inference will be drawn using both maximum likelihood (see Section 6.2) and Bayesian estimation (see Section 7.2.3), whereby the dent parameter posterior distributions are sampled.

9.1.2   Non-Contact Results Using Maximum Likelihood

In our first approach, we use the method of maximum likelihood described in Section 7.1. Given the deterministic model given in the previous section, we begin, as always, by forming a probabilistic model for the observed data. As briefly described in Section 9.1.1, the test problem we are considering is that of a geometrically imperfect barrel vault shell with an acoustic fluid volume on one side of the shell. The shell has a single dent-like imperfection that we have modeled using the RBF described in Eq. (9.5).

To infer the presence of an initial geometric imperfection, only pressure sensors in the fluid are used to measure the pressure waves resulting from the response of the solid domain as illustrated in Figure 9.5. In other words, we will indirectly identify and characterize a dent by "listening" to the shell's response within the fluid. Assume that the fluid pressure can be measured at eight uniformly spaced locations (see Figure 9.5) and at times $t_n = n\Delta_t, n = 1 \cdots N$ with constant sampling frequency $1/\Delta_t$. So as not to confuse "pressure" with "probability," we will use the generic observable **y** to describe our sensor measurements and **z** to serve as the model response vector. In this first study, our model response is pressure, that is, $\mathbf{z} \equiv \mathbf{p}$.

As we have done throughout this book, our discrete signal model is written as

$$\mathbf{y} \equiv y_j(n) = z_j(n, \boldsymbol{\theta}) + \eta_j(n), \quad n = 1 \cdots N, \ j = 1 \cdots M \tag{9.7}$$

where each $\eta_j(n)$ is assumed to be chosen independently from a zero-mean Gaussian distribution with variance σ^2. The signal $z_j(n, \boldsymbol{\theta})$ denotes the pressure measurement for sensor j at discrete time index n, and depends entirely on the structural model parameters $\boldsymbol{\theta}$. The model response is obtained by time marching Eq. (9.4), where again we use the Newmark scheme from Section (5.4) to numerically integrate (9.4) with parameters, $\delta = \frac{1}{2}$ and $\alpha = \frac{1}{4}$, chosen in accordance with the literature [18].

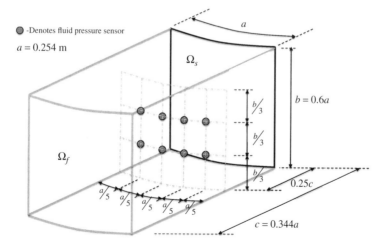

Figure 9.5   Schematic of sensor placement within the fluid domain. *Source*: Reproduced from [5], Figure 3, with permission of Elsevier

Under these assumptions, we have already shown (see development of Eq. 7.5) that the probability (likelihood) of observing the vector **y** is given by

$$p_H(\mathbf{y}|\theta) = \frac{1}{(2\pi\sigma)^{NM/2}} \exp\left[-\frac{1}{\sigma^2}\sum_{n=1}^{N}\sum_{j=1}^{M}(y_j(n) - z_j(n,\theta))^2\right] \tag{9.8}$$

The parameters θ that maximize (9.8) are the maximum likelihood estimations (MLE)s. We have also previously shown that maximizing (9.8) is equivalent to minimizing

$$C(\mathbf{y},\theta) = \sum_{n=1}^{N}\sum_{j=1}^{M}(y_j(n) - z_j(n,\theta))^2 \tag{9.9}$$

where $C(\mathbf{y},\theta)$ is our familiar cost function. While this is not precisely the same as the log-likelihood, the only difference is a scaling constant, hence both functions are minimized by the same parameter vector. The identification problem consists of finding the parameter vector $\theta \equiv \{\vec{c}, \tilde{\sigma}, A\}^{\mathrm{T}}$ (all dimensions in meters) that minimizes Eq. (9.9).

The primary difficulty in structural parameter estimation lies directly in the form of this function. The authors have observed that in many structural system identification problems of interest, the function $C(\mathbf{y},\theta)$ is multimodal, with many similarly valued minima occurring for parameter values that are significantly different from the true (desired) parameter values [19, 20]. This is particularly true in problems such as damage identification, where the damage parameters have a small influence on the observed data. Thus, the practitioner is confronted with minimizing a very challenging function, a problem we discussed at length in Chapter 7.

To better illustrate the problem, consider the "ground truth" plot of Figure 9.3, where the dent has been magnified 500×. To gain a better understanding of the global minimum we are trying to locate during the optimization, plots of the cost function landscape are provided in Figure 9.6. These landscapes illustrate the mean-squared error (essentially our cost, 9.9) between the observed data and the measured data for the parameter values on the axes. The minimum represents the smallest possible error between the ground truth and the measured case, thus it identifies the true parameter combination. As it is not possible to visualize the true minimum for four parameter variables, the landscapes demonstrate how the cost function varies, while two parameters are fixed to their true value. From Figure 9.6, one can observe that there are several "valleys," and thus a number of local minima; typical of this type of structural imperfection identification problem [20]. The observed multiple minima motivate the use of the differential evolution (DE) estimator described in Section 7.1.3, over more efficient classical approaches (e.g., gradient descent).

As can be seen in Figure 9.6a, for a dent amplitude and standard deviation fixed at the true values, the location of the dent is relatively easy to find. However, even if the location of the dent is known, Figure 9.6b, identifying the amplitude and the standard deviation is not nearly as easy to do. In particular, the likelihood is hardly influenced at all by changing the dent amplitude A. For parameters that do not play a large role in model fitting, we can expect a large uncertainty (large variance) in the resulting parameter probability density functions (PDFs); this will be explored using the Bayesian estimation approach of the next section.

To obtain the MLEs we therefore turn to the DE algorithm described in Section 7.1.3. Recall that the strength of this particular implementation of DE is that it allows for the systematic increase of the probability, Pr_d, of choosing parameter vectors that have a difference pointing

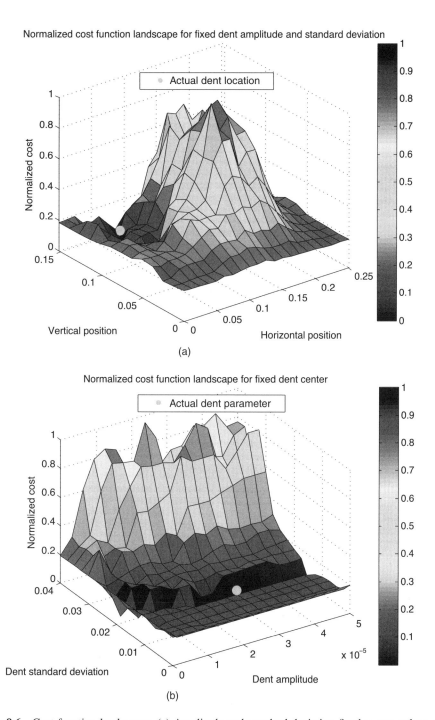

Figure 9.6 Cost function landscapes. (a) Amplitude and standard deviation fixed to true values. (b) Dent center fixed to true value

in the direction of lower cost (see Eqs. 7.41 and 7.42), as well as simultaneously changing the probability of accepting parameters from the mutated vector during the crossover (Pr_{cr}) versus retaining parameters from the initial target vector. As we mentioned in Section 7.1.3, this flexibility provides faster convergence for structural estimation problems.

In this example, we choose to increase the probabilities linearly: the $Pr_d(0)$ is set to 0.5 at the first iteration and then linearly increases, so that at the last iteration $Pr_d(K) = 1$. This can be expressed in software as simply $Pr_d(k) = 0.5 + \frac{k}{2K}$, where k is our usual iteration index, i.e., $k = 1 \ldots K$. While a linear increase is used here, a logarithmic or exponential increase could have also been applied.

To generate the observed data from the ground truth parameters, the fluid-structure forward model formulated in Section 9.1.1 was iterated for the true dent parameters, and then zero-mean Gaussian noise ($\mathcal{N}(0, \sigma_{SNR})$) was added to the "measured" data from the eight sensors, to simulate sensor noise. That is to say, the measured data are given by Eq. (9.7) with the noise amplitude dictated by the SNR, defined as

$$SNR = \frac{\mu}{\sigma_{SNR}}, \text{ where } \mu \equiv \text{ mean amplitude of the sensor response} \qquad (9.10)$$

A signal-to-noise ratio (SNR) of 20 was used to determine the standard deviation in the likelihood function (Eq. 9.8). In what follows, MLEs for the dent parameters, as obtained via different implementations of DE, are provided.

9.1.2.1 Linearly increasing Pr_d and constant Pr_{cr}

First, the proposed DE is compared to the DE from the literature (in which both the Pr_d and the Pr_{cr} remained constant, both equal to 0.5). In the first set of comparison results, the Pr_d are varied linearly from $0.5 \leq Pr_d \leq 1.0$, while the Pr_{cr} remains constant at 0.5. Allowing the Pr_d to increase with iteration increases the probability of choosing a negative cost differential, $d_{qq'}$ (see Eq. 7.41), and leads to faster convergence as we showed in Section 7.1.3. Twenty trials of both the DE from the literature and the modified algorithm were run, in order to assess the variability in the estimates. The average cost of the best fit parameter vectors (i.e., the parameter vector that yielded the lowest cost at the final iteration) from the 20 trials was used as a measure of convergence, as can be seen from curve 2 in Figure 9.7.

9.1.2.2 Linearly Increasing Pr_d and Linearly Increasing Pr_{cr}

Comparable gains in convergence rate can be achieved by allowing the $Pr_{cr}(k)$ to increase with iteration, versus keeping it constant at 0.5; Allowing the $Pr_{cr}(k)$ to increase permits a greater probability that we will accept parameters from the mutated vector, versus retaining parameters from the original target vector. In this set of experiments, both the $Pr_d(k)$ and the $Pr_{cr}(k)$ are varied linearly from 0.5 to 1.0. The average cost of the best fit parameter vectors from the 20 trials is depicted by curve 3 in Figure 9.7.

Table 9.1 shows the mean parameter values at the converged iterations associated with the best fit parameter vectors from the 20 trials, as compared to the ground truth values. While the horizontal and vertical dent location parameters converge to values close to the true values, the dent amplitude parameter is relatively difficult to estimate for all three versions of the DE algorithm.

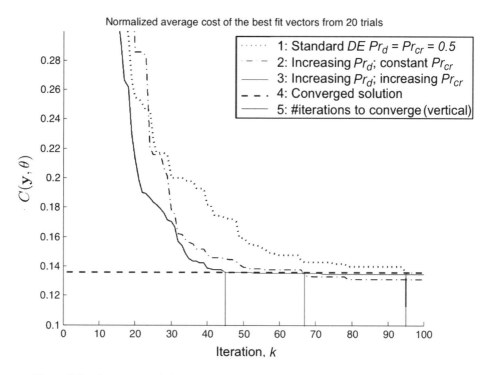

Figure 9.7   Average population cost values for the best fit parameter vectors over 20 trials

Table 9.1   Average identified parameter estimates collected over 20 trials.

	Mean values (m)		
	c_1	c_2	A
True parameters	0.06213	0.12700	3.0e−5
Standard DE $Pr_d = Pr_{cr} = 0.5$	0.06251	0.12703	3.170e−5
Increasing Pr_d; constant Pr_{cr}	0.06263	0.12793	3.536e−5
Increasing Pr_d; increasing Pr_{cr}	0.06206	0.13093	3.764e−5

* Both modified DE and traditional DE-based MLEs for the dent location (c_1, c_2) and dent amplitude parameters. While the traditional DE provides slightly better estimates, this comes at a large computational cost. The modified DE estimates are obtained in half the time as the traditional DE approach.

While the costs associated with all three curves converge to approximately the same value, the behavior and rate of this convergence vary among the methods. All three variations of the algorithm behave similarly for the first approximately 20 iterations. This is to be expected as the Pr_d and Pr_{cr} at the beginning of the algorithm are the same as the traditional DE algorithm. They are slowly increased so that around iteration $k = 20$, the probability has been significantly increased such that the convergence of the parameter vectors is affected. The traditional DE algorithm (curve 1 in Figure 9.7) takes approximately twice as long to converge to the solution as does curve 3. This suggests that filling the parameter vector population with diverse values throughout the duration of the algorithm (as is done in curve 1) is not the fastest way to arrive at an optimal parameter value. Faster convergence is observed in curves 2 and 3 when the algorithm is allowed to narrow in on the global minimum versus continuing to search throughout the parameter space for the true parameter values.

This numerical study has demonstrated the DE algorithm introduced in section 7.1.3. The approach can be used to provide efficient maximum likelihood estimates of parameters related to structural damage. However, as we have mentioned in Chapter 7, the problem can also be approached from a Bayesian point of view.

9.1.3   Bayesian Identification of Dents

In this section we consider the same identification problem, but use the Bayesian estimation approach described in Sections 7.2.3, and 7.3.2. Recall that in this approach we treat the unknown damage parameters as random variables, each modeled with its own PDF. The job of the Bayesian estimator is to obtain these PDFs from which we can extract the final parameter estimate as the mean, median, or some other property of the distribution.

Recall also that the key ingredient in Bayesian estimation is the same challenging likelihood we just finished navigating in the previous section (see Figure 9.6) using the MLE approach. Moreover, because we cannot obtain an analytical solution, we must resort to sampling this function using the Markov chain Monte Carlo (MCMC) algorithm of Section 7.2.3.

As we have discussed in Section 7.2, we can use Bayes' theorem, to relate the joint parameter distribution to the likelihood via

$$p_\theta(\theta \mid \mathbf{y}) = C^{-1} p_H(\mathbf{y} \mid \theta) p_\pi(\theta) \qquad (9.11)$$

wherein $p_\pi(\theta)$ is the prior, embodying any prior understanding concerning the parameter vector $\theta \in \mathbb{R}^P$. The normalizing constant C is given by the multidimensional integral $\int_{\mathbb{R}^P} p_\eta(\mathbf{y} \mid \theta) p_\pi(\theta) d\theta$. Also recall that while Eq. (9.11) provides the joint (PDF), we are typically interested in the *marginal* distribution associated with the individual parameters (e.g., $p_{\theta_1}(\theta_1)$). Analytically, this requires integrating Equation (9.11) over each of the other parameters in the vector. Denoting the parameter vector with the ith parameter removed as θ_{-i}, the marginal distributions are found by

$$p_{\theta_i}(\theta_i) = \int_{\mathbb{R}^{P-1}} p_\theta(\theta \mid \mathbf{y}) d\theta_{-i}. \qquad (9.12)$$

Either the posterior mean or the median are commonly employed choices for the final parameter estimate. In addition, we may use the posterior to form a credible interval for each parameter (i.e., the probability that the parameter falls in a specific interval). This is precisely the information that we are interested in. The goal of the stochastic inverse problem is therefore to solve (9.12). We have already shown in Section 7.2.3 how to numerically obtain the parameter posteriors without having to analytically solve Eq. (9.12) using the MCMC algorithm.

Recall that the MCMC method numerically generates K samples from the posterior distribution of interest, that is, draws

$$\theta_i(k) \sim p_{\theta_i}(\theta_i), \quad k = 1 \cdots K \qquad (9.13)$$

via a generating distribution function $q(\cdot)$, and subsequently evaluating the ratio (see again, section 7.2.3)

$$r = \frac{p_{\theta_i}(\theta_i^*)q(\theta_i(k)|\theta_i^*)}{p_{\theta_i}(\theta_i(k))q(\theta_i^*|\theta_i(k))} = \frac{p_\eta(\mathbf{y}|\theta_i^*)p_{\pi_i}(\theta_i^*)}{p_\eta(\mathbf{y}|\theta_i)p_{\pi_i}(\theta_i)}\frac{q(\theta_i(k)|\theta_i^*)}{q(\theta_i^*|\theta_i(k))}. \tag{9.14}$$

The candidate value θ_i^* is accepted with probability $\min(r, 1)$ as $\theta_i(k+1) = \theta_i^*$; otherwise, we retain the previous value $\theta_i(k+1) = \theta_i(k)$. The stationary Markov chain that results contains samples from the desired posterior distribution. Typically, however, there is a nonstationary transient period during the start of the chain, as the values converge toward their stationary distribution. Particularly, if the prior $\theta_i(0) \sim p_{\pi_i}(\theta)$ is poor, it may take some number of "burn-in" samples, B, before the influence of the data (influence of the likelihood in evaluating r) overwhelms the bad initial guess (i.e., "you can twist perception, reality won't budge" [21]). Provided that there are enough data, even a poorly chosen prior will often not significantly impact the result. The process repeats for $K > B$ samples, and the final $K - B$ values are taken to be truly representative samples from $p(\theta_i)$. Moreover, we know from Section 7.2.4 that for multivariate parameter vectors we can use the Gibbs sampling strategy whereby each parameter is taken in turn, holding the others fixed. In other words, for iteration k in the Markov chain, for the ith parameter, we are sampling from the conditional posterior

$$p_{\theta_{-i}}(\theta_i(k+1)|\theta_1(k+1), \cdots, \theta_{i-1}(k+1), \theta_{i+1}(k), \cdots, \theta_P(k)) \tag{9.15}$$

Each of the parameters are considered sequentially, until we have sampled the entire vector $\theta(k+1)$, at which point we increment k and begin the next iteration.

For the candidate generating distribution we use Eqn. 7.119 so that

$$q(\theta_i^*|\theta_i(k)) = \frac{1}{2A} \text{ if } |\theta_i^* - \theta_i(k)| < A \tag{9.16}$$

and zero otherwise, i.e., the uniform distribution with width $2A$. Clearly this distribution is symmetric in exchange of its arguments, so that the ratio of candidate-generating distributions in Eq. (9.14) cancel out, and r simply becomes the ratio of posterior distributions, evaluated at both the candidate and original values of the parameter. One can think of the proposal as applying a perturbation to the existing solution, and subsequent acceptance or rejection of that perturbation is dictated via the ratio r. The tuning parameter A is adjusted during the burn-in phase according to the rules specified in Eqns. 7.124 and 7.125

The MCMC approach is computationally expensive as it requires running the forward model N times at each iteration in the Markov chain and either accepting or rejecting the model parameter according to Eq. (9.14). This is unavoidable for most parameters in structural dynamics problems. A slight computational savings can be enjoyed for the noise variance parameter σ via clever choice of prior.

As we have pointed out in Section 7.2.5, a Gamma prior can be used with a joint Gaussian likelihood to produce an analytical posterior distribution for the variance parameter. In this work, we choose a diffuse Gamma prior with $\alpha = 1$ and $\beta = 0$, which yields

$$\sigma^2 \sim \frac{1}{\Gamma\left(\frac{(K+N)}{2}+1, \frac{2}{C(\mathbf{y},\theta)}\right)} \tag{9.17}$$

for the variance posterior distribution. At each step k in the Markov chain, we simply draw a sample from the inverse Gamma distribution (9.17) as $\sigma^2(k)$. In the end, we not only have our model parameter posterior distributions but also an estimate of the noise in our system.

9.1.3.1 MCMC Results

To illustrate the algorithm performance, we consider two cases. In the first, the "true" dent matches exactly the assumed functional form, that is, an RBF with unknown location, amplitude, and standard deviation. That is to say, there is no modeling error and only additive measurement noise is considered. In the second case, modeling error is introduced in the form of perturbations to the true dent conditions. Specifically, at each discrete location in the simulated plate response, we add perturbations $\Delta_y \sim U[-0.005t, 0.005t]$ to the out-of-plane response where t is the plate thickness.

For both cases, we collect a single, noise-corrupted time-series response $y(n), n = 1 \cdots 16$, sampled at an interval of $\Delta_t = 2.2e^{-6}$ s. This sampling time and number of points are sufficient to resolve the pressure wave as it moves across the sensors. The actual dent profile was defined by an amplitude and standard deviation of $A = 5e^{-5}$ m and $\tilde{\sigma} = 0.018$ m, respectively, while the dent was located at $c_1 = 0.0990$ m and $c_2 = 0.1016$ m. The noise level was set to $+100$ dB below the level of the signal. After discarding the first 10,000 iterations of the MCMC algorithm for "burn-in," we retain the next 40,000 as samples from the damage parameter posterior distributions. These distributions are shown in Figures 9.8 through 9.11 for both of the examined cases (modeling error and no modeling error). Each of the PDFs has been normalized to unit amplitude at the maximum value for comparison purposes. It is apparent that, depending on the parameter being estimated, that modeling error can significantly alter the ability to alter the resulting PDF. Generally speaking, this type of error causes an increase in the variance of the posterior PDF. This can be seen clearly in each of the parameter estimates. Sometimes this type of error can also result in additional "peaks" in the posterior PDF (see, e.g., Figure 9.9).

However, even with an increased variance, the maximum *a posterior* estimate (parameter value at which PDF is maximized) are close to the true values and, moreover, the 90% credible interval always contains the true values.

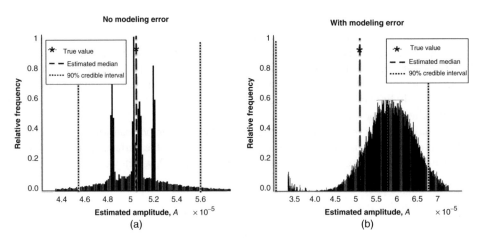

Figure 9.8 Estimated PDFs for the dent amplitude for the case with no modeling error (a) and modeling error (b)

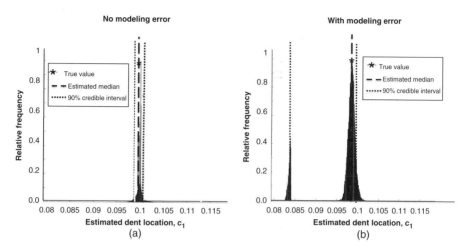

Figure 9.9   Estimated PDFs for the dent location parameter c_1 for the case with no modeling error (a) and modeling error (b).

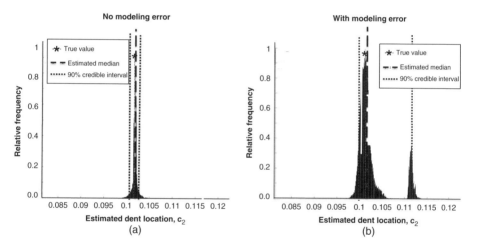

Figure 9.10   Estimated PDFs for the dent location parameter c_2 for the case with no modeling error (a) and modeling error (b).

An obvious criticism of this basic approach is that it does not include the possibility of multiple instances of damage. This is a shortcoming we have discussed in theory in Section 7.3 and one we will address in the next section in the context of the dent identification problem.

9.1.3.2   RJMCMC Results

To this point, our identification of dents has focused on a single imperfection. In reality, we may also need to identify the number of imperfections. Attempting to identify the number and/or type of damage places the identification problem squarely in the realm of model selection,

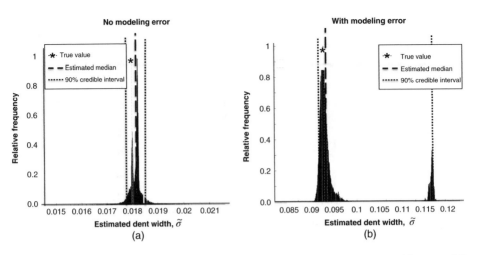

Figure 9.11 Estimated PDFs for the dent width (standard deviation) for the case with no modeling error (a) and modeling error (b)

discussed in Section 7.3. In this section, we use the "reversible jump" MCMC approach of Section 7.3.2 to identify both the number of imperfections and the parameters associated with those imperfections.

Here we consider two types of moves in constructing the RJMCMC sampler:

- **Birth/Death** (see Section 7.3.2 for a description of this pair of moves).
- **Split/Merge** As the name implies, a split is when a single damage instance is divided into two instances, while a merge results when two nearby damage instances combine to form one. As with Death/Birth moves, the probability of a split is zero when n_k is some maximum value (user defined), while a merge cannot occur for $n_k < 2$.

  Merge: To achieve good acceptance ratios, merge moves between any two kernels (j_1 and j_2, $j_1 \neq j_2$), in the parameter vector θ of dimension n_k, are only allowed if the following two conditions are met:

$$\frac{||c_{j_1} - c_{j_2}||}{\sqrt{2\sigma_{j_1}^2 + 2\sigma_{j_2}^2}} \leq \delta_c \qquad (9.18)$$

and

$$||A_{j_1} - A_{j_2}|| \leq \delta_A \qquad (9.19)$$

where the values of $\delta_c = 2$ and $\delta_A = 0.3$ were used. During a merge move, Eqs. (9.18) and (9.19) are evaluated for every possible combination of kernel pairs in the parameter vector. If both conditions are met for a particular pair, then that pair is eligible to be merged. This ensures that the average of Eq. (9.5) (average dent displacement) will be the same as that resulting from the "split" move, defined next. Given the conditions have been met, kernels

j_1 and j_2 are deleted and replaced with a new kernel possessing parameters

$$\tilde{\sigma}_j = \sqrt{\tilde{\sigma}_{j_1}^2 + \tilde{\sigma}_{j_2}^2}$$

$$A_j = \frac{A_{j_1}\sigma_{j_1} + \omega_{j_2}\sigma_{j_2}}{\sigma_j^2}$$

$$\vec{c}_j = \frac{\vec{c}_{j_1} + \vec{c}_{j_2}}{2} \tag{9.20}$$

Split: The split move must be the reverse of the merge move (i.e., an application of the split move must return a merged kernel (j) back to the two kernels j_1, j_2), and thus the parameters of the two new kernels must be consistent with Eqs. (9.20). In this example, a kernel j is selected randomly with uniform probability from the existing n_k kernels and is substituted by two new kernels j_1, j_2 as follows:

– A scalar u_σ is drawn from $U[0, 1]$ and we set

$$\sigma_{j_1}^2 = u_\sigma \sigma_j^2, \sigma_{j_2}^2 = (1 - u_\sigma)\sigma_j^2 \tag{9.21}$$

– We similarly draw a two-dimensional vector $\mathbf{u}_c$ from a circle of radius $R = \frac{\sqrt{2}}{2}\delta_c\sigma_j$ and set

$$\vec{c}_{j_1} = \vec{c}_j - \mathbf{u}_c, \vec{c}_{j_2} = \vec{c}_j + \mathbf{u}_c \tag{9.22}$$

– Finally, a scalar u_A is drawn uniformly from $U\left[-\frac{\delta_A}{2}, \frac{\delta_A}{2}\right]$ and

$$A_{j_1} = A - u_A, A_{j_2} = A + u_A \tag{9.23}$$

where

$$\hat{A} = \frac{A_j + u_A\left(\sqrt{u_\sigma} - \sqrt{1-u_\sigma}\right)}{\sqrt{u_\sigma} + \sqrt{1 - u_\sigma}}. \tag{9.24}$$

The new parameter vector therefore becomes

$$\mathbf{u} = \{u_\sigma, \mathbf{u}_c, u_A\} \tag{9.25}$$

and $g(\mathbf{u})$ (the mapping function given by Eqn. 7.162) becomes the product of the uniform distributions used in generating this move so that $g(\mathbf{u}) = \frac{1}{\pi R^2}\frac{1}{\delta_A}$. The Jacobian of this move is therefore

$$|J(g(\mathbf{u}))| = \frac{4\sigma_j}{\sqrt{u_\sigma}\left(\sqrt{1-u_\sigma} + \sqrt{u_\sigma}\right)\sqrt{1 - u_\sigma}}. \tag{9.26}$$

Both types of moves (split and merge) are shown graphically in Figure 9.12.

As a test of the RJMCMC estimator, assume the two-dent damage profile given in Figure 9.13. The location of these dents are given by $\mathbf{c}_1 = (0.06, 0.19)$ and $\mathbf{c}_2 = (0.12, 0.07)$, respectively. The amplitudes of the two dents are $A_1 = 7e - 5$ and $A_2 = 4.5e - 5$, while both dents are given a standard deviation of $\tilde{\sigma}_{1,2} = 0.012$. In this, and likely many other damage

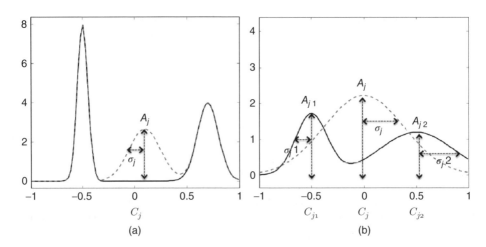

Figure 9.12 Illustration of both "merge" (a) and "split" (b) moves. In the former, two damaged kernels are combined to form a single kernel located at the midpoint and with a standard deviation and amplitude governed by (9.20). For the split move, a single kernel is divided into two with the parameters chosen according to the rules (9.21–9.23)

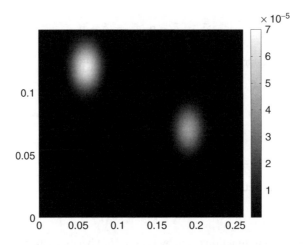

Figure 9.13 Two-dent damage configuration considered in the RJMCMC example given in this section

identification applications, a prior probability mass function (PMF) on the "model" variable should enforce a limited damage scenario, that is, a few damage instances are considered more likely than many.

Consider first a discrete Poisson PMF as describing the probability of observing model n_k dents, given an available Poisson rate parameter λ

$$p_{\mathcal{M}}(n_k|\lambda) = \frac{1}{n_k!}\lambda^{n_k}e^{-\lambda}. \qquad (9.27)$$

If we wish to eliminate the dependency on λ, we may note that

$$p_{\mathcal{M}}(n_k|\lambda) = \frac{p_{\mathcal{M},\Lambda}(n_k, \lambda)}{p_\Lambda(\lambda)} \tag{9.28}$$

so that

$$p_{\mathcal{M},\Lambda}(n_k, \lambda) = p_{\mathcal{M}}(n_k|\lambda)p_\Lambda(\lambda) \tag{9.29}$$

giving the desired marginal distribution

$$p_{\mathcal{M}}(n_k) = \int_\Lambda p_{\mathcal{M}}(n_k|\lambda)p_\Lambda(\lambda)d\lambda. \tag{9.30}$$

To simplify this integral, we may once again use the concept of conjugacy (see Section 7.2.1). We note that if we choose for our hyperparameter distribution $p_\Lambda(\lambda) \sim Gamma(\alpha, \beta)$, and assume no prior data was available in choosing α, β, then this integral can be accomplished in closed form to yield

$$p_{\mathcal{M}}(n_k) = NB\left(\alpha, \frac{\beta}{1+\beta}\right) \tag{9.31}$$

where $NB(\cdot, \cdot)$ is the negative binomial distribution. Choosing $\alpha = \beta = 1$ yields

$$p_{\mathcal{M}}(n_k) = NB\left(1, \frac{1}{2}\right) = 2^{-(n_k+1)} \tag{9.32}$$

for our model prior distribution.

This distribution is shown in Figure 9.14 along with the posterior PMF obtained by running the RJMCMC algorithm of Section 7.3.2. Essentially, zero probability was assigned to models of order $> n_k = 3$ due to a combination of the problem physics (only two dents present) and the prior, which favored fewer dents. In addition to providing information about the number of damage instances, the RJMCMC procedure provides estimates of the model parameters.

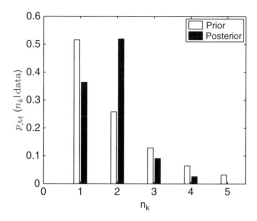

Figure 9.14 Prior and posterior probability distributions for the model order parameter n_k. The posterior distribution, obtained via RJMCMC, selects the model with two dents as the most probable given the observed acoustic pressure field

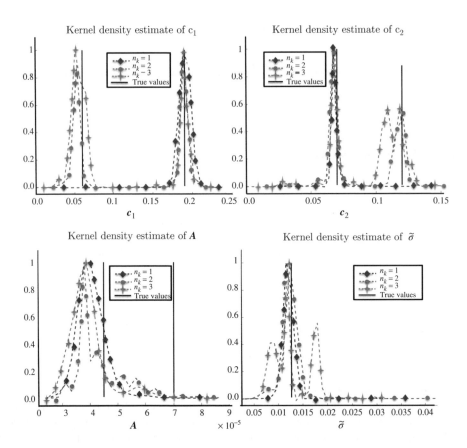

Figure 9.15 Posterior probability distribution associated with each of the four parameters used to describe the damage, plotted as a function of the number of damage instances identified. For $n_k = 1$, the distributions are unimodal with the probability aligning with the larger of the two damage instances. For the multiple dent model the locations are correctly identified. The amplitude parameter presents the biggest challenge and no single damage is assigned the correct value for the larger dent. However, consider that two or more collocated dents with the same, lower amplitude can sum to model a dent with the correct amplitude

Under each model ($n_k = 1, 2, 3, \cdots$), we obtain Markov chains associated with the parameters $c_1, c_2, A, \tilde{\sigma}$. The posterior probability distributions for each of these parameters are shown in Figure 9.15 as a function of the number of damage instances. For $n_k = 1$, the distributions are unimodal, with only the dent at location $c_1 = 0.19, c_2 = 0.07$ (in meters) clearly identified. The fact that this dent, as opposed to the second dent, was identified is likely a consequence of the particular draw from the prior distribution used in seeding the Markov chains. As we show in a subsequent example (see Section 9.3), when there are multiple peaks in the likelihood, the MCMC approach can sometimes become "stuck" in one particular peak. Nonetheless, for $n_k = 1$, the algorithm correctly identifies one of the two dents. We know from Figure 9.14, however, that the $n_k = 2, 3$ cases are higher probability. For both $n_k = 2$ and $n_k = 3$, two dents are clearly identified by the algorithm at the correct locations. The algorithm further correctly identifies the damage width ($\tilde{\sigma} = 0.12$) shared by both dents. The approach struggles, however, in correctly discerning the two different damage amplitudes. While the smaller amplitude damage is identified at slightly less than the true value, the larger amplitude does not appear to be. However, an explanation can be found by considering again our damage model (9.5). In the instance that the Markov chains associated with c_1, c_2 predict similar locations (two dents on top of one another), the amplitudes simply sum. Taking the maximum *a posterior* estimate from Figure 9.15 for amplitude $\sim 3.5e-5m$, we see that two collocated dents with this amplitude yields, from the perspective of the likelihood, a single dent with twice the amplitude, $\sim 7e-5m$, that is, the true amplitude. In fact, this highlights one of the challenges of the RJMCMC process, namely, interpreting parameters from different models. Although using multiple models comes with both computational and interpretation challenges, it clearly provides information relevant to the damage identification process and ultimately decisions that are made regarding structural maintenance.

Identifying geometric imperfections in shell structures is a difficult, yet widespread problem within the structural mechanics community. The reduction in component strength due to such imperfections can be large, necessitating that the component be removed from the surface and repaired. Condition-based maintenance for such structures therefore requires the accurate identification of the presence, size, and location of this type of damage as well as a measure of the uncertainty in the estimate. For example, the uncertainty associated with the inferred model parameters can now be propagated forward into the instability problem discussed in [10], allowing for a better characterization of the buckling strength of the imperfect shell structure.

In this section, we have considered different approaches to identifying this type of damage, using both maximum likelihood and Bayesian estimation methods. Regardless of the particulars of the approach, the challenge in these types of problems is to successfully navigate a complicated likelihood function for which the damage parameters have only a mild influence. This is a recurring theme in structural identification and further highlights the importance of advanced numerical methods for exploring such functions.

9.2   Modeling and Identification of Delamination

We have already discussed the modeling of delamination in a composite beam structure (see Section 5.2) and the detection of the delamination using higher order spectral analysis (8.5). In this section, we go one step further and attempt to estimate the parameters associated with the delamination (location, length, and depth).

Recall the dynamic beam model shown schematically in Figure 9.16. This model is low dimensional (only three independent coordinates need to be specified), yet was shown experimentally to accurately capture the localized buckling that occurs because of the presence of the delamination under static loading in Section 5.2. In this section we must first take the static analysis of section 5.2 and extend it to a dynamic regime, where our vibration-based analysis and estimation techniques apply. The global beam motion is assumed to be dictated by the first mode of the response; thus, the global displacement assumes the form

$$y_1(x_1, t) = q_1(t)\psi_1(x_1)$$

where

$$\psi_1(x) = \frac{3}{2}\left[\left(\frac{x}{L}\right)^2 - \frac{1}{3}\left(\frac{x}{L}\right)^3\right]$$

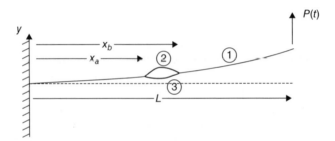

Figure 9.16 Schematic of the dynamic delaminated beam model. Region 1 is simply modeled as a linear cantilevered beam whose motion is governed by the first mode of vibration. Regions 2 and 3 are modeled as nonlinear beams where axial stretching is permitted. The dynamic vibration problem depicted here is different than that considered in the static case of Section 5.2. *Source*: Reproduced from [25], Figure 3, with permission of Elsevier

is a normalized shape function describing the vertical beam deflection at any point x, measured from the left end of the beam. The other two coordinates, describing the time-dependent motion of the upper (region 2) and lower (region 3) laminates, respectively, are assumed to be of the form

$$y_2(x_2, t) = y_1(x_2 + x_a, t) + q_2(t)\Psi_2(x_2) + \frac{1}{2}(1 - a)h$$

$$y_3(x_3, t) = y_1(x_3 + x_a, t) + q_3(t)\Psi_3(x_3) - \frac{ah}{2}$$

$$\Psi_{2,3} = 1 - \cos^2\left(\pi\frac{x_{2,3} - x_a}{x_b - x_a}\right) \qquad (9.33)$$

where the shape functions $\Psi_{2,3}$ describe the deflected shape of regions 2 and 3 in figure 9.16. The constant terms in Eq. (9.33) denote the neutral axis offsets, measured from the global neutral axis, of each of the laminates.

Following the procedure described in Chapter 4, we take an energy-based approach to deriving the dynamic equations of motion. First, the bending energies associated with the three sections of the composite are given by

$$U_b = \frac{EI_1}{2}\int_0^{x_a}\left(\frac{\partial^2 y_1}{\partial x_1^2}\right)^2 dx_1 + \frac{EI_1}{2}\int_{x_b}^L\left(\frac{\partial^2 y_1}{\partial x_1^2}\right)^2 dx_1 + \sum_{i=2}^3\frac{EI_i}{2}\int_0^{L_i}\left(\frac{\partial^2 y_i}{\partial x_i^2}\right)^2 dx_i \quad (9.34)$$

where $I_1 = \frac{bh^3}{12}$, $I_2 = \frac{b(ah)^3}{12}$, and $I_3 = \frac{b((1-a)h)^3}{12}$ are the area moments of inertia. The lengths of these beams are $L_1 = x_a$, $L_2 = L_3 = (x_b - x_a)$. The nonlinearity in the equations of motion result from the midline stretching that can occur in regions 2 and 3 of the structure. To first order, the stretching of the two neutral axes of these beams is

$$u_i = \frac{1}{2}\int_0^{L_i}\left(\frac{\partial y_i}{\partial x_i}\right)^2 dx_i. \qquad (9.35)$$

In addition, the nonzero slopes at the left and right ends of the delaminated region results in additional stretching/compression of the two neutral axes. The result of both sources of stretching leads to the end deflections

$$\delta_2 = \frac{1}{2}\int_0^{L_2}\left(\frac{\partial y_2}{\partial x_2}\right)^2 dx_2 + \frac{1}{2}(1-a)h\frac{\partial y_1}{\partial x_1}(x_a) - \frac{1}{2}(1-a)h\frac{\partial y_1}{\partial x_1}(x_b)$$

$$\delta_3 = \frac{1}{2}\int_0^{L_3}\left(\frac{\partial y_3}{\partial x_3}\right)^2 dx_3 - \frac{ah}{2}\frac{\partial y_1}{\partial x_1}(x_a) + \frac{ah}{2}\frac{\partial y_1}{\partial x_1}(x_b). \tag{9.36}$$

The net stretching energy then becomes

$$U_s = \frac{ahbE}{2L_2}\delta_2^2 + \frac{(1-a)hbE}{2L_3}\delta_3^2. \tag{9.37}$$

The kinetic energy is given by

$$T = \frac{\rho bh}{2}\int_0^{x_a}\dot{y}_1^2\,dx_1 + \frac{\rho bh}{2}\int_{x_b}^L\dot{y}_1^2\,dx_1 + \frac{\rho bah}{2}\int_0^{L_2}\dot{y}_2^2\,dx_2 + \frac{\rho b(1-a)h}{2}\int_0^{L_3}\dot{y}_3^2\,dx_3 \tag{9.38}$$

while the external work done by the time-varying point load + dissipative forces is given by

$$W = P(t)L - \sum_{i=1}^3 (c_i\dot{y}_i)y_i \tag{9.39}$$

leading to the three generalized forces $Q_1 = P(t)L - c_1\dot{q}_1$, $Q_2 = -c_2\dot{q}_2$, and $Q_3 = -c_3\dot{q}_3$. Denote this generalized forcing vector $F(t) = (Q_1, Q_2, Q_3)$. Here, it is assumed that the load is applied at the beam tip and that all damping is viscous. Forming the Lagrangian $L = T - U_b - U_s$ (see Section 4.3.4) and substituting into

$$\frac{\partial}{\partial t}\left(\frac{\partial L}{\partial \dot{q}_i}\right) - \frac{\partial L}{\partial q_i} = Q_i \tag{9.40}$$

results in a set of three coupled, nonlinear differential equations in terms of the time-dependent vector $\mathbf{q}(t) \equiv (q_1(t), q_2(t), q_3(t))$

$$[M]\ddot{\mathbf{q}}_t + [C]\dot{\mathbf{q}}_t + [K_L]\mathbf{q}_t + [K_M]q_1(t)\mathbf{q}_t + [K_Q]\mathbf{q}_t^2 + [K_C]\mathbf{q}_t^3 = F(t) \tag{9.41}$$

The damping for both the global beam motion and laminates is assumed to follow a viscous model such that we have

$$[C] = \begin{pmatrix} c_1 & 0 & 0 \\ 0 & c_{2,3} & 0 \\ 0 & 0 & c_{2,3} \end{pmatrix} \tag{9.42}$$

The equations of motion are therefore described in terms of the material properties of the beam, the dimensions of the beam, and the parameters associated with the damage. These damage parameters are x_a, the delamination starting location, x_b, the delamination end point, and a, the delamination depth specified as a fraction of the overall beam thickness h. In addition, we have chosen to add a viscous damping model for both the global vibration of the beam and the local vibrations of the delaminated portions of the beam.

Specifically, the linear stiffness matrix $[K_L]$ is given by

$$[K_L] = \begin{pmatrix} \frac{bh^3(-4L^3+3(a-1)a(x_b-x_a)^3)E}{16L^6} & 0 & 0 \\ 0 & \frac{a^3bh^3\pi^4E}{6(x_b-x_a)^3} & 0 \\ 0 & 0 & \frac{(1-a)^3bh^3\pi^4E}{6(x_b-x_a)^3} \end{pmatrix} \quad (9.43)$$

The two matrices involving quadratic terms are defined as

$$[K_M] = \begin{pmatrix} 0 & 0 & 0 \\ 0 & -\frac{3(1-a)abh^2\pi^2(2L-x_a-x_b)E}{8L^3(x_b-x_a)} & 0 \\ 0 & 0 & \frac{3(1-a)abh^2\pi^2(2L-x_a-x_b)E}{8L^3(x_b-x_a)} \end{pmatrix}$$

$$[K_Q] = \begin{pmatrix} 0 & -\frac{3(1-a)abh^2\pi^2(2L-x_a-x_b)E}{16L^3(x_b-x_a)} & \frac{3(1-a)abh^2\pi^2(2L-x_a-x_b)E}{8L^3(x_b-x_a)} \\ 0 & 0 & 0 \\ 0 & 0 & 0 \end{pmatrix} \quad (9.44)$$

The matrix multiplying the cubic term is given by

$$[K_C] = \begin{pmatrix} 0 & 0 & 0 \\ 0 & \frac{abh\pi^4E}{8(x_b-x_a)^3} & 0 \\ 0 & 0 & \frac{(1-a)bh\pi^4E}{8(x_b-x_a)^3} \end{pmatrix} \quad (9.45)$$

The mass matrix is slightly more complicated. Defining

$$\alpha = \frac{-1}{16L^3\pi^2}bh(x_b-x_a)[(x_a+x_b)\rho((-3+\pi^2)x_a^2 + 6x_ax_b + (-3+\pi^2)x_b^2)$$
$$-2L((-3+2\pi^2)x_a^2 + 2(3+\pi^2)x_ax_b + (-3+2\pi^2)x_b^2)] \quad (9.46)$$

the mass matrix is given by

$$[M] = \begin{pmatrix} \frac{33}{140}bhL\rho & a\alpha & (1-a)\alpha \\ a\alpha & \frac{3}{8}abh\rho(x_b-x_a) & 0 \\ (1-a)\alpha & 0 & \frac{3}{8}(1-a)bh\rho(x_b-x_a) \end{pmatrix} \quad (9.47)$$

thus completing the model.

In the following analysis, it is assumed that there is no uncertainty in the beam length or material properties. We therefore set $L = 0.24$ (m), $h = 2.25$ (mm) $EI = 75,889,600,000$ (GPa), and $\rho = 1234.0$ (kg/m). The damage parameters, x_a, x_b, a are assumed to be unknown as are the viscous damping coefficients $c_1, c_{2,3}$ (see Eq. 9.42) and the initial global beam deflection $y_1(0)$. Rather than assume the ability to apply a dynamic load, we assume the more likely scenario of observing the free-decay response of the beam to impulse excitation; hence the need to identify the initial condition. Again we assume that we will have access to a single, noise-corrupted signal and that the noise is additive, i.i.d. Gaussian distributed. Thus, the likelihood function our familiar Eqn. 7.5, assuming that only one signal (in this case, the global motion) can be recorded. In Section 8.5 we demonstrated that global, vibration-based detection was not possible unless the measurement was recorded from near the delamination site. This stems from the fact that the only coupling between the laminate vibrations and the rest of the beam is inertial. Thus, for relatively small delaminations (what we are interested in), there is little in the way of influence with respect to the global motion away from the delamination site. In what follows, we therefore must assume that we are able to record the beam response from the delamination site. Thus, the model data are given by $y_n \equiv y_2((x_b - x_a)/2, n\Delta t)$ (Eq. 9.33). Again we assume a transient response to an initial displacement and use $N = 512$ sampled points at a sampling interval of $\Delta_t = 0.001$ (s). The variance of the additive Gaussian noise was fixed such that the signal-to-noise ratio was $SNR = +20\,\mathrm{dB}$.

Figure 9.17 shows the observed (noise-corrupted) data along with the true underlying free-decay signal. Also shown are the estimated structure parameter distributions, initial condition distributions, and the distribution of the noise variance along with the true values for each parameter. Despite the *very* small influence of the nonlinearity on the response, the algorithm is still able to identify the delamination start and end points as well as the depth. The delamination depth appears easier to identify as evidenced by the very narrow confidence interval around the correct value of $a = 0.1$. The delamination start and end points have very little influence on the global response as was illustrated in [26], hence the confidence intervals for these two parameters are large. The Bayesian approach is clearly giving us an indication of the degree to which we can trust our estimates. In this case, we can be fairly certain of the delamination depth, but much less certain about our ability to estimate both the beginning and end points of the delamination. If our application requires a tighter confidence interval, we might need to try a larger excitation (allowing the nonlinearity to more strongly influence the response) or maybe even change to a local damage detection method. Regardless, the information about our faith in the ability to estimate these parameters is clearly valuable information. Both the linear damping c_1, the initial beam deflection $y_1(0)$, and the noise variance σ_G^2 are easily estimated with a high degree of confidence. These parameters clearly have a large influence on the global response, thus one might expect them to be easily obtained.

It turns out that shallow delaminations, such as the $a = 0.1$ case just presented, are more difficult to identify than thicker ones. Consider the case of a delamination of depth $a = 0.2$ with the start and end points of the delamination fixed at $x_a = 0.05$ (m) and $x_b = 0.2$ (m), respectively. As in the previous example, the global damage parameters, $c_1, y_1(0)$ are trivial to identify. The difficult-to-identify parameters, $x_a, x_b, a, c_{2,3}$, are shown in Figure 9.18. The variance associated with the estimates of each of the parameters is reduced from the previous case. This example points out that certain combinations of parameters are more easily identified than others. It depends to a large extent on the degree to which the estimated model parameters influence the observed data. Parameters with little effect on the observed response will be hard to identify, whereas the converse is also true.

As a final example, we demonstrate how the approach can be used to track damage in a structure. For this example, the delamination starting point and depth were fixed to the values $x_a = 0.05$ (m) and $a = 0.2$, respectively. The delamination end point was slowly varied from $x_b = 0.1$ (m) up to $x_b = 0.2$ (m). Again, using only the noisy free-decay response, the goal was to estimate and track the delamination end point. Figure 9.19 shows the progression of estimated delamination length along with the associated 95% credible interval. Also shown are the "true" values for x_b used in generating the time series. For each successive

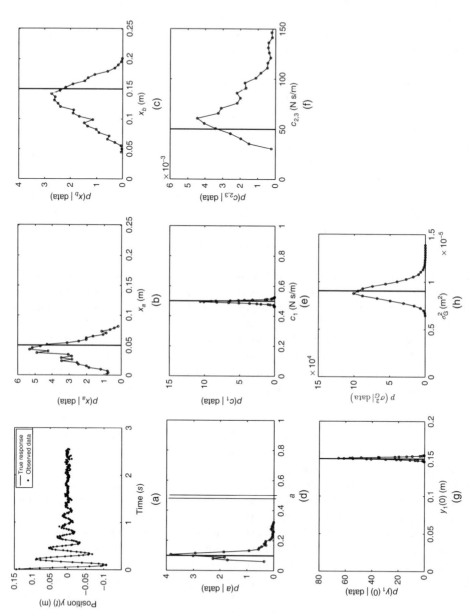

Figure 9.17 Actual and observed (i.e., noise corrupted) impulse response data followed by posterior distributions of, respectively, delamination start point x_a, delamination end point x_b, delamination depth a, global coefficient of viscous damping c_1, local damping coefficient $c_{2,3}$, initial beam tip deflection $y_1(0)$, and noise variance σ_G^2. *Source:* reproduced from [25], Figure 4, with permission of Elsevier

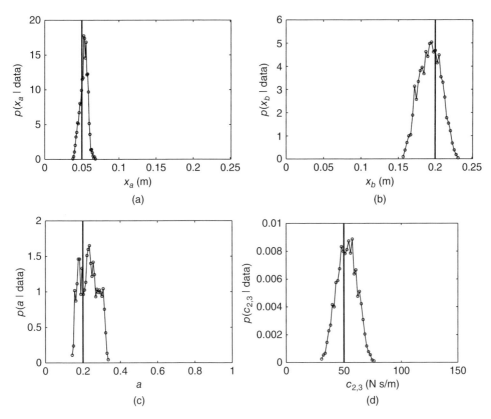

Figure 9.18 Estimated posterior distributions of, respectively, delamination start point x_a, delamination end point x_b, delamination depth a, and local damping coefficient $c_{2,3}$. *Source*: Reproduced from [25], Figure 5, with permission of Elsevier

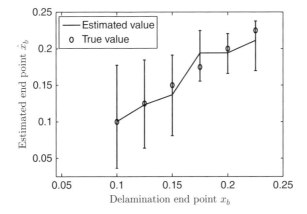

Figure 9.19 Estimated delamination end point $\hat{x}_b$ as a function of the actual delamination length. The intervals of confidence were created as the central 95% of the sampled values from $p(x_b)$ obtained using the MCMC algorithm. *Source*: Reproduced from [25], Figure 6, with permission of Elsevier

damage case, we use a prior PDF that is based on the posterior PDF from the previous case. Specifically, we select the prior from a uniform distribution centered at the previous posterior mean value with a width of roughly 2 standard deviations. For the first damage case ($x_b = 0.1$ (m)), we also assumed a uniform prior on the range $x_b \sim U(x_a = 0.05, 0.1)$ (i.e., assuming an initial delamination between 0 and 0.05 m). Again, for this particular example, the damage parameters x_a, x_b, a have very little influence on the global vibrational response, thus the confidence intervals tend to be relatively large. However, the algorithm is clearly tracking the progression of the delamination with reasonable accuracy. By the time the delamination is large with respect to the beam length, the confidence intervals narrow considerably. The large nonlinearities produce a more readily identifiable signature in the global response, hence the associated nonlinearity parameters are more easily identified. The information provided in Figure 9.19 is precisely the information that the owner of a structure would be interested in: the estimated damage extent and associated confidence in that estimate. Given this information and the cost associated with the structural damage, one can make optimal decisions regarding how best to maintain that structure, a subject we tackle in the next chapter.

9.3   Modeling and Identification of Cracked Structures

In this section we focus on the identification of the parameters that describe a crack in an aluminum plate. This particular type of damage is ubiquitous in the structural health monitoring literature and numerous works have been devoted to detection and identification of cracks in structures. This particular problem is one of some importance to the U.S. Navy, for example. Certain classes of ship possess an aluminum superstructure; that superstructure is experiencing severe degradation to stress corrosion cracking, a problem alluded to in the introduction (see Figure 1.1). In what follows we propose a crack model that faciliates fast, efficient identification of the crack parameters.

9.3.1   Cracked Plate Model

Unfortunately, analytical solutions for the dynamic response of a rectangular plate with an arbitrary crack (position, orientation, and size) do not presently exist to our knowledge. Solecki [27] produced a solution for the natural frequencies of a plate with a single crack, but did not extend it to the more general description we require for damage identification. A second possible family of models are FE solutions. These clearly have the desired flexibility, but involve time marching potentially large systems of equations and so are computationally intense.

As we have described in Section 5.1.1 (see also section 4.4.2), we can capture the physics of a crack using a simple model whereby we tailor an FE description of the crack according to the analytic stress field (given by the Mitchell solution [28]) at the crack tip. Specifically, the elements away from the crack tip are standard eight-noded quadrilateral Mindlin serendipity elements, with nodes at each corner and at the middle of each side (see Figure 9.20a and b for the element description). Adjacent to the crack tip, the eight-noded quads are modified as described in Section 5.1.1. Because these augmented triangular elements capture the crack tip behavior, they are placed around the crack tip in a pinwheel fashion as shown in Figure 9.20d,

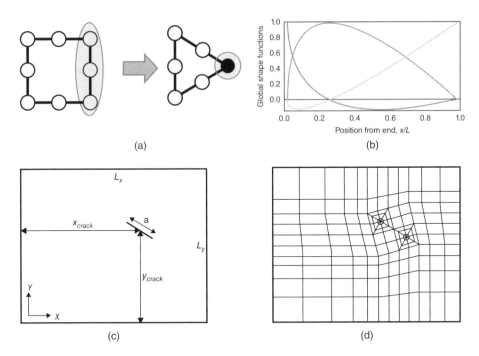

Figure 9.20 (a) Nodal configuration of the Mindlin serendipity element, (b) resulting interpolation functions, (c) schematic of plate with crack parameters, and (d) resulting course FEM mesh. *Source:* Reproduced from [29], Figure 4, with permission of Elsevier

The end result is an element with a stress field that varies as $\frac{1}{\sqrt{r}}$, where r is the distance from the collapsed node (to be placed on the crack tip). This permits a more sparse mesh near the crack tip than would otherwise be necessary and reduces the number of degrees of freedom required and, hence, the computation time.

In describing the crack, we take the model parameters to be the location of the center of the crack (x_c, y_c), the crack length (a), and the orientation of the crack measured from the positive x-axis (α). It is these parameters we will be interested in identifying based on measurements of the plate displacement, $w(x, y, t)$. These parameters are shown schematically in Figure 9.20c.

One of the chief reasons for developing a low-dimensional model is that the estimation techniques required of structural damage identification problems can be computationally intensive. In this section, we use the population-based MCMC (pop-MCMC)algorithm of Section 7.2.7 to estimate the damage parameters. Efficiency of the forward model is essential as the pop-MCMC algorithm requires a very large number of model-to-date comparisons in the evaluation of r_{MH}, r_{swap}, r_{DE}.

On the basis of our description of the plate, an FE eigensolution (see 4.5.1.2 for a description of how to build such a solution from the FE matrices) is used to build in the features of the crack singularity. The eigenvectors and eigenvalues can then be used to describe the system response in terms of the nodal displacements $w(x, y, t)$. The values at the actual locations at which sensors are placed are then simply interpolated via the natural neighbor method. This model is many times faster than time marching a large system of equations, yet retains the flexibility inherent in the FE approach.

The cracked plate model involves several standard, additional assumptions. First, the model is linear and it is assumed that material properties are known exactly (although these could be left as unknown parameters and found via the Bayesian/MCMC approach). The deflections are presumed small and crack growth is not considered; the latter is reasonable under small deflections. The crack is also assumed to remain open, such that impacts at the crack interface are ignored. Any mass lost due to the crack is presumed negligible. Finally, there is only one crack and the tips are at least one-half of the crack length from the edges of the plate. Because finding smaller cracks is the more interesting and useful problem, the allowable zone is still a very large fraction of the plate's area.

As with any FE analysis, convergence of the mesh must also be verified. Owing to the large number of iterations, it is not practical to verify convergence for every perturbed set of parameters. Instead, because all the meshes used are qualitatively similar, convergence was checked on several representative parameter vectors.

9.3.2 Crack Parameter Identification

The structure of interest is assumed to be a clamped rectangular plate measuring 1.25 (m) × 1 (m) with thickness $h = 0.01$ (m) with material properties $E = 209$ (GPa) (Young's modulus), $v = 0.3$ (Poisson's ratio), and $\rho = 7850$ (kg/m^3) (density). It will also be assumed that the plate has been instrumented with $j = 1 \cdots M$ displacement sensors capable of sampling the plate's response to an input at $n = 1 \cdots N$ equally spaced points in time. In terms of notation, we will use these discrete indices to define the points in space and time at which we will generate the model response, that is, $w_j(n, \boldsymbol{\theta}) \equiv w(x_j, y_j, t_n)$. Note, as in previous sections, that we have lumped the parameters of interest (in this case, the crack parameters) into the vector $\boldsymbol{\theta}$. The observed signal model is therefore written as per our convention

$$s_j(n) = w_j(n, \boldsymbol{\theta}) + \eta_j(n), \quad n = 1 \cdots N, \ j = 1 \cdots M \tag{9.48}$$

where the $\eta_j(n)$ are taken as realizations of an i.i.d., zero-mean, Gaussian random process, that is, each $\eta_j(n) \sim N(0, \sigma^2)$. Under this noise model, the likelihood function for the data is our familiar expression (7.5)

$$p_H(\mathbf{s}|\boldsymbol{\theta}) = \frac{1}{(2\pi\sigma^2)^{NM/2}} e^{-\frac{1}{2\sigma^2} \sum_{n=1}^{N} \sum_{j=1}^{M} (s_j(n) - w_j(n,\boldsymbol{\theta}))^2}. \tag{9.49}$$

Again, using our prior convention, we denote the sum-squared error over all sensors by the cost

$$C(\mathbf{s}, \boldsymbol{\theta}) = \sum_{n=1}^{N} \sum_{j=1}^{M} (s_j(n) - w_j(n, \boldsymbol{\theta}))^2. \tag{9.50}$$

As we discussed in Section 7.2.7, the cost for many damage identification problems possesses many minima which can "trap" numerical estimators designed to explore the minima. We therefore proposed the pop-MCMC estimator as one possible means of navigating such cost functions and providing the estimates $\hat{\boldsymbol{\theta}}$.

To this end, we form the sequence of tempered posterior distributions, as we did with Eq. (7.138), as

$$p_q(\boldsymbol{\theta}^{(q)}) = p_H(\mathbf{s}|\boldsymbol{\theta}^{(q)})^{\zeta_q} p_\pi(\boldsymbol{\theta}^{(q)})$$

$$= \frac{p_\pi(\boldsymbol{\theta}^{(q)})}{(2\pi\sigma_q^2)^{\zeta_q NM/2}} e^{-\frac{\zeta_q}{2\sigma_q^2} C(\mathbf{s}, \boldsymbol{\theta}^{(q)})}, \quad q = 1 \cdots Q \tag{9.51}$$

using the sequence $\zeta_1 = 1.0$,

$$\zeta_{q+1} = \zeta_q - \frac{1}{Q} = 1 \cdots Q \tag{9.52}$$

as suggested by Jasra *et al.* [30]. The idea here, as in the toy example of Section 7.2.7, is to explore a composite posterior where the "smoothed" marginal posteriors $q = 2, \cdots, Q$ are related to the true posterior of interest, $p_1(\boldsymbol{\theta}^{(1)})$. To do so, we will therefore generate "Q" different Markov chains, each exploring a slightly different version of the posterior distribution (again, see Section 7.2.7).

This particular implementation of the pop-based algorithm proceeds as follows. The parameter vectors for each of the Q chains are initialized by drawing samples from the priors. Then, for each iteration in the Markov chain, one of the chains $q \in [1, Q]$ is selected with uniform probability, and a standard Metropolis–Hastings (MH) update is performed for each of the P parameters in the qth vector $\boldsymbol{\theta}^{(q)}$ using the Gibbs sampling strategy of Section 7.2.4. Thus, for each parameter $i \in [1, P]$ one generates a candidate value using Eq. (7.136), and evaluates the ratio

$$r_{MH} = \text{Exp}\left[-\frac{\zeta_q}{2\sigma_q^2} \left(C(\mathbf{s}, \theta_i^{*(q)}|\boldsymbol{\theta}_{-i}^{(q)}) - C(\mathbf{s}, \theta^{(q)}|\boldsymbol{\theta}_{-i}^{(q)}) \right) \right] \frac{p_\pi(\theta_i^{*(q)})}{p_\pi(\theta_i^{(q)})} \tag{9.53}$$

and accepts with probability $\min(r_{MH}, 1)$. The (unknown) noise variance associated with the qth chain, σ_q, also needs to be sampled. As we already showed in Section 7.2.5, by choosing a vague Gamma prior for this parameter, one may directly sample from the posterior via

$$\sigma_q^2 \sim \frac{1}{\Gamma(MN/2, 2/(\zeta_q C(\mathbf{s}, \boldsymbol{\theta}^{(q)}))}. \tag{9.54}$$

Once each of the parameters (including the noise variance) has been updated for chain q, two different chains $u, v \in [1, Q]$ are selected at random with uniform probability. With 50% probability, either a "swap" move or a "DE" move between these chains is performed. For the swap move, the prior ratios will cancel, thus one evaluates

$$r_{swap} = \text{Exp}\left[-\zeta_u \left(\frac{1}{2\sigma_v^2} C(\mathbf{s}, \theta^{(v)}) - \frac{1}{2\sigma_u^2} C(\mathbf{s}, \theta^{(u)}) \right) - \zeta_v \left(\frac{1}{2\sigma_u^2} C(\mathbf{s}, \theta^{(u)}) - \frac{1}{2\sigma_v^2} C(\mathbf{s}, \theta^{(v)}) \right) \right] \tag{9.55}$$

accepting the move with probability $\min(r_{swap}, 1)$. For the DE move one requires three randomly drawn chains (see Eq. 7.143) to generate the trial vector $\boldsymbol{\theta}^{(u')}$. The ratio of priors in

the acceptance criteria is also required as they do not cancel out for this type of move. One
therefore evaluates

$$r_{DE} = \text{Exp}\left[-\frac{\zeta_u}{2\sigma_u^2}\left(C(\mathbf{s}, \boldsymbol{\theta}^{(u')}) - C(\mathbf{s}, \boldsymbol{\theta}^{(u)})\right)\right] \times \frac{\prod_{i=1}^{P} p_{\pi}(\theta_i^{(u')})}{\prod_{i=1}^{P} p_{\pi}(\theta_i^{(u)})} \qquad (9.56)$$

and accepts the move with probability $\min(r_{DE}, 1)$. To summarize, the algorithm picks a chain
at random and performs a standard MH update on each of the parameters, including the noise
variance. The algorithm then selects chains at random and performs either a swap move (Eq.
9.55) or crossover (Eq. 9.56) with 50% probability. This procedure repeats for some number
of iterations until enough samples have been drawn from the posterior distribution.

The algorithm is fairly simple to implement in software, however it is clearly computationally intensive. Each evaluation of a ratio (r_{MH}, r_{swap} or r_{DE}) requires solving the forward
model described in the previous section. The efficiency of the forward model is therefore of
great importance in using MCMC in structural system identification problems.

9.3.2.1 Numerical Identification of a Cracked Plate

There are four parameters that determine the state of damage in the plate model: crack length a,
crack location (x_c, y_c), and crack orientation θ_c. These damage parameters are fixed to the values $a = 0.1$ (m), $x_c = 0.3$ (m), $y_c = 0.5$ (m), and $\theta_c = 30°$. Further assume that four displacement sensors have been placed on the surface of the plate at the x–y locations $(0.375, 0.375)$,
$(0.375, 0.862)$, $(0.862, 0.375)$, $(0.862, 0.862)$. The acquired data will consist of each sensor's
response to four separate impacts (hammer strikes) at locations $(0.29, 0.275)$, $(0.29, 0.725)$,
$(0.96, 0.275)$, and $(0.96, 0.275)$. These sensor locations were chosen so as to maximize the
sum of the first four modes in the response. No claims of optimality are made regarding this
choice, in fact finding sensor locations that produce a more well-defined likelihood is an active
area of research we attempt to address in Section 9.3.3.

The data used in the identification procedure consist of four simulated impulse response
signals, sampled at 4 kHz for a duration of 2 s ($N = 8000$ observations). The first nine modes
were used in generating the solution, all well below the Nyquist frequency of 2 kHz. The SNR
was set at 17 dB ($50{:}1$) corresponding to the level of noise observed in previous experiments.

Given this response, the goal is to estimate the posterior distributions associated with each
of the crack parameters. Using the standard MCMC Algorithm, we were unable to consistently
identify the parameters due to the aforementioned problem by exploring complicated likeli-
hood functions. Figure 9.21 shows the argument of the log-likelihood, $C(\mathbf{s}, \boldsymbol{\theta})$ (Eq. 9.50), as a
function of different parameter combinations (holding the others fixed at their true values) As
with the simple 1-D example (Figure 7.22), one can immediately see the difficulty presented by
this estimation problem. Consider Figure 9.21a. The location parameter x_c has a minimum at
the true location, $x_c = 0.3$, however there is also a "trough" of minima with a particularly low
value at $(x_c = 0.85, y_c = 0.28)$. A similarly complicated likelihood is shown for the parameters
a, θ_c (Figure 9.21b). These plots illustrate again a fundamental aspect of damage identification,
namely the fact that multiple damage states can yield very similar structural response data.

This, of course, is precisely the reason the population-based approach to sampling the
posterior is so valuable. In the example that follows, $Q = 10$ chains were used for each
of the parameters a, θ_c, x_c, y_c. The prior for the crack parameter a is taken to be a Gamma

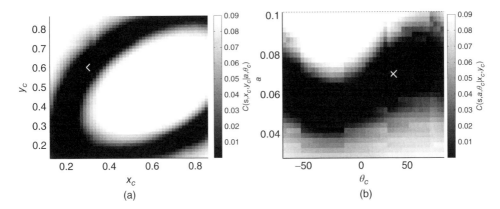

Figure 9.21 Main argument of the likelihood function, $C(\mathbf{s}, \boldsymbol{\theta})$, plotted as a function of (a) crack location on the plate and (b) the parameters a (crack length) and θ_c, crack orientation. For each plot, the remaining parameters were held fixed at their true values. True parameter values are denoted with an "X". *Source*: Reproduced from [29], Figure 5, with permission of Elsevier

distribution, $p_{\pi_a} = a^{\alpha-1} \frac{\mathrm{Exp}[-a/\beta]}{\Gamma(\alpha)\beta^{\alpha}}$ with parameters $\alpha = 1.25, \beta = 0.025$. Thus, essentially no damage is assumed at the outset. Certainly this distribution can be altered to reflect a known level of damage, however the more typical case is to assume the plate is healthy and allow the data (likelihood) to drive the posterior. The form of this prior (Gamma) was chosen on the basis of the fact that the crack cannot be negative. One might also have chosen a Beta prior which has finite support on the plate, however for the prior parameters chosen, there is essentially no probability of the crack extending off the end of the plate. For the crack location parameters x_c, y_c, uniform priors were chosen over the span of the crack, that is, $x_c \sim U(0, 1.25), y_c \sim U(0, 1)$ reflecting the fact that no *a priori* knowledge about the crack location exists. Similarly, the prior on crack angle was taken as $\theta_c \sim U(-90, 90)$ because, in general, one will not know the crack orientation *a priori* either. The prior distributions are displayed in Figure 9.22. Initializing the Markov chains using these priors, the pop-MCMC algorithm was run for 60,000 iterations with a burn-in of $B = 50,000$ iterations. The remaining 10,000 iterations were stored and used to form the posterior densities shown in Figure 9.22.

In this simulation, and in others the authors have looked at, crack length is perhaps the most easily identified parameter. The Markov chains in all populations tend to the true value ($a = 0.1$) after only a few thousand iterations. Likewise, the location parameter y_c exhibits a unimodal posterior distribution. The other parameters, x_c, θ_c are significantly more challenging to estimate as both show the clear presence of multiple maxima in the likelihood. The location parameter x_c exhibits multimodal behavior. In particular, one can see the multiple minima observed in Figure 9.21. The two peaks located at $x_c = 0.3, x_c = 0.4$ presented a significant challenge for the regular MCMC algorithm. Because crack lengths were initially assumed small, the first minimum encountered by the Markov chain was the one located at $x_c = 0.4$ and the solution would often remain here. Running multiple chains easily overcomes this problem.

Perhaps the greatest utility of the approach, however, can be seen in the identified orientation θ_c. Regardless of crack geometry, there are nearly always multiple solutions for θ_c that come close to maximizing the likelihood (hence the posterior). The results are multiple well-defined local maxima that easily trap the standard MCMC algorithm. By

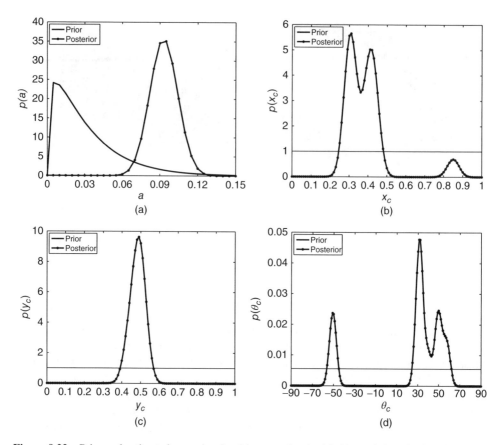

Figure 9.22 Prior and estimated posterior densities associated with (a) crack length, (b and c) crack location, and (d) crack orientation. The true parameter values are $a = 0.1$, $(x_c, y_c) = (0.3, 0.5)$, and $\theta_c = 30$. *Source*: Reproduced from [29], Figure 6, with permission of Elsevier

contrast, Figure 9.22 captures the relative heights of these maxima, indicating that the highest probability for crack orientation is at the true value, $\theta_c = 30°$.

As a second example, Figure 9.23 shows the identified posterior distributions for the case where the true crack parameters were set to the values $a = 0.1, x_c = 0.6, y_c = 0.35, \theta_c = -20$. In addition, the Gamma prior (biased toward no crack) was changed to a uniform prior to demonstrate the insensitivity of the approach to this choice. It could be that the practitioner has no *a priori* information regarding the presence and length of a crack, thus a uniform prior would be appropriate. All parameters are again correctly identified, provided that the final estimate is taken as the maximum *a posteriori* value. Again, one sees multiple "good" solutions that can often trap the standard MCMC algorithm. For example, the crack location parameter y_c has a fairly high probability of being $y_c = 0.7$ despite the fact that the true value is a factor of two different. Similarly, multiple peaks for the orientation parameter θ_c can be seen. We note that the standard MCMC algorithm was not able to sample from these complex posterior distributions.

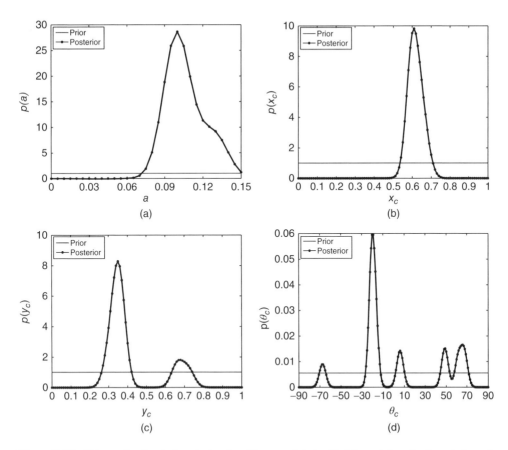

Figure 9.23 Prior and estimated posterior densities associated with (a) crack length, (b and c) crack location, and (d) crack orientation. The true parameter values are $a = 0.1$, $(x_c, y_c) = (0.6, 0.35)$, and $\theta_c = -20$. *Source*: Reproduced from [29], Figure 7, with permission of Elsevier

The complexity of the posterior distributions observed in these types of problems is simply a consequence of trying to identify parameters that have minimal affect on the global vibrations. It is obvious from the likelihood plots (Figure 9.21) that varying crack configuration can lead to nearly the same vibrational response. This is simply part of the physics of structural damage identification problems and motivates our development and usage of estimation tools such as the pop-MCMC algorithm of Section 7.2.7. The sampling in this approach is done in such a way as to avoid becoming stuck in locally optimal solutions. Thus, rather than producing point estimates which may or may not coincide with the true parameter value, the pop-MCMC estimator allows for a full accounting of the probability associated with different crack configurations.

Of course, the downside to this more sophisticated estimation approach is that it requires repeated iterations of the forward model. Thus, great care should be taken in developing an efficient (low-dimensional) model. In this section, we used the tailored "serendipity" elements developed in Section 5.1.1 (see also Section 4.4.2) to describe the stress field near the crack tip. This allows for good model convergence with many fewer elements than are used in standard FE codes. The combination of efficient modeling and effective parameter identification routines can provide a wealth of information about the state of a structure. Both the parameter estimates and the credible intervals associated with those estimates are obtained. This allows for confidence-based decisions regarding the maintenance of a structure and also provides information needed in prognostics models for damage evolution in structures.

9.3.2.2 Experimental Implementation

We now turn our attention to an experimental implementation of the approach to estimation outlined in the previous section. The test specimen was a 0.76 m by 0.60 m (30″ by 24″, 5:4 aspect ratio), 1.55 mm (1/16 in.) thick 6061-T6 aluminum plate. The density was calculated using a digital scale and a vernier caliper. Normalized dimensions are given in Table 9.2.

Normalized values of the true crack parameters are given in Table 9.3. The crack was created by first machining a 0.813 mm (1/32 in.) wide slot. The slot was extended by a laser cutting tool so that the tips of the crack had an effective diameter of 0.20 mm (0.008 in.). One crack tip is shown in Figure 9.24 along with a small coin for reference. The crack was located at a point away from modal nodes and the edges of the plate, and rotated through an angle of −60° from horizontal. The authors' earlier simulation study, see Ref. [31], suggested that this would be a relatively easy crack to identify.

The plate was clamped in a bolted fixture, as shown in Figure 9.24. Each of the 28 × 12 mm bolts were tightened to 135 N-m (100 ft-lbs). Lock washers were used on each fastener. Young's modulus and Poisson's ratio were both determined experimentally. Tension tests following ASTM E8 were performed on six samples cut from the same sheet of aluminum as the test plate, and were determined to have an average Young's modulus of approximately 52.5 GPa. This differs significantly from the handbook value of 68.9 GPa. The experimentally determined Poisson's ratio was 0.11, significantly smaller than the 0.33 suggested by handbooks. Calibrating the model with these values was critical to the success of the algorithm.

Table 9.2 Plate dimensions

Normalized dimensions	Value
Width (L_x/L_y)	1.25
Length (L_y/L_y)	1
Thickness (h/L_y)	0.0025

Table 9.3 True crack parameters

Normalized parameter	Value
Crack Center $\left(\dfrac{x_{crack}}{L_x}, \dfrac{y_{crack}}{L_y}\right)$	(0.64, 0.60)
Crack Length, $\dfrac{a}{L_y}$	0.153
Crack Orientation	−60°

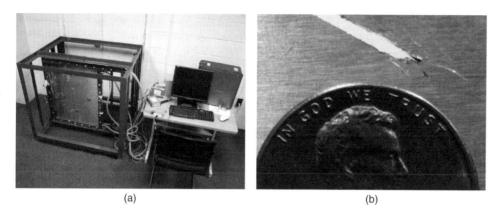

(a)   (b)

Figure 9.24 The experimental test setup (a) and a close-up of the machined crack tip (b). *Source*: Reproduced from [32], Figure 2, with permission of Elsevier

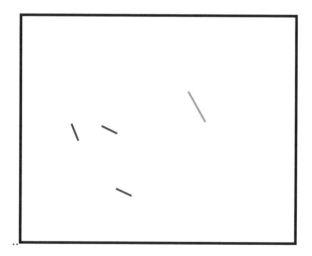

Figure 9.25 The arbitrarily selected locations and orientations of the three strain gages (short lines) and the true location of the crack (long line). *Source*: Reproduced from [32], Figure 3, with permission of Elsevier

Three single-axis Vishay Micro Measurements model EA-13-125AD-120 strain gages were located in arbitrary positions and orientations in the third quadrant of the plate, away from the crack, as shown in Figure 9.25. The sampled strain response at each of the three gage locations is denoted $s_j(n) \equiv s_j(n\Delta_t), n = 1\cdots N, \ j = 1\cdots 3$ in accordance with our discrete model (9.48) where Δ_t is our usual sampling interval. The signals were filtered with a passband filter between the frequencies of 13 and 66 Hz before acquisition as this is the frequency range over which our model is predicting. To see this, we can look at the estimated Fourier amplitudes of the unfiltered signal, showing that the first three frequencies would be well inside the band, and higher frequencies would be well outside it. The length of time modeled is selected to maximize the quantity of data used, while avoiding using data with a low SNR. Because the magnitude of the response decays, the SNR decreases over approximately 0.75 s from about 25 to about 15. In other words, there is no point retaining data that contains little in the way of plate dynamics.

The plate was excited from rest by the impact of a simple rubber-tipped mallet. The impact point was selected to be an arbitrary point along the diagonal of the plate. There are two phenomena not captured by the model that are dominant immediately after impact. The first is nonlinearity of the response due to the large transverse deflection experienced. For deflections larger than about one plate thickness, membrane stresses are significant, whereas the FE model includes only the effects of bending. The second unmodeled phenomenon is the power in non-modal frequencies transmitted by impact. This "ringing" in the response dissipated quickly, but must not be compared to the model prediction by the MCMC process, as the behavior cannot be captured by the linear model. A more complex model may be used, but at the cost of computational speed. Once these two effect subside, free bending vibration dominates the response.

This transition can be seen in the magnitude of the estimated Fourier transforms of each gage's signal, shown in Figure 9.26. In Figure 9.26a, the 0.75 s of the response immediately following impact are analyzed. In contrast, Figure 9.26b shows the estimated magnitude Fourier transform of 0.75 s of the response beginning 0.55 s after impact, and here the frequencies of the three free vibration modes dominate with sharp peaks. The importance of allowing the early portion of the response to decay is readily apparent. The unfortunate side effect of this windowing, however, is that the initial conditions of the plate are unknown at the moment modeling begins.

Recall that to estimate the parameters, we require the time-domain comparison in the argument of Eq. (9.50). The initial conditions of the plate are therefore needed to generate model time series (as a function of the system parameters) that can be compared to the (now windowed) data. The solution is to simply solve a system of nonlinear equations in modal space for the amplitude and phase of each of the modes, which would produce the strain and strain rate measured at each gage location at the moment modeling begins (i.e., a nonlinear least-squares estimation problem). Testing showed that a unique solution is found regardless of the seed values given to the solver. The strain can be read directly, and the strain rate can be approximated by the central difference method. In this way, each gage provides two initial conditions, while each mode had two independent variables: magnitude and phase. Thus, it is possible to model as many modes as there are gages in use.

Given the observed data $s_j(n)$ and the initial conditions required of the model $w_j(n, \theta)$ we can now estimate the crack parameters just as we did in the preceding numerical example. Using the same pop-MCMC approach, the posterior parameter distributions for each of the four parameters x_c, y_c, a, θ_c were estimated; these are shown in Figure 9.27. We note that the kernel density estimation approach described in section 6.3 was used to smooth the final posterior estimate the true values of the parameters are shown on the distributions as vertical lines. Within each mode of the posterior, the 90% high-probability density (HPD) region appears shaded. The HPD represents the smallest region of the parameter space where the true crack parameters are likely to reside, within some given probability level. It is similar to a confidence interval, but does not have to be a continuous region, and is thus more applicable to multimodal distributions similar to the ones we encounter here. A more detailed discussion of the HPD region can be found in the book by Gelman *et al.* [33].

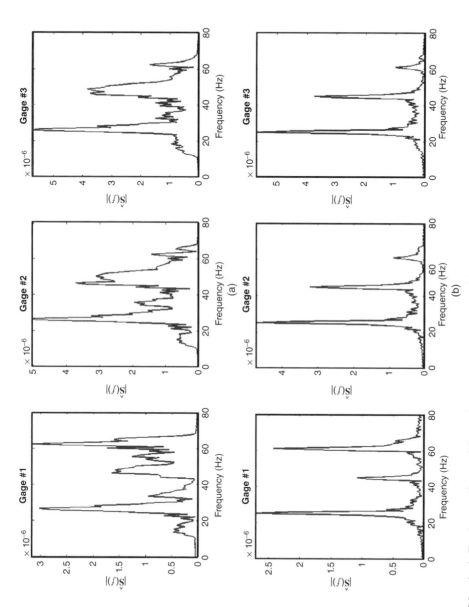

Figure 9.26   Magnitude Fourier transforms of the strain response $s_j(t)$, $j = 1 \cdots 3$ immediately after impact (a) and after 0.55 s settling time elapsed (b).

Source: Reproduced from [32], Figure 4, with permission of Elsevier

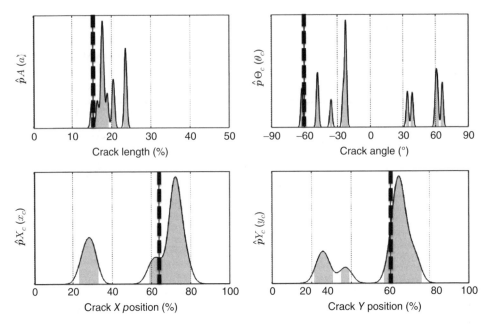

Figure 9.27 Estimated posterior distribution for each of the crack parameters as obtained using the pop-MCMC estimation algorithm along with the HPD confidence interval (shaded). Solid vertical lines show the true values of the parameters. *Source*: Reproduced from [32], Figure 5, with permission of Elsevier

The length of the crack, consistent with numerical results, is the easiest parameter to estimate given the impulse response data. In this experiment, the crack length was overestimated by a small margin. The position of the center of the crack was located correctly with a moderate to high level of certainty. Locations symmetric to the correct solution are shown to be lower probability possibilities. Again, as with the numerical example, the orientation of the crack is easily the most difficult parameter to estimate. For detecting, localizing, and estimating damage extent, it appears the impulse response measurements are largely sufficient. However, if one were to predict crack direction, either additional measurements would likely be required, or perhaps a different (more complex) form of excitation (e.g., swept sine-wave) would be needed.

In Figure 9.28, the experimentally measured strain response from the various gages is compared to the strain predicted by the identified model. The model predicted results were obtained using optimal damage parameters, taken as those that maximized the posterior distribution. Using the posterior mean, although common, clearly will not produce good estimates for multimodal posterior PDFs.

Each row in Figure 9.28 represents the data from a single gage, with the time domain response on (a), and the estimated magnitude Fourier transform of the signals on (b). In general, there is very good agreement. The consequences of locating the gages arbitrarily are immediately apparent: there is relatively little contribution from the third mode in the response of the gages. The relative amplitude of the third mode is also adversely affected by the windowing of the data, as higher modes decay more quickly. Lacking significant input from the third mode

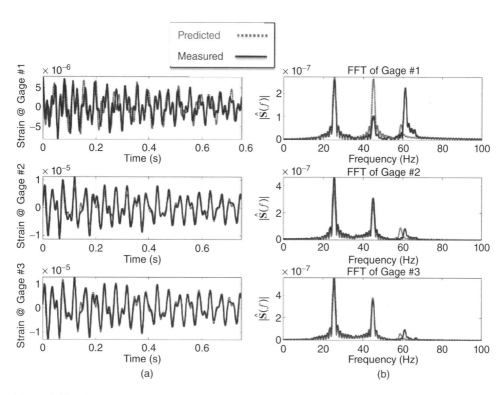

Figure 9.28 Comparison between measured response and identified response in both the time domain (a) and frequency domain (b). *Source*: Reproduced from [32], Figure 6, with permission of Elsevier

makes it more difficult to identify the crack location. In particular, it is difficult to discriminate the correct solution from solutions "mirrored" about an axis of symmetry. Nonetheless, the maximum *a posteriori* estimate correctly identifies the crack location. In the next section, we propose a more principled choice of sensor placement.

It is also worth repeating that for this test only three strain gages were used, thus limiting the model response to a combination of only three modes. It is likely that adding sensors, and the commensurate increase in modal resolution, would lower the probabilities associated with the spatially symmetric solutions (multiple modes in the posterior). This would increase the resources required for sensors and data acquisition equipment, however, although computation time would only increase marginally. These types of trades necessarily depend on how one chooses to balance system cost versus performance.

As a second estimation approach, the method of maximum likelihood was also used for comparison to the Bayesian approach. For this approach we used the DE estimator of section (7.1.3) to maximize the familiar cost function (9.50). To generate confidence intervals, the estimate was repeated 50 times, and the MLE associated with each data set was retained. The combined results were then used to generate a smoothed histogram using the same kernel density estimation settings as used to generate the Bayesian posteriors in Figure 9.27. The results of this comparison are shown in Figure 9.29. Perhaps not surprisingly, the two approaches produce qualitatively similar distributions. Given that the likelihood function lies at the heart

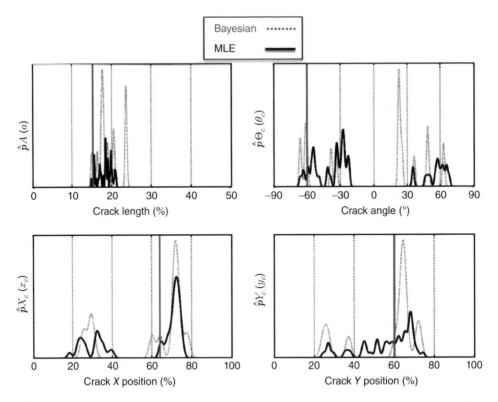

Figure 9.29 Comparison of posterior generated with Bayesian methods and the distribution of repeated frequentist genetic algorithm solutions. True values are indicated by vertical lines. *Source*: Reproduced from [32], Figure 7, with permission of Elsevier

of both estimation approaches one would expect algorithms that are attempting to sample from the maximum would lead to similar sets of samples, even though the confidence intervals are to be interpreted in a slightly different manner. The Bayesian posterior has the formal advantage of being a mathematically rigorous probability distribution, however in this example, the practical difference between the two estimation approaches is small.

As with the Bayesian estimation approach, the MLE results are also very much conditional on the type of data being acquired. This study focused on a single system impulse response measured at three different locations. This type of data is easily acquired in a practical setting as it does not require much in the way of specialized equipment. For example, a single wave-slam event on a ship hull produces these types of free-decay responses. However, there is a trade-off in using such simple methods for obtaining data. As seen from the damage parameter PDFs, there exists a fair amount of uncertainty in the estimates. Had we used, for example, a controlled, swept-sine input, our confidence in the resulting parameter estimates would likely increase because of the increased number and quality of data. Nonetheless, this approach to damage identification does a reasonable job of identifying the damage magnitude and a reasonable job of locating the damage as well. Only the crack orientation proved too difficult to reliably estimate with reasonable fidelity.

9.3.3   Optimization of Sensor Placement

Having established a well-defined approach for the estimation of crack parameters, an interesting question becomes: might we place the strain sensors in such a way as to improve our estimates? The answer, of course, requires us to define what we mean by "optimal" sensor placement. From our perspective, the optimal placement would be the one that provides us with damage parameter estimates that are both unbiased and have as small a variance as possible (a reasonable goal considering Eqn. 6.4). This is a difficult question to answer analytically, however is one that we can approach in a sensible manner.

We know from our work to this point that achieving unbiased, minimum variance estimates goes hand-in-hand with a well-defined likelihood that possesses a unique minimum corresponding to the true parameters to be identified. In this study we consider the position and orientation of the sensors, as well as the impact location, to be free parameters that we can adjust to create a better defined likelihood and hence a more easily identified damage state. Although it is not generally known how to construct a cost function that accomplishes this, we can suggest heuristic arguments for how one might proceed. Consider the first work of Papadimitriou *et al.* [34] and others (see also Trendafilova *et al.* [35] and Kripakaran and Smith [36]), where it was suggested that a good goal might be for each sensor to provide as much unique information as possible when compared to the other sensors. The idea, of course, is that less redundancy in sensor information will lead to greater identifiability of the unknown parameters. We too will use "minimal redundancy" as a surrogate for "identifiability" in what follows.

Begin by recalling the cross-correlation coefficient (see Eq. 3.25) between model time series $s_i(t)$ and $s_j(t)$ as

$$\rho^2_{S_i S_j}(\tau) = \frac{\int_{-\infty}^{\infty} (s_i(t) - \mu_{S_i})(s_j(t+\tau) - \mu_{S_j}) dt}{\sigma_{S_i} \sigma_{S_j}} \tag{9.57}$$

where $\mu_{S_{i,j}}, \sigma_{S_{i,j}}$ are the signal means and standard deviations, respectively, and τ is a measure of time delay. In some sense, Eq. (9.57) quantifies the amount of information common to signals $s_i(t), s_j(t)$. In fact, we have already shown in Section 3.5.1, Eq. (3.127) that for Gaussian-distributed random processes $S_i(t), S_j(t)$, the mutual information is given by [37]

$$I_{S_i S_j}(s_i, s_j; \tau) = -\frac{1}{2}\log_2(1 - \rho^2_{S_i S_j}(\tau)). \tag{9.58}$$

Thus, a small cross-correlation leads to a correspondingly small mutual information between the two signals. To quantify the common information, Eq. (9.58) is typically maximized over the range $\tau \in [0, N]$, that is to say, we take the measure of similarity to be that associated with the maximum degree of commonality. In practice, this means we generate the discrete model signals $s_i(n), s_j(n), n = 1 \cdots N$ and estimate the cross-correlation using the methods of Section 6.4.2. The result will be the function $\hat{\rho}_{S_i S_j}(n)$ and we simply record the maximum (over all n) as the amount of common information between the two signals. The goal in what follows is to minimize this redundant information and therefore maximize the unique information being provided by each sensor.

However, one would also like this redundancy minimized between a wide range of possible damage cases and the undamaged case, that is, make the damaged and undamaged signatures different. As we have pointed out in previous chapters, identifiability of the damage is of large importance. By mandating that the measurements produce a "damaged" signature that is different from an "undamaged" one, we can help ensure this property.

One way to accomplish all of the above-stated goals is to focus on the *difference* between damaged and undamaged response signals (as opposed to the signals themselves). For a given damage case, "q" may denote this difference as $\Delta_{jq} = s_{jq} - s_{ju}$, where $q = 1 \cdots Q$ denotes one of Q particular damage cases and u is the undamaged case. The index j again selects which of our M sensor data is being accessed. This difference is a function of each of the desired damage parameters and the sensor/excitation parameters so that

$$\Delta_{jq} \equiv \Delta_{jq}(\theta_q, x_j, y_j, \phi_j, h_x, h_y), \quad j = 1 \cdots M, \quad q = 1 \cdots Q.$$

The independent variables x_j, y_j, ϕ_j represent the location and orientation of the jth strain gage, while h_x, h_y is the location of the impact excitation. The crack parameter vector for the "qth" damage scenario is again given by $\theta_q \equiv (a^{(q)}, x_c^{(q)}, y_c^{(q)}, \theta_c^{(q)})$

To minimize redundant information over all possible crack locations, orientations, and impact locations over all Q damage scenarios considered, a reasonable cost function to be minimized is

$$C(\theta_q, x_j, y_j, \phi_j, h_x, h_y) = \sum_{q=1}^{Q} \sum_{j=1}^{M-1} \sum_{k=j+1}^{M} \max_\tau \rho^2_{\Delta_{jq}\Delta_{kq}}(\tau) \tag{9.59}$$

which says that we want to minimize the largest (over all τ) correlation between sensor data.

In choosing the subset of damage scenarios to model, we take the worst case scenario where the sensors are restricted to the third quadrant and the cracks in the subset of damaged plates are restricted to the first quadrant. Specifically, we used a set of 64 "test cracks," varying in position and orientation, but all sharing the same length. The crack length will affect the magnitude of the change in strain but it should not significantly impact the relative changes between modes. Plate-centered cracks were also included. Ref. [31] suggests that this last geometry would be a particularly difficult crack location to identify because the plate center (i.e., the crack location) coincides with a number of vibration nodes. Having the crack center and multiple vibration nodes coincide means that, if the impact energy is distributed uniformly through the first few modes, then very little of the total energy will be devoted to exciting the crack. This has the effect of masking the presence of the crack or at least significantly downplaying its effect on the response. In addition, to avoid having a hammer strike very near a sensor or very near the crack, the impact location was restricted to the second quadrant.

The parameter space for the optimization problem implied by (9.59) is fairly complicated. For each of the "M" gages, three variables are needed to specify its location and orientation. Two additional variables specify the location of the impact. Formally, this reduces to solving the problem

$$\min_{x_j, y_j, \phi_j, h_x, h_y} C(\theta_q, x_j, y_j, \phi_j, h_x, h_y), \quad j = 1 \cdots M, \quad q = 1 \cdots Q. \tag{9.60}$$

To this end, we used a genetic algorithm (similar to the DE algorithm of Section 7.1.3) to obtain rough estimates for the optimal sensor locations. A steepest descent algorithm is then used to "fine-tune" the sensor locations. As with our DE approach to producing MLEs, we are using the output of the genetic algorithm as a starting point and then switching over to a

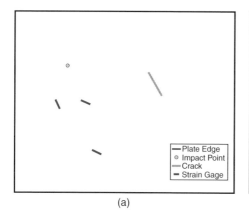
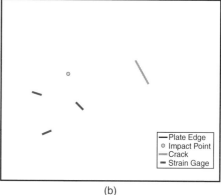

(a)   (b)

Figure 9.30   Locations of arbitrarily located and oriented gages and impact point (a) and locations and orientation of carefully chosen sensor locations and impact point (b). *Source*: Reproduced from [39], Figure 3, with permission of SAGE

Table 9.4   Value of objective function for each set

Set Name	Cost
Optimum	134.7
Arbitrary	186.5
Worst Possible	384.0

more classical search routine. A succinct review of genetic algorithms is provided in Ref. [38], while Ref. [32] provides the specifics of the algorithm used.

The sensor and impact locations that resulted from the optimization procedure (for the single crack shown) are given in Figure 9.30b. The randomly located gages and impact sites used in the example of the previous section are shown for comparison in Figure 9.30a. The *cost* of these two designs is presented in Table 9.4. It turns out that the arbitrarily placed sensors and excitation of the previous section performed reasonably well, however we predict that the optimized set of sensors and excitation will do better at identifying the damage.

Normalized values of the true crack parameters were already given in Table 9.3 (as shown in Figure 9.24) The crack was located at a point away from modal nodes and the edges of the plate, and rotated through an angle of −60° from horizontal. It should be noted that this crack parameter set (i.e., "truth") was not included as one of the subset of cracks used to select sensor locations.

Given the new sensor configuration, the experiment of the previous section was repeated. That is to say, the plate was again excited from rest by the impact of a simple rubber-tipped mallet and the same amount of time was allowed to pass for the "ringing" to die down before recording began. The initial conditions of the plate at the start of the recorded data were also estimated identically to the previous experiment and all signals were again sampled at a frequency of 25 kHz.

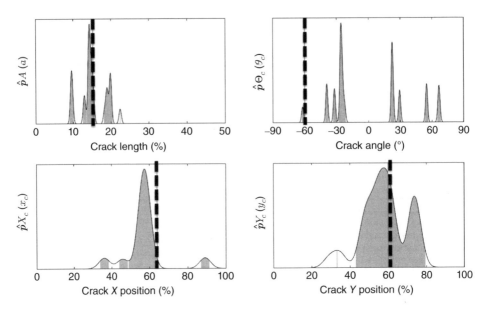

Figure 9.31 Posterior distribution of crack parameters found with optimized sensor locations. The 90% HPD interval is shaded. Solid vertical lines show the true values of the parameters. *Source*: Adapted from [39], Figure 5, reproduced with permission of SAGE

From this data, we repeated the population-based Bayesian estimation of the crack parameter posterior PDFs. These distributions are shown in Figure 9.31; the shaded regions again indicating the 90% HPD interval. The true crack parameters are shown with vertical lines. These results can be compared to those of the arbitrarily located/oriented gages given earlier in Figure 9.27. It is evident that the x-position estimate is improved greatly. The y-position is also improved, as the dominant peak is now closer to the actual value, although a strong secondary peak now appears at a slightly higher y-position. The crack length estimate, a measure of the severity of the damage, is also slightly improved as the dominant peak nearly coincides with the actual crack length and the neighboring peaks are at or below the levels of the nonoptimized results. The crack orientation remains the most difficult to identify with no real improvement over the arbitrary sensor arrangement.

While the number of local minima is not significantly reduced by changing the sensor placement, the depth of the global minimum in the likelihood is increased slightly so that the dominant peaks are closer to the true answer, with somewhat lower secondary peaks. In particular, the mirrored solution in the x-position (mirrored meaning symmetric about the center of the plate) is less pronounced with the optimized sensor set. This would suggest that the sensor placement strategy is helping to eliminate this additional peak from the likelihood function.

The MLEs for each crack parameter are also compared in Table 9.5. The error of the estimate dropped for both the x and y crack location parameters and the crack length. This is important as the center position of the crack (given by the x and y position) locates the damage, while the crack length indicates its severity. Thus, from this repeated experiment, we again conclude that we can reliably detect the presence of the crack and estimate its length and location, albeit with some degree of uncertainty. Nonetheless, we can quantify this uncertainty and factor it into

Table 9.5 Comparison of maximum likelihood estimate and errors of that estimate from both the arbitrarily located sensors and the optimized sensor suite

Parameter	Arbitrary Set		Optimized Set	
	MLE	Error (%)	MLE	Error (%)
Crack Center $\left(\dfrac{x_{crack}}{L_x}, \dfrac{y_{crack}}{L_y} \right)$	(0.724, 0.642)	(8.4, 4.2)	(0.574, 0.577)	(6.6, 2.3)
Crack Length, $\dfrac{a}{L_y}$	0.176	15.5	0.143	6.2
Crack Orientation	$-22.6°$	41.5	$-25.7°$	38.1

Table 9.6 Comparison of 90% HPD size for arbitrarily located sensors and impact point with that of the optimized sensors and impact point

Parameter	HPD: Arbitrary set (%)	HPD: Optimized set (%)
Crack center $\left(\dfrac{x_{crack}}{L_x}, \dfrac{y_{crack}}{L_y} \right)$	(30.6, 32.1)	(28.4, 36.4)
Crack length, $\dfrac{a}{L_y}$	14.5	15.6
Crack orientation	14.2	14.6

* The percentage shown is that of the size of the HPD region as a fraction of the total parameter space

our decisions regarding how best to maintain the structure. The crack orientation remains the parameter that is most difficult to identify, thereby limiting our ability to predict the direction of crack growth. It is possible that different numbers of sensors, different types of excitation, or even different types of measurements (e.g., acceleration instead of strain) would help in this regard (Table 9.6).

While the proposed approach to sensor placement improves the estimates, the gains are not dramatic. Both Bayesian and MLE techniques demonstrated that the optimized sensor locations/orientations improved the final parameter estimates, but did not help much in reducing the size of the highest probability density region of the posteriors of the crack parameters. In addition, while the approach did not reduce the number of local maximum in the likelihood, the depth of the true minimum (and hence our ability to find it) was improved, for example, identification of the x location of the crack center.

9.4 Modeling and Identification of Corrosion

Corrosion comprises a number of processes by which a chemical reaction, usually due to environmental stressors, causes a degradation in the integrity of a structural component. A prime example is a maritime environment which, owing to the salinity of the water, can cause the metal in maritime structures (e.g., ships) to degrade over time.

Past efforts toward the identification of corrosion damage on plates have proceeded in a number of directions, including approaches based on Lamb wave propagation [40, 41, 42], measurements of refracted light [43] and [44], magnetic field measurements [45], and thermal radiation-based approaches [46]. In each of these approaches, one monitors the system for changes in the characteristics of the associated signals. As we have stated several times, our preference is to instead cast the problem as one of modeling and estimation.

Moreover, one of our goals has been to use structural vibration as the means of generating the observations we use in estimation. While our modeling and estimation framework could easily be applied to different types of acquired data (thermal, magnetic, etc.), using simple strain or acceleration sensors to measure global structural vibrations is practically more attractive. The trade-off, of course, is that the estimation of localized damage-related parameters is challenging using global vibration measurements.

In this section we quantify the extent of corrosion damage (presumed uniform) in a homogeneous, isotropic metal plate exposed to a corrosive salt water environment. The plate we study here is the same as that used for the experiments used earlier in this chapter. As with these earlier experiments, we again rely on the pop-MCMC algorithm described earlier in Section 7.2.7 to estimate the entire parameter posterior PDFs.

For corrosion damage, the key damage parameter captured by the model will be the *average* plate thickness. As a result, the model we develop cannot successfully identify highly localized phenomena, such as pitting corrosion and intergranular corrosion. While we could develop a localized damage model, our aforementioned preference for analytical models with as few parameters as possible drives us toward this more global approach. Nonetheless, we demonstrate that one can still reliably infer the presence and extent of corrosion damage.

To begin, consider a collection of measured strain time series, $\mathbf{s} \equiv (s_j(0), s_j(\Delta_t), \cdots, s_j(n\Delta_t))$, consisting of the usual N observations at each of $j = 1 \cdots M$ sensors. As per convention, we will leave out the sampling interval and simply denote the sampled time with the discrete index n, that is, our data are $s_j(n)$. Along with that observed data, we again postulate a physics-based model for the observations $\mathbf{w}(n, \theta) \equiv (w_j(0, \theta), w_j(1, \theta), \cdots, w_j(N, \theta))$, $j = 1 \cdots M$. The parameter vector $\theta \equiv (\theta_1, \theta_2, \cdots, \theta_P)$ is comprised of the unknown model parameters, for example, θ_1 could be the plate thickness. The objective is to estimate the model parameter vector θ given the observations $\mathbf{s}$. A good estimate is therefore one that renders a model signal $\mathbf{w}(\theta)$ that accurately describes the experimental data.

As we have done throughout, we assume that our sensor measurements follow the additive noise model (9.48) where the noise values $\eta_j(n), n = 1 \cdots N$, $j = 1 \cdots M$ are assumed to consist of independent draws from a normal (Gaussian) distribution. Given this noise model, the joint probability distribution for all $N \times M$ observations is again of the familiar form

$$p_H(\mathbf{s}|\theta) = \frac{1}{(2\pi\sigma^2)^{NM/2}} e^{-\frac{1}{2\sigma^2} \sum_{n=1}^{N} \sum_{j=1}^{M} (s_j(n) - w_j(n,\theta))^2} \tag{9.61}$$

Each of the model parameters θ_i is modeled as a random variable Θ_i with probability distribution function: $p_{\Theta_i}(\theta_i), i = 1 \cdots P$. The final parameter estimates are then taken as the mode of these parameter PDFs and the width of the distributions indicates the confidence associated with that estimate: narrow (wide) peaks imply more (less) confidence in the estimates. To obtain these distributions, we again turn to the pop-MCMC approach to estimation.

Before proceeding, we require (as always) a physics-based model that predicts the observed strain response, $w_j(n, \theta)$. Recall from Section 5.5 that we can use a Galerkin approach to form the dynamic strain response for a corroded plate subject to impact excitation. Specifically, it was shown how one could expand the displacement and rotation at any spatial location as a linear combination of admissible functions. The specific weights (coefficients) associated with this expansion were determined via an eigenvalue problem (5.51).

Once the mode shapes and natural frequencies are determined, it is a simple matter to calculate strain time series at any particular point on the plate using the definition of strain in Mindlin plates as

$$\epsilon_x = z\frac{\partial \Psi_y}{\partial x} \qquad \epsilon_y = z\frac{\partial \Psi_x}{\partial y} \qquad \epsilon_{xy} = z\left(\frac{\partial \Psi_y}{\partial y} - \frac{\partial \Psi_x}{\partial x}\right) \tag{9.62}$$

Recall that in Eq. (9.62) Ψ_x and Ψ_y are the rotation of the cross-section of the plate at any spatial location, and at any time. More specifically,

$$\Psi_x(x, y, t) = \sum_{i=1}^{S} A_i \psi_x(x, y) e^{i\omega_i t}/a \tag{9.63a}$$

$$\Psi_y(x, y, t) = \sum_{k=1}^{T} A_k \psi_y(x, y) e^{i\omega_k t}/b \tag{9.63b}$$

where the constants A_i, and A_k are found via analytical methods and are related to the initial conditions of the plate, and $S = T = 3$ based on the convergence study mentioned earlier in section 5.5.

Given this construction, the strain field at a particular location (x, y) can be written (as before, Eq. 5.57)

$$w(x, y) = \epsilon_x \cos^2\phi + \epsilon_y \sin^2\phi + \epsilon_{xy} \sin\phi \cos\phi \tag{9.64}$$

which is, implicitly, a function of the model parameters θ. After specifying the angle of orientation of the gage on the plate, ϕ, as well as the specific position of the sensor (x_j, y_j) and the time of interest $t = n\Delta_t$, we have for our model generated data $w_j(n, \theta) \equiv w(t_n, x_j, y_j)$ which can be used in the likelihood (9.61).

With each of the other model parameters being largely known from previous work, or easily measurable, the parameter vector contains the unknowns, $\theta = \{h, \tau, \sigma^2\}$. We are particularly interested in the thickness parameter h as it captures the degree of corrosion damage. However, we have also included the standard deviation of the additive noise (see Eq. 9.61) and rotational spring constants (see Fig. 5.23 and the associated modeling discussion) as additional unknowns that must be estimated. Of course, we have already shown in Eq. (7.133) that through appropriate choice of parameter prior distribution (namely, the inverse Gamma distribution) we may sample from this noise parameter PDF directly at each stage of the Markov chain when it comes to estimating our parameter vector.

With regard to the rotation parameter, Figure 9.32 shows the change in the first three normalized natural frequencies plotted as the normalized rotational spring constants are varied from zero to 500 N/rad. It is clear that the natural frequencies rapidly approach the fully clamped values as the rotational spring constant increases; this bounds the domain of reasonable torsional spring stiffnesses and will be used to set the prior PDF for this parameter.

We therefore have a physics-based model which uses the eigenfrequencies and mode shapes as building blocks for the solution. The model is further expressed in terms of only two key unknown parameters h and τ. To estimate these parameters, we require a means of predicting the strain response for a given set of parameters, and then forming the likelihood ratio used in the MCMC algorithm (see Eq. (9.53).

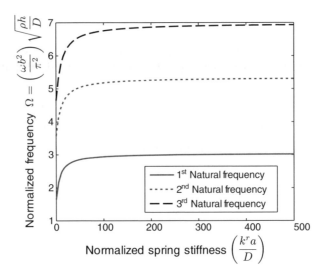

Figure 9.32 Normalized natural frequencies of plates increase as the rotational spring stiffness increases. For a particular plate, w is the natural frequency, b is the width, ρ is the density, h is the plate thickness, k^r is the rotational spring stiffness, a is the length of the plate, and D is the flexural rigidity. Source [47], Fig. 2, reproduced with permission of Elsevier.

Because the true initial conditions (a hammer strike) are not known, an arbitrary time shortly after the impact is selected as time zero. The model will use the strain at this time, as well as the rate of change of the strain, as the initial conditions for the time-series prediction. This limited amount of information about the initial conditions of the plate means that only one mode per gage can be modeled (without under-determining the problem). Thus, with three gages we can only estimate the first three plate modes in the identification process. Systems with more gages may therefore produce more accurate estimates, however one of our goals was to provide a good damage identification capability with limited resources. All other modes present in the recorded data are filtered out by a bandpass filter. The high-pass portion of the filter removes drift in the experimentally recorded strain from temperature change or other external sources.

A population of $Q = 16$ Markov chains was used in both the numerical study and experimental testing that follow. For each parameter, the Markov chains consisted of $K = 64{,}000$ points, however the initial 16,000 points were considered part of the "burn-in" process and were therefore discarded. Stationarity of the resulting Markov chains was assessed by repeating the analysis several times, making sure that the same posterior PDF was estimated. On an i7-2600k (3.4 GHz)-based desktop PC, the analysis required two minutes of processing time. This was one of the chief motivations for the modeling effort as a typical FE model would have required at least two orders of magnitude more time, potentially limiting the range of applications of this approach to damage assessment.

9.4.1 Experimental Setup

The test specimen was a 0.76 m by 0.60 m (30 in. by 24 in., 5:4 aspect ratio) 6061-T6 aluminum plate. The density was calculated using a digital scale and a vernier caliper to measure the

Table 9.7 Plate dimensions and information about thickness variability

True normalized plate dimension	Mean value	$\hat{c}_v$	Increase in $\hat{c}_v$
Width	$\dfrac{a}{b} = 1.25$	—	—
Length	$\dfrac{b}{b} = 1$	—	—
Thickness (Healthy)	$\dfrac{h_0}{h_0} = 1$	0.021	—
Thickness after 7 days of corrosion	$\dfrac{h_1}{h_0} = 0.953$	0.030	47%
Thickness after 15 days of corrosion	$\dfrac{h_2}{h_0} = 0.908$	0.059	187%

* Note that the healthy plate thickness normalized by plate length is $h_0/b = 0.00252$, indicating that this is, in fact, a thin plate. The increase in the coefficient of variation indicates that the plate becomes less uniform in thickness as corrosion progresses

mass and volume, respectively. Normalized dimensions are given in Table 9.2. The thickness of the plate reported three times, first the original, undamaged thickness, and then the thickness after each of the two corrosion cycles. Thickness is normalized by the original thickness so that the extent of the corrosion damage is clear. No plate has truly uniform thickness, and the plate used in this study is no exception. The thicknesses given in Table 9.2 are average values of nine separate measurements taken around the perimeter of the plate. To show how the variation in thickness increased as the plate corroded, the coefficient of variation is also given. The coefficient of variation is defined as

$$\hat{c}_v = \left(1 + \frac{1}{4N}\right) \cdot \frac{\sigma_x}{\bar{x}} \tag{9.65}$$

where $N = 9$ is the number of samples, σ_x is the standard deviation of the samples, and $\bar{x}$ is the sample mean of the measurements. The coefficient of variation is, for large N, roughly the standard deviation normalized by the mean, which allows for easy comparisons between samples that have different means. From examining the coefficient of variation, we can see that the variability of the plate thickness increases with the amount of corrosion. Hence the "uniform thickness" assumption is clearly violated (see Table 9.7). Nonetheless, we will show that the basic physics of the corroded plate are still captured by the model.

An initial set of vibration data was acquired using a new (i.e., uncorroded) plate. Then, to induce corrosion, the plate was submerged in a saltwater bath as shown in Figure 9.33. The saltwater had the same salt concentration as common seawater (3.5% salinity). The corrosion was accelerated by applying a voltage source to the submerged plate. After substantive corrosion, another set of vibration data was acquired. A new saltwater bath was prepared and the process was repeated.

To acquire the strain response data, the plate was mounted into the same test rig shown in Figure 9.24. The plate was clamped in a bolted fixture, where each of the 28, 12 mm bolts were tightened to 121 N-m (90 ft-lbs). Lock washers were used on each fastener. As in the preceding section on crack identification, the plate was excited with a rubber-tipped mallet and the same steps were taken to eliminate the nonlinear, large-amplitude response before recording. The

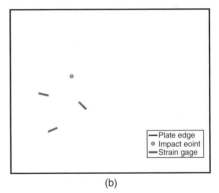

(a)   (b)

Figure 9.33 (a) Accelerated corrosion environment used to degrade the plate under test, shown in Figure 9.24 and (b) gage and impact location on the plate

goal, of course, is to collect observations in the regime over which we expect the model to accurately predict the response.

Metal foil strain gages, of the same type used in the crack experiments of the previous section, were again selected as a simple, inexpensive, means of acquiring the strain response data **s**. Other types of sensors (vibrometers, fiber-optic strain sensors, etc.) could just as easily be used in the identification process in the event that they are affordable for a given platform.

For these tests, three single-axis Vishay Micro Measurements model EA-13-125AD-120 strain gages were located in positions and orientations in the third quadrant of the plate, as shown at the right of Figure 9.33. The signals were sampled at a frequency of 10 kHz, well above the frequencies of interest. The locations of the gages were simply retained from the previous section in crack identification and no claims are made that these are preferred/optimal locations for detecting uniform corrosion damage.

9.4.2 Results and Discussion

9.4.2.1 Noise Study

Before proceeding to the experiment, a series of numerical experiments were conducted to better understand the performance of the estimator under varying levels of additive noise. To vary the amount of noise in a controlled manner, an artificial experimental signal **s** was created using the (noise-free) model $\mathbf{w}(n, \boldsymbol{\theta})$ with a prescribed thickness and torsional stiffness; various levels of Gaussian white noise were then superimposed on this signal. The Bayesian procedure was then used in an attempt to recover the specified plate thickness. This was repeated with five different noise signals at each noise level.

The final parameter estimates were taken as those associated with the largest peak in the posterior PDF. Although the mean or median of the posterior PDF is sometimes used in Bayesian estimation problems, this would have been a poor choice here as the distributions for h were typically multimodal, a fact we have repeatedly pointed out in the estimation problems of the previous sections.

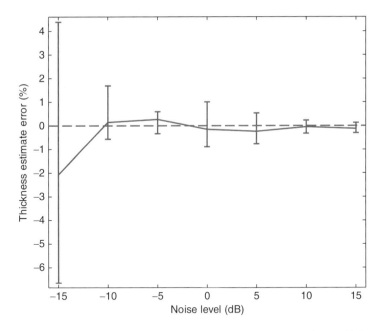

Figure 9.34 Range of the error associated with five independent estimates of plate thickness as a function of noise level. *Source*: Reproduced from [47], Figure 5, with permission of Elsevier

Seven different noise levels were considered; at each level, five test cases (each corresponding to a different noise signal) were run. Figure 9.34 shows the range of the error, over five separate thickness estimates as a function of noise level. The high noise level (−15 dB) range of error is most significant; the range of error goes from approximately −6.5% to 4.5%. As the noise level goes down (and the decibels go up), the range of error diminishes and the results converge toward zero error, as one would expect. As for practical implications, even at high noise levels (−15 dB) the thickness error range is less than 10%. For the experimental signals encountered in this study, the noise levels were close to 10 dB. This suggests that estimation errors related to additive sensor noise should be quite low. On the basis of past experience, we expect the dominant source of error to be modeling error.

To quantify how dissimilar the two signals (data and model) appear to be, one may compute the sum-squared difference of the signals at each instance and average the differences across the life of the signal. Again, this corresponds to the driving argument in the likelihood function (9.49) which, for uniform prior distribution, governs the form of the parameter posterior distributions as well. The trend in the sum-squared difference between the noise-corrupted data and the model predicted strain is shown in Figure 9.35. Again, as one might expect, higher noise levels lead to a larger response error (greater variance in the posterior distribution). To give context to the magnitude of sum-squared error values, the simulated response and the model predictions are plotted against each other for large and small levels of noise in Figure 9.35b and c.

It is interesting when contrasted with the trend of the thickness estimate in Figure 9.34. If the sum-squared error is large, as it is when the noise is great, then the time series of the

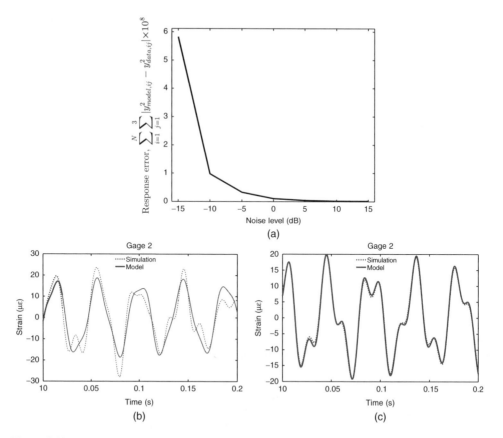

Figure 9.35 (a) The average of the sum-squared error of the model predicted strain response versus the filtered simulated data drops rapidly as noise decreases. Examples of the model prediction versus simulated data for gage 2 with noise of -15 dB (b) and 10 dB (c). *Source*: Reproduced from [47], Figure 6, with permission of Elsevier

model prediction is not very similar to the experimentally gathered data. It can be seen that the sum-squared error between the data and model response increases more quickly than does the error in the estimate of the plate thickness. Thus, an increase in the difference between data and model in the time domain does not necessarily imply an equal increase in the error of the thickness parameter estimate.

9.4.2.2 Experimental Study

The next step is to test the modeling and estimation approach using experimentally acquired data. The plate was tested first in the new, or healthy, condition before any corrosion. The plate was struck four separate times and the vibration response data (measured as strain vs. time) for each sensor was used to infer the thickness of the plate using the pop-MCMC algorithm. The PDFs that result from the Bayesian process are shown in Figure 9.36. For each parameter, a kernel density estimation approach (see Section 6.3, Eq. 6.40) was applied to the Markov

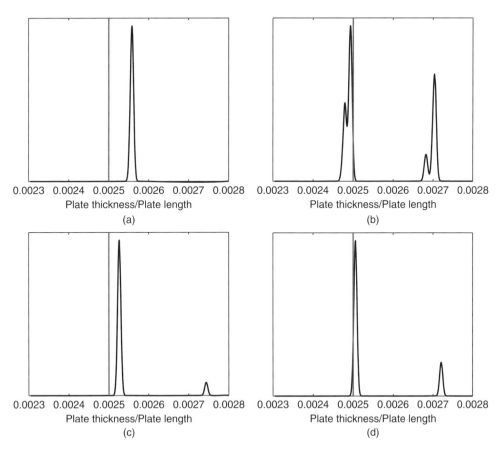

Figure 9.36 PDFs of the plate thickness estimate for the four experiments with the healthy, undamaged plate. The correct value is shown as a solid vertical line. *Source*: Reproduced from [47], Figure 7, with permission of Elsevier

Table 9.8 Mean and error range of the estimated plate thickness from experimental measurements

Experiment	Mean error (%)	Error range (min, max) (%)
Healthy plate	0.67	(−1.42, 2.58)
After 7 days of corrosion	0.71	(−0.97, 2.82)
After 15 days of corrosion	−1.51%	(−14.78, 14.35)

* A negative error implies that the estimate is too low

chain samples. Specifically, each of the PDFs in Figure 9.36 were calculated using a normal kernel function and a bandwidth found using Shimazaki's variable bandwidth method [48]. It can be seen that the estimates for the thickness of the healthy, undamaged plate are very good (Table 9.8).

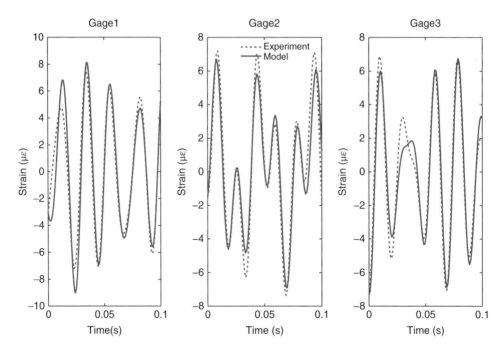

Figure 9.37 Comparison of the strain response predicted by the model, and the actual recorded data for the three gages on the undamaged plate. *Source*: Reproduced from [47], Figure 8, with permission of Elsevier

An example of the response predicted by the model versus the actual recorded strain data is shown in Figure 9.37. This data is the same as that produced by the PDF in Figure 9.36a. There is generally very good agreement between the experimentally recorded strain data and the model predicted response.

After corrosion was induced over seven days, the plate was tested again. It was again struck four separate times, and the strain from the resulting vibrations recorded after each strike. Again, the plate thickness was inferred from that data. The original recorded signal, the filtered signal, the model-based strain prediction, and the resulting thickness estimate PDF can be seen in Figure 9.38. The estimates of plate thickness are still reasonably good. However, the quality of the estimates degraded slightly after the first corrosion cycle as compared with the healthy plate. On the basis of the results of our numerical simulation, this degradation is far more likely to be the result of modeling error than experimental noise.

An example of the response predicted by the model versus the actual recorded strain data is shown in Figure 9.39. This data is the same as that which produced the PDF in Figure 9.38a. Here, in Figure 9.39 there is generally good agreement between the experimentally recorded strain data and the model predicted response, but not quite as good as with the healthy plate data in Figure 9.37.

After further corrosion was induced over an additional eight days, the plate was tested again. It was again struck four separate times, and the strain from the resulting vibrations recorded after each strike, and the plate thickness was again inferred from these datasets.

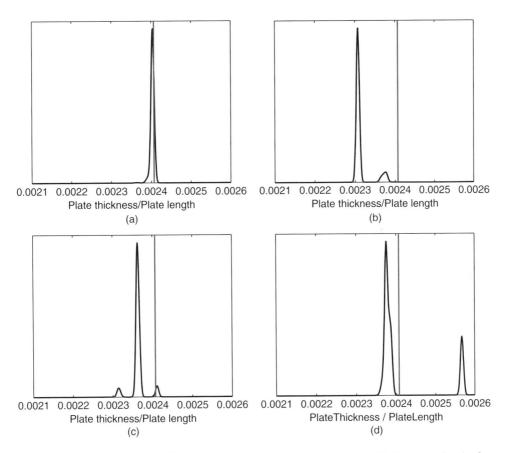

0.0021 0.0022 0.0023 0.0024 0.0025 0.0026   0.0021 0.0022 0.0023 0.0024 0.0025 0.0026
Plate thickness/Plate length   Plate thickness/Plate length
(a)   (b)

0.0021 0.0022 0.0023 0.0024 0.0025 0.0026   0.0021 0.0022 0.0023 0.0024 0.0025 0.0026
Plate thickness/Plate length   PlateThickness / PlateLength
(c)   (d)

Figure 9.38 PDFs of the plate thickness estimate for the four experiments with the plate after the first corrosion cycle. The correct value is shown as a solid vertical line. *Source* [47], Figure 9, reproduced with permission of Elsevier.

The original recorded signal, the filtered signal, the model-based strain prediction, and the resulting thickness estimate PDF can be seen in Figure 9.40 and 9.41. The accuracy and consistency of plate thickness estimates degraded significantly as additional corrosion cycles were applied. This coincided with a decrease in the uniformity of the thickness of the plate, as indicated in Table 9.2. The standard deviation of the plate thickness is now almost triple that of the undamaged plate. Despite efforts to agitate the corrosion media, the high-flow velocities that typically lead to uniform corrosion in industry were not present, and variations in the plate thickness became evident. This seems to be the most likely source of the decrease in accuracy of the thickness estimates.

An example of the response predicted by the model versus the actual recorded strain data taken from the heavily corroded plate is shown in Figure 9.40. This data is the same as that which produced the PDF shown in Figure 9.41a. The relationship between the model prediction and the recorded strain data is very weak. As the plate no longer satisfies the model assumption that the plate is uniform, this is again unsurprising.

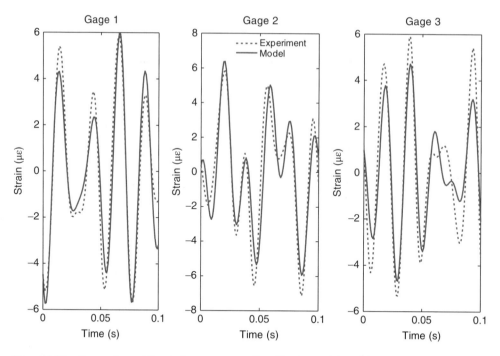

Figure 9.39 Comparison of the strain response predicted by the model, and the actual recorded data for the three gages on the plate after one corrosion cycle had been applied. *Source* [47], Figure 10, reproduced with permission of Elsevier.

What is interesting is that even in the event that the model begins to break down, we can learn something valuable about the physics of the problem. The method was intended to estimate the thickness of plates of uniform thickness, and it succeeds admirably when this condition is satisfied. However, when this condition is not satisfied, it alerts the practitioner by producing PDFs with many widely distributed peaks. Moreover, the width (scatter) of the distribution appears to increase with the magnitude of the corrosion. At this point, we have two choices. We can either remain content with the ability to reliably detect the localized damage as a breakdown of our global model or we could choose to move to a more localized approach as we alluded to at the start of this section.

9.5 Chapter Summary

The goal of this chapter was to formally combine deterministic and probabilistic models to predict the data we observe with our sensors. We then use the estimation methods of Chapter 7 to find those parameters that are most likely in either the Bayesian or maximum likelihood sense. For structural systems, where data are plentiful, these two approaches often yield similar

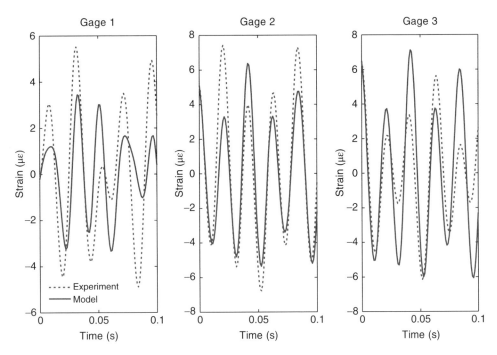

Figure 9.40 Comparison of the strain response predicted by the model, and the actual recorded data for the three gages on the plate after two corrosion cycles had been applied. *Source*: Reproduced from [47], Figure 12, with permission of Elsevier

results as we have shown in our experimental examples. Our preference for the Bayesian approach stems mainly for its ability to also provide us credible intervals by way of estimating the entire damage parameter posterior distributions. While the physics of the different damage models changed (e.g., cracking vs. corrosion vs. dents), the basic construction of our estimation problem did not. That is to say, we used the same additive, jointly Gaussian, i.i.d. noise probability model. While the tools of the preceding chapters would allow us to handle other noise models, in practice the Gaussian assumption is a good one, again as we have shown in experiment.

At this point, we have a means of obtaining damage parameter estimates. While useful, this does not solve our initial problem which concerned how best to maintain a structure. To take this final step, we have to chart a path that connects what we know about a structure now to the future integrity of that structure. We also have to formally assign costs to our maintenance decisions so as to properly chart a maintenance plan that is optimal in the sense that we define it. While this last, but most important piece, of the puzzle is far from complete, the next and final chapter attempts to combine all we have learned to this point and lay out the framework by which optimal maintenance may be accomplished.

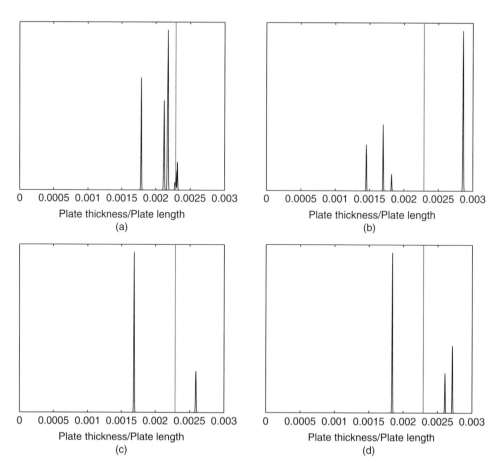

Figure 9.41 PDFs of the plate thickness estimate for the four experiments with the plate after the second corrosion cycle. The correct value is shown as a solid vertical line. *Source* [47], Figure 11, reproduced with permission of Elsevier.

References

[1] Z. P. Bazant, L. Cedolin, Stability of Structures, Oxford University Press, Oxford, 1991.

[2] C. A. Featherston, Imperfection sensitivity of curved panels under combined compression and shear, International Journal of Non-Linear Mechanics 38 (2003) 225–238.

[3] J. Singer, J. Arbocz, T. Weller, Buckling Experiments, Vol. 2, John Wiley & Sons, Inc., San Diego, CA, 2002.

[4] B. L. O. Edlund, Buckling of metallic shells: buckling and postbuckling behaviour of isotropic shells, especially cylinders, Structural Control and Health Monitoring 14 (2007) 693–713.

[5] H. M. Reed, J. M. Nichols, C. J. Earls, A modified differential evolution algorithm for damage identification in submerged shell structures, Mechanical Systems and Signal Processing 39 (2013) 396–408.

[6] H. Langtangen, Computational Partial Differential Equations: Numerical Methods and Diffpack Programming, Springer-Verlag, Berlin, 1999.

[7] G. Everstine, Structural analogies for scalar field problems, International Journal for Numerical Methods in Engineering 17 (1981) 471–476.

[8] G. Everstine, Finite element formulations of structural acoustics problems, Computers and Structures 65 (1997) 307–321.

 [9] R. L. Hardy, Theory and applications of the multiquadratic-biharmonic method, Computers and Mathematics with Applications 19 (8/9) (1990) 163–208.

[10] C. J. Stull, C. J. Earls, W. Aquino, A posteriori initial imperfection identification in shell buckling problems, Computer Methods in Applied Mechanics and Engineering 198 (2008) 260–268.

[11] C. J. Stull, J. M. Nichols, C. J. Earls, Stochastic inverse identification of geometric imperfections in shell structures, Computer Methods in Applied Mechanics and Engineering 200 (1–4) (2011) 2256–2267.

[12] E. N. Dvorkin, K. J. Bathe, A continuum mechanics based four-node shell element for general nonlinear analysis, Engineering Computations 1 (1983) 77–88.

[13] K. J. Bathe, E. N. Dvorkin, A four - node plate bending element based on Mindlin/Reissner plate theory and a mixed interpolation, International Journal for Numerical Methods in Engineering 21 (1985) 367–383.

[14] K. J. Bathe, E. N. Dvorkin, A formulation of general shell elements - the use of mixed interpolation of tensorial components, International Journal for Numerical Methods in Engineering 22 (1986) 697–722.

[15] H. P. Langtangen, Computational Partial Differential Equations, Springer-Verlag, 2000.

[16] K. J. Bathe, E. N. Dvorkin, On the automatic solution of nonlinear finite element equations, Computers and Structures 17 (5-6) (1983) 871–879.

[17] V. Papadopoulos, I. Pavlos, The effect of non-uniformity of axial loading on the buckling behavior of shells with random imperfections, International Journal of Solids and Structures 44 (2007) 6299–6317.

[18] K. Bathe, Finite Element Procedures, Prentice-Hall, New Jersey, 1996.

[19] C. J. Stull, C. J. Earls, P. Koutsourelakis, Model-based structural health monitoring of naval ships, Computer Methods in Applied Mechanics and Engineering 200 (2011) 1137–1149.

[20] C. J. Stull, J. Nichols, C. J. Earls, Stochastic inverse identification of geometric imperfections in shell structures, Computer Methods in Applied Mechanics and Engineering 200 (2011) 2256–2267.

[21] Rush, Show don't tell, Presto (1989).

[22] R. M. Neal, Probabilistic inference using Markov chain Monte Carlo methods, Tech. Rep. CRG-TR-93-1, Department of Computer Science, University of Toronto, Ontario, Canada (1993).

[23] R. M. Neal, Suppressing random walks in Markov chain Monte Carlo using ordered overrelaxation, Tech. Rep. 9508, Department of Statistics, University of Toronto, Ontario, Canada (1995).

[24] W. A. Link, R. J. Barker, Bayesian Inference with Ecological Examples, Academic Press, San Diego, CA, 2010.

[25] J. M. Nichols, W. A. Link, K. D. Murphy, C. C. Olson, A Bayesian approach to identifying structural nonlinearity using free-decay response: application to damage detection in composites, Journal of Sound and Vibration 329 (15) (2010) 2995–3007.

[26] J. M. Nichols, K. D. Murphy, Modeling and detection of delamination in a composite beam: a polyspectral approach, Mechanical Systems and Signal Processing 24 (2) (2009) 365–378.

[27] R. Solecki, Bending vibration of a rectangular plate with arbitrarily located rectilinear crack, Engineering Fracture Mechanics 22 (4) (1985) 687–695.

[28] J. Mitchell, Elementary distributions of plane stress, Proceedings of the London Mathematical Society 32 (1901) 35–61.

[29] J. M. Nichols, E. Z. Moore, K. D. Murphy, Bayesian identification of a cracked plate using a population-based Markov chain Monte Carlo method, Computers and Structures 89 (13–14) (2011) 1323–1332.

[30] A. Jasra, D. A. Stephens, C. C. Holmes, On population-based simulation for static inference, Statistics and Computing 17 (2007) 263–279.

[31] E. Z. Moore, J. M. Nichols, K. D. Murphy, Crack identification in a freely vibrating plate using Bayesian parameter estimation, Mechanical Systems and Signal Processing 25 (6) (2011) 2125–2134.

[32] E. Z. Moore, J. M. Nichols, K. D. Murphy, Model-based SHM: demonstration of identification of a crack in a thin plate using free vibration data, Mechanical Systems and Signal Processing 29 (2012) 284–295.

[33] A. Gelman, J. B. Carlin, H. S. Stern, D. B. Rubin, Baysian Data Analysis, 2nd ed., Chapman & Hall / CRC, 2004.

[34] C. Papadimitriou, M. Christodoulou, M. Pavlidou, S. Karamanos, Optimal sensor and actuator configuration for structural identification, in: Proceedings of the ASME Design Engineering Technical Conference, Vol. 6 A, Pittsburgh, PA, 2001, pp. 829–838.

[35] I. Trendafilova, W. Heylen, H. V. Brussel, Measurement point selection in damage detection using the mutual information concept, Smart Materials and Structures 10 (3) (2001) 528–533.

[36] P. Kripakaran, I. F. Smith, Configuring and enhancing measurement systems for damage identification, Advanced Engineering Informatics 23 (4) (2009) 424–432.

[37] J. M. Nichols, Examining structural dynamics using information flow, Probabilistic Engineering Mechanics 21 (2006) 420–433.

[38] H. Y. Guo, L. Zhang, L. Zhang, J. X. Zhou, Optimal placement of sensors for structural health monitoring using improved genetic algorithms, Smart Materials and Structures 13 (2004) 528–538.

[39] E. Z. Moore, K. D. Murphy, J. M. Nichols, Optimized sensor placement for damage parameter estimation: experimental results for a cracked plate, Structural Health Monitoring 12 (3) (2013) 197–206.

[40] P. Fromme, P. D. Wilcox, M. J. S. Lowe, P. Cawley, On the development and testing of a guided ultrasonic wave array for structural integrity monitoring, IEEE Transactions on Ultrasonics, Ferroelectrics, and Frequency Control 53 (4) (2006) 777–784. doi: 10.1109/TUFFC.2006.1611037.

[41] V. Giurgiutiu, J. Bao, W. Zhao, Active sensor wave propagation health monitoring of beam and plate structures, Proceedings of SPIE - The International Society for Optical Engineering 4327 (2001) 234–245, sensor wave propagation; Structural health monitoring. doi: 10.1117/12.436535.

[42] K. R. Lohr, J. L. Rose, Ultrasonic guided wave and acoustic impact methods for pipe fouling detection, Journal of Food Engineering 56 (4) (2003) 315–324. doi: 10.1016/S0260-8774(02)00156-5.

[43] J. H. Ali, W. B. Wang, P. P. Ho, R. R. Alfano, Detection of corrosion beneath a paint layer by use of spectral polarization optical imaging, Optics Letters 25 (17) (2000) 1303–1305, optical imaging.

[44] E. Jin, F. Chiang, ESPI and digital speckle correlation applied to inspection of crevice corrosion on aging aircraft, Research in Nondestructive Evaluation 10 (2) (1998) 63–73.

[45] N. Kasai, Y. Fujiwara, K. Sekine, T. Sakamoto, Evaluation of back-side flaws of the bottom plates of an oil-storage tank by the RFECT, NDT and E International 41 (7) (2008) 525–529. doi: 10.1016/j.ndteint.2008.05.002.

[46] V. Vavilov, E. Grinzato, P. Bison, S. Marinetti, M. Bales, Surface transient temperature inversion for hidden corrosion characterisation: theory and applications, International Journal of Heat and Mass Transfer 39 (2) (1996) 355–371. doi: 10.1016/0017-9310(95)00126-T.

[47] E. Z. Moore, K. D. Murphy, E. G. Rey, J. M. Nichols, Modeling and identification of uniform corrosion damage on a thin late sing a Bayesian framework, Journal of Sound and Vibration 340 (2015) 112–125.

[48] H. Shimazaki, S. Shinomoto, Kernel bandwidth optimization in spike rate estimation, Journal of Computational Neuroscience 29 (1–2) (2010) 171–182. doi: 10.1007/s10827-009-0180-4.

10

Decision Making in Condition-Based Maintenance

The previous chapters have been focused on modeling physical processes associated with damage and estimating the parameters of those models. This development has been motivated implicitly by the idea that it will provide us with the information needed to make good decisions about the maintenance and usage of a given structure. In fact, we argued in the introduction that the goal of any structural monitoring system is to make decisions that improve safety, reduce maintenance costs, and/or increase the operational envelope.

The topic of optimal decision making is an important one, yet it is often overlooked in many areas of science. In the field of structural health monitoring (SHM), researchers define their role as to provide information about current damage state, and perhaps projections of future damage state, to a decision maker. For example, a recent paper by Clauss *et al.* nicely integrates the components required of "decision support" for a ship transit problem [1]. Specifically, they predict both wave and corresponding ship motion and present the information to an operator. Other works are similarly devoted to decision support, whereby the information relevant to decision making is provided to the end user (see, e.g., [2–5] in the context of maintaining ship structures). In some cases, the system may even provide the decision maker with a set of possible actions he/she might take (see, e.g., [6] in the context of "fire control").

The role of the analyst currently ends here. Instead, it is implicitly assumed that the decision maker has access to the relevant information and uses intuition to make good decisions. Presumably, this "intuition" includes the rough outlines of a cost–benefit analysis, although such analyses are seldom cited in the damage detection literature. We view this "heuristic" approach as unsatisfying for at least two reasons. First, there is no objective algorithm by which structural models and associated parameter estimates are translated into decisions. Instead, decisions are made subjectively and are neither repeatable nor transparent. Second, such decisions are extremely unlikely to be optimal for any but the simplest problems.

In this chapter, we argue for a different approach that represents an attempt to provide optimal decisions using the same kind of rigor used in modeling and estimation. As we show, the mathematics of decision making can be formalized using many of the tools we have already

Modeling and Estimation of Structural Damage, First Edition. Jonathan M. Nichols and Kevin D. Murphy.
© 2016 John Wiley & Sons, Ltd. Published 2016 by John Wiley & Sons, Ltd.

developed. All that is required is a conceptual framework for (i) describing our space of possible decisions mathematically, (ii) a formal definition of what we want in a "good" decision, and (iii) a means of optimizing over our space of possibilities to find the best decision.

Approaches that produce an actual decision or maintenance plan as the output are few. A notable exception is the recent work of Huynh *et al.* [7], where the authors developed an approach that produces an optimal maintenance plan for monitoring a system degraded by fatigue cracks. An important feature of [7] is that it is dynamic in the sense that it incorporates newly available information in the decision-making process. We too view this as an essential component of decision making. The approach we develop in this chapter dynamically integrates information relevant to structural fatigue of a Naval vessel to reduce the probability of failure.

10.1   Structured Decision Making

Here we describe a decision-analytic approach that we refer to as structured decision making (SDM). The key steps of SDM are depicted in Figure (1.9): acquire data, identify and predict system state via modeling and estimation, and maximize an objective function defined on a space of possible decisions or actions. Numerous works are devoted to the various methods of collecting structural vibration data. In fact, just in the experimental examples provided in this book we have used accelerometers, strain gages (resistive and fiber-optic), and load cells, to name a few. In this chapter, as we have done throughout this book, we assume that an appropriately chosen data acquisition system is installed on the structure of interest.

The second piece of the process is perhaps the most challenging: obtaining the current state of the structure and then using that information to predict the future state. The former has largely been the topic of this book, while the latter comprises the field of *prognostics*. Research in predicting the evolution of structural damage is ongoing and many useful models are available. Some are general, relating cycles of observed stress to the remaining life of a structure or structural component. One such model is used in the example of this chapter. Other models attempt to predict the evolution of specific damage mechanisms such as crack growth (see, e.g., [8]). We do not attempt to cover the diversity of such models, rather our focus in this chapter is on one particular model and how it can be used in an automated decision-making framework.

Finally, if we are to make an optimal decision, we must first define a space of possible decisions on which to optimize. Often, our decisions can be described as a set of possible *actions* we might take. The actions may be discrete, as when the decision is to replace a potentially damaged structure or continue to use it. The actions may also be continuous as when the decision is the speed (within some specified interval) at which to propel a damaged ship through rough seas. In a simple example, let's consider the maintenance of a bridge on which we have installed a sensing system. The sensors record the bridge vibrations and these data are used to identify the bridge condition and predict the future condition. In this example, our space of possible actions can be quite simple: do we send a bridge inspector to the site, or not? It is a binary decision (only two possible actions) and is a likely one we would want our system to make at regular temporal intervals. In fact, we stress that any practical decision-making process is likely to be recurrent in time, that is to say, we revisit our decision at certain intervals. The reason, of course, is that as new information becomes available, the optimal decision may change. The recurrent nature of the process is highlighted in Figure 10.1. As part of the decision, we would weigh the cost of sending out an inspector versus the cost of a possible

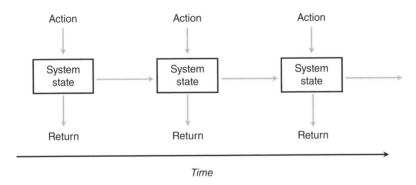

Figure 10.1 Any decision-making process is likely to be recurrent. As new information becomes available, the optimal decision will likely change. Our decision-making process must be able to accommodate such information

structural failure. This latter cost requires the probability of bridge failure, which is estimated from our model. Thus, the final ingredient combines model predictions and any other information relevant to the decision (e.g., cost, time, etc.) into an *objective function*, denoted $C(\mathbf{A})$. The objective function captures mathematically what it is that we want in a "good" decision and is defined as a function of a series of actions (decisions) we might make, $\mathbf{A} \equiv (A_{t_1}, A_{t_2}, \ldots,)$. The optimal decision can therefore be defined as the particular sequence of actions that minimizes

$$\mathbf{A}^* = \min_{\mathbf{A}} C(\mathbf{A}) \tag{10.1}$$

This framework therefore allows us to declare the optimal decision as the one that minimizes this cost and, by definition, best meets our stated objectives.

In what follows we describe the SDM process in the context of a fairly general structural monitoring problem, namely, how to balance usage of a structure with the cost of structural failure. As we will see, even in a seemingly simple problem there are many ingredients that need to be combined to produce a truly automated SDM system.

10.2 Example: Ship in Transit

The problem we consider as an illustrative example is that of a ship transiting from point A to point B. The "health" of ship structures is a point of some concern for the United States Navy. The superstructure on most ships is constructed from aluminum which, while lightweight, has proved sensitive to both stress corrosion cracking (due to the salty environment), and fatigue failure at weld lines. Both forms of damage can result in the failure of load-bearing components and require the ship to be docked and repaired. In addition to the monetary cost of the repair, time out-of-service is also a concern as it clearly reduces fleet readiness.

Damage to naval vessels is exacerbated by the continual loading placed on the structure by waves. Ocean waves are modeled as a continuous random process with frequency content dependent on weather conditions (to be explained shortly). The speed at which the vessel is traveling and the height and frequency of the waves govern the severity of the forcing and hence the rate at which fatigue damage is accumulated. As will be shown, faster ship speeds result in harsher loading conditions and faster rates of degradation.

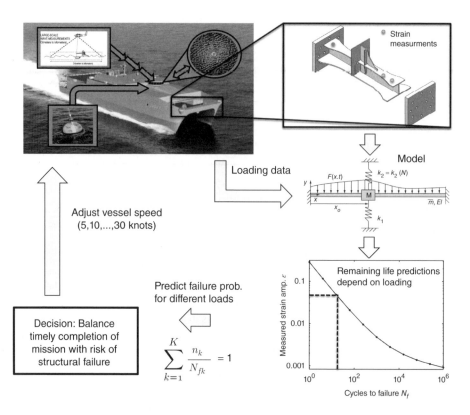

Figure 10.2 Schematic of the ship transit problem. The structural integrity of the vessel is governed by an aluminum support "stringer," which has been instrumented with strain sensors. A model of this stringer that allows one to estimate the current state and predict the future state using the available data (both strain data and wave predictions) and Miner's rule has been developed. The decisions in this example are setting the ship velocity at different points in transit *Source*: Reproduced from [9], Figure 2, with permission of Elsevier

Hence, we arrive at an interesting trade-off balancing time to completion of a particular mission and the integrity of the structure. It is this problem that is illustrated pictorially in Figure 10.2. A ship is required to travel a certain distance through the ocean within a fixed amount of time. Along the way the vessel is subject to wave loading, the severity of which increases with vessel speed. The structural integrity of the ship is assumed to be governed by the steel "stringers" used in certain naval vessels for hull support. We assume that the stringer is instrumented with several strain sensors such that we may record the strain response of this particular component. Using this data, we use a simple model for predicting future damage states and estimating the probability of component failure (Section 10.2.3). Our decisions will be based on a well-defined objective function that balances failure probability with time to completion. The space of possible actions will be the space of possible vessel speeds; the decision on what speed to use will be revisited at regular intervals. Although this example is relatively simple, it involves every aspect of SHM research: sensing, estimation, prediction, and decision making. From this example it should be clear how one might approach more sophisticated problems.

10.2.1 *Loading Data*

One of the interesting aspects of this problem is that the loading is a combination of wind, waves, and vessel speed. The subject of ocean wave modeling is well established by now and holds that the ocean wave height $\eta(y, t)$, as a function of both space and time, is modeled as a random process that is specified by its power spectral density (PSD) $S_{\eta\eta}(f)$ and marginal distribution $p_{\mathcal{N}}(\eta)$. The assumption of stationarity is expected to hold over the time scale at which a typical observation is made. For this study we use the spectral density given by Soares [10], which contains both wind-driven and "swell" components

$$S_{\eta\eta}^{(wind)}(f) = 0.11 H_w^2 T_w (T_w f)^{-0.5} e^{-0.44(T_w f)^{-4}} \gamma^{q_w}$$

$$S_{\eta\eta}^{(swell)}(f) = 0.11 H_s^2 T_s (T_s f)^{-0.5} e^{-0.44(T_s f)^{-4}} \gamma^{q_s} \tag{10.2}$$

so that

$$S_{\eta\eta}(f) = S_{\eta\eta}^{(wind)}(f) + S_{\eta\eta}^{(swell)}(f) \tag{10.3}$$

A number of parameters are required to completely specify this spectrum; these are

H_s Significant wave height

H_r Ratio of swell component energy to wind component energy

T_z Average wave period

T_r Ratio of swell period to wind component period $\qquad(10.4)$

from which the individual swell and wind component parameters can be defined [10]

$$H_{sw} = H_s \sqrt{\frac{1}{1 + H_r^2}}$$

$$H_{ss} = H_s \sqrt{\frac{H_r^2}{1 + H_r^2}}$$

$$T_{zw} = T_z (1 + H_r^2 / T_r) / (1 + H_r^2)$$

$$T_{zs} = T_z (T_r + H_r^2) / (1 + H_r^2) \tag{10.5}$$

From these, we may finally obtain the normalized wave heights and wave periods

$$H_w = \frac{H_{sw}}{\sqrt{F_1}}$$

$$H_s = \frac{H_{ss}}{\sqrt{F_1}}$$

$$T_w = \frac{T_{zw}}{F_2}$$

$$T_s = \frac{T_{zs}}{F_2} \tag{10.6}$$

for the wind (subscript w) and swell (subscript s) portions of the spectrum, respectively. The normalization constants F_1, F_2 depend on the empirically derived parameter γ. Using the data from Soares [10], we fit the following second-order polynomials

$$F_1 = -0.0067\gamma^2 + 0.2544\gamma + 0.7545$$

$$F_2 = 0.0044\gamma^2 - 0.0512\gamma + 1.0427 \tag{10.7}$$

so that the specification of γ uniquely defines both normalization constants. Valid ranges for this parameter were given in Ref. [10] as $\gamma \in [1, 6]$; in this example we use $\gamma = 3$. Finally, this spectrum requires the exponents

$$q_w = e^{-(1.296T_w f - 1)^2 / (2\sigma^2)}$$

$$q_s = e^{-(1.296T_s f - 1)^2 / (2\sigma^2)} \tag{10.8}$$

which are largely determined by T_w, T_s. The remaining parameter σ was suggested to be in the range $\sigma \in [0.07, 0.09]$, hence we use $\sigma = 0.08$ here. This particular spectral model was found to fit several thousand acquired spectra in the North Atlantic and North Sea and showed good agreement. Hence, we use it here to provide realistic loading scenarios for the transit problem being considered.

On the basis of this model, we see that the loading is determined largely by a significant wave height H_s and a significant wave period T_z. These parameters will vary over both time and space as weather conditions change. It is common to capture both in a single number—the "sea state." Sea state is an integer value designed to capture the severity of a particular sea condition. Table 10.1 gives the parameters H_s, T_z as a function of sea state. To construct our spatiotemporal loading, we assign a sea state to each point in space and time, thereby defining a loading spectral density. Also required are the ratios T_r, H_r which can be used to obtain the wind-driven component parameters. Following Ref. [10], we use $T_r = 2.1$ and $H_r = 0.49$ in our examples. Sample spectral densities for different sea states are provided in Figure 10.3. A more severe sea state results, in general, in lower frequencies but with far greater energy.

We know from Chapter (3) that by specifying the PSD we are equivalently specifying the covariance matrix for this stationary random process. We may use the approach described in Section 6.7 to create loading time histories with this spectral density and a particular marginal

Table 10.1 Sea states and associated wave heights and periods

Sea state	Mean wave height, H_s, in m (ft)	Mean wave period, T_z
0–1	0.05 (0.16)	—
2	0.3 (0.98)	6.9
3	0.87 (2.87)	7.5
4	1.87 (6.15)	8.8
5	3.25 (10.66)	9.7
6	5.00 (16.40)	12.4
7	7.50 (24.61)	15.0
8	11.50 (37.73)	16.4
>8	> 13.99 (> 45.90)	20.0

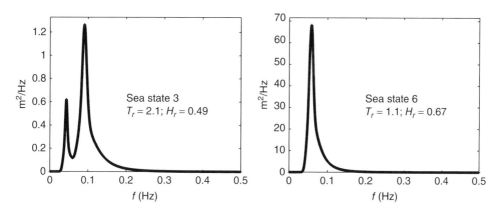

Figure 10.3 One-sided power spectral density functions associated with sea states 3 and 6 *Source*: Reproduced from [9], Figure 3, with permission of Elsevier

PDF. In the example presented in this section, we are implicitly assuming a Gaussian marginal density despite the fact that we know this is not quite correct. The loading will always be positive and centered around some mean value. However, as we see over the next two sections, our model for fatigue damage depends on the number of loading cycles and their associated amplitudes. This is precisely the information the spectral density function provides, thus the specific form of the marginal PDF is of diminished importance.

On the basis of a spectral representation, we seek an expression for the wave height $\eta(y, t)$ as a function of both space and time. Spectral theory for stationary random processes is well established (see the development in Section 3.3), but requires some care in practical implementation. Begin by modeling the wave height at a given spatial location as a zero-mean random process $\eta(t)$, $-\infty < t < \infty$. For such a process, we may use the Fourier–Stieltjes integral of Section (3.3.2)

$$\eta(t) = \int_{-\infty}^{\infty} e^{i\omega t} \, dZ(\omega) \tag{10.9}$$

to represent $\eta(t)$ as an infinite sum of sines and cosines with complex coefficients $dZ(\omega)$. In this representation, $Z(\omega)$ is itself a complex random process and, therefore, so too is $dZ(\omega) = Z(\omega + d\omega) - Z(\omega)$. As we showed in Section (3.3.3), this implies that for each ω, the quantity $dZ(\omega)$ is a random variable, defined so that $E[dZ(\omega)] = 0$ and

$$E[dZ(\omega)dZ^*(\omega)] = E[|dZ(\omega)|e^{i\phi(\omega)}|dZ(\omega)|e^{-i\phi(\omega)}]$$
$$= E[|dZ(\omega)|^2]$$
$$= S_{\eta\eta}(\omega)d\omega \tag{10.10}$$

that is, the complex random amplitudes $|dZ(\omega)|$ can be related (in expectation) to the PSD.

The challenge in being given $S_{\eta\eta}(\omega)$ and extracting $dZ(\omega)$ is that any phase information is lost through conjugation, that is, we cannot know what $\phi(\omega)$ is. We can, however, treat $\phi(\omega)$ as a random variable and consider the amplitudes $|dZ(\omega)|$ as fixed, known quantities. Using this approach, we see that

$$E[(|dZ(\omega)| \cos(\phi(\omega)) + i|dZ(\omega)| \sin(\phi(\omega))) \, (|dZ(\omega)| \cos(\phi(\omega)) - i|dZ(\omega)| \sin(\phi(\omega)))]$$

$$= |dZ(\omega)|^2 E[\cos^2(\phi(\omega))] + |dZ(\omega)|^2 E[\sin^2(\phi(\omega))] \qquad (10.11)$$

If we take the phases as uniformly distributed on the interval $[0, 2\pi)$, these expectations simplify to $\frac{1}{2}$ and we are left with

$$|dZ(\omega)|^2 = S_{\eta\eta}(\omega)d\omega \qquad (10.12)$$

or

$$|dZ(\omega)| = \sqrt{S_{\eta\eta}(\omega)d\omega}. \qquad (10.13)$$

Now, given that the random process is real, the only way for the imaginary components of (10.9) to vanish is if $dZ(\omega) = dZ^*(-\omega)$, which allows us to write (10.9) [11]

$$\eta(t) = \int_{-\infty}^{\infty} \cos(\omega t)dU(\omega) + \int_{-\infty}^{\infty} \sin(\omega t)dV(\omega) \qquad (10.14)$$

where $dU(\omega) = \text{Re}[dZ(\omega)] = |dZ(\omega)| \cos(\phi)$ and $dV(\omega) = -\text{Im}(dZ(\omega)) = -|dZ(\omega)| \sin(\phi)$. Substituting these expressions into Eq. (10.14) yields

$$\eta(t) = \int_{-\infty}^{\infty} |dZ(\omega)| \cos(\omega t) \cos(\phi) - |dZ(\omega)| \sin(\omega t) \sin(\phi)$$

$$= \int_{-\infty}^{\infty} |dZ(\omega)| \cos(\omega t + \phi)$$

$$= 2\int_{0}^{\infty} \cos(\omega t + \phi(\omega)) \sqrt{S_{\eta\eta}(\omega)d\omega} \qquad (10.15)$$

where in the last line we have made use of the fact that both cosine and $S_{\eta\eta}(\omega)$ are even functions. Note that the expression (10.15) is a factor of $\sqrt{2}$ larger than that given in Ref. [12] (page 104). This is simply due to our defining $S_{\eta\eta}(\omega)$ as the two-sided (as opposed to one-sided) autospectral density.

At this point, we have a random process $\eta(t)$ that is consistent with the PSD $S_{\eta\eta}(\omega)$. However, we also require a spatial dependence for the transit problem. To this end, we make use of linear wave theory [12] (specifically see Section 3.1 of the cited work) and assume the spatial dependence is also periodic. In terms of this development, this periodic spatial dependency will simply show up as a constant (with regard to time) phase difference so that we may represent the wave height

$$\eta(y, t) = 2\int_{0}^{\infty} \cos(k(\omega)y - \omega t - \phi(\omega)) \sqrt{S_{\eta\eta}(\omega)d\omega} \qquad (10.16)$$

The wavenumber $k(\omega)$ defines the spatial wave period $\lambda = 2\pi/k(\omega)$ and is, in general, a function of the temporal wave frequency ω. In this example, we use the common "deep-water" assumption [12] in which case $k(\omega) = \omega^2/g$ where g is the familiar gravitational constant. Finally, our location in space will depend on both the elapsed time and our velocity, $v \equiv dy(t)/dt$. Hence, we arrive at the final expression for the forcing function

$$\eta(y, t) = 2 \int_0^\infty \cos \left(\frac{\omega^2}{g} vt - \omega t - \phi(\omega) \right) \sqrt{S_{\eta\eta}(\omega)d\omega} \qquad (10.17)$$

where we have implicitly assumed the wave train is encountering the ship head-on. Other wave directions, that is, heading information, can easily be incorporated by multiplying the velocity by the cosine of the heading angle as in Ref. [1].

Thus, we have that an $N-$point realization of a random process with a known, two-sided autospectral density function can be generated by the Fourier series approximation

$$\eta(y, i\Delta_t) = 2 \sum_{k=0}^{K-1} \cos \left(\frac{\omega_k^2}{g} vi\Delta_t + \omega_k i\Delta_t + \phi_k \right) \sqrt{S_{\eta\eta}(\omega_k)\Delta_\omega}, i = 0 \cdots N - 1 \quad (10.18)$$

where Δ_t is the desired temporal resolution of the process, $\Delta_\omega = \frac{2\pi}{N\Delta_t}$, and $\omega_k = k\Delta_\omega$ for all nonnegative frequencies up to the Nyquist frequency $K = N/2$ ($\omega_K = \pi/\Delta_t$).

In practice, the sea-state data defining our loading will be predicted over space and time using weather service models. We do not use such models in this example, but rely on the fact that such models exist. Let's say we are traveling in a straight path from point A to point B. Divide the total spatial distance into D evenly spaced spatial intervals and the maximum allowable transit time into T temporal intervals. A sea-state map, such as the one shown in Figure 10.4, will be denoted by the $D \times T$ matrix $\mathbf{W}$ and is assumed to be made available to the ship's captain by the weather service.

This map shows the predicted sea state at any point in the path as a function of time. This particular sea-state map was generated by first creating a $D \times T$ matrix of independent, identically distributed (i.i.d.), normally distributed entries. Using the procedure described in Section (6.7.2), this matrix was filtered in both space and time to produce a matrix with a strong covariance in each direction (e.g., highly correlated in both space and time). Specifically, we use

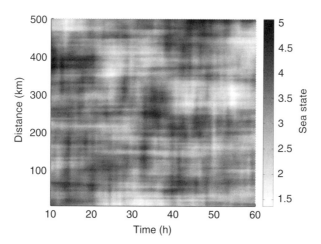

Figure 10.4 Sea-state prediction matrix $\mathbf{W}$ over a 1 km journey and spanning a prediction horizon of 60 h. These predictions will change in time so that the matrix $\mathbf{W}$ is but one in a collection of predictions $\mathbf{W} = (\mathbf{W}^{(0)}, \mathbf{W}^{(1)}, \ldots, \mathbf{W}^{(T)})$ that will be issued during transit *Source*: Reproduced from [9], Figure 4, with permission of Elsevier

covariance matrices of the form

$$C_{ij}^{(TT)} = \max(1.0 - |(i-j)\alpha_T|, 0) = \mathbf{H_T}\mathbf{H_T^T}$$

$$C_{ij}^{(YY)} = \max(1.0 - |(i-j)\alpha_Y|, 0) = \mathbf{H_S}\mathbf{H_S^T} \qquad (10.19)$$

where α_T, α_Y quantify the linear decay in covariance in the temporal and spatial dimensions, respectively. Factoring these matrices by Cholesky decomposition yields the filters $\mathbf{H}_T, \mathbf{H}_S$ (as in Eq. 6.175). Denote as $\tilde{\mathbf{W}}$ an $D \times T$ matrix of i.i.d., normally distributed random variables with unit variance and zero mean. A matrix $\mathbf{W}$ with spatiotemporal covariance specified by (10.19) is therefore given by

$$\mathbf{W} = \mathbf{H_S}^T \tilde{\mathbf{W}} \mathbf{H_T}. \qquad (10.20)$$

Of course, during a given journey we would expect a *sequence* of weather predictions $W = (\mathbf{W}^{(0)}, \mathbf{W}^{(1)}, \ldots, \mathbf{W}^{(T)})$. This sequence will also possess strong temporal correlations (predicted weather at time t will be similar to the predicted weather a short time into the future), specified by yet a third covariance matrix.

$$C_{ij}^{(PP)} = \max(1.0 - |(i-j)\alpha_P|, 0) = \mathbf{H_P}\mathbf{H_P}^T. \qquad (10.21)$$

Thus, using the entries of $\mathbf{W}^{(0)}$ as the starting values, we can generate the temporal evolution of the sea-state predictions for each space/time point as it evolves in time by performing another filtering procedure using $\mathbf{H}_P$. The end result is the $D \times T \times T$ matrix of sea-state predictions $\mathbf{W}$, where each value can be simply rescaled to the appropriate sea-state range, in this case a real number on the interval $[1, 6]$. Figure 10.4, therefore, shows one slice of the full matrix $\mathbf{W}$, obtained with covariance parameters $\alpha_T = \alpha_Y = \alpha_P = 0.05$. These values produce reasonable spatiotemporal correlations in sea state and have been used throughout the remainder of this chapter in generating weather predictions.

This section has described the steps associated with generating a realistic loading model for the ship-in-transit problem and serves as a reminder of why so much time was spent in the earlier chapters establishing random process models and spectral theory. Many problems in SHM require one to model forcing functions that result from naturally occurring processes. Therefore, we consider the ability to do so accurately to be of great importance to those seeking to use the resulting random vibration data in condition-based maintenance.

10.2.2   Ship "Stringer" Model

Rather than attempt to gage the structural integrity of an entire ship, we analyze one of the key load-bearing members in a mono-hull naval vessel. A failure of this component is considered a structural failure of the ship and immediate repairs would be required. The particular component of interest is the "stringer" shown in Figure 10.5a and is essentially an I-beam with a "stiffener" placed at the center of the beam. As we have done throughout, the goal is to capture the basic physics of a typical damage scenario with the simplest possible model.

To study the transverse vibrations, the beam is modeled as an Euler–Bernoulli beam with a concentrated mass (M) and stiffness (k_{eff}). This model is shown in Figure 10.5b. The mass replicates the added inertia and the springs the desired additive stiffness. The reason for the two

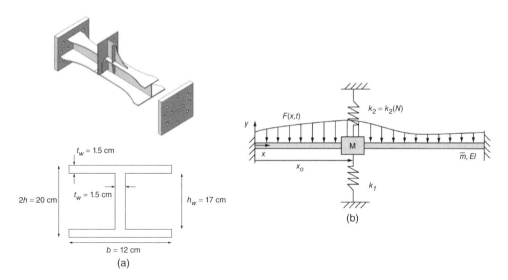

Figure 10.5 Load-bearing stringer (a) and associated model (b) *Source*: Reproduced from [9], Figure 5, with permission of Elsevier

springs is so that even if a weld degrades and fails on the stiffener, much of the stiffener will remain intact, that is, all welds will not likely fail simultaneously. Thus, although the stiffener will be degraded in failure, it will not reduce to zero.

As this system is forced, one would expect the stiffness k_2 to degrade in time owing to fatigue damage, while k_1 will remain fixed. To derive the equations of motion for this model, we use the energy methods presented in Chapter (4). The elastic strain energy is characterized by the product of elastic modulus E and the mass moment of inertia I, while the potential energy is stored in the discrete springs. Thus, we have for the strain energy

$$U = \int_0^L \left[\frac{1}{2} EI w_{,xx}^2 + \frac{1}{2} k_{eff} w^2 \delta(x - x_o) \right] dx \tag{10.22}$$

where $\delta(x - x_o)$ is the Dirac delta function centered at x_o. For the cross-section shown in Figure 10.5, the moment of inertia is given by

$$I = \frac{bh^3 - (b - t_w)h_w^3}{12} \tag{10.23}$$

The lateral deflection of the beam, as a function of both space and time, is denoted $w(x, t)$; however, in the development we simply use w and w_x, $\dot{w}$ to denote spatial and temporal derivatives, respectively. Note that the Dirac delta function has units of one over length; however, it is multiplied by the differential element dx, and hence the spring constant has the traditional units of force/length. The kinetic energy is given by

$$T = \int_0^L \left[\frac{1}{2} \bar{m} \dot{w}^2 + \frac{1}{2} M w^2 \delta(x - x_o) \right] dx. \tag{10.24}$$

Here, $\bar{m}$ is the mass per unit length of the beam and M is the concentrated mass of the stiffener. As with the concentrated spring, the concentrated mass has the traditional units (mass- and not mass per unit length, as $\bar{m}$ does).

Taking the variation of these expressions and substituting into Hamilton's principle ($\int_{t_1}^{t_2}[\delta T - \delta U]dt$, see Chapter 4, Section 4.3.5), one gets

$$\int_{t_1}^{t_2} \left\{ \int_0^L \left(\left[\bar{m}\dot{w}\ \delta\dot{w} + M\dot{w}\ \delta\dot{w}\delta\left(x - x_o \right) \right] - \left[EIw_{,xx}\delta w_{,xx} + k_{eff}w\ \delta w\delta(x - x_o) \right] \right) dx \right\} dt = 0$$

(10.25)

where the Dirac delta function should not be confused with the variation symbol (also given by δ). Integrating by parts with respect to space (for the strain energy) and time (for the kinetic energy) leads to the boundary conditions and the following governing linear partial differential equation for displacement

$$\bar{m}\ddot{w} + M\ddot{w}\delta(x - x_o) + EIw_{,xxx} + k_{eff}w\delta(x - x_o) = 0.$$

(10.26)

This equation may be spatially discretized using Galerkin's method. For the assumed solution, we use the model

$$w(x, t) = a_1(t)\Psi_1(x) + a_2(t)\Psi_2(x)$$

(10.27)

where the shape functions are taken as

$$\Psi_1(x) = \frac{1}{2}\left[1 - \cos\left(\frac{2\pi x}{L}\right)\right]$$

$$\Psi_2(x) = \frac{1}{2}\left[1 - \cos\left(\frac{4\pi x}{L}\right)\right].$$

(10.28)

These functions were chosen because they satisfy the geometric boundary conditions (as required by Galerkin) of no displacement and no slope (the stringer is fixed at both ends). Another reason for this choice is that in the absence of the central spring (i.e., $k_{eff} \rightarrow 0$), one gets the standard "mode one" type response of a clamped–clamped beam. Our reasons for focusing only on the first mode have to do with the expected wave loading. The frequency content of the ocean waves is $O(10^{-1})$ Hz, while the natural frequencies of the beam are well above 10 Hz; in this example, mode 1 occurs at 26 Hz. As a result, one would not expect significant contributions from the higher modes but will rather see the quasi-static beam bending in the first mode only.

The associated displacement (Eq. 10.27) is shown in Figure 10.6. Conversely, for a stiff spring ($k_{eff} \rightarrow \infty$), there will not be any center displacement. In this limiting case, one gets the mode shape of Figure 10.6d. Figure 10.6b and c shows the transitional cases between these extremes. Substituting Eq. (10.27) into (10.26) and performing a Galerkin procedure results in two linear ordinary differential equations. We further assume a modal damping model, specified by coefficients c_{11}, c_{22} (see Eq. 4.25), so that the governing equations are given in matrix form by

$$[m]\ddot{\mathbf{a}} + [c]\dot{\mathbf{a}}(t) + [k]\mathbf{a} = \mathcal{F}$$

(10.29)

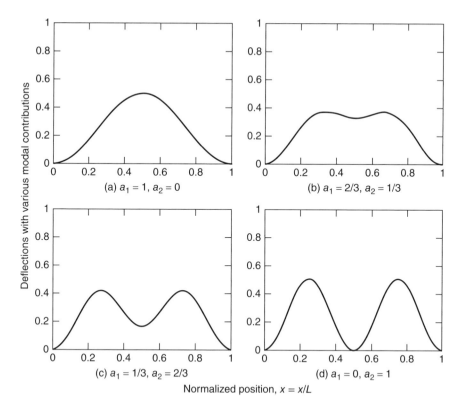

Figure 10.6 Mode shapes associated with the model given in Eq. (10.27). For zero central stiffness, the mode shape for a clamped–clamped beam is recovered (a). For an infinitely stiff central portion, we recover two clamped-clamped beams (d). Other contributions (b and d) yield mixed results. This model captures the basic physics of the problem *Source*: Reproduced from [9], Figure 6, with permission of Elsevier

where

$$[m] = \begin{bmatrix} \left(\frac{3}{8}\bar{m}L + M\left[\frac{1}{2}\left(1 - \cos\left[\frac{2\pi x_o}{L}\right]\right)\right]^2\right) & \left(\frac{1}{4}\bar{m}L + \frac{M}{4}\left(1 - \cos\left[\frac{2\pi x_o}{L}\right]\right)\left(1 - \cos\left[\frac{4\pi x_o}{L}\right]\right)\right) \\ \left(\frac{1}{4}\bar{m}L + \frac{M}{4}\left(1 - \cos\left[\frac{2\pi x_o}{L}\right]\right)\left(1 - \cos\left[\frac{4\pi x_o}{L}\right]\right)\right) & \left(\frac{3}{8}\bar{m}L + M\left[\frac{1}{2}\left(1 - \cos\left[\frac{4\pi x_o}{L}\right]\right)\right]^2\right) \end{bmatrix}$$

(10.30)

$$[c] = \begin{bmatrix} c_{11} & 0 \\ 0 & c_{22} \end{bmatrix}$$

(10.31)

and

$$[k] = \begin{bmatrix} \left(2\pi^4\frac{EI}{L^3} + k_{eff}\left[\frac{1}{2}\left(1 - \cos\left[\frac{2\pi x_o}{L}\right]\right)\right]^2\right) & \left(\frac{k_{eff}}{4}\left(1 - \cos\left[\frac{2\pi x_o}{L}\right]\right)\left(1 - \cos\left[\frac{4\pi x_o}{L}\right]\right)\right) \\ \left(\frac{k_{eff}}{4}\left(1 - \cos\left[\frac{2\pi x_o}{L}\right]\right)\left(1 - \cos\left[\frac{4\pi x_o}{L}\right]\right)\right) & \left(32\pi^4\frac{EI}{L^3} + k_{eff}\left[\frac{1}{2}\left(1 - \cos\left[\frac{4\pi x_o}{L}\right]\right)\right]^2\right) \end{bmatrix}$$

(10.32)

If we further assume that the stiffener is located at the midpoint, $x_o = L/2$, the mass and stiffness matrices reduce to

$$[m] = \begin{bmatrix} \left[\frac{3\bar{m}L}{8} + M \right] & \left[\frac{\bar{m}L}{4} \right] \\ \left[\frac{\bar{m}L}{4} \right] & \left[\frac{3\bar{m}L}{8} \right] \end{bmatrix} \tag{10.33}$$

$$[k] = \begin{bmatrix} \left[2\pi^4 \frac{EI}{L^3} + k_{eff} \right] & [0] \\ [0] & 32\pi^4 \frac{EI}{L^3} \end{bmatrix}. \tag{10.34}$$

The applied load vector in this case is found to be

$$\mathcal{F} = \left\{ \begin{array}{c} \int_0^L f(x,t)\Psi_1(x)dx \\ \int_0^L f(x,t)\Psi_2(x)dx \end{array} \right\}. \tag{10.35}$$

If it is assumed that the excitation is harmonic, and spatially constant (with a force per unit length of F), we have $f(x,t) = F(\omega)\cos(\omega t + \phi(\omega))$ and the load vector becomes simply

$$\mathcal{F} = \frac{F(\omega)L}{2} \left\{ \begin{array}{c} 1 \\ 1 \end{array} \right\} \cos(\omega t + \phi(\omega)). \tag{10.36}$$

At this point, the matrix equation (10.29), along with the forcing (10.36), can be used to solve for the time-dependent coefficients $\mathbf{a}(t) = (a_1(t), a_2(t))$. This is accomplished using the familiar solution procedure for linear, multiple-input, multiple-output differential equations covered in Chapter 4, Section (4.6). For such systems we know that the response of a particular DOF to the dual input (10.36) will be the superposition of responses to inputs at each DOF taken separately. Thus, we write for the general solution

$$a_1(t) = \frac{F(\omega)L}{2}[H_{11}(i\omega)e^{i\omega t} + H_{12}(i\omega)e^{i\omega t}]$$

$$a_2(t) = \frac{F(\omega)L}{2}[H_{21}(i\omega)e^{i\omega t} + H_{22}(i\omega)e^{i\omega t}] \tag{10.37}$$

where the notation $H_{ij}(i\omega)$ denotes the linear transfer function relating a unit amplitude, harmonic input $e^{i\omega t}$ at DOF j to output DOF i. To find these transfer functions, we can substitute (10.37) and (10.36) into (10.29). The result is a system of four equations that can be solved for the four unknown transfer functions. Using this approach we may compactly write

$$\begin{bmatrix} k_{eff} + \frac{2\pi^4 EI}{L^3} + i8c_{11}\omega - (M + \frac{3Lm}{8})\omega^2 & -\frac{Lm}{4}\omega^2 \\ -\frac{Lm}{4}\omega^2 & \frac{32\pi^4 EI}{L^3} + ic_{22}\omega - \frac{3Lm}{8}\omega^2 \end{bmatrix}$$

$$\times \begin{bmatrix} H_{11}(i\omega) & H_{12}(i\omega) \\ H_{21}(i\omega) & H_{22}(i\omega) \end{bmatrix} = \begin{bmatrix} 1 & 0 \\ 0 & 1 \end{bmatrix} \tag{10.38}$$

or, more compactly,

$$\mathbf{Z}(i\omega)\mathbf{H}(i\omega) = \mathbf{I} \tag{10.39}$$

Thus, we see that a simple inversion of the coefficient matrix yields the needed transfer functions which can be substituted into Eq. (10.37) and combined with Eq. (10.27) to establish the displacement at any point in space and time. In practice, however, it is more common to measure the strain response, $\epsilon(x, t)$, using either resistive or fiber-optic strain sensors. Both have been used to monitor the vibration of ship structures in various locations during transit (see, e.g., [13–15]).

We know from the basic Euler–Bernoulli beam theory that the strain is related to displacement via $\epsilon(x, t) = -h\frac{\partial^2 w(x,t)}{\partial x^2}$, where h is the distance from the beam neutral axis to the surface on which the strain is being measured. In this case, we may finally write our observed response

$$\epsilon(x, t) = -\frac{FLh}{2}\left\{\frac{\partial^2 \Psi_1(x)}{\partial x^2}\left[H_{11}(i\omega) + H_{12}(i\omega)\right]e^{i\omega t} + \frac{\partial^2 \Psi_1(x)}{\partial x^2}[H_{21}(i\omega)e + H_{22}(i\omega)]e^{i\omega t}\right\} \quad (10.40)$$

where

$$\frac{\partial^2 \Psi_1(x)}{\partial x^2} = \frac{2\pi^2}{L^2}\cos\left(\frac{2\pi x}{L}\right)$$

$$\frac{\partial^2 \Psi_2(x)}{\partial x^2} = \frac{8\pi^2}{L^2}\cos\left(\frac{4\pi x}{L}\right). \quad (10.41)$$

Of course, in simulating (10.40) we consider the real part of the expression only so that writing $H_{ij}(i\omega) = |H_{ij}(i\omega)|e^{i\phi_{ij}}$, where $\phi_{ij} = \tan^{-1}(\text{Im}\{H_{ij}(i\omega)\}/\text{Re}\{H_{ij}(i\omega)\})$. We also will cease to write the phases as frequency dependent for notational convenience, although such a dependence implicitly exists. Using this notation, the strain response at any point on the beam becomes

$$\epsilon(x, t) = -\frac{FLh}{2}\left\{\frac{2\pi^2}{L^2}\cos\left(\frac{2\pi x}{L}\right)[|H_{11}(i\omega)|\cos(\omega t + \phi_{11}) + |H_{12}(i\omega)|\cos(\omega t + \phi_{12})]\right.$$

$$\left. + \frac{8\pi^2}{L^2}\cos\left(\frac{4\pi x}{L}\right)[|H_{21}(i\omega)|\cos(\omega t + \phi_{21}) + |H_{22}(i\omega)|\cos(\omega t + \phi_{22})]\right\}. \quad (10.42)$$

This is the structure strain response to a harmonic forcing with constant amplitude $\frac{F(\omega)L}{2}$. To simplify the analysis further, we assume that the strain response is measured at the midpoint, $x = L/2$. With this assumption, the Eq. (10.42) becomes

$$\epsilon(L/2, t) = -\frac{FLh}{2}\left\{\frac{8\pi^2}{L^2}\left[|H_{21}(i\omega)|\cos(\omega t + \phi_{21}) + |H_{22}(i\omega)|\cos(\omega t + \phi_{22})\right]\right.$$

$$\left. - \frac{2\pi^2}{L^2}\left[|H_{11}(i\omega)|\cos(\omega t + \phi_{11}) + |H_{12}(i\omega)|\cos(\omega t + \phi_{12})\right]\right\}. \quad (10.43)$$

However, as we have discussed in the previous section, our forcing function is a realization of a random process with known PSD, spread across a broad array of frequencies. More specifically, the force felt by the structure due to the pressure of a surface wave at frequency ω can be expressed as the Froude–Krylov force

$$F_p(t) = \frac{1}{2}A\rho_w g\eta(y, t)$$

$$= A\rho_w g\int_0^\infty \sqrt{S_{\eta\eta}(\omega)d\omega}\cos\left(\frac{\omega^2}{g}v + \omega t + \phi(\omega)\right) \quad (10.44)$$

where $\rho_w = 1030 \quad \text{kg/m}^3$ is the density of seawater and A is the area over which the pressure force acts. Because the stringer model requires a force per unit length, the "area" in (10.44) possesses units of meters, not meters-squared and is more appropriately thought of as the characteristic dimension of the problem, that is, a length scale over which the pressure wave acts.

There will also be an inertial force due to the wave slamming that increases as a function of velocity. A summary of modeling approaches for this type of loading is provided in Ref. [16]. More specifically, for a plate subject to a wave with impact velocity v_I, the force per unit length is

$$F_s = \frac{1}{2}\rho_w A C_1 v_I^2 \tag{10.45}$$

where C_1 is a dimensionless constant that depends on the geometry of the structure under study [16], and A is again a characteristic length scale (in meters) over which the slamming event occurs. Following the experimental results in Ref. [16], we use $C_1 = 0.25$ in this work. An impact in the time domain can be expected to possess broad-band spectral content. We therefore model the effect of the impacting force as an increase in the amplitude of the pressure loading proportional to the square of the ship's velocity. Thus, combining with (10.44) the loading may be written as

$$F_{tot}(t) = \sum_{k=1}^{K} A\rho_w \left(g\sqrt{S_{\eta\eta}(\omega_k)d\omega_k} + \frac{C_1}{2K}v^2 \right) \cos\left(\frac{\omega_k^2}{g}v + \omega_k t + \phi(\omega_k) \right)$$

$$= \sum_{k=1}^{K} F(\omega_k) \cos\left(\frac{\omega_k^2}{g}v + \omega_k t + \phi(\omega_k) \right) \tag{10.46}$$

where we have spread the energy in the impact loading evenly across all K frequencies. Under this model, a stationary vessel would experience no impact load, while rapidly moving vessels would experience large amplitude loading. The loading model (10.46) correctly captures the expected increase in loading due to increasing sea state and ship velocity.

Given that the structural response to a forcing $F(\omega)\cos(\omega t + \phi)$ is given by Eq. (10.43), we may write the response to the wave loading across all frequencies ω using the Fourier series

$$\epsilon(L/2, t) = -\frac{A\rho_w Lh}{2} \sum_{k=0}^{N_{modes}} \left(g\sqrt{S_{\eta\eta}(\omega_k)\Delta_\omega} + \frac{C_1 v^2}{2K} \right)$$

$$\times \left\{ \frac{8\pi^2}{L^2} \left[|H_{21}(i\omega_k)| \cos\left(\frac{\omega_k^2}{g}vt + \omega_k t + \phi_{21} + \phi \right) + |H_{22}|(i\omega_k) \cos\left(\frac{\omega_k^2}{g}vt + \omega_k t + \phi_{22} + \phi \right) \right] \right.$$

$$\left. -\frac{2\pi^2}{L^2} \left[|H_{11}(i\omega_k)| \cos\left(\frac{\omega_k^2}{g}vt + \omega_k t + \phi_{11} + \phi \right) + |H_{12}(i\omega_k)| \cos\left(\frac{\omega_k^2}{g}v + \omega_k t + \phi_{12} + \phi \right) \right] \right\}. \tag{10.47}$$

To implement (10.47), the problem was discretized temporally giving the simulated response at times $t_i = i\Delta_t, i = 1\cdots N$, where, in this example, we use $\Delta_t = T_z/20$, that is, a small fraction of the dominant wave period. This temporal discretization implies the (positive) discrete frequencies $\omega_k = k\Delta_\omega$, $k = 0\cdots N/2$, where $\Delta_\omega = \frac{2\pi}{N\Delta_t}$ and $N_{modes} = N/2$ are the number of available modes to use in the expansion.

At this point, our model is sufficient to capture the basic physics governing the problem. For a given velocity and a given sea state, the vibrational strain response of the structure over some spatial interval can be predicted. The final ingredient is a model that forecasts the amount of fatigue damage accumulated by the structure given both speed and wave loading. This topic is discussed next.

10.2.3 Cumulative Fatigue Model

Most prognostic models specify a number of cycles to failure N_f at a given level of loading. The higher the loading, the fewer cycles it takes for a structure to fail. Different models exist for predicting N_f; here we use the Manson–Coffin fatigue model

$$\bar{e} = b(2N_f)^c \tag{10.48}$$

where $\bar{e}$ is the amplitude of the strain response of the specimen under periodic loading and b and c are constants related to the material under study [17]. In this example, the material is structural steel for which we have that $b = 0.45$ and $c = -0.6$. A plot of Eq. (10.48) is provided in Figure 10.7 for these parameters. Again, however, the structural response we will encounter is best modeled as a random process with no clearly defined period. What is required is therefore a means of decomposing the response $\epsilon(L/2, t)$ into a number of cycles n_k at response amplitudes $\epsilon_k, k = 1 \cdots N_A$. The most popular approach for accomplishing this is the so-called rainflow analysis [18], a version of which is implemented here.

Now, for each identified load level ϵ_k, there is a corresponding number of cycles to failure N_{f_k}, found by inverting Eq. (10.48)

$$N_{f_k} = \frac{1}{2} b^{-1/c} \epsilon_k^{1/c}. \tag{10.49}$$

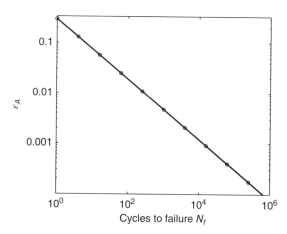

Figure 10.7 The Manson–Coffin equation governing material fatigue for parameters $b = 0.45$ and $c = -0.6$. These parameters were chosen on the basis of the assumption that the material in question was structural steel [17] *Source*: Reproduced from [9], Figure 7, with permission of Elsevier

Given the number of cycles observed at a given load level n_k and N_{f_k}, we can predict the cumulative amount of fatigue damage. The Palmgren–Miner rule is a linear model for cumulative fatigue given by

$$P_f = \sum_{k=1}^{K} \frac{n_k}{N_{f_k}} \tag{10.50}$$

for some number K of different amplitude loadings as determined by the aforementioned rainflow counting method (see, e.g., [19]). The quantity P_f may be interpreted as a probability of failure and is typically compared to a threshold, that is, $P_f \gtrsim C$ to assess the severity of the degradation. Typically, the constant C is set to a number near unity, suggesting that a structure for which $P_f > C$ is in imminent danger of failure. It is noted that in the framework presented here, there is nothing preventing the insertion of a different fatigue model in lieu of (10.50). The only requirement is that there exists some mechanism for predicting fatigue failure as a function of observed stress cycles; the Palmgren–Miner rule is one frequently used means of accomplishing this [19]. For example, this particular fatigue model has been used recently in developing aircraft maintenance and usage plans [20]. Other possibilities include adding a material-dependent exponent to the ratio in (10.50), that is, $P_f = \sum_{k=1}^{K} \left(\frac{n_k}{N_{f_k}} \right)^{C_k}$, or making the failure probability dependent on both number of cycles at a load and the load level itself [21].

It is also worth mentioning that there exists a potentially large uncertainty associated with the number of cycles to failure, N_{f_k}. In fact, for a typical fatigue model (such as 10.49), the reported cycles to failure is better thought of as the median, denoted $\bar{N}_{f_k}$, of a log-normal distribution with shape parameter σ_L and location parameter $\mu = \ln(\bar{N}_{f_k})$ [22, 23]. It therefore makes sense to consider $\bar{P}_f \equiv E[P_f]$ as opposed to P_f in the optimization to be described in Section (10.3). We first note that the mean of the reciprocal of our log-normally distributed random variable is $E[N_{f_k}^{-1}] = e^{-\ln(\bar{N}_{f_k})+\sigma_L^2/2} = \frac{1}{\bar{N}_{f_k}} e^{\sigma_L^2/2}$. If we then consider σ_L constant for all load levels, by the linearity of the expectation operator we have that

$$\bar{P}_f = \sum_{k=1}^{K} n_k E\left[\frac{1}{N_{f_k}} \right]$$

$$= e^{\sigma_L^2/2} \sum_{k=1}^{K} \frac{n_k}{\bar{N}_{f_k}} = e^{\sigma_L^2/2} P_f \tag{10.51}$$

From the standpoint of optimization, maximizing P_f with regard to ship speed and sea state is therefore equivalent to maximizing $\bar{P}_f$, because σ_L is assumed constant for all load levels (i.e., it is independent of ship speed and sea state). The only difference is that Eq. (10.49) is now interpreted as the median value $\bar{N}_{f_k}$ of a random variable N_{f_k}. For notational convenience, we will continue to refer to the probability of failure as simply P_f with the understanding that we really mean expected probability of failure.

Figure 10.8 shows a sample strain response time series assuming a ship speed of $v = 20$ knots and sea state 5. Also shown is the corresponding rainflow histogram detailing the distribution of cycles n_k at load level ϵ_k. On the basis of these results, we may now relate both ship speed, v, and sea-state conditions, W, to the rate at which damage is accumulated. This can be accomplished in the following manner. For each sea state (1–6) and for each possible ship speed (1–40 knots), we determine the appropriate loading signal (10.46). On the

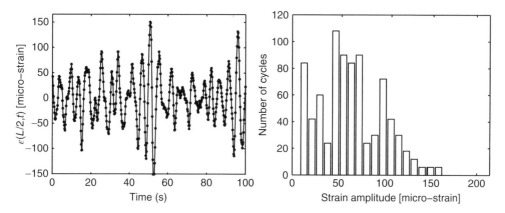

Figure 10.8 Sample response time series (sea state 4, traveling 20 knots) and the associated rainflow plot. The rainflow plot is based on 1 h of transit time *Source*: Reproduced from [9], Figure 8, with permission of Elsevier

basis of the Fourier series model, we determine the strain response at the stringer midpoint over some predefined time interval using Eq. (10.47). In the simulations to follow, we choose a time of 1 h. The strain response is analyzed using the aforementioned rainflow counting algorithm to determine the damage incurred, P_f, over that time. Thus, we have a precomputed function $P_f(W, v)$ having units 1/s that provides the accumulated probability of failure per hour. This function is shown in Figure 10.9. Note that this function is similar to the damage

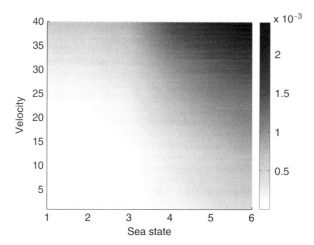

Figure 10.9 Plot of the function $P_f(W, v)$. Damage accumulated per hour as a function of both sea state W and ship velocity v. This function relates the decision variable v and sea conditions W to the cost function (10.52) to be minimized in generating the optimal decision. The function is based solely on the physics of the problem, and hence can be precomputed. In principle, this would allow for wave-structure models of far greater complexity than used here to be implemented *Source*: Reproduced from [9], Figure 9, with permission of Elsevier

rate function described in Ref. [4] and results in similar damage accumulation rates. We point out that wave-structure interaction models of far greater complexity than the one described here could also be used to generate the needed failure function $P_f(W, v)$. Our focus on this relatively simple structural model stems from our goal of highlighting the decision-making process, as opposed to an advanced structural modeling capability.

10.3 Optimal Transit

At this point, we have all of the needed modeling components to address the problem of optimal transit. Our goal will be to travel a distance of Y_{tot} km within some maximum allowable time T_{max}. These two values are assumed given at the outset. In this example, we are balancing the probability of structural failure against the time it takes to complete a given mission. We would like to be able to minimize both quantities, that is, arrive safely in as little time as possible. However, as seen by our model, the faster we travel (particularly in bad sea conditions) the more severe the loading, and hence the larger the accumulated value for $P_f(W, v)$. The job of the decision engine is to therefore balance these competing outcomes in the manner of our choosing. Such trade-offs are common to many real-world decisions.

10.3.1 Problem Statement

As alluded to earlier, the decisions we make are to be selected from a space of possible actions. These actions are typically adjustments to a problem variable that impact our definition of what it means to make a "good" decision. In this problem, the ship's captain has control over the vessel speed. Hence, we choose for our space of possible action the discrete set of velocities $v = \{0, 1, ..., v_{max}\}$ knots, where v_{max} is the vessel's top speed. To find the optimal velocities, it is first required that we discretize the problem space into fixed temporal or spatial intervals on which we will define our decisions. For this problem we choose to revisit our decision every Δ h., and thus we have $T = (T_{max} - T_{min})/\Delta$ decision points where T_{min}, T_{max} are the minimum and maximum allowable transit times. The former is set by the maximum ship velocity as $T_{min} = Y_{tot}/v_{max}$, while the latter can be infinite; although given the premium on transit time, never becomes very large.

At each time t, we will be solving for the spatial location we need to reach at decision point $t + 1, t + 2, \cdots$, that is, we solve for the vector $Y_t \equiv \{Y_{t+1}, Y_{t+2}, ..., Y_{t+T}\}$. These spatial points will therefore define the optimal velocity vector (our action) $v_{obt} \equiv (Y_{t+1} - Y_t)/\Delta$.

As we have mentioned, we seek the actions that balance the objectives of minimal transit time and minimal structural damage. With these discretizations in mind, we choose the cost function

$$C(Y_t|T) = \beta T + \sum_{t'=t}^{T} P_f(W(Y_{t'}, Y_{t'+1}), Y_{t'}, Y_{t'+1}) \tag{10.52}$$

where β is a scalar that controls the relative weighting between our desired outcomes (fast transit with minimal structural damage). The sum is taken from the current decision time, t, to the end of the trip, specified by the total transit time T. In minimizing (10.52), we are therefore choosing the optimal sequence of spatial locations we need to reach at times $t + 1, ..., T$ which defines the optimal velocity vector v_{opt}. Note that system state is characterized only by ship location at time t, with no information about current damage state (damage accumulated before time t). This characterization is adequate for our specification of this problem because damage accumulation between t and $t+1$ depends only on sea state and velocity during t to $t+1$, and not on damage state at t (see Figure 10.9). We also point out the units of Eq. (10.52). Recall the function $P_f(\cdot)$ has units of % damage (structural life) per unit time, hence the summation yields % structural life consumed over the period T. The constant β therefore also has units of % structural life per unit time (i.e., how much we value a decrement in remaining life vs. a unit of transit time Δ).

It is worth mentioning that the damage rate function $P_f(W(Y_t, Y_{t+1}), Y_t, Y_{t+1})$ is not necessarily defined at the desired spatiotemporal points required of the optimization. Thus, in practice, we simply use linear interpolation of this function to obtain the needed values. In doing so, we are implicitly assuming that the sea height is roughly constant over a given trip segment (i.e., between t and $t+1$). This can be approximated as the beginning or ending sea height or the average height of these two (in this work, we use the average). With small enough trip segments, it is unlikely that this choice makes much difference.

10.3.2 Solutions via Dynamic Programming

In this section we briefly describe the mechanics of solving (10.52). Unlike a standard optimization problem, the optimal solution at one point in time is likely not optimal at future points. We therefore formulate the problem as a Markov decision process (MDP; [24]) and solve using stochastic dynamic programming [24]. MDPs are models for sequential decision making when outcomes are uncertain and consist of four essential elements: states, actions, transition probabilities to determine how each state evolves given the different actions, and rewards (i.e., utility or objective function). Each of these elements is defined shortly for the ship transit problem.

Here, we are interested in exploring the consequences of choosing different actions in a state, which generates a reward (through our objective function) and determines the future state at the next period through our model. The goal of solving an MDP is to obtain an optimal policy or strategy that provides a prescription of which action to choose given any possible state at a future time and is optimal in that it maximizes our utility or objective function. For instance, with a finite-time horizon, we could be interested in maximizing our expected total reward (our expected reward at the end of the time horizon) or we could be interested in maximizing our average reward at each decision point. These are two very different problems and depend on the specific goals of the project. Here, we were interested in maximizing our expected total reward, which was to reach our final destination in the least amount of time while incurring the least amount of damage (10.52).

Setting up sequential decision problems with uncertainty as MDPs is the most common framework to handle these problems, and stochastic dynamic programming is the standard solution approach. It allows one to explore and understand the implications of choosing specific actions when one is uncertain about how the actions may influence the system. Therefore, MDPs provide a logical and transparent approach to making optimal decisions in the face of uncertainty, that is, to solving (10.52).

Here, we set the problem up as a finite-time horizon problem where, as a result of choosing and implementing a policy, the decision maker receives rewards in each of the decision periods $1,\ldots,T$. The reward sequence (a time-specific reward is given at each decision period) is random because it is not known before implementing the policy. The objective is to choose a policy so that the corresponding random reward sequence is optimal (e.g., as large as possible). To this end, we solve the Bellman equation (sometimes referred to as the optimality equations) defined as

$$V_t(Y_t | T) = \max_{Y_{t+1}} R(Y_t, Y_{t+1}) + \delta E_t[V_{t+1}(Y_{t+1})]. \qquad (10.53)$$

We can see that the Bellman equation is made up of two pieces, the first being the current rewards for time t given the current state Y_t and action Y_{t+1}. Because the state is the current location and the action is the location at time $t + 1$, the next period's state *is* the current action and thus the transition probability for each state given an action is 1.

The second piece is the discounted (δ) sum of the expected future rewards $E_t[V_{t+1}(Y_{t+1})]$. To solve the problem, we use backward iteration (i.e., dynamic programming) to recursively evaluate expected rewards. Specifically, we start with the final time period, and given a terminal value $V_T(Y_{tot})$ (often set to 0), we can evaluate the maximal current reward (and thus the set of actions to take) and continue to work through the problem backward and solving recursively to obtain an optimal policy.

We now need to define a per-period reward given the states, actions, transition probabilities, and the uncertain sea-state conditions. We have already postulated $P_f(W(Y_t, Y_{t+1}), Y_t, Y_{t+1})$ to be the expected damage cost of traveling in one time period from Y_t to Y_{t+1} when the average sea state between locations Y_t and Y_{t+1} and between times t and $t + 1$ is $W(Y_t, Y_{t+1})$. Therefore, for times 0 through $T - 2$, the per-period reward function is simply

$$R_t(Y_t, Y_{t+1}) = -P_f(W(Y_t, Y_{t+1}), Y_t, Y_{t+1}) \qquad (10.54)$$

so that the smaller the probability of failure, the larger the reward. At time $T - 1$, the only allowable action is Y_{tot} and the cost of the remaining trip is

$$V_{T-1}(Y_{T-1}) = P_f(W(Y_{T-1}, Y_{tot}), Y_{tot} - X_{T-1}). \qquad (10.55)$$

Furthermore, we need to place restrictions on the per-period reward function so that we do not overshoot the total distance of the trip Y_{tot}. For any values of $Y_{t+1} - Y_t > Y_{max}$, where $Y_{max} = v_{max}\Delta$ is the maximal distance that can be traveled in one time period, the cost is set to a large number, M, to ensure that only trip plans that can reach the destination are considered. To prevent premature arrival, however, any reward values associated with actions such as the $Y_{t+1} \geq Y_{tot}$ are set to M. Thus, the use of the penalties ensures that paths that fail to arrive or that arrive too soon at the destination are not considered.

We need two final pieces of information to fully specify our MDP: the terminal value, V_T, and the discount factor, δ. We defined the terminal value to be 0, $V_T = 0$ and the discount factor to be 1, $\delta = 1$. This implies that there is no value in making decisions after we have arrived at our destination and because the total time frame of the trip is small there is no need to discount the future.

The Bellman equation can be solved for varying trip durations T which, when multiplied by β as in (10.52), provides the additional term needed to rank the associated trip plans. In the results that follow, the results obtained using the optimization (10.53) are compared to those obtained from a "constant speed" strategy whereby the vessel operator sets the velocity based on final arrival time only and does not account for structural damage.

10.3.3   Transit Examples

As a first example, assume the ship is to transit $Y_{tot} = 1000$ km and is given a maximum allowable time to make the journey of $T_{max} = 50$ h. We consider the maximum ship speed to be $v_{max} = 40$ knots and the maximum sea state during the journey to be sea state 6. Before the journey, we receive a weather prediction describing how the sea state is expected to evolve as a function of time. In the examples to follow, the sea-state model time and space variables were discretized to 1 h and 10 km, respectively. On the basis of this predicted loading, and the maximum allowable transit time, the job of the decision engine is to produce the optimal velocity vector $\mathbf{v}^* = (\mathbf{Y}_{t+1} - \mathbf{Y}_t)/\Delta$ using the approach described in the previous section. We consider the characteristic length scale over which the loading acts on the structure to be $A = 10$ m.

There are two (related) ways that the optimization framework can be used. First, the total trip time can be assumed to be fixed and the trip strategy can found that minimizes the damage cost. Second, a set of trip strategies for a range of alternative travel times can be computed and the optimal trip time determined using a utility function such as (10.52). Given a maximum ship speed of 40 knots, the least cost trip plans are shown for alternative trip lengths of 15, 20, … ,50 h. Constant speed trips (not shown) would be represented in such a plot as straight lines. Figure 10.10a illustrates a series of optimal paths, each one for a different fixed transit time. Paths are compared using a cost function parameter of $\beta = 1/50$.

By definition, the optimal strategy will achieve a lower cost than the constant speed approach. Using SDM, it is straightforward to assess exactly how much better one strategy is than another. Figure 10.10b plots the accumulated % damage, defined as $100 \times P_f$, over the entire trip for both constants, and optimal strategies as a function of the time taken to complete the trip. The solid line in Figure 10.10b represents the total expected accumulated

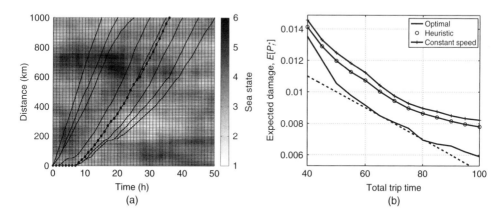

(a)   (b)

Figure 10.10   (a) Optimal decisions for different (fixed) transit times. The path is shown superimposed on top of the loading conditions that are predicted by the weather model. For larger allowable trip times, the optimal decision is the one that effectively "dodges" the predicted areas of rough seas. Shorter transit times provide the practitioner less flexibility in avoiding challenging seas. Each path reflects the decisions (velocities) that minimize (10.52). The optimal trip path for all possible times is indicated with a *. This path is, by definition, the path for which a line of constant cost is tangent to the cost versus time plot shown in (b) *Source*: Reproduced from [9], Figure 10, with permission of Elsevier

damage (over the entire trip) for a line of constant cost

$$C - \sum_{t=1}^{T_{\min}} P_f(W(Y_t, Y_{t+1}), Y_t, Y_{t+1}) - \beta T \tag{10.56}$$

thus, the optimal trip time (assuming this quantity is allowed to vary) is given by the point at which the optimal strategy is tangent to this line. As we pointed out in Eq. (10.52), the units of this constant cost are in % structural life. On the basis of this simulation, the optimal trip strategy is one that completes the mission in 30-40 h. The optimal trip length using this value of β is a 35 h trip with an expected damage cost of 0.5%. This trip is shown in Figure 10.10a marked with "*"s. In contrast, a constant speed trip is associated with an expected damage cost of 0.9%. Thus, the dynamic programming approach yields a sizable reduction in damage costs. These results are typical of the improvement realized by employing the SDM procedure.

Depending on the time allotted for transit, the optimal strategy shows varying degrees of improvement over the fixed speed strategies. As might be expected, for trips that must be completed in a small amount of time, we have little choice in how the trip is planned. In these cases, we are required to move at near top speed for the entire journey, and hence there is little improvement over the "fixed speed" strategy. However, if the allotted time is large, the algorithm has a greater freedom to find a path that can avoid conditions that degrade the structure. In this case, the improvement in accumulated damage can be large, registering a large improvement for many different trip durations.

Plots such as Figure 10.10 could be used by trip planners to select a preferred trip speed based on their assessment of the relative value placed on speed versus damage. It is clear from the figure that large reductions in damage are possible by reducing speed, Alternatively, a utility function such as (10.52) could be used to make that assessment. The β parameter represents

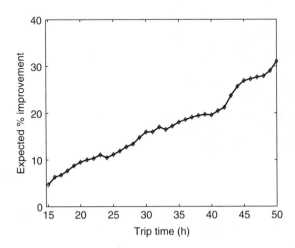

Figure 10.11 Expected % improvement obtained by using the optimal approach over the constant speed strategy as a function of time. Not surprisingly, the more time we allow to complete the trip, the more freedom the algorithm has to chart a course around challenging sea conditions *Source*: reproduced from [9], Figure 11, with permission of Elsevier

the relative weight placed on speed, and the optimal trip strategy can be found graphically by locating the point of tangency between the damage possibility curves in Figure 10.10 and a line with slope β. The tangent line using a value of $\beta = 1/50$ is shown in the figure and yields an optimal trip length of between 30 and 40 h. This means that each additional hour of trip time is valued the same as an additional 0.01% increase in the damage index. If more weight were placed on speed, the dashed line would be steeper and the tangency point would occur at a lower speed/higher damage combination.

We have run the optimal and constant speed strategies over a number of different sea states to determine the expected gain in performance. Figure 10.11 plots the expected % improvement over the constant velocity approach, averaged over 20 different realizations of sea state (any individual trip could produce results that are slightly better or slightly worse).

The above-described results are static in the sense that they assume a particular weather forecast at the beginning of the trip and plan accordingly. In reality, the predicted sea conditions will change as the weather models are continually updated. A better decision can therefore be made by incorporating this new information as it becomes available. Rather than holding the weather predictions static at each decision point, we simply solve the optimization problem at each decision point using the updated forecast. In situations where the weather is rapidly changing or where the conditions make forecasting difficult, this adjustment can yield substantial gains as quantified by a reduction in the cost. This is known as an open loop with feedback strategy. We will denote the strategy that uses initial weather forecasts only as the "original" approach, while the adaptive approach that continually updates predictions will be referred to as "actual." The "perfect foresight" case gives the optimal strategy had we known the precise sea-state conditions at the trip outset. This plan is unattainable, of course, but represents a good yardstick for comparing adaptive versus static strategies.

An example of this is shown in Figure 10.12. The left panel shows the original sea-state predictions, whereas the right panel shows the sea states that actually occurred. Three trip plans are shown superimposed on each panel, the original optimal plan, requiring 37 h, the actual trip taken, requiring 33 h and the perfect foresight trip plan, requiring 23 h.

It can be seen that the strategy with revision led to a shorter trip (37 vs. 33 h). Although it need not be the case, here the actual trip (with revisions) was associated with slightly less damage (0.83 vs. 0.74). In general, however, the ability to respond to the changing sea states will result in a trip with higher utility than if revisions are not performed. The example shown is typical in that the strategy with revisions is closer to the ideal trip strategy, taking 27 h to complete. Nonetheless, we repeated this numerical experiment, comparing the "original" versus "actual" (adaptive) plans, for 10 different realizations of sea state (10 different trips). On average, taking the changing weather predictions into account resulted in a 17% reduction in cumulative damage over the plan that uses initial weather predictions only.

The proposed framework trivially allows for new information to be incorporated into the decision-making process. We could just as easily have incorporated updated ship models if such information were available. This would result in updates to the damage rate function used in calculating our cost. In addition, if an on-board monitoring system were to determine that the damage/hour were different than the precomputed function (see, e.g., Figure 10.9), one simply updates that function accordingly. This possibility is increasingly likely as condition monitoring systems are beginning to see use in the field.

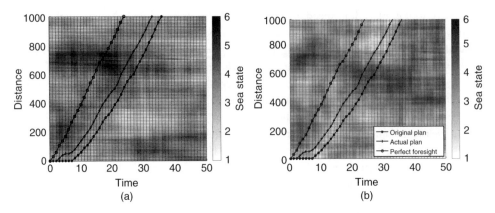

Figure 10.12   Effect of revision on trip plans. The paths are shown superimposed on top of the loading conditions that are predicted by the weather model. The left panel shows the conditions predicted at the beginning of the trip. The right panel shows the actual conditions that occurred. Superimposed are the optimal paths based on the initial weather predictions only ("Original plan"), the path based on updated weather predictions ("Actual plan"), and the best possible path ("Perfect foresight") *Source*: Reproduced from [9], Figure 12, with permission of Elsevier

10.4   Summary

The goal of this book was to provide the reader with the ingredients necessary to implement the approach to structural maintenance described pictorially in Figure (1.9). In fact, this last chapter serves as a good summary, demonstrating each of the basic components of the physics-based approach to damage identification and subsequent decision making. While conceptually attractive, the approach requires a breadth of understanding, borrowing from solid mechanics, probability theory, random processes, and decision theory (among others). We have done our best to do each of these subjects justice although we freely acknowledge that more in-depth treatment of each component can be found elsewhere. Our goal was to take only those pieces we have found particularly useful for the damage identification problem.

Perhaps the most useful contribution of this work is to cast the problem of damage identification as one of modeling and estimation. Given some assumptions about the physics of the structure and the sources of uncertainty, we have shown how one can both detect and identify structural damage. A rather large amount of background material was presented on the modeling of damage physics and on the probabilistic modeling of uncertain events. We have used the former to model those aspects of the problem we can accurately predict, while the latter describes those we cannot. Both kinds of models are essential to the estimation of structural damage, the key ingredient to the physics-based approach to structural maintenance.

We devoted a number of pages to estimation theory, focusing on both Bayesian and maximum likelihood methods. Both provide a powerful set of tools for drawing inferences about the state of a structure from observed data. The strength of viewing damage identification as an estimation problem is that we can leverage all available information we have about the problem in a principled manner. In fact, as we progressed through the later chapters we have described approaches requiring increasing levels of *a priori* knowledge about the structure. To begin, we showed how basic assumptions about the general form of the structural model could be used to detect damage-induced nonlinearities via the method of surrogate data. Expanding on this general theme, we also showed how a nonlinear structural response would manifest itself in estimates of the higher order spectra. We then moved toward more specific damage models, including dented structures, delamination in composites, corrosion damage, and cracked plates. In these cases, our knowledge of damage-based physics was leveraged to produce estimates of specific damage-related parameters. A number of case studies were provided, both numerical and experimental, to highlight both the modeling and estimation methods.

In the last chapter we provided an example of what we feel should be the ultimate output of an SHM system: a decision that minimizes cost of ownership. Interestingly, while a great deal of research has gone into developing the ingredients required of SDM, this final step has gone largely unnoticed. Yet, if one is to make a business case for implementing an SHM system, the impact of the system on maintenance decisions must be quantified. While many have made heuristic arguments for an SHM system, SDM places these arguments on a firm mathematical foundation. Moreover, "SDM" goes a step further and informs the structure's owner of the optimal way in which to use the SHM system. In short, SDM provides a clear path to integrating all available sources of data to produce smart decisions. The utility of SDM clearly extends far beyond problems in structural dynamics and is certainly a tool worth understanding.

It is our hope that this book is a useful reference for both the practitioner and theoretician as we have made every effort to accommodate both in our treatment. However, if nothing else, we hope that the reader follows the basic philosophical approach toward the problem. That is to say, we hope to have made a convincing argument that this problem is best treated as one of modeling and estimation as opposed to one of pattern recognition. Again, we feel the uncertainty present in many real-world systems will prevent the latter from being used in practice. Only time will tell. However, basic philosophy of science aside, we hope that this work will also prove useful to those studying general structural vibration problems. We have attempted to provide a single source for both modeling and analysis techniques that we have found useful over the years, but that do not appear together in many structural vibrations texts. This has no doubt come at the expense of other important topics in vibrations, however we felt that the excluded topics have seen enough attention in other places so as not to be missed here. Thanks to those of you who have read this far, and best of luck in your research!

References

[1] G. F. Class, S. Kosleck, D. Testa, Critical situations of vessel operations in short crested seas–forecast and decision support system, Journal of Offshore Mechanics and Arctic Engineering 134 (2012) 031601.
[2] P. Lacey, H. Chen, Improved passage planning using weather forecasting maneuvering guidance, and instrumentation feedback, Marine Technology and SNAME News 32 (1) (1995) 1–19.
[3] E. M. Bitner-Gregersen, R. Skjong, Concept for a risk based navigation decision assistant, Marine Structures 22 (2009) 275–286.
[4] U. D. Nielsen, J. J. Jensen, P. T. Pedersen, Y. Ito, Onboard monitoring of fatigue damage rates in the hull girder, Marine Structures 24 (2011) 182–206.
[5] U. D. Nielsen, Z. Lajic, J. J. Jensen, Towards fault-tolerant decision support systems for ship operator guidance, Reliability Engineering & System Safety 104 (2012) 1–14.
[6] F. Calabrese, A. Corallo, A. Margherita, A. A. Zizzari, A knowledge-based decision support system for shipboard damage control, Expert Systems with Applications 39 (2012) 8204–8211.
[7] K. T. Huynh, A. Barros, C. Bérenguer, Maintenance decision-making for systems operating under indirect condition monitoring: value of online information and impact of measurement uncertainty, IEEE Transactions on Reliability 61 (2) (2012) 410–425.

[8] H. Sehitoglu, K. Gall, A. M. Garcia, Recent advances in fatigue crack growth, International Journal of Fracture 80 (1996) 165–192.

[9] J. M. Nichols, P. L. Fackler, K. Pacifici, K. D. Murphy, J. D. Nichols, Reducing fatigue damage for ships in transit through structured decision making, Marine Structures 38 (2014) 18–43.

[10] C. G. Soares, Representation of double-peaked sea wave spectra, Ocean Engineering 11 (2) (1984) 185–207.

[11] M. B. Priestly, Spectral Analysis and Time Series, Probability and Mathematical Statistics, Elsevier Academic Press, London, 1981.

[12] S. M. Han, H. Benaroya, Nonlinear and Stochastic Dynamics of Compliant Offshore Structures, Kluwer Academic Publishers, Dordrecht, Netherlands, 2002.

[13] D. J. Witmer, J. W. Lewis, The BP oil tanker structural monitoring system, Marine Technology and SNAME News 32 (4) (1995) 277–296.

[14] G. Wang, K. Pran, G. Sagvolden, G. B. Havsgard, A. E. Jensen, G. A. Johnson, S. T. Vohra, Ship hull structure monitoring using fibre optic sensors, Smart Materials and Structures 10 (2001) 472–478.

[15] J. M. Nichols, M. Seaver, S. T. Trickey, K. Scandell, L. W. Salvino, E. Aktaş, Real-time strain monitoring of a navy vessel during open water transit, Journal of Ship Research 54 (4) (2010) 225–230.

[16] A. Mizoguchi, K. Tanizawa, Impact wave loads due to slamming - a review, Ship Technology Research 43 (1996) 139–151.

[17] V. T. Troshchenko, L. A. Khamaza, Strain-life curves of steels and methods for determining the curve parameters. Part 1. Conventional methods, Strength of Materials 42 (6) (2010) 647–659.

[18] S. D. Downing, D. F. Socie, Simple rainflow counting algorithms, International Journal of Fatigue 4 (1) (1982) 31–40.

[19] Y. Liu, S. Mahadevan, Stochastic fatigue damage modeling under variable amplitude loading, International Journal of Fatigue 29 (2007) 1149–1161.

[20] X. Wang, M. Rabiei, J. Hurtado, M. Modarres, P. Hoffman, A probabilistic-based airframe integrity management model, Reliability Engineering & System Safety 94 (2009) 932–941.

[21] S. Zengah, A. Aid, M. Benguediab, Comparative study of fatigue damage models using different number of classes combined with the rainflow method, Engineering, Technology & Applied Science Research 3 (3) (2013) 446–451.

[22] F. G. Pascual, W. Q. Meeker, Estimating fatigue curves with the random fatigue-limit model, Technometrics 41 (1999) 277–302.

[23] T. Lassen, P. Darcis, N. Recho, Fatigue behavior of welded joints - Part 1: statistical methods for fatigue life prediction, Welding Journal 84 (2005) 183s–187s.

[24] M. L. Puterman, Markov Decision Processes: Discrete Stochastic Dynamic Programming, John Wiley & Sons, Inc., Hoboken, NH, 2005.

A

Useful Constants and Probability Distributions

Table A.1 Mathematical definition of useful functions and constants

Glossary of key functions and constants	
Gamma function	$\Gamma(x) = \int_0^\infty t^{x-1}e^{-t}dt$ for $x > 0$
Digamma function	$\Psi(x) = \frac{d}{dx}\ln[\Gamma(x)] = \frac{\Gamma'(x)}{\Gamma(x)}$
Beta function	$B(x, y) = \int_0^1 t^{x-1}(1-t)^{y-1}dt = \frac{\Gamma(x)\Gamma(y)}{\Gamma(x+y)}$
Euler–Mascheroni constant	$\gamma = \lim_{m\to\infty}\left[1 + \frac{1}{2} + \frac{1}{3} + \cdots + \frac{1}{m} - \ln(m)\right] = 0.5772156649$
	$= -\Gamma'(1) = -\Psi(1)$

Table A.2 Probability density functions and the associated mean and variance

Distribution type	Probability density function	Mean, Variance
Beta	$p(x) = \begin{cases} \frac{\Gamma(\eta+\lambda)}{\Gamma(\eta)\Gamma(\lambda)}x^{\lambda-1}(1-x)^{\eta-1} & : \quad 0 \le x \le 1 \\ 0 & : \quad \text{otherwise} \end{cases}$ where $\lambda > 0, \eta > 0$	$\mu = \frac{\eta}{\eta+\lambda}$ $\sigma^2 = \frac{\eta\lambda}{(\eta+\lambda)^2(\eta+\lambda+1)}$
Cauchy	$p(x) = \frac{1}{\pi b\left[1 + \frac{(x-a)^2}{b^2}\right]} \qquad -\infty < x < \infty$ where $b > 0$	$\nexists\mu$ $\nexists\sigma^2$
Chi	$p(x) = \begin{cases} \frac{2(n/2)^{n/2}}{\sigma^n\Gamma(n/2)}x^{n-1}e^{-(n/(2\sigma^2))x^2} & : \quad 0 \le x < \infty \\ 0 & : \quad \text{otherwise} \end{cases}$ where $\sigma > 0$ and n is a positive integer	$\mu = \sqrt{2}\frac{\Gamma((n+1)/2)}{\Gamma(n/2)}$ $\sigma^2 = n - \mu^2$

(continued)

Modeling and Estimation of Structural Damage, First Edition. Jonathan M. Nichols and Kevin D. Murphy.
© 2016 John Wiley & Sons, Ltd. Published 2016 by John Wiley & Sons, Ltd.

Table A.2   (*Continued*)

Distribution type	Probability density function	Mean, Variance

Chi-squared

$$p(x) = \begin{cases} \frac{1}{2^{n/2}\Gamma(n/2)}x^{n/2-1}e^{-x/2} & : \quad 0 < x < \infty \\ 0 & : \quad \text{otherwise} \end{cases}$$

where n is a positive integer

$\mu = n$

$\sigma^2 = 2n$

Dirac delta function

$p(x) = \delta(x)$

$\mu = 0, \sigma^2 = 0$

Exponential

$$p(x) = \begin{cases} \lambda e^{-\lambda x} & : \quad x \geq 0 \\ 0 & : \quad \text{otherwise} \end{cases}$$

where $\lambda > 0$

$\mu = \lambda^{-1}$

$\sigma^2 = \lambda^{-2}$

F-distribution

$$p(x) = \begin{cases} \frac{v^{\frac{v}{2}}w^{\frac{w}{2}}}{B(\frac{v}{2},\frac{w}{2})}\frac{x^{\frac{v}{2}-1}}{(w+vx)^{(v+w)/2}} & : \quad 0 \leq x < \infty \\ 0 & : \quad \text{otherwise} \end{cases}$$

where v, w are positive integers

$\mu = \frac{w}{w-2}$ if $w > 2$

$\sigma^2 = \frac{2w^2(v+w-2)}{v(w-2)^2(w-4)}$ if $w > 4$

Gamma

$$p(x) = \begin{cases} \frac{\lambda^\eta}{\Gamma(\eta)}x^{\eta-1}e^{-\lambda x} & : \quad 0 < x < \infty \\ 0 & : \quad \text{otherwise} \end{cases}$$

where $\lambda > 0, \eta > 0$

$\mu = \frac{\eta}{\lambda}$

$\sigma^2 = \frac{\eta}{\lambda^2}$

Generalized beta

$$p(y) = \begin{cases} \frac{1}{b-a}\frac{\Gamma(\eta+\lambda)}{\Gamma(\eta)\Gamma(\lambda)}\left(\frac{y-a}{b-a}\right)^{\lambda-1}\left(\frac{b-y}{b-a}\right)^{\eta-1} & : \quad a \leq y \leq b \\ 0 & : \quad \text{otherwise} \end{cases}$$

where $\lambda > 0, \eta > 0, a \geq 0$

$\mu = a + \frac{(b-a)\lambda}{\eta+\lambda}$

$\sigma^2 = \frac{(b-a)^2\eta\lambda}{(\eta+\lambda)^2(\eta+\lambda+1)}$

Generalized normal

$$p(x) = \frac{\beta}{2\alpha\Gamma\left(\frac{1}{\beta}\right)}e^{-(|x-\mu|/\alpha)^\beta} \quad -\infty < x < \infty$$

where $\alpha > 0, \beta > 0$

$\mu = \mu$

$\sigma^2 = \alpha^2\frac{\Gamma(3/\beta)}{\Gamma(1/\beta)}$

Kumaraswamy

$$p(x) = \begin{cases} abx^{a-1}(1-x^a)^{b-1} & : \quad 0 \leq x \leq 1 \\ 0 & : \quad \text{otherwise} \end{cases}$$

where $a > 0, b > 0$

$\mu = bB(1+\frac{1}{a},b)$

$\sigma^2 = bB(1+\frac{2}{a},b) - \mu^2$

Laplace

$$p(x) = \frac{1}{2}\lambda e^{-\lambda|x|} \quad -\infty < x < \infty$$

where $\lambda > 0$

$\mu = 0$

$\sigma^2 = \frac{2}{\lambda^2}$

Log normal

$$p(x) = \begin{cases} \frac{1}{\sqrt{2\pi}\sigma x}e^{-(\ln(x)-\ln(m))^2/2\sigma^2} & : \quad 0 < x < \infty \\ 0 & : \quad \text{otherwise} \end{cases}$$

where $m > 0, \sigma > 0$

$\mu = me^{\sigma^2/2}$

$\sigma^2 = m^2e^{\sigma^2}(e^{\sigma^2}-1)$

Logistic

$$p(x) = \frac{e^{-(x-\mu)/s}}{s(1+e^{-(x-\mu)/s})^2} \quad -\infty < x < \infty$$

where $s > 0$

$\mu = \mu$

$\sigma^2 = \frac{\pi^2 s^2}{3}$

Log-logistic

$$p(x) = \begin{cases} \frac{\frac{\beta}{\alpha}(\frac{x}{\alpha})^{\beta-1}}{[1+(\frac{x}{\alpha})^\beta]^2} & : \quad 0 \leq x < \infty \\ 0 & : \quad \text{otherwise} \end{cases}$$

where $\alpha > 0, \beta > 0$

$\mu = \frac{\alpha\pi}{\beta\sin(\pi/\beta)}$ if $\beta > 1$

$\sigma^2 = \alpha^2\left(\frac{2\pi}{\beta\sin(2\pi/\beta)} - \frac{\pi^2}{\beta^2\sin^2(\pi/\beta)}\right)$ if $\beta > 2$

Maxwell

$$p(x) = \begin{cases} \frac{4}{\sqrt{\pi}}\frac{x^2e^{-x^2/(\alpha^2)}}{\alpha^3} & : \quad 0 \leq x < \infty \\ 0 & : \quad \text{otherwise} \end{cases}$$

where $\alpha > 0$

$\mu = \frac{2\alpha}{\sqrt{\pi}}$

$\sigma^2 = \left(\frac{3}{2} - \frac{4}{\pi}\right)\alpha^2$

Table A.2 (*Continued*)

Distribution type	Probability density function	Mean, Variance
Mixed Gaussian	$p(x) = \dfrac{1}{2\sigma\sqrt{2\pi}}[e^{-(x-\mu)^2/2\sigma^2} + e^{-(x+\mu)^2/2\sigma^2}]$ where $\mu \geq 0, \sigma > 0 \qquad -\infty < x < \infty$	$\mu = 0$ $\sigma^2 = \mu^2 + \sigma^2$
Nakagami	$p(x) = \begin{cases} \dfrac{2}{\Gamma(m)}\left(\dfrac{m}{\Omega}\right)^m x^{2m-1}e^{-mx^2/\Omega} & : \quad x > 0 \\ 0 & : \quad \text{otherwise} \end{cases}$ where $\Omega > 0$ and $m \geq \dfrac{1}{2}$	$\mu = \dfrac{\Gamma(m+1/2)}{\Gamma(m)}\left(\dfrac{\Omega}{m}\right)^{1/2}$ $\sigma^2 = \Omega - \mu^2$
Normal	$p(x) = \dfrac{1}{\sigma\sqrt{2\pi}}e^{-(x-\mu)^2/2\sigma^2} \qquad -\infty < x < \infty$ where $\sigma > 0$	$\mu = \mu$ $\sigma^2 = \sigma^2$
Pareto	$p(x) = \begin{cases} cx_o^c x^{-(c+1)} & : \quad x_o \leq x < \infty \\ 0 & : \quad \text{otherwise} \end{cases}$ where $c > 0, \; x_o \geq 1$	$\mu = \dfrac{cx_o}{c-1}$ if $c > 1$ $\sigma^2 = [\dfrac{c}{c-2} - \dfrac{c^2}{(c-1)^2}]x_o^2$ if $c > 2$
Rayleigh	$p(x) = \begin{cases} \dfrac{x}{\sigma^2}e^{-x^2/2\sigma^2} & : \quad 0 \leq x < \infty \\ 0 & : \quad \text{otherwise} \end{cases}$ where $\sigma > 0$	$\mu = \sigma\sqrt{\pi/2}$ $\sigma^2 = \sigma^2(2 - \pi/2)$
Simpson	$p(x) = \begin{cases} \dfrac{a-|x|}{a^2} & : \quad -a \leq x \leq a \\ 0 & : \quad \text{otherwise} \end{cases}$	$\mu = 0, \; \sigma^2 = \dfrac{a^2}{6}$
Sine wave	$p(x) = \begin{cases} \dfrac{1}{\pi\sqrt{A^2-x^2}} & : \quad -A < x < A \\ 0 & : \quad \text{otherwise} \end{cases}$ where $A > 0$	$\mu = 0$ $\sigma^2 = \dfrac{A^2}{2}$
Students-t	$p(x) = \dfrac{1}{\sqrt{n}B(\frac{1}{2},\frac{n}{2})}(1 + \dfrac{x^2}{n})^{-(n+1)/2} \qquad -\infty < x < \infty$ where $n = $ positive integer	$\mu = 0$ $\sigma^2 = \dfrac{n}{n-2}$
Uniform	$p(x) = \begin{cases} \dfrac{1}{a} & : \quad 0 < x \leq a \\ 0 & : \quad \text{otherwise} \end{cases}$	$\mu = \dfrac{a}{2}, \sigma^2 = \dfrac{a^2}{12}$
Weibull	$p(x) = \begin{cases} \dfrac{\eta}{\sigma}\left(\dfrac{x}{\sigma}\right)^{\eta-1}e^{-\left(\frac{x}{\sigma}\right)^{\eta}} & : \quad 0 \leq x < \infty \\ 0 & : \quad \text{otherwise} \end{cases}$ where $\eta > 0, \sigma > 0$	$\mu = \sigma\Gamma\left(1 + \dfrac{1}{\eta}\right)$ $\sigma^2 = \sigma^2\left(\Gamma(1 + \dfrac{2}{\eta}) - [\Gamma(1 + \dfrac{1}{\eta})]^2\right)$

B

Contour Integration of Spectral Density Functions

In a number of places in this book, we have been required to integrate functions possessing singularities on the range of integration. For example, we know that the power spectral density (PSD) function associated with a spring-mass system with mass, damping, and stiffness parameters m, c, k is given by

$$S_{YY}(\omega) = \frac{S_{XX}(\omega)}{m^2(-\omega^2 + i\omega c/m + \omega_n^2)^2} \tag{B.1}$$

where we have defined $\omega_n^2 = k/m$. We also recall that from Parseval's relationship that $E[Y(t)]^2 \approx \int_{-\infty}^{\infty} S_{YY}(\omega)d\omega$ so that an expression for the signal variance can be obtained directly by integrating the PSD. The problem is that this function possesses singularities in the complex plane, rendering the integral intractable by standard methods.

To see this, we may recall from Section 4.5.1 that a convenient reparametrization gives $c = 2\zeta m\omega_n$ and $\omega_d \equiv \omega_n\sqrt{1 - \zeta^2}$. If we further prescribe that the input PSD is a constant, given by $S_{XX}(\omega) = P$, we can write the output PSD in the form

$$S_{YY}(\omega) = P|H_1(\omega)|^2 = \frac{P}{m^2} \frac{1}{(\omega - p_1)(\omega - p_2)(\omega - p_3)(\omega - p_4)} d\omega \tag{B.2}$$

where $p_{1,2} = \pm\omega_d + i\zeta\omega_n$ and $p_{3,4}$ are the complex conjugates of $p_{1,2}$ obtained by simply reversing the sign on the imaginary component. In general, ω is a complex variable so that the integral clearly does not exist at $p_{1,2}$, hence the traditional calculus of integration cannot be applied. An approach for integrating functions of the form (B.2) is possible, however, and requires several useful formula from complex integral calculus. In what follows, we present only that which is necessary for the integrals we encounter in this book. The more general topic of integrating complex functions is discussed in [1].

The first of the needed formula was developed by Augustin-Louis Cauchy and is referred to as the Cauchy integral formula. Consider a function of the form $f(z)/(z - p)$, where z is complex and the complex constant p is a (single) fixed point in the domain of integration.

Modeling and Estimation of Structural Damage, First Edition. Jonathan M. Nichols and Kevin D. Murphy.
© 2016 John Wiley & Sons, Ltd. Published 2016 by John Wiley & Sons, Ltd.

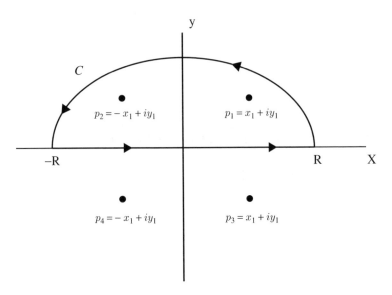

Figure B.1 Depiction of a contour integral around the upper half of the complex plane. This integral, defined on the closed curve C, can be decomposed into two integrals, one on the semicircle spanning $[-R, R]$ and the other along the real axis

Specifically, we will require integrals of this type of function along a closed, piecewise smooth path C in the complex plane. For this type of integral, the Cauchy integral formula states

$$\oint_C \frac{f(z)}{z - p} dz = 2\pi i f(p) \tag{B.3}$$

where the integral is defined as positive for a clockwise-oriented path (if C is drawn counter-clockwise, the integral is equal to $-2\pi i f(p)$). Using the Cauchy formula, it can be shown [1] that for a function with *multiple* poles, $p_1, p_2, \cdots, p_K$

$$\oint_C \frac{f(z)}{\prod_{k=1}^{K}(z - p_k)} dz = 2\pi i \sum_{k=1}^{K} \lim_{z \to p_k} (z - p_k) \frac{f(z)}{\prod_{k=1}^{K}(z - p_k)}$$

$$= 2\pi i \sum_{k=1}^{K} \lim_{z \to p_k} \frac{f(z)}{\prod_{k' \neq k}^{K}(z - p_{k'})}. \tag{B.4}$$

This expression is a special case of the *residue theorem* for functions of the form (B.2). Because this is the only version of the theorem we require, it is presented here in the less general form. A further simplification is possible if we consider these types of integrals where the limits of integration are along the real axis, $-\infty, +\infty$. That is to say, we are really interested in the integral $\int_{-\infty}^{\infty} \frac{f(z)}{\prod_{k=1}^{K}(z - p_k)} dz$. To approach this type of integral using complex analysis, consider the diagram shown in Figure B.1.

This figure depicts a complex function with four complex poles (p_1, p_2, p_3, p_4) where we have drawn a closed curve around the top half of the complex plane. We know from (B.4) how to solve this integral around the $K = 2$ poles. We could also break this integral up into two pieces, the first along the real axis and the second along the semicircle which we will define as C_R. Thus, we may write:

$$\oint_C \frac{f(z)}{\prod_{k=1}^{K}(z - p_k)} dz = \oint_{C_R} \frac{f(z)}{\prod_{k=1}^{K}(z - p_k)} dz + \int_{-R}^{R} \frac{f(z)}{\prod_{k=1}^{K}(z - p_k)} dz \qquad (B.5)$$

However, it can be shown [1] that in the limit as $R \to \infty$, the integral over C_R goes to zero, hence we have that

$$\int_{-\infty}^{\infty} \frac{f(z)}{\prod_{k=1}^{K}(z - p_k)} dz = \oint_C \frac{f(z)}{\prod_{k=1}^{K}(z - p_k)} dz$$

$$= 2\pi i \sum_{k=1}^{K} \lim_{z \to p_k} \frac{f(z)}{\prod_{k' \neq k}^{K}(z - p_{k'})} \qquad (B.6)$$

where K is now read as the number of poles in the half of the complex plane we are considering. This remarkable result states that to perform the desired infinite integral along the real axis, we may choose a closed curve around either half of the complex plane and perform the integral in that half only. For some types of problems, one half of the complex plane will possess more poles than another, hence choosing to integrate around the half with the fewer number of poles (smaller K) can be advantageous. For most spectral density functions the number of poles in the positive and negative halves of the complex plane is equal, hence there is no advantage to choosing one half over the other. Our convention in this book is to integrate over the upper half of the complex plane. Application of the formula is straightforward provided the integrand can be put in pole-zero form.

The formula (B.6) provides a convenient way to solve the type of integrals we repeatedly face when integrating spectral density functions (e.g., B.2). Returning to the problem (B.2) and noting that $f(z) = 1$ we can apply the above-described method of residues over the positive half of the complex plane by integrating around p_1, p_2 $(K = 2)$ yielding

$$\int_{-\infty}^{\infty} S_{YY}(\omega) d\omega = P \left\{ 2\pi i \frac{1}{m^2} \frac{1}{(p_1 - p_2)(p_1 - p_3)(p_1 - p_4)} + 2\pi i \frac{1}{m^2} \frac{1}{(p_2 - p_1)(p_2 - p_3)(p_2 - p_4)} \right\}$$

$$= P \frac{\pi}{2\zeta m^2 \omega_d^2 \omega_n + 2\zeta^3 m^2 \omega_n^3}$$

$$= P \frac{\pi}{2m^2 \omega_n^3 \zeta}$$

$$= P \frac{\pi}{ck} \qquad (B.7)$$

Checking units, we have $ck \sim (\mathrm{N}^2 \cdot \mathrm{s/m}^2)$ and $P \sim (\mathrm{N}^2 \cdot \mathrm{s})$, so that the variance will have units m^2 as expected. Note that we could also have handled a nonconstant input PSD provided we knew the functional form. In fact, many "colored" noise PSD functions can also be described in the same pole-zero form we have used for the response. In this case, the input signal simply contributes additional poles to the expression that can be easily integrated around.

This approach is also used in the book to perform integrals over higher order Volterra kernels, for example, $H_2(\omega_1, \omega_2)$. For example, the expression for the PSD of the response of a quadratically nonlinear system is given by Eq. (6.82). The second term in this expression contains the integral of the magnitude squared of a second-order Volterra kernel. For a spring-mass system, we have shown this kernel to be expressable as a product of first-order (linear) kernels such that we can write

$$\frac{1}{\pi} \int |H_2(\omega - \xi, \xi)|^2 S_{XX}(\omega - \xi) S_{XX}(\xi) d\xi$$

$$= \frac{1}{\pi} |H_1(\omega)|^2 \int_{-\infty}^{\infty} (-k2 + c2\xi(\omega - \omega_\xi))^2 |H_1(\omega - \xi)|^2 |H_1(\xi)|^2 S_{XX}(\omega - \xi) S_{XX}(\xi) d\xi. \quad \text{(B.8)}$$

In what follows, we again assume that the input PSD is constant, that is, the random process $x(t)$ is modeled as an i.i.d. sequence of random variables. We will denote this constant $S_{XX}(\omega) = P$. With this assumption, the integral is of the required form with eight separate poles; these we denote $p_{1,2} = \pm \omega_d + i\zeta\omega_n, p_{3,4} = p_{1,2}^*, p_{5,6} = \omega - p_{1,2}$ and $p_{7,8} = \omega - p_{5,6}$. Pulling the (constant) PSDs outside the integral, we may express the integral term as

$$F(\omega) = \int_{-\infty}^{\infty} (-k2 + c2\omega_2(\omega - \omega_2))^2 |H_1(\omega_2)|^2 |H_1(\omega - \omega_2)|^2 d\omega_2$$

$$= \int_{-\infty}^{\infty} \frac{1}{m^4} \left\{ \frac{(-k2 + c2\omega_2(\omega - \omega_2))^2}{(\omega_2 - p_1)(\omega_2 - p_2)(\omega_2 - p_3)(\omega_2 - p_4)(\omega_2 - p_5)(\omega_2 - p_6)(\omega_2 - p_7)(\omega_2 - p_8)} \right\} d\omega_2$$

$$= \frac{2\pi i}{m^4} \left(\frac{(-k_2 + c_2 p_1(\omega - p_1))^2}{(p_1 - p_2)(p_1 - p_3)(p_1 - p_4)(p_1 - p_5)(p_1 - p_6)(p_1 - p_7)(p_1 - p_8)} \right.$$

$$\frac{(-k_2 + c_2 p_2(\omega - p_2))^2}{(p_2 - p_1)(p_2 - p_3)(p_2 - p_4)(p_2 - p_5)(p_2 - p_6)(p_2 - p_7)(p_2 - p_8)}$$

$$\frac{(-k_2 + c_2 p_7(\omega - p_7))^2}{(p_7 - p_1)(p_7 - p_2)(p_7 - p_3)(p_7 - p_4)(p_7 - p_5)(p_7 - p_6)(p_7 - p_8)}$$

$$\left. \frac{(-k_2 + c_2 p_8(\omega - p_8))^2}{(p_8 - p_1)(p_8 - p_2)(p_8 - p_3)(p_8 - p_4)(p_8 - p_5)(p_8 - p_6)(p_8 - p_7)} \right) \quad \text{(B.9)}$$

where we have chosen to integrate over the four poles residing in the positive half of the complex plane (imaginary component is greater than zero). This expression can be simplified to

$$F(\omega)$$

$$= \frac{2\sqrt{k_1}m\pi(2c_2 k_1 k_2 m(4k_1 - 3m\omega^2) + c_2^2 k_1(4k_1^2 + (c_1^2 - 3k_1 m)\omega^2 + m^2\omega^4) + k_2^2 m(4c_1^2 + m(4k_1 + m\omega^2)))}{c_1(k_1/m)^{3/2} m^3(c_1^2 + m^2\omega^2)(4c_1^2\omega^2 + (-4k_1 + m\omega^2)^2)}$$

$$\text{(B.10)}$$

from which we have for the final expression for the nonlinear PSD function

$$P_{YY}(\omega) = |H_1(\omega)|^2 P + |H_1(\omega)|^2 F(\omega) P^2. \quad \text{(B.11)}$$

If we would rather have the response accelerance PSD or mobility PSD, we simply change the kernels used in the expressions. We first note that for acceleration $|H_1^{(accel)}(\omega)|^2 = \omega^4 |H_1^{(disp)}(\omega)|^2$ and that $|H_2^{(accel)}(\omega_1, \omega_2)|^2 = (\omega_1 + \omega_2)^4 |H_2^{(disp)}(\omega_1, \omega_2)|^2$.

Similarly we have for mobility $|H_1^{(vel)}(\omega)|^2 = \omega^2|H_1^{(disp)}(\omega)|^2$ and that $|H_2^{(vel)}(\omega_1,\omega_2)|^2 = |(\omega_1 + \omega_2)^2 H_2^{(disp)}(\omega_1,\omega_2)|^2$. Reconsidering the given derivation we have for accelerance and mobility PSD

$$P_{\ddot{Y}\ddot{Y}}(\omega) = \omega^4|H_1(\omega)|^2 P + \omega^4|H_1(\omega)|^2 F(\omega)P^2$$

$$P_{\dot{Y}\dot{Y}}(\omega) = \omega^2|H_1(\omega)|^2 P + \omega^2|H_1(\omega)|^2 F(\omega)P^2. \tag{B.12}$$

Figure 6.8 shows a comparison between theoretical and estimated accelerance PSD for a quadratically nonlinear spring-mass system.

A more complicated example occurs in the derivation of the bispectral density function. Recall from Section 6.5, Eq. (6.119) that the bispectral density contains a term that is $O(P^3)$ for input PSD of level P. For the systems we are interested in, this term is typically higher order as $P \sim O(10^{-2} - 10^{-1})$, however as we showed in Section 6.5.2, specifically Figure 6.15, this term can be observed in the estimate. We therefore present the derivation of this term in what follows.

Assuming a single degree-of-freedom (DOF) system, for constant PSD P Eq. (6.119) requires the integral

$$P^3 \frac{8}{2\pi} \int_{-\infty}^{\infty} H_2(\omega_1 + \xi, \omega_2 - \xi)H_2(-\omega_1 - \xi, \xi)H_2(-\omega_2 + \xi, -\xi)d\xi$$

$$= P^3 \frac{8}{2\pi}|H_1(\omega_1 + \omega_2)|^2|H_1(\omega_1)|^2|H_1(\omega_2)|^2$$

$$\times \int_{-\infty}^{\infty} \{[-k_2 - c_2\xi(\xi + \omega_1)][-k_2 + c_2(-\xi - \omega_1)(\xi - \omega_2)][-k_2 + c_2\xi(-\xi + \omega_2)]$$

$$|H_1(\omega_1 + \xi)|^2|H_1(\omega_2 - \xi)|^2|H_1(\omega_1 + \xi)|^2|H_1(\xi)|^2|H_1(-\omega_2 + \xi)|^2|H_1(\xi)|^2\}d\xi$$

$$\equiv P^3 \frac{8}{2\pi}|H_1(\omega_1 + \omega_2)|^2|H_1(\omega_1)|^2|H_1(\omega_2)|^2 G(\omega_1,\omega_2) \tag{B.13}$$

where we have noted that $|H_1(-\omega)|^2 = |H_1(\omega)|^2$. The challenge lies in the integral term $G(\omega_1,\omega_2)$, which can be placed in the needed pole-zero form and evaluated using (B.6) to yield

$G(\omega_1,\omega_2) =$

$$\frac{2\pi i}{m^6}\Bigg[\frac{(-k_2 - c_2 p_1(p_1 + \omega_1))(-k_2 + c_2(-p_1 - \omega_1)(p_1 - \omega_2))(-k_2 + c_2 p_1(-p_1 + \omega_2))}{(p_1 - p_2)(p_1 - p_3)(p_1 - p_4)(p_1 - p_5)(p_1 - p_6)(p_1 - p_7)(p_1 - p_8)(p_1 - p_9)(p_1 - p_{10})(p_1 - p_{11})(p_1 - p_{12})}$$

$$+ \frac{(-k_2 - c_2 p_2(p_2 + \omega_1))(-k_2 + c_2(-p_2 - \omega_1)(p_2 - \omega_2))(-k_2 + c_2 p_2(-p_2 + \omega_2))}{(p_2 - p_1)(p_2 - p_3)(p_2 - p_4)(p_2 - p_5)(p_2 - p_6)(p_2 - p_7)(p_2 - p_8)(p_2 - p_9)(p_2 - p_{10})(p_2 - p_{11})(p_2 - p_{12})}$$

$$+ \frac{(-k_2 - c_2 p_7(p_7 + \omega_1))(-k_2 + c_2(-p_7 - \omega_1)(p_7 - \omega_2))(-k_2 + c_2 p_7(-p_7 + \omega_2))}{(p_7 - p_1)(p_7 - p_2)(p_7 - p_3)(p_7 - p_4)(p_7 - p_5)(p_7 - p_6)(p_7 - p_8)(p_7 - p_9)(p_7 - p_{10})(p_7 - p_{11})(p_7 - p_{12})}$$

$$+ \frac{(-k_2 - c_2 p_8(p_8 + \omega_1))(-k_2 + c_2(-p_8 - \omega_1)(p_8 - \omega_2))(-k_2 + c_2 p_8(-p_8 + \omega_2))}{(p_8 - p_1)(p_8 - p_2)(p_8 - p_3)(p_8 - p_4)(p_8 - p_5)(p_8 - p_6)(p_8 - p_7)(p_8 - p_9)(p_8 - p_{10})(p_8 - p_{11})(p_8 - p_{12})}$$

$$+ \frac{(-k_2 - c_2 p_9(p_9 + \omega_1))(-k_2 + c_2(-p_9 - \omega_1)(p_9 - \omega_2))(-k_2 + c_2 p_9(-p_9 + \omega_2))}{(p_9 - p_1)(p_9 - p_2)(p_9 - p_3)(p_9 - p_4)(p_9 - p_5)(p_9 - p_6)(p_9 - p_7)(p_9 - p_8)(p_9 - p_{10})(p_9 - p_{11})(p_9 - p_{12})}$$

$$+ \frac{(-k_2 - c_2 p_{10}(p_{10} + \omega_1))(-k_2 + c_2(-p_{10} - \omega_1)(p_{10} - \omega_2))(-k_2 + c_2 p_{10}(-p_{10} + \omega_2))}{(p_{10} - p_1)(p_{10} - p_2)(p_{10} - p_3)(p_{10} - p_4)(p_{10} - p_5)(p_{10} - p_6)(p_{10} - p_7)(p_{10} - p_8)(p_{10} - p_9)(p_{10} - p_{11})(p_{10} - p_{12})}\Bigg]$$

$$\tag{B.14}$$

where the poles are given by

$$p_{1,2} = \pm\omega_d + i\zeta\omega_n$$

$$p_{3,4} = \pm\omega_d - i\zeta\omega_n$$

$$p_{5,6} = \omega_2 - p_{1,2}$$

$$p_{7,8} = \omega_2 - p_{3,4}$$

$$p_{9,10} = -\omega_1 + p_{1,2}$$

$$p_{11,12} = -\omega_1 + p_{3,4} \tag{B.15}$$

Evaluating these types of integrals is clearly tedious but not difficult. Use of symbolic manipulation software is of considerable help in dealing with calculations such as the one we have just done.

Reference

[1] M. D. Greenberg, Advanced Engineering Mathematics, Prentice-Hall, Inc., Englewood Cliffs, NJ, 1988.

C

Derivation of Terms for the Trispectrum of an MDOF Nonlinear Structure

In this Appendix we carry out the integrals required in generating an analytical expression for the full trispectral density function. This derivation continues the work outlined in section 6.5.4. Specifically, we return to the terms required of Eqn. 6.153. Substitution of (6.150) into the first term in Eq. (6.153) gives

$$
C^l_{pijk}(\tau_1, \tau_2, \tau_3) = E\left[\int_\mathbb{R} h_{p,1}(\tau_4)X(t-\tau_4)d\tau_4 \int_\mathbb{R} h_{i,1}(\tau_5)X(t-\tau_5+\tau_1)d\tau_5 \right.
$$
$$
\left. \times \int_\mathbb{R} h_{j,1}(\tau_6)X(t-\tau_6+\tau_2)d\tau_6 \int_\mathbb{R} h_{k,1}(\tau_7)X(t-\tau_7+\tau_3)d\tau_7 \right]
$$
$$
= \int_{\mathbb{R}^4} h_{p,1}(\tau_4)h_{i,1}(\tau_5)h_{j,1}(\tau_6)h_{k,1}(\tau_7)
$$
$$
\times E[X(t-\tau_4)X(t-\tau_5+\tau_1)X(t-\tau_6+\tau_2)X(t-\tau_7+\tau_3)]d\tau_{4\to7}
$$
$$
= \int_{\mathbb{R}^4} h_{p,1}(\tau_4)h_{i,1}(\tau_5)h_{j,1}(\tau_6)h_{k,1}(\tau_7)
$$
$$
\times \left\{ E[X(t-\tau_4)X(t-\tau_5+\tau_1)]E[X(t-\tau_6+\tau_2)X(t-\tau_7+\tau_3)] \right.
$$
$$
+ E[X(t-\tau_4)X(t-\tau_6+\tau_2)]E[X(t-\tau_5+\tau_1)X(t-\tau_7+\tau_3)
$$
$$
\left. \times E[X(t-\tau_4)X(t-\tau_7+\tau_3)]E[X(t-\tau_5+\tau_1)X(t-\tau_6+\tau_2)] \right\} d\tau_{4\to7} \quad (C.1)
$$

where the delays $\tau_{4\to7}$ are all dummy variables of integration (introduced via convolution) and the expectation and integration operators have been interchanged. Again using Isserlis' theorem, the product of normally distributed random variables is factored into three products of autocorrelation functions associated with the input $x(t)$. Using the definition of the Fourier

---

Modeling and Estimation of Structural Damage, First Edition. Jonathan M. Nichols and Kevin D. Murphy.
© 2016 John Wiley & Sons, Ltd. Published 2016 by John Wiley & Sons, Ltd.

transform (FT) of the kernels, for example, $H_{p,1}(\omega) = \int_{\mathbb{R}} h_{p,1}(\tau)e^{-i\omega\tau}d\tau$, and also making the substitution $E[X(t+a)X(t+b)] = \frac{1}{2\pi}\int_{\mathbb{R}} S_{XX}(\omega)e^{i\omega(b-a)}d\omega$ yields

$$
\begin{aligned}
C^I_{pijk}(\tau_1,\tau_2,\tau_3) = \frac{1}{8\pi^3}\int_{\mathbb{R}^3} \Big\{ & \delta(\omega_2+\omega_3)H_{p,1}(-\omega_1)H_{i,1}(\omega_1)H_{j,1}(\omega_2)H_{k,1}(-\omega_2)S_{XX}(\omega_1)S_{XX}(\omega_2) \\
& + \delta(\omega_1+\omega_3)H_{p,1}(-\omega_2)H_{i,1}(-\omega_3)H_{j,1}(\omega_2)H_{k,1}(\omega_3)S_{XX}(\omega_2)S_{XX}(\omega_3) \\
& + \delta(\omega_1+\omega_2)H_{p,1}(-\omega_3)H_{i,1}(-\omega_2)H_{j,1}(\omega_2)H_{k,1}(\omega_3)S_{XX}(\omega_3)S_{XX}(\omega_2)\Big\} \\
& \times e^{i(\omega_1\tau_1+\omega_2\tau_2+\omega_3\tau_3)}d\omega_1\,d\omega_2\,d\omega_3.
\end{aligned} \tag{C.2}
$$

Equation (C.2) can immediately be recognized as the inverse FT of the corresponding portion of the trispectrum

$$
\begin{aligned}
S^I_{pijk}(\omega_1,\omega_2,\omega_3) = & \\
& \times \delta(\omega_2+\omega_3)H_{p,1}(-\omega_1)H_{i,1}(\omega_1)H_{j,1}(\omega_2)H_{k,1}(-\omega_2)S_{XX}(\omega_1)S_{XX}(\omega_2) \\
& + \delta(\omega_1+\omega_3)H_{p,1}(-\omega_2)H_{i,1}(-\omega_3)H_{j,1}(\omega_2)H_{k,1}(\omega_3)S_{XX}(\omega_2)S_{XX}(\omega_3) \\
& + \delta(\omega_1+\omega_2)H_{p,1}(-\omega_3)H_{i,1}(-\omega_2)H_{j,1}(\omega_2)H_{k,1}(\omega_3)S_{XX}(\omega_3)S_{XX}(\omega_2)
\end{aligned} \tag{C.3}
$$

These terms are exactly canceled by the linear terms associated with the three submanifolds in Eq. (3.104), thus the trispectrum for a linear system is zero everywhere. Expressions for the submanifold terms are given shortly. Note that by "submanifold" we mean those terms that exist only on restricted subsets of the full frequency space i.e., those pre-multiplied by delta functions involving combinations of frequency variables.

C.1 Simplification of $C^{VIII}_{pijk}(\tau_1,\tau_2,\tau_3)$

Consider the expected value

$$
\begin{aligned}
C^{VIII}_{pijk}(\tau_1,\tau_2,\tau_3) &= E[Y_{i,1}(t+\tau_1)Y_{j,1}(t+\tau_2)Y_{k,1}(t+\tau_3)Y_{p,3}(t)] \\
&= \int_{\mathbb{R}^6} h_{p,3}(\tau_7,\tau_8,\tau_9)h_{i,1}(\tau_4)h_{j,1}(\tau_5)h_{k,1}(\tau_6)E[X(t-\tau_4)X(t-\tau_5) \\
&\quad \times X(t-\tau_6)X(t-\tau_7+\tau_1)X(t-\tau_8+\tau_2)X(t-\tau_9+\tau_3)]d\tau_{4\to9}
\end{aligned} \tag{C.4}
$$

Expanding the expected values using Isserlis' theorem, exploiting the symmetry of the kernels, and combining terms reduces this expression to

$$
\begin{aligned}
C^{VIII}_{pijk}(\tau_1,\tau_2,\tau_3) = \frac{1}{8\pi^3}\int_{\mathbb{R}^6} h_{i,1}(\tau_4)h_{j,1}(\tau_5)h_{k,1}(\tau_6)h_{p,3}(\tau_7,\tau_8,\tau_9)\Big\{ & 6\phi^{1,4,7}_{XX}\phi^{2,5,8}_{XX}\phi^{3,6,9}_{XX} \\
\times\, 3\phi^{1,4,2,5}_{XX}\phi^{3,6,7}_{XX}\phi^{8,9}_{XX} + 3\phi^{1,4,3,6}_{XX}\phi^{2,5,7}_{XX}\phi^{8,9}_{XX} + 3\phi^{1,4,7}_{XX}\phi^{2,5,3,6}_{XX}\phi^{8,9}_{XX} & \Big\}d\tau_{4\to9}
\end{aligned} \tag{C.5}
$$

thus, there are four terms that require simplification. The first term is simplified as

$$
\begin{aligned}
S^{VIIIa}_{pijk}(\tau_1,\tau_2,\tau_3) = \frac{1}{8\pi^3}\int_{\mathbb{R}^9} & 6h_{i,1}(\tau_4)h_{j,1}(\tau_5)h_{k,1}(\tau_6)h_{p,3}(\tau_7,\tau_8,\tau_9)S_{XX}(\omega_1)e^{i\omega_1(-\tau_1+\tau_4-\tau_7)} \\
& \times S_{XX}(\omega_2)^{i\omega_2(-\tau_2+\tau_5-\tau_8)}S_{XX}(\omega_3)^{i\omega_3(-\tau_3+\tau_6-\tau_9)}d\tau_{4\to9}
\end{aligned}
$$

$$= \frac{1}{8\pi^3} \int_{\mathbb{R}^3} 6H_{i,1}(\omega_1)H_{j,1}(\omega_2)H_{k,1}(\omega_3)H_{p,3}(-\omega_1,-\omega_2,-\omega_3)$$

$$\times S_{XX}(\omega_1)S_{XX}(\omega_2)S_{XX}(\omega_3)e^{i(\omega_1\tau_1+\omega_2\tau_2+\omega_3\tau_3)}d\omega_{1\to3} \tag{C.6}$$

where, in the second step, we have made use of the fact that changing the sign of the frequency arguments does not affect the integral. The given expression is in the correct form of an inverse triple FT. The desired component of the trispectrum is therefore

$$S_{pijk}^{VIIIa}(\omega_1,\omega_2,\omega_3) = 6H_{i,1}(\omega_1)H_{j,1}(\omega_2)H_{k,1}(\omega_3)H_{p,3}(-\omega_1,-\omega_2,-\omega_3)$$

$$\times S_{XX}(\omega_1)S_{XX}(\omega_2)S_{XX}(\omega_3) \tag{C.7}$$

The other three terms are only valid on submanifolds of the complete trispectrum. To see this, consider

$$C_{pijk}^{VIIIb}(\tau_1,\tau_2,\tau_3) = \frac{1}{8\pi^3} \int_{\mathbb{R}^9} 3h_{i,1}(\tau_4)h_{j,1}(\tau_5)h_{k,1}(\tau_6)h_{p,3}(\tau_7,\tau_8,\tau_9)S_{XX}(\omega_2)e^{i\omega_2(\tau_2-\tau_5-\tau_1+\tau_4)}$$

$$\times S_{XX}(\omega_3)e^{i\omega_3(-\tau_7-\tau_3+\tau_6)}S_{XX}(\omega_1)^{i\omega_1(-\tau_9+\tau_8)}d\tau_{4\to9}d\omega_{1\to3}$$

$$= \frac{1}{8\pi^3} \int_{\mathbb{R}^3} 3H_{i,1}(-\omega_2)H_{j,1}(\omega_2)H_{k,1}(\omega_3)H_{p,3}(-\omega_3,-\omega_1,\omega_1)$$

$$\times S_{XX}(\omega_1)S_{XX}(\omega_2)S_{XX}(\omega_3)e^{i(-\omega_2\tau_1+\omega_2\tau_2+\omega_3\tau_3)}d\omega_{1\to3} \tag{C.8}$$

This term can only be expressed as an inverse triple FT if $\omega_1 = -\omega_2$. Thus, the needed term for the trispectrum becomes

$$S_{pijk}^{VIIIb}(\omega_1,\omega_2,\omega_3) = \delta(\omega_1+\omega_2)3H_{i,1}(\omega_1)H_{j,1}(\omega_2)H_{k,1}(\omega_3)H_{p,3}(-\omega_3,\omega_2,\omega_1)$$

$$\times S_{XX}(\omega_1)S_{XX}(\omega_2)S_{XX}(\omega_3) \tag{C.9}$$

Similarly, the other two integrals become

$$S_{pijk}^{VIIIc}(\omega_1,\omega_2,\omega_3) = \delta(\omega_1+\omega_2)3H_{i,1}(\omega_1)H_{j,1}(\omega_2)H_{k,1}(\omega_3)H_{p,3}(-\omega_2,\omega_1,\omega_3)$$

$$\times S_{XX}(\omega_1)S_{XX}(\omega_2)S_{XX}(\omega_3)$$

$$S_{pijk}^{VIIId}(\omega_1,\omega_2,\omega_3) = \delta(\omega_1+\omega_2)3H_{i,1}(\omega_1)H_{j,1}(\omega_2)H_{k,1}(\omega_3)H_{p,3}(-\omega_1,\omega_2,\omega_3)$$

$$\times S_{XX}(\omega_1)S_{XX}(\omega_2)S_{XX}(\omega_3) \tag{C.10}$$

respectively. The terms S_{pijk}^{VIIIb}, S_{pijk}^{VIIIc}, and S_{pijk}^{VIIId} are therefore only supported on submanifolds of the trispectrum (these three terms are the first three given in Eq. C.16).

C.2   Submanifold Terms in the Trispectrum

To determine the contributions to the trispectrum produced by Eqs. (6.154), a Third-order Volterra series model is used. The PSD $S_{ji}(\omega)$ can be expressed as the FT of the expected value $E[Y_j(t)Y_i(t+\tau)]$. Substituting Eq. (6.108) into this expectation for both $Y_i(t)$, $Y_j(t)$ yields nine separate terms. Four of the terms result in odd products of the input $x(t)$ which, for Gaussian excitation, are zero, while one of the terms (involving products of two $h_3(\cdot)$ kernels) is higher order with regard to the input PSD. The four remaining terms in the expectation are

$$C^I_{ji}(\tau_1) = E[Y_{j,1}(t)Y_{i,1}(t+\tau_1)]$$

$$C^{II}_{ji}(\tau_1) = E[Y_{j,1}(t)Y_{i,3}(t+\tau_1)]$$

$$C^{III}_{ji}(\tau_1) = E[Y_{i,1}(t)Y_{j,3}(t+\tau_1)]$$

$$C^{IV}_{ji}(\tau_1) = E[Y_{j,2}(t)Y_{i,2}(t+\tau_1)]. \qquad (C.11)$$

The first term is a purely linear term and can be simplified in a straightforward manner:

$$C^I_{ji}(\tau_1) = \int_{\mathbb{R}^2} h_{j,1}(\tau_2)h_{i,1}(\tau_3)E[X(t-\tau_2)X(t+\tau_1-\tau_3)]d\tau_2 d\tau_3$$

$$= \frac{1}{2\pi}\int_{\mathbb{R}^3} h_{j,1}(\tau_2)h_{i,1}(\tau_3)S_{XX}(\omega_1)e^{i\omega_1(\tau_1-\tau_3+\tau_2)}d\tau_2 d\tau_3 d\omega_1$$

$$= \frac{1}{2\pi}\int_{\mathbb{R}} H_{j,1}(-\omega_1)H_{i,1}(\omega_1)S_{XX}(\omega_1)e^{i\omega_1\tau_1}d\omega_1 \qquad (C.12)$$

so that we have

$$S_{ji}(\omega_1) = H_{j,1}(-\omega_1)H_{i,1}(\omega_1)S_{XX}(\omega_1). \qquad (C.13)$$

Substituting this expression into each of the PSD terms in Eq. (6.154) exactly produces the negative of the linear terms given earlier in this Appendix, thus the cumulant trispectrum function has no linear terms.

The same general procedure can be followed for terms $s^{II \to IV}_{ji}(\tau)$, leading to the complete expression for the PSD

$$S_{ji}(\omega_1) = H_{j,1}(-\omega_1)H_{i,1}(\omega_1)S_{XX}(\omega_1)$$

$$+ \frac{3}{2\pi}H_{j,1}(-\omega_1)S_{XX}(\omega_1)\int_{\mathbb{R}} H_{i,3}(\omega_1,-\omega,\omega)S_{XX}(\omega)d\omega$$

$$+ \frac{3}{2\pi}H_{i,1}(\omega_1)S_{XX}(\omega_1)\int_{\mathbb{R}} H_{j,3}(-\omega_1,-\omega,\omega)S_{XX}(\omega)d\omega$$

$$+ \frac{1}{\pi}\int_{\mathbb{R}} H_{j,2}(-\omega_1+\omega,-\omega)H_{i,2}(\omega_1-\omega,\omega)S_{XX}(\omega_1-\omega)S_{XX}(\omega)d\omega \qquad (C.14)$$

Substituting into the first term of Eqs. (6.154) and ignoring all terms greater than $O(S_{XX}(\omega))^3$ gives the associated portion of the trispectrum,

$$-\delta(\omega_1+\omega_2)S_{ji}(-\omega_2)S_{pk}(\omega_3) = -\delta(\omega_1+\omega_2)S_{XX}(\omega_2)S_{XX}(\omega_3)$$

$$\times \left\{ H_{p,1}(-\omega_3)H_{k,1}(\omega_3)H_{j,1}(\omega_2)H_{i,1}(-\omega_2) \right.$$

$$+ \frac{3}{2\pi}H_{p,1}(-\omega_3)H_{k,1}(\omega_3)H_{j,1}(\omega_2)\int_{\mathbb{R}} H_{i,3}(-\omega_2,-\omega,\omega)S_{XX}(\omega)d\omega$$

$$+ \frac{3}{2\pi} H_{p,1}(-\omega_3) H_{k,1}(\omega_3) H_{i,1}(-\omega_2) \int_{\mathbb{R}} H_{j,3}(\omega_2, -\omega, \omega) S_{XX}(\omega) d\omega$$

$$+ \frac{3}{2\pi} H_{j,1}(\omega_2) H_{i,1}(-\omega_2) H_{p,1}(-\omega_3) \int_{\mathbb{R}} H_{j,3}(\omega_3, -\omega, \omega) S_{XX}(\omega) d\omega$$

$$+ \frac{3}{2\pi} H_{j,1}(\omega_2) H_{i,1}(-\omega_2) H_{k,1}(\omega_3) \int_{\mathbb{R}} H_{p,3}(-\omega_3, -\omega, \omega) S_{XX}(\omega) d\omega \bigg\}. \quad \text{(C.15)}$$

Repeating this process for each of the other two terms in Eq. (6.154) gives the remaining sub-manifold terms for the trispectrum. It turns out, however, that for structural systems the integral terms are of higher order and can be ignored. The only term that significantly contributes is just the linear PSD term which, as we have shown, cancels with the linear term from the moment trispectrum.

C.3   Complete Trispectrum Expression

The final expression for the trispectrum includes both the moment spectrum terms (i.e., $S_{pijk}(\omega_1, \omega_2, \omega_3)$) plus the additional terms related to the products of second moment spectra (e.g., Eq. C.15). For the moment spectrum, the main portion, valid everywhere in the $\omega_1, \omega_2, \omega_3$ plane, is already given by Eq. (6.155). However, the additional terms resulting from the moment spectrum (e.g., Eq. C.9) also must be accounted for. These terms are given by

$$T_{ijkp}^{sm}(\omega_1 + \omega_2 + \omega_3) = 3S_{XX}(\omega_1)S_{XX}(\omega_2)S_{XX}(\omega_3)$$

$$\times \{ \delta(\omega_1 + \omega_2) H_{i,1}(\omega_1) H_{j,1}(\omega_2) H_{k,1}(\omega_3) H_{p,3}(-\omega_3, \omega_1, \omega_2)$$

$$+ \delta(\omega_1 + \omega_3) H_{i,1}(\omega_1) H_{j,1}(\omega_2) H_{k,1}(\omega_3) H_{p,3}(-\omega_2, \omega_1, \omega_3)$$

$$+ \delta(\omega_2 + \omega_3) H_{i,1}(\omega_1) H_{j,1}(\omega_2) H_{k,1}(\omega_3) H_{p,3}(-\omega_1, \omega_2, \omega_3)$$

$$+ \delta(\omega_1 + \omega_2) H_{p,1}(-\omega_3) H_{j,1}(\omega_2) H_{k,1}(\omega_3) H_{i,3}(\omega_1, \omega_1, \omega_2)$$

$$+ \delta(\omega_1 + \omega_3) H_{p,1}(-\omega_2) H_{j,1}(\omega_2) H_{k,1}(\omega_3) H_{i,3}(\omega_1, \omega_1, \omega_3)$$

$$+ \delta(\omega_2 + \omega_3) H_{p,1}(-\omega_1) H_{j,1}(\omega_2) H_{k,1}(\omega_3) H_{i,3}(\omega_1, \omega_2, \omega_3)$$

$$+ \delta(\omega_1 + \omega_2) H_{p,1}(-\omega_3) H_{i,1}(\omega_1) H_{k,1}(\omega_3) H_{j,3}(\omega_2, \omega_1, \omega_2)$$

$$+ \delta(\omega_1 + \omega_3) H_{p,1}(-\omega_2) H_{i,1}(\omega_1) H_{k,1}(\omega_3) H_{j,3}(\omega_2, \omega_1, \omega_3)$$

$$+ \delta(\omega_2 + \omega_3) H_{p,1}(-\omega_1) H_{i,1}(\omega_1) H_{k,1}(\omega_3) H_{j,3}(\omega_2, \omega_2, \omega_3)$$

$$+ \delta(\omega_1 + \omega_2) H_{p,1}(-\omega_3) H_{i,1}(\omega_1) H_{j,1}(\omega_2) H_{j,3}(\omega_3, \omega_1, \omega_2)$$

$$+ \delta(\omega_1 + \omega_3) H_{p,1}(-\omega_2) H_{i,1}(\omega_1) H_{j,1}(\omega_2) H_{j,3}(\omega_3, \omega_1, \omega_3)$$

$$+ \delta(\omega_2 + \omega_3) H_{p,1}(-\omega_1) H_{i,1}(\omega_1) H_{j,1}(\omega_2) H_{j,3}(\omega_3, \omega_2, \omega_3) \}$$

$$+ 2\delta(\omega_1 + \omega_2) H_{i,1}(-\omega_2) H_{j,1}(\omega_2) H_{p,2}(-\omega_3 + \omega_1, -\omega_1) H_{k,2}(\omega_3 - \omega_1, \omega_1) S_{XX}(\omega_3 - \omega_1)$$

$$\times S_{XX}(\omega_1) S_{XX}(\omega_2)$$

$$+ 2\delta(\omega_1 + \omega_3)H_{i,1}(-\omega_3)H_{k,1}(\omega_3)H_{p,2}(-\omega_2 + \omega_1, -\omega_1)H_{j,2}(\omega_2 - \omega_1, \omega_1)S_{XX}(\omega_2 - \omega_1)$$
$$\times S_{XX}(\omega_1)S_{XX}(\omega_3)$$

$$+ 2\delta(\omega_2 + \omega_3)H_{j,1}(-\omega_3)H_{k,1}(\omega_3)H_{p,2}(-\omega_1 + \omega_2, -\omega_2)H_{i,2}(\omega_1 - \omega_2, \omega_2)S_{XX}(\omega_1 - \omega_2)$$
$$\times S_{XX}(\omega_2)S_{XX}(\omega_3)$$

$$+ 2\delta(\omega_1 + \omega_2)H_{p,1}(-\omega_3)H_{k,1}(\omega_3)H_{i,2}(\omega_1 - \omega_2, -\omega_1)H_{j,2}(\omega_2 - \omega_1, \omega_1)S_{XX}(\omega_2 - \omega_1)$$
$$\times S_{XX}(\omega_1)S_{XX}(\omega_3)$$

$$+ 2\delta(\omega_1 + \omega_3)H_{p,1}(-\omega_2)H_{j,1}(\omega_2)H_{i,2}(\omega_1 - \omega_3, -\omega_1)H_{k,2}(\omega_3 - \omega_1, \omega_1)S_{XX}(\omega_3 - \omega_1)$$
$$\times S_{XX}(\omega_1)S_{XX}(\omega_2)$$

$$+ 2\delta(\omega_2 + \omega_3)H_{p,1}(-\omega_1)H_{i,1}(\omega_1)H_{k,2}(\omega_3 - \omega_2, -\omega_3)H_{j,2}(\omega_2 - \omega_3, \omega_3)S_{XX}(\omega_2 - \omega_3)$$
$$\times S_{XX}(\omega_1)S_{XX}(\omega_3) \tag{C.16}$$

where the superscript *sm* is simply used to denote "submanifold." Noting that to leading order the contribution of Eq. (6.154) is simply to cancel out the linear term of the moment trispectrum, and the complete trispectrum expression is given by combining Eq. (C.16) with Eq. (6.155).

Index

Acceleration
 Frequency response, 136, 230
 Of spring-mass system, 92, 122, 251
Accelerance (bispectrum), 263
Airy stress function, 146
Algebra, 24–25
Amplitude
 Forcing, 125–126, 251, 340, 395, 422,
  468, 557–558
 Fourier, 63, 71, 121–123, 292, 307, 317,
  549
 Imperfection, 147–150, 154, 350, 484
Arrival time estimation, 352
Autocorrelation (Autocovariance) function
 Analytical expression, 237, 451
 Description, 55–59
 Estimation of, 235
 Relationship to power spectrum, 72–73
Average (*see* mean)

Bayes estimation, 22, 364–374, 492, 521,
 526–527
Bayes rule, 314
Beam
 Buckled, 115
 Damage in, 347–351, 383–386
 Element, 173, 348, 383
 Euler-Bernoulli, 96–99, 153, 162,
  172–173, 184, 383, 557
 Static & dynamic modeling of, 108–113,
  154, 196

Timoshenko, 97, 100
Beta distribution, 372
Bias, 206, 229, 251, 363, 423–425
Bispectral density
 Analytical expression, 246
 Damage detection, 410–411, 423–444
 Definition, 76
 Estimation, 249, 423
Boundary conditions, 99, 101, 109–113,
  132–133, 162, 175, 179–180, 185,
  198, 353, 482–483, 554

Calculus of variations, 102–103
Cauchy integration, 575–580
Central limit theorem, 45–46
Characteristic function, 43–47
Choleski decomposition, 287, 552
Coefficient of variation, 531
Coherence functions
 Damage detection, 411, 432–433,
  445
 Definition, 247–248
 Estimation, 250
Compliance, 118, 194
Conditional probability (*see* probability,
  conditional)
Conjugate distribution, 365–366, 381–382
Convolution, 123, 133, 236, 404
Corrosion (*see* damage
  modeling/identification)
Covariance, 40, 75, 87, 287–290, 552

Modeling and Estimation of Structural Damage, First Edition. Jonathan M. Nichols and Kevin D. Murphy.
© 2016 John Wiley & Sons, Ltd. Published 2016 by John Wiley & Sons, Ltd.

Cracking (*see* damage
    modeling/identification)
Cramer-Rao bound, 207, 356–357, 363
Cross Correlation
    Coefficient, 40–41, 60, 275, 523
    Estimation of, 235
    Function, 59–60, 238–241
    In information theory, 87–90
Cross-spectral density
    Definition, 70–75
    Estimation of, 230
Cumulant
    Definition, 57, 59
    In higher-order spectra, 78–79
    In damage detection, 408

D' Alembert's principle, 129
Damage modeling
    Boundary slippage, 145–151
    Corrosion, 178–182
    Cracking, 117, 143–145, 189–198,
        508
    Delamination, 9–10, 151–160, 438
    Imperfections in plates, 481
    Loose joint, 160–178
Damage identification
    Corrosion, 535–539
    Cracking, 403, 510–522
    Delamination, 404, 501
Damage detection
    Data driven, 11–13
    Local vs. global, 7
    Loose joints, 403, 418, 467–470
    Motivation for, 3–7
    Nonlinearity induced, 403–478
    Surrogate approach (*see* surrogate
        testing)
Damping, 108, 125, 138, 149, 172, 174, 239
Decision making, 15, 544
Deflection of beams & plates, 98–100, 108,
    145, 153, 157–158, 179, 184, 189,
    195
Deflection coefficient, 434
Delay (estimation of), 352
Density (*see* probability density function)
Derivative of random process, 65

Deterministic, 2–3, 17–18
Distribution function (*see* probability
    density function)
Duffing's equation, 340

Efficiency (of estimator, 208)
Eigenvalues & eigenvectors, 125, 151, 164,
    174, 198, 486, 509, 529
Energy methods, 98–105, 153–156, 161,
    178–179, 197, 502–503, 553
Ensemble, 55, 212, 215, 276
Ergodic, 213–215
Error (Type I & Type II, *see also* ROC
    curve), 315, 409, 423–431, 436
Events (probability), 24
Expectation, 39

False alarm (*see also* Error, Type-I), 315,
    319–320
Fatigue model, 559
Finite element, 117–118, 131, 143, 152,
    160, 172, 341, 483
Flexible foundation, 115, 182–184
Forcing vector, 126, 174, 503, 556
Forced vibration, 119, 125, 554–559
Fourier analysis, 61–74
    Fourier coefficient estimation, 221, 224
    Fourier series, 62
    Fourier transform, 63
Free decay (free vibration), 120, 336, 506

Galerkins method, 133, 198, 528, 554
Gamma function, 371, 571
Gaussian random variable, 27, 32, 45, 92
Gaussian random process, 92, 211, 335,
    493, 504, 510, 523, 532
Gibbs sampling, 379, 493, 511

Harmonic
    Analysis, 61 (*see also* Fourier series)
    Excitation, 92, 119, 121, 124
Henon integration, 116
Higher-order spectra (*see also* bispectral
    density, trispectral density), 74–81,
    410
Hypothesis testing, 312–313, 406, 446

Impulse response, 123, 133, 236, 404, 506, 512, 520
Information theoretics
  Definition, 82–91
  Estimation of, 275
  Use in damage detection, 414, 453–456
Initial conditions, 125, 130–131, 181, 336
Interval, confidence, 312, 327–328, 337, 364, 371, 459, 505–507, 518–522
Isserlis' Theorem, 56, 232, 242–246, 256–258, 402, 410

Jacobian, 34, 497
Joint probability, 28–31, 52, 60, 86, 91, 209, 279, 285, 313, 404, 528

Kernel, 217–218, 374, 415, 453–456, 466, 496–500, 534–535
Kinetic energy, 103–107, 109, 111, 178–179, 196–197, 503, 553–554
Kurtosis, 177–178, 409, 417–418

Lagrange's equations, 105–107
Least squares, 227, 336, 345, 488, 518
Likelihood function, 209, 211, 313, 335, 349, 353, 370, 396, 488, 510, 528
Likelihood ratio, 315–317, 378, 529
Linear system, 92, 122, 174, 177

Marginal distribution, 29, 287, 492, 499, 511
Markov chain Monte Carlo, 374–379, 386–389 494, 500, 511–513, 529–530
Markov process, 81–82, 412
Matched filter, 317
Maximum a posteriori criteria, 315
Maximum likelihood estimation, 209, 336, 350, 487–488, 521, 527
Mean, 39
Mean square error, 206
Minimum variance estimator (*see* CR bound)
Modal coordinates, 126, 147, 236–239, 342, 451, 554
Model-based damage detection, 14–15

Model selection, 392–400
Mode shape, 130, 147, 154, 161, 165–172, 342, 451, 462, 554
Multivariate distribution (*see* probability distribution, multivariate)
Mutual Information
  Definition of, 85–87
  Estimation of, 276
  Uses in damage detection, 355, 414–415, 453–454, 523

Natural frequency (*see* resonance)
Negative Binomial distribution, 499
Newmark method of integration, 174, 487
Newton's laws, 96, 101–110, 113, 145
Neyman-Pearson detector, 315, 320, 433
Non-central Chi Squared distribution, 424
Nonlinearity
  Damage induced, 172, 177, 403, 485, 502, 518
  System definition, 124, 405–406
  Volterra modeling of, 133–140
Normal distribution (*see* Gaussian distribution)

Optimal sensor placement, 523–527
Optimal transit, 562
Orthogonality, 125, 130, 226

Parseval's relation, 68, 250, 260, 292
Periodic motion, 92, 108, 118, 122, 127
Phase, 119–120
Poincare (recurrence theorem), 214
Potential energy, 103, 109, 112, 553–554
Power spectral density
  Constant, 74
  Damage detection, 411–412, 420, 448
  Definition, 67–68
  Estimation of, 228, 416
  Excitation, 409, 416, 423, 467, 549
Probability
  Conditional, 30, 88
  Detection, 315, 323
  Failure, 560
  False alarm, 315, 319–320
  Marginal, 29, 211

Probability density function
    Definition of, 27
    Joint, 28–31, 335
    Multivariate, 31
    Transformation of, 32
    Types of, 571–573

Quadratic nonlinearity, 242, 422, 438

Random process (*see also* Gaussian random
        process)
    Definition of, 51–52
    Generation of, 285
    Properties of, 54–57
    Realization of, 51, 203
    Spectral, 67, 549
Random variable, 25
Receiver Operating Characteristic (ROC),
        323, 419, 421, 427–428, 473–474
Resonance (natural) frequency, 95–96, 119,
        130–131, 198, 423, 431, 439–440,
        471, 530, 554
Rician distribution, 442
Rotating machinery, 132, 465–466

Sets, 23–25
Sigma algebra, 24–25
Signal-to-noise ratio, 320, 322, 349, 357,
        490, 505, 533
Significance (statistical level of), 305, 312
Spectral density (*see* cross- power- spectral
        density)
Spring-mass system, 92, 119, 230, 308, 430,
        451, 458
Standard deviation, 36
Stationarity
    Definition, 58
    Testing of, 302–312, 475, 530
Statistics, 23, 61
Stochastic process (*see* random process)
Strain energy, 97–99, 109, 118, 192,
        196–197, 503, 553–554
Sufficient statistic, 318
Superposition, 122–123, 127, 556
Surrogate testing, 446, 449–451, 472–476

Test statistic, 205, 309–312, 318, 409–415

Time-delay (*see* Arrival time)
Timoshenko beam theory (*see* Beam,
        Timoshenko)
Torsional vibration, 108, 117, 128, 180–181
Transfer function, 127–128, 247, 258, 271,
        432, 556
Transfer entropy
    Definition of, 88–91
    Estimation of, 280
    Use in damage detection, 414–415,
        455–457, 466
Transient response (*see* free-decay)
Travelling wave solution, 129
Triangular matrix, 287
Trispectrum
    Analytical expression, 266
    Damage detection, 410–411, 439–4434
    Definition, 79
    Estimation, 269–270
Two degree-of-freedom system, 278,
        451–452

Undamped vibration, 105, 125
Univariate distribution (*see* probability
        density function)

Variance
    Definition, 27, 35, 40
    Estimation of, 206
    Noise, 230, 321, 353, 380–383, 485,
        487, 493, 505, 511
    Of estimator, 206–207, 523
    Of spectral estimators, 229, 250, 270
    Relation to spectral density, 74, 292
Variational mechanics, 103–105
Vibration
    Beams, 131, 154, 161, 438, 467, 553
    Plate, 181
    Rods, 132, 465
    Strings and bars, 128–129, 190
Viscous damping (*see also* damping), 106,
        149, 172, 422, 503, 506
Volterra series, 133–134, 232, 264, 424

Wiener-Khintchin theorem, 72–73,
        232–233, 411, 423–424, 427, 437